An introduction to dynamical systems

To

VLADIMIR IGOREVICH ARNOLD

and

STEPHEN SMALE

for their inspirational work

An introduction to

DYNAMICAL SYSTEMS

D. K. ARROWSMITH
Lecturer, School of Mathematical Sciences,
Queen Mary College & Westfield College, University of London

C. M. PLACE
Lecturer (formerly Department of Mathematics,
Westfield College, University of London)

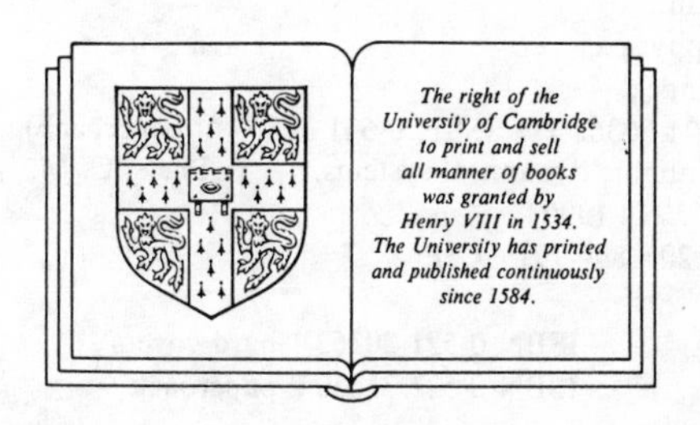
The right of the
University of Cambridge
to print and sell
all manner of books
was granted by
Henry VIII in 1534.
The University has printed
and published continuously
since 1584.

CAMBRIDGE UNIVERSITY PRESS
Cambridge
New York Port Chester
Melbourne Sydney

Published by the Press Syndicate of the University of Cambridge
The Pitt Building, Trumpington Street, Cambridge CB2 1RP
40 West 20th Street, New York NY 10011, USA
10 Stamford Road, Oakleigh, Melbourne 3166, Australia

First published 1990

Printed in Great Britain at the University Press, Cambridge

British Library cataloguing in publication data
Arrowsmith, D.K.
An introduction to dynamical systems.
1. Differentiable dynamical systems
I. Title II. Place, C.M.
514'.7

Library of Congress cataloguing in publication data
Arrowsmith, D. K.
An introduction to dynamical systems / D.K. Arrowsmith and C.M. Place
p. cm.
Bibliography: p.
Includes index.
ISBN 0 521 30362 1. – ISBN 0 521 31650 2 (paperback)
1. Differentiable dynamical systems. I. Place, C. M. II. Title.
QA614.8.A77 1990
515.'.352–dc20 89-7191 CIP

ISBN 0 521 30362 1 hard covers
ISBN 0 521 31650 2 paperback

MP

CONTENTS

PREFACE

In recent years there has been a marked increase of research interest in dynamical systems and a number of excellent postgraduate texts have been published. This book is specifically aimed at the interface between undergraduate and postgraduate studies. It is intended both to stimulate the interest of final year undergraduates and to provide a solid foundation for postgraduates who intend to embark on research in the field. For example, a challenging third-year undergraduate course can be constructed by selecting topics from the first four chapters. Indeed, lecture courses taught by one of us (CMP) provided the basis for Chapters 1, 2 and 4. On the other hand, Chapter 6 is directed at first-year postgraduate students. It contains a selection of current research topics that illustrate the interaction between superficially different research problems.

A major feature of the book is its extensive set of exercises; more than 300 in all. These exercises not only illustrate the topics discussed in the text, but also guide the reader in the completion of technical details omitted from the main discussion. Detailed model solutions have been prepared and hints to their construction are provided.

The reader is assumed to have attended courses in analysis and linear algebra to second-year undergraduate standard. Prior knowledge of dynamical systems is not necessary; however, some familiarity with the qualitative theory of differential equations and Hamiltonian dynamics might be an advantage.

We would like to thank Martin Casdagli for sharpening our understanding of Birkhoff attractors, David Knowles and Chris Norman for helpful discussions and Carl Murray for steering some awkward diagrams to a laser printer. We are grateful to the *Quarterly Journal of Applied Mathematics* and Springer-Verlag for allowing us to use diagrams from some of their publications and our thanks go to Sandra Place for her fast and accurate typing of much of the manuscript. One of us (CMP) would like to acknowledge the Brayshay Foundation for its financial support throughout this project. Finally, we must both pay tribute to the patience and support of our families during the long, and often difficult, gestation period of the manuscript.

1

Diffeomorphisms and flows

1.1 Introduction

A dynamical system is one whose state changes with time (t). Two main types of dynamical system are encountered in applications: those for which the time variable is discrete ($t \in \mathbb{Z}$ or $\mathbb{N}$) and those for which it is continuous ($t \in \mathbb{R}$).

Discrete dynamical systems can be presented as the iteration of a function, i.e.

$$\mathbf{x}_{t+1} = \mathbf{f}(\mathbf{x}_t), \qquad t \in \mathbb{Z} \text{ or } \mathbb{N}. \tag{1.1.1}$$

When t is continuous, the dynamics are usually described by a differential equation

$$\frac{d\mathbf{x}}{dt} = \dot{\mathbf{x}} = \mathbf{X}(\mathbf{x}). \tag{1.1.2}$$

In (1.1.1 and 2), $\mathbf{x}$ represents the state of the system and takes values in the *state* or *phase space*. Sometimes the phase space is Euclidean space or a subset thereof, but it can also be a non-Euclidean structure such as a circle, a sphere, a torus or some other *differentiable manifold*.

In this chapter we will consider two special cases of the above equations, namely when:

(i) $\mathbf{f}$ in (1.1.1) is a *diffeomorphism*; and

(ii) the solutions of (1.1.2) can be described by a *flow* with velocity given by the *vector field* $\mathbf{X}$.

These two cases have been widely studied and they are fundamental to our understanding of dynamical systems. Smale, in his definitive work (Smale, 1967), pointed out that (i) and (ii) are closely related and our discussion emphasises this connection.

Any description of the theory of (i) and (ii) involves differentiable maps so let us begin by recalling some definitions. Let U be an open subset of $\mathbb{R}^n$. Then a function $g: U \to \mathbb{R}$ is said to be of *class* C^r if it is r-fold continuously differentiable, $1 \leqslant r \leqslant \infty$. Let V be an open subset of $\mathbb{R}^m$ and $\mathbf{G}: U \to V$. Given coordinates

$(x_1, \ldots, x_n)$ in U and $(y_1, \ldots, y_m)$ in V, $\mathbf{G}$ may be expressed in terms of component functions g_i: $U \to \mathbb{R}$, where

$$y_i = g_i(x_1, \ldots, x_n), \qquad i = 1, \ldots, m. \tag{1.1.3}$$

The map $\mathbf{G}$ is called a *C^r-map* if g_i is C^r for each $i = 1, \ldots, m$. $\mathbf{G}$ is said to be *differentiable* if it is a C^r-map for some $1 \leqslant r \leqslant \infty$ and to be *smooth* if it is C^∞. Maps that are continuous but not differentiable are, conventionally, referred to as *C^0-maps*.

Definition 1.1.1 $\mathbf{G}$ *is said to be a* diffeomorphism *if it is a bijection and both* $\mathbf{G}$ *and* $\mathbf{G}^{-1}$ *are differentiable mappings.* $\mathbf{G}$ *is called a* C^k-diffeomorphism *if both* $\mathbf{G}$ *and* $\mathbf{G}^{-1}$ *are C^k-maps.*

Observe that the bijection $\mathbf{G}$: $U \to V$ is a diffeomorphism if and only if $m = n$ and the matrix of partial derivatives

$$\mathbf{DG}(x_1, \ldots, x_n) = \left[\frac{\partial g_i}{\partial x_j}\right]_{i,j=1}^{n} \tag{1.1.4}$$

is non-singular at every $\mathbf{x} \in U$. Thus $\mathbf{G}(x, y) = (\exp(y), \exp(x))^{\mathrm{T}}$ with $U = \mathbb{R}^2$ and $V = \{(x, y) | x, y > 0\}$ is a diffeomorphism because Det $\mathbf{DG}(x, y) = -\exp(x + y) \neq 0$ for each $(x, y) \in \mathbb{R}^2$.

If $\mathbf{G}$ satisfies Definition 1.1.1 with $\mathbf{G}$ and $\mathbf{G}^{-1}$ continuous, rather than differentiable, maps then $\mathbf{G}$ is said to be a *homeomorphism*. As we shall see, such maps play a central role in the topological theory of flows and diffeomorphisms.

The above definitions are adequate provided phase space is Euclidean, but, as we have already mentioned, the natural setting for dynamics is a *differentiable manifold*. The important point here is that manifolds have the property that they are 'locally Euclidean' and this allows us to extend the idea of differentiability to functions defined on them. If M is a manifold of dimension n then, for any $\mathbf{x} \in M$, there is a neighbourhood $W \subseteq M$ containing $\mathbf{x}$ and a homeomorphism $\mathbf{h}$: $W \to \mathbb{R}^n$ which maps W onto a neighbourhood of $\mathbf{h}(\mathbf{x}) \in \mathbb{R}^n$. Since we can define coordinates in $U = \mathbf{h}(W) \subseteq \mathbb{R}^n$ (the coordinate curves of which can be mapped back onto W), we can think of $\mathbf{h}$ as defining local coordinates on the patch W of M (see Figure 1.1).

The pair $(U, \mathbf{h})$ is called a *chart* and we can use it to give meaning to differentiability on W. Let us assume, for simplicity, that $\mathbf{f}$: $W \to W$, then $\mathbf{f}$ induces a map $\tilde{\mathbf{f}} = \mathbf{h} \cdot \mathbf{f} \cdot \mathbf{h}^{-1}$: $U \to U$ (see Figure 1.2). We say that $\mathbf{f}$ is a C^k-map on W if $\tilde{\mathbf{f}}$ is a C^k-map on U. This construction allows us to give a definition of a *local* diffeomorphism on M.

In order to obtain a global description of the manifold, we cover it with a family of open sets, W_α, each with its associated chart $(U_\alpha, \mathbf{h}_\alpha)$ (predictably, the set of all charts is called an *atlas*). If $W_\alpha \cap W_\beta$ is not empty, then either $(U_\alpha, \mathbf{h}_\alpha)$ or $(U_\beta, \mathbf{h}_\beta)$ can be used to provide local coordinates for $W_\alpha \cap W_\beta$. This possibility induces overlap maps, $\mathbf{h}_{\alpha\beta}$ and $\mathbf{h}_{\beta\alpha}$ between $\mathbf{h}_\alpha(W_\alpha \cap W_\beta) \subseteq U_\alpha$ and $\mathbf{h}_\beta(W_\alpha \cap W_\beta) \subseteq U_\beta$ (see

Figure 1.1 Examples of differentiable manifolds and some 'patches' of local coordinates. Several open sets based on patches of this kind may be required in order to cover the whole manifold.

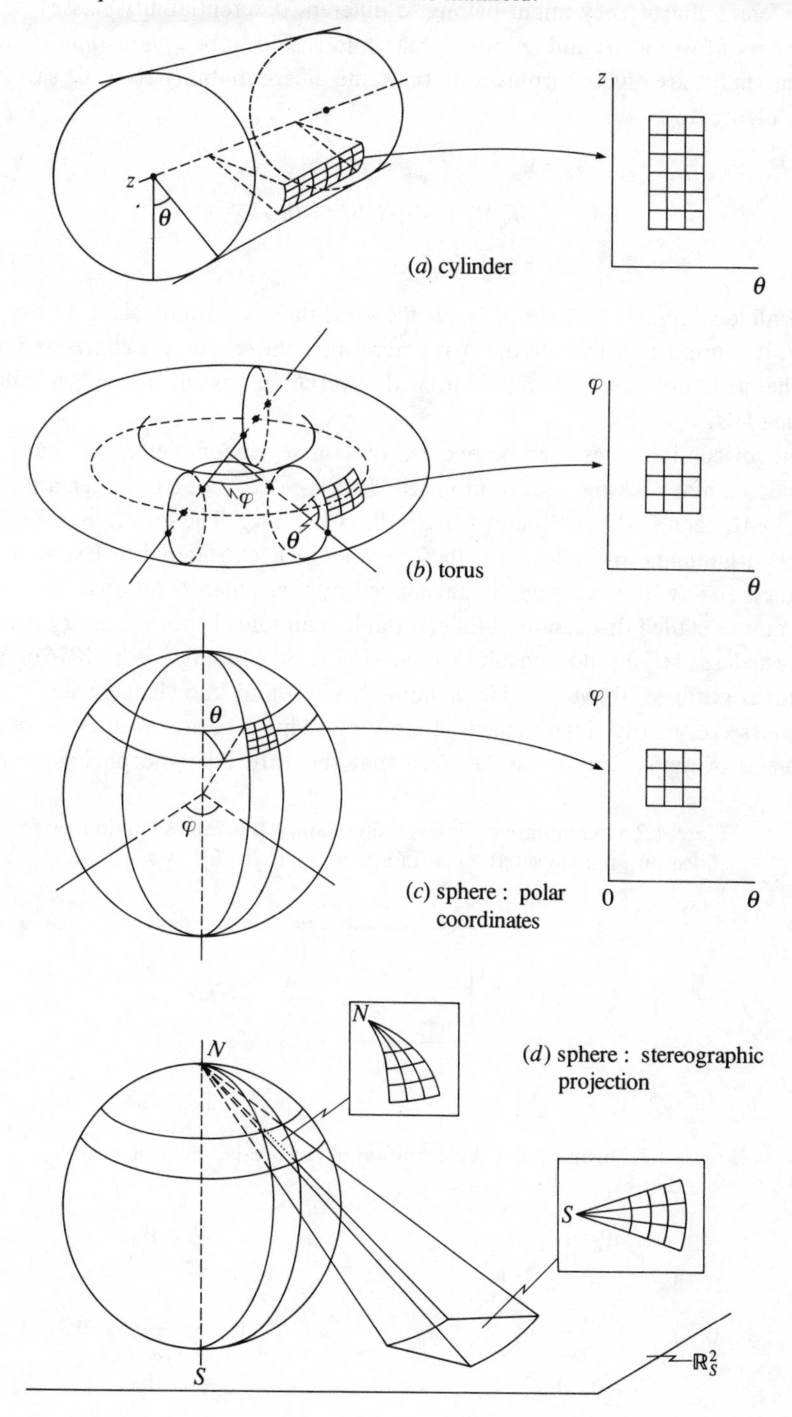

Figure 1.3). If we now consider $\mathbf{f}$: $W_\alpha \cap W_\beta \to W_\alpha \cap W_\beta$, we have two alternative representatives $\tilde{\mathbf{f}}_\alpha = \mathbf{h}_\alpha \cdot \mathbf{f} \cdot \mathbf{h}_\alpha^{-1}$ and $\tilde{\mathbf{f}}_\beta = \mathbf{h}_\beta \cdot \mathbf{f} \cdot \mathbf{h}_\beta^{-1}$ for $\mathbf{f}$. Since $\tilde{\mathbf{f}}_\alpha$ and $\tilde{\mathbf{f}}_\beta$ are determined by different charts, they might belong to different differentiability classes, so that the class of $\mathbf{f}$ would be ambiguous. A manifold is said to be *differentiable* if all the overlap maps are diffeomorphisms of the same differentiability class, C^r say. Now, from Figure 1.3,

$$\begin{aligned} \tilde{\mathbf{f}}_\beta &= \mathbf{h}_\beta \cdot \mathbf{f} \cdot \mathbf{h}_\beta^{-1} \\ &= (\mathbf{h}_\beta \cdot \mathbf{h}_\alpha^{-1}) \cdot (\mathbf{h}_\alpha \cdot \mathbf{f} \cdot \mathbf{h}_\alpha^{-1}) \cdot (\mathbf{h}_\alpha \cdot \mathbf{h}_\beta^{-1}) \\ &= \mathbf{h}_{\alpha\beta} \cdot \tilde{\mathbf{f}}_\alpha \cdot \mathbf{h}_{\alpha\beta}^{-1}. \end{aligned} \tag{1.1.5}$$

Thus all local representatives of $\mathbf{f}$ have the same differentiability class, C^k say, with $k \leqslant r$. It is important to note that r is determined entirely by the charts and hence by the structure of M. A manifold with overlap maps of class C^r is called a C^r-*manifold*.

The discussion presented above is, of course, incomplete. We have only considered maps taking a chart into itself. This is clearly not true in general. Given $\mathbf{f}$: $M \to M$, then $\mathbf{f}$: $W_\alpha \to W_\beta$ and $\mathbf{f}$: $W_\alpha \cap W_\gamma \to W_\beta \cap W_\delta$. The generalisation of our simple arguments that allows for these omissions is considered in Exercise 1.1.2. Needless to say, the 'message' is unchanged by these manipulations.

A more detailed discussion of differentiable manifolds is not necessary here (the interested reader should consult Arnold (1973) or Chillingworth (1976)). While the ideas outlined above provide valuable background knowledge, we will rarely find ourselves involved with charts, atlases, etc. This is because our concern is the *dynamics* of maps defined on M *given* that they are diffeomorphisms or flows.

Figure 1.2 Commutative diagram illustrating the representation of $\mathbf{f}$ defined on an open set W of M in a local shart $(U, \mathbf{h})$.

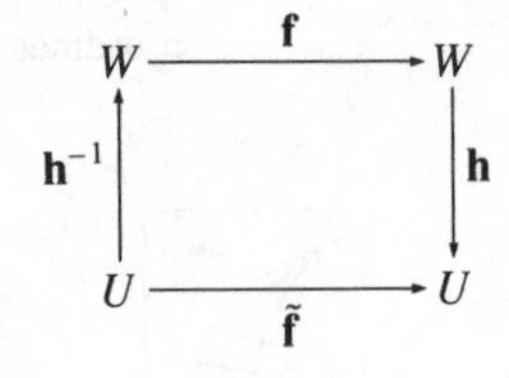

Figure 1.3 Illustration of the definition of the overlap maps $\mathbf{h}_{\alpha\beta}$ and $\mathbf{h}_{\beta\alpha}$. Note that $\mathbf{h}_{\beta\alpha} = \mathbf{h}_{\alpha\beta}^{-1}$.

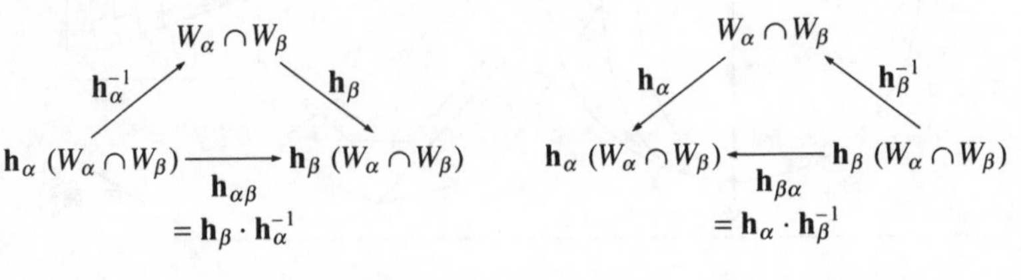

$W_\alpha \cap W_\beta$

$\mathbf{h}_\alpha^{-1}$ $\quad$ $\mathbf{h}_\beta$

$\mathbf{h}_\alpha\,(W_\alpha \cap W_\beta) \longrightarrow \mathbf{h}_\beta\,(W_\alpha \cap W_\beta)$

$\mathbf{h}_{\alpha\beta}$

$= \mathbf{h}_\beta \cdot \mathbf{h}_\alpha^{-1}$

$W_\alpha \cap W_\beta$

$\mathbf{h}_\alpha$ $\quad$ $\mathbf{h}_\beta^{-1}$

$\mathbf{h}_\alpha\,(W_\alpha \cap W_\beta) \longleftarrow \mathbf{h}_\beta\,(W_\alpha \cap W_\beta)$

$\mathbf{h}_{\beta\alpha}$

$= \mathbf{h}_\alpha \cdot \mathbf{h}_\beta^{-1}$

These maps are usually presented to us in local coordinates so that the manifold structure does not appear explicitly.

1.2 Elementary dynamics of diffeomorphisms

1.2.1 Definitions

Let M be a differentiable manifold and suppose $\mathbf{f}: M \to M$ is a diffeomorphism. For each $\mathbf{x} \in M$, the iteration (1.1.1) generates a sequence, the distinct points of which define the *orbit* or *trajectory* of $\mathbf{x}$ under $\mathbf{f}$. More precisely, the orbit of $\mathbf{x}$ under $\mathbf{f}$ is $\{\mathbf{f}^m(\mathbf{x}) | m \in \mathbb{Z}\}$. For $m \in \mathbb{Z}^+$, $\mathbf{f}^m$ is the composition of $\mathbf{f}$ with itself m times. Since $\mathbf{f}$ is a diffeomorphism $\mathbf{f}^{-1}$ exists and $\mathbf{f}^{-m} = (\mathbf{f}^{-1})^m$. Finally, $\mathbf{f}^0 = \mathbf{id}_M$, the identity map on M. Typically, the orbit of $\mathbf{x}$ is a bi-infinite sequence of distinct points of M. However, there are two important exceptions to this state of affairs.

Definition 1.2.1 *A point* $\mathbf{x}^* \in M$ *is called a* fixed point *of* $\mathbf{f}$ *if* $\mathbf{f}^m(\mathbf{x}^*) = \mathbf{x}^*$ *for all* $m \in \mathbb{Z}$.

Definition 1.2.2 *A point* $\mathbf{x}^* \in M$ *is a* periodic point *of* $\mathbf{f}$ *if* $\mathbf{f}^q(\mathbf{x}^*) = \mathbf{x}^*$, *for some integer* $q \geqslant 1$.

The least value of q satisfying Definition 1.2.2 is called the *period* of the point $\mathbf{x}^*$ and the orbit of $\mathbf{x}^*$, i.e.

$$\{\mathbf{x}^*, \mathbf{f}(\mathbf{x}^*), \ldots, \mathbf{f}^{q-1}(\mathbf{x}^*)\}, \tag{1.2.1}$$

is said to be a *periodic orbit of period* q or a *q-cycle* of $\mathbf{f}$. Clearly, since $\mathbf{f}^q(\mathbf{x}^*) = \mathbf{x}^*$, it is the sequence $\{\mathbf{f}^m(\mathbf{x}^*)\}_{m=-\infty}^{\infty}$ which is q-periodic. Notice that a fixed point is a periodic point of period one and a periodic point of $\mathbf{f}$ with period q is a fixed point of $\mathbf{f}^q$. Morever, if $\mathbf{x}^*$ is a periodic point of period q for $\mathbf{f}$ then so are all of the other points in the orbit of $\mathbf{x}^*$. For example, if $\mathbf{f}^q(\mathbf{x}^*) = \mathbf{x}^*$ then $\mathbf{f}(\mathbf{f}^q(\mathbf{x}^*)) = \mathbf{f}(\mathbf{x}^*) = \mathbf{f}^q(\mathbf{f}(\mathbf{x}^*))$ and $\mathbf{f}(\mathbf{x}^*)$ is therefore a periodic point of period q, and so on for $\mathbf{f}^2(\mathbf{x}^*), \ldots, \mathbf{f}^{q-1}(\mathbf{x}^*)$.

Fixed and periodic points can be classified according to the behaviour of the orbits of points in their vicinity. The following ideas are due to Liapunov.

Definition 1.2.3 *A fixed point,* $\mathbf{x}^*$, *is said to be* stable *if, for every neighbourhood* N *of* $\mathbf{x}^*$, *there is a neighbourhood* $N' \subseteq N$ *of* $\mathbf{x}^*$ *such that if* $\mathbf{x} \in N'$ *then* $\mathbf{f}^m(\mathbf{x}) \in N$ *for all* $m > 0$.

Essentially, Definition 1.2.3 says that iterates of points 'near to' a stable fixed point, remain 'near to' it for $m \in \mathbb{Z}^+$. If a fixed point $\mathbf{x}^*$ is stable and $\lim_{m \to \infty} \mathbf{f}^m(\mathbf{x}) = \mathbf{x}^*$, for all $\mathbf{x}$ in some neighbourhood of $\mathbf{x}^*$, then the fixed point is said to be *asymptotically stable*. Trajectories of points near to an asymptotically stable fixed point move toward it as m increases. Fixed points that are stable, but not asymptotically stable, are said to be *neutrally* or *marginally* stable and those that are not stable in the sense of Definition 1.2.3 are *unstable*.

1.2.2 Diffeomorphisms of the circle

The circle (S^1) is arguably the simplest non-Euclidean differentiable manifold. It is compact (see Chillingworth, 1976, p. 143) so 'behaviour at infinity' is not a problem; it has no boundary so that dynamics can be studied without the complication of boundary conditions on the functions concerned and it is one-dimensional. The dynamics of diffeomorphisms on the circle therefore provide an ideal opportunity for us to illustrate the definitions given in §1.2.1.

Some of the simplest examples of diffeomorphisms on S^1 are the pure rotations. They are easily defined in terms of the angular displacement (θ) at the centre of the circle relative to a reference radius (see Figure 1.4). In terms of this local coordinate, an anticlockwise rotation by α may be written as

$$R_\alpha(\theta) = (\theta + \alpha) \bmod 1. \tag{1.2.2}$$

Here we have assumed that θ is measured in units of 2π. If $\alpha = p/q$, $p, q \in \mathbb{Z}$ and relatively prime, then

$$R_\alpha^q(\theta) = (\theta + p) \bmod 1 = \theta \tag{1.2.3}$$

and we conclude (cf. Definition 1.2.2) that every point of the circle is a periodic point of period-q, i.e. the orbit of any point is a q-cycle (see Figure 1.4). If α is irrational then

$$R_\alpha^m(\theta) = (\theta + m\alpha) \bmod 1 \neq \theta, \tag{1.2.4}$$

for any θ and, in fact, the orbit of any point fills the circle densely (see Exercise 1.2.1).

Obviously more general diffeomorphisms of S^1 do not simply rotate all points uniformly. Crudely speaking they compress some arcs of the circle and stretch

Figure 1.4 Typical orbit of the pure rotation R_α for $\alpha = p/q = 2/5$. Observe that the orbit of θ winds around the circle $p = 2$ times before returning to θ on the fifth iteration.

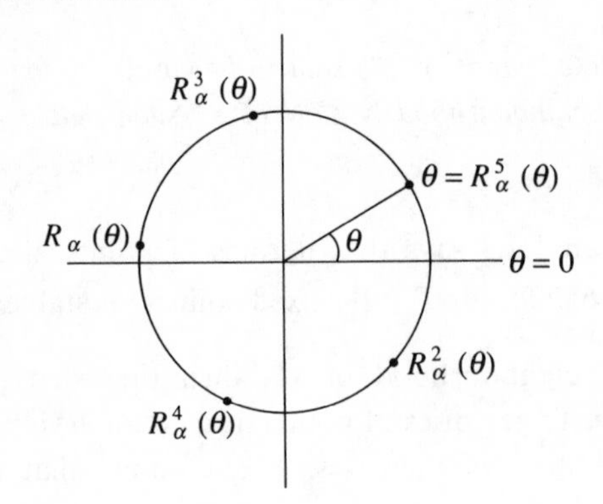

others. It is then difficult to recognise fixed or periodic points from the representation of orbits on the circle itself. This is a problem for any map (f) of the circle, whether it is a diffeomorphism or not, and it is solved by considering a *lift* of f.

The natural setting for introducing the lift of $f: S^1 \to S^1$ is when f is a homeomorphism rather than a diffeomorphism and it would be perverse to artifically confine our discussion to the differentiable case. Moreover, by taking f to be a homeomorphism at this point we can better appreciate the consequences of imposing differentiability on f and f^{-1}. Thus, let $f: S^1 \to S^1$ be a homeomorphism and suppose there is a continuous function $\bar{f}: \mathbb{R} \to \mathbb{R}$ such that

$$\pi(\bar{f}(x)) = f(\pi(x)) \tag{1.2.5}$$

(see Figure 1.5), where

$$\pi(x) = x \bmod 1 = \theta. \tag{1.2.6}$$

Then $\bar{f}$ is called a *lift of* $f: S^1 \to S^1$ onto $\mathbb{R}$.

Proposition 1.2.1 *Let $\bar{f}$ be a lift of the orientation-preserving homeomorphism $f: S^1 \to S^1$. Then*

$$\bar{f}(x+1) = \bar{f}(x) + 1 \tag{1.2.7}$$

for every $x \in \mathbb{R}$.

Proof. Observe that

$$f(\pi(x)) = f(\pi(x+1)) \tag{1.2.8}$$

because $\pi(x) = \pi(x+1)$ by (1.2.6). If we substitute for $f \cdot \pi$ from (1.2.5), (1.2.8) becomes

$$\pi(\bar{f}(x)) = \pi(\bar{f}(x+1)) \tag{1.2.9}$$

and it follows that

$$\bar{f}(x+1) = \bar{f}(x) + k(x), \tag{1.2.10}$$

where $k(x)$ is an integer possibly depending on x. However, since $\bar{f}$ is continuous, $k(x)$ must be continuous and this is only possible if $k(x) \equiv k \in \mathbb{Z}$.

Figure 1.5 Commutative diagram illustrating the definition of the lift of a circle homeomorphism f. The map π takes infinitely many equivalent points of $\mathbb{R}$ onto a single point of S^1.

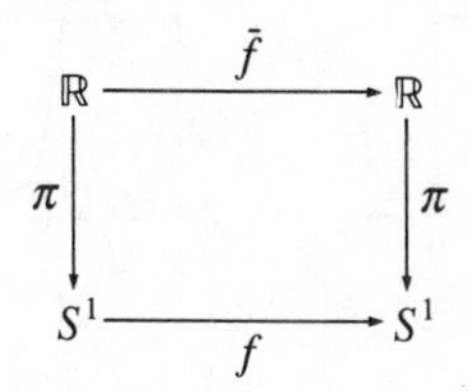

Suppose $k \geqslant 2$, then $\bar{f}(x)$ and $\bar{f}(x+1)$ differ by more than two and $\bar{f}$ takes the form shown schematically in Figure 1.6(a). Clearly, the points x_0 and x_1 satisfying $\bar{f}(x_0)=1$ and $\bar{f}(x_1)=2$ are both less than unity. This means that π maps them to distinct points on S^1. However, $\bar{f}(x_0)$ and $\bar{f}(x_1)$ differ by unity and therefore represent the same point on S^1. This contradicts the hypothesis that f is a homeomorphism. Hence $k \leqslant 1$.

If $k=0$, $\bar{f}(0)=\bar{f}(1)$ and $\bar{f}$ fails to be injective on (0, 1) (see Figure 1.6(b)). Again this contradicts the fact that f is a homeomorphism.

If $k<0$ then continuity of $\bar{f}$ can only be maintained if f is orientation-reversing in contradiction to hypothesis. Moreover, it is clear that similar arguments would lead to a minus sign in the right hand side of (1.2.7) for orientation-reversing f.

Finally, we conclude that $k=1$ and (1.2.7) follows. □

It is important to realise that not every continuous function satisfying (1.2.7) is the lift of some homeomorphism. The function shown in Figure 1.7 is continuous and satisfies (1.2.7) but fails to be the lift of a homeomorphism because it is not injective. Figure 1.7 also highlights the geometrical significance of (1.2.7); namely that the graph of $\bar{f}$ in the interval $[k, k+1]$ is obtained by shifting the graph of $\bar{f}$ in [0, 1] vertically by k units. In this way any continuous function g, defined on [0, 1], that is injective, and such that $g(1)=g(0)+1$, can be used to construct a lift $\bar{f}$ for some homeomorphism $f\colon S^1 \to S^1$. The function f is given by (1.2.5). A simple example of this construction is given in Figure 1.8(a) where

$$g(x) = -x^2 + 2x + \tfrac{1}{2}, \tag{1.2.11}$$

$x \in [0, 1]$. In this case, $\bar{f}$ is a continuous bijection but it is not differentiable at $x=1, 2, \ldots$. This reflects on the corresponding f which is a homeomorphism,

Figure 1.6 Schematic forms for $\bar{f}$ when (1.2.10) has (a) $k=2$; (b) $k=0$. In both cases, the hypothesis that f is a homeomorphism is contradicted.

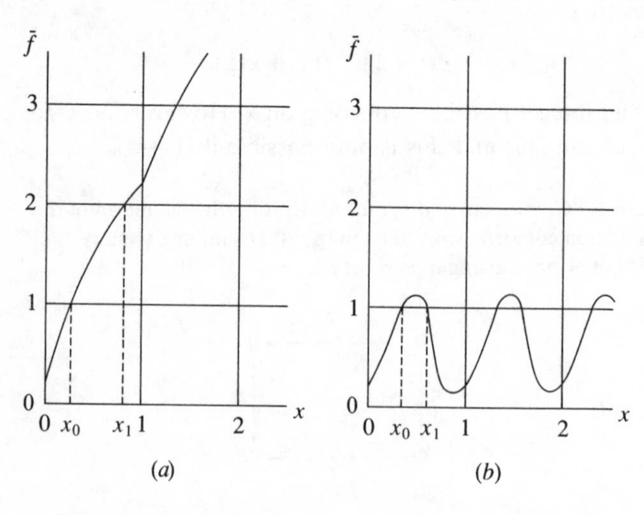

but not a diffeomorphism of S^1. To obtain the latter, $\bar{f}$ must be a bijection and differentiable for all $x \in \mathbb{R}$. An example of this type is shown in Figure 1.8(b) where

$$g(x) = x + \tfrac{1}{2} + \tfrac{1}{10}\sin 2\pi x, \tag{1.2.12}$$

$x \in [0, 1]$.

Notice, we have, without loss of generality, taken $\bar{f}(0) \in [0, 1)$ in both of the above examples. Observe that, $\pi(\bar{f}(x) + k) = \pi(\bar{f}(x))$, for any $k \in \mathbb{Z}$. Thus if $\bar{f}(x)$ is a lift of f then so is $\bar{f}_k(x) = \bar{f}(x) + k$, $k \in \mathbb{Z}$. Therefore, unless otherwise stated, we will assume that $\bar{f}$ is the member of this family of lifts satisfying $\bar{f}(0) \in [0, 1)$.

Figure 1.7 The function $\bar{f}$ shown here cannot be the lift of a homeomorphism $f: S^1 \to S^1$ because it is not injective.

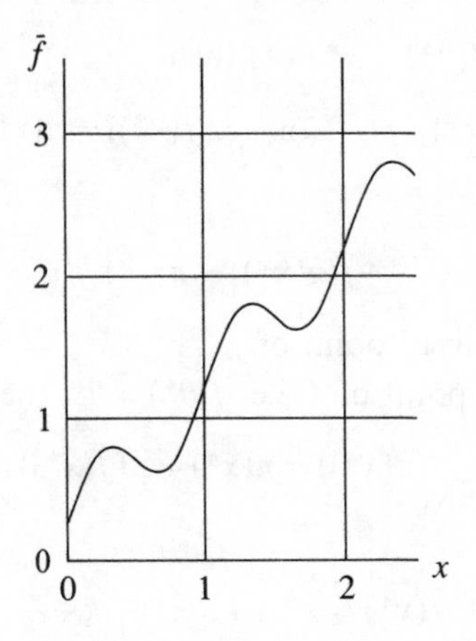

Figure 1.8 The function shown in (a) is the lift of a homeomorphism, but not of a diffeomorphism, of the circle. Lifts of diffeomorphisms are differentiable functions of x, see (b) for example, where $\bar{f}$ is obtained from (1.2.12).

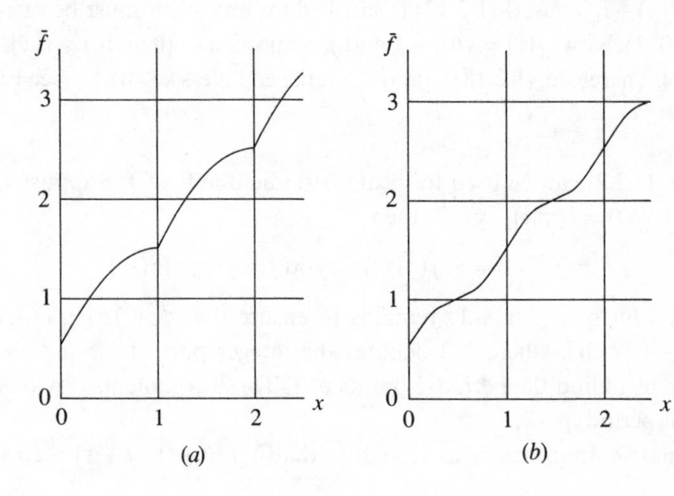

How are the fixed or periodic points of $f: S^1 \to S^1$ related to the properties of the lift $\bar{f}$?

Proposition 1.2.2 *Let $f: S^1 \to S^1$ be an orientation-preserving homeomorphism and suppose that $\bar{f}$ is the lift of f with $\bar{f}(0) \in [0, 1)$. Then $\pi(x^*)$ is a fixed point of f if and only if either*

$$\bar{f}(x^*) = x^* \tag{1.2.13a}$$

or

$$\bar{f}(x^*) = x^* + 1. \tag{1.2.13b}$$

Proof. If $\bar{f}(x^*) = x^*$ (or $\bar{f}(x^*) = x^* + 1$) then

$$\pi(\bar{f}(x^*)) = \pi(x^*) \qquad (\text{or} \quad \pi(\bar{f}(x^*)) = \pi(x^* + 1) = \pi(x^*)). \tag{1.2.14}$$

In either case,

$$f(\pi(x^*)) = \pi(x^*) \tag{1.2.15}$$

by (1.2.5) and $\pi(x^*)$ is a fixed point of f.

If $\theta^* = \pi(x^*)$ is a fixed point of f, i.e. $f(\theta^*) = \theta^*$, then

$$f(\pi(x^*)) = \pi(x^*) = \pi(\bar{f}(x^*)) \tag{1.2.16}$$

by (1.2.5). Thus

$$\bar{f}(x^*) = x^* + k, \qquad k \in \mathbb{Z}. \tag{1.2.17}$$

Let $x^* = y^* + l$, $l \in \mathbb{Z}$, $y^* \in [0, 1)$ then (1.2.17) becomes

$$\bar{f}(y^*) + l = y^* + l + k. \tag{1.2.18}$$

Here we have noted that a simple induction on $\bar{f}(x+1) = \bar{f}(x) + 1$ gives $\bar{f}(x + l) = \bar{f}(x) + l$. Thus, if (1.2.17) is satisfied for any x^*, it must be satisfied for a point $y^* \in [0, 1)$. Now, $\bar{f}(1) = \bar{f}(0) + 1$ and $\bar{f}$ is injective so that $\bar{f}(0) \leqslant \bar{f}(y) < \bar{f}(0) + 1$ for $y \in [0, 1)$. Therefore, (1.2.18) cannot be satisfied unless $k = 0$ or 1 (see Figure 1.9).

□

Proposition (1.2.2) can be used to locate periodic points of f. Suppose that f has lift $\bar{f}$, i.e. $\pi(\bar{f}(x)) = f(\pi(x))$, $x \in \mathbb{R}$, then

$$\pi(\bar{f}^2(x)) = \pi(\bar{f}(\bar{f}(x))) = f(\pi(\bar{f}(x)) = f^2(\pi(x)). \tag{1.2.19}$$

Thus $\bar{f}^2$ is a lift of f^2. It only remains to ensure that $\bar{f}^2(0) \in [0, 1)$ (i.e. choose the lift $\bar{f}^2 - [\bar{f}^2(0)]$, where $[\cdot]$ denotes the integer part of $\cdot$), and Proposition 1.2.2 allows us to find the period-2 points of f. These arguments obviously extend to points of period $q > 2$.

An alternative approach is to recognise that if $\bar{f}^q(0) \in [l, l + 1)$ then (1.2.13) is replaced by

$$\bar{f}^q(x^*) = x^* + l \qquad \text{or} \qquad \bar{f}^q(x^*) = x^* + l + 1. \tag{1.2.20}$$

This point of view often has the advantage that $\bar{f}, \bar{f}^2, \ldots, \bar{f}^q, \ldots$ can be presented on the same diagram (see Figure 1.10) without ending up with a confusion of curves in the vicinity of $y = x$ and $y = x + 1$.

The lift $\bar{f}$ of $f: S^1 \to S^1$ not only provides a means of conveniently finding fixed and periodic points, it can also allow us to determine their stability. If (1.2.13a) is satisfied at x^*, then the orbits of points near to x^* under $\bar{f}$ can be obtained by moving between $y = \bar{f}(x)$ and $y = x$ as in Figure 1.11. The fixed point x^* is stable (unstable) if

$$|D\bar{f}(x^*)| < 1 \qquad (>1), \tag{1.2.21}$$

(see any first course in Numerical Analysis). The stability of $\theta^* = \pi(x^*)$ is clearly the same as that of x^*. When (1.2.13b) is satisfied, we can *either* replace $\bar{f}$ by $\bar{f} - 1$, so that x^* is then represented by an intersection with $y = x$, and proceed as above *or* construct paths for the orbits of $\bar{f}$ by using $y = \bar{f}(x)$ and $y = x + 1$. The stability of the fixed point is still given by (1.2.21).

1.3 Flows and differential equations

The iteration problem (1.1.1) for a diffeomorphism $\mathbf{f}: M \to M$ given different $\mathbf{x}_0 \in M$ is equivalent to the study of the set of functions $\{\mathbf{f}^m | m \in \mathbb{Z}\}$. This set has the property

Figure 1.9 Examples illustrating why (1.2.18) can only be satisfied if $k = 0$ or 1 for the case when f is an orientation-preserving homeomorphism. As $\bar{f}_3$ shows, if $k > 1$ then f fails to be injective. Notice (1.2.17) has a countable infinity of solutions for each solution to (1.2.18).

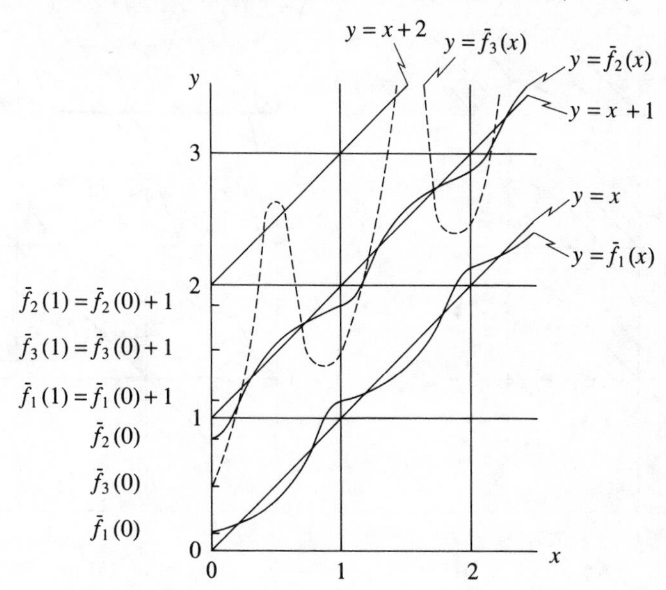

that:

$$\mathbf{f}^0 = \mathbf{id}_M \qquad \text{and} \qquad \mathbf{f}^i \cdot \mathbf{f}^j = \mathbf{f}^{i+j}, \tag{1.3.1}$$

for each $i, j \in \mathbb{Z}$. It is said to be an *action* of the group $\mathbb{Z}$ on M or, more precisely, the $\mathbb{Z}$-action generated by $\mathbf{f}$ (see Chillingworth, 1976). In this section we consider the action of the group $\mathbb{R}$ on M; such $\mathbb{R}$-actions are called *flows* on M.

Definition 1.3.1 *A flow on M is a continuously differentiable function* $\boldsymbol{\varphi}: \mathbb{R} \times M \to M$ *such that, for each* $t \in \mathbb{R}$, *the restriction* $\boldsymbol{\varphi}(t, \cdot) = \boldsymbol{\varphi}_t(\cdot)$ *satisfies*

(a) $$\boldsymbol{\varphi}_0 = \mathbf{id}_M; \tag{1.3.2a}$$

(b) $$\boldsymbol{\varphi}_t \cdot \boldsymbol{\varphi}_s = \boldsymbol{\varphi}_{t+s}, \qquad t, s \in \mathbb{R}. \tag{1.3.2b}$$

Observe that (1.3.2a and b) imply that $(\boldsymbol{\varphi}_t)^{-1}$ exists and is given by $\boldsymbol{\varphi}_{-t}$. Since $\boldsymbol{\varphi} \in C^1$, it follows (see Exercise 1.3.1) that $\boldsymbol{\varphi}_t: M \to M$ is a diffeomorphism for each $t \in \mathbb{R}$.

Let us pursue the analogy with diffeomorphisms a little further. We define the *orbit* or *trajectory* of $\boldsymbol{\varphi}$ through $\mathbf{x}$ to be $\{\boldsymbol{\varphi}_t(\mathbf{x}) | t \in \mathbb{R}\}$ oriented in the sense of

Figure 1.10 Plots of $\bar{f}^q(x)$ vs x for (*a*) $\bar{f}(x) = -x^2 + 2x + \frac{1}{2}$; (*b*) $\bar{f}(x) = x + \frac{1}{2} + \frac{3}{20} \sin 2\pi x$. Observe that case (a) corresponds to a homeomorphism with a 3-cycle but no fixed points or 2-cycles. On the other hand, case (*b*) is the lift of a diffeomorphism with 2-, 4- and 6-cycles but no 1-, 3- or 5-cycles.

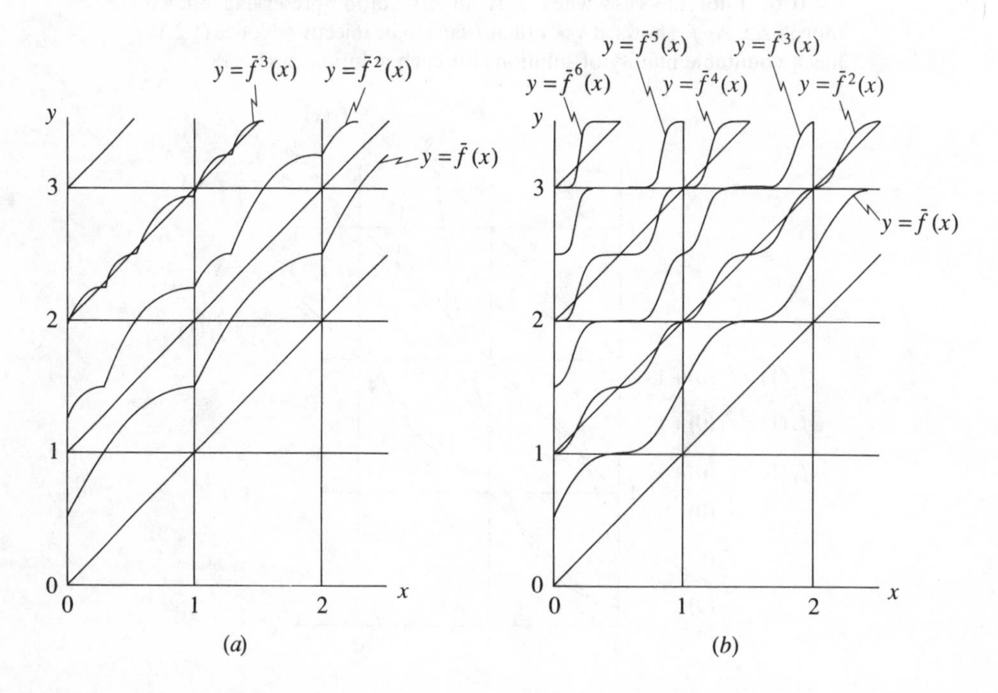

increasing t. It can be shown (see Exercise 1.3.2) that there is one and only one trajectory of $\boldsymbol{\varphi}$ passing through each point $\mathbf{x} \in M$. If $\boldsymbol{\varphi}_t(\mathbf{x}^*) = \mathbf{x}^*$ for all $t \in \mathbb{R}$ then $\mathbf{x}^*$ is said to be a *fixed point* of the flow. Fixed points of flows can be stable, asymptotically stable, neutrally stable or unstable in the sense of Liapunov. Precise definitions are obtained by the transcription $\mathbf{f}^m \mapsto \boldsymbol{\varphi}_t$ and $m \in \mathbb{Z} \mapsto t \in \mathbb{R}$ in Definition 1.2.3 and the comments following it.

The orbit of a fixed point is just the point itself. If $\mathbf{x}$ is not a fixed point it is said to be *ordinary* or *regular*. The trajectory through an ordinary point gives rise to an oriented curve on M and $\boldsymbol{\varphi}$ has *periodic points* if this curve is closed.

Definition 1.3.2 *A* closed orbit *of a flow is a trajectory, γ, which is not a fixed point but is such that $\boldsymbol{\varphi}_\tau(\mathbf{x}) = \mathbf{x}$ for some $\mathbf{x} \in \gamma$ and $\tau \neq 0$.*

Clearly, if $\boldsymbol{\varphi}_\tau(\mathbf{x}) = \mathbf{x}$ the orbit returns to $\mathbf{x}$ after time τ. If T is the least, positive time for which this occurs, $\mathbf{x}$ is a *periodic point* with period T. It is easily shown (see Exercise 1.3.3) that if a closed orbit has one point with period T, then every point of γ is periodic with period T. Thus, T is also called the period of γ.

The set of all trajectories of a flow is called its *phase portrait*. Since each trajectory corresponds geometrically to an oriented curve or point on M, a valuable pictorial representation of the flow is obtained by sketching or plotting typical trajectories. Some examples are shown in Figure 1.12. Notice that the caption to this figure does not specify $\boldsymbol{\varphi}_t$, instead a differential equation is given. How are flows related

Figure 1.11 Graphical illustration of the iteration $x_{n+1} = \bar{f}(x_n)$ showing the stability of $x_1^*, x_1^* + 1, \ldots$. Note that $|\mathrm{D}f(x^*)| < 1$ for all these points. The remaining fixed points, $x_2^*, x_2^* + 1, \ldots$, satisfy $|\mathrm{D}f(x^*)| > 1$ and are unstable. Observe that the graphical representation of the iteration can still give the stability of a fixed point x^* even when $|\mathrm{D}f(x^*)| = 1$.

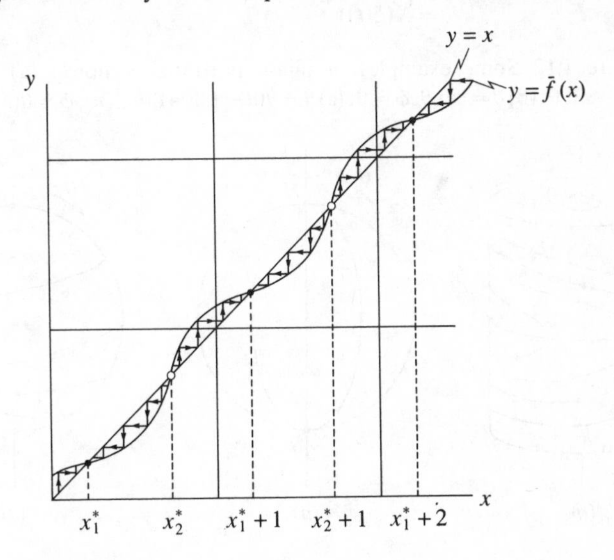

to differential equations? We define the *velocity* or *vector field*, $\mathbf{X}$, of a flow $\boldsymbol{\varphi}$ by

$$\mathbf{X}(\mathbf{x}) = \frac{\mathrm{d}\boldsymbol{\varphi}_t}{\mathrm{d}t}(\mathbf{x})\Big|_{t=0} = \operatorname*{Lim}_{\varepsilon \to 0}\left\{\frac{\boldsymbol{\varphi}(\varepsilon, \mathbf{x}) - \boldsymbol{\varphi}(0, \mathbf{x})}{\varepsilon}\right\} \tag{1.3.3}$$

for each $\mathbf{x} \in M$. Geometrically, $\{\boldsymbol{\varphi}_t(\mathbf{x}) | t \in \mathbb{R}\}$ defines a curve on M passing through $\mathbf{x}$. The vector $\mathbf{X}(\mathbf{x})$ is directed along the tangent to this curve at $\mathbf{x}$ and has magnitude equal to the speed of description of the curve under the parametrisation by t. It is important to realise that, in contrast to vector fields defined on $\mathbb{R}^n$, $\mathbf{X}(\mathbf{x}) \notin M$. For each $\mathbf{x} \in M$, the set, $TM_{\mathbf{x}}$, of all vectors tangent to M at $\mathbf{x}$ is called the tangent space to M at $\mathbf{x}$ and $\mathbf{X}(\mathbf{x}) \in TM_{\mathbf{x}}$. Figure 1.13 illustrates $TM_{\mathbf{x}}$ for a typical point $\mathbf{x} \in S^2$. If M is an n-dimensional manifold, then $TM_{\mathbf{x}}$ is isomorphic to $\mathbb{R}^n$ for all $\mathbf{x} \in M$. Each element of $TM_{\mathbf{x}}$ corresponds to an equivalence class of curves on M having the same tangent vector at $\mathbf{x}$ (see Chillingworth, 1976, p. 164).

Proposition 1.3.1 *$\boldsymbol{\varphi}_t(\mathbf{x}_0)$ is the solution of $\dot{\mathbf{x}} = \mathbf{X}(\mathbf{x})$ which passes through $\mathbf{x}_0$ at $t = 0$.*

Proof. Let $\xi(t) = \boldsymbol{\varphi}_t(\mathbf{x}_0)$. Then

$$\begin{aligned}
\dot{\xi}(t) &= \operatorname*{Lim}_{\varepsilon \to 0}\left\{\frac{\xi(t+\varepsilon) - \xi(t)}{\varepsilon}\right\} \\
&= \operatorname*{Lim}_{\varepsilon \to 0}\left\{\frac{\boldsymbol{\varphi}_{t+\varepsilon}(\mathbf{x}_0) - \boldsymbol{\varphi}_t(\mathbf{x}_0)}{\varepsilon}\right\} \\
&= \operatorname*{Lim}_{\varepsilon \to 0}\left\{\frac{\boldsymbol{\varphi}_\varepsilon \cdot \boldsymbol{\varphi}_t(\mathbf{x}_0) - \boldsymbol{\varphi}_t(\mathbf{x}_0)}{\varepsilon}\right\} \\
&= \operatorname*{Lim}_{\varepsilon \to 0}\left\{\frac{\boldsymbol{\varphi}(\varepsilon, \xi(t)) - \boldsymbol{\varphi}(0, \xi(t))}{\varepsilon}\right\} \\
&= \mathbf{X}(\xi(t)).
\end{aligned} \tag{1.3.4}$$

Figure 1.12 Some examples of phase portraits of flows: (a) $\dot{\theta} = z$, $\dot{z} = -\sin\theta$; (b) $\dot{\theta} = \sin\theta$, $\dot{\varphi} = 0$; (c) $\dot{\theta} = \theta(\theta - (3\pi/4))(\theta - \pi)$, $\dot{\varphi} = \theta(\pi - \theta)$.

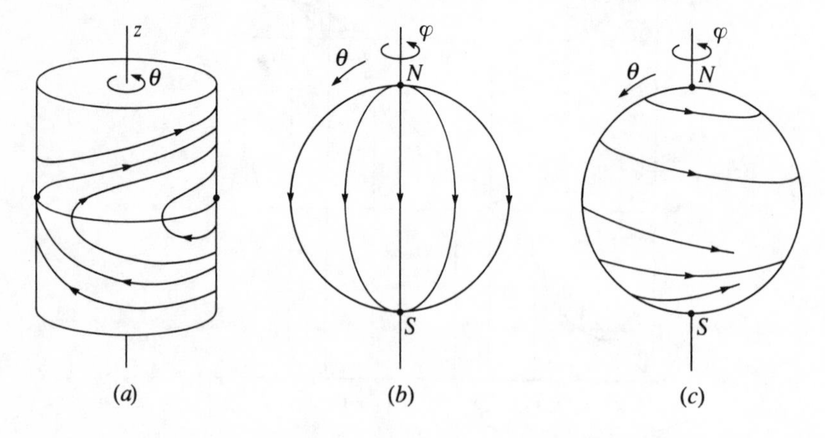

Thus, $\xi(t)$ is a solution of $\dot{\mathbf{x}} = \mathbf{X}(\mathbf{x})$ and, since $\boldsymbol{\varphi}_0 = \mathbf{id}_M$, $\xi(0) = \boldsymbol{\varphi}_0(\mathbf{x}_0) = \mathbf{x}_0$, as required. □

Notice that if $\mathbf{X}(\mathbf{x}^*) = \mathbf{0}$ then $\boldsymbol{\varphi}_t(\mathbf{x}^*) \equiv \mathbf{x}^*$ is the solution of $\dot{\mathbf{x}} = \mathbf{X}(\mathbf{x})$ passing through $\mathbf{x}^*$. Moreover, if $\boldsymbol{\varphi}_t(\mathbf{x}^*) = \mathbf{x}^*$ for all t then (1.3.3) implies $\mathbf{X}(\mathbf{x}^*) = \mathbf{0}$. We conclude, therefore, that $\mathbf{x}^*$ is a fixed point of $\boldsymbol{\varphi}_t$ if and only if $\mathbf{X}(\mathbf{x}^*) = \mathbf{0}$. Such points are referred to as *singular points* of the vector field $\mathbf{X}$.

Proposition 1.3.1 means that every flow on M corresponds to an autonomous differential equation. Unfortunately the converse is not true. This is because there are autonomous differential equations with solutions that cannot be extended indefinitely in t. For example, $\dot{x} = x^2$ has general solution

$$x(t) = \begin{cases} (C-t)^{-1}, & t \in (-\infty, C); \\ 0, & t \in \mathbb{R}; \\ (C'-t)^{-1}, & t \in (C', \infty), \end{cases} \tag{1.3.5}$$

$C, C' \in \mathbb{R}$. Only the trivial solution has domain $\mathbb{R}$. In such cases, *local flows* can still be defined. For example, the function

$$\varphi_t(x_0) = \frac{x_0}{1 - x_0 t}, \tag{1.3.6}$$

provides a local flow for $\dot{x} = x^2$. When $x_0 > 0$, (1.3.5) implies $t \in (-\infty, x_0^{-1})$ in (1.3.6). It is easy to verify that φ_t satisfies (1.3.2) provided t, s and $t+s$ *all* belong to $(-\infty, x_0^{-1})$. The same function φ_t can be used when $x_0 < 0$ provided t is restricted

Figure 1.13 Illustration of the tangent space, $TM_{\mathbf{x}}$, to the sphere S^2 at $\mathbf{x}$. Let the circles a and b define the latitude and longitude of $\mathbf{x}$. If $\mathbf{a}$ and $\mathbf{b}$ are tangent to a and b, respectively, at $\mathbf{x}$ then $TM_{\mathbf{x}} = Sp\{\mathbf{a}, \mathbf{b}\}$. Observe that b, c and d are all curves on the sphere having tangent vector $\mathbf{b}$.

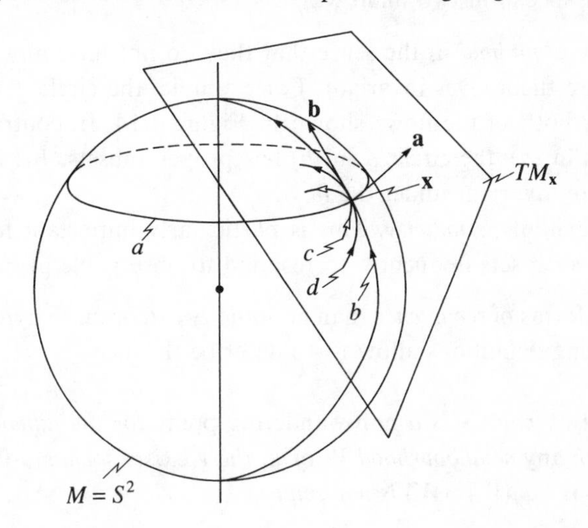

to the interval (x_0^{-1}, ∞). Equation (1.3.6) obviously provides the trivial solution when $x_0 = 0$. This local flow is sufficient to characterise the solutions of $\dot{x} = x^2$ in a neighbourhood of the origin of the t, x-plane. For example, for $|x_0| < \varepsilon$, (1.3.6) certainly gives the solutions to $\dot{x} = x^2$ for $t \in (-\varepsilon^{-1}, \varepsilon^{-1})$. Flows of this type are frequently used implicitly when local properties are discussed (e.g. the saddle-node singularity in Example 2.7.4).

With the above proviso in mind, differential equations, vector fields and flows merely provide alternative ways of presenting the same dynamics. These alternatives have arisen for historical reasons; applications frequently lead to differential equations; local analysis is usually presented in terms of vector fields; and global analysis uses the language of flows. We hope the reader will become familiar with all three possibilities.

1.4 Invariant sets

Sometimes the orbit of a point under $\mathbf{f}$ or φ remains within a particular region of phase space for all $m \in \mathbb{Z}$ or $t \in \mathbb{R}$. A set $\Lambda \subseteq M$ is said to be *invariant* under the diffeomorphism $\mathbf{f}$ (or flow φ) if $\mathbf{f}^m(\mathbf{x}) \in \Lambda$ ($\varphi_t(\mathbf{x}) \in \Lambda$) for each $\mathbf{x} \in \Lambda$ and all $m \in \mathbb{Z}$ ($t \in \mathbb{R}$). We write

$$\mathbf{f}^m(\Lambda) \subseteq \Lambda \qquad \text{for all } m \in \mathbb{Z} \tag{1.4.1}$$

or

$$\varphi_t(\Lambda) \subseteq \Lambda \qquad \text{for all } t \in \mathbb{R}. \tag{1.4.2}$$

Invariant sets are said to be *positively* (*negatively*) *invariant* if the orbits of their elements remain within them for $m \in \mathbb{Z}^+$ ($\mathbb{Z}^-$) or $t \geqslant 0$ ($t \leqslant 0$).

Clearly, the orbit of any point is an example of an invariant set. It follows therefore that fixed points, cycles and closed orbits are all invariant sets. However, they are rather special in two main ways.

(i) They are *minimal* in the sense that they do not have any proper subsets that are themselves invariant. For example, the circle $\mathscr{C}$ is an invariant set for both of the flows shown in Figure 1.14. In contrast to the flow shown in (*a*), the circle $\mathscr{C}$ in (*b*) has proper subsets, P_1, P_2, Γ_1 and Γ_2, that are invariant under the flow.

(ii) They exhibit *periodicity*. This is particularly important for applications where such sets frequently correspond to observable phenomena.

More subtle forms of *recurrence* than periodicity can occur in dynamical systems and the following definitions allow us to describe them.

Definition 1.4.1 *A point* $\mathbf{x}$ *is a* non-wandering point *for the diffeomorphism* $\mathbf{f}$ (or flow φ) *if, given* any *neighbourhood* W *of* $\mathbf{x}$, *there exists some* $m > 0$ ($t > t_0 > 0$) *for which* $f^m(W) \cap W$ ($\varphi_t(W) \cap W$) *is not empty.*

The set of non-wandering points for $\mathbf{f}$ (φ) is called the *non-wandering set*, $\Omega(\mathbf{f})$ ($\Omega(\varphi)$). It is easy to see that fixed points and periodic orbits lie in Ω (see Exercises 1.4.2 and 1.4.3), however, points exhibiting milder forms of recurrence are also present. For example, consider an irrational rotation of the circle, S^1. No point of the circle is periodic, but the orbit of any point $\mathbf{x}$ ultimately approaches $\mathbf{x}$ arbitrarily closely. Thus, every point of S^1 is a non-wandering point and $\Omega = S^1$.

The structure of Ω will be examined more closely in § 3.6, but we can recognise some important subsets of it by formalising the idea that fixed points and closed orbits frequently attract or repel the trajectories of phase points not contained in them.

Definition 1.4.2 *A point* $\mathbf{y} \in M$ *is said to be an* $\begin{Bmatrix} \alpha\text{-} \\ \omega\text{-} \end{Bmatrix}$ limit point *of the trajectory of* $\mathbf{f}$ (φ) *through* $\mathbf{x}$ *if there is a sequence* m_i (t_i) $\to \begin{Bmatrix} -\infty \\ +\infty \end{Bmatrix}$ *such that*

$$\lim_{i\to\infty} \mathbf{f}^{m_i}(\mathbf{x}) = \mathbf{y} \quad (\lim_{i\to\infty} \varphi_{t_i}(\mathbf{x}) = \mathbf{y}).$$

The set of all $\begin{Bmatrix} \alpha\text{-} \\ \omega\text{-} \end{Bmatrix}$ limit points of $\mathbf{x}$ is known as the $\begin{Bmatrix} \alpha\text{-} \\ \omega\text{-} \end{Bmatrix}$ limit set of $\mathbf{x}$, denoted by $\begin{Bmatrix} L_\alpha(\mathbf{x}) \\ L_\omega(\mathbf{x}) \end{Bmatrix}$. These sets are invariant under $\mathbf{f}$ (φ). Let $\mathbf{z} = \mathbf{f}^m(\mathbf{y})$, $m \in \mathbb{Z}$ ($\mathbf{z} = \varphi_t(\mathbf{y})$, $t \in \mathbb{R}$), where $\mathbf{y}$ satisfies Definition 1.4.2. Then $\lim_{i\to\infty} \mathbf{f}^{m_i+m}(\mathbf{x}) = \mathbf{z}$ ($\lim_{i\to\infty} \varphi_{t_i+t}(\mathbf{x}) = \mathbf{z}$) so that $\mathbf{z}$ and $\mathbf{y}$ belong to the same limit set of $\mathbf{x}$.

Notice that α- and ω-limits sets are subsets of Ω for any $\mathbf{x}$. Recall if $\mathbf{y} \notin \Omega$ then

Figure 1.14 The circle $\mathscr{C}$ is an invariant set for both of the flows shown. However, in (*a*) $\mathscr{C}$ has no proper subsets that are themselves invariant; while in (*b*) $\mathscr{C}$ is the disjoint union of the invariant sets P_1, P_2, Γ_1, Γ_2.

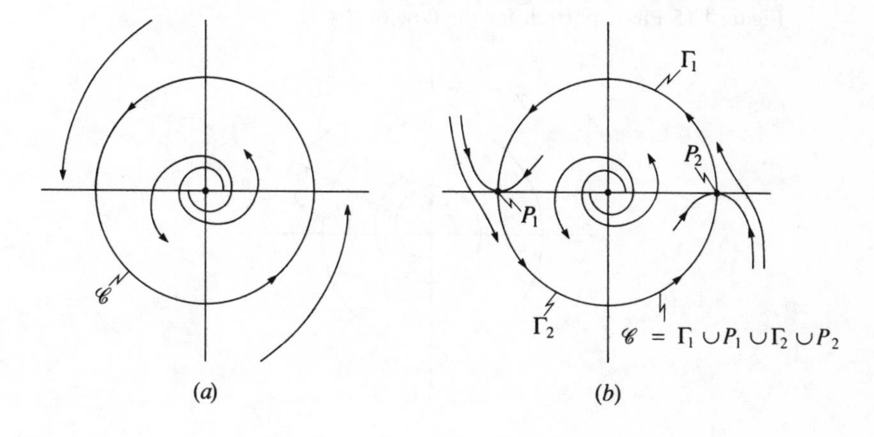

there exists a neighbourhood $V \ni \mathbf{y}$ such that $\mathbf{f}^m(V) \cap V$ is empty for all $m > 0$. However, $\mathbf{y} \in L_\omega(\mathbf{x})$ implies $\mathbf{f}^{m_i}(\mathbf{x}) \in V$ for $i \geqslant N$, say, and hence there is $\mathbf{z} = \mathbf{f}^N(\mathbf{x}) \in V$ such that $\mathbf{f}^{m_i - N}(\mathbf{z}) \in V$ for $i > N$. Thus $\mathbf{f}^m(V) \cap V$ cannot be empty for all m and $\mathbf{y}$ must lie in Ω.

Example 1.4.1 Find $L_\alpha(\mathbf{x})$ and $L_\omega(\mathbf{x})$ for *(a)* $\mathbf{x} = \mathbf{0}$; *(b)* $\mathbf{x} \neq \mathbf{0}$, when $\boldsymbol{\varphi}$ is the flow on $\mathbb{R}^2$ induced by

$$\dot{r} = r(1 - r), \qquad \dot{\theta} = 1, \tag{1.4.3}$$

where (r, θ) are plane polar coordinates.

Solution. $\boldsymbol{\varphi}$ has a unique, attracting closed orbit γ given by $r(t) \equiv 1$, with period $T = 2\pi$, and an unstable fixed point at the origin (see Figure 1.15).

(a) $\mathbf{x} = \mathbf{0}$
Note $\boldsymbol{\varphi}_t(\mathbf{0}) = \mathbf{0}$ for all t therefore

$$L_\omega(\mathbf{0}) = L_\alpha(\mathbf{0}) = \{\mathbf{0}\}. \tag{1.4.4}$$

(b) $\mathbf{x} \neq \mathbf{0}$
Let $\mathbf{y} = (\cos\theta_0, \sin\theta_0) \in \gamma$ and let t_i be the sequence of $t > 0$ at which the orbit of $\mathbf{x}$ crosses the radial line from $\mathbf{0}$ through $\mathbf{y}$. Then $\lim_{i \to \infty} \boldsymbol{\varphi}_{t_i}(\mathbf{x}) = \mathbf{y}$ and $\mathbf{y}$ is an ω-limit point of $\mathbf{x}$. This argument is valid for any $\mathbf{y} \in \gamma$ and any $\mathbf{x} \neq \mathbf{0}$. Therefore, $L_\omega(\mathbf{x}) = \gamma$ for any $\mathbf{x} \neq \mathbf{0}$.

A similar argument allows us to show that

$$L_\alpha(\mathbf{x}) = \begin{cases} \{\mathbf{0}\} & |\mathbf{x}| < 1, \\ \gamma & |\mathbf{x}| = 1. \end{cases} \tag{1.4.5}$$

Figure 1.15 Phase portrait for the flow of (1.4.3).

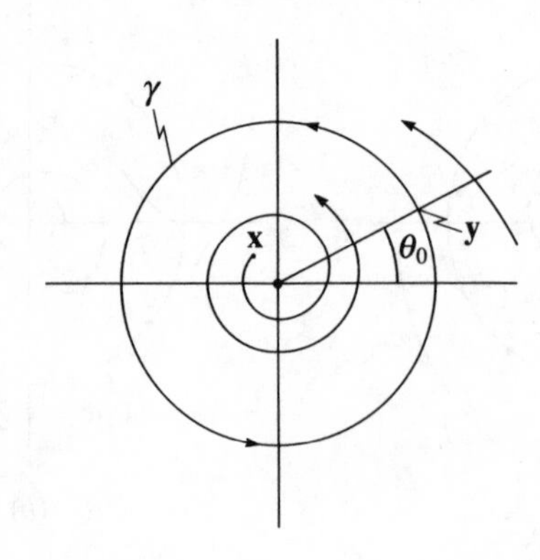

However, for $|\mathbf{x}| > 1$, the $\operatorname*{Lim}_{i \to \infty} \varphi_{t_i}(\mathbf{x})$ does not exist for any sequence t_i such that $t_i \to -\infty$ as $i \to \infty$ and therefore $L_\alpha(\mathbf{x})$ is empty. □

Example 1.4.2 Let the flow φ have the phase portrait shown in Figure 1.16. What are $L_\omega(\mathbf{x})$ and $L_\alpha(\mathbf{x})$ for $\mathbf{x} \in A, B, C$ respectively? What feature do all three ω-limit sets have in common?

Solution. Sequences $\{t_i\}_{i=0}^{\infty}$ can be constructed as in Example 1.4.1 to show that

$$\mathbf{x} \in A\colon L_\alpha(\mathbf{x}) = \{P_1\};\ L_\omega(\mathbf{x}) = \partial A$$

$$\mathbf{x} \in B\colon L_\alpha(\mathbf{x}) = \{P_2\};\ L_\omega(\mathbf{x}) = \partial B$$

$$\mathbf{x} \in C\colon L_\alpha(\mathbf{x}) = \text{empty set};\ L_\omega(\mathbf{x}) = \partial A \cup \partial B.$$

Let Γ_A and Γ_B be the trajectories of the flow which form the separatrices of the saddle point P_0. Observe that

$$\partial A = \Gamma_A \cup P_0, \qquad \partial B = \Gamma_B \cup P_0, \tag{1.4.6}$$

and it follows that all three ω-limit sets are unions of fixed points and the trajectories joining them. □

Example 1.4.2 illustrates an important theorem concerning the global properties of *planar* flows.

Theorem 1.4.1 (Poincaré–Bendixson) *A non-empty, compact limit set of a flow on the plane, which contains no fixed point, is a closed orbit.*

This theorem states that the types of limit sets illustrated in Examples 1.4.1 and 1.4.2 are the only compact ones that can occur in flows on the plane. It is one of the few theorems which gives the existence of a global feature of a phase portrait.

Figure 1.16 Phase portrait of the flow required for Example 1.4.2. The points $P_{0,1,2}$ are fixed points. The open sets A, B have boundaries ∂A, ∂B, respectively. C is the complement of the closure of $A \cup B$.

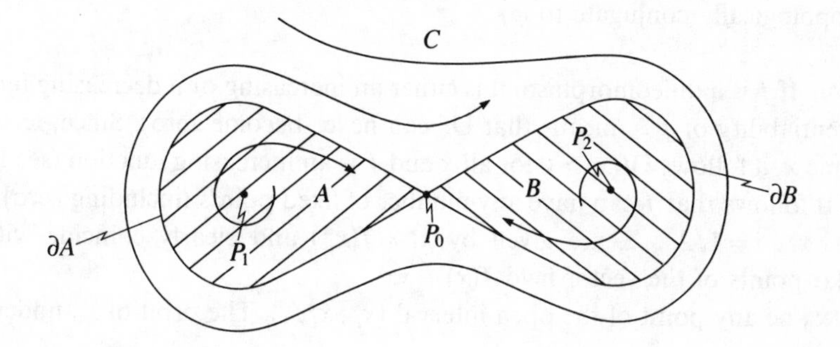

Definition 1.4.3 *A* limit cycle *is a closed orbit* γ *such that either* $\gamma \subseteq L_\omega(\mathbf{x})$ *or* $\gamma \subseteq L_\alpha(\mathbf{x})$ *for some* $\mathbf{x} \notin \gamma$.

Theorem 1.4.1 has the important corollary that a non-empty, compact set Λ which is positively or negatively invariant contains either a limit cycle or a fixed point. This result can be useful in demonstrating the existence of limit cycles (Arrowsmith & Place, 1982, pp. 147–51).

1.5 Conjugacy

We now turn to the equivalence relations which allow us to recognise when two diffeomorphisms or two flows exhibit the 'same' behaviour. These equivalence relations lie at the heart of topological or qualitative theory.

Definition 1.5.1 *Two diffeomorphisms* $\mathbf{f}, \mathbf{g}: M \to M$ *are said to be* topologically (*or* C^0-) conjugate *if there is a homeomorphism,* $\mathbf{h}: M \to M$, *such that*

$$\mathbf{h} \cdot \mathbf{f} = \mathbf{g} \cdot \mathbf{h}. \tag{1.5.1}$$

Topological conjugacy of two flows $\boldsymbol{\varphi}_t, \boldsymbol{\psi}_t: M \to M$ is defined in the same way with (1.5.1) replaced by $\mathbf{h} \cdot \boldsymbol{\varphi}_t = \boldsymbol{\psi}_t \cdot \mathbf{h}$ for all $t \in \mathbb{R}$.

Definition 1.5.1 means that $\mathbf{h}$ takes each orbit of $\mathbf{f}$ (or $\boldsymbol{\varphi}_t$) onto an orbit of $\mathbf{g}$ ($\boldsymbol{\psi}_t$) preserving the parameter m (t), i.e.

$$\mathbf{f}^m(\mathbf{x}) \xrightarrow{\mathbf{h}} \mathbf{g}^m(\mathbf{h}(\mathbf{x})), \qquad \text{for each } m \in \mathbb{Z}, \tag{1.5.2}$$

or

$$\boldsymbol{\varphi}_t(\mathbf{x}) \xrightarrow{\mathbf{h}} \boldsymbol{\psi}_t(\mathbf{h}(\mathbf{x})), \qquad \text{for each } t \in \mathbb{R}. \tag{1.5.3}$$

The significance of (1.5.2 and 1.5.3) is illustrated in Figure 1.17. Notice, by uniqueness of the trajectories of each flow, a given trajectory of $\boldsymbol{\varphi}_t$ is mapped onto one and only one trajectory of $\boldsymbol{\psi}_t$ and vice versa.

Example 1.5.1 Let $f: \mathbb{R} \to \mathbb{R}$ be a diffeomorphism with $\mathrm{D}f(x) > 0$ for some $x \in \mathbb{R}$. Given that the differential equation $\dot{x} = f(x) - x$ defines a flow $\varphi_t: \mathbb{R} \to \mathbb{R}$, show that f is topologically conjugate to φ_1.

Solution. If f is a diffeomorphism it is either an increasing or a decreasing function (differentiability of f^{-1} means that $\mathrm{D}f$ can never become zero). Since, $\mathrm{D}f(x) > 0$ for some x, it follows $\mathrm{D}f(x) > 0$ for all x and f is an increasing function (see Figure 1.18). It follows that f can have any number of fixed points (including zero). Such points, x_i^*, $i = 1, 2, \ldots$, are given by $x_i^* = f(x_i^*)$ and clearly coincide with the singular points of the vector field $f(x) - x$.

Let x_0 be any point of the open interval (x_i^*, x_{i+1}^*). The orbit of x_0 under both

Figure 1.17 Diagram illustrating conjugacy of: (*a*) diffeomorphisms; (*b*) flows. Note that (*b*) is valid for all $t \in \mathbb{R}$ and (1.5.1) implies that $\mathbf{h}(\mathbf{f}^m(\mathbf{x})) = \mathbf{g}^m(\mathbf{h}(\mathbf{x}))$ for all $m \in \mathbb{Z}$.

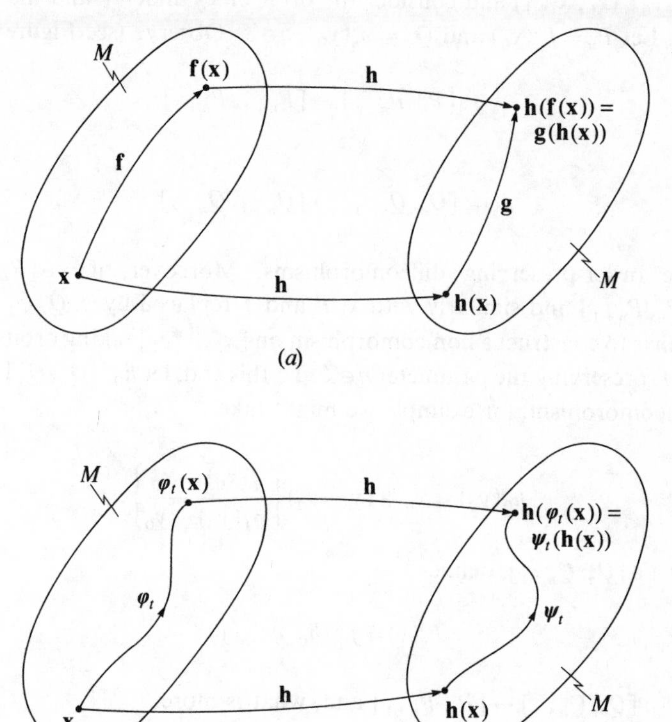

Figure 1.18 Typical graph of a diffeomorphism $f: \mathbb{R} \to \mathbb{R}$ for which $\mathrm{D}f(x) > 0$ for some $x \in \mathbb{R}$. The fixed points of f are given by the intersections of the curve $y = f(x)$ and the straight line $y = x$.

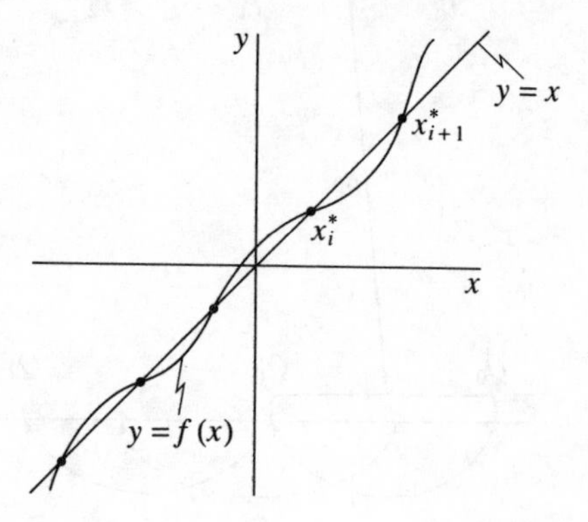

f and φ_1 is confined to this interval and has the same orientation for both maps (N.B. $\operatorname{sign}(x_{n+1}-x_n)=\operatorname{sign}(f(x_n)-x_n)=\operatorname{sign}(\dot{x})$, $n\in\mathbb{Z}$).

Let $x_0, y_0 \in (x_i^*, x_{i+1}^*)$ and consider the orbit of x_0 under f and the orbit of y_0 under φ_1. Let $P_n = f^n(x_0)$ and $Q_n = \varphi_n(y_0)$, $n\in\mathbb{Z}$. Observe (see Figure 1.19) that

$$f\colon [P_n, P_{n+1}] \to [P_{n+1}, P_{n+2}]$$

and (1.5.4)

$$\varphi_1\colon [Q_n, Q_{n+1}] \to [Q_{n+1}, Q_{n+2}],$$

$n\in\mathbb{Z}$, are order-preserving diffeomorphisms. Moreover, if $x\in[P_0, P_1]$ then $f^n(x)\in[P_n, P_{n+1}]$ and similarly with x, P and f replaced by y, Q, φ_1.

Our aim is to construct a homeomorphism on $[x_i^*, x_{i+1}^*]$ taking orbits of φ_1 onto orbits of f, preserving the parameter $n\in\mathbb{Z}$. To this end, let $h_0\colon [Q_0, Q_1]\to[P_0, P_1]$ be a homeomorphism, for example we might take

$$h_0(y_0) = x_0 + (y - y_0)\left\{\frac{f(x_0)-x_0}{\varphi_1(y_0)-y_0}\right\}. \tag{1.5.5}$$

Now, for $y\in[Q_n, Q_{n+1}]$, define

$$h_n(y) = f^n \cdot h_0 \cdot \varphi_{-n}(y). \tag{1.5.6}$$

Clearly, $h_n\colon [Q_n, Q_{n+1}] \to [P_n, P_{n+1}]$ and, what is more,

$$h_n(Q_{n+1}) = h_{n+1}(Q_{n+1}) = P_{n+1}, \qquad n\in\mathbb{Z}. \tag{1.5.7}$$

Figure 1.19 Orbits of the points x_0 and y_0 under f and φ_1, respectively. It is convenient to define $P_n = f^n(x_0)$ and $Q_n = \varphi_n(y_0)$, for $n\in\mathbb{Z}$.

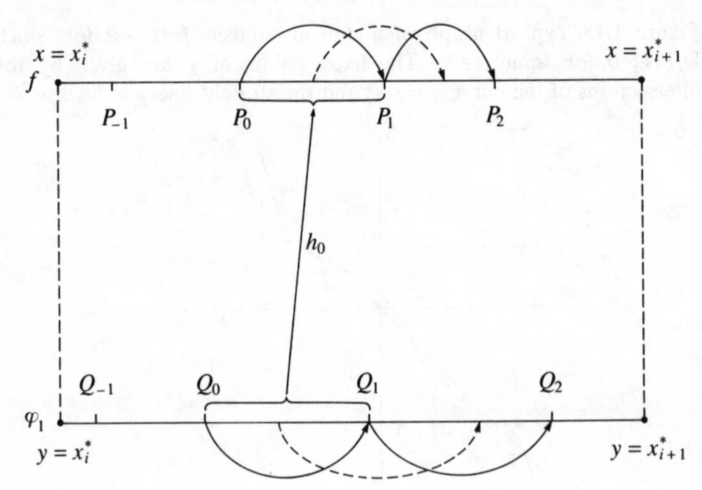

It follows that $h: [x_i^*, x_{i+1}^*] \to [x_i^*, x_{i+1}^*]$ defined by

$$h(y) = \begin{cases} x_i^* & y = x_i^* \\ h_n(y) \quad \text{for} & y \in [Q_n, Q_{n+1}], \quad n = \mathbb{Z} \\ x_{i+1}^* & y = x_{i+1}^* \end{cases} \tag{1.5.8}$$

is a homeomorphism.

Finally, it is easy to verify that h exhibits the conjugacy of f and φ_1. If $x \in [x_i^*, x_{i+1}^*]$ then $x \in [Q_n, Q_{n+1}]$ for some n and

$$\begin{aligned} h \cdot \varphi_1(x) &= h_{n+1} \cdot \varphi_1(x) \\ &= f^{n+1} \cdot h_0 \cdot \varphi_{-n-1} \cdot \varphi_1(x) \\ &= f \cdot h_n(x) \\ &= f \cdot h(x) \end{aligned} \tag{1.5.9}$$

as required. □

It is important to note that Example 1.5.1 highlights a *special* property of some increasing diffeomorphisms of the line. Not all diffeomorphisms on $\mathbb{R}$ are topologically conjugate to the time-one map of some flow. For example, if f is a decreasing diffeomorphism on $\mathbb{R}$, the orbits of f oscillate about its fixed point (see Figure 1.20). Such behaviour is impossible for the time-one map of any flow on $\mathbb{R}$.

If $\mathbf{h}$, in Definition 1.5.1, is a C^k-diffeomorphism with $k \geqslant 1$, rather than a homeomorphism then $\mathbf{f}$ and $\mathbf{g}$ (or $\boldsymbol{\varphi}_t$ and $\boldsymbol{\psi}_t$) are said to be *C^k-conjugate*. This kind of conjugacy is far more restrictive than topological conjugacy. For example, the real valued functions $f(x) = 2x$ and $g(x) = 8x$, $x \in \mathbb{R}$, are topologically conjugate but they are not C^k-conjugate for any $k \geqslant 1$ (see Exercise 1.5.1). C^k-conjugacy of

Figure 1.20 Graphical derivation of a typical orbit of a decreasing diffeomorphism $f: \mathbb{R} \to \mathbb{R}$ in the neighbourhood of its fixed point. The orbit clearly oscillates from one side of the fixed point to the other.

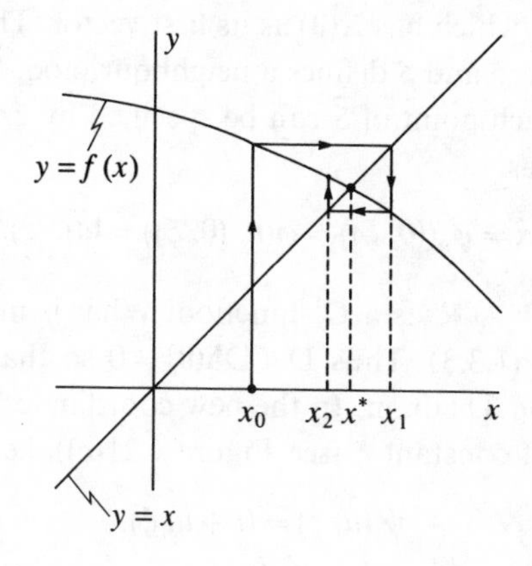

φ_t and ψ_t corresponds to there being a k-times differentiable change of coordinates, $\mathbf{h}$, which transforms the differential equation, $\dot{\mathbf{x}} = \mathbf{X}(\mathbf{x})$, of φ_t into that, $\dot{\mathbf{y}} = \mathbf{Y}(\mathbf{y})$ say, of ψ_t. Recall C^k-conjugacy of φ_t and ψ_t means that there is a function $\mathbf{h} \in C^k$ such that $\mathbf{h}(\varphi_t(\mathbf{x})) = \psi_t(\mathbf{h}(\mathbf{x}))$. Differentiate this equation with respect to t and evaluate at $t = 0$, to obtain

$$\left\{\mathbf{Dh}(\varphi_t(\mathbf{x}))\frac{\mathrm{d}\varphi_t}{\mathrm{d}t}(\mathbf{x})\right\}\bigg|_{t=0} = \frac{\mathrm{d}\psi_t}{\mathrm{d}t}(\mathbf{h}(\mathbf{x}))\bigg|_{t=0} \tag{1.5.10}$$

or

$$\mathbf{Dh}(\mathbf{x})\mathbf{X}(\mathbf{x}) = \mathbf{Y}(\mathbf{h}(\mathbf{x})), \tag{1.5.11}$$

since $\varphi_0 = \mathbf{id}_M$. Now consider the change of coordinates $\mathbf{y} = \mathbf{h}(\mathbf{x})$ applied to $\dot{\mathbf{x}} = \mathbf{X}(\mathbf{x})$. With the aid of (1.5.11), we find

$$\dot{\mathbf{y}} = \mathbf{Dh}(\mathbf{x})\dot{\mathbf{x}} = \mathbf{Dh}(\mathbf{x})\mathbf{X}(\mathbf{x}) = \mathbf{Y}(\mathbf{h}(\mathbf{x})) = \mathbf{Y}(\mathbf{y}) \tag{1.5.12}$$

as required. Thus, when $\mathbf{h}$ exhibits the conjugacy of φ and ψ, the *derivative map*, $\mathbf{Dh}$, transforms the vector field $\mathbf{X}(\mathbf{x})$ into $\mathbf{Y}(\mathbf{y})$ with $\mathbf{y} = \mathbf{h}(\mathbf{x})$.

An important example of C^1-conjugacy of flows occurs in the qualitative study of local phase portraits in the neighbourhood of an ordinary point. Let $\mathbf{x}_0$ be an ordinary point of the flow $\varphi\colon \mathbb{R} \times \mathbb{R}^n \to \mathbb{R}^n$ of the vector field $\mathbf{X}\colon \mathbb{R}^n \to \mathbb{R}^n$.

Definition 1.5.2 *A* local (cross) section at $\mathbf{x}_0$ *is an open set, S, containing $\mathbf{x}_0$, in a hyperplane $H \subseteq \mathbb{R}^n$ which is transverse to $\mathbf{X}(\mathbf{x}_0)$.*

For convenience, we will assume that H has normal $\mathbf{X}(\mathbf{x}_0)$ in the following discussion. Observe, (see Figure 1.21) that there is a neighbourhood, V, of $\mathbf{x}_0$ such that any point $\mathbf{x} \in V$ can be written as $\mathbf{x} = \varphi_u(\mathbf{y})$, where $\mathbf{y} \in S$. In other words, we can use the trajectories of the flow to define new coordinates on V.

These new coordinates are best related to *local coordinates* at $\mathbf{x}_0$, therefore, let $\mathbf{x} \mapsto \mathbf{x} - \mathbf{x}_0$ so that $\mathbf{x}_0$ is at the origin of both sets of coordinates. Now suppose we choose a basis in $\mathbb{R}^n$ which has $\mathbf{X}(\mathbf{0})$ as its first vector. Then the first coordinate of every point $\mathbf{y} \in S$ is zero and S defines a neighbourhood, $\tilde{S}$, of the origin in $\mathbb{R}^{n-1}$ (see Figure 1.21(b)). Each point of $\tilde{S}$ can be specified by $\xi \in \mathbb{R}^{n-1}$ and every point $\mathbf{x}$ of V can be written as

$$\mathbf{x} = \varphi_u((0, \xi)) = \varphi(u, (0, \xi)) = \mathbf{h}(u, \xi). \tag{1.5.13}$$

By definition of φ, $\mathbf{h}\colon \mathbb{R}^n \to \mathbb{R}^n$ is a C^1-function. What is more, $\mathbf{h}|\tilde{S}$ is the identity and $\mathrm{D}_u\mathbf{h}(\mathbf{0}) = \mathbf{X}(\mathbf{0})$, by (1.3.3). Thus Det $\mathbf{Dh}(\mathbf{0}) \neq 0$ so that $\mathbf{h}^{-1}$ exists and is C^1 by the Inverse Function Theorem. In the new coordinates, the trajectories of the flow are simply lines of constant ξ (see Figure 1.21(c)), i.e.

$$\psi_t(u, \xi) = (t + u, \xi). \tag{1.5.14}$$

Figure 1.21 Various representations of the 'flow-box' containing the ordinary point $\mathbf{x}_0$: (*a*) in the original coordinates; (*b*) in local coordinates at $\mathbf{x}_0$ and (*c*) using local coordinates defined by the flow lines.

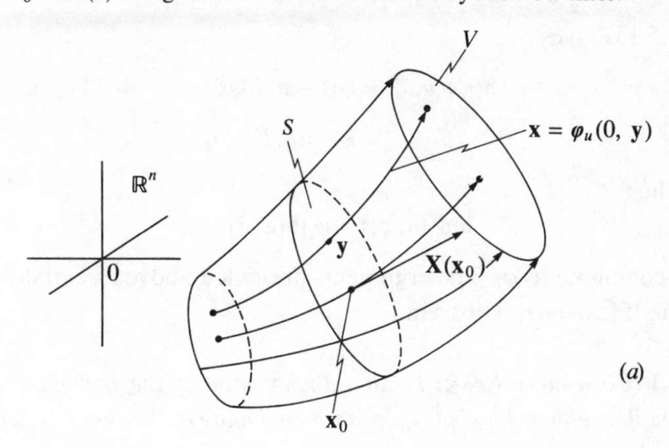

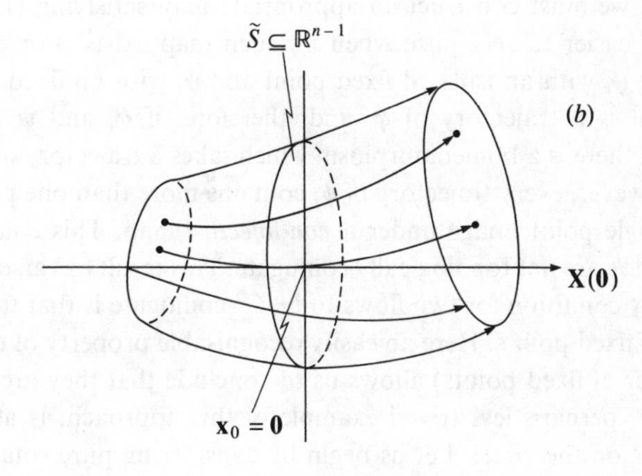

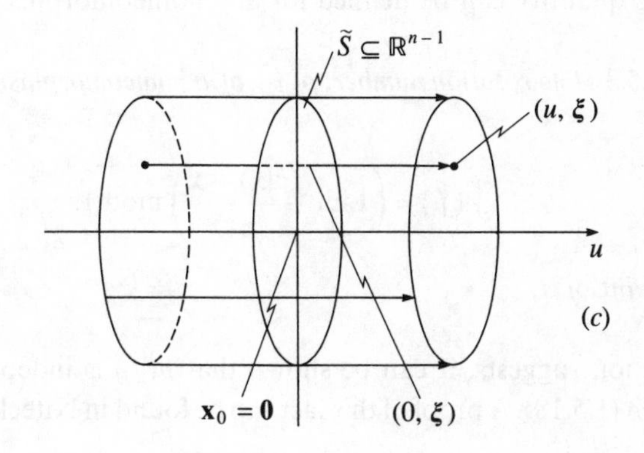

To show that ψ_t and φ_t are conjugate, observe that (1.5.14) implies

$$\mathbf{h}(\psi_t(u, \xi)) = \mathbf{h}(t+u, \xi). \tag{1.5.15}$$

However, (1.5.13) gives

$$\begin{aligned} \mathbf{h}(t+u, \xi) &= \varphi(t+u, (0, \xi)) \\ &= \varphi_t(\varphi(u, (0, \xi)) \end{aligned} \tag{1.5.16}$$

by (1.3.2). Thus

$$\mathbf{h}(\psi_t(u, \xi)) = \varphi_t(\mathbf{h}(u, \xi)) \tag{1.5.17}$$

and ψ_t is C^1-conjugate to φ_t. The arguments presented above essentially constitute a proof of the 'Flow-box' Theorem.

Theorem 1.5.1 (Flow-box) *Let $\mathbf{x}_0$ be an ordinary point of the flow $\boldsymbol{\varphi}$. Then in every sufficiently small neighbourhood of $\mathbf{x}_0$, $\boldsymbol{\varphi}$ is C^1-conjugate to the flow $\boldsymbol{\psi}(t, \mathbf{x}) = \mathbf{x} + t\mathbf{e}_1$, where $\mathbf{e}_1$ is a unit vector parallel to the x_1-axis.*

The above examples emphasise that in order to prove two flows or diffeomorphisms conjugate, we must construct an appropriate map satisfying (1.5.1). It is often a great deal easier to recognise when no such map exists. For example, consider two flows: φ_t with an isolated fixed point and ψ_t with no fixed points at all. The fixed point is a trajectory of φ_t and, therefore, if φ_t and ψ_t are topologically conjugate, there is a homeomorphism which takes a trajectory of ψ_t onto the fixed point. However, every trajectory of ψ_t contains more than one point and can only have a single point image under a *non-injective* map. This contradiction proves that φ_t and ψ_t are not topologically conjugate. This result has an obvious extension: a necessary condition for two flows to be C^0-conjugate is that they have the same number of fixed points. Here an easily recognisable property of the flows (namely, the number of fixed points) allows us to conclude that they are not conjugate.

Another, perhaps less trivial example of this approach, is afforded by diffeomorphisms on the circle. Let us begin by considering pure rotations.

A property that distinguishes rational and irrational rotations is their *rotation number*. This quantity can be defined for any homeomorphism $f: S^1 \to S^1$.

Definition 1.5.3 *The rotation number, $\rho(f)$, of a homeomorphism $f: S^1 \to S^1$ is given by*

$$\rho(f) = \left(\operatorname*{Lim}_{n \to \infty} \frac{\bar{f}^n(x) - x}{n} \right) \bmod 1, \tag{1.5.18}$$

where $\bar{f}$ is a lift of f.

As our notation suggests, it can be shown that $\rho(f)$ is independent of the point x occurring in (1.5.18). A proof of this fact can be found in Nitecki (1971, pp. 33–4).

As Figure 1.22 shows

$$\bar{R}_\gamma(x) = x + \gamma \tag{1.5.19}$$

is a lift of the pure rotation $R_\gamma(\theta) = (\theta + \gamma) \bmod 1$. Thus $\bar{R}_\gamma^n(x) = x + n\gamma$ and $\rho(R_\gamma) = \gamma$, i.e. the rotation number of R_γ is simply γ itself. A rational rotation, R_α, $\alpha = p/q \in \mathbb{Q}$, cannot be topologically conjugate to an irrational rotation, R_β, $\beta \in \mathbb{R}\backslash\mathbb{Q}$. We saw in §1.2.2 that the orbit of any point θ under R_α was periodic with period q, i.e. $R_\alpha^q(\theta) = \theta$, while $R_\beta^m(\theta) \neq \theta$ for any $\theta \in [0, 2\pi)$ or $m \in \mathbb{Z}$. Clearly, any map taking an orbit of R_β onto an orbit of R_α would fail to be injective. Therefore, the pure rotations with rational rotation number are topologically distinct from (i.e. not C^0-conjugate to) those with irrational rotation number. Now, the pure rotations are diffeomorphisms on S^1 and $\rho(f)$ is defined for any diffeomorphism $f: S^1 \to S^1$. To what extent, therefore, can the above result for pure rotations be carried over to general diffeomorphisms on S^1?

Proposition 1.5.1 *A diffeomorphism $f: S^1 \to S^1$ has periodic points if and only if its rotation number, $\rho(f)$, is rational.*

Proof. If f has a periodic point then, given a lift, $\bar{f}$, of f, there exists $x^* \in \mathbb{R}$ such that

$$\bar{f}^q(x^*) = x^* + p, \tag{1.5.20}$$

for some integers p and q. It follows that $\bar{f}^{nq}(x^*) = x^* + np$, and therefore

$$\frac{\bar{f}^{nq}(x^*) - x^*}{nq} = \frac{np}{nq} = \frac{p}{q}. \tag{1.5.21}$$

Hence $\rho(f)$ is rational.

To prove the converse, suppose f has no periodic points then,

$$\bar{f}^q(x) \neq x + p, \tag{1.5.22}$$

or

$$\bar{f}^q(x) - x \neq p, \tag{1.5.23}$$

for any integers p, q and any $x \in \mathbb{R}$. Since $g_q(x) = \bar{f}^q(x) - x$ satisfies $g_q(x+1) = g_q(x)$,

Figure 1.22 Commutative diagram illustrating the connection between the pure rotation R_γ and its lift $\bar{R}_\gamma$.

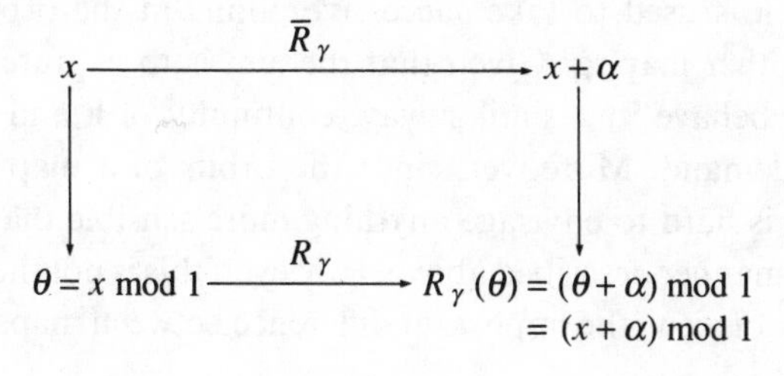

for each x, (1.5.23) means that there exists $\varepsilon > 0$ such that either

$$g_q(x) < p - \varepsilon, \qquad \text{for all } x; \tag{1.5.24}$$

or

$$g_q(x) > p + \varepsilon, \qquad \text{for all } x. \tag{1.5.25}$$

Suppose (1.5.24) holds, then $\bar{f}^q(x) < x + p - \varepsilon$, for all x, and therefore

$$\begin{aligned} \bar{f}^{nq}(x) &= \bar{f}^q(\bar{f}^{(n-1)q}(x)) < \bar{f}^{(n-1)q}(x) + p - \varepsilon \\ &= \bar{f}^q(\bar{f}^{(n-2)q}(x)) + p - \varepsilon < \cdots < x + n(p - \varepsilon). \end{aligned} \tag{1.5.26}$$

Similarly, when (1.5.25) is valid

$$\bar{f}^{nq}(x) > x + n(p + \varepsilon). \tag{1.5.27}$$

Thus, $\lim_{n\to\infty}[\bar{f}^{nq}(x) - x]/nq$ is either greater than $(p+\varepsilon)/q$ or less than $(p-\varepsilon)/q$, for any integers p and q, and so $\rho(f) \neq (p/q) \bmod 1$. □

Typically, circle diffeomorphisms with rational rotation number, $p/q \in \mathbb{Q}$, have an *even* number of period-q cycles. A sketch of $\bar{f}^q(x)$ (see Figure 1.23) not only reveals why the number of cycles is even, but it also shows that the stable and unstable points alternate around the circle.

The following result shows that circle diffeomorphisms with irrational rotation number can behave like irrational rotations.

Theorem 1.5.2 (Denjoy) *If an orientation-preserving diffeomorphism $f: S^1 \to S^1$ is of class C^2 and $\rho(f) = \beta \in \mathbb{R}\backslash\mathbb{Q}$, then it is topologically conjugate to the pure rotation R_β.*

For a proof of Denjoy's Theorem the interested reader should consult Arnold (1983, pp. 105–6) or Nitecki (1971, pp. 45–9). This important result means that every orbit of f is dense in the circle provided $f \in C^2$ and $\rho(f)$ is irrational. If $f \notin C^2$, then more complicated phenomena, such as invariant Cantor sets, can occur (see §6.4.1 and Nitecki, 1971).

1.6 Equivalence of flows

Topological conjugacy is arguably the natural equivalence relation for maps. A homeomorphism **h** is used to take successive points in the orbit of one map, **f**, onto those of another map, **g**. Given that the aim is to capture the fact that the orbits of **f** and **g** behave in a similar way, continuity of **h** and its inverse is the least we should demand. Moreover, since the orbits of a map are sequences of discrete points, it is hard to envisage anything more sensible than mapping orbits onto orbits in the manner described above. However, this is not the case for flows.

From this point of view, the important difference between maps and flows is that the orbits of the latter are parametrised by a *continuous* variable t. This allows us some additional freedom in the mapping of orbits onto orbits.

Definition 1.6.1 *Two flows, $\boldsymbol{\varphi}_t$ and $\boldsymbol{\psi}_t$, are said to be* topologically (or C^0) equivalent *if there is a homeomorphism, $\mathbf{h}$, taking orbits of $\boldsymbol{\varphi}_t$ onto those of $\boldsymbol{\psi}_t$, preserving their orientation.*

Since equivalence only demands that *orientation* be preserved, we allow $\mathbf{h}(\boldsymbol{\varphi}_t(\mathbf{x})) = \boldsymbol{\psi}_{\tau_{\mathbf{y}}(t)}(\mathbf{y})$, with $\mathbf{y} = \mathbf{h}(\mathbf{x})$, where $\tau_{\mathbf{y}}$ is an increasing function of t for every $\mathbf{y}$ (see Figure 1.24). This relaxation of the requirement that the parameter t be preserved, provides more satisfactory equivalence classes for flows. For example, the planar differential

Figure 1.23 (*a*) Sketch of $\bar{f}^q(x)$. Observe that, since $\bar{f}^q(1) = \bar{f}^q(0) + 1$, if a fixed point, x_0^*, occurs then there must be at least one further fixed point x_1^*. Moreover, if x_0^* is stable then x_1^* must be unstable. (*b*) Example of $\bar{f}^3(x)$ for a circle diffeomorphism with a stable 3-cycle. Note that an unstable 3-cycle must also occur. (*c*) Illustration of periodic points of f on the circle for the lift shown in (*b*).

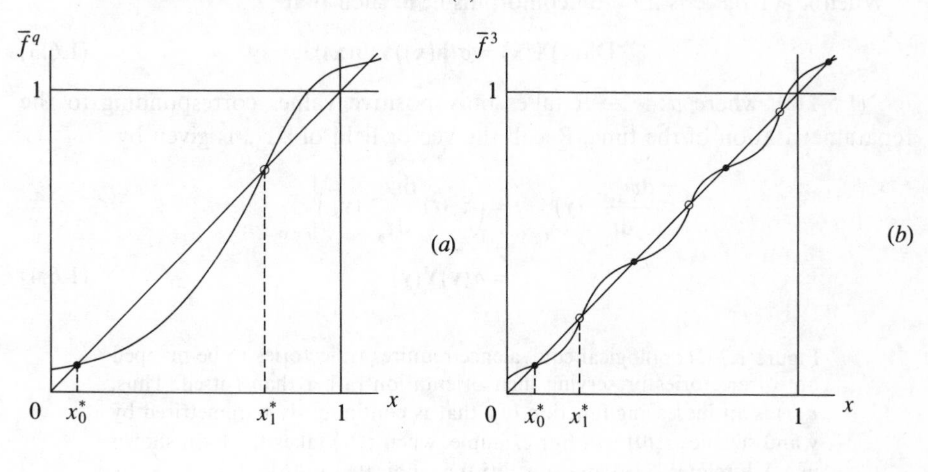

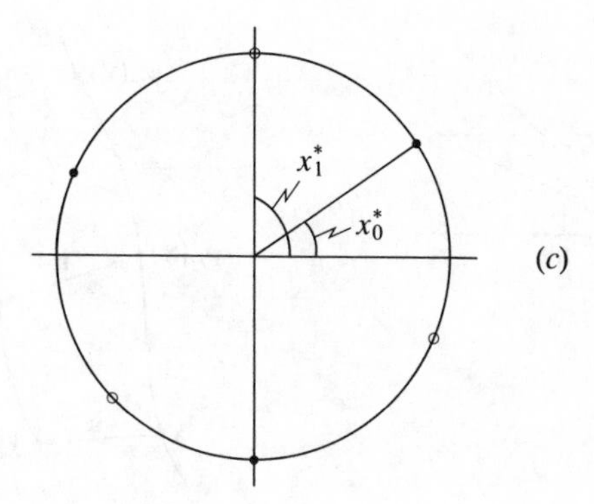

equations

$$\dot{r} = \tfrac{1}{2}r(1-r), \qquad \dot{\theta} = 1, \tag{1.6.1}$$

$$\dot{r} = r(1-r), \qquad \dot{\theta} = 2, \tag{1.6.2}$$

where (r, θ) are polar coordinates, have similar phase portraits. Both have an attractive closed orbit γ with $r(t) \equiv 1$ and an unstable focus at the origin. However, the closed orbit has period-2π in (1.6.1) and period-π in (1.6.2). Thus, if $\mathbf{h}: \gamma \to \gamma$ preserves the parameter t, it must fail to be a bijection. Thus (1.6.1) and (1.6.2) *are not* topologically *conjugate*, but they *are* topologically *equivalent*. Observe that the time rescaling $t \mapsto 2t$ transforms (1.6.2) into (1.6.1).

If Definition 1.6.1 is satisfied with $\mathbf{h} \in C^k$, $k \geqslant 1$, then the stronger relationship between φ and ψ can be emphasised by saying that they are *C^k-equivalent*. If two flows φ and ψ are C^k-equivalent ($k \geqslant 0$) then their vectors fields $\mathbf{X}(\mathbf{x})$ and $\mathbf{Y}(\mathbf{y})$ are also said to be C^k-equivalent. This terminology is frequently used because flows are often described implicitly in terms of their vector fields. For example, in applications one is often provided with a model differential equation but no explicit form for its solutions.

When $k \geqslant 1$ there is a C^k-diffeomorphism, $\mathbf{h}$, such that

$$\mathbf{Dh}(\mathbf{x})\mathbf{X}(\mathbf{x}) = \sigma(\mathbf{h}(\mathbf{x}))\mathbf{Y}(\mathbf{h}(\mathbf{x})) \tag{1.6.3}$$

(cf. (1.5.11)), where $\sigma: \mathbb{R}^n \to \mathbb{R}$ takes only positive values corresponding to the reparametrisation of the time. Recall the vector field of $\psi_{\tau_{\mathbf{y}}(t)}$ is given by

$$\begin{aligned}\left.\frac{\mathrm{d}\psi_{\tau_{\mathbf{y}}(t)}}{\mathrm{d}t}(\mathbf{y})\right|_{t=0} &= \left.\left(\dot{\tau}_{\mathbf{y}}(t)\,\frac{\mathrm{d}\psi_{\tau_{\mathbf{y}}}}{\mathrm{d}\tau_{\mathbf{y}}}(\mathbf{y})\right)\right|_{t=0} \\ &= \sigma(\mathbf{y})\mathbf{Y}(\mathbf{y})\end{aligned} \tag{1.6.4}$$

Figure 1.24 Topological equivalence requires trajectories to be mapped onto trajectories preserving their orientation rather than t itself. Thus, $\tau_{\mathbf{y}}(t)$ is an increasing function of t that is continuously parametrised by $\mathbf{y}$ and satisfies $\tau_{\mathbf{y}}(0) = 0$. For example, when $\tau_{\mathbf{y}}(t)$ takes the form shown in (a), $\mathbf{h}$ relates $\varphi_t(\mathbf{x})$ and $\psi_{\tau_{\mathbf{y}}(t)}(\mathbf{h}(\tilde{\mathbf{x}}))$ as indicated in (b).

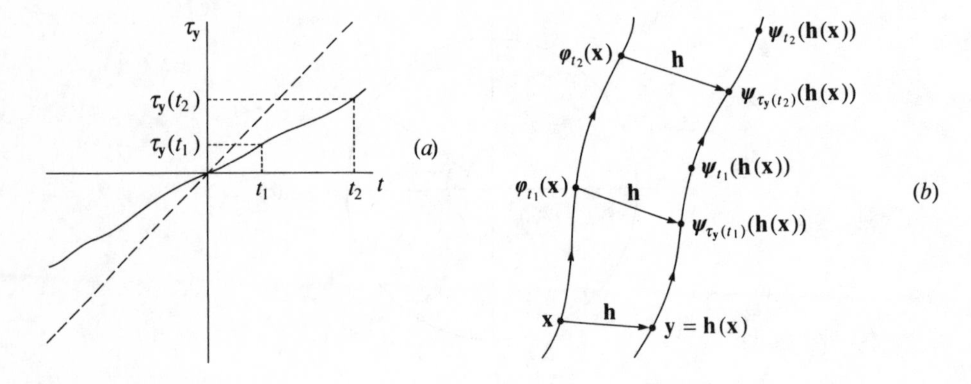

where $\sigma(\mathbf{y}) = \dot{\tau}_{\mathbf{y}}(0)$ is a positive scale factor altering the magnitude but not the direction of $\mathbf{Y}(\mathbf{y})$.

Example 1.6.1 Show that the vector fields $\mathbf{J}\mathbf{x}$ and $\mathbf{J}_0\mathbf{x}$, with

$$\mathbf{J} = \begin{pmatrix} \alpha & -\beta \\ \beta & \alpha \end{pmatrix} \quad \text{and} \quad \mathbf{J}_0 = \begin{pmatrix} 1 & -1 \\ 1 & 1 \end{pmatrix}, \tag{1.6.5}$$

where $\alpha, \beta > 0$, are topologically equivalent.

Solution. The differential equations $\dot{\mathbf{x}} = \mathbf{J}\mathbf{x}$ and $\dot{\mathbf{x}} = \mathbf{J}_0\mathbf{x}$ are easily solved using plane polar coordinates. We find $\dot{\mathbf{x}} = \mathbf{J}\mathbf{x}$ gives $\dot{r} = \alpha r$, $\dot{\theta} = \beta$ with solutions

$$r(t) = r_0 \exp(\alpha t), \qquad \theta = \beta t + \theta_0. \tag{1.6.6}$$

The equation $\dot{\mathbf{x}} = \mathbf{J}_0\mathbf{x}$ becomes $\dot{R} = R$, $\dot{\Theta} = 1$ and its solutions are

$$R(t) = R_0 \exp(t), \qquad \Theta = t + \Theta_0. \tag{1.6.7}$$

If we let $t \mapsto \beta t$ in (1.6.6), we obtain

$$r(t) = r_0 \exp(\alpha t/\beta), \qquad \theta = t + \theta_0. \tag{1.6.8}$$

Since $\beta > 0$, the flows defined by (1.6.6) and (1.6.8) are topologically equivalent with $\mathbf{h}$ equal to the identity. In other words, they have identical trajectories and differ only in the speed at which they are described.

Elimination of t from (1.6.7) and (1.6.8) gives

$$\frac{R}{R_0} = \left(\frac{r}{r_0}\right)^{\beta/\alpha}, \qquad \Theta = (\theta - \theta_0) + \Theta_0. \tag{1.6.9}$$

Equation (1.6.9) defines a map taking the trajectory of (1.6.8) through (r_0, θ_0) onto the trajectory of (1.6.7) through (R_0, Θ_0) (see Figure 1.25). For $r, r_0 > 0$, this map is 1:1, continuous and preserves orientation (indeed it preserves t itself);

Figure 1.25 Illustration of the effect of the map (1.6.9) on the orbit of (1.6.8) through (r_0, θ_0). The result is the orbit of (1.6.7) passing through (R_0, Θ_0).

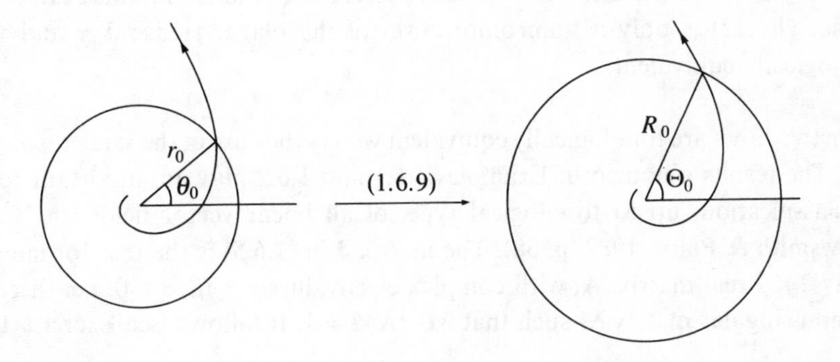

however, it involves four parameters. In fact, (1.6.9) represents a family of maps of the plane onto itself. We require a single homeomorphism taking each trajectory of (1.6.8) onto an orbit of (1.6.7) and, therefore, we must choose values for the parameters.

Observe, every trajectory of (1.6.7) and (1.6.8) crosses the unit circle once and only once. Let us choose to map the orbit of (1.6.8) that crosses the unit circle at angular coordinate θ_0 onto the orbit of (1.6.7) that crosses the unit circle with angular coordinate $\Theta_0 = \theta_0$. The map $\mathbf{h}$ obtained in this way is given by setting $r_0 = R_0 = 1$ and $\theta_0 = \Theta_0$ in (1.6.9), i.e.

$$R = r^{\beta/\alpha}, \qquad \Theta = \theta, \tag{1.6.10}$$

with $r > 0$, $0 \leqslant \theta < 2\pi$. Thus, if we define $\mathbf{h}(\mathbf{0}) = \mathbf{0}$, we have constructed a homeomorphism which exhibits the topological equivalence of (1.6.7) and (1.6.8). Since we have already established the equivalence of (1.6.6) and (1.6.8), we finally conclude that $\mathbf{Jx}$ and $\mathbf{J}_0\mathbf{x}$ are topologically equivalent. □

Example 1.6.2 Use the map $r' = r$, $\theta' = \theta - \ln r$ $(r > 0)$ to demonstrate that the vector fields $\mathbf{J}_0\mathbf{x}$, where $\mathbf{J}_0$ is given in (1.6.5), and $\mathbf{x}$ are topologically equivalent.

Solution. Let $\mathbf{h}$ be given by

$$\mathbf{h}(\mathbf{x}) = \mathbf{h}(r\cos\theta, r\sin\theta) = \begin{cases} r\cos(\theta - \ln r), r\sin(\theta - \ln r), & r > 0 \\ (0, 0), & r = 0. \end{cases} \tag{1.6.11}$$

The map $\mathbf{h}$: $\mathbb{R}^2 \to \mathbb{R}^2$ is continuous and has continuous inverse, $r = r'$, $\theta = \theta' + \ln r'$, $r' > 0$. Since $\mathbf{h}(\mathbf{0}) = \mathbf{0}$, $\mathbf{h}$ takes the fixed point trajectory of the flow of $\mathbf{J}_0\mathbf{x}$ onto that of $\mathbf{x}$. For $\mathbf{x} \neq \mathbf{0}$, $\mathbf{h}$ is differentiable so we can check its effect on the flow by transforming the differential equation $\dot{\mathbf{x}} = \mathbf{J}_0\mathbf{x}$ or, in polar coordinates, $\dot{r} = r$, $\dot{\theta} = 1$. We find

$$\dot{r}' = \dot{r} = r = r' \quad \text{and} \quad \dot{\theta}' = \dot{\theta} - \frac{\dot{r}}{r} = \dot{\theta} - 1 = 0, \tag{1.6.12}$$

which is just the polar form of $\dot{\mathbf{x}} = \mathbf{x}$. Of course, $\mathbf{h}$ is not differentiable at the origin so that (1.6.11) is only a homeomorphism of the plane. Hence $\mathbf{J}_0\mathbf{x}$ and $\mathbf{x}$ are topologically equivalent. □

When two flows are topologically equivalent we say they are of the same *topological type*. The results obtained in Examples 1.6.1 and 1.6.2 play an important role in the classification, up to topological type, of all linear vector fields on $\mathbb{R}^2$ (see Arrowsmith & Place, 1982, p. 58). The matrix $\mathbf{J}$ in (1.6.5) is the real Jordan form of any 2×2 real matrix, $\mathbf{A}$, with complex eigenvalues $\alpha \pm i\beta$, $\alpha > 0$, i.e. there is a real non-singular matrix $\mathbf{M}$ such that $\mathbf{M}^{-1}\mathbf{AM} = \mathbf{J}$. It follows (see Exercise 1.5.6) that the flows of **Ax** and **Jx** are linearly conjugate. Examples 1.6.1 and 1.6.2 show that all such vector fields are topologically equivalent to the vector field **x**.

The complete classification of linear vector fields on $\mathbb{R}^2$ is summarised in Figure 1.26. Each point of the (Tr **A**, Det **A**)-plane represents a similarity class of real, 2×2 matrices. The striking feature is that the vast majority of points in Figure 1.26 correspond to vector fields of stable, unstable or saddle type. Such linear vector fields are said to be *hyperbolic* (see § 2.1) and Figure 1.26 suggests that hyperbolic behaviour is 'typical' for linear vector fields on $\mathbb{R}^2$. The point to note is that, without a suitable equivalence relation, the idea of what is typical has no meaning. We will return to the question of typical or *generic* properties of flows and diffeomorphisms in § 3.1.

1.7 Poincaré maps and suspensions

We have already noted that the flow map $\varphi_t: M \to M$ is a diffeomorphism for each fixed t. Thus, one way of obtaining a diffeomorphism from a flow is to take its time-τ map, $\varphi_\tau: M \to M$, $\tau > 0$. Clearly, the orbits of φ_τ are constrained to follow the trajectories of the flow because $\{\varphi_\tau^m(\mathbf{x}) | m \in \mathbb{Z}\} = \{\varphi_{m\tau}(\mathbf{x}) | m \in \mathbb{Z}\} \subseteq \{\varphi_t(\mathbf{x}) | t \in \mathbb{R}\}$. This means that the dynamics of φ_τ are strongly influenced by the flow φ and they are not typical of those of diffeomorphisms on M. It is perhaps worth stressing that, while the orbits of $\mathbf{x}$ under φ_{τ_1} and φ_{τ_2}, $\tau_1 \neq \tau_2$, behave in a similar way for any $\mathbf{x} \in M$ (because both are subsets of the same trajectory of φ), the two maps are not necessarily of the same topological type. For example, suppose φ_t has a closed orbit γ of period T and that $\tau_1 = \alpha T$, $\alpha \in \mathbb{Q}$, whilst $\tau_2 = \beta T$, $\beta \in \mathbb{R} \backslash \mathbb{Q}$. It follows that φ_{τ_1} has an invariant circle γ consisting entirely of periodic points (of period q if $\alpha = p/q$). The same closed curve γ is invariant for φ_{τ_2} but the orbit of any point $\mathbf{x} \in \gamma$ under φ_{τ_2} fills out γ densely. Therefore, φ_{τ_1} cannot be topologically conjugate

Figure 1.26 Topological types of all linear vector fields on the plane. Each point in the (Tr **A**, Det **A**)-plane corresponds to an equivalence class of linear vector fields. Details of the derivation of this diagram are given in Chapter 2 of Arrowsmith & Place (1982). The differential equation $\dot{\mathbf{x}} = \mathbf{x}$ has Tr **A** = 2, Det **A** = 1 and is therefore unstable.

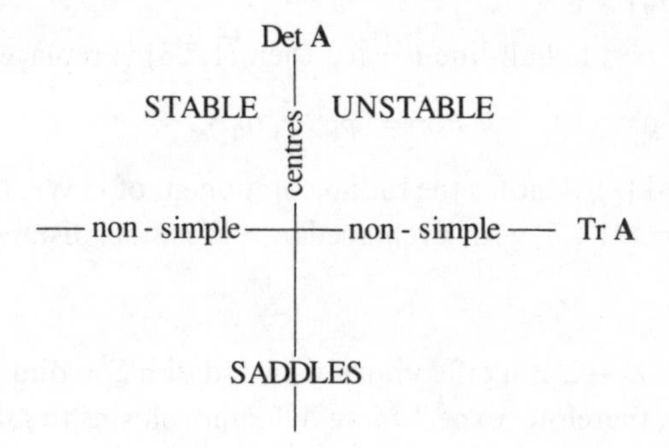

to φ_{τ_2}, i.e. the maps are of different topological type. We will have cause to return to time-τ maps of flows in Chapter 5.

Another, more significant, way of obtaining a diffeomorphism from a flow is to construct its *Poincaré map*. Let φ be a flow on M with vector field $\mathbf{X}$ and suppose that Σ is a co-dimension one submanifold of M satisfying:

(i) every orbit of φ meets Σ for arbitrarily large positive and negative times;
(ii) if $\mathbf{x} \in \Sigma$ then $\mathbf{X}(\mathbf{x})$ is not tangent to Σ.

Then Σ is said to be a *global* (*cross*) *section* of the flow. Let $\mathbf{y} \in \Sigma$ and $\tau(\mathbf{y})$ be the least, positive time for which $\varphi_{\tau(\mathbf{y})}(\mathbf{y}) \in \Sigma$.

Definition 1.7.1 *The* Poincaré (*or first return*) map *for* Σ *is defined to be*

$$\mathbf{P}(\mathbf{y}) = \varphi_{\tau(\mathbf{y})}(\mathbf{y}), \qquad \mathbf{y} \in \Sigma. \tag{1.7.1}$$

Example 1.7.1 Obtain the Poincaré map, P, of the flow defined by

$$\dot{r} = r(1 - r); \qquad \dot{\theta} = 1, \qquad r > 0, \tag{1.7.2}$$

where (r, θ) are plane polar coordinates, taking Σ to be the half-line $\theta = 0$. How does P change if Σ is taken to be the half-line $\theta = \theta_0$?

Solution. The phase portrait of (1.7.2) is shown in Figure 1.15. Σ is the positive x-axis in the plane and (1.7.1) can be written

$$P(x) = (\varphi_{\tau(x)}(x, 0))_x, \qquad x > 0, \tag{1.7.3}$$

where $\varphi_t(r, \theta) \in \mathbb{R}^2$ is the flow of (1.7.2), $(\cdot)_x$ denotes the x-component of $\cdot$, and $\tau(x)$ is the time taken for a phase point at $x \in \Sigma$ to make one complete revolution about the origin. Since $\dot{\theta} = 1$, $\tau(x) = 2\pi$.

The radial equation, $\dot{r} = r(1 - r)$, has solution

$$r(t) = r_0/\{r_0 + (1 - r_0)\exp(-t)\}, \tag{1.7.4}$$

with $r(0) = r_0$, so that

$$P(x) = x/\{x + (1 - x)\alpha\}, \qquad x > 0, \tag{1.7.5}$$

where $\alpha = \exp(-2\pi) < 1$.

If Σ is taken to be the half-line $\theta = \theta_0$, then (1.7.3) is replaced by

$$P(r) = (\varphi_{\tau(r)}(r, \theta_0))_r, \tag{1.7.6}$$

where $\tau(r) = 2\pi$ and $(\cdot)_r$ denotes the radial component of $\cdot$. We, therefore, conclude that P takes the form (1.7.5) with x replaced by r, the radial distance along $\theta = \theta_0$.

□

By construction $\mathbf{P}: \Sigma \to \Sigma$ is a diffeomorphism and $\dim \Sigma = \dim M - 1$. In contrast to time-τ maps we, therefore, expect these diffeomorphisms to reflect the properties of flows in one higher dimension. For example, the Poincaré map $P(x)$ in (1.7.5) has a fixed point at $x = 1$ (observe $x^* = P(x^*)$ implies $(1 - x^*)(1 - \alpha) = 0$, which is only satisfied for $x^* = 1$). Furthermore, if $x \gtrless 1$, then $P(x) \lessgtr x$ so that $x = 1$ is an attracting fixed point. This *fixed point* in P clearly corresponds to the stable *limit cycle* in the phase portrait of $\boldsymbol{\varphi}$ (see Figure 1.15).

Another example is afforded by the flow on the torus, T^2, defined by

$$\dot{\theta} = \alpha, \qquad \dot{\varphi} = \beta, \qquad \alpha, \beta > 0, \tag{1.7.7}$$

where θ and φ are as shown in Figure 1.27. The equations (1.7.7) have solutions

$$\theta = \alpha t + \theta_0 \qquad \text{and} \qquad \varphi = \beta t + \varphi_0 \tag{1.7.8}$$

reduced mod 2π, so $\begin{cases} \varphi \\ \theta \end{cases}$ first returns to $\begin{cases} \varphi_0 \\ \theta_0 \end{cases}$ when $\begin{cases} t = t_\varphi \\ t = t_\theta \end{cases}$, where $\begin{matrix} \beta t_\varphi = 2\pi \\ \alpha t_\theta = 2\pi \end{matrix}$. Thus if $\alpha/\beta = p/q$, $p, q \in \mathbb{Z}^+$ and relatively prime, then $q t_\varphi = p t_\theta$ and the orbit through (θ_0, φ_0) returns to this point after q revolutions around the torus in the φ-sense and p revolutions in the θ-sense. It follows that if α and β are rationally related then every point of T^2 is a periodic point of the flow, i.e. every point lies on a closed orbit. If on the other hand α and β are not rationally related then the orbit through (θ_0, φ_0) never returns to that point although it approaches it arbitrarily closely.

A global section of the torus is obtained by taking $\varphi = \varphi_0$, a constant, when Σ is a circle, S^1, with coordinate θ. Since the orbit of the flow first returns to $\varphi = \varphi_0$ after time $t_\varphi = 2\pi/\beta$ and $\theta = \alpha t + \theta_0$, we conclude that the Poincaré map, $P: S^1 \to S^1$, is a rotation by $2\pi\alpha/\beta$. The properties of pure rotations (see § 1.2.2) obviously reflect the behaviour of the flow described above.

There are flows for which there is no global section (see Exercise 1.7.2). Therefore, it is not true to say that every flow corresponds to a diffeomorphism by taking Poincaré maps. However, the converse is true, i.e. every diffeomorphism $\mathbf{f}$ is the Poincaré map of a flow – called the *suspension of* $\mathbf{f}$. This is a very important

Figure 1.27 Diagram showing how the coordinates, θ and φ, used in (1.7.7) are defined.

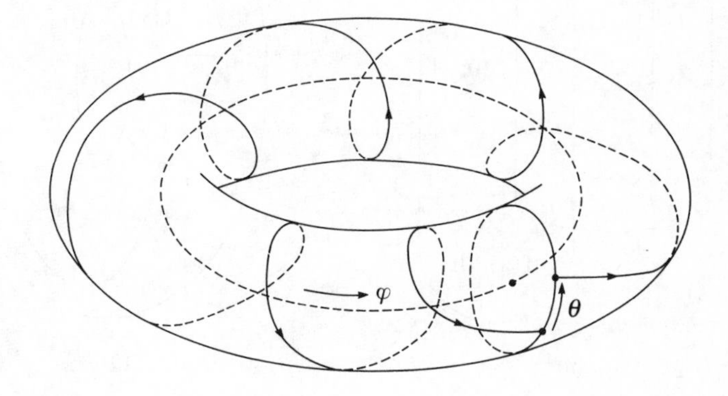

observation. It means that any result that can be proved for diffeomorphisms should have a counterpart for flows in one higher dimension (see Smale, 1967). The following explicit definition is given on p. 59 of Arnold & Avez, 1968.

Definition 1.7.2 *The flow*

$$\psi_t(\mathbf{x}, \theta) = (\mathbf{f}^{[t+\theta]}(\mathbf{x}), t + \theta - [t + \theta]), \tag{1.7.9}$$

where $\mathbf{x} \in M$, $\theta \in [0, 1]$ *and* $[\cdot]$ *denotes the integer part of* $\cdot$, *defined on a compact manifold* $\tilde{M}$ *by identification of* $(\mathbf{x}, 1)$ *and* $(\mathbf{f}(\mathbf{x}), 0)$ *in the topological product* $M \times [0, 1]$, *is called the* suspension *of the diffeomorphism* $\mathbf{f}: M \to M$.

It is easy to verify that $\psi_t(\mathbf{x}, \theta)$ in (1.7.9) formally satisfies the requirements of Definition 1.3.1. Geometrically, (1.7.9) corresponds to considering the product $M \times [0, 1]$ and taking a unit vector field in the $[0, 1]$-direction. Now imagine identifying the 1-end and the 0-end in such a way that $(\mathbf{x}, 1)$ is attached to $(\mathbf{f}(\mathbf{x}), 0)$ for each $\mathbf{x} \in M$ (see Figure 1.28).

It must be pointed out that the manifold $\tilde{M}$ is not always $M \times S^1$ as Figure 1.28 suggests. $M \times S^1$ is obtained if $\mathbf{f}$ is continuously deformable, through diffeomorphisms, to the identity. For example, if we let $M = S^1$ and $\mathbf{f}$ be a rotation then $\tilde{M}$ is the torus $T^2 = S^1 \times S^1$. However, if $\mathbf{f}$ is a reflection in a diameter of the circle then $\tilde{M}$ must be a Klein bottle to achieve the identification of $(\mathbf{x}, 1)$ and $(\mathbf{f}(\mathbf{x}), 0)$. Another, perhaps simpler example is to let $M = (0, 1)$ and f be reflection in $x = \frac{1}{2}$. As Figure 1.29 shows, $\tilde{M}$ is a Möbius strip.

Figure 1.28 Schematic illustration of the construction of the suspension of a diffeomorphism $\mathbf{f}$ that is continuously deformable into the identity: (*a*) before; (*b*) after; identification of $(\mathbf{x}, 1)$ and $(\mathbf{f}(\mathbf{x}), 0)$.

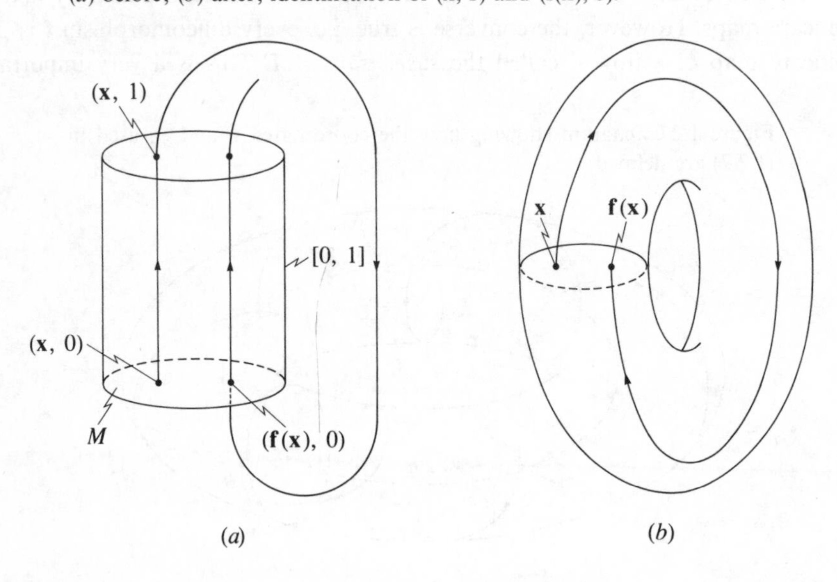

An alternative way of viewing Definition 1.7.2 is to think of linking $(\mathbf{x}, 1)$ and $(\mathbf{f}(\mathbf{x}), 0)$ by a smooth 'fibre' of unit length along which the suspension is considered to flow. This must be done for each $\mathbf{x} \in M$. Since $\mathbf{f}$ is a diffeomorphism, if $\mathbf{y} \in M$ is close to $\mathbf{x}$ then $\mathbf{f}(\mathbf{y})$ is close to $\mathbf{f}(\mathbf{x})$ and the fibres of the identification lie close to each other. If we were to take a finite sample of these fibres we should obtain something resembling unit length of a, possibly twisted, multicored electrical flex.

Obviously, this procedure does not define the precise shape of the identifying fibre or, in other words, it does not uniquely determine the suspended flow. What is important is that the component of the flow in the new dimension is never zero. It then follows that all admissible shapes of the identifying fibres give rise to topologically equivalent suspended flows. The flow given in (1.7.9) is a particular

Figure 1.29 The suspension of the diffeomorphism f: $(0, 1) \to (0, 1)$ given by reflection in $x = \frac{1}{2}$ is defined on a Möbius band. The twist in the manifold on which the suspension is defined arises because $(x, 1)$ must be identified with $(f(x), 0)$.

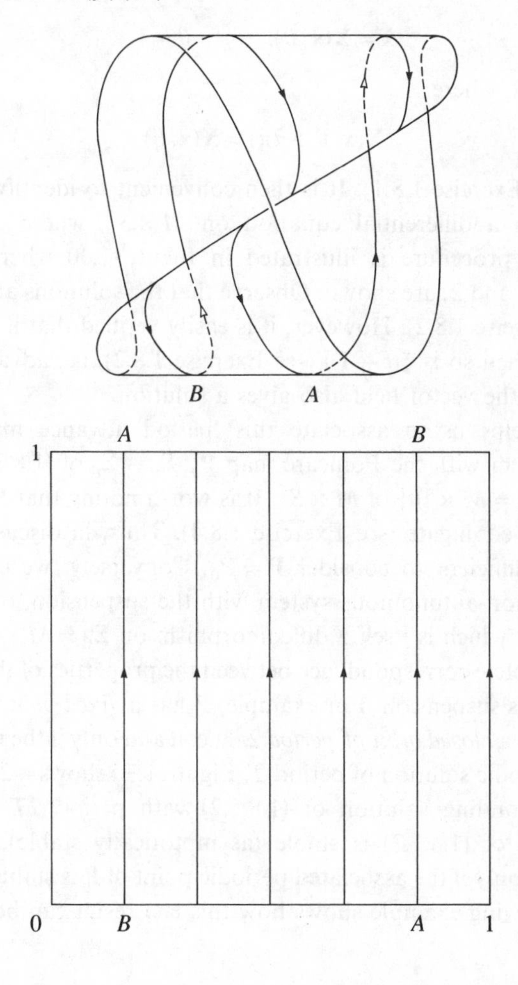

representative of this equivalence class which clearly exhibits the connection with the diffeomorphism **f**. When looked at from this point of view it is easier to understand how the nature of **f** (whether or not it is deformable to $\mathbf{id}_M$) affects the resultant manifold on which the suspension is defined.

1.8 Periodic non-autonomous systems

An important application of the ideas developed in § 1.7 is in the analysis of differential equations of the form

$$\dot{\mathbf{x}} = \mathbf{X}(\mathbf{x}, t), \qquad \mathbf{x} \in M, \tag{1.8.1}$$

where

$$\mathbf{X}(\mathbf{x}, t + T) = \mathbf{X}(\mathbf{x}, t), \tag{1.8.2}$$

for all $t \in \mathbb{R}$. The transformation $t \mapsto t/\gamma$, $\mathbf{X}(\mathbf{x}, t) \mapsto \gamma \mathbf{X}(\mathbf{x}, \gamma t)$, with $\gamma = T/2\pi$, allows (1.8.1) to be written as the autonomous system

$$\dot{\mathbf{x}} = \mathbf{X}(\mathbf{x}, \theta), \qquad \dot{\theta} = 1, \tag{1.8.3}$$

defined on $M \times \mathbb{R}$, where

$$\mathbf{X}(\mathbf{x}, \theta + 2\pi) = \mathbf{X}(\mathbf{x}, \theta) \tag{1.8.4}$$

for all $\theta \in \mathbb{R}$ (see Exercise 1.8.1). It is then convenient to identify $\theta + 2\pi m$, $m \in \mathbb{Z}$, with θ to obtain a differential equation on $M \times S^1$, where θ is the circular coordinate. This procedure is illustrated in Figure 1.30 where some possible solutions of (1.8.1 and 2) are shown. Observe that the solutions are not necessarily periodic (see Exercise 1.8.2). However, it is easily verified that if $\xi(t)$ is a solution of (1.8.1 and 2) then so is $\xi(t + T)$ (see Exercise 1.8.2). i.e. advancing a solution by one period of the vector field also gives a solution.

Figure 1.30 helps us to associate this 'period advance map' of the non-autonomous system with the Poincaré map $\mathbf{P}_\theta: \Sigma_\theta \to \Sigma_\theta$ of (1.8.3) defined on the global section, $\Sigma_\theta = M \times \{\theta\}$ of $M \times S^1$. It is worth noting that $\mathbf{P}_\theta$ and $\mathbf{P}_{\theta'}$, $\theta \neq \theta'$ are topologically conjugate (see Exercise 1.8.4). Thus, in discussing topological properties it is sufficient to consider $\mathbf{P} = \mathbf{P}_0$. Conversely, we can associate the solutions of the non-autonomous system with the suspension, on $M \times S^1$, of the Poincaré map, $\mathbf{P}$, which is itself a diffeomorphism on $\Sigma_0 = M \times \{0\}$.

There is a complete correspondence between the properties of the Poincaré map, $\mathbf{P}$, and those of its suspension. For example, $\mathbf{P}$ has a *fixed point* $\mathbf{x}^*$ if and only if its suspension has a *closed orbit of period* 2π, i.e. if and only if the non-autonomous system has a periodic solution of period T. Figure 1.30 shows a 2-cycle of $\mathbf{P}$ along with the corresponding solution of (1.8.1,2) with period $2T$. Furthermore, a periodic solution of (1.8.1,2) is stable (asymptotically stable), in the sense of Liapunov, if and only if the associated periodic point of $\mathbf{P}$ is stable (asymptotically stable). The following example shows how this last result can be applied.

Figure 1.30 (*a*) Schematic representation, in the extended phase space, $M \times \mathbb{R}$, of some possible solutions of the non-autonomous system (1.8.1), (1.8.2). (*b*) Corresponding solutions of the autonomous equation (1.8.3), (1.8.4) on $M \times S^1$. $\Sigma_\theta = M \times \{\theta\}$ is a global section for the flow of (1.8.3), (1.8.4) and this allows us to define the Poincaré map $\mathbf{P}_\theta: \Sigma_\theta \to \Sigma_\theta$.

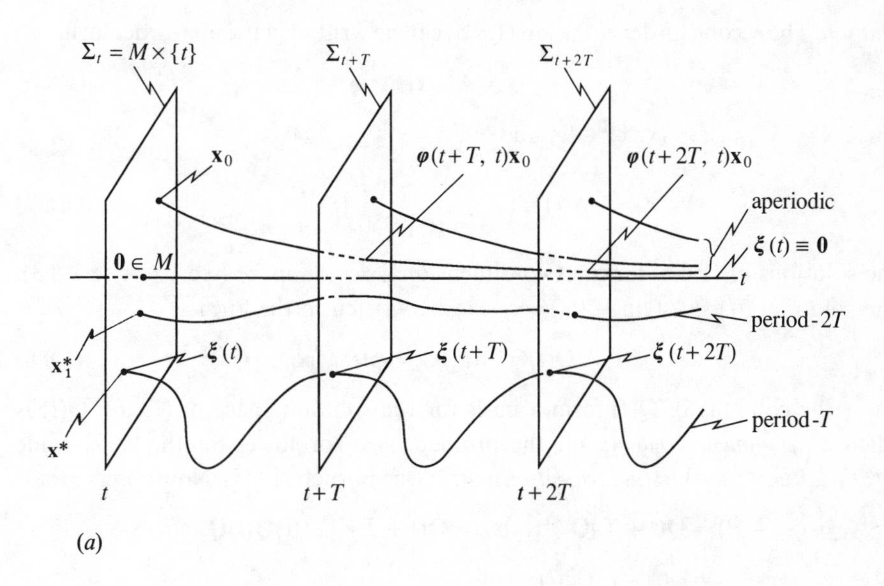

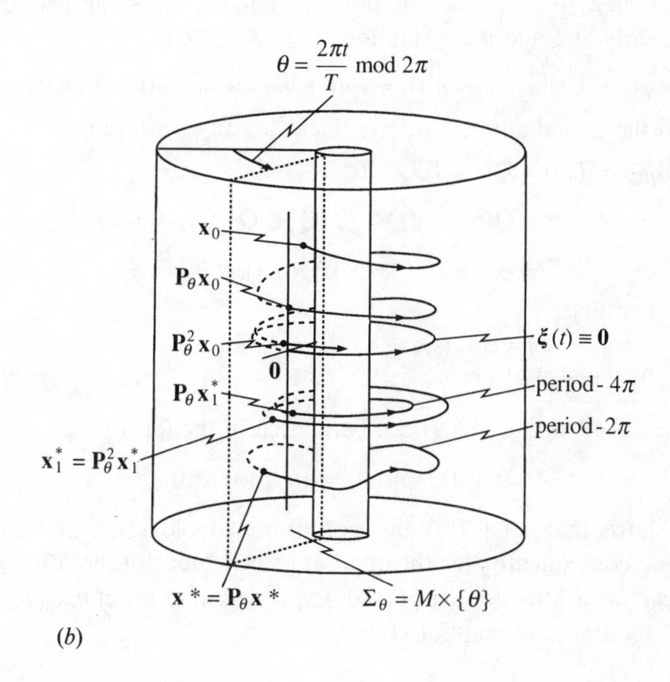

Example 1.8.1 Find the period advance map for the non-autonomous system

$$\ddot{x} + [\omega(t)]^2 x = 0, \tag{1.8.5}$$

where $\omega(t) = \omega(t+T)$, $t \in \mathbb{R}$. Obtain the Poincaré map $\mathbf{P}$ and show that $\operatorname{Det} \mathbf{P} = 1$. Hence, deduce that the null solution of (1.8.5) is stable (in the sense of Liapunov) if $|\operatorname{Tr} \mathbf{P}| < 2$ and unstable if $|\operatorname{Tr} \mathbf{P}| > 2$.

Solution. The second-order equation (1.8.5) can be written in the first-order form

$$\dot{\mathbf{x}} = \mathbf{A}(t)\mathbf{x}, \tag{1.8.6}$$

where $\mathbf{x} = (x_1, x_2)^{\mathrm{T}} = (x, \dot{x})^{\mathrm{T}} \in \mathbb{R}^2$ and

$$\mathbf{A}(t) = \begin{pmatrix} 0 & 1 \\ -[\omega(t)]^2 & 0 \end{pmatrix}. \tag{1.8.7}$$

The solutions of (1.8.6) form a two-dimensional vector space (see Exercise 1.8.3). The solution $\xi(t)$ satisfying $\xi(t_0) = \mathbf{x}_0$ can be written in the form

$$\xi(t) = \mathbf{Q}(t)\mathbf{Q}^{-1}(t_0)\mathbf{x}_0 = \varphi(t, t_0)\mathbf{x}_0, \tag{1.8.8}$$

where the columns of $\mathbf{Q}(t)$ form a basis for the solution space of (1.8.6). $\mathbf{Q}(t)$ is called a *fundamental matrix* for the problem (see Jordan & Smith, 1977) while $\varphi(t, t_0)$ is known as the *state transition matrix* (see Barnett, 1975). Now observe that

$$\begin{aligned} \xi(t+T) &= \mathbf{Q}(t+T)\mathbf{Q}^{-1}(t_0)\mathbf{x}_0 = \mathbf{Q}(t+T)\mathbf{Q}^{-1}(t)\mathbf{Q}(t)\mathbf{Q}^{-1}(t_0)\mathbf{x}_0 \\ &= \varphi(t+T, t)\xi(t). \end{aligned}$$

Thus $\varphi(t+T, t)\colon \Sigma_t \to \Sigma_{t+T}$ (see notation in Figure 1.30) is the period advance map at t. Clearly, if ξ and η are solutions of (1.8.6) then

$$\varphi(t+T,t)(a\xi(t) + b\eta(t)) = a\varphi(t+T,t)\xi(t) + b\varphi(t+T,t)\eta(t),$$

$a, b \leqslant \mathbb{R}$, and the period advance map is linear for any t. Moreover,

$$\begin{aligned} \varphi(t+T, t) &= \mathbf{Q}(t+T)\mathbf{Q}^{-1}(t) \\ &= \mathbf{Q}(t+T)[\mathbf{Q}^{-1}(T)\mathbf{Q}(T)\mathbf{Q}^{-1}(0)\mathbf{Q}(0)]\mathbf{Q}^{-1}(t) \\ &= \varphi(t+T, T)\varphi(T, 0)\varphi(0, t). \end{aligned} \tag{1.8.9}$$

It can be shown that:

(i) $\varphi(t+T, t_0+T) = \varphi(t, t_0)$;

(ii) $\varphi(t, 0)^{-1} = \varphi(0, t)$;

(see Exercise 1.8.4) so that (1.8.9) can be written in the form

$$\varphi(t+T, t)\varphi(t, 0) = \varphi(t, 0)\varphi(T, 0). \tag{1.8.10}$$

This result shows that $\varphi(t+T, t)$ and $\varphi(T, 0)$ are topologically (indeed linearly) conjugate and consequently, for the qualitative behaviour of the solutions of (1.8.6), we can focus attention on $\varphi(T, 0)$. Expressed in terms of θ, $\varphi(T, 0) = \mathbf{P}_0 = \mathbf{P}\colon \mathbb{R}^2 \to \mathbb{R}^2$, the Poincaré map for (1.8.6).

To show that Det $\mathbf{P} = 1$, note that (1.8.8) implies that

$$\frac{\mathrm{d}}{\mathrm{d}t}(\boldsymbol{\varphi}(t, 0)) = \mathbf{A}(t)\boldsymbol{\varphi}(t, 0), \tag{1.8.11}$$

since $\dot{\mathbf{Q}}(t) = \mathbf{A}(t)\mathbf{Q}(t)$. It follows (see Exercise 1.8.6) that, if $W(t) = \mathrm{Det}(\boldsymbol{\varphi}(t, 0))$, then $\dot{W}(t) = \mathrm{Tr}(\mathbf{A}(t))W(t) = 0$ for (1.8.6). Hence $W(t) = W(0) = \mathrm{Det}(\boldsymbol{\varphi}(0, 0)) = 1$ and, in particular,

$$W(T) = \mathrm{Det}(\boldsymbol{\varphi}(T, 0)) = \mathrm{Det}\,\mathbf{P} = 1. \tag{1.8.12}$$

The null solution of (1.8.6) corresponds to the fixed point of $\mathbf{P}$ at the origin. The stability type of the null solution is the same as that of the fixed point and the latter is determined by the eigenvalues, $\lambda_{1,2}$, of $\mathbf{P}$. Since Det $\mathbf{P} = 1$, the characteristic equation of $\mathbf{P}$ is $\lambda^2 - (\mathrm{Tr}\,\mathbf{P})\lambda + 1 = 0$ and

$$\lambda_{1,2} = \tfrac{1}{2}\{\mathrm{Tr}\,\mathbf{P} \pm [(\mathrm{Tr}\,\mathbf{P})^2 - 4]^{1/2}\}. \tag{1.8.13}$$

If $|\mathrm{Tr}\,\mathbf{P}| < 2$ then $(\mathrm{Tr}\,\mathbf{P})^2 < 4$ and the eigenvalues are complex with $\lambda_1 = \lambda_2^* = \exp(\mathrm{i}\beta)$ (since $\lambda_1\lambda_2 = 1$), where $\tan\beta = [4 - (\mathrm{Tr}\,\mathbf{P})^2]^{1/2}/\mathrm{Tr}\,\mathbf{P}$. Let $\mathbf{u} + \mathrm{i}\mathbf{v}$, $\mathbf{u}, \mathbf{v} \in \mathbb{R}^2$, be the eigenvector of $\mathbf{P}$ with eigenvalue λ_1. Then the matrix $\mathbf{K} = (\mathbf{v} \vdots \mathbf{u})$ is such that

$$\mathbf{K}^{-1}\mathbf{P}\mathbf{K} = \begin{pmatrix} \cos\beta & -\sin\beta \\ \sin\beta & \cos\beta \end{pmatrix}, \tag{1.8.14}$$

i.e. $\mathbf{P}$ is conjugate to a rotation about $\mathbf{x} = \mathbf{0}$. It follows that the orbit of $\mathbf{x} \neq \mathbf{0}$ under $\mathbf{P}$ lies on an ellipse and, consequently, the fixed point at $\mathbf{x} = \mathbf{0}$ is *stable* in the sense of Liapunov (see Figure 1.31).

If $|\mathrm{Tr}\,\mathbf{P}| > 2$, then $\lambda_{1,2}$ are real with $\lambda_1 = \lambda$ ($|\lambda| > 1$) $\lambda_2 = \lambda^{-1}$. In this case, there is a non-singular $\mathbf{K}$ such that

$$\mathbf{K}^{-1}\mathbf{P}\mathbf{K} = \mathbf{D} = \begin{pmatrix} \lambda & 0 \\ 0 & \lambda^{-1} \end{pmatrix}. \tag{1.8.15}$$

Figure 1.31 When $|\mathrm{Tr}\,\mathbf{P}| < 2$, the orbits of points $\mathbf{x} \neq \mathbf{0}$ under $\mathbf{P}$ lie on ellipses as shown. Observe that, for any $\mathbf{x} \in N'$, $\mathbf{P}^m\mathbf{x} \in N$ for all $m \in \mathbb{Z}$. Thus, $\mathbf{x} = \mathbf{0}$ is stable in the sense of Liapunov (see Definition 1.2.3).

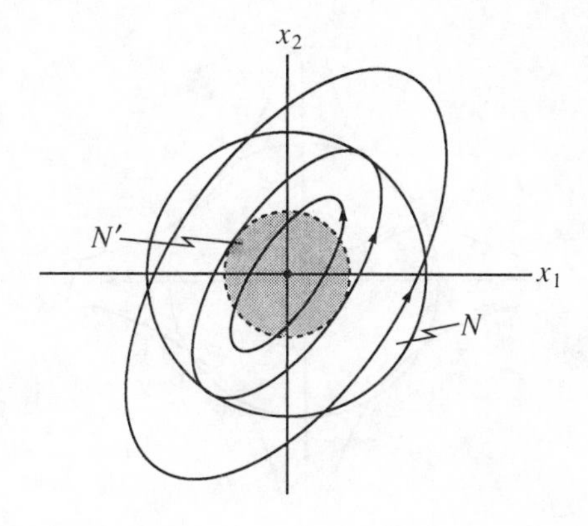

Here the orbits of **P** lie on hyperbolae and, as Figure 1.32 shows, the $\mathbf{x} = \mathbf{0}$ is an *unstable* fixed point. □

Example 1.8.1 suggests that periodic perturbations of the frequency, ω, of a harmonic oscillator can de-stabilise the equilibrium point with $x = 0$. This is essentially what a child on a swing achieves by appropriate movements of weight, in order to build up the amplitude of the oscillations of the swing. A simple example illustrating how this instability can be achieved is given in Arnold (1973, pp. 205–6). This phenomenon is known as *parametric resonance*.

1.9 Hamiltonian flows and Poincaré maps

Another application of Poincaré maps, that is of current research interest, lies in the study of *non-integrable*, conservative Hamiltonian systems. While the reader will no doubt have encountered the integrable case in a Classical Mechanics course, it will be useful to review the basic ideas emphasising the connection with flows.

Definition 1.9.1 *Let U be an open subset of $\mathbb{R}^{2n}$ and $H: U \to \mathbb{R}$ be a twice continuously differentiable function. The system of differential equations $\dot{\mathbf{x}} = \mathbf{X}_H(\mathbf{x})$, $\mathbf{X}_H: U \to \mathbb{R}^{2n}$ given by*

$$\dot{q}_i = \frac{\partial H}{\partial p_i}, \qquad \dot{p}_i = -\frac{\partial H}{\partial q_i}, \qquad i = 1, \ldots, n, \tag{1.9.1}$$

where $\mathbf{x} = (q_1, \ldots, q_n, p_1, \ldots, p_n)^{\mathrm{T}}$ is said to be a conservative Hamiltonian system with *n*-degrees of freedom.

Figure 1.32 For $\lambda > 1$ the hyperbolae $x_1 x_2 = c$, $c \neq 0$, are invariant curves for the map $\mathbf{Dx}$, where $\mathbf{D}$ is given by (1.8.15). The origin is a hyperbolic saddle point and therefore for every $N' \subseteq N$, there exists $\mathbf{x} \in N'$ for which $\mathbf{P}^m \mathbf{x} \notin N$, for some $m \in \mathbb{Z}^+$. Hence the saddle point is unstable in the sense of Liapunov.

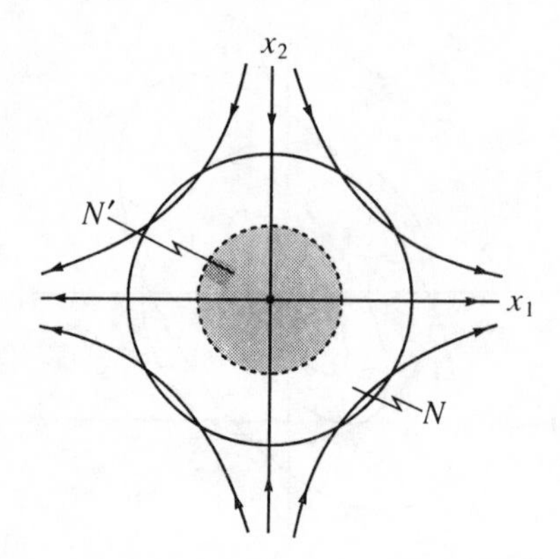

$H = H(\mathbf{q}, \mathbf{p})$ is the Hamiltonian for the system and the equations (1.9.1) are known as *Hamilton's equations*. The state of the system at time t is specified by

$$\mathbf{x}(t) = \begin{pmatrix} \mathbf{q}(t) \\ \mathbf{p}(t) \end{pmatrix}.$$

The *configuration*, $\mathbf{q}(t)$, of the system is given by the n *generalised coordinates* $q_i(t)$ and $\mathbf{p}(t)$ consists of the n *conjugate generalised momenta* $p_i(t)$. A system with n degrees of freedom is often called an *n-F system.*

In general, q_i and p_i change with t but H does not. Observe

$$\dot{H} = \sum_{i=1}^{n} \left(\frac{H}{\partial q_i} \dot{q}_i + \frac{\partial H}{\partial p_i} \dot{p}_i \right) = 0, \tag{1.9.2}$$

for all t, by (1.9.1). Thus, H is a *conserved* quantity or a constant of the motion. Alternatively, (1.9.1) is an autonomous system of differential equations which defines a *Hamiltonian flow*, $\boldsymbol{\varphi}_t^H \colon U \to \mathbb{R}^{2n}$. Equation (1.9.2) means that H is constant on the trajectories of $\boldsymbol{\varphi}_t^H$ i.e. H is a *first integral* for (1.9.1) (see Arrowsmith & Place, 1982, pp. 101–6).

In general, Hamiltonian flows occur on differentiable manifolds and Definition 1.9.1 is valid for each chart $(U_\alpha, \mathbf{h}_\alpha)$. Thus (see Figure 1.33) $H_\alpha \colon U_\alpha \to \mathbb{R}$ gives rise to a vector field $\mathbf{X}_{H_\alpha}$, via (1.9.1), for each α. Moreover, when $W_\alpha \cap W_\beta$ $(\alpha \neq \beta)$ is non-empty, the two sets of local coordinates on U_α and U_β are related by the

Figure 1.33 Illustration of the way in which a Hamiltonian function defined on a manifold M gives rise to Hamiltonians, H_α and H_β on the charts $(U_\alpha, \mathbf{h}_\alpha)$ and $(U_\beta, \mathbf{h}_\beta)$, respectively.

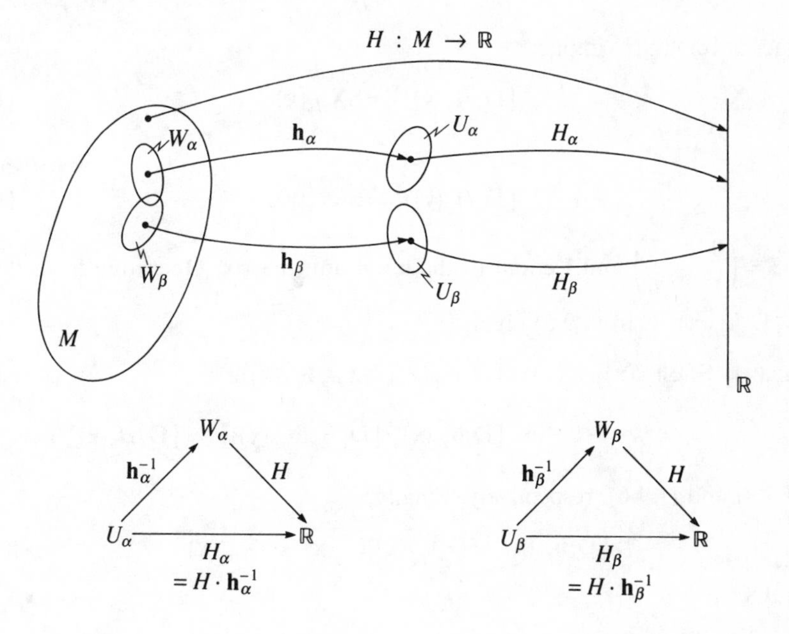

overlap map $\mathbf{h}_{\alpha\beta}$ (see Figure 1.3). Thus, if $\mathbf{x} = (q_1, \ldots, q_n, p_1, \ldots, p_n)^{\mathrm{T}}$ in U_α and $\mathbf{y} = (Q_1, \ldots, Q_n, P_1, \ldots, P_n)^{\mathrm{T}}$ in U_β represent the same point on M, then

$$H_\alpha(\mathbf{x}) = H_\beta(\mathbf{h}_{\alpha\beta}(\mathbf{x})), \tag{1.9.3}$$

and

$$H_\alpha(\mathbf{h}_{\alpha\beta}^{-1}(\mathbf{y})) = H_\beta(\mathbf{y}). \tag{1.9.4}$$

Of course, we require that the vector fields $\mathbf{X}_{H_\alpha}$ and $\mathbf{X}_{H_\beta}$ give rise to the same dynamics on the overlap between two charts and this imposes constraints on the manifold itself. To make the dynamics on W_α and W_β agree on $W_\alpha \cap W_\beta$, we demand that

$$\mathrm{D}_{\mathbf{x}}\mathbf{h}_{\alpha\beta}(\mathbf{x})\mathbf{X}_{H_\alpha}(\mathbf{x}) = \mathbf{X}_{H_\beta}(\mathbf{h}_{\alpha\beta}(\mathbf{x})) \tag{1.9.5}$$

(see (1.5.12)). Now, differentiation of (1.9.3) gives

$$\mathrm{D}_{\mathbf{x}}H_\alpha(\mathbf{x}) = \mathrm{D}_{\mathbf{y}}H_\beta(\mathbf{h}_{\alpha\beta}(\mathbf{x}))\mathrm{D}_{\mathbf{x}}\mathbf{h}_{\alpha\beta}(\mathbf{x}). \tag{1.9.6}$$

Equation (1.9.6) looks more familiar in component form, i.e.

$$\left(\frac{\partial H_\alpha}{\partial q_1}, \ldots, \frac{\partial H_\alpha}{\partial q_n}, \frac{\partial H_\alpha}{\partial p_1}, \ldots, \frac{\partial H_\alpha}{\partial p_n}\right)$$

$$= \left(\frac{\partial H_\beta}{\partial Q_1}, \ldots, \frac{\partial H_\beta}{\partial Q_n}, \frac{\partial H_\beta}{\partial P_1}, \ldots, \frac{\partial H_\beta}{\partial P_n}\right) \begin{pmatrix} \frac{\partial Q_1}{\partial q_1}, \ldots, \frac{\partial Q_1}{\partial q_n}, \frac{\partial Q_1}{\partial p_1}, \ldots, \frac{\partial Q_1}{\partial p_n} \\ \vdots \qquad\qquad\qquad\qquad \vdots \\ \frac{\partial P_n}{\partial q_1} \ldots\ldots\ldots\ldots\ldots \frac{\partial P_n}{\partial p_n} \end{pmatrix}. \tag{1.9.7}$$

Furthermore, (1.9.1) implies

$$[\mathrm{D}_{\mathbf{x}}H_\alpha(\mathbf{x})]^{\mathrm{T}} = \mathbf{S}\mathbf{X}_{H_\alpha}(\mathbf{x}) \tag{1.9.8}$$

and

$$[\mathrm{D}_{\mathbf{y}}H_\beta(\mathbf{y})]^{\mathrm{T}} = \mathbf{S}\mathbf{X}_{H_\beta}(\mathbf{y}), \tag{1.9.9}$$

with $\mathbf{S} = \begin{pmatrix} \mathbf{0} & -\mathbf{I} \\ \mathbf{I} & \mathbf{0} \end{pmatrix}$ and $\mathbf{I}$ equal to the $n \times n$ unit matrix. Operating from the left with $[\mathrm{D}_{\mathbf{x}}\mathbf{h}_{\alpha\beta}(\mathbf{x})]^{\mathrm{T}}\mathbf{S}$ in (1.9.5) gives

$$[\mathrm{D}_{\mathbf{x}}\mathbf{h}_{\alpha\beta}(\mathbf{x})]^{\mathrm{T}}\mathbf{S}\mathrm{D}_{\mathbf{x}}\mathbf{h}_{\alpha\beta}(\mathbf{x})\mathbf{X}_{H_\alpha}(\mathbf{x}) = [\mathrm{D}_{\mathbf{x}}\mathbf{h}_{\alpha\beta}(\mathbf{x})]^{\mathrm{T}}\mathbf{S}\mathbf{X}_{H_\beta}(\mathbf{h}_{\alpha\beta}(\mathbf{x})), \tag{1.9.10}$$

$$= [\mathrm{D}_{\mathbf{x}}\mathbf{h}_{\alpha\beta}(\mathbf{x})]^{\mathrm{T}}[\mathrm{D}_{\mathbf{y}}H_\beta(\mathbf{h}_{\alpha\beta}(\mathbf{x}))]^{\mathrm{T}} = [\mathrm{D}_{\mathbf{x}}H_\alpha(\mathbf{x})]^{\mathrm{T}}, \tag{1.9.11}$$

by (1.9.9) and (1.9.6), respectively. Finally,

$$[\mathrm{D}_{\mathbf{x}}\mathbf{h}_{\alpha\beta}(\mathbf{x})]^{\mathrm{T}}\mathbf{S}\mathrm{D}_{\mathbf{x}}\mathbf{h}_{\alpha\beta}(\mathbf{x})\mathbf{X}_{H_\alpha}(\mathbf{x}) = \mathbf{S}\mathbf{X}_{H_\alpha}(\mathbf{x}), \tag{1.9.12}$$

by (1.9.8).

Clearly, (1.9.5) is satisfied if and only if the overlap map $\mathbf{h}_{\alpha\beta}$ is such that

$$[\mathrm{D}_{\mathbf{x}}\mathbf{h}_{\alpha\beta}(\mathbf{x})]^{\mathrm{T}}\mathbf{S}\mathrm{D}_{\mathbf{x}}\mathbf{h}_{\alpha\beta}(\mathbf{x}) = \mathbf{S}, \qquad (1.9.13)$$

for each $\mathbf{x} \in \mathbf{h}_{\alpha}(W_{\alpha} \cap W_{\beta})$.

Definition 1.9.2 *A diffeomorphism* $\mathbf{h}: U \to \mathbb{R}^{2n}$, $U \subseteq \mathbb{R}^{2n}$, *is said to be* symplectic *if*

$$[\mathrm{D}\mathbf{h}(\mathbf{x})]^{\mathrm{T}}\mathbf{S}\mathrm{D}\mathbf{h}(\mathbf{x}) = \mathbf{S} \qquad (1.9.14)$$

for all $\mathbf{x} \in \mathbb{R}^{2n}$, *with* $\mathbf{S} = \begin{pmatrix} 0 & -\mathbf{I} \\ \mathbf{I} & 0 \end{pmatrix}$ *where* $\mathbf{I}$ *is the* $n \times n$ *identity matrix.*

A differentiable manifold for which all the overlap maps satisfy (1.9.13) is said to be a *symplectic manifold*. The theory of symplectic manifolds provides a coordinate free approach to Hamiltonian mechanics (Abraham & Marsden, 1978; Arnold, 1968).

It is important to realise that (1.9.13) is sufficient to ensure that the form (1.9.1) of Hamilton's equations is valid on both U_{α} and U_{β} (see (1.9.8,9)). The arguments involved in obtaining (1.9.13) are not confined to overlap maps. Consider the effect of a coordinate transformation, $\mathbf{h}$, on a Hamiltonian system defined on $\mathbb{R}^{2n}$. If we demand that the equations of motions of the new coordinates be derived from the transformed Hamiltonian by applying (1.9.1), then we can conclude, by precisely the same steps as we have used above, that $\mathbf{h}$ must be symplectic. However, preservation of Hamilton's equations in this sense is the property that defines *canonical* transformations in Classical Mechanics. Thus symplectic and canonical transformations are one and the same thing.

A property that distinguishes a Hamiltonian flow, φ_t^H, from other flows of even dimension is that φ_t^H preserves volumes of phase space.

Theorem 1.9.1 (Liouville) *Let* φ_t *be the flow induced by* $\dot{\mathbf{x}} = \mathbf{X}(\mathbf{x})$ *and* $\Omega(t)$ *be the volume of the image,* $\varphi_t(D)$, *of any region* D *of its phase space. If* $\operatorname{div}\mathbf{X} \equiv 0$, *then* φ_t *preserves volume, i.e.* $\Omega(t) = \Omega(0)$ *for all* t.

To illustrate the ideas behind the proof of Theorem 1.9.1 we will assume that D and $\varphi_t(D)$ both lie in the same chart. Since φ_t is a diffeomorphism for each t, we can regard it as a change of coordinates in phase space. With notation in Figure 1.34,

$$\Omega(t) = \int_{\varphi_t(D)} \mathrm{d}q'_1, \ldots, \mathrm{d}p'_n. \qquad (1.9.15)$$

Since $\mathbf{x}' = \varphi_t(\mathbf{x})$, this can be written as

$$\Omega(t) = \int_D \operatorname{Det}(\mathrm{D}\varphi_t(\mathbf{x}))\mathrm{d}^{2n}x, \qquad (1.9.16)$$

where $d^{2n}x = dq_1, \ldots, dp_n$. Now,

$$\varphi_t(\mathbf{x}) = \mathbf{x} + t\mathbf{X}(\mathbf{x}) + O(t^2), \tag{1.9.17}$$

and therefore

$$D\varphi_t(\mathbf{x}) = \mathbf{I} + tD\mathbf{X}(\mathbf{x}) + O(t^2). \tag{1.9.18}$$

Thus

$$\left.\frac{d\Omega}{dt}\right|_{t=0} = \dot{\Omega}(0) = \lim_{t\to 0} \int_D \left(\frac{\mathrm{Det}(D\varphi_t(\mathbf{x})) - 1}{t}\right) d^{2n}x, \tag{1.9.19}$$

since $\Omega(0) = \int_D dq_1, \ldots, dp_n$. However, observe that if $D\mathbf{X}(\mathbf{x})$ has eigenvalues $\eta_i(\mathbf{x})$ then

$$\begin{aligned} \mathrm{Det}(D\varphi_t(\mathbf{x})) &= \mathrm{Det}(\mathbf{I} + tD\mathbf{X}(\mathbf{x}) + O(t^2)), \\ &= \prod_i (1 + t\eta_i(\mathbf{x}) + O(t^2)), \\ &= 1 + t\sum_i \eta_i(\mathbf{x}) + O(t^2), \\ &= 1 + t\,\mathrm{Tr}\, D\mathbf{X}(\mathbf{x}) + O(t^2). \end{aligned} \tag{1.9.20}$$

Of course,

$$\mathrm{Tr}\, D\mathbf{X}(\mathbf{x}) = \mathrm{div}\, \mathbf{X}(\mathbf{x}) \tag{1.9.21}$$

and substitution in (1.9.19) gives

$$\dot{\Omega}(0) = \int_D \mathrm{div}\, \mathbf{X}(\mathbf{x}) d^{2n}x. \tag{1.9.22}$$

Figure 1.34 The flow map φ_t takes D at time zero (see (*a*)) to $\varphi_t(D)$ at time t (see (*b*)). Since φ_t is a diffeomorphism, this transformation can be regarded as a change of coordinates from $(q_1, \ldots, q_n, p_1, \ldots, p_n) = \mathbf{x}^T$ to $(q'_1, \ldots, q'_n, p'_1, \ldots, p'_n) = \mathbf{x}'^T$.

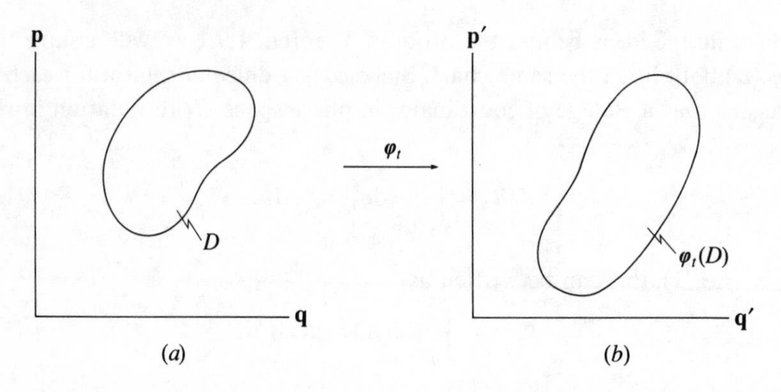

The above arguments do not depend on the initial time being zero and (1.9.22) can be generalised to

$$\dot{\Omega}(t) = \int_{D(t)} \operatorname{div} \mathbf{X}(\mathbf{x}) \mathrm{d}^{2n}x. \tag{1.9.23}$$

Clearly, if div $\mathbf{X}(\mathbf{x}) \equiv 0$ then $\dot{\Omega}(t) \equiv 0$ and $\Omega(t) = \Omega(0)$ for all t.

Let us apply Theorem 1.9.1 to a Hamiltonian flow φ_t^H. The vector field $\mathbf{X}$ is given by (1.9.1) and

$$\operatorname{div} \mathbf{X}(\mathbf{x}) = \sum_{i=1}^{n} \frac{\partial}{\partial q_i}\left(\frac{\partial H}{\partial p_i}\right) - \frac{\partial}{\partial p_i}\left(\frac{\partial H}{\partial q_i}\right) \equiv 0. \tag{1.9.24}$$

Hence φ_t^H preserves phase space volumes. This result highlights, in a geometrical way, the very special nature of Hamiltonian flows. In general, even dimensional flows may expand volumes in some parts of phase space and contract them in others. Clearly, (1.9.24) imposes a global restriction on φ_t^H. The volume-preserving nature of Hamiltonian flows is also reflected in the nature of the transformations that relate them to one another. It can be shown (see Arnold, 1968, p. 222 and Exercise 1.9.5) that (1.9.14) implies $\mathrm{Det}(\mathrm{D}\mathbf{h}(\mathbf{x})) \equiv 1$ so that symplectic transformations preserve volumes of phase space. However, it is perhaps worth noting that $\mathrm{Det}(\mathrm{D}\mathbf{h}(\mathbf{x})) \equiv 1$ only implies $\mathbf{h}$ is symplectic when $\mathbf{h}\colon \mathbb{R}^2 \to \mathbb{R}^2$ (see Exercises 1.9.5 and 1.9.6).

It is reasonable to consider to what extent Hamilton's equations can be simplified by symplectic transformations. Let $\mathbf{h}\colon (\mathbf{q}, \mathbf{p}) \to (\mathbf{Q}, \mathbf{P})$ and $\tilde{H}(\mathbf{Q}, \mathbf{P}) = H(\mathbf{h}^{-1}(\mathbf{Q}, \mathbf{P}))$. In particular, the transformed equations will be simpler if the new Hamiltonian is independent of one of the generalised coordinates. For example, suppose $\tilde{H}$ does not depend on Q_n, then

$$\dot{P}_n = -\frac{\partial \tilde{H}}{\partial Q_n} = 0 \tag{1.9.25}$$

and

$$P_n(t) = P_n(0) = I_n \tag{1.9.26}$$

The constant value I_n can be thought of as a parameter. For a given value of I_n, H now depends on only $(n-1)$ pairs of conjugate variables; the number of degrees of freedom has been reduced by one and the order of Hamilton's equations has decreased by two.

Ideally, one would like $\tilde{H}$ to be independent of all Q_i, $i = 1, \ldots, n$. Then

$$P_i(t) = P_i(0) = I_i \tag{1.9.27}$$

and

$$\dot{Q}_i = \frac{\partial \tilde{H}}{\partial P_i} = \omega_i(I_1, \ldots, I_n), \tag{1.9.28}$$

$i = 1, \ldots, n$. Notice $\dot{Q}_i$ depends only on the parameters $I_1, \ldots, I_n$ and is therefore independent of t. Thus (1.9.28) can be trivially integrated to give

$$Q_i(t) = \omega_i t + K_i, \tag{1.9.29}$$

$i = 1, \ldots, n$, $K_i \in \mathbb{R}$. Systems for which such a reduction is possible are said to be *integrable* and the system defined by (1.9.27 and 28) is referred to as their *normal form*. The variables $(\mathbf{Q}, \mathbf{P})$ displaying this form are called *action-angle variables*; the P_i (or I_i) being the 'actions' and the Q_i being the 'angles' (or cyclic variables). The latter name arises because (1.9.27 and 28) is the polar form of a simple harmonic oscillator with radial coordinate I_i and angular coordinate Q_i.

Traditional courses in Classical Mechanics focus attention on the integrable case. For example, 1-F systems with analytic H, linear equations of motion (i.e. normal modes), non-linear systems that are separable into 1-F systems are commonly discussed. However, these systems are not typical. In general, Hamiltonian systems are non-integrable and they can exhibit much more exotic dynamics. To illustrate this we must consider systems with at least two degrees of freedom and Poincaré maps play a key role in making such problems manageable.

A system with two-degrees of freedom has a four-dimensional phase space and it is, therefore, not feasible to picture its flow directly. Since the system is conservative, (generically) its trajectories lie in three-dimensional submanifolds or 'shells' on which the Hamiltonian $H(\mathbf{q}, \mathbf{p})$ is constant. Thus, by choosing a particular value for $H(\mathbf{q}, \mathbf{p})$ we can reduce the dimensionality of the problem by one. Now, we are frequently interested in systems exhibiting some kind of recurrence. For example, non-integrable perturbations of an integrable system or the behaviour of a non-integrable system in the neighbourhood of a closed orbit. In such cases, we can reduce our problem to one in two dimensions by constructing an appropriate Poincaré map. Of course, we have lost some detail of the dynamics in this process. After all we are only sampling the orbit periodically. However, the interesting point is that sufficient information is retained to show that the dynamics of 2-F, conservative systems can be very complicated. Moreover, since this information is in two-dimensions it is quite easy to present and appreciate in graphical form.

To show how the Poincaré map is constructed, let us first examine an *integrable* case, where solutions can be written down explicitly. Consider the biharmonic oscillator

$$\begin{aligned} \dot{q}_1 &= p_1, & \dot{p}_1 &= -\omega_1^2 q_1, \\ \dot{q}_2 &= p_2, & \dot{p}_2 &= -\omega_2^2 q_2. \end{aligned} \tag{1.9.30}$$

The Hamiltonian $H(\mathbf{q}, \mathbf{p})$ is given by

$$H(q_1, q_2; p_1, p_2) = \tfrac{1}{2} \sum_{i=1}^{2} (p_i^2 + \omega_i^2 q_i^2) \tag{1.9.31}$$

and (1.9.30) has solutions of the form

$$q_i = A_i \cos(\omega_i t + \eta_i),$$
$$p_i = -\omega_i A_i \sin(\omega_i t + \eta_i), \tag{1.9.32}$$

$\eta_i \in \mathbb{R}$, $i = 1, 2$. The aim is to construct the Poincaré map in such a way that one pair of conjugate variables (q_2, p_2, say) are removed. Thus we argue that by restricting to the Hamiltonian shell $H(\mathbf{q}, \mathbf{p}) = h_0 > 0$ we can express p_2 in terms of q_1, p_1 and q_2. Since q_2 is periodic with period $2\pi/\omega_2$, the orbit of a phase point in the plane $q_2 = 0$ returns to $q_2 = 0$ after time $2\pi/\omega_2$ (see Figure 1.35). Therefore, the Poincaré map $\mathbf{P}$ defined on the section $q_2 = 0$ is given by

$$\mathbf{P}\begin{pmatrix} q_1 \\ p_1 \end{pmatrix} = \begin{pmatrix} A_1 \cos\left(\omega_1\left[t + \dfrac{2\pi}{\omega_2}\right] + \eta_1\right) \\ -\omega_1 A_1 \sin\left(\omega_1\left[t + \dfrac{2\pi}{\omega_2}\right] + \eta_2\right) \end{pmatrix}$$
$$= \begin{pmatrix} \cos 2\pi\mu & -\dfrac{1}{\omega_1}\sin 2\pi\mu \\ \omega_1 \sin 2\pi\mu & \cos 2\pi\mu \end{pmatrix}\begin{pmatrix} q_1 \\ p_1 \end{pmatrix}, \tag{1.9.33}$$

with $\mu = \omega_1/\omega_2$. Clearly, $\mathbf{P}$ represents a rotation for which the ellipses

$$p_1^2 + \omega_1^2 q_1^2 = C, \tag{1.9.34}$$

with $0 < C < 2h_0$, are invariant curves. These closed invariant curves correspond to invariant tori in the flow on the $H = h_0$ shell.

The important thing to notice about (1.9.33) is that Det $\mathbf{P} = 1$. This means (see Exercise 1.9.5) that the Poincaré map, constructed in the manner described above, is *area-preserving*. That this is also the case when the system is non-integrable follows from the *Poincaré–Cartan invariant* (Arnold, 1968, pp. 233–40 or Arnold & Avez, 1968, pp. 230–2). A derivation of this invariant for 2-F systems requires

Figure 1.35 The Poincaré map defined on the section $q_2 = 0$. It is clear from (1.9.32) that q_2 returns to zero periodically with period $2\pi/\omega_2$.

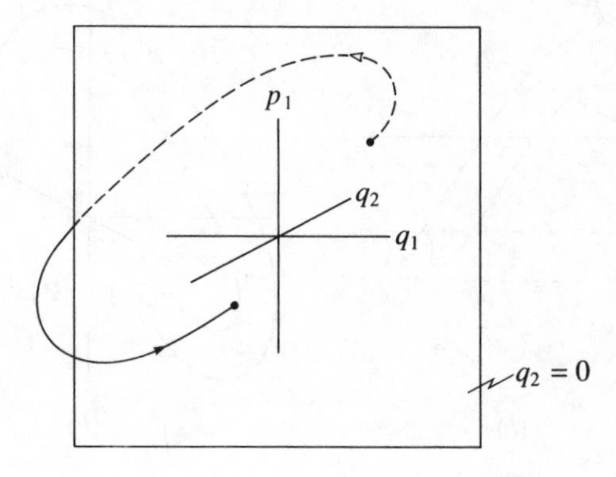

a knowledge of differential forms, however, for 1-F systems it can be obtained in the familiar notation of vector analysis. Consider the extended phase space for a 1-F system with coordinates (q, p, t). Let $\mathbf{v} = (-p, 0, H)$, then $\operatorname{curl} \mathbf{v} = (\partial H/\partial p, -\partial H/\partial q, 1)$ is the vector field of the Hamiltonian H in extended phase space (see (1.9.1)). Now apply Stokes Theorem to the tubular region shown in Figure 1.36(a). Here the sides of the tube consist of flow lines of curl $\mathbf{v}$. Dissecting

Figure 1.36 (a) Tubular region to which Stoke's Theorem is applied for 1-F systems. The vector field curl $\mathbf{v}$ is tangent to the surface at every point of the tube so that $\operatorname{curl} \mathbf{v} \cdot d\mathbf{S} \equiv 0$. The closed curves γ_1 and γ_2 are obtained by taking sections transverse to the tube of flow flines. (b) Dissection of the tube shown in (a) used to obtain (1.9.35).

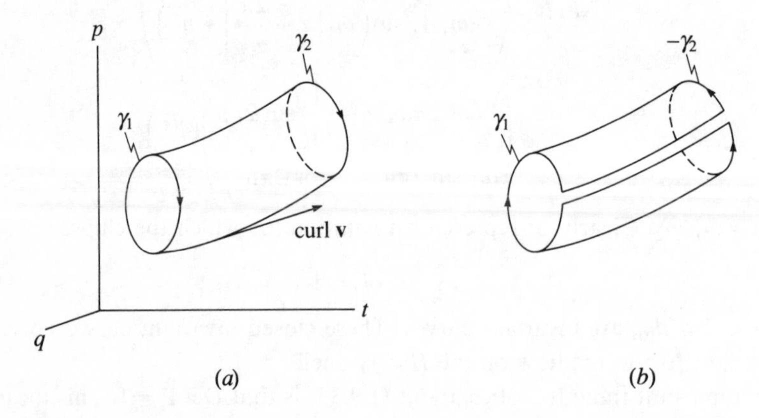

Figure 1.37 If γ_2 is given by $\mathbf{r} = \mathbf{r}(u)$, then $\mathbf{r}(u) = (q_1(u), 0, p_1(u), p_2(u), t(u))$, where $p_2(u)$ is determined by $H(\mathbf{q}, \mathbf{p}) = h_0$. The curve, $\bar{\gamma}_2$, obtained by projecting γ_2 onto $t = 0$ (see (a)), is the image of γ_1 under the Poincaré map $\mathbf{P}$ (see (b)).

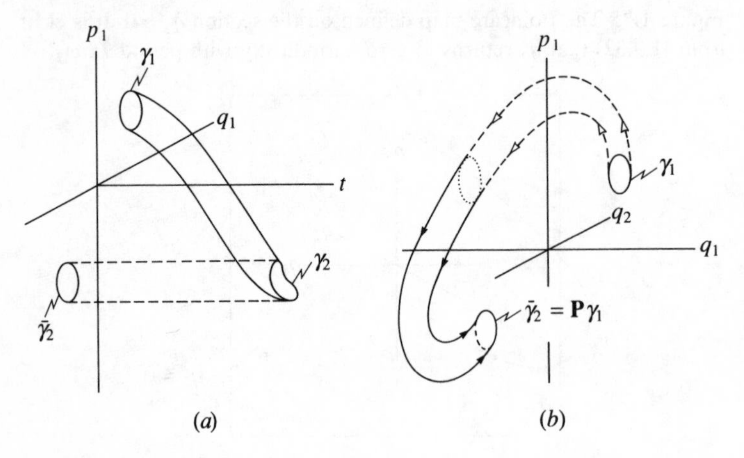

the tube as shown in Figure 1.36(*b*), we observe that

$$\int_S \text{curl}\,\mathbf{v}\cdot d\mathbf{S} = 0 = \int_{\gamma_1} \mathbf{v}\cdot d\mathbf{r} - \int_{\gamma_2} \mathbf{v}\cdot d\mathbf{r}, \tag{1.9.35}$$

where $d\mathbf{r} = (dq, dp, dt)$. Thus

$$\int_{\gamma_1} p\,dq - H\,dt = \int_{\gamma_2} p\,dq - H\,dt \tag{1.9.36}$$

and it follows that $\int_\gamma p\,dq - H\,dt$ is invariant under the flow.

With the aid of differential two forms (see Arnold, 1968, pp. 234–6), we can obtain Stokes' Theorem in five dimensions and derive the corresponding result for 2-F systems; namely

$$\int_{\gamma_1} p_1\,dq_1 + p_2\,dq_2 - H\,dt = \int_{\gamma_2} p_1\,dq_1 + p_2\,dq_2 - H\,dt, \tag{1.9.37}$$

Figure 1.38 Some typical orbits of the Hénon map (1.9.40) for $\cos\alpha = 0.8$. Two fixed points can be seen: one elliptic (see § 6.5) and one saddle-like. What appear to be closed curves are each the orbit of a single point, i.e. the orbit is confined to what is topologically an invariant circle. For small numbers of iterations of (1.9.40) individual points of these orbits can be distinguished moving around the origin (cf. Exercise 1.9.9). As the number of iterations increases, the plotted points merge into what looks like a closed curve. Individual orbit points are more apparent in the vicinity of the saddle point. (After Hénon, 1969.)

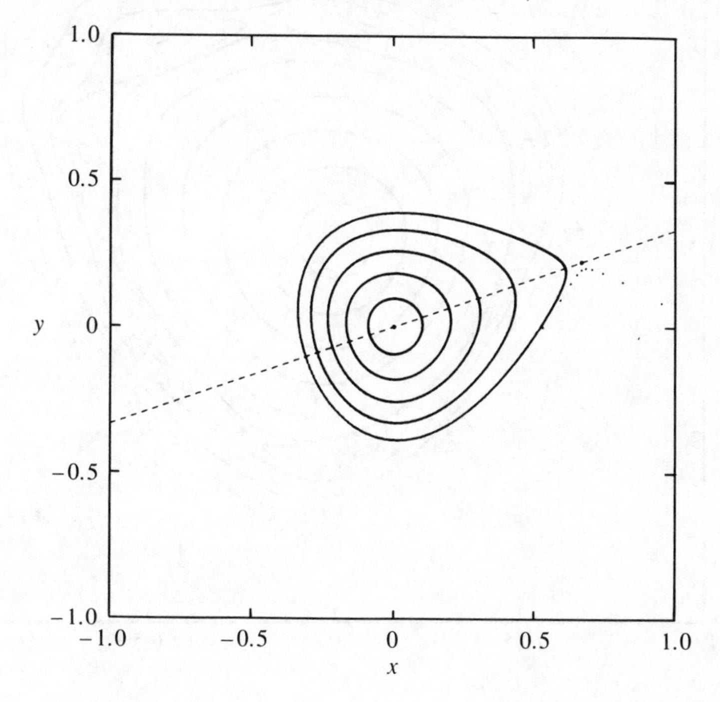

where γ_1 and γ_2 are closed curves bounding a tube of the flow in the five-dimensional extended phase space with coordinates (q_1, q_2, p_1, p_2, t). Now let γ_1 consist entirely of points such that $H = h_0$, $q_2 = t = 0$. Suppose we follow the lines of the flow φ_t^H until we return to $q_2 = 0$. Although H remains at h_0 and q_2 returns to zero, the

Figure 1.39 A selection of plotted orbits of (1.9.40) for $\cos\alpha = 0.4$. Orbits of points near to the saddle point become highly irregular. Successive iterates still move around the fixed point at (0, 0) but they are no longer confined to a closed curve. Instead they appear to spread over a two-dimensional region in an erratic manner. Eventually, they are pulled away along the unstable manifold of the saddle and, left to themselves, will cause an overflow error in the computer doing the plotting. On the other hand, orbits of points near the origin still appear to be confined to invariant circles. Between these extremes, a new feature called an *island chain* can be seen. The 'islands' themselves are formed around the points of an elliptic periodic orbit, here of period six. The 'straits' between successive islands contain a hyperbolic periodic orbit also of period six. The orbits of points near to the elliptic periodic points move from island to island, returning to an invariant circle surrounding the initial elliptic point at every sixth iteration. Some information to help the reader to observe island chains is given in Exercise 1.9.9. (After Hénon, 1969.)

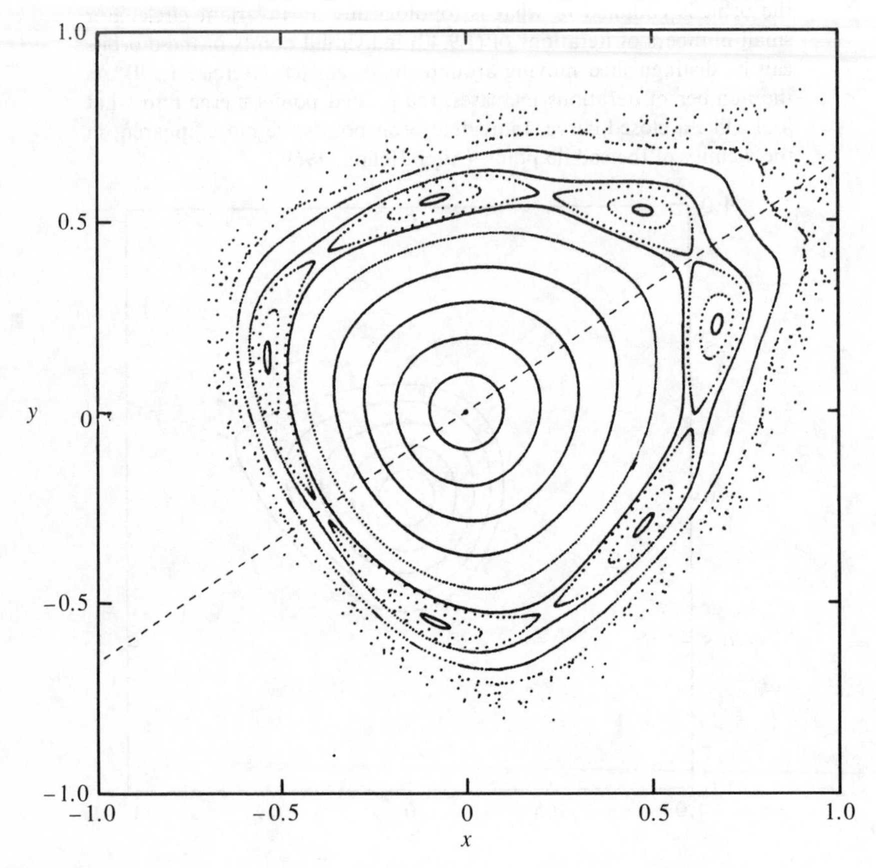

time required to reach $q_2 = 0$ will, in general, be different for different points of γ_1. Thus, in extended phase space, points of the image, γ_2, of γ_1 do not all have the same t coordinate. Let us put γ_1 and γ_2 defined in this way into (1.9.37). Since H and q_2 are constant on both curves, we have

$$\int_{\gamma_i} H \, dt = \int_{\gamma_i} p_2 \, dq_2 = 0 \tag{1.9.38}$$

for $i = 1, 2$, so that (1.9.37) becomes

$$\int_{\gamma_1} p_1 \, dq_1 = \int_{\gamma_2} p_1 \, dq_1 = \int_{\bar{\gamma}_2} p_1 \, dq_1, \tag{1.9.39}$$

where $\bar{\gamma}_2$ is the projection of γ_2 onto $t = 0$. Now $\bar{\gamma}_2$ is the image of γ_1 under the

Figure 1.40 Analogous plots to those shown in Figure 1.39 but with $\cos \alpha = 0.24$. Observe that a five-fold island chain is the dominant feature here. In fact, (see § 6.5) island chains of all periods occur but only a few are easily visible. The orbits looking like separatrices of the hyperbolic periodic points are deceptive (see Figure 1.41). (After Hénon, 1969.)

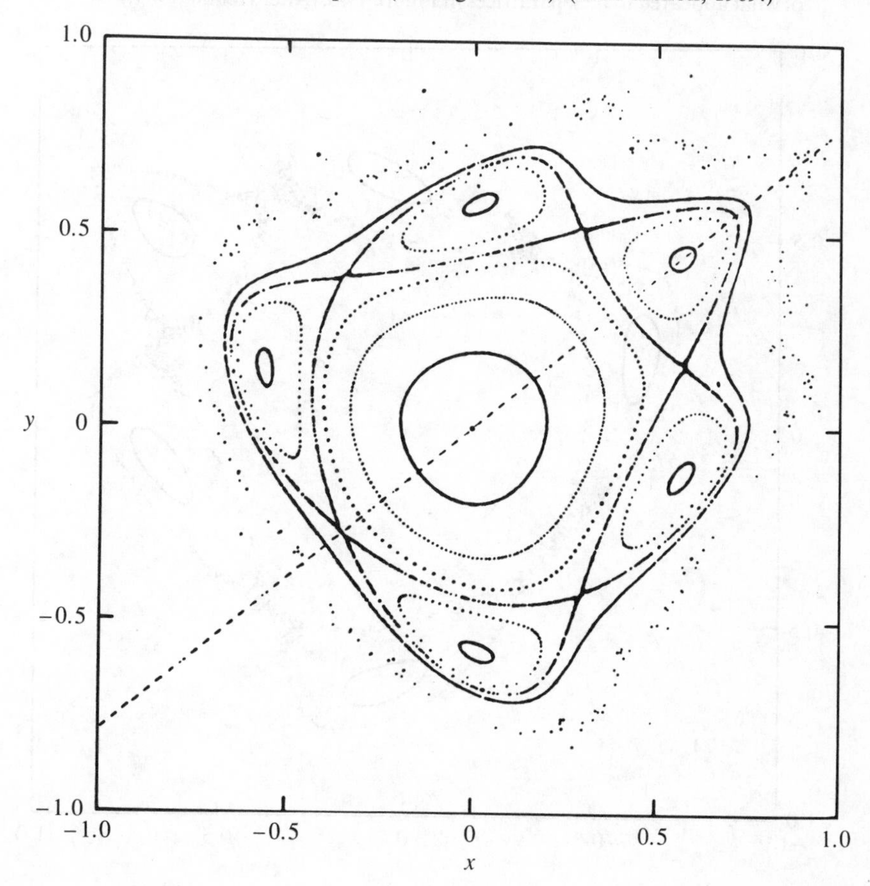

Poincaré map **P** (see Figure 1.37). Hence **P** is an area-preserving map on the section $H(\mathbf{q}, \mathbf{p}) = h_0$ and $q_2 = 0$ in the phase space of the system.

Numerical experiment has shown that Poincaré maps constructed in the manner described above exhibit complicated dynamics (Hénon, 1983, pp. 84–95; Lichtenberg & Lieberman, 1982). This complexity is a feature of area-preserving maps of the plane and it is typified by the quadratic mapping of Hénon: namely

$$\begin{aligned} x_{t+1} &= x_t \cos \alpha - y_t \sin \alpha + x_t^2 \sin \alpha, \\ y_{t+1} &= x_t \sin \alpha + y_t \cos \alpha - x_t^2 \cos \alpha, \end{aligned} \tag{1.9.40}$$

where α is a real parameter and $t \in \mathbb{Z}$ (see Hénon, 1969). This map is not constructed

Figure 1.41 Two orbits of (1.9.40) for $\cos \alpha = 0.22$. The first is the orbit of a point near an island centre giving invariant circles around the five elliptic periodic points. In the present context, it serves only to indicate the position of the islands. The remaining points are all generated by iterating a single initial point. Once again, the iterates spread out, in a stochastic manner, over a two-dimensional region in the neighbourhood of what appeared to be separatrices in Figure 1.40. (After Hénon, 1969.)

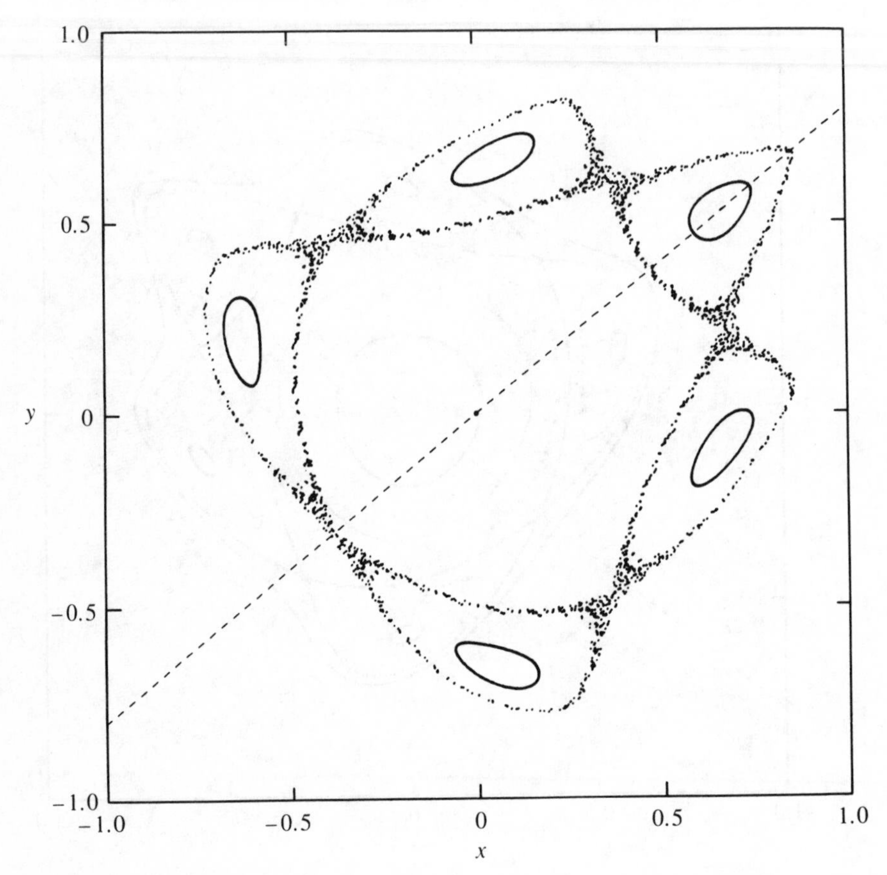

as the Poincaré map of a Hamiltonian system. Instead it represents the most general quadratic planar map that is area-preserving and has a pure rotation for its linear part. Some striking features of the dynamics of (1.9.40) are illustrated in Figures 1.38–1.42 but the reader cannot do better than to consult Hénon's excellent review (1983) for more details. Figures 1.38–1.42 show invariant circles, islands chains, chaotic orbits and their repetition on all scales. All this leads to a picture of immense complexity that is by no means fully understood. We will return to such matters in Chapters 3 and 6.

Figure 1.42 The result of magnifying a detail of Figure 1.40 containing one of the hyperbolic periodic points. A two-dimensional orbit like that shown in Figure 1.41 is apparent. However, not only are more island chains visible around the fixed point (0, 0), but also analogous islands can be seen around the adjacent elliptic periodic points. As we shall see, if these islands were again magnified, then we should find more two-dimensional orbits and more island chains and so on. Thus the complexity of the map is repeated on all scales. (After Hénon, 1969.)

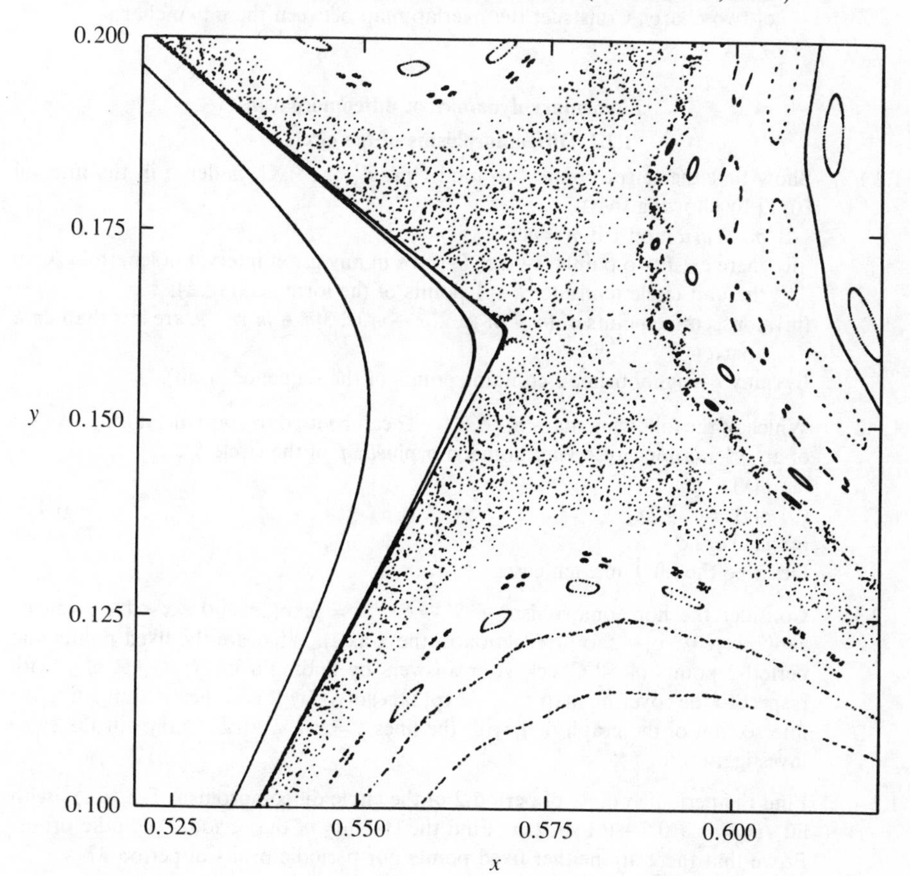

Exercises

1.1 Introduction

1.1.1 Let M be the unit circle in the complex plane. Explain how the map $\pi: \mathbb{R} \to S^1$ given by $x \mapsto \exp(ix)$ can be used to define a set of charts on the circle S^1. Define explicitly two charts which form an atlas on S^1. What is the differentiability of the overlap maps for this atlas?

1.1.2 Let M be a C^r-manifold. Show that if $\mathbf{f}: M \to M$ is a C^k-map, $k \leqslant r$, and the points $\mathbf{x}_0$ and $\mathbf{f}(\mathbf{x}_0)$ are in overlapping charts $(U_\alpha, \mathbf{h}_\alpha)$, $(U_\gamma, \mathbf{h}_\gamma)$ and $(U_\beta, \mathbf{h}_\beta)$, $(U_\delta, \mathbf{h}_\delta)$ respectively, then the map

$$\mathbf{h}_\beta \cdot \mathbf{f} \cdot \mathbf{h}_\alpha^{-1} \tag{E1.1}$$

is C^k at $\mathbf{h}_\alpha(\mathbf{x}_0)$ if and only if

$$\mathbf{h}_\delta \cdot \mathbf{f} \cdot \mathbf{h}_\gamma^{-1} \tag{E1.2}$$

is C^k at $\mathbf{h}_\gamma(\mathbf{x}_0)$. What does this imply about the differentiability of $\mathbf{f}$ at $\mathbf{x}_0$?

1.1.3 (a) Find an atlas of the torus $T^2 = \{(x \bmod 1, y \bmod 1) | (x, y) \in \mathbb{R}^2\}$ containing four charts by using the local diffeomorphism $\pi: \mathbb{R}^2 \to T^2$ given by $(x, y) \mapsto (x \bmod 1, y \bmod 1)$.
(b) Use stereographic projection on the unit sphere to obtain an atlas consisting of two charts. Construct the overlap map between these two charts.

1.2 Elementary dynamics of diffeomorphisms

1.2.2 Diffeomorphisms of the circle

1.2.1 Show that the set of points $\{m\alpha \bmod 1 | m \in \mathbb{Z}\}$, $\alpha \in \mathbb{R} \backslash \mathbb{Q}$, is dense in the interval $[0, 1]$ by showing that:
(i) $m\alpha \neq m'\alpha \bmod 1$ if $m \neq m'$;
(ii) there exist two points $\theta + m\alpha$, $\theta + m'\alpha$ in any given interval of length $2\pi/k$ on the unit circle (consider $k+1$ points of the form $m\alpha$, $m \in \mathbb{Z}$);
(iii) consecutive points of $(m-m')\alpha$, $2(m-m')\alpha$, $3(m-m')\alpha, \ldots$ are less than $2\pi/k$ apart;
(iv) any ε-neighbourhood contains points of the sequence in (iii).

1.2.2 Which of the following maps $g: [0, 1] \to \mathbb{R}$ can be used to construct a lift $\bar{f}: \mathbb{R} \to \mathbb{R}$ of an orientation-preserving homeomorphism f of the circle S^1.
(a) $g(x) = x^2$,
(b) $g(z) = x^2 - 2x$, (E1.3)
(c) $g(x) = 2x^2 - x$.
Describe the lift $\bar{f}$ in each case.

1.2.3 Consider the homeomorphism $f: S^1 \to S^1$, $S^1 = \{\exp(2\pi ix) | 0 \leqslant x < 1\}$, given by $\exp(2\pi ix) \mapsto \exp(-2\pi ix)$ (reflection in the x-axis). What are the fixed points and period-2 points of f? Check your answers by finding a lift $\bar{f}: \mathbb{R} \to \mathbb{R}$ of f with respect to the covering map $\pi: x \mapsto \exp(2\pi ix)$ of S^1 by $\mathbb{R}$ and then investigating the intersection of the graph of $\bar{f}$ with the lines $y = x + n$, $n \in \mathbb{Z}$. Carry out the same investigation for $\bar{f}^2$.

1.2.4 Find the periodic points of period 2 of the circle diffeomorphism $f: S^1 \to S^1$ with lift $\bar{f}(x) = x + 0.5 + 0.1 \sin 2\pi x$. Find the stability of the period-2 periodic orbits. Prove that there are neither fixed points nor periodic orbits of period 3?

1.2.5 Consider an orientation-*reversing* homeomorphism $f: S^1 \to S^1$. Show that any lift $\bar{f}: \mathbb{R} \to \mathbb{R}$ of f will have the property that $\bar{f}$ is strictly decreasing and $\bar{f}(x+1) = \bar{f}(x) - 1$. Show that f always has two fixed points. Is f^2 orientation-reversing or -preserving?

1.3 Flows and differential equations

1.3.1 Use (1.3.2) to show that $(\varphi_t)^{-1} = \varphi_{-t}$ for each $t \in \mathbb{R}$. Hence show that it follows from Definition 1.3.1 that $\varphi_t: M \to M$ is a diffeomorphism for every real t.

1.3.2 Let φ be a flow on the manifold M and suppose that the orbits $\{\varphi_t(\mathbf{x}_0)\}$ and $\{\varphi_t(\mathbf{x}_1)\}$ intersect. Prove that the orbits coincide.

1.3.3 Let γ be a closed orbit of the flow φ on the manifold M and suppose there exists $T > 0$ and $\mathbf{x}_0 \in \gamma$ such that $\varphi_T(\mathbf{x}_0) = \mathbf{x}_0$. Prove that $\varphi_T(\mathbf{x}) = \mathbf{x}$, for every $\mathbf{x} \in \gamma$.

Locate two closed orbits γ_1 and γ_2 and positive periods T_1 and T_2 for the flow of

$$\dot{r} = r(r-1)(r-2); \qquad \dot{\theta} = r^2. \tag{E1.4}$$

1.3.4 Verify that

$$\varphi(t, x) = x \exp(t)/[x \exp(t) - x + 1] \tag{E1.5}$$

is a flow on [0, 1] and find its associated vector field. Why is it not a flow on $\mathbb{R}$?

1.3.5 Find the differential equations associated with the following (local) flows:

(a) $\varphi_t(x) = \dfrac{x}{(1-2x^2t)^{\frac{1}{2}}}$ on $\mathbb{R}$; (E.1.6)

(b) $\varphi_t(x, y) = \left(x \exp(t), \dfrac{y}{1-yt} \right)$ on $\mathbb{R}^2$. (E1.7)

Convince yourself that, for any interval of t of the form $(-a, a)$, there is a neighbourhood of origin in which (E1.6) and (E1.7), respectively, correctly describe the solutions of the differential equations you have obtained.

1.4 Invariant sets

1.4.1 Find the minimal closed invariant sets for (a) irrational and (b) rational rotations of the circle. What is the most general closed invariant set in both cases?

1.4.2 Prove that the fixed and periodic points of a diffeomorphism $\mathbf{f}$ on a manifold M lie in its non-wandering set Ω. Show that Ω is (a) closed and (b) invariant under $\mathbf{f}$.

1.4.3 Prove that the non-wandering set of a flow φ on a manifold M is (a) closed and (b) invariant. Show that a closed orbit of φ is a subset of the non-wandering set.

1.4.4 Describe the behaviour of the diffeomorphism $f(x) = ax$, $a \in \mathbb{R}$, when

(i) $a < -1$;
(ii) $a = -1$;
(iii) $-1 < a < 0$;
(iv) $0 < a < 1$; (E1.8)
(v) $a = 1$;
(vi) $1 < a$.

Give $L_\alpha(x)$ and $L_\omega(x)$, $x \neq 0$, for each case.

1.4.5 Sketch examples of phase portraits on the plane with a non-empty, compact limit set containing (a) one; (b) two; (c) three; (d) four fixed points. Suggest how differential equations with flows of these types might be constructed.

1.4.6 Use the Poincaré–Bendixson Theorem to show that the Van der Pol oscillator

$$\ddot{x} + \varepsilon(x^2 - 1)\dot{x} + x = 0, \qquad \varepsilon < 0, \tag{E1.9}$$

has at least one stable limit cycle for sufficiently small values of ε.

1.5 Conjugacy

1.5.1 Show that:
(i) $f(x) = 2x$ and $g(x) = 8x$ are topologically conjugate on $\mathbb{R}$ but not differentiably conjugate;
(ii) $f(x) = 2x$ and $g(x) = -2x$ are not topologically conjugate by showing conjugacy preserves orientation of a map;
(iii) $f(x) = 2x$ and $g(x) = \frac{1}{2}x$ are not topologically conjugate by investigating the nature of the fixed point at the origin in the two cases.

1.5.2 Prove that $\varphi: x \mapsto x^{2n+1}$, $n \in \mathbb{N}$, is a topological conjugacy of the diffeomorphisms $f(x) = 2x$ and $g(y) = 2^{2n+1}y$ on $\mathbb{R}$. Why is there no differentiable conjugacy when $n > 0$.

1.5.3 Let $\mathbf{f}, \mathbf{g}: \mathbb{R}^n \to \mathbb{R}^n$ be C^k-conjugate $(k \geqslant 1)$ diffeomorphisms by $\mathbf{h}: \mathbb{R}^n \to \mathbb{R}^n$ and suppose $\mathbf{f}(\mathbf{0}) = \mathbf{0}$. Prove that the Jacobian matrices of $\mathbf{f}$ at $\mathbf{0}$ and $\mathbf{g}$ at $\mathbf{h}(\mathbf{0})$ are similar. What does this imply about the eigenvalues of $\mathbf{Df}(\mathbf{0})$ and $\mathbf{Dg}(\mathbf{h}(\mathbf{0}))$?

Show that $f(x) = \alpha x$ and $g(x) = \beta x$ are not C^k-conjugate for $\alpha \neq \beta$. When are f and g not C^0-conjugate?

1.5.4 Let f, g be diffeomorphisms on $\mathbb{R}$ given by

$$f(x) = x + \sin x \tag{E1.10}$$

and

$$g(x) = x + h_k(x), \tag{E1.11}$$

where

$$h_k(x) = x - \frac{x^3}{3!} + \frac{x^5}{5!} + \cdots + (-1)^k \frac{x^{2k+1}}{(2k+1)!}, \tag{E1.12}$$

$k \in \mathbb{Z}^+$. Prove that f and g are not topologically conjugate for any k.

1.5.5 Find the number of period-2 points of the diffeomorphisms f and g of the circle S^1, where the lifts are:

$$\bar{f}(x) = x + 0.5 + 0.1 \sin 2\pi x; \tag{E1.13}$$

$$\bar{g}(x) = x + 0.3 + 0.1 \sin 2\pi x. \tag{E1.14}$$

Show that f and g are not topologically conjugate.

1.5.6 Prove that the two linear systems $\dot{\mathbf{x}} = \mathbf{A}\mathbf{x}$, $\dot{\mathbf{y}} = \mathbf{B}\mathbf{y}$, $\mathbf{x}, \mathbf{y} \in \mathbb{R}^n$, have flows which are linearly conjugate if and only if the matrices $\mathbf{A}$ and $\mathbf{B}$ are similar.

1.5.7 Let φ, ψ be flows on $\mathbb{R}^n$ and suppose that φ has a fixed point at the origin. If φ and ψ are C^k-conjugate, $k > 0$, by $\mathbf{h}: \mathbb{R}^n \to \mathbb{R}^n$, show that the vector fields $\mathbf{X}$ and $\mathbf{Y}$ of φ and ψ are such that the matrices $\mathbf{DX}(\mathbf{0})$ and $\mathbf{DY}(\mathbf{h}(\mathbf{0}))$ are similar. Compare your answer with that of Exercise 1.5.6 and comment on why the converse of the above is not true in the non-linear case.

1.5.8 Find the rotation number of the circle homeomorphism f with lift $\bar{f}: \mathbb{R} \to \mathbb{R}$ given by:

(i) $\bar{f}(x) = x + \frac{2}{3}$;

(ii) $\bar{f}(x) = x^3 + \frac{3}{4}$, $0 \leqslant x < 1$, $\bar{f}(x+1) = \bar{f}(x) + 1$;

(iii) $\bar{f}(x) = x + \frac{1}{2} + \dfrac{1}{2\pi} \sin 2\pi x$; (E1.15)

by locating fixed or periodic points of f.

1.6 Equivalence of flows

1.6.1 Show that flows

$$\exp(\mathbf{J}t),\ \mathbf{J} = \begin{pmatrix} 0 & -\beta \\ \beta & 0 \end{pmatrix} \text{ and } \exp(\mathbf{J}_0 t),\ \mathbf{J}_0 = \begin{pmatrix} 0 & -1 \\ 1 & 0 \end{pmatrix}$$

are topologically equivalent. Hence prove that all linear flows, $\exp(\mathbf{A}t)$, for which $\mathbf{A}$ has pure imaginary eigenvalues are topologically equivalent to $\exp(\mathbf{J}_0 t)$. Why are these flows not all topologically conjugate?

1.6.2 It is often easy to recognise why two flows are *not* topologically equivalent by noting key features in their phase portraits. Describe, for each diagram in Figure E1.1, a distinguishing feature preserved by topological equivalence which is not shared by the others. Explain your answers. Why is the invariant circle in (*b*) not a limit cycle?

Figure E1.1

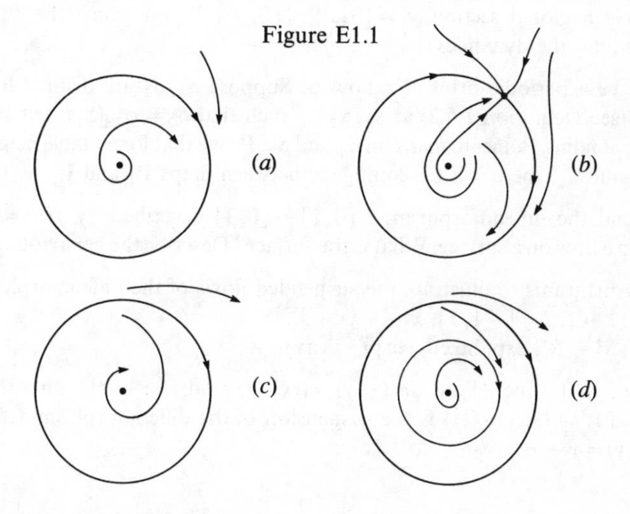

1.6.3 Show that the topological types of a saddle and node are different by considering the separatrices of the saddle.

1.6.4 Consider the system $\dot{\mathbf{x}} = \mathbf{X}_\alpha(\mathbf{x})$, $\alpha \in \mathbb{R}$,

$$\dot{x} = 1, \qquad \dot{y} = \alpha \tag{E1.16}$$

on $\mathbb{R}^2$ and the flow on T^2 induced by the map $\pi: \mathbb{R}^2 \to T^2$ where $\pi(x, y) = (x \bmod 1, y \bmod 1)$. Show that the phase portraits of the system (E1.16) for α (a) rational and (b) irrational are not topologically equivalent.

1.6.5 Let φ and φ' be topologically conjugate flows on the manifold M and ψ and ψ' be topologically conjugate flows on the manifold N. Prove that the product flow

$\varphi \times \psi$ is topologically conjugate to $\varphi' \times \psi'$ on $M \times N$. Show that this result does not extend to topological equivalence of flows by considering flows on $M = N = S^1$.

1.6.6 The topological types of linear flow, $\exp(\mathbf{A}t)$, given in Figure 1.26 can be sub-divided into *algebraic types*. Sketch phase portraits for $\exp(\mathbf{A}t)$ when:
(a) $[\mathrm{Tr}\,\mathbf{A}]^2 > 4\,\mathrm{Det}\,\mathbf{A}$ (nodes);
(b) $[\mathrm{Tr}\,\mathbf{A}]^2 = 4\,\mathrm{Det}\,\mathbf{A}$ (improper nodes); and
(c) $[\mathrm{Tr}\,\mathbf{A}]^2 < 4\,\mathrm{Det}\,\mathbf{A}$, $\mathrm{Tr}\,\mathbf{A} \neq 0$ (foci).
Modify Figure 1.26 to show these algebraic types on the ($\mathrm{Tr}\,\mathbf{A}$, $\mathrm{Det}\,\mathbf{A}$)-plane.

1.7 Poincaré maps and suspensions

1.7.1 Show that a map $f: \mathbb{R} \to \mathbb{R}$ of the form $f(x) = \alpha x$, $\alpha \in \mathbb{R}$, has a fixed point at $x = 0$ and that it is stable or unstable according as $|\alpha| < 1$ or $|\alpha| > 1$ respectively.
Consider the time-one map φ_1 of the flow of $\dot{x} = x - x^2$. Deduce that φ_1 has fixed points at $x = 0$ and $x = 1$. Obtain the linear approximations to φ_1 at these points and deduce their stability. Sketch the behaviour of the map φ_1. Compare your results with the phase portrait of the flow.

1.7.2 Show that flows with fixed points do not satisfy the requirements for the existence of a global cross section. In what sense can the system $\dot{r} = r - r^3$, $\dot{\theta} = 1$, be said to have a global section, $S = \{(r, \theta) | r \geqslant 0, \theta = 0\}$, on which the Poincaré map determines the dynamics?

1.7.3 Let γ be a periodic orbit of a flow φ. Suppose S_1, S_2 are distinct local sections for γ (see Definition 1.5.2) at $\mathbf{x}_1, \mathbf{x}_2 \in \gamma$ such that $S_2 = \varphi_{T_0}(S_1)$. Let $\mathbf{P}_1, \mathbf{P}_2$ be the corresponding Poincaré maps on S_1 and S_2. Prove that for suitable neighbourhoods of $\mathbf{x}_1$ and $\mathbf{x}_2$, there is a C^1-conjugacy between maps $\mathbf{P}_1$ and $\mathbf{P}_2$.

1.7.4 Suspend the diffeomorphism $f: [0, 1] \to [0, 1]$ described by $f(x) = \frac{1}{2}(x + x^2)$ to obtain a flow on a surface. What is the surface? Describe the behaviour of the flow.

1.7.5 Draw diagrams to illustrate the suspended flows of the diffeomorphisms
(a) $f: I \to I$, $I = [-1, 1]$, $x \mapsto -\frac{3}{2}x + \frac{1}{2}x^3$,
(b) $f: S^1 \to S^1$, $\exp(2\pi i x) \mapsto \exp(-2\pi i x)$. (E1.17)

1.7.6 Show that the flow $\varphi_t(x, y) = ((x + t) \bmod 1,\, y + \alpha t)$ on the cylinder $(x \bmod 1, y) | (x, y) \in \mathbb{R}^2\}$ is the suspension of the diffeomorphism $f: \mathbb{R} \to \mathbb{R}$ given by $f(y) = y + \alpha$.

1.8 Periodic non-autonomous systems

1.8.1 Let $\dot{\mathbf{x}} = \mathbf{X}(\mathbf{x}, t)$, $(\mathbf{x}, t) \in \mathbb{R}^n \times \mathbb{R}$, be a periodic differential equation with $\mathbf{X}(\mathbf{x}, t) = \mathbf{X}(\mathbf{x}, t + T)$ for some $T > 0$. Show that the transformations $t' = (2\pi/T)t$ and $\mathbf{X}'(\mathbf{x}, t') = (T/2\pi)\mathbf{X}(\mathbf{x}, t)$ give a 2π-periodic system $\mathrm{d}\mathbf{x}/\mathrm{d}t' = \mathbf{X}'(\mathbf{x}, t')$.

1.8.2 Prove that if $\mathbf{x} = \xi(t)$ is a solution of $\dot{\mathbf{x}} = \mathbf{X}(\mathbf{x}, t)$, with $\mathbf{X}(\mathbf{x}, t + T) = \mathbf{X}(\mathbf{x}, t)$ and $(\mathbf{x}, t) \in \mathbb{R}^n \times \mathbb{R}$, then $\mathbf{x} = \xi(t + T)$ is also a solution. Show that such systems can have non-periodic solutions by considering $\dot{x} = (1 + \sin t)x$ with $(\mathbf{x}, t) \in \mathbb{R} \times \mathbb{R}$.

1.8.3 (a) Show that the set of solutions of $\dot{\mathbf{x}} = \mathbf{A}(t)\mathbf{x}$, $(\mathbf{x}, t) \in \mathbb{R}^n \times \mathbb{R}$, form a vector space.
(b) Let $\xi_i(t)$, $i = 1, \ldots, n + 1$ be a set of solutions. Show that there exist

$\alpha_1, \ldots, \alpha_{n+1}$, not all zero, such that $\sum_{i=1}^{n+1} \alpha_i \xi_i(0) = \mathbf{0}$. Use uniqueness of solution to show that $\sum_{i=1}^{n+1} \alpha_i \xi_i(t) \equiv \mathbf{0}$.

(c) Show that the vector space of solutions is n-dimensional.

1.8.4 Define the state transition matrix $\boldsymbol{\varphi}(t, t_0)$ of the system $\dot{\mathbf{x}} = \mathbf{A}(t)\mathbf{x}$, $(\mathbf{x}, t) \in \mathbb{R}^n \times \mathbb{R}$, $\mathbf{A}(t+T) = \mathbf{A}(t)$. Prove that

(i) $\boldsymbol{\varphi}(t+T, t_0+T) = \boldsymbol{\varphi}(t, t_0)$,
(ii) $\boldsymbol{\varphi}(t, 0)^{-1} = \boldsymbol{\varphi}(0, t)$. (E1.18)

Use these results to show that

$$\boldsymbol{\varphi}(t+T, t) = \boldsymbol{\varphi}(0, t)^{-1}\boldsymbol{\varphi}(T, 0)\boldsymbol{\varphi}(0, t). \tag{E1.19}$$

What does this imply about the family of Poincaré maps $\mathbf{P}_{\theta_0}: \Sigma_{\theta_0} \to \Sigma_{\theta_0}$, $\theta_0 \in [0, 2\pi]$ of the system

$$\dot{\mathbf{x}} = \mathbf{A}(\theta)\mathbf{x}, \quad \dot{\theta} = 1, \tag{E1.20}$$

where $\mathbf{A}(\theta + 2\pi) = \mathbf{A}(\theta)$, and $\Sigma_{\theta_0} = \{(\mathbf{x}, \theta_0) | \mathbf{x} \in \mathbb{R}^2\}$?

1.8.5 The state transition matrix is useful where $\mathbf{A}$ is independent of t or aperiodic.

(a) Verify that $\xi_1(t) = \begin{pmatrix} \exp(\lambda_1 t) \\ 0 \end{pmatrix}$ and $\xi_2(t) = \begin{pmatrix} 0 \\ \exp(\lambda_2 t) \end{pmatrix}$ are linearly independent solutions of $\dot{\mathbf{x}} = \mathbf{A}\mathbf{x}$, $\mathbf{A} = \begin{pmatrix} \lambda_1 & 0 \\ 0 & \lambda_2 \end{pmatrix}$. Hence find $\boldsymbol{\varphi}(t, t_0)$ for this system.

(b) Verify that $\xi_1(t) = \begin{pmatrix} 2 \\ \exp(t) \end{pmatrix}$ and $\xi_2(t) = \begin{pmatrix} \exp(t) \\ 1 \end{pmatrix}$ are a basis for the solution space of $\dot{\mathbf{x}} = \mathbf{A}(t)\mathbf{x}$ with $\mathbf{A}(t) = \begin{pmatrix} 1 & -2\exp(-t) \\ \exp(t) & -1 \end{pmatrix}$. Hence construct $\boldsymbol{\varphi}(t, 0)$ and find $\mathbf{x}(t)$ given that $\mathbf{x}(0) = \begin{pmatrix} 3 \\ 1 \end{pmatrix}$.

1.8.6 Let $\boldsymbol{\varphi}(t, t_0)$ be the state transition matrix of the system $\dot{\mathbf{x}} = \mathbf{A}(t)\mathbf{x}$ and define $W(t) = \mathrm{Det}(\boldsymbol{\varphi}(t, 0))$. Prove

(i) $\boldsymbol{\varphi}(t+h, 0) = \boldsymbol{\varphi}(t, 0) + h\mathbf{A}(t)\boldsymbol{\varphi}(t, 0) + O(h^2)$; (E1.21)

(ii) $W(t+h) = \mathrm{Det}(\mathbf{I} + h\mathbf{A}(t) + O(h^2))W(t)$

$$= \prod_{i=1}^{n} (1 + h\lambda_i(t) + O(h^2))W(t), \tag{E1.22}$$

where λ_i, $i = 1, \ldots, n$ are the eigenvalues of $\mathbf{A}$;

(iii) $\dot{W}(t) = \mathrm{Tr}(\mathbf{A}(t))W(t)$. (E1.23)

1.8.7 Find solutions for the system

$$\begin{pmatrix} \dot{x}_1 \\ \dot{x}_2 \end{pmatrix} = \begin{pmatrix} \cos t & -\sin t \\ \sin t & \cos t \end{pmatrix} \begin{pmatrix} x_1 \\ x_2 \end{pmatrix} \tag{E1.24}$$

and hence, or otherwise, obtain the period advance map $\boldsymbol{\varphi}(2\pi, 0) = \mathbf{P}$. Calculate the eigenvalues of $\mathbf{P}$ and determine the stability of the null solution of (E1.24).

1.8.8 If $\boldsymbol{\varphi}(t, t_0)$ is the state transition matrix of the system

$$\dot{\mathbf{x}} = \mathbf{A}(t)\mathbf{x}, \tag{E1.25}$$

show that

$$\dot{\mathbf{x}} = \mathbf{A}(t)\mathbf{x} + \mathbf{B}(t) \tag{E1.26}$$

has solution

$$\mathbf{x}(t) = \varphi(t, t_0)\left(\mathbf{x}_0 + \int_{t_0}^{t} \varphi(t_0, \tau)\mathbf{B}(\tau)\,\mathrm{d}\tau\right) \tag{E1.27}$$

when $\mathbf{x} = \mathbf{x}_0$ at $t = t_0$.

Find the solution of

$$\dot{\mathbf{x}} = \begin{pmatrix} 0 & -1 \\ 1 & 0 \end{pmatrix}\mathbf{x} + \begin{pmatrix} \sin t \\ \cos t \end{pmatrix} \tag{E1.28}$$

when $\mathbf{x}(0) = \mathbf{x}_0$.

1.9 Hamiltonian flows and Poincaré maps

1.9.1 Find the phase portraits of the flows in $\mathbb{R}^2$ with Hamiltonian:
(a) $H(x_1, x_2) = x_1^2 + x_2^2$;
(b) $H(x_1, x_2) = x_1^2 + x_2^3$; (E1.29)
(c) $H(x_1, x_2) = x_1^2 + x_2^4 - x_2^2$.

1.9.2 Sketch the phase portraits for Hamilton's equations when $H(x_1, x_2) = x_1^2 + x_2^3 + \mu x_2$ and
(i) $\mu < 0$;
(ii) $\mu = 0$;
(iii) $\mu > 0$.

1.9.3 Prove that the fixed point of a planar flow with a quadratic Hamiltonian is generically either a centre or a hyperbolic saddle.

1.9.4 Show that, for $\mu > 0$, the Hamiltonian vector field given by $H = -\mu r^2 + r^4 + r^5 \cos 5\theta$, where (r, θ) are plane polar coordinates, has 11 fixed points consisting of 6 centres and 5 saddles. Show that the separatrices of the saddles form a chain of 5 islands around the origin such that each island contains a centre.

1.9.5 Prove that a change of variables on $\mathbb{R}^2$ is symplectic if and only if it is orientation- and area-preserving. Show that:
(i) the change from Cartesian to plane polar coordinates, $(x, y) \mapsto (r, \theta)$, is not symplectic;
(ii) the transformation $(x, y) \mapsto (\tau, \theta)$, where $\tau = r^2/2$, is symplectic.
Illustrate the fact that symplectic transformations preserve the form of Hamilton's equations by applying the transformations in (i) and (ii) to the system.

$$\dot{x} = y - y(x^2 + y^2); \qquad \dot{y} = -x + x(x^2 + y^2). \tag{E1.30}$$

1.9.6 Let $\mathbf{h} \in L(\mathbb{R}^4)$ be given by $\mathbf{h}(\mathbf{x}) = \mathbf{P}\mathbf{x}$ with

$$\mathbf{P} = \begin{pmatrix} \mathbf{A} & \mathbf{B} \\ \mathbf{C} & \mathbf{D} \end{pmatrix}, \tag{E1.31}$$

where $\mathbf{A}, \mathbf{B}, \mathbf{C}, \mathbf{D}$ are 2×2 matrices. Show that $\mathbf{h}$ is symplectic if and only if

$$\mathbf{D}^{\mathrm{T}}\mathbf{A} - \mathbf{B}^{\mathrm{T}}\mathbf{C} = \mathbf{I}, \qquad \mathbf{D}^{\mathrm{T}}\mathbf{B} = \mathbf{B}^{\mathrm{T}}\mathbf{D} \qquad \text{and} \qquad \mathbf{C}^{\mathrm{T}}\mathbf{A} = \mathbf{A}^{\mathrm{T}}\mathbf{C}, \tag{E1.32}$$

where $\mathbf{I}$ is the 2×2 unit matrix. Hence construct a counterexample to show that $\mathrm{Det}(\mathbf{Dh}(\mathbf{x})) \equiv 1$ does not imply $\mathbf{h}$ is symplectic.

1.9.7 Show that the Volterra–Lotka equations

$$\dot{x} = (a - by)x, \qquad \dot{y} = -(c - fx)y \tag{E1.33}$$

do *not* form a Hamiltonian system. Show that the change of variable $x = \exp(q)$, $y = \exp(p)$ allows (E1.33) to be written as a Hamiltonian system with Hamiltonian

$$H(q, p) = ap - b\exp(p) + cq - f\exp(q). \tag{E1.34}$$

Why can it be concluded (without calculation) that the transformation $(q, p) \mapsto (x, y)$ is not symplectic?

1.9.8 (a) Find action-angle variables for the Hamiltonian system

$$\dot{x} = \omega y, \qquad \dot{y} = -\omega x \tag{E1.35}$$

by using $I = x^2 + y^2$.

(b) Consider the pendulum equations

$$\dot{x} = y, \qquad \dot{y} = -\sin x \tag{E1.36}$$

and show that the transformation Ψ: $(x, y) \mapsto (I, \theta)$, where $I = (y^2/2) - \cos x$ and θ is the polar angle, has Jacobian $1 + O(I)$. Show that the transformed vector field is

$$\dot{I} = 0, \qquad \dot{\theta} = -1 + O(|I|). \tag{E1.37}$$

In view of Exercise 1.9.5 what do these results imply about Ψ?

1.9.9 Write a computer program to plot orbits of the Hénon area-preserving map (1.9.40).

(a) For $\cos\alpha = 0.8$, plot several orbits with initial points $(x, 0)$ for $x \in (0, \frac{1}{2}]$. Compare your results with Figure 1.38.

(b) For $\cos\alpha = 0.4$, plot orbits with initial points $(x, 0)$ for $x \in (\frac{1}{2}, 1]$. Observe a six-fold island chain and chaotic orbits.

2

Local properties of flows and diffeomorphisms

In this chapter we consider the topological behaviour of diffeomorphisms and flows in the neighbourhood of an isolated fixed point. Let $\mathbf{x}_i^*$ be an isolated fixed point of the diffeomorphism $\mathbf{f}_i$ for $i = 1, 2$. Then the fixed points $\mathbf{x}_1^*$ and $\mathbf{x}_2^*$ are said to be of the same *topological type* if $\mathbf{f}_1$ and $\mathbf{f}_2$ are topologically conjugate when restricted to sufficiently small neighbourhoods of their respective fixed points. A similar definition holds for flows $\boldsymbol{\varphi}_1$ and $\boldsymbol{\varphi}_2$ but topological equivalence replaces topological conjugacy. The central result is that the topological type of a fixed point $\mathbf{x}^*$ is determined by the linear approximation to $\mathbf{f}$ or $\boldsymbol{\varphi}$ at $\mathbf{x}^*$ *provided* the fixed point is *hyperbolic*. We begin by defining this term for linear diffeomorphisms and flows.

2.1 Hyperbolic linear diffeomorphisms and flows

Definition 2.1.1 *A linear diffeomorphism* $\mathbf{A}: \mathbb{R}^n \to \mathbb{R}^n$ *is said to be* hyperbolic *if it has no eigenvalues with modulus equal to unity.*

The main properties of such diffeomorphisms are summarised in the following theorem.

Theorem 2.1.1 *If* $\mathbf{A}: \mathbb{R}^n \to \mathbb{R}^n$ *is a hyperbolic linear diffeomorphism, then there are subspaces* E^s *and* $E^u \subseteq \mathbb{R}^n$*, invariant under* $\mathbf{A}$*, such that* $\mathbf{A}|E^s$ *is a contraction,* $\mathbf{A}|E^u$ *is an expansion and* $E^s \oplus E^u = \mathbb{R}^n$.

A linear diffeomorphism $\mathbf{L}: \mathbb{R}^l \to \mathbb{R}^l$ is a *contraction* (*expansion*) if all its eigenvalues have modulus *less than* (*greater than*) unity. For example, if $\mathbf{L}$ has distinct eigenvalues then, for any $\mathbf{x} \in \mathbb{R}^l$ and $k \in \mathbb{Z}^+$,

$$\mathbf{L}^k\mathbf{x} = \mathbf{M}^{-1}\mathbf{D}^k\mathbf{M}\mathbf{x}, \tag{2.1.1}$$

where

$$\mathbf{D} = [\lambda_i \delta_{ij}]_{i,j=1}^l \qquad \text{with} \qquad \delta_{ij} = \begin{cases} 1 & i = j \\ 0 & i \neq j. \end{cases}$$

The ith column of $\mathbf{M}$ is an eigenvector of $\mathbf{L}$ with eigenvalue λ_i. If $|\lambda_i| < 1$ for every $i = 1, \ldots, l$, then $\mathbf{D}^k \to \mathbf{0}$ as $k \to \infty$ and the orbit under $\mathbf{L}$ of any point in $\mathbb{R}^l$ ultimately approaches the origin; hence the term 'contraction'. Similarly, if $|\lambda_i| > 1$ for every $i = 1, \ldots, l$ the orbit 'expands' away from the origin. Corresponding results when $\mathbf{L}$ has eigenvalues with multiplicity greater than unity are considered in Exercise 2.1.1

The subspaces E^s and E^u of Theorem 2.1.1 are easily identified as the eigenspaces of eigenvalues with modulus less than 1 and greater than 1, respectively. Indeed, they are usually called the *stable* and *unstable eigenspaces* of $\mathbf{A}$. The direct sum of E^s and E^u gives the whole of $\mathbb{R}^n$, since, for hyperbolic $\mathbf{A}$, there is no eigenvalue that is not associated with either E^s or E^u. Of course, expansions ($E^u = \mathbb{R}^n$) and contractions ($E^s = \mathbb{R}^n$) are hyperbolic. When $E^s, E^u \neq \mathbb{R}^n$ the diffeomorphism is said to be of 'saddle-type' (see Figure 2.1).

Given any linear transformation $\mathbf{A}: \mathbb{R}^n \to \mathbb{R}^n$, there is a corresponding linear flow on $\mathbb{R}^n$, given by

$$\varphi_t(\mathbf{x}) = \exp(\mathbf{A}t)\mathbf{x}, \tag{2.1.2}$$

where $\exp(\mathbf{A}t)$ is the *exponential matrix* defined by

$$\exp(\mathbf{A}t) = \sum_{k=0}^{\infty} \frac{(\mathbf{A}t)^k}{k!}. \tag{2.1.3}$$

It is easy to show that $\exp(\mathbf{A}t)\mathbf{x}$ has vector field $\mathbf{A}\mathbf{x}$ and thus $\exp(\mathbf{A}t)\mathbf{x}_0$ is the solution of the linear differential equation $\dot{\mathbf{x}} = \mathbf{A}\mathbf{x}$ passing through $\mathbf{x}_0$ at $t = 0$.

Definition 2.1.2 *The linear flow* $\exp(\mathbf{A}t)\mathbf{x}$ *is said to be* hyperbolic *if* $\mathbf{A}$ *has no eigenvalues with zero real part.*

If $\exp(\mathbf{A}t)\mathbf{x}$ is hyperbolic then $\mathbf{A}$ must be non-singular and $\mathbf{A}\mathbf{x} = \mathbf{0}$ has only the trivial solution $\mathbf{x} = \mathbf{0}$. It follows that the origin is the only fixed point of the flow. In such cases, both the fixed point and the vector field $\mathbf{A}\mathbf{x}$ are said to be hyperbolic.

If the linear flow $\exp(\mathbf{A}t)\mathbf{x}$ is hyperbolic then the exponential matrix, $\exp(\mathbf{A}t)$, represents a hyperbolic linear diffeomorphism for each $t \neq 0$. Equation (2.1.3) implies that an eigenvector of $\mathbf{A}$ with eigenvalue λ is also an eigenvector of $\exp(\mathbf{A}t)$ with eigenvalue $\exp(\lambda t)$. If $\exp(\mathbf{A}t)\mathbf{x}$ is a hyperbolic flow, $\mathrm{Re}\,\lambda_i \neq 0$ for each i, hence $|\exp(\lambda_i t)| = \exp(\mathrm{Re}\,\lambda_i t) \neq 1$, for all $t \neq 0$, so that $\exp(\mathbf{A}t)$ is hyperbolic. It follows that Theorem 2.1.1 applies and $\mathbb{R}^n$ can be decomposed into stable (E^s) and unstable (E^u) subspaces associated with the flow. Clearly, $E^s(E^u)$ is the direct sum of eigenspaces associated with eigenvalues of $\mathbf{A}$ for which $\mathrm{Re}\,\lambda_i < 0\,(>0)$. A linear flow $\exp(\mathbf{L}t)\mathbf{x}$ on $\mathbb{R}^l$ is said to be a *contraction* (*expansion*) if all the eigenvalues of $\mathbf{L}$ have *negative* (*positive*) real parts. Thus $\exp(\mathbf{A}t)|E^s$ is a contraction and $\exp(\mathbf{A}t)|E^u$ is an expansion.

We have already suggested, in § 1.6, that hyperbolic linear flows can be classified into a finite number of types by using topological equivalence. The following theorem makes a precise statement of this result.

Theorem 2.1.2 *Let* $\dot{\mathbf{x}} = \mathbf{A}\mathbf{x}$ *define a hyperbolic linear flow on* $\mathbb{R}^n$ *with* $\dim E^s = n_s$. *Then* $\dot{\mathbf{x}} = \mathbf{A}\mathbf{x}$ *is topologically equivalent to the system*

$$\begin{aligned} \dot{\mathbf{x}}_s &= -\mathbf{x}_s, & \mathbf{x}_s &\in \mathbb{R}^{n_s}, \\ \dot{\mathbf{x}}_u &= \mathbf{x}_u, & \mathbf{x}_u &\in \mathbb{R}^{n_u}, \end{aligned} \tag{2.1.4}$$

where $n_u = n - n_s$.

Note that (2.1.4) is completely characterised by n_s (or n_u). Thus, two hyperbolic linear flows $\exp(\mathbf{A}t)\mathbf{x}$ and $\exp(\mathbf{B}t)\mathbf{x}$ are topologically equivalent if $\mathbf{A}$ and $\mathbf{B}$ have the same number of eigenvalues with negative real part. This follows from Theorem 2.1.2 because both are topologically equivalent to the flow of (2.1.4).

Figure 2.1 Some typical orbits for a selection of hyperbolic, linear diffeomorphisms on the plane. The position of the eigenvalues of each diffeomorphism relative to the unit circle in the complex plane is also shown. The problem of classification up to C^0-conjugacy is considered in Exercise 2.1.8.

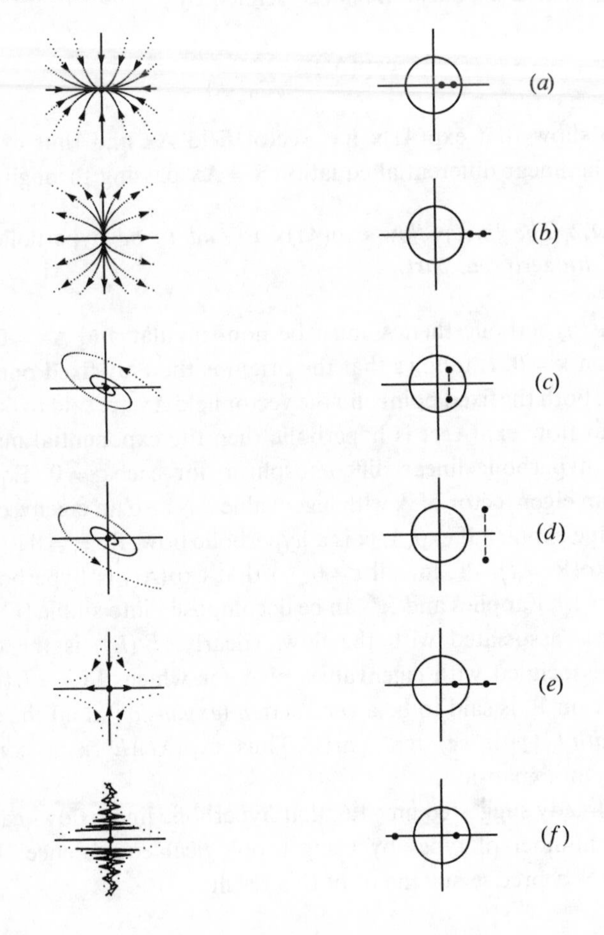

We have already illustrated Theorem 2.1.2 for $n = 2$ in Figure 1.26. The three topological types stable, saddle and unstable correspond to $n_s = 2, 1, 0$; respectively. For $n > 2$, equation (2.1.4) with $0 < n_s < n$ is said to have a *multidimensional* saddle point at $\mathbf{x} = \begin{pmatrix} \mathbf{x}_s \\ \mathbf{x}_u \end{pmatrix} = \mathbf{0}$. In general, (2.1.4) gives rise to $n + 1$ distinct topological types on $\mathbb{R}^n$.

An analogous result to Theorem 2.1.2 exists for hyperbolic linear diffeomorphisms. Figure 2.1 shows some orbits of typical planar examples. It is apparent that the nature of the orbits is not determined by whether the diffeomorphism is a contraction, expansion or is of saddle-type. Indeed, the saddle-type diffeomorphisms shown in Figure 2.1(e) and Figure 2.1(f) behave quite differently; the orbit of the latter repeatedly moving across the y-axis. The following theorem (Robbin, 1972) allows for such possibilities and provides a characterisation of the topological type of a hyperbolic linear diffeomorphism.

Let $\mathbf{A}$ be a hyperbolic linear diffeomorphism with stable (unstable) eigenspace $E^s_{\mathbf{A}}(E^u_{\mathbf{A}})$. Define $\mathbf{A}_i = \mathbf{A} | E^i_{\mathbf{A}}$, $i = \mathrm{s, u}$. Then $\mathbf{A}_i$ is said to be orientation-preserving (reversing) if $\mathrm{Det}\, \mathbf{A}_i > 0$ ($\mathrm{Det}\, \mathbf{A}_i < 0$).

Theorem 2.1.3 *Let* **A**, **B**: $\mathbb{R}^n \to \mathbb{R}^n$ *be hyperbolic linear diffeomorphisms. Then* **A** *and* **B** *are topologically conjugate if and only if*:

(i) $\dim E^s_{\mathbf{A}} = \dim E^s_{\mathbf{B}}$ (*or equivalently* $\dim E^u_{\mathbf{A}} = \dim E^u_{\mathbf{B}}$);

(ii) *for* $i = \mathrm{s, u}$, $\mathbf{A}_i$ and $\mathbf{B}_i$ *are either both orientation-preserving or both orientation-reversing.*

This result highlights an interesting difference between general diffeomorphisms and flows. The hyperbolic linear flow $\exp(\mathbf{A}t)$ has the property that $\exp(\mathbf{A}t)|E^i$, $i = \mathrm{s, u}$, is orientation-preserving for all t. Thus, for example, there is, up to topological equivalence, only one planar flow of saddle-type (see Theorem 2.1.2). On the other hand, hyperbolic linear diffeomorphisms of the plane are not so restricted. As Theorem 2.1.3 predicts, there are four topologically distinct saddle-types corresponding to $\mathbf{A}_s$ and $\mathbf{A}_u$ each being either orientation-preserving or reversing. In general, Theorem 2.1.3 can be used (see Exercise 2.1.8) to show that there are $4n$ topological types of hyperbolic linear diffeomorphism on $\mathbb{R}^n$.

2.2 Hyperbolic non-linear fixed points

A non-linear dynamical system is, in general, defined on a differentiable manifold. However, the topological type of a fixed point is determined by the restriction of the system to sufficiently small neighbourhoods of the point. These neighbourhoods can be chosen to lie in a single chart so that only one representative of the system is involved. It is therefore convenient to take our diffeomorphisms and flows to be defined on open sets in $\mathbb{R}^n$. This simplifies the language required to state the results we require and provides a more practical approach to calculations.

2.2.1 Diffeomorphisms

Let U be an open subset of $\mathbb{R}^n$ and $\mathbf{f}: U \to \mathbb{R}^n$ be a non-linear diffeomorphism with an isolated fixed point at $\mathbf{x}^* \in U$. The *linearisation* of $\mathbf{f}$ at $\mathbf{x}^*$ is given by

$$\mathrm{D}\mathbf{f}(\mathbf{x}^*) = \left[\frac{\partial f_i}{\partial x_j}\right]_{i,j=1}^{n}\Bigg|_{\mathbf{x}=\mathbf{x}^*}, \tag{2.2.1}$$

where $x_1, \ldots, x_n$ are coordinates on U.

Definition 2.2.1 *A fixed point* $\mathbf{x}^*$ *of a diffeomorphism* $\mathbf{f}$ *is said to be* hyperbolic *if* $\mathrm{D}\mathbf{f}(\mathbf{x}^*)$ *is a hyperbolic, linear diffeomorphism.*

The following theorems allow us to obtain valuable information from $\mathrm{D}\mathbf{f}(\mathbf{x}^*)$.

Theorem 2.2.1 (Hartman–Grobman) *Let* $\mathbf{x}^*$ *be a hyperbolic fixed point of the diffeomorphism* $\mathbf{f}: U \to \mathbb{R}^n$*. Then there is a neighbourhood* $N \subseteq U$ *of* $\mathbf{x}^*$ *and a neighbourhood* $N' \subseteq \mathbb{R}^n$ *containing the origin such that* $\mathbf{f}|N$ *is topologically conjugate to* $\mathrm{D}\mathbf{f}(\mathbf{x}^*)|N'$.

It follows from Theorem 2.2.1 and Theorem 2.1.3 that there are $4n$ topological types of hyperbolic fixed point for diffeomorphisms $\mathbf{f}: U \to \mathbb{R}^n$.

Theorem 2.2.2 (Invariant Manifold) *Let* $\mathbf{f}: U \to \mathbb{R}^n$ *be a diffeomorphism with a hyperbolic fixed point at* $\mathbf{x}^* \in U$*. Then on a sufficiently small neighbourhood* $N \subseteq U$ *of* $\mathbf{x}^*$*, there exist local stable and unstable manifolds,*

$$W^{\mathrm{s}}_{\mathrm{loc}}(\mathbf{x}^*) = \{\mathbf{x} \in U \,|\, \mathbf{f}^n(\mathbf{x}) \to \mathbf{x}^* \text{ as } n \to \infty\}, \tag{2.2.2}$$

$$W^{\mathrm{u}}_{\mathrm{loc}}(\mathbf{x}^*) = \{\mathbf{x} \in U \,|\, \mathbf{f}^n(\mathbf{x}) \to \mathbf{x}^* \text{ as } n \to -\infty\}, \tag{2.2.3}$$

of the same dimensions as E^{s} *and* E^{u} *for* $\mathrm{D}\mathbf{f}(\mathbf{x}^*)$ *and tangent to them at* $\mathbf{x}^*$.

Theorem 2.2.2 has global repercussions because it allows us to define *global* stable and unstable manifolds at $\mathbf{x}^*$ by

$$W^{\mathrm{s}}(\mathbf{x}^*) = \bigcup_{m \in \mathbb{Z}^+} \mathbf{f}^{-m}(W^{\mathrm{s}}_{\mathrm{loc}}(\mathbf{x}^*)) \tag{2.2.4}$$

$$W^{\mathrm{u}}(\mathbf{x}^*) = \bigcup_{m \in \mathbb{Z}^+} \mathbf{f}^{m}(W^{\mathrm{u}}_{\mathrm{loc}}(\mathbf{x}^*)). \tag{2.2.5}$$

The behaviour of $W^{\mathrm{s}}(\mathbf{x}^*)$ and $W^{\mathrm{u}}(\mathbf{x}^*)$ has a profound effect on the complexity of the dynamics of $\mathbf{f}$. In particular, if $W^{\mathrm{s}}(\mathbf{x}^*)$ and $W^{\mathrm{u}}(\mathbf{x}^*)$ meet transversely at one point, they must do so infinitely many times (see §3.7) and a *homoclinic tangle* (see Figure 2.2) results. The relationship between such tangles and 'chaotic orbits' of $\mathbf{f}$ is discussed in Chapter 3.

The ideas described above have a natural extension to the periodic points of $\mathbf{f}$. Let $\mathbf{x}^*$ belong to a q-cycle of $\mathbf{f}$ then it is said to be a *hyperbolic periodic point* of $\mathbf{f}$ if it is a hyperbolic fixed point of $\mathbf{f}^q$. The orbit of $\mathbf{x}^*$ under $\mathbf{f}$ is referred to as a *hyperbolic periodic orbit* and its topological type is determined by that of the corresponding fixed point of $\mathbf{f}^q$. Moreover, information about stable and unstable manifolds at each point of the q-cycle can be obtained by applying Theorem 2.2.2 to $\mathbf{f}^q$ (see Exercise 2.2.4).

2.2.2 Flows

There are analogous results to Theorems 2.2.1 and 2.2.2 for non-linear flows. Let $\dot{\mathbf{x}} = \mathbf{X}(\mathbf{x})$, $\mathbf{x} \in U$, U an open subset of $\mathbb{R}^n$, be such that $\mathbf{X}(\mathbf{x}^*) = \mathbf{0}$, $\mathbf{x}^* \in U$. Then the *linearisation* of $\dot{\mathbf{x}} = \mathbf{X}(\mathbf{x})$ at $\mathbf{x}^*$ is the linear differential equation

$$\dot{\mathbf{y}} = \mathrm{D}\mathbf{X}(\mathbf{x}^*)\mathbf{y},$$

where

$$\mathrm{D}\mathbf{X}(\mathbf{x}^*) = \left[\frac{\partial X_i}{\partial x_j}\right]_{i,j=1}^{n}\Bigg|_{\mathbf{x}=\mathbf{x}^*} \tag{2.2.6}$$

and $\mathbf{y} = (y_1, \ldots, y_n)^{\mathrm{T}}$ are local coordinates at $\mathbf{x}^*$.

Definition 2.2.2 *A singular point $\mathbf{x}^*$ of a vector field $\mathbf{X}$ is said to be* hyperbolic *if no eigenvalue of $\mathrm{D}\mathbf{X}(\mathbf{x}^*)$ has zero real part.*

If $\mathbf{x}^*$ is a singular point of $\mathbf{X}$, then it is a fixed point of the flow of $\dot{\mathbf{x}} = \mathbf{X}(\mathbf{x})$. Definition 2.2.2 means that $\mathbf{x}^*$ is a hyperbolic singular point of $\mathbf{X}$ if the flow, $\exp(\mathrm{D}\mathbf{X}(\mathbf{x}^*)t)\mathbf{x}$, of the linearisation of $\dot{\mathbf{x}} = \mathbf{X}(\mathbf{x})$ is hyperbolic in the sense of Definition 2.1.2. It is sometimes convenient to distinguish those non-hyperbolic singular points for which $\mathrm{D}\mathbf{X}(\mathbf{x}^*)$ has at least one zero eigenvalue, such points are said to be *non-simple* (Arrowsmith & Place, 1982, p. 52).

Theorem 2.2.3 (Hartman–Grobman) *Let $\mathbf{x}^*$ be a hyperbolic fixed point of $\dot{\mathbf{x}} = \mathbf{X}(\mathbf{x})$ with flow $\boldsymbol{\varphi}_t \colon U \subseteq \mathbb{R}^n \to \mathbb{R}^n$. Then there is a neighbourhood N of $\mathbf{x}^*$ on which $\boldsymbol{\varphi}$ is topologically conjugate to the linear flow* $\exp(\mathrm{D}\mathbf{X}(\mathbf{x}^*)t)\mathbf{x}$.

Figure 2.2 Sketch of a homoclinic tangle. The stable and unstable manifolds of the saddle point $\mathbf{x}^*$ intersect infinitely many times.

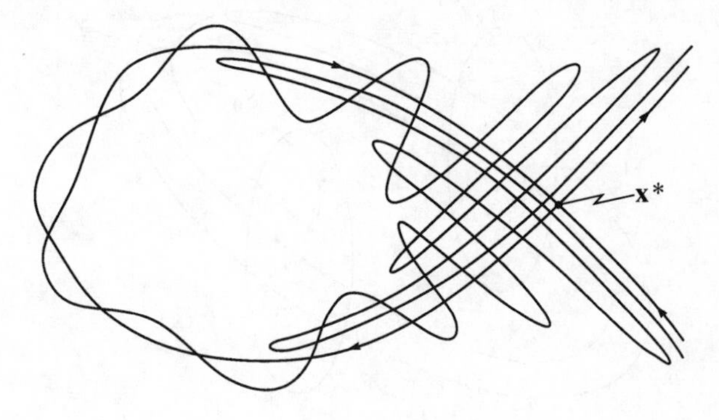

The Invariant Manifold Theorem is also valid for flows. The statement is the same as that in Theorem 2.2.2 except that $\mathbf{f} \mapsto \boldsymbol{\varphi}_t$, $\mathbf{Df}(\mathbf{x}^*) \mapsto \dot{\mathbf{y}} = \mathbf{DX}(\mathbf{x}^*)\mathbf{y}$ and $W^{\mathrm{s(u)}}_{\mathrm{loc}} = \{\mathbf{x} \in U \mid \boldsymbol{\varphi}_t(\mathbf{x}) \to \mathbf{x}^* \text{ as } t \to \infty(-\infty)\}$. Global stable and unstable manifolds, W^{s} and W^{u}, are defined (see (2.2.4 and 5)) by taking a union over real $t > 0$ of $\boldsymbol{\varphi}_{-t}(W^{\mathrm{s}}_{\mathrm{loc}})$ and $\boldsymbol{\varphi}_t(W^{\mathrm{u}}_{\mathrm{loc}})$, respectively. Simple examples of the above theorems are discussed in Chapter 3 of Arrowsmith & Place (1982).

The Hartman Theorem (2.2.3) coupled with Theorem 2.1.2 provides the following classification of hyperbolic fixed points.

Theorem 2.2.4 *Let* $\mathbf{x}^*$ *be a hyperbolic fixed point of* $\dot{\mathbf{x}} = \mathbf{X}(\mathbf{x})$ *with flow* $\boldsymbol{\varphi}_t \colon U \subseteq \mathbb{R}^n \to \mathbb{R}^n$. *Then there is a neighbourhood* N *of* $\mathbf{x}^*$ *on which* $\boldsymbol{\varphi}_t$ *is topologically equivalent to the flow of the linear differential equation*

$$\begin{aligned} \dot{\mathbf{x}}_{\mathrm{s}} &= -\mathbf{x}_{\mathrm{s}}, \qquad && \mathbf{x}_{\mathrm{s}} \in \mathbb{R}^{n_{\mathrm{s}}}, \\ \dot{\mathbf{x}}_{\mathrm{u}} &= \mathbf{x}_{\mathrm{u}}, && \mathbf{x}_{\mathrm{u}} \in \mathbb{R}^{n_{\mathrm{u}}}, \end{aligned} \tag{2.2.7}$$

where $n_{\mathrm{u}} = n - n_{\mathrm{s}}$. *Here* n_{s} *is the dimension of the stable eigenspace of* $\exp(\mathbf{DX}(\mathbf{x}^*)t)$.

The idea of hyperbolicity can be extended to closed orbits of flows. Let γ be a closed orbit in the flow $\boldsymbol{\varphi}$ of the vector field $\mathbf{X}$ and suppose S_0 is a local section at $\mathbf{x}_0 \in \gamma$. If $\mathbf{x} \in S_0$ is sufficiently close to $\mathbf{x}_0$ then the trajectory through $\mathbf{x}$ will return to S_0 at $\mathbf{P}_0(\mathbf{x})$ (see Figure 2.3). This defines a local Poincaré map $\mathbf{P}_0 \colon S_0 \to S_0$ for which $\mathbf{x}_0$ is a fixed point. The closed orbit γ is said to be a *hyperbolic closed orbit* if $\mathbf{x}_0$ is a hyperbolic fixed point of the diffeomorphism $\mathbf{P}_0$. This definition is independent of $\mathbf{x}_0 \in \gamma$ because a Poincaré map $\mathbf{P}_1 \colon S_1 \to S_1$ obtained by taking a local section at $\mathbf{x}_1 \in \gamma$ is C^1-conjugate to $\mathbf{P}_0$ (see Exercise 2.2.6). It follows that for each $\mathbf{x} \in \gamma$, the associated Poincaré map defines local stable and unstable manifolds

Figure 2.3 Schematic illustration of Poincaré maps, $\mathbf{P}_0$ and $\mathbf{P}_1$, constructed on local sections at different points of a closed orbit γ in a flow. These maps are C^1-conjugate.

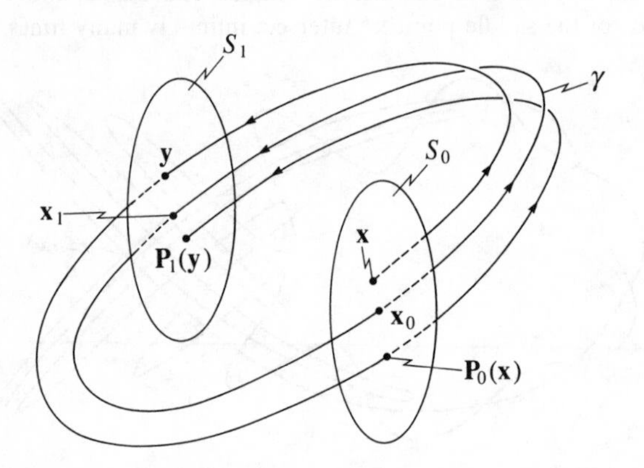

with dimensions that are independent of $\mathbf{x}$. The totality of these manifolds (i.e. $\bigcup_{\mathbf{x}\in\gamma} W^{s,u}_{loc}(\mathbf{x})$) defines stable and unstable manifolds for γ (see Figure 2.4).

In contrast, the problem of attributing a form of hyperbolicity to a closed invariant curve of a diffeomorphism, $\mathbf{f}$, is much more subtle. There is no analogue of the Poincaré map of the flow in this case. Indeed, if S_0 is a local section at a point, $\mathbf{x}_0$, of such an 'invariant circle', then, in general, $\mathbf{f}^n(\mathbf{x}_0)$ may not belong to S_0 for any integer $n \neq 0$ (e.g. consider an irrational rotation on the plane). An alternative way of associating the invariant circle with the fixed point of a map must be found. To illustrate the ideas involved, we will consider the planar case.

Let $\mathscr{C}$ be an invariant circle of $\mathbf{f}: \mathbb{R}^2 \to \mathbb{R}^2$ and let $\mathscr{C} \times I$, I an interval of $\mathbb{R}$, be an annular neighbourhood of $\mathscr{C}$. Consider the set of all closed curves in $\mathscr{C} \times I$ that can be represented by the graph of a function $\sigma: \mathscr{C} \to I$. The set, Σ, of all functions σ can be given a Banach space structure. It is in this space that $\mathscr{C}$ can be associated with a fixed point of a map. This is achieved as follows. Since $\mathbf{f}$ is a diffeomorphism, it must map closed curves onto closed curves. Thus, if $\mathscr{C}_0$ is a closed curved near to $\mathscr{C}$, represented by σ_0, then $\mathscr{C}_1 = \mathbf{f}(\mathscr{C}_0)$ is also a closed curve. If $\mathscr{C}_1 \subseteq \mathscr{C} \times I$ and is represented by the graph of a function σ_1, then we can think of σ_1 as the image of σ_0 under a map $\mathscr{F}: \Sigma \to \Sigma$, i.e. $\sigma_1 = \mathscr{F}(\sigma_0)$. Now, $\mathbf{f}(\mathscr{C}) = \mathscr{C}$ so that if σ^* represents $\mathscr{C}$, then $\mathscr{F}(\sigma^*) = \sigma^*$ and σ^* is a fixed point of $\mathscr{F}$. If this fixed point is hyperbolic then the invariant circle of $\mathbf{f}$ is said to be *normally hyperbolic*. To ensure that $\mathscr{F}(\sigma) \in \Sigma$, the diffeomorphism $\mathbf{f}$ must be such that the contraction (expansion) it produces normal to $\mathscr{C}$ is stronger than its tendency to move points along the

Figure 2.4 Sketch illustrating the stable and unstable manifolds of a hyperbolic closed orbit γ: (a) stable and unstable manifolds of Poincaré maps constructed on local sections at $\mathbf{x} \in \gamma$ with normal vectors $\mathbf{X}(\mathbf{x})/|\mathbf{X}(\mathbf{x})|$; ($b$) trajectories of the flow $\boldsymbol{\varphi}$ on the stable and unstable manifolds of γ.

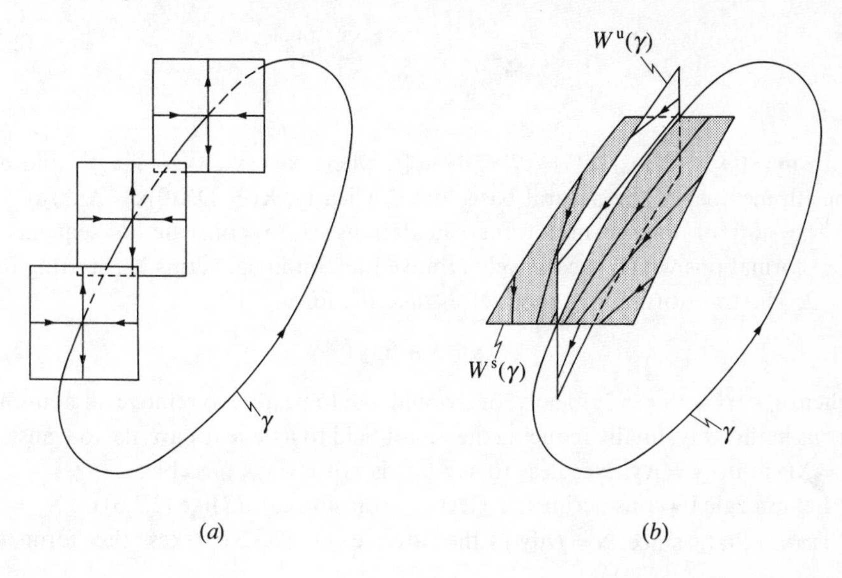

tangent to the invariant circle. What is more, the statement that the fixed point of $\mathscr{F}: \Sigma \to \Sigma$ is hyperbolic assumes that Σ can be equipped with a metric in terms of which the characteristic exponential contraction (expansion) (see Exercise 2.1.2), associated with hyperbolic fixed points, can be demonstrated. The technical details involved here are beyond the scope of the present text and the interested reader should consult Hirsch, Pugh & Shub (1977).

The theorems described above, along with the known topological behaviour of linear systems, provide a clear picture of the behaviour of flows and diffeomorphisms near to isolated hyperbolic fixed points. Let us now consider how to obtain similar information when the fixed points involved are non-hyperbolic. In such cases, higher order terms in the Taylor expansion of **f** or **X** play the determining role. The calculations of topological type consist of two stages:

(i) the construction of a normal form in which the non-linear terms take their 'simplest' form;

(ii) the determination of the topological type of the fixed point from the normal form.

These two steps will occupy us for the remainder of this chapter.

2.3 Normal forms for vector fields (Arnold, 1983, §§ 22 and 23)

Let $\mathbf{X}: \mathbb{R}^n \to \mathbb{R}^n$ be a smooth vector field satisfying $\mathbf{X}(\mathbf{0}) = \mathbf{0}$. We can formally construct the Taylor expansion of $\mathbf{X}$ about $\mathbf{0}$, namely

$$\mathbf{X}(\mathbf{x}) = \mathbf{X}_1 + \mathbf{X}_2 + \cdots + \mathbf{X}_k + O(|\mathbf{x}|^{k+1}), \tag{2.3.1}$$

where $\mathbf{X}_r \in H^r$, the real vector space of vector fields whose components are homogeneous polynomials of degree r. For $r = 1$ to k we write

$$\mathbf{X}_r = \underset{\sum_i m_i = r}{\sum_{m_1=0}^{r} \cdots \sum_{m_n=0}^{r}} \sum_{j=1}^{n} a_{\mathbf{m}}^{(j)} \mathbf{x}^{\mathbf{m}} \mathbf{e}_j \tag{2.3.2}$$

with $\mathbf{m} = (m_1, \ldots, m_n)$, $\mathbf{x}^{\mathbf{m}} = x_1^{m_1} x_2^{m_2} \cdots x_n^{m_n}$, where $\mathbf{x} = (x_1, x_2, \ldots, x_n)^{\mathrm{T}}$, and $\mathbf{e}_j$ is the jth member of the natural basis for $\mathbb{R}^n$. Clearly, $\mathbf{X}_1 = \mathbf{DX}(\mathbf{0})\mathbf{x} = \mathbf{Ax}$, say.

The aim of the normal form calculations is to construct a sequence of transformations which successively remove the non-linear terms $\mathbf{X}_r$, starting from $r = 2$. The transformations themselves take the form

$$\mathbf{x} = \mathbf{y} + \mathbf{h}_r(\mathbf{y}), \tag{2.3.3}$$

where $\mathbf{h}_r \in H^r$ with $r \geqslant 2$. Ideally, one would like to be able to remove all non-linear terms in this way, finally reducing the vector field to its linear part, i.e. to transform $\dot{\mathbf{x}} = \mathbf{X}(\mathbf{x})$ into $\dot{\mathbf{y}} = \mathbf{Ay}$. Needless to say this is not always possible.

Let us begin by considering the effect of a transformation like (2.3.3) on $\mathbf{X}_1 = \mathbf{Ax}$. Observe that, since $\mathbf{x} = O(|\mathbf{y}|)$, the inverse of (2.3.3) takes the form (see

Exercise 2.3.1)

$$\mathbf{y} = \mathbf{x} - \mathbf{h}_r(\mathbf{x}) + O(|\mathbf{x}|^{r+1}). \tag{2.3.4}$$

Thus

$$\begin{aligned}\dot{\mathbf{y}} &= [\mathbf{I} - \mathbf{Dh}_r(\mathbf{x}) + O(|\mathbf{x}|^r)]\dot{\mathbf{x}} \\ &= [\mathbf{I} - \mathbf{Dh}_r(\mathbf{y} + O(|\mathbf{y}|^r)) + O(|\mathbf{y}|^r)]\mathbf{A}(\mathbf{y} + \mathbf{h}_r(\mathbf{y})) \\ &= [\mathbf{I} - \mathbf{Dh}_r(\mathbf{y}) + O(|\mathbf{y}|^r)]\mathbf{A}(\mathbf{y} + \mathbf{h}_r(\mathbf{y})) \\ &= \mathbf{Ay} - \{\mathbf{Dh}_r(\mathbf{y})\mathbf{Ay} - \mathbf{Ah}_r(\mathbf{y})\} + O(|\mathbf{y}|^{r+1}).\end{aligned} \tag{2.3.5}$$

Note that $\mathbf{Dh}_r(\mathbf{y} + O(|\mathbf{y}|^r)) = \mathbf{Dh}_r(\mathbf{y}) + O(|\mathbf{y}|^{2(r-1)})$ but $2(r-1) \geqslant r$ for $r \geqslant 2$. The quantity

$$\{\mathbf{Dh}_r(\mathbf{y})\mathbf{Ay} - \mathbf{Ah}_r(\mathbf{y})\} \tag{2.3.6}$$

is known as the *Lie* (or *Poisson*) *bracket* of the vector fields $\mathbf{Ay}$ and $\mathbf{h}_r(\mathbf{y})$. It is convenient to introduce an operator, $L_{\mathbf{A}}$, which acts on vector fields in such a way as to produce their Lie bracket with $\mathbf{Ay}$. Thus (2.3.6) can be written as $L_{\mathbf{A}}\mathbf{h}_r(\mathbf{y})$.

Now suppose $\dot{\mathbf{x}} = \mathbf{Ax} + \mathbf{X}_r(\mathbf{x})$, $\mathbf{X}_r \in H^r$, $r \geqslant 2$. Since $\mathbf{X}_r(\mathbf{y} + \mathbf{h}_r(\mathbf{y})) = \mathbf{X}_r(\mathbf{y}) + O(|\mathbf{y}|^{r+1})$ for $r \geqslant 2$, (2.3.5) becomes

$$\dot{\mathbf{y}} = \mathbf{Ay} - L_{\mathbf{A}}\mathbf{h}_r(\mathbf{y}) + \mathbf{X}_r(\mathbf{y}) + O(|\mathbf{y}|^{r+1}). \tag{2.3.7}$$

Notice that both $L_{\mathbf{A}}\mathbf{h}_r$ and $\mathbf{X}_r \in H^r$, so that the deviation of the right hand side of (2.3.7) from $\mathbf{Ay}$ has no terms of order less than r in $|\mathbf{y}|$. This means that if $\mathbf{X}$ is such that $\mathbf{X}_2, \ldots, \mathbf{X}_{r-1} = \mathbf{0}$, they will remain zero under a transformation of the form (2.3.3). Furthermore, (2.3.7) makes clear how we may be able to remove $\mathbf{X}_r$ by a suitable choice of $\mathbf{h}_r$.

Proposition 2.3.1 *Given that the inverse of* $L_{\mathbf{A}}$ *exists, the differential equation*

$$\dot{\mathbf{x}} = \mathbf{Ax} + \mathbf{X}_r(\mathbf{x}) + O(|\mathbf{x}|^{r+1}), \tag{2.3.8}$$

with $\mathbf{X}_r \in H^r$, $r \geqslant 2$, *is transformed into*

$$\dot{\mathbf{y}} = \mathbf{Ay} + O(|\mathbf{y}|^{r+1}) \tag{2.3.9}$$

by the transformation $\mathbf{x} = \mathbf{y} + \mathbf{h}_r(\mathbf{y})$, *where*

$$\mathbf{h}_r(\mathbf{y}) = L_{\mathbf{A}}^{-1}\mathbf{X}_r(\mathbf{y}). \tag{2.3.10}$$

Equation (2.3.7) shows that $\mathbf{X}_r$ can be eliminated if $\mathbf{h}_r$ can be chosen such that

$$L_{\mathbf{A}}\mathbf{h}_r(\mathbf{y}) = \mathbf{X}_r(\mathbf{y}). \tag{2.3.11}$$

This is known as the *homological equation* associated with the linear vector field $\mathbf{Ay}$. Proposition 2.3.1 simply asserts that $\mathbf{h}_r$ satisfying (2.3.11) can be found if the inverse operator $L_{\mathbf{A}}^{-1}$ exists. To find out when this is the case we must examine $L_{\mathbf{A}}$ more closely.

The Lie bracket $L_{\mathbf{A}}: H^r \to H^r$ is a linear map (see Exercise 2.3.3) and its eigenvalues can be expressed in terms of those of $\mathbf{A} = \mathrm{D}\mathbf{X}(\mathbf{0})$. For example, when $\mathbf{A}$ has distinct eigenvalues λ_i, $i = 1, \ldots, n$, its eigenvectors form a basis for $\mathbb{R}^n$. Relative to this basis $\mathbf{A}$ is diagonal, i.e. $\mathbf{A} = [\lambda_i \delta_{ij}]_{i,j=1}^n$, with eigenvectors $\mathbf{e}_i$, $i = 1, \ldots, n$. Let $x_1, \ldots, x_n$ be the coordinates of $\mathbf{x}$ relative to this eigenbasis and consider $L_{\mathbf{A}}\mathbf{x}^{\mathbf{m}}\mathbf{e}_i$. Equation (2.3.6) implies

$$\begin{aligned} L_{\mathbf{A}}\mathbf{x}^{\mathbf{m}}\mathbf{e}_i &= \mathrm{D}(\mathbf{x}^{\mathbf{m}}\mathbf{e}_i)\mathbf{A}\mathbf{x} - \mathbf{A}(\mathbf{x}^{\mathbf{m}}\mathbf{e}_i) \\ &= \left\{ \sum_{j=1}^{n} \lambda_j x_j \frac{m_j}{x_j} \mathbf{x}^{\mathbf{m}}\mathbf{e}_i \right\} - \lambda_i \mathbf{x}^{\mathbf{m}}\mathbf{e}_i \\ &= \{(\mathbf{m} \cdot \boldsymbol{\lambda}) - \lambda_i\} \mathbf{x}^{\mathbf{m}}\mathbf{e}_i. \end{aligned} \tag{2.3.12}$$

Thus, for each $i = 1, \ldots, n$, the monomial $\mathbf{x}^{\mathbf{m}}\mathbf{e}_i$ is an eigenvector of $L_{\mathbf{A}}$ with eigenvalue $\Lambda_{\mathbf{m},i} = (\mathbf{m} \cdot \boldsymbol{\lambda}) - \lambda_i$, where $\boldsymbol{\lambda} = (\lambda_1, \ldots, \lambda_n)$. Even when $\mathbf{A}$ cannot be diagonalised, it can be shown (see Exercise 2.3.8 and 2.3.9) that the eigenvalues of $L_{\mathbf{A}}$ are still given by the above expression. Therefore, the inverse operator, $L_{\mathbf{A}}^{-1}$, exists if and only if

$$\Lambda_{\mathbf{m},i} = (\mathbf{m} \cdot \boldsymbol{\lambda}) - \lambda_i \neq 0 \tag{2.3.13}$$

for every allowed $\mathbf{m}$ and $i = 1, \ldots, n$.

Definition 2.3.1 *The n-tuple of eigenvalues $\boldsymbol{\lambda} = (\lambda_1, \ldots, \lambda_n)$ is said to be* resonant of order r *if*

$$\lambda_i = \sum_{j=1}^{n} m_j \lambda_j, \tag{2.3.14}$$

for some $m_j \in \mathbb{N}$ (recall $\mathbb{N}$ is the set of non-negative integers), with $\sum_{j=1}^{n} m_j = r$, and some $i = 1, \ldots, n$.

Using the eigenvectors of $L_{\mathbf{A}}$ as a basis for H^r, we can write

$$\mathbf{h}_r(\mathbf{x}) = \sum_{\substack{\mathbf{m},i \\ \sum m_j = r}} h_{\mathbf{m},i} \mathbf{x}^{\mathbf{m}}\mathbf{e}_i \tag{2.3.15}$$

and

$$\mathbf{X}_r(\mathbf{x}) = \sum_{\substack{\mathbf{m},i \\ \sum m_j = r}} X_{\mathbf{m},i} \mathbf{x}^{\mathbf{m}}\mathbf{e}_i. \tag{2.3.16}$$

Substituting into (2.3.11), we see that if the eigenvalues of $\mathbf{A}$ are not resonant of order r, then

$$h_{\mathbf{m},i} = \frac{X_{\mathbf{m},i}}{(\mathbf{m} \cdot \boldsymbol{\lambda} - \lambda_i)}, \tag{2.3.17}$$

gives $\mathbf{h}_r$ explicitly in terms of $\mathbf{X}_r$.

Example 2.3.1 Let $\mathbf{x} = \mathbf{y} + \mathbf{h}_2(\mathbf{y})$. Choose $\mathbf{h}_2(\mathbf{y})$ so as to eliminate terms of order $|\mathbf{x}|^2$ from the following vector fields:

$$\text{(a) } \mathbf{X}(\mathbf{x}) = \begin{pmatrix} 3x_1 - x_2^2 \\ x_2 \end{pmatrix}; \quad \text{(b) } \mathbf{X}(\mathbf{x}) = \begin{pmatrix} 3x_1 + 3x_1^2 \\ x_2 \end{pmatrix}. \tag{2.3.18}$$

Verify your results by transforming the differential equation $\dot{\mathbf{x}} = \mathbf{X}(\mathbf{x})$.

Solution

(a)
$$\mathbf{X}_2(\mathbf{x}) = -\begin{pmatrix} x_2^2 \\ 0 \end{pmatrix} = \sum_{\substack{\mathbf{m}, i \\ m_1 + m_2 = 2}} X_{\mathbf{m},i} \mathbf{e}_i, \tag{2.3.19}$$

where

$$X_{\mathbf{m},i} = \begin{cases} -1 & \mathbf{m} = (0, 2), i = 1 \\ 0 & \text{otherwise.} \end{cases} \tag{2.3.20}$$

Therefore (2.3.17) implies

$$\mathbf{h}_2(\mathbf{y}) = \begin{pmatrix} y_2^2 \\ 0 \end{pmatrix}. \tag{2.3.21}$$

To verify this result, note that $\mathbf{x} = \mathbf{y} + \mathbf{h}_2(\mathbf{y})$, i.e.

$$x_1 = y_1 + y_2^2, \qquad x_2 = y_2, \tag{2.3.22}$$

means that

$$y_1 = x_1 - x_2^2, \qquad y_2 = x_2. \tag{2.3.23}$$

Hence

$$\dot{y}_1 = \dot{x}_1 - 2x_2\dot{x}_2, \qquad \dot{y}_2 = \dot{x}_2, \tag{2.3.24}$$

with $\dot{x}_1 = 3x_1 - x_2^2$ and $\dot{x}_2 = x_2$. Thus

$$\begin{aligned} \dot{y}_1 &= 3x_1 - x_2^2 - 2x_2^2 = 3(y_1 + y_2^2) - 3y_2^2 = 3y_1 \\ \dot{y}_2 &= \dot{x}_2 = x_2 = y_2 \end{aligned} \tag{2.3.25}$$

so that the transformed vector field $\tilde{\mathbf{X}}(\mathbf{y}) = \begin{pmatrix} 3y_1 \\ y_2 \end{pmatrix}$ and the quadratic terms have been removed.

(b) Here

$$X_{\mathbf{m},i} = \begin{cases} 3 & \mathbf{m} = (2, 0), i = 1 \\ 0 & \text{otherwise} \end{cases}$$

and (2.3.17) gives $\mathbf{h}_2(\mathbf{y}) = \begin{pmatrix} y_1^2 \\ 0 \end{pmatrix}$. Differentiation of $\mathbf{x} = \mathbf{y} + \mathbf{h}_2(\mathbf{y})$ with respect to time leads to

$$\dot{x}_1 = \dot{y}_1(1 + 2y_1), \qquad \dot{x}_2 = \dot{y}_2, \tag{2.3.26}$$

where $\dot{x}_1 = 3x_1 + 3x_1^2$, $\dot{x}_2 = x_2$. Clearly, $\dot{y}_2 = y_2$ as in (a), but

$$\begin{aligned}\dot{y}_1 &= \frac{3x_1 + 3x_1^2}{1 + 2y_1} = 3y_1(1 + 2y_1 + O(y_1^2))(1 - 2y_1 + O(y_1^2)) \\ &= 3y_1 + O(y_1^3).\end{aligned} \tag{2.3.27}$$

Thus, the transformed vector field is $\tilde{\mathbf{X}}(\mathbf{y}) = \begin{pmatrix} 3y_1 + O(y_1^3) \\ y_2 \end{pmatrix}$. □

Example 2.3.1 highlights the fact that the transformation $\mathbf{x} = \mathbf{y} + \mathbf{h}_r(\mathbf{y})$ that removes $\mathbf{X}_r$ *may* change $\mathbf{X}_{r+1}, \mathbf{X}_{r+2}, \ldots$ etc. For example, in case (b) $\mathbf{X}_3(\mathbf{x}) \equiv \mathbf{0}$ before transformation but the cubic term in (2.3.27) can be shown (see Exercise 2.3.10) to be $6y_1^3$. However, in spite of these changes, it is clear that, provided we do not encounter resonant terms, we can successively remove quadratic, cubic, quartic, etc. terms, thereby forcing the non-linearity of the transformed vector field to higher and higher orders.

Now let us consider the case when resonance occurs. Recall that if $\mathbf{A}$ can be diagonalised, the eigenvectors of $L_{\mathbf{A}}$ form a basis for H^r. The subset of eigenvectors of $L_{\mathbf{A}}$ with non-zero eigenvalues then form a basis for the image, B^r, of H^r under $L_{\mathbf{A}}$. It follows that the component of $\mathbf{X}_r$ in B^r can be expanded in terms of these eigenvectors and $\mathbf{h}_r$ chosen as in (2.3.17) to ensure the removal of these terms. The component, $\mathbf{w}_r$, of $\mathbf{X}_r$ lying in a complementary subspace, G^r, of B^r in H^r will be unchanged by the transformation $\mathbf{x} = \mathbf{y} + \mathbf{h}_r(\mathbf{y})$ obtained from B^r. These *resonant* terms therefore remain in the transformed vector field. Of course, since $\mathbf{X}_r(\mathbf{y} + \mathbf{h}_{r+k}(\mathbf{y})) = \mathbf{X}_r(\mathbf{y}) + O(|\mathbf{y}|^{r+k+1})$, $r \geqslant 2$, $k = 1, 2, \ldots$, these terms are not changed by subsequent transformations to remove non-resonant terms of higher order. Ultimately, we arrive at the following theorem.

Theorem 2.3.1 (Normal Form Theorem) *Given a smooth vector field* $\mathbf{X}(\mathbf{x})$ *on* $\mathbb{R}^n$ *with* $\mathbf{X}(\mathbf{0}) = \mathbf{0}$, *there is a polynomial transformation to new coordinates*, $\mathbf{y}$, *such that the differential equation* $\dot{\mathbf{x}} = \mathbf{X}(\mathbf{x})$ *takes the form*

$$\dot{\mathbf{y}} = \mathbf{J}\mathbf{y} + \sum_{r=2}^{N} \mathbf{w}_r(\mathbf{y}) + O(|\mathbf{y}|^{N+1}), \tag{2.3.28}$$

where $\mathbf{J}$ *is the real Jordan form of* $\mathbf{A} = D\mathbf{X}(\mathbf{0})$ *and* $\mathbf{w}_r \in G^r$, *a complementary subspace in* H^r *of* $B^r = L_{\mathbf{A}}(H^r)$.

Notice we have assumed that $\mathbf{A}$ has been reduced to real Jordan form before attempting to eliminate non-linear terms. Moreover, Theorem 2.3.1 is not dependent on $\mathbf{J}$ being diagonal. When Jordan blocks occur in $\mathbf{J}$, $L_{\mathbf{J}}$ cannot be diagonalised but the form (2.3.28) can still be achieved provided $L_{\mathbf{J}}^{-1}$ exists (see Exercises 2.3.8 and 2.3.9). The right hand side of (2.3.8) is called the *normal form* of $\mathbf{X}(\mathbf{x})$. The polynomial transformation referred to in Theorem 2.3.1 is simply the composition of a sequence of coordinate changes each of the form $\mathbf{x} = \mathbf{y} + \mathbf{h}_r(\mathbf{y})$, $\mathbf{h}_r \in H^r$, for $2 \leqslant r \leqslant N$.

Example 2.3.2 Let $\mathbf{X}(\mathbf{x})$ be a smooth vector field with $\mathbf{X}(\mathbf{0}) = \mathbf{0}$ and $\mathbf{X}_1 = \begin{pmatrix} 2x_1 - \frac{1}{2}x_2 \\ x_2 \end{pmatrix}$. Show the normal form of $\mathbf{X}$ is

$$\begin{pmatrix} 2y_1 + Ky_2^2 \\ y_2 \end{pmatrix} + O(|\mathbf{y}|^{N+1}), \tag{2.3.29}$$

where K is a real constant and N is any integer greater than or equal to 2. Find K when

$$\mathbf{X}(\mathbf{x}) = \begin{pmatrix} 2x_1 - \frac{1}{2}x_2 + x_1^2 - \frac{1}{2}x_1x_2 + x_2^2 \\ x_2 + x_1x_2 - \frac{1}{2}x_2^2 \end{pmatrix} + O(|\mathbf{x}|^3). \tag{2.3.30}$$

Solution. Observe $\mathbf{X}_1 = \mathbf{DX}(\mathbf{0})\mathbf{x} = \mathbf{Ax}$ with $\mathbf{A} = \begin{pmatrix} 2 & -\frac{1}{2} \\ 0 & 1 \end{pmatrix}$. The matrix $\mathbf{A}$ has eigenvalues $\lambda_1 = 2$ and $\lambda_2 = 1$ and consequently its eigenvectors form a basis for $\mathbb{R}^2$. Let x_1', x_2' be coordinates relative to this basis. The Jordan form of $\mathbf{A}$ is $\mathbf{J} = \begin{pmatrix} 2 & 0 \\ 0 & 1 \end{pmatrix}$ and $L_{\mathbf{J}}: H^r \to H^r$ has eigenvectors $\mathbf{x}'^{\mathbf{m}}\mathbf{e}_i$ with eigenvalues $\Lambda_{\mathbf{m},i} = m_1\lambda_1 + m_2\lambda_2 - \lambda_i$, $i = 1, 2$, $m_1, m_2 \geqslant 0$, $m_1 + m_2 = r$. To obtain the normal form of $\mathbf{X}$ we must look for resonant terms, i.e. those $\mathbf{m}$, i for which $\Lambda_{\mathbf{m},i} = 0$. For $r = 2$, the values of $\Lambda_{\mathbf{m},i}$ are given in Table 2.3.1. Only $\Lambda_{(0,2),1} = 0$ so there is only one resonant term of order two: namely $y_2^2\mathbf{e}_1$. For $r \geqslant 3$, the resonance condition $\Lambda_{\mathbf{m},i} = 0$ takes the form

$$(m_1 + r) = \lambda_i, \qquad i = 1, 2. \tag{2.3.31}$$

Now, $m_1 + r \geqslant r \geqslant 3$, while $\max \lambda_i = 2$ and so we conclude that there are no

Table 2.3.1. $\Lambda_{\mathbf{m},i} = \mathbf{m} \cdot \boldsymbol{\lambda} - \lambda_i$ for $m_1 + m_2 = 2$

m_1	m_2	$\Lambda_{\mathbf{m},1}$	$\Lambda_{\mathbf{m},2}$
2	0	2	3
1	1	1	2
0	2	0	1

resonances of order $r \geqslant 3$. Thus $\mathbf{w}_r(\mathbf{y}) \equiv \mathbf{0}$ for $r \geqslant 3$ and $\mathbf{X}$ has normal form

$$\begin{pmatrix} 2y_1 + Ky_2^2 \\ y_2 \end{pmatrix} + O(|\mathbf{y}|^{N+1}) \tag{2.3.32}$$

for some K and $N \geqslant 2$.

In order to find K, we must take account of the linear transform $\mathbf{x} = \mathbf{M}\mathbf{x}'$ used to reduce $\mathbf{A}$ to Jordan form. The eigenvectors of $\mathbf{A}$ are scalar multiples of $\mathbf{u}_1 = \begin{pmatrix} 1 \\ 0 \end{pmatrix}$, $\mathbf{u}_2 = \begin{pmatrix} 1 \\ 2 \end{pmatrix}$ so, for example, we can take $\mathbf{M} = (\mathbf{u}_1 \vdots \mathbf{u}_2)$. Hence

$$\begin{aligned} \dot{\mathbf{x}}' &= \mathbf{M}^{-1}\mathbf{X}(\mathbf{M}\mathbf{x}') \\ &= \mathbf{M}^{-1}\mathbf{X}_1(\mathbf{M}\mathbf{x}') + \mathbf{M}^{-1}\mathbf{X}_2(\mathbf{M}\mathbf{x}') + O(|\mathbf{x}'|^3). \end{aligned} \tag{2.3.33}$$

Now, (2.3.30) gives

$$\mathbf{X}_1(\mathbf{x}) = \mathbf{A}\mathbf{x} \qquad \text{and} \qquad \mathbf{X}_2(\mathbf{x}) = \begin{pmatrix} x_1^2 - \frac{1}{2}x_1x_2 + x_2^2 \\ x_1x_2 - \frac{1}{2}x_2^2 \end{pmatrix}$$

so that (2.3.33) becomes

$$\dot{\mathbf{x}}' = \mathbf{X}'(\mathbf{x}') = \begin{pmatrix} 2 & 0 \\ 0 & 1 \end{pmatrix} \begin{pmatrix} x_1' \\ x_2' \end{pmatrix} + \begin{pmatrix} x_1'^2 + 4x_2'^2 \\ x_1'x_2' \end{pmatrix} + O(|\mathbf{x}'|^3). \tag{2.3.34}$$

In view of the discussion preceding Theorem 2.3.1, we note that the transformations required to remove $x_1'\mathbf{e}_1$ and $x_1'x_2'\mathbf{e}_2$, or indeed any higher-order resonance term, will not affect the coefficient of $x_2'^2\mathbf{e}_1$. Hence $K = 4$. □

In a field where so many changes of variables are involved it is not usual to persist with primed variables. Normally, we say that

'$\mathbf{X}$ becomes $\begin{pmatrix} 2x_1 + x_1^2 + 4x_2^2 \\ x_2 + x_1x_2 \end{pmatrix} + O(|\mathbf{x}|^3)$ when $\mathbf{x} \to \mathbf{M}^{-1}\mathbf{x}$'.

This means

'let $\mathbf{x}' = \mathbf{M}^{-1}\mathbf{x}$, do the transformation and then drop primes'.

Example 2.3.2 exhibits (albeit in a rather simple form) another feature typical of normal form calculations. For given $\mathbf{X}$, it is often quite easy to find out which terms are resonant by simply examining $\mathbf{m}\cdot\boldsymbol{\lambda} - \lambda_i$ for each $(\mathbf{m}, i)$. However, if the coefficient of a particular resonant term is required this can involve much more work. The latter problem frequently occurs when checking non-degeneracy conditions at non-hyperbolic fixed points (see Propositions 2.4.1–2.4.3). In Example 2.3.2, only the linear transformation had to be followed through in order to find K, but if, for example, a resonant cubic coefficient is needed, then linear and quadratic transforms must be performed in order to find it in terms of the coefficients in the original Taylor expansion of **X**. Such calculations can become rather lengthy.

2.4 Non-hyperbolic singular points of vector fields

A singular point $\mathbf{x}^*$ of a vector field $\mathbf{X}$ is non-hyperbolic if at least one of the eigenvalues of $D\mathbf{X}(\mathbf{x}^*)$ has zero real part. These zeroes imply the existence of algebraic constraints to be satisfied by the elements of $D\mathbf{X}(\mathbf{x}^*)$. The planar case is particularly important.

Let $\mathbf{X}: \mathbb{R}^2 \to \mathbb{R}^2$ be a smooth vector field with a non-hyperbolic singular point at $\mathbf{x} = \mathbf{0}$. Then there are four possibilities:

(i) $D\mathbf{X}(\mathbf{0})$ has real eigenvalues and one of them is zero, i.e.

$$\operatorname{Det} D\mathbf{X}(\mathbf{0}) = 0, \qquad \operatorname{Tr} D\mathbf{X}(\mathbf{0}) \neq 0; \tag{2.4.1}$$

(ii) $D\mathbf{X}(\mathbf{0})$ has pure imaginary eigenvalues, i.e.

$$\operatorname{Tr} D\mathbf{X}(\mathbf{0}) = 0, \qquad \operatorname{Det} D\mathbf{X}(\mathbf{0}) > 0; \tag{2.4.2}$$

(iii) both eigenvalues of $D\mathbf{X}(\mathbf{0})$ are zero but $D\mathbf{X}(\mathbf{0})$ is not the null matrix, i.e.

$$\operatorname{Tr} D\mathbf{X}(\mathbf{0}) = \operatorname{Det} D\mathbf{X}(\mathbf{0}) = 0, \qquad D\mathbf{X}(\mathbf{0}) \neq \mathbf{0}; \tag{2.4.3}$$

(iv)

$$D\mathbf{X}(\mathbf{0}) = \mathbf{0}. \tag{2.4.4}$$

The equalities (inequalities) in (2.4.1—2.4.4) are called *degeneracy* (*non-degeneracy*) *conditions*. Indeed, non-hyperbolic (hyperbolic) singularities are frequently referred to as *degenerate* (*non-degenerate*) *singularities*. As we shall see, in general, degeneracy/non-degeneracy conditions can involve higher-order coefficients in the Taylor expansion of **X** about $\mathbf{x}^*$ (see Example 2.7.4), but (2.4.1–4) are sufficient for our present purpose.

The number of degeneracy conditions that are satisfied at a particular singularity indicates its *level of degeneracy* (or *codimension*, see Dumortier, 1978; Guckenheimer & Holmes, 1983). Thus, if either $\operatorname{Det} D\mathbf{X}(\mathbf{x}^*) = 0$ or $\operatorname{Tr} D\mathbf{X}(\mathbf{x}^*) = 0$ is the only degeneracy condition satisfied at $\mathbf{x}^*$, then $\mathbf{x}^*$ is at the first level of degeneracy. However, a fixed point $\mathbf{x}^*$ for which (2.4.3) is valid is, at least, at the second level of degeneracy.

Figure 2.5 shows the phase portraits for the linear vector fields $D\mathbf{X}(\mathbf{0})\mathbf{x}$ when (i)–(iii) are satisfied. Observe that all are sensitive to non-linear perturbations (see Exercise 2.4.2). Clearly, the perturbations appearing in the normal form of **X** are most important. The Normal Form Theorem (2.3.1) is valid for non-hyperbolic fixed points and the resonant terms depend upon the degeneracy/non-degeneracy conditions that are satisfied. For example, when $n = 2$, we have listed the ways in which non-hyperbolic fixed points can occur when $D\mathbf{X}(\mathbf{0}) \neq \mathbf{0}$ (see (2.4.1)—(2.4.3)). We will now examine their normal forms.

Proposition 2.4.1 *Let* $\mathbf{X}(\mathbf{x})$ *be a smooth vector field with a singularity satisfying* (2.4.1) *at the origin. Then the normal form of* $\mathbf{X}$ *is given by*

$$\begin{pmatrix} \lambda_1 & 0 \\ 0 & 0 \end{pmatrix}\begin{pmatrix} y_1 \\ y_2 \end{pmatrix} + \sum_{r=2}^{N} \begin{pmatrix} a_r y_1 y_2^{r-1} \\ b_r y_2^r \end{pmatrix} + O(|\mathbf{y}|^{N+1}), \tag{2.4.5}$$

where a_r, $b_r \in \mathbb{R}$.

Proof. Let $\mathbf{DX}(\mathbf{0}) = \mathbf{A}$, then (2.4.1) allows us to take $\lambda_1 = \operatorname{Tr} \mathbf{A} \neq 0$ and $\lambda_2 = 0$. As in Example 2.3.2, the transformation $\mathbf{x} \mapsto \mathbf{M}^{-1}\mathbf{x}$, with $\mathbf{M} = (\mathbf{u}_1 \vdots \mathbf{u}_2)$ takes $\mathbf{X}_1$ into the form

$$\mathbf{X}_1(\mathbf{x}) = \begin{pmatrix} \lambda_1 & 0 \\ 0 & 0 \end{pmatrix}\begin{pmatrix} x_1 \\ x_2 \end{pmatrix} = \mathbf{Jx}$$

and the coefficients of terms of order two or more will in general change.

Figure 2.5 Phase portraits for linear vector fields $\mathbf{Jx}$, where $\mathbf{J}$ is the real Jordan form of $\mathbf{DX}(\mathbf{0})$, when (*a*) (2.4.1) with $\operatorname{Tr}(\mathbf{DX}(\mathbf{0})) > 0$; (*b*) (2.4.2); (*c*) (2.4.3), are satisfied.

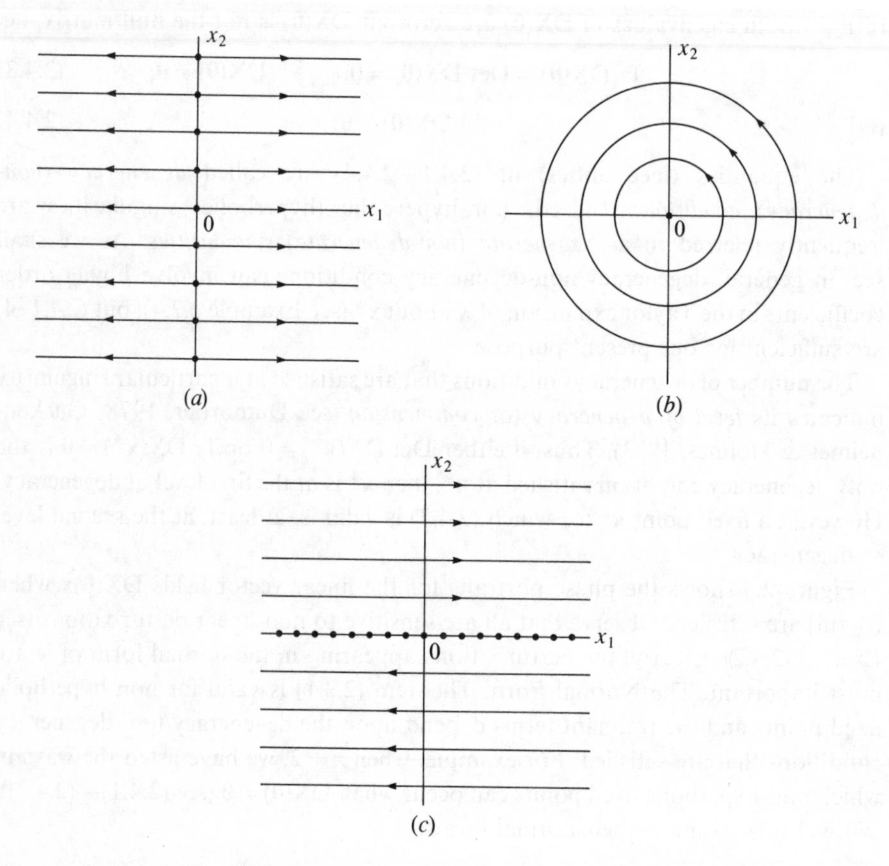

We seek to simplify terms $O(|\mathbf{x}|^r)$, $r \geqslant 2$, by transformations of the form $\mathbf{x} = \mathbf{y} + \mathbf{h}_r(\mathbf{y})$, $\mathbf{h}_r \in H^r$. Since $\mathbf{J}$ is diagonal, the only terms which cannot be removed are resonant monomials $\mathbf{x}^{\mathbf{m}}\mathbf{e}_i$, $i = 1, 2$. Observe that, since $\lambda_2 = 0$,

$$\lambda_1 + (r-1)\lambda_2 = \lambda_1,$$

and

$$(0)\lambda_1 + r\lambda_2 = \lambda_2,$$

for any $r \geqslant 2$. Thus, the resonance condition $\mathbf{m} \cdot \boldsymbol{\lambda} - \lambda_i = 0$ is satisfied for $\mathbf{m} = (1, r-1)$ when $i = 1$ and $\mathbf{m} = (0, r)$ when $i = 2$. It follows that

$$y_1 y_2^{r-1}\mathbf{e}_1 \qquad \text{and} \qquad y_2^r \mathbf{e}_2$$

are resonant monomials. In fact, it is not difficult to see that these are the only resonant terms. Thus, $\mathbf{X}(\mathbf{x})$ has normal form (2.4.5) by Theorem 2.3.1. □

Proposition 2.4.2 *Let $\mathbf{X}(\mathbf{x})$ be a smooth vector field with a singularity satisfying* (2.4.2) *at the origin. Then the normal form of $\mathbf{X}$ is given by*

$$\begin{pmatrix} 0 & -\beta \\ \beta & 0 \end{pmatrix}\begin{pmatrix} y_1 \\ y_2 \end{pmatrix} + \sum_{k=1}^{[\frac{1}{2}(N-1)]} (y_1^2 + y_2^2)^k \left\{ a_k \begin{pmatrix} y_1 \\ y_2 \end{pmatrix} + b_k \begin{pmatrix} -y_2 \\ y_1 \end{pmatrix} \right\} + O(|\mathbf{y}|^{N+1}), \quad (2.4.6)$$

where $\beta = (\mathrm{Det}\, \mathbf{DX}(\mathbf{0}))^{1/2}$, $N \geqslant 3$, $[\cdot]$ denotes the integer part of $\cdot$ and $a_k, b_k \in \mathbb{R}$.

Proof. The conditions (2.4.2) imply $\mathbf{A} = \mathbf{DX}(\mathbf{0})$ has eigenvalues $\lambda_1 = \bar{\lambda}_2 = \mathrm{i}\beta$ ($\bar{\ }$ denotes complex conjugate) with complex eigenvectors $\mathbf{u} \pm \mathrm{i}\mathbf{v}$. It follows that the real linear transformation $\mathbf{x} \mapsto \mathbf{M}^{-1}\mathbf{x}$ with $\mathbf{M} = (\mathbf{v} \vdots \mathbf{u})$ reduces $\mathbf{X}_1$ to the real Jordan form

$$\mathbf{X}_1(\mathbf{x}) = \begin{pmatrix} 0 & -\beta \\ \beta & 0 \end{pmatrix}\begin{pmatrix} x_1 \\ x_2 \end{pmatrix} = \mathbf{Jx}$$

possibly changing second and higher order terms. Unfortunately, $\mathbf{J}$ is not diagonal so that the monomials $\mathbf{x}^{\mathbf{m}}\mathbf{e}_i$ are not eigenvectors of $L_{\mathbf{J}}$.

To put $\mathbf{J}$ into diagonal form we will use the complex linear transformation $\mathbf{x} = \mathbf{M}_{\mathbb{C}}\mathbf{z}$ where $\mathbf{M}_{\mathbb{C}} = \frac{1}{2}\begin{pmatrix} 1 & 1 \\ -\mathrm{i} & \mathrm{i} \end{pmatrix}$. The columns of $\mathbf{M}_{\mathbb{C}}$ are simply the normalised eigenvectors of $\mathbf{J}$, so that $\mathbf{M}_{\mathbb{C}}^{-1}\mathbf{J}\mathbf{M}_{\mathbb{C}} = \begin{pmatrix} \mathrm{i}\beta & 0 \\ 0 & -\mathrm{i}\beta \end{pmatrix} = \mathbf{J}_{\mathbb{C}}$ and $\mathbf{z} = \begin{pmatrix} z_1 \\ z_2 \end{pmatrix} = \begin{pmatrix} x_1 + \mathrm{i}x_2 \\ x_1 - \mathrm{i}x_2 \end{pmatrix} = \begin{pmatrix} z \\ \bar{z} \end{pmatrix}$. Now the monomials $\mathbf{z}^{\mathbf{m}}\mathbf{e}_i$, $i = 1, 2$, are eigenvectors of $L_{\mathbf{J}_{\mathbb{C}}}$ and the usual resonance condition (2.3.14) applies.

Clearly $\lambda_1 = -\lambda_2$, so that for any positive integer k,

$$(k+1)\lambda_1 + k\lambda_2 = \lambda_1 \quad (2.4.7)$$

and

$$k\lambda_1 + (k+1)\lambda_2 = \lambda_2. \tag{2.4.8}$$

These are the only resonant terms. In both (2.4.7) and (2.4.8) $r = m_1 + m_2 = 2k+1$, so there are no resonant terms of even order in $|\mathbf{z}|$. The resonant monomials are given by

$$z^{k+1}\bar{z}^k \mathbf{e}_1 \qquad \text{and} \qquad z^k \bar{z}^{k+1} \mathbf{e}_2,$$

where $k \in \mathbb{Z}^+$. Transforming these vector fields back into the real coordinates x_1, x_2, we obtain

$$\tfrac{1}{2}|\mathbf{x}|^{2k}\left[\begin{pmatrix} x_1 \\ x_2 \end{pmatrix} - \mathrm{i}\begin{pmatrix} -x_2 \\ x_1 \end{pmatrix}\right] \qquad \text{and} \qquad \tfrac{1}{2}|\mathbf{x}|^{2k}\left[\begin{pmatrix} x_1 \\ x_2 \end{pmatrix} + \mathrm{i}\begin{pmatrix} -x_2 \\ x_1 \end{pmatrix}\right],$$

respectively. Notice $(z\bar{z})^k = |\mathbf{x}|^{2k}$. Thus, for each $k \in \mathbb{Z}^+$,

$$\left\{|\mathbf{x}|^{2k}\begin{pmatrix} x_1 \\ x_2 \end{pmatrix}, |\mathbf{x}|^{2k}\begin{pmatrix} -x_2 \\ x_1 \end{pmatrix}\right\}$$

spans a subspace, G^{2k+1}, of H^{2k+1} complementary to $L_{\mathbf{J}}(H^{2k+1})$. Finally, Theorem 2.3.1 implies that the normal form of $\mathbf{X}$ is given by (2.4.6). □

Thus far we have been able to diagonalise $\mathbf{A} = \mathbf{DX}(\mathbf{0})$ and use the resonance condition (2.3.13) to find a complementary subspace G^r. However, if $\mathbf{DX}(\mathbf{0})$ satisfies (2.4.3) this is no longer the case and we must find G^r directly.

Proposition 2.4.3 *Let $\mathbf{X}(\mathbf{x})$ be a smooth vector field with a singularity satisfying (2.4.3) at the origin. Then the normal form of $\mathbf{X}$ is given by*

$$\begin{pmatrix} 0 & 1 \\ 0 & 0 \end{pmatrix}\begin{pmatrix} y_1 \\ y_2 \end{pmatrix} + \sum_{r=2}^{N}\begin{pmatrix} a_r y_1^r \\ b_r y_1^r \end{pmatrix} + O(|\mathbf{y}|^{N+1}), \tag{2.4.9}$$

where $a_r, b_r \in \mathbb{R}$.

Proof. The condition (2.4.3) implies both eigenvalues of $\mathbf{A} = \mathbf{DX}(\mathbf{0})$ are zero but $\mathbf{A} \neq \mathbf{0}$. It follows that there is only one eigenvector and (see Arrowsmith & Place, 1982, pp. 43–4) there is a linear transformation $\mathbf{x} \mapsto \mathbf{M}^{-1}\mathbf{x}$ reducing the linear part of $\mathbf{X}$ to the form

$$\mathbf{X}_1(\mathbf{x}) = \begin{pmatrix} 0 & 1 \\ 0 & 0 \end{pmatrix}\begin{pmatrix} x_1 \\ x_2 \end{pmatrix} = \mathbf{J}\mathbf{x}.$$

In the absence of eigenvectors for $L_{\mathbf{A}}$, we focus attention on finding a basis for $B^r = L_{\mathbf{J}}(H^r)$. Clearly,

$$\{x_1^{m_1} x_2^{m_2} \mathbf{e}_i \,|\, m_1, m_2 \in \mathbb{N}, m_1 + m_2 = r, i = 1, 2\}$$

forms a basis for H^r, so we consider

$$L_{\mathbf{J}} x_1^{m_1} x_2^{m_2} \mathbf{e}_1 = m_1 x_1^{m_1-1} x_2^{m_2+1} \mathbf{e}_1$$

and

$$L_{\mathbf{J}} x_1^{m_1} x_2^{m_2} \mathbf{e}_2 = -x_1^{m_1} x_2^{m_2} \mathbf{e}_1 + m_1 x_1^{m_1-1} x_2^{m_2+1} \mathbf{e}_2$$

with $m_1 + m_2 = r$, and conclude that

$$\left\{ \begin{pmatrix} x_2^r \\ 0 \end{pmatrix}, \begin{pmatrix} x_1 x_2^{r-1} \\ 0 \end{pmatrix}, \begin{pmatrix} x_1^2 x_2^{r-2} \\ 0 \end{pmatrix}, \ldots, \begin{pmatrix} x_1^{r-1} x_2 \\ 0 \end{pmatrix}, \right.$$
$$\left. \begin{pmatrix} -x_1 x_2^{r-1} \\ x_2^r \end{pmatrix}, \begin{pmatrix} -x_1^2 x_2^{r-2} \\ 2x_1 x_2^{r-1} \end{pmatrix}, \ldots, \begin{pmatrix} -x_1^r \\ r x_1^{r-1} x_2 \end{pmatrix} \right\} \qquad (2.4.10)$$

form a basis for B^r. Notice the set (2.4.10) has $2r$ elements, while H^r has dimension $2r+2$, so that $\dim G^r = 2$. Since $G^r \subseteq H^r$ such that $G^r \cap B^r = \{\mathbf{0}\}$ and $G^r \oplus B^r = H^r$, inspection of (2.4.10) shows that we could choose

$$G^r = Sp\left\{ \begin{pmatrix} x_1^r \\ 0 \end{pmatrix}, \begin{pmatrix} 0 \\ x_1^r \end{pmatrix} \right\}, \qquad (2.4.11)$$

when the normal form (2.4.9) follows by Theorem 2.3.1. □

It should be emphasised that (2.4.11) is not unique. For example, the alternative choice

$$G^r = Sp\left\{ \begin{pmatrix} 0 \\ x_1^{r-1} x_2 \end{pmatrix}, \begin{pmatrix} 0 \\ x_1^r \end{pmatrix} \right\} \qquad (2.4.12)$$

is favoured by Arnold and his co-workers (Arnold, 1983, p. 296; Bogdanov, 1981b). The form (2.4.9) is due to Takens (1974a). Of course, these two forms are completely equivalent or, more precisely, smoothly conjugate (see Exercise 2.4.4).

2.5 Normal forms for diffeomorphisms (Arnold, 1983, §25)

A similar analysis to that described in §§2.3 and 4 for differential equations can be carried out for diffeomorphisms. The important difference is in the condition for resonance.

Let $\mathbf{f}: \mathbb{R}^n \to \mathbb{R}^n$ be a diffeomorphism, satisfying $\mathbf{f}(\mathbf{0}) = \mathbf{0}$, whose Taylor expansion about $\mathbf{x} = \mathbf{0}$ is

$$\mathbf{f}(\mathbf{x}) = \mathbf{f}_1 + \mathbf{f}_2 + \cdots + \mathbf{f}_k + O(|\mathbf{x}|^{k+1}), \qquad (2.5.1)$$

where $\mathbf{f}_r \in H^r$ and $\mathbf{f}_1 = D\mathbf{f}(\mathbf{0})\mathbf{x} = \mathbf{A}\mathbf{x}$. Suppose

$$\mathbf{f}(\mathbf{x}) = \mathbf{A}\mathbf{x} + \mathbf{f}_r(\mathbf{x}) + O(|\mathbf{x}|^{r+1}), \qquad (2.5.2)$$

$r \geqslant 2$, and consider the effect of the transformation

$$\mathbf{x} = \mathbf{y} + \mathbf{k}_r(\mathbf{y}) = \mathbf{K}(\mathbf{y}), \tag{2.5.3}$$

with $\mathbf{k}_r \in H^r$. If $\mathbf{K}$ transforms $\mathbf{f}$ into $\tilde{\mathbf{f}}$, then

$$\tilde{\mathbf{f}}(\mathbf{y}) = \mathbf{K}^{-1}(\mathbf{f}(\mathbf{K}(\mathbf{y}))).$$

Thus

$$\begin{aligned}\tilde{\mathbf{f}}(\mathbf{y}) &= \mathbf{K}^{-1}(\mathbf{f}(\mathbf{y} + \mathbf{k}_r(\mathbf{y}))),\\ &= \mathbf{K}^{-1}(\mathbf{A}\mathbf{y} + \mathbf{A}\mathbf{k}_r(\mathbf{y}) + \mathbf{f}_r(\mathbf{y}) + O(|\mathbf{y}|^{r+1})),\\ &= \mathbf{A}\mathbf{y} + \mathbf{A}\mathbf{k}_r(\mathbf{y}) + \mathbf{f}_r(\mathbf{y}) - \mathbf{k}_r(\mathbf{A}\mathbf{y}) + O(|\mathbf{y}|^{r+1}).\end{aligned}$$

Clearly, if $\mathbf{k}_r(\mathbf{y})$ is chosen so that

$$\mathbf{k}_r(\mathbf{A}\mathbf{y}) - \mathbf{A}\mathbf{k}_r(\mathbf{y}) = \mathbf{f}_r(\mathbf{y}) \tag{2.5.4}$$

then

$$\tilde{\mathbf{f}}(\mathbf{y}) = \mathbf{A}\mathbf{y} + O(|\mathbf{y}|^{r+1}). \tag{2.5.5}$$

Equation (2.5.4) is the homological equation associated with $\mathbf{A} = \mathbf{Df}(\mathbf{0})$. The operator $M_{\mathbf{A}} : H^r \to H^r$, defined by

$$M_{\mathbf{A}}\mathbf{k}_r(\mathbf{x}) = \mathbf{k}_r(\mathbf{A}\mathbf{x}) - \mathbf{A}\,\mathbf{k}_r(\mathbf{x}), \tag{2.5.6}$$

can be shown to be linear (see Exercise 2.5.1) with eigenvectors $\mathbf{x}^{\mathbf{m}}\mathbf{e}_i$. When $\mathbf{A} = [\lambda_i \delta_{ij}]_{i,j=1}^n$, the eigenvalues of $M_{\mathbf{A}}$ are given by

$$M_{\mathbf{A}}\mathbf{x}^{\mathbf{m}}\mathbf{e}_i = (\boldsymbol{\lambda}^{\mathbf{m}} - \lambda_i)\mathbf{x}^{\mathbf{m}}\mathbf{e}_i, \tag{2.5.7}$$

$i = 1, \ldots, n$, where $\boldsymbol{\lambda}^{\mathbf{m}} = \lambda_1^{m_1}\lambda_2^{m_2}, \ldots, \lambda_n^{m_n}$. Eigenvector expansions of $\mathbf{k}_r(\mathbf{x})$ and $\mathbf{f}_r(\mathbf{x})$, analogous to (2.3.15 and 16), lead to

$$k_{\mathbf{m},i} = \frac{f_{\mathbf{m},i}}{(\boldsymbol{\lambda}^{\mathbf{m}} - \lambda_i)} \tag{2.5.8}$$

for the coefficients in the expansion of $\mathbf{k}_r(\mathbf{x})$. In this case, therefore, the n-tuple of eigenvalues $\boldsymbol{\lambda} = (\lambda_1, \ldots, \lambda_n)$ is said to be resonant of order r if

$$\lambda_i = \boldsymbol{\lambda}^{\mathbf{m}} \tag{2.5.9}$$

for some allowed $\mathbf{m}$ and $i = 1, \ldots, n$. Recall $\mathbf{m} = (m_1, \ldots, m_n)$, $m_i \in \mathbb{N}$ and $\sum_i m_i = r$.

An analogous result to Theorem 2.3.1 can be proved for a diffeomorphism $\mathbf{f}$. More specifically, $\mathbf{f}$ can be reduced to the form

$$\mathbf{J}\mathbf{y} + \sum_{r=2}^{N} \mathbf{w}_r(\mathbf{y}) + O(|\mathbf{y}|^{N+1}), \tag{2.5.10}$$

where $\mathbf{w}_r$ is resonant of order r (in the sense of (2.5.9)), the reduction being brought about by a polynomial transformation. The expression (2.5.10) is the *normal form of* **f**.

A fixed point, $\mathbf{x}^*$, of **f** is non-hyperbolic if $\mathbf{Df}(\mathbf{x}^*)$ has at least one eigenvalue on the unit circle in the complex plane. As in the case of vector fields, characteristic classes of resonant terms occur at non-hyperbolic fixed points. The nature of these terms again depends on how the non-hyperbolicity occurs.

Example 2.5.1 Show that any smooth diffeomorphism $f: \mathbb{R} \to \mathbb{R}$, satisfying $f(0) = 0$ and $Df(0) = -1$, can be written in the form

$$f(x) = -x + \sum_{r=1}^{N} a_{2r+1} x^{2r+1} + O(x^{2N+3}), \tag{2.5.11}$$

for any integer $N \geqslant 1$. Is any analogous simplification possible when $\mathrm{D}f(0) = +1$?

Solution. In this case, $\mathrm{D}f(0) = \lambda_1 = -1$ and $\boldsymbol{\lambda}^{\mathbf{m}} = \lambda_1^{m_1}$ with $m_1 = r$ for terms of order r. It follows that the resonance condition (2.5.9) is satisfied only if r is odd. Thus, there are transformations of the form (2.5.3) allowing the terms of order x^r, with r even, to be eliminated from the Taylor expansion of f about $x = 0$. Such transformations only change terms of order x^k for $k \geqslant r+1$. Therefore, successive application of these transformations for $r = 2, 4, \ldots, 2N+2$ leads to (2.5.11).

When $\mathrm{D}f(0) = +1$, $\boldsymbol{\lambda}^{\mathbf{m}} = \lambda_1^r = 1 = \lambda_1$ for all r, so that no analogous simplification of the Taylor expansion of f is possible in this case. □

It is perhaps worth noting that if $|\mathrm{D}f(0)| = |\lambda_1| \neq 1$ then $\boldsymbol{\lambda}^{\mathbf{m}} = \lambda_1^r \neq 1$ for any $r \geqslant 2$ and there are no resonant terms in the normal form of $f: \mathbb{R} \to \mathbb{R}$.

Example 2.5.2 Let $\mathbf{f}: \mathbb{R}^2 \to \mathbb{R}^2$ be a smooth map with $\mathbf{f}(\mathbf{0}) = \mathbf{0}$ and

$$\mathbf{Df}(\mathbf{0}) = \begin{pmatrix} \cos\alpha & -\sin\alpha \\ \sin\alpha & \cos\alpha \end{pmatrix}, \tag{2.5.12}$$

where α is an irrational multiple of 2π. Show that the normal form of **f** involves only odd order terms. Does this result remain valid if $\alpha = 2\pi p/q$, $p, q \in \mathbb{Z}^+$ and relatively prime?

Solution. $\mathbf{Df}(\mathbf{0})$ given in (2.5.12) can be diagonalised, by the complex change of variable $z_1 = z = x + \mathrm{i}y$, $z_2 = \bar{z} = x - \mathrm{i}y$, to obtain the Jordan form

$$\mathbf{J}_{\mathbb{C}} = \begin{pmatrix} \exp(\mathrm{i}\alpha) & 0 \\ 0 & \exp(-\mathrm{i}\alpha) \end{pmatrix} = \begin{pmatrix} \lambda_1 & 0 \\ 0 & \lambda_2 \end{pmatrix}. \tag{2.5.13}$$

The resonance condition (2.5.9) becomes

$$\exp[\mathrm{i}(m_1 - m_2)\alpha] = \begin{cases} \exp(\mathrm{i}\alpha) & i = 1 \quad (2.5.14\mathrm{a}) \\ \exp(-\mathrm{i}\alpha) & i = 2. \quad (2.5.14\mathrm{b}) \end{cases}$$

Clearly, (2.5.14) gives resonant terms of the form

$$\begin{pmatrix} z^{m+1}\bar{z}^m \\ 0 \end{pmatrix} \quad \text{and} \quad \begin{pmatrix} 0 \\ z^m\bar{z}^{m+1} \end{pmatrix}, \tag{2.5.15}$$

$m \in \mathbb{Z}^+$. Moreover, these are the only resonant terms when α is an irrational multiple of 2π, so that the normal form of **f** involves *only* odd order terms.

The terms appearing in (2.5.15) are called *inevitable* resonant terms since they occur for all values of α. When $\alpha = 2\pi p/q$, the above result is no longer valid because *additional* resonant terms (of both odd and even order) can occur. Observe that in this case (2.5.14a) is satisfied by

$$m_1 - m_2 = 1 - lq, \qquad l \in \mathbb{Z}, \tag{2.5.16}$$

giving rise to resonances of the form

$$\begin{pmatrix} z^{m-lq+1}\bar{z}^m \\ 0 \end{pmatrix}. \tag{2.5.17}$$

In general, $2m - lq + 1$, $l \neq 0$, need not be odd and the normal form of **f** may contain even order terms. For example, from (2.5.16) the lowest order additional resonance term occurs for $l = +1$ and $m = q - 1$, i.e. it takes the form

$$\begin{pmatrix} \bar{z}^{q-1} \\ 0 \end{pmatrix}. \tag{2.5.18}$$

Clearly, $q - 1$ can be even. □

When using the complex notation employed in Example 2.5.2, it is, in fact, sufficient to consider only the first component of the map. Notice, (2.5.13) shows that the second component of the linearisation is simply the complex conjugate of the first. Moreover, it is clear from (2.5.14) that if (m_1, m_2) satisfies (a) then (m_2, m_1) satisfies (b). This means that the resonant terms in the second component are the complex conjugate of those in the first. This is apparent in (2.5.15). These observations also apply to the additional resonances. Note that (2.5.14b) leads to (2.5.16) with m_1 and m_2 interchanged. To avoid duplication, it is usual to consider only the first component and to refer to this as the *complex form* of $\mathbf{f}: \mathbb{R}^2 \to \mathbb{R}^2$, denoted by $f(z)$. This notation is convenient but slightly confusing as it suggests that f depends only on z rather than on both z and $\bar{z}$. Indeed, some authors (see Moser, 1968) prefer to emphasise this point by saying that $\mathbf{f}: \mathbb{R}^2 \to \mathbb{R}^2$ is 'represented by a function of z and $\bar{z}$'. Of course, it is true to say that if z is known then so is $\bar{z}$, however the map $\kappa: \mathbb{C} \to \mathbb{C}$ such that $\bar{z} = \kappa(z)$ is not analytic. Thus in the present context it must be emphasised that f is not an analytic function of z (see Exercise 2.5.3).

Mappings of the kind considered in Example 2.5.2 play a major role in Chapters 5 and 6. The following result emphasises the symmetry properties of their normal forms.

Proposition 2.5.1 *Let* $\mathbf{f}$: $\mathbb{R}^2 \to \mathbb{R}^2$ *have* $\mathbf{f}(\mathbf{0}) = \mathbf{0}$ *and* $\mathbf{Df}(\mathbf{0}) = \mathbf{R}_\alpha$, *a rotation through angle* α. *Then* $\mathbf{f}$ *is in normal form if and only if*

$$\mathbf{f} \cdot \mathbf{R}_\alpha = \mathbf{R}_\alpha \cdot \mathbf{f}. \tag{2.5.19}$$

Proof. In complex notation, (2.5.19) can be written

$$f(\exp(i\alpha)z) = \exp(i\alpha)f(z), \tag{2.5.20}$$

where

$$f(z) = \exp(i\alpha)z + \sum_{r=2}^{\infty} \sum_{\substack{\mathbf{m} \\ m_1+m_2=r}} a_\mathbf{m} z^{m_1}\bar{z}^{m_2}.$$

Let $w = \exp(i\alpha)z$ so that

$$\begin{aligned} f(\exp(i\alpha)z) = f(w) &= \exp(i\alpha)w + \sum_{r=2}^{\infty} \sum_{\substack{\mathbf{m} \\ m_1+m_2=r}} a_\mathbf{m} w^{m_1}\bar{w}^{m_2} \\ &= \exp(2i\alpha)z + \sum_{r=2}^{\infty} \sum_{\substack{\mathbf{m} \\ m_1+m_2=r}} a_\mathbf{m} \exp[i\alpha(m_1 - m_2)]z^{m_1}\bar{z}^{m_2} \\ &= \exp(i\alpha)\left\{\exp(i\alpha)z + \sum_{r=2}^{\infty} \sum_{\substack{\mathbf{m} \\ m_1+m_2=r}} a_\mathbf{m} \exp[i\alpha(m_1 - m_2 - 1)]z^{m_1}\bar{z}^{m_2}\right\}. \end{aligned} \tag{2.5.21}$$

Now, if f is in normal form only resonant terms are present and therefore either:

$$m_1 - m_2 - 1 = 0, \text{ if } \alpha \text{ is irrational}; \tag{2.5.22}$$

or

$$m_1 - m_2 - 1 = -lq, \; l \in \mathbb{Z}, \text{ if } \alpha = 2\pi p/q. \tag{2.5.23}$$

In either case, $\exp[i\alpha(m_1 - m_2 - 1)] = 1$ and (2.5.20) follows. Conversely, if (2.5.20) is to be satisfied, then $a_\mathbf{m} = 0$ unless $\mathbf{m} = (m_1, m_2)$ is such that

$$\exp[i\alpha(m_1 - m_2 - 1)] = 1. \tag{2.5.24}$$

However, (2.5.24) is precisely the same as the resonance condition (2.5.14a) and consequently f must be in normal form. □

Examples 2.5.1 and 2 have not allowed us to illustrate the mechanics of the removal of the non-resonant terms. If the coefficients of the resonant terms in the normal form are required (as they often are, see Chapters 4 and 5) then details of the transformations giving the normal form must be considered.

Example 2.5.3 Let $\mathbf{f}: \mathbb{R}^2 \to \mathbb{R}^2$ be as described in Example 2.5.2, with α an irrational multiple of 2π. Given that $\mathbf{f}$ has the complex form

$$f(z) = \exp(i\alpha)z + \sum_{\substack{\mathbf{m} \\ m_1+m_2=2}} a_{\mathbf{m}} z^{m_1}\bar{z}^{m_2} + O(|z|^{m_1+m_2+1}), \tag{2.5.25}$$

find a transformation of the form

$$z = K(w) = w + \sum_{\substack{\mathbf{m} \\ m_1+m_2=2}} k_{\mathbf{m}} w^{m_1}\bar{w}^{m_2} \tag{2.5.26}$$

such that $\tilde{f}(w) = K^{-1}(f(K(w)))$ has no quadratic terms.

Show that $\mathbf{f}$ can be written in the form

$$\begin{pmatrix} \cos\alpha & -\sin\alpha \\ \sin\alpha & \cos\alpha \end{pmatrix}\begin{pmatrix} x \\ y \end{pmatrix} + (x^2+y^2)\begin{pmatrix} c & -d \\ d & c \end{pmatrix}\begin{pmatrix} x \\ y \end{pmatrix} + O(|\mathbf{x}|^4) \tag{2.5.27}$$

for some constants $c, d \in \mathbb{R}$. Describe, without doing any calculations, how you would find c and d from (2.5.25).

Solution. For $m_1 + m_2 = 2$, $\boldsymbol{\lambda}^{\mathbf{m}} - \lambda_1$, with $\lambda_1 = \lambda = \exp(i\alpha)$, is given by

$$[\lambda^{m_1}\bar{\lambda}^{m_2} - \lambda] = \begin{cases} \lambda(\lambda - 1), & \mathbf{m} = (2,0), \\ \lambda(\bar{\lambda} - 1), & \mathbf{m} = (1,1), \\ \lambda(\bar{\lambda}^3 - 1), & \mathbf{m} = (0,2). \end{cases} \tag{2.5.28}$$

Since α is an irrational multiple of 2π, the right hand sides of (2.5.28) are non-zero for all $\mathbf{m}$. Hence the choice

$$k_{20} = \frac{a_{20}}{\lambda^2 - \lambda}, \qquad k_{11} = \frac{a_{11}}{1 - \lambda}, \qquad k_{02} = \frac{a_{02}}{\bar{\lambda}^2 - \lambda} \tag{2.5.29}$$

ensures that $\tilde{f}(w) = K^{-1}(f(K(w)))$ has no quadratic terms.

To obtain (2.5.27), observe that $\tilde{f}$ now takes the form (2.5.25) with $m_1 + m_2 = 3$ and $a_{\mathbf{m}}$ replaced by $\tilde{a}_{\mathbf{m}}$, say. Of course, the coefficients $\tilde{a}_{\mathbf{m}}$ in this expression will, in general, be different from the cubic terms of (2.5.25) but they can be expressed in terms of the original quadratic and cubic coefficients. We now consider the effect of a transformation like (2.5.26) with $m_1 + m_2 = 3$. In this case,

$$[\lambda^{m_1}\bar{\lambda}^{m_2} - \lambda] = \begin{cases} \lambda(\lambda^2 - 1) & \mathbf{m} = (3,0) \\ 0 & \mathbf{m} = (2,1) \\ \lambda(\bar{\lambda}^2 - 1) & \mathbf{m} = (1,2) \\ \lambda(\bar{\lambda}^4 - 1) & \mathbf{m} = (0,3) \end{cases} \tag{2.5.30}$$

and $k_{\mathbf{m}}$, $m_1 + m_2 = 3$, can be chosen to remove all but the $z^2\bar{z}$-term. By combining the $m_1 + m_2 = 2$ and $m_1 + m_2 = 3$ transformations, we conclude that f can be reduced to

$$\lambda w + \tilde{a}_{21} w^2 \bar{w} + O(|w|^4). \tag{2.5.31}$$

If we write $\tilde{a}_{21} = c + id$, $\lambda = \exp(i\alpha) = \cos\alpha + i\sin\alpha$ and $w = x + iy$, (2.5.31) becomes

$$[x\cos\alpha - y\sin\alpha + cx(x^2+y^2) - dy(x^2+y^2)] + i[x\sin\alpha + y\cos\alpha + cy(x^2+y^2) + dx(x^2+y^2)] \tag{2.5.32}$$

and the real form (2.5.27) follows.

To find c and d, we must express $\tilde{a}_{21}$ in terms of $a_{\mathbf{m}}$, $m_1 + m_2 = 2$ and 3, occurring in (2.5.25). To do this we use the $k_{\mathbf{m}}$ given in (2.5.29) to compute the cubic terms in $\tilde{f}$. This, somewhat lengthy, calculation is considered in [Iooss, 1979, pp. 28–32]. □

2.6 Time-dependent normal forms (Arnold, 1983, §26)

The normal form calculations of §2.3 can be extended to a time-dependent vector field, $\mathbf{X}(\mathbf{x}, t)$, satisfying

$$\mathbf{X}(\mathbf{x}, t + 2\pi) = \mathbf{X}(\mathbf{x}, t), \tag{2.6.1}$$

$t \in \mathbb{R}$, $\mathbf{x} \in \mathbb{R}^n$, $\mathbf{X}(\cdot, t)\colon \mathbb{R}^n \to \mathbb{R}^n$ for each t. In this case, given that $\mathbf{X}(\mathbf{0}, t) = \mathbf{0}$ for all t, Taylor expansion gives

$$\mathbf{X}(\mathbf{x}, t) = \mathbf{A}(t)\mathbf{x} + \sum_{r=2}^{k} \left\{ \sum_{\substack{\mathbf{m} \\ \sum_i m_i = r}} \sum_{j=1}^{n} a_{\mathbf{m}j}(t)\mathbf{x}^{\mathbf{m}}\mathbf{e}_j \right\} + O(|\mathbf{x}|^{k+1}; t), \tag{2.6.2}$$

where $\mathbf{A}(t)$, $a_{\mathbf{m}j}(t)$ and $O(|\mathbf{x}|^{k+1}; t)$ are 2π-periodic functions of t.

Let the non-autonomous linear system

$$\dot{\mathbf{x}} = \mathbf{A}(t)\mathbf{x}, \qquad \mathbf{A}(t + 2\pi) = \mathbf{A}(t), \qquad \text{for all } t \in \mathbb{R}, \tag{2.6.3}$$

have state transition matrix $\boldsymbol{\varphi}(t, t_0)$ and period advance map $\mathbf{P} = \boldsymbol{\varphi}(2\pi, 0)$ (see §1.8).

Theorem 2.6.1 (Floquet) *If we write* $\mathbf{P} = \exp(2\pi\mathbf{\Lambda})$, *then there exists a change of variable* $\mathbf{x} = \mathbf{B}(t)\mathbf{y}$, *with* $\mathbf{B}(t + 2\pi) = \mathbf{B}(t)$ *for all* $t \in \mathbb{R}$, *such that* (2.6.3) *becomes*

$$\dot{\mathbf{y}} = \mathbf{\Lambda}\mathbf{y}. \tag{2.6.4}$$

It should be noted that not every real linear transformation has a real logarithm. As we shall see, the occurrence of complex $\mathbf{\Lambda}$ does not present any problems in the normal form calculations. However, if a real matrix $\mathbf{\Lambda}$ is desired for other reasons then this can be achieved with at worst a 4π-periodic transformation $\mathbf{B}(t)$. The background to these remarks is examined in Exercises 2.6.1–2.6.4.

If we assume that Theorem 2.6.1 applies to the linear part of (2.6.2), then the change of variables $\mathbf{x} \mapsto \mathbf{B}(t)^{-1}\mathbf{x}$ reduces $\mathbf{X}(\mathbf{x}, t)$ to the form

$$\mathbf{X}(\mathbf{x}, t) = \mathbf{\Lambda x} + \sum_{r=2}^{k} \left\{ \sum_{\substack{\mathbf{m} \\ \sum_i m_i = r}} \sum_{j=1}^{n} a_{\mathbf{m}j}(t)\mathbf{x}^{\mathbf{m}}\mathbf{e}_j \right\} + O(|\mathbf{x}|^{k+1}; t), \tag{2.6.5}$$

In (2.6.5), $a_{\mathbf{m}j}(t)$ are still 2π-periodic functions of t, but they will, in general, be different from those in (2.6.2).

The following result generalises (2.3.5) (see Exercise 2.6.6).

Proposition 2.6.1 *The linear equation* $\dot{\mathbf{x}} = \mathbf{\Lambda x}$, $\mathbf{\Lambda}$ *independent of* t, *is transformed into*

$$\dot{\mathbf{y}} = \mathbf{\Lambda y} - \left\{ (\mathrm{D}_{\mathbf{x}}\mathbf{h}_r)\mathbf{\Lambda y} - \mathbf{\Lambda h}_r + \frac{\partial \mathbf{h}_r}{\partial t} \right\} + O(|\mathbf{y}|^{r+1}; t) \tag{2.6.6}$$

by the change of coordinates

$$\mathbf{x} = \mathbf{y} + \mathbf{h}_r(\mathbf{y}, t), \tag{2.6.7}$$

where $\mathbf{h}_r \in H^r$ *in* $\mathbf{y}$, $r \geqslant 2$, *and is* 2π-*periodic in* t.

Equation (2.6.6) leads us to define the homological equation

$$L_{\mathbf{\Lambda}}\mathbf{h}_r + \frac{\partial \mathbf{h}_r}{\partial t} = \mathbf{X}_r(\mathbf{x}, t)$$

$$= \sum_{\substack{\mathbf{m} \\ \sum_i m_i = r}} \sum_{j=1}^{n} a_{\mathbf{m}j}(t)\mathbf{x}^{\mathbf{m}}\mathbf{e}_j. \tag{2.6.8}$$

Let us assume, as in § 2.3, that $\mathbf{\Lambda} = [\lambda_i \delta_{ij}]_{i,j=1}^{n}$, where, in general, $\lambda_i \in \mathbb{C}$. To solve (2.6.8) we Fourier analyse the time dependence in $a_{\mathbf{m}j}(t)$ and $\mathbf{h}_r(\mathbf{x}, t)$, i.e.

$$\mathbf{X}_r(\mathbf{x}, t) = \sum_{\substack{\mathbf{m} \\ \sum_i m_i = r}} \sum_{j=1}^{n} \sum_{\nu=-\infty}^{\infty} a_{\mathbf{m},j,\nu} \exp(i\nu t)\mathbf{x}^{\mathbf{m}}\mathbf{e}_j \tag{2.6.9}$$

and

$$\mathbf{h}_r(\mathbf{x}, t) = \sum_{\substack{\mathbf{m} \\ \sum_i m_i = r}} \sum_{j=1}^{n} \sum_{\nu=-\infty}^{\infty} h_{\mathbf{m},j,\nu} \exp(i\nu t)\mathbf{x}^{\mathbf{m}}\mathbf{e}_j. \tag{2.6.10}$$

In (2.6.9 and 10), $h_{\mathbf{m},j,\nu}, a_{\mathbf{m},j,\nu} \in \mathbb{C}$ and $h_{\mathbf{m},j,-\nu} = \bar{h}_{\mathbf{m},j,\nu}$, $a_{\mathbf{m},j,-\nu} = \bar{a}_{\mathbf{m},j,\nu}$, since $\mathbf{X}_r, \mathbf{h}_r \in \mathbb{R}^n$. Substituting (2.6.9) and (2.6.10) into (2.6.8), we obtain

$$h_{\mathbf{m},j,\nu} = \frac{a_{\mathbf{m},j,\nu}}{i\nu + \mathbf{m} \cdot \boldsymbol{\lambda} - \lambda_j}, \tag{2.6.11}$$

provided that

$$(\mathrm{i}\nu + \mathbf{m}\cdot\boldsymbol{\lambda} - \lambda_j) \neq 0 \tag{2.6.12}$$

(cf. 2.3.15–17). It follows that (2.6.5) can be reduced to the time-dependent normal form

$$\dot{\mathbf{x}} = \mathbf{\Lambda x} + \sum_{r=2}^{N} \mathbf{w}_r(\mathbf{x}, t) + O(|\mathbf{x}|^{N+1}; t). \tag{2.6.13}$$

Here the time dependence of the resonant term $\mathbf{w}_r(\mathbf{x}, t)$ involves only a finite number of Fourier harmonics because for given $\mathbf{m}$ and j the resonance condition

$$\mathrm{i}\nu + \mathbf{m}\cdot\boldsymbol{\lambda} - \lambda_j = 0, \qquad \sum_i m_i = r \tag{2.6.14}$$

determines $\nu \in \mathbb{Z}$ uniquely. The case when $\mathbf{\Lambda}$ cannot be diagonalised is considered in Exercise 2.6.6.

Example 2.6.1 Consider a time-dependent vector field $\mathbf{X}(\cdot, t)$: $\mathbb{R}^2 \to \mathbb{R}^2$ satisfying (2.6.1). Let the period advance map $\mathbf{P}$ of its non-autonomous linearisation have complex eigenvalues $\mu_{1,2} = \exp(\pm 2\pi \mathrm{i}\omega)$.

Assume that ω is *irrational* and show that $\dot{\mathbf{x}} = \mathbf{X}(\mathbf{x}, t)$ can be written (see Exercise 2.6.7) in the complex form

$$\dot{z} = \mathrm{i}\omega z + \sum_{s=1}^{k-1} c_s z|z|^{2s} + O(|z|^{2k+1}; t), \tag{2.6.15}$$

where $O(|z|^{2k+1}; t)$ is 2π-periodic in t.

Given that ω is *rational* and equal to p/q, $p, q \in \mathbb{Z}^+$ and relatively prime, show that the resonant term of order r involves Fourier harmonics given by

$$\nu = lp, \tag{2.6.16}$$

where $l \in \mathbb{Z}$ and satisfies

$$\left\lceil \frac{1-r}{q} \right\rceil \leqslant l \leqslant \left\lfloor \frac{r+1}{q} \right\rfloor. \tag{2.6.17}$$

In (2.6.17), $\lfloor(\cdot)\rfloor$ ($\lceil(\cdot)\rceil$) is the largest integer less than or equal to $(\cdot)$ (smallest integer greater than or equal to $(\cdot)$). Hence deduce that when ω is rational (2.6.15) contains *additional* resonant terms. Show that the lowest order additional resonant term is proportional to $\bar{z}^{q-1} \exp(\mathrm{i}pt)$.

Solution. Since the eigenvalues of $\mathbf{P}$ are $\exp(\pm 2\pi \mathrm{i}\omega)$, there exists a (complex) linear change of coordinates such that

$$\mathbf{P} = \begin{pmatrix} \exp(2\pi \mathrm{i}\omega) & 0 \\ 0 & \exp(-2\pi \mathrm{i}\omega) \end{pmatrix}, \tag{2.6.18}$$

which is of the form $\mathbf{P} = \exp(2\pi\mathbf{\Lambda})$ with

$$\mathbf{\Lambda} = \begin{pmatrix} \mathrm{i}\omega & 0 \\ 0 & -\mathrm{i}\omega \end{pmatrix}. \tag{2.6.19}$$

The coordinates (z_1, z_2) corresponding to these forms can be chosen to satisfy $z_1 = \bar{z}_2 = z$ (see Exercise 2.6.7) and therefore it is sufficient to consider the differential equation for z. This takes the form

$$\dot{z} = \mathrm{i}\omega z + \sum_{r=2}^{N} \sum_{\substack{\mathbf{m} \\ m_1+m_2=r}} b_{\mathbf{m}}(t) z^{m_1} \bar{z}^{m_2} + O(|z|^{N+1}; t), \tag{2.6.20}$$

where $b_{\mathbf{m}}(t) \in \mathbb{C}$. The resonant terms are determined by (2.6.14) with $\lambda_j = \mathrm{i}\omega$ (see Exercise 2.6.10), i.e.

$$\nu + (m_1 - m_2 - 1)\omega = 0, \tag{2.6.21}$$

with $m_1 + m_2 = r \geqslant 2$.

If ω is irrational then (2.6.21) can only be satisfied if $\nu = 0$ and $m_1 - m_2 = 1$. Thus, for $m_1 + m_2 = r$, we have

$$2m_1 = r + 1 \qquad \text{and} \qquad 2m_2 = r - 1. \tag{2.6.22}$$

Since $m_1, m_2 \in \mathbb{N}$, resonance can only occur if r is odd. We conclude that (2.6.20), with $N = 2k$, can be reduced to the form

$$\dot{z} = \mathrm{i}\omega z + \sum_{s=1}^{k-1} c_s z|z|^{2s} + O(|z|^{2k+1}; t). \tag{2.6.23}$$

When $\omega = p/q$, (2.6.21) is satisfied if

$$\nu = lp \quad (a) \qquad \text{and} \qquad (m_1 - m_2 - 1) = -lq \quad (b) \tag{2.6.24}$$

when $l \in \mathbb{Z}$. Now $l = 0$ gives $\nu = 0$, $m_1 - m_2 = 1$ and the terms of the form $z|z|^{2s}$ are resonant for rational or irrational ω. Such resonances are 'inevitable', however, when $l \neq 0$ additional resonances occur. For $m_1 + m_2 = r$, (2.6.24b) gives

$$2m_1 = r + 1 - lq, \tag{2.6.25}$$

and, since $0 \leqslant m_1 \leqslant r$, we conclude that

$$\left\lceil \frac{1-r}{q} \right\rceil \leqslant l \leqslant \left\lfloor \frac{r+1}{q} \right\rfloor. \tag{2.6.26}$$

Notice that not every integer l satisfying (2.6.26) gives a solution for (2.6.21); only those for which m_1 (from (2.6.25)) is a non-negative integer are acceptable.

For given q, (2.6.26) implies that no additional resonant terms occur until $r = q - 1$ when $l = \dfrac{r+1}{q} = 1$. (N.B. $r \geqslant 2$ implies $q \geqslant 3$, see §5.5.) Observe that $l = -1$ is not involved until $r = q + 1$ and $|l| \geqslant 2$ can only occur for $r \geqslant 2q - 1$. Thus, the lowest order additional resonance occurs for $r = q - 1$ and $l = +1$. Equation (2.6.25) then gives $m_1 = 0$ so that the corresponding resonant term is proportional to $\bar{z}^{q-1} \exp(\mathrm{i}pt)$. □

As we shall see, the calculations described above play a central role in the construction of vector field approximations to the planar maps treated in Chapter 5.

2.7 Centre manifolds (Carr, 1981; Guckenheimer & Holmes, 1983)

Let us return to the problem of the determination of the topological type of a non-hyperbolic fixed point. Having constructed the normal form, with its minimal set of non-linear terms, we must consider whether or not its topological type can be obtained. For flows, the existence of a centre manifold sometimes allows us to obtain the desired topological type from a lower dimensional problem. Here we can only outline the main ideas.

Let $\mathbf{A}$ be a non-hyperbolic, linear diffeomorphism on $\mathbb{R}^n$. Then the Jordan Form Theorem implies that there is a basis for $\mathbb{R}^n$ such that

$$\mathbf{A} = \begin{pmatrix} \mathbf{J}_1 & & \mathbf{0} \\ & \ddots & \\ \mathbf{0} & & \mathbf{J}_k \end{pmatrix}, \tag{2.7.1}$$

where

$$\mathbf{J}_j = \begin{pmatrix} \lambda_j & 1 & & \mathbf{0} \\ & \ddots & \ddots & \\ & & \ddots & 1 \\ \mathbf{0} & & & \lambda_j \end{pmatrix},$$

or

$$\mathbf{J}_j = \begin{pmatrix} \mathbf{R}_j & \mathbf{I} & & \mathbf{0} \\ & \ddots & \ddots & \\ & & \ddots & \mathbf{I} \\ \mathbf{0} & & & \mathbf{R}_j \end{pmatrix}, \tag{2.7.2}$$

with

$$\mathbf{R}_j = \begin{pmatrix} \alpha_j & -\beta_j \\ \beta_j & \alpha_j \end{pmatrix}, \qquad \mathbf{I} = \begin{pmatrix} 1 & 0 \\ 0 & 1 \end{pmatrix}, \tag{2.7.3}$$

$1 \leqslant \dim \mathbf{J}_j \leqslant n$ and $\sum_{j=1}^{k} \dim \mathbf{J}_j = n$. Each Jordan block can be uniquely associated with an eigenvalue $\lambda_j \in \mathbb{R}$ or $\alpha_j + i\beta_j \in \mathbb{C}$ of $\mathbf{A}$. As in the hyperbolic case (see §2.1), unstable and stable eigenspaces, E^u and E^s, are associated with the eigenvalues having modulus not equal to unity; the new feature is the *centre eigenspace*, E^c, formed by the direct sum of the eigenspaces of eigenvalues with modulus equal to one. It is E^c that replaces the fixed point as the entity about which hyperbolic motion takes place. If we are interested in the non-hyperbolic flow, $\exp(\mathbf{A}t)$, then it is convenient to group the Jordan blocks of $\mathbf{A}$ according to whether the associated eigenvalue has positive (E^u), negative (E^s) or zero (E^c) real part. It is important to note that the exponential attraction/repulsion off E^c results in recurrence being possible only on E^c itself (see Exercises 2.7.1, 2.7.2 and Figure 2.6).

The Centre Manifold Theorem extends the above ideas to the local behaviour at non-hyperbolic fixed points of non-linear flows. In this sense, it represents a generalisation of the Invariant Manifold Theorem for flows given in §2.2.2. Let $\mathbf{X}: \mathbb{R}^n \to \mathbb{R}^n$ be a smooth vector field with a non-hyperbolic singularity at the origin (i.e. $\mathbf{X}(\mathbf{0}) = \mathbf{0}$ and $D\mathbf{X}(\mathbf{0})$ has some eigenvalues with zero real part) and suppose that E^u, E^s and E^c are the unstable, stable and centre eigenspaces of $D\mathbf{X}(\mathbf{0})$.

Theorem 2.7.1 (Centre manifold) *Let φ be the flow of $\mathbf{X}$. Then there exists locally a centre manifold, W^c_{loc}, containing the origin and invariant under φ such that W^c_{loc} has tangent space E^c at $\mathbf{x} = \mathbf{0}$. This manifold can be chosen to be C^k for any $k \in \mathbb{N}$, but its domain of definition can depend on k. Furthermore, there are locally smooth stable and unstable manifolds, W^s_{loc} and W^u_{loc}, which contain $\mathbf{x} = \mathbf{0}$, are invariant under φ, have tangent spaces E^s and E^u, respectively, and are such that $\varphi_t | W^s_{loc}$ is a contraction while $\varphi_t | W^u_{loc}$ is an expansion.*

It is important to realise that, unlike W^s_{loc} and W^u_{loc}, the centre manifold is not necessarily unique.

Example 2.7.1 Show that the differential equation

$$\dot{x} = x^2, \qquad \dot{y} = -y \tag{2.7.4}$$

has infinitely many smooth centre manifolds.

Solution. The linearisation of (2.7.4) at the origin is

$$\dot{\mathbf{x}} = D\mathbf{X}(\mathbf{0})\mathbf{x} = \begin{pmatrix} 0 & 0 \\ 0 & -1 \end{pmatrix} \begin{pmatrix} x \\ y \end{pmatrix}. \tag{2.7.5}$$

The eigenvalues of $D\mathbf{X}(\mathbf{0})$ are 0 and -1, with eigenvectors along the x- and y-axes, respectively. The Centre Manifold Theorem predicts the existence of a curve invariant under the flow and tangent to the x-axis at $\mathbf{x} = \mathbf{0}$. The x-axis itself is clearly such a centre manifold because $y \equiv 0$ implies $\dot{y} = 0$.

However, there are other such centre manifolds. The phase portrait of (2.7.4) is shown in Figure 2.7. The orbit passing through a typical point (x_0, y_0) with

Figure 2.6 Some examples of the eigenspaces E^u, E^s and E^c for flows on $\mathbb{R}^3$: *(a)* $\dot{x} = -x$, $\dot{y} = 0$, $\dot{z} = z$: at $\mathbf{x} = \mathbf{0}$, $E^u = z$-axis, $E^s = x$-axis, $E^c = y$-axis; *(b)* $\dot{x} = -y$, $\dot{y} = x$, $\dot{z} = -z$: $E^u = \{\mathbf{0}\}$, $E^s = z$-axis, $E^c = xy$-plane; *(c)* $\dot{x} = y$, $\dot{y} = 0$, $\dot{z} = -z$: at $\mathbf{x} = \mathbf{0}$, $E^u = \{\mathbf{0}\}$, $E^s = z$-axis, $E^c = xy$-plane. Observe the connection with Figure 2.5.

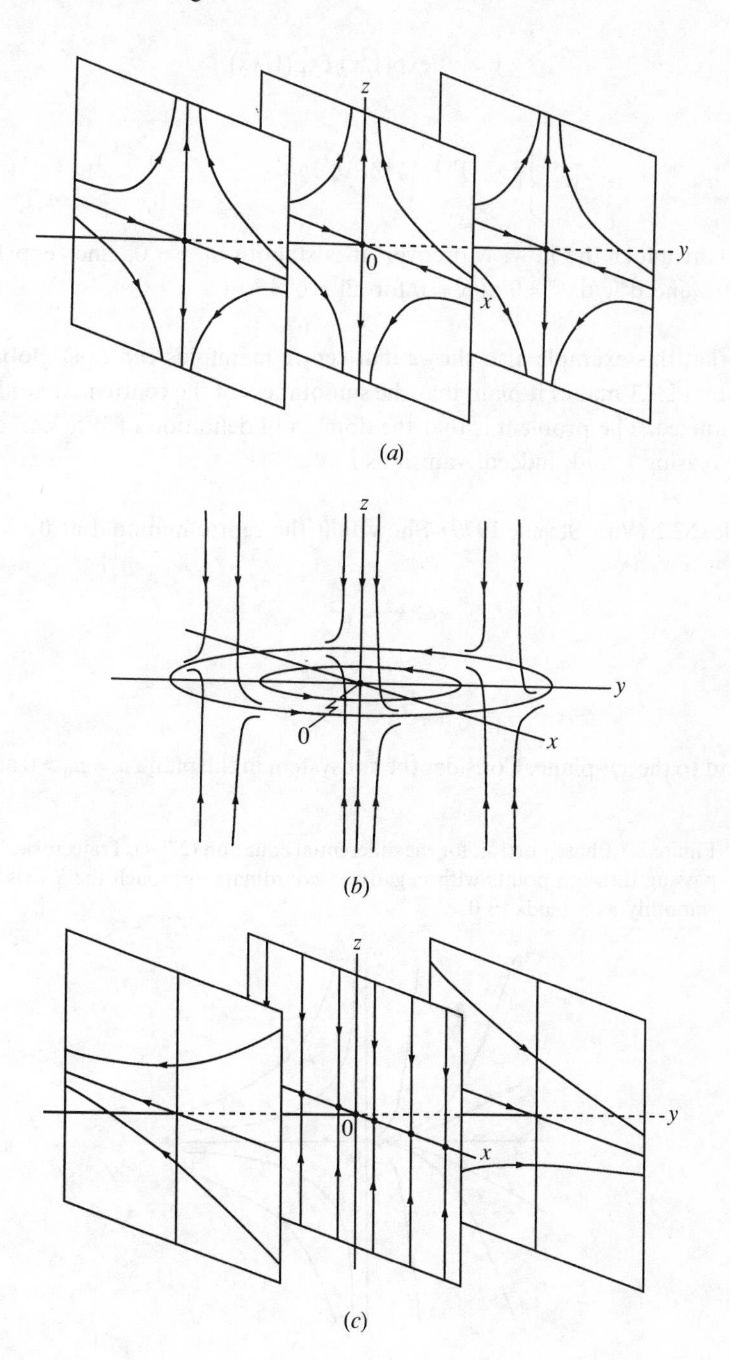

$x_0 < 0$ is given by a particular solution of

$$\frac{dy}{dx} = -\frac{y}{x^2}, \tag{2.7.6}$$

namely,

$$y = y_0 \exp(1/x)/\exp(1/x_0). \tag{2.7.7}$$

The curve

$$y = \begin{cases} y_0 \exp(1/x)/\exp(1/x_0), & x < 0, \\ 0, & x \geqslant 0, \end{cases} \tag{2.7.8}$$

is invariant under the flow. Moreover, it is smooth at $x = 0$, since $\exp(1/x) \to 0$ as $x \to 0^-$ and $d^m y/dx^m = 0$ at $x = 0$ for all $m \in \mathbb{N}$. □

Notice that this example also shows that centre manifolds can exist globally.

Theorem 2.7.1 makes it plain that the smoothness of the centre manifold cannot be guaranteed. The problem is that the domain of definition of W^c_{loc} can decrease with increasing k and, indeed, vanish as $k \to \infty$.

Example 2.7.2 (Van Strien, 1979) Show that the centre manifold at $\mathbf{0} \in \mathbb{R}^3$ of the system

$$\begin{aligned} \dot{x} &= x^2 - \mu^2, \\ \dot{y} &= y + x^2 - \mu^2, \\ \dot{\mu} &= 0, \end{aligned} \tag{2.7.9}$$

is tangent to the $x\mu$-plane. Consider the subsystem in the plane $\mu = \mu_0 > 0$ and show

Figure 2.7 Phase portrait for the differential equation (2.7.4). Trajectories passing through points with negative x-coordinate approach the x-axis smoothly as x tends to 0.

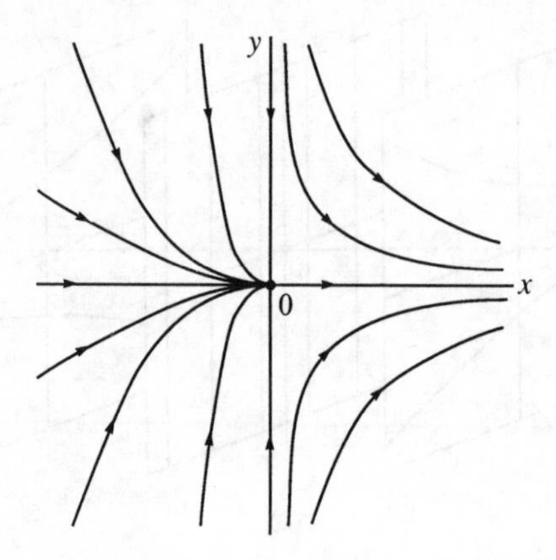

that:

(i) $(x, y) = (\mu_0, 0)$ is an unstable node;
(ii) $(x, y) = (-\mu_0, 0)$ is a saddle.

Let $\mathscr{C}_{\mu_0}$ be the curve obtained by restricting the centre manifold to the plane $\mu = \mu_0$. Show that $\mathscr{C}_{\mu_0}$ is a union of trajectories containing the node and saddle. Moreover, if $\mathscr{C}_{\mu_0}$ is given by $y = g_{\mu_0}(x)$, prove that $g_{1/2m}$ is $(m-1)$-times differentiable at $x = 1/2m$ for $m \geqslant 3$.

Solution. The centre eigenspace E^c of the linearisation of (2.7.9) at the origin is the $x\mu$-plane. Therefore, there exists a C^k centre manifold, W^c_{loc}, on some neighbourhood of $(x, y, \mu) = \mathbf{0}$ which is tangent to the $x\mu$-plane, by Theorem 2.7.1.

The subsystem on the $\mu = \mu_0$ plane, $\dot{x} = x^2 - \mu_0^2$, $\dot{y} = y + x^2 - \mu_0^2$, has fixed points at $(x, y) = (\pm\mu_0, 0)$. The linearisations have eigenvalues $\pm 2\mu_0, 1$ and therefore, for $\mu_0 > 0$, $(\mu_0, 0)$ is an unstable node and $(-\mu_0, 0)$ is a saddle.

Observe that the lines $(x, y, \mu) = (\pm\mu, y, \mu)$, $y \in \mathbb{R}$, parallel to the y-axis, are invariant sets for (2.7.9) so that $M = \{((x, y, \mu) | x^2 = \mu^2, \mu > 0\}$ is an invariant surface (see Figure 2.8). The centre manifold W^c_{loc} is invariant and tangent to the $x\mu$-plane. Therefore, $W^c_{loc} \cap M$ must be an invariant curve of (2.7.9) passing through the origin and contained in M. The only possible choice for such a curve on M is the set of fixed points. This, in turn, implies (see Figure 2.9) that $\mathscr{C}_{\mu_0}$ contains the stable manifold of the saddle $(-\mu_0, 0)$. This follows because only points on this manifold approach the fixed point with increasing time. For $x > \mu_0$, $\mathscr{C}_{\mu_0}$ is a trajectory of the unstable node $(\mu_0, 0)$. Since the stable manifold of the saddle is smooth, the finite differentiability of $g_{\mu_0}(x)$ arises at the node $(\mu_0, 0)$ (see Exercise 2.7.4).

Figure 2.8 The surface $M = \{(x, y, \mu): x^2 = \mu^2, \mu > 0\}$ is an invariant set for (2.7.9). For $x < 0$, it contains the line $x = -\mu$ of saddle points and, for $x > 0$, it contains the line $x = \mu$ of nodes.

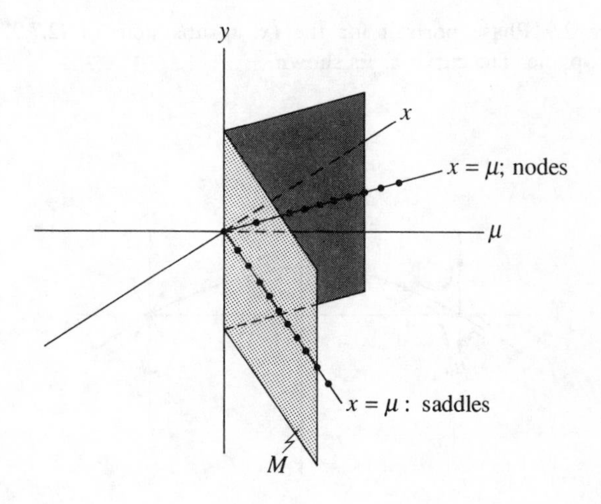

Assume that g_{μ_0} is C^k-differentiable in x, then it can be written in the form

$$g_{\mu_0}(x) = \sum_{i=1}^{k} a_i(\mu_0)(x-\mu_0)^i + o(x-\mu_0)^k. \tag{2.7.10}$$

Since $y = g_{\mu_0}(x)$ is also an invariant curve of (2.7.9) it satisfies

$$(x^2 - \mu_0^2)\frac{dy}{dx} = y + x^2 - \mu_0^2. \tag{2.7.11}$$

Substituting (2.7.10) in (2.7.11) and writing $x^2 - \mu_0^2$ as $(x-\mu_0)^2 + 2\mu_0(x-\mu_0)$, we obtain

$$\sum_{i=1}^{k} ia_i(x-\mu_0)^{i+1} + 2\mu_0 \sum_{i=1}^{k} ia_i(x-\mu_0)^i$$
$$= \sum_{i=1}^{k} a_i(x-\mu_0)^i + (x-\mu_0)^2 + 2\mu_0(x-\mu_0) + o(x-\mu_0)^k. \tag{2.7.12}$$

Comparing coefficients of $(x-\mu_0)^i$ in (2.7.12), we obtain

$$a_1 = \frac{-2\mu_0}{1-2\mu_0}, \qquad a_2 = \frac{a_1 - 1}{1-4\mu_0} \tag{2.7.13}$$

and for $i = 3, \ldots, k$

$$(1 - 2i\mu_0)a_i = (i-1)a_{i-1}. \tag{2.7.14}$$

Now set $\mu_0 = 1/2m$, $m \geqslant 3$, and assume that g_{μ_0} is C^k-differentiable with $k \geqslant m$. Using (2.7.14) we have $a_{m-1} = 0$ and hence $a_{m-2} = \cdots = a_2 = 0$. However, (2.7.13) implies that $a_2 = -m^2/(m-1)(m-2) \neq 0$ and this gives a contradiction. Thus $g_{1/2m}(x)$ is at most C^{m-1}-differentiable at $x = 1/2m$. On examining (2.7.13) and (2.7.14), we see that the coefficients a_i are well-defined for $i = 1, 2, \ldots, m-1$ when $\mu_0 = 1/2m$, and so $g_{1/2m}(x)$ is C^{m-1}-differentiable at $x = 1/2m$. □

Figure 2.9 Phase portrait for the (x, y)-subsystem of (2.7.9) in the $\mu = \mu_0$-plane. The curve $\mathscr{C}_{\mu_0}$ is shown.

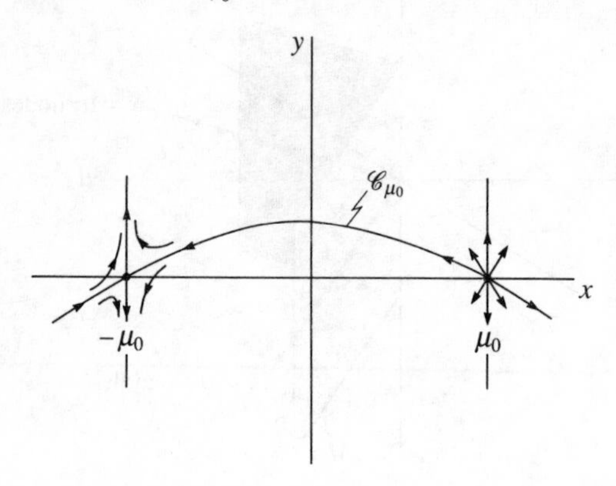

The following result generalises Theorem 2.2.4 for hyperbolic fixed points.

Theorem 2.7.2 *Let* $\mathbf{X}$ *be a smooth vector field with a non-hyperbolic fixed point at* $\mathbf{x}=\mathbf{0}$. *Let* $\tilde{\mathbf{X}}=\mathbf{X}|W^c_{loc}$ *be the restriction of* $\mathbf{X}$ *to its centre manifold. Then the flow of* $\dot{\mathbf{x}}=\mathbf{X}(\mathbf{x})$ *is topologically equivalent, in some neighbourhood of the origin, to that of the partitioned system*

$$\begin{aligned} \dot{\mathbf{x}}_c &= \tilde{\mathbf{X}}(\mathbf{x}_c), & \mathbf{x}_c &\in W^c_{loc}, \\ \dot{\mathbf{x}}_u &= \mathbf{x}_u, & \mathbf{x}_u &\in W^u_{loc}, \\ \dot{\mathbf{x}}_s &= -\mathbf{x}_s, & \mathbf{x}_s &\in W^s_{loc}. \end{aligned} \tag{2.7.15}$$

Theorem 2.7.2 implies that the topological type of a non-hyperbolic singularity is determined by the topological behaviour of the flow restricted to its centre manifold. It is this result that allows us to use the Centre Manifold Theorem to determine the topological type of a non-hyperbolic singularity.

Example 2.7.3 Find the topological type of the singularity at $\mathbf{x}=\mathbf{0}$ in the differential equation

$$\dot{x}=xy+y^2, \qquad \dot{y}=y-x^2. \tag{2.7.16}$$

Solution. The linearisation of (2.7.16) at the origin shows that its centre manifold, W^c_{loc}, is tangent to the x-axis at $(x, y)=\mathbf{0}$. Assume that the equation of W^c_{loc} can be written in the form

$$y=g(x)=\sum_{m=0}^{r} a_m x^m + O(|x|^{r+1}) \tag{2.7.17}$$

for some $r\in\mathbb{N}$. Substituting for y, $\dot{y}$ in (2.7.16), we obtain

$$\left(\sum_{m=0}^{r} m a_m x^{m-1}\right)\left[x\sum_{m=0}^{r} a_m x^m + \left(\sum_{m=0}^{r} a_m x^m\right)^2\right] = \left(\sum_{m=0}^{r} a_m x^m\right) - x^2 + O(|x|^{r+1}). \tag{2.7.18}$$

Observe that, tangency of W^c_{loc} to the x-axis at $x=0$ implies $a_0=a_1=0$. Comparing coefficients of x^2 gives $a_2=1$ (left hand side (2.7.18) is $O(x^4)$) and therefore

$$y=x^2+O(|x|^3). \tag{2.7.19}$$

It follows that on W^c_{loc}, (2.7.16) takes the form

$$\begin{aligned} \dot{x} &= x^3 + O(|x|^4), \\ \dot{y} &= 2x^4 + O(|x|^5). \end{aligned} \tag{2.7.20}$$

Here we have used $\dot{y}=[2x+O(|x|^2)]\dot{x}$. Therefore, the system is weakly repelling on W^c_{loc} (see Figure 2.10(*a*)). Consequently, we conclude that the origin is a weak (i.e. non-hyperbolic) node as shown in Figure 2.10(*b*). □

Example 2.7.4 Find the topological type of the singularity at $\mathbf{x}=\mathbf{0}$ for a differential equation with normal form

$$\dot{x}=x\left(\lambda+\sum_{i=1}^{N-1} a_i y^i\right)+O(|\mathbf{x}|^{N+1}),$$
$$\dot{y}=\sum_{i=2}^{N} b_i y^i+O(|\mathbf{x}|^{N+1}), \tag{2.7.21}$$

where $N\geqslant 2$, given that $b_2\neq 0$.

Solution. Let the centre manifold, W^c_{loc}, tangent to the y-axis, be given by

$$x=g(y)=\sum_{m=2}^{N} c_m y^m+O(|y|^{N+1}). \tag{2.7.22}$$

Differentiating (2.7.22) with respect to t and substituting from (2.7.21) yields

$$\left(\sum_{m=2}^{N} mc_m y^{m-1}\right)\left(\sum_{i=2}^{N} b_i y^i\right)=\left(\sum_{m=2}^{N} c_m y^m\right)\left(\lambda+\sum_{i=1}^{N-1} a_i y^i\right)+O(|\mathbf{x}|^{N+1}). \tag{2.7.23}$$

Comparing coefficients of y^2 gives $\lambda c_2=0$, since, given that $b_2\neq 0$, the left hand side of (2.7.23) is $O(y^3)$. It follows that $c_2=0$ and summations over m in (2.7.23) can be started at $m=3$. Now compare coefficients of y^3 to obtain $\lambda c_3=0$, since the left hand side of (2.7.23) is now $O(y^4)$, and so on. Thus we conclude that $c_m=0$, $0\leqslant m\leqslant N$, and W^c_{loc} is given by $x=0$ up to terms of order N.

On W^c_{loc}, (2.7.21) takes the form

$$\dot{x}=O(|y|^N)\dot{y}=O(|y|^{N+2}),$$
$$\dot{y}=b_2y^2+O(|y^3|), \tag{2.7.24}$$

Figure 2.10 (*a*) Centre manifold for (2.7.16) in the neighbourhood of the origin. The vector field restricted to the centre manifold is as indicated. (*b*) Sketch of the local phase portrait of (2.7.16) at the origin.

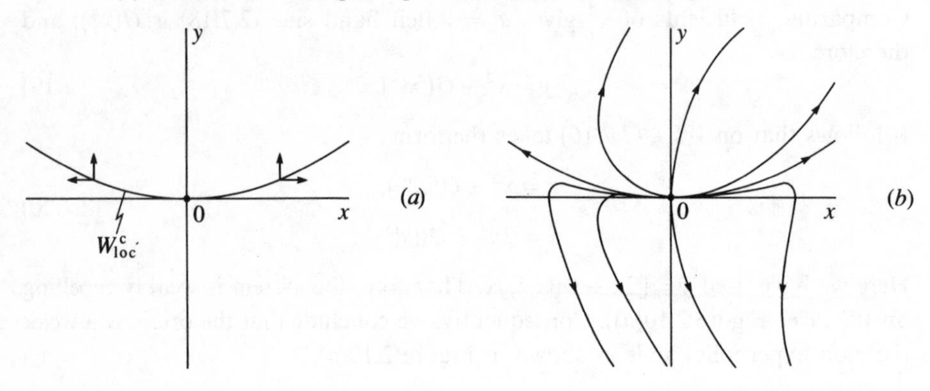

and the topological type of the flow restricted to the centre manifold is shown in Figure 2.11.

The hyperbolic expansion or contraction associated with the non-zero eigenvalue, λ, combines with this centre manifold behaviour to give local phase portraits of the type shown in Figure 2.12 (or its time reversal). □

The reader will recognise the similarity between (2.7.21) and the normal form obtained in Proposition 2.4.1. A non-hyperbolic singularity with normal form (2.7.21) is said to be of *saddle-node* type. Such a singularity $\mathbf{x}^*$ is characterised by the degeneracy condition $\operatorname{Det} \mathbf{DX}(\mathbf{x}^*) = 0$ and the non-degeneracy conditions $\operatorname{Tr} \mathbf{DX}(\mathbf{x}^*) \neq 0$ and $b_2 \neq 0$. This means that the topological type of this singularity is determined by $\mathbf{X}_1 + \mathbf{X}_2$ in (2.3.1).

Definition 2.7.1 *The truncation of the Taylor expansion of* $\mathbf{X}$ *about* $\mathbf{x}^*$ *obtained by excluding terms of order* $k+1$ *and higher is called the* k-jet *of* $\mathbf{X}$ *in* $\mathbf{x}^*$, *denoted by* $j^k(\mathbf{X})(\mathbf{x}^*)$.

Figure 2.11 Topological type (up to order-preserving homeomorphism) of the fixed point at the origin of (2.7.21) restricted to its centre manifold: (a) $b_2 > 0$; (b) $b_2 < 0$.

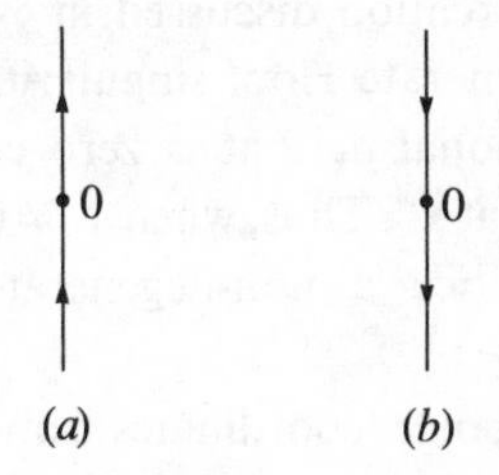

Figure 2.12 Local phase portrait at the origin of (2.7.21) when both λ and b_2 are greater than zero. The fixed point at the origin is said to be a *saddle-node*.

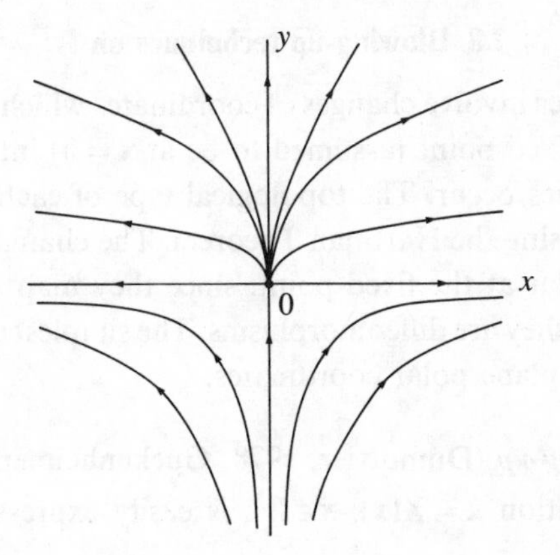

Example 2.7.4 shows that the terms of order greater than 2 in (2.7.21) do not change the local phase portrait obtained. We say that the topological type is *2-jet determined*. Indeed, one could think of the Hartman Theorem as the statement that the topological type of a hyperbolic fixed point is 1-jet determined.

Theorem 2.7.2 is helpful in the above examples because the centre manifold is one-dimensional and the topological behaviour of the flow restricted to it is immediately obvious. This is not the case for the normal forms given in Propositions 2.4.2 and 2.4.3. Both of these forms have two-dimensional centre manifolds.

For proposition 2.4.2, this is not a problem since (2.4.6) has the polar form

$$\dot{r} = \sum_{k=1}^{[\frac{1}{2}(N-1)]} a_k r^{2k+1} + O(r^{N+1}; \theta), \tag{2.7.25}$$

$$\dot{\theta} = \beta + \sum_{k=1}^{[\frac{1}{2}(N-1)]} b_k r^{2k} + O(r^{N+1}; \theta), \tag{2.7.26}$$

where $\beta = (\text{Det}\,\mathbf{DX}(\mathbf{0}))^{1/2} > 0$, $N \geqslant 3$. If $a_k = 0$ for $k < l$ and $a_l \neq 0$, it follows that $\mathbf{x} = \mathbf{0}$ is a weak focus: stable if $a_l < 0$; unstable if $a_l > 0$. The simplest singularity of this type is characterised by the degeneracy condition $\text{Tr}\,\mathbf{DX}(\mathbf{x}^*) = 0$ and the non-degeneracy conditions $\text{Det}\,\mathbf{DX}(\mathbf{x}^*) > 0$ and $a_1 \neq 0$. It is known as a *non-degenerate, Hopf singularity*. Here, 'Hopf' refers to the connection between this singularity and the Hopf bifurcation discussed in §4.2.2, while 'non-degenerate' distinguishes it from the degenerate Hopf singularities for which $a_k = 0$ for $k < l$ and $a_l \neq 0$, $l > 1$. Each additional a_k that is zero corresponds to an additional degeneracy condition satisfied at $\mathbf{x}^*$. Thus, when $a_1 \neq 0$ we have the least degenerate singularity of this type. Notice a non-degenerate Hopf singularity is 3-jet determined.

While transformation to polar coordinates solves the problem for (2.4.6), something rather more sophisticated is required to cope with (2.4.9) or, indeed, with the case where $\mathbf{DX}(\mathbf{0}) = \mathbf{0}$ for which we cannot construct a normal form in the manner described in §2.3.

2.8 Blowing-up techniques on $\mathbb{R}^2$

Blowing-up techniques involve changes of coordinates which expand, or 'blow-up', the non-hyperbolic fixed point (assumed to be at $\mathbf{x} = \mathbf{0}$) into a curve on which a number of singularities occur. The topological type of each of these singularities is then investigated using the Hartman Theorem. The changes of coordinates used are, of course, singular at the fixed point, since they map a curve onto a point. Elsewhere, however, they are diffeomorphisms. The simplest example is well known to everyone, namely plane polar coordinates.

2.8.1 Polar blowing-up (Dumortier, 1978; Guckenheimer & Holmes, 1983)
The differential equation $\dot{\mathbf{x}} = \mathbf{X}(\mathbf{x})$, $\mathbf{x} \in \mathbb{R}^2$, is easily expressed in terms of polar

coordinates (r, θ) to obtain

$$\dot{r} = X_r(r, \theta), \qquad \dot{\theta} = r^{-1}X_\theta(r, \theta), \tag{2.8.1}$$

where $\mathbf{X} = X_r\mathbf{e}_r + \mathbf{X}_\theta\mathbf{e}_\theta$ and $\mathbf{e}_r$, $\mathbf{e}_\theta$ are, respectively, radial and angular unit vectors. This form can be regarded as defining a differential equation on a half cylinder or, equivalently, on a punctured plane (see Figure 2.13). In these terms, the usual polar coordinates correspond to the singular case in which the $r = 0$ circle is mapped onto the origin and the half-cylinder, $\mathbb{R}^+ \times S^1$, is mapped diffeomorphically onto $\mathbb{R}^2\backslash\{\mathbf{0}\}$. In Cartesian coordinates, the map which achieves this is, of course,

$$\mathbf{\Phi}(r, \theta) = (r\cos\theta, r\sin\theta) \tag{2.8.2}$$

and $\mathbf{\Phi}^{-1}$ can be regarded as 'blowing-up' the origin of the plane into the $r = 0$ circle.

It is important to realise that we cannot always study the $r = 0$ circle directly. Observe that if $j^l(\mathbf{X})(\mathbf{0}) \equiv \mathbf{0}$, $l \leqslant k$ and $j^{k+1}(\mathbf{X})(\mathbf{0}) \neq \mathbf{0}$ then (cf. (2.3.16))

$$\begin{aligned} r\dot{r} = x_1\dot{x}_1 + x_2\dot{x}_2 &= \sum_{\substack{\mathbf{m} \\ m_1+m_2=k+1}} \{X_{\mathbf{m},1}\mathbf{x}^{\mathbf{m}}x_1 + X_{\mathbf{m},2}\mathbf{x}^{\mathbf{m}}x_2\} + \cdots \\ &= r^{k+2}R(r, \theta). \end{aligned} \tag{2.8.3}$$

Similarly,

$$\begin{aligned} r^2\dot{\theta} = x_1\dot{x}_2 - x_2\dot{x}_1 &= \sum_{\substack{\mathbf{m} \\ m_1+m_2=k+1}} \{X_{\mathbf{m},2}\mathbf{x}^{\mathbf{m}}x_1 - X_{\mathbf{m},1}\mathbf{x}^{\mathbf{m}}x_2\} + \cdots \\ &= r^{k+2}\Theta(r, \theta). \end{aligned} \tag{2.8.4}$$

Figure 2.13 Illustration of the equivalence of the half-cylinder and the punctured plane: $\mathbf{g}$ is a diffeomorphism, for example $\mathbf{g}(r, \theta)$ could have cartesian coordinates $e^r(\cos\theta, \sin\theta)$. Plane polar coordinates correspond to the non-injective case $\mathbf{\Phi}(r, \theta) = (r\cos\theta, r\sin\theta)$.

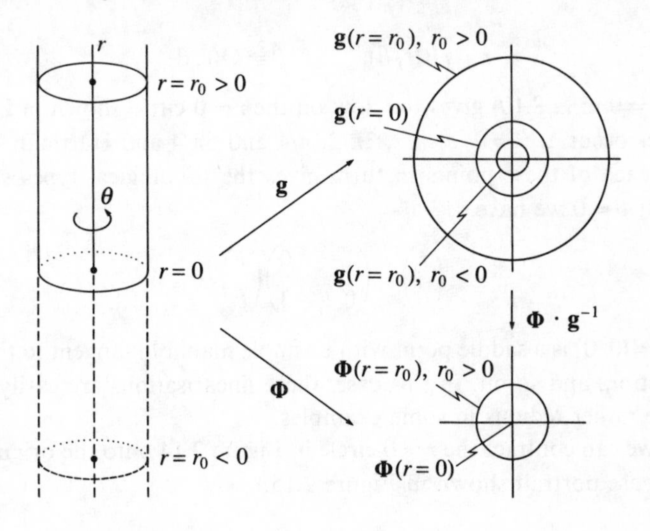

Therefore, $\dot{\mathbf{x}} = \mathbf{X}(\mathbf{x})$ has polar form

$$\dot{r} = r^{k+1}R(r, \theta), \qquad \dot{\theta} = r^k\Theta(r, \theta), \tag{2.8.5}$$

and no information is obtained by setting $r = 0$ directly. Instead, notice that the phase curves of (2.8.5) are given by

$$\frac{\mathrm{d}r}{\mathrm{d}\theta} = \frac{rR(r, \theta)}{\Theta(r, \theta)} \tag{2.8.6}$$

for $r > 0$. In other words, the phase curves of (2.8.5) are the same as those of the system

$$\dot{r} = rR(r, \theta), \qquad \dot{\theta} = \Theta(r, \theta) \tag{2.8.7}$$

obtained by dividing (2.8.5) by r^k. Moreover, since $r^k > 0$, the orientation of the trajectories of (2.8.5) and (2.8.7) are the same. However, in general $\Theta(0, \theta) \not\equiv 0$ in (2.8.7) and we can use the Hartman Theorem to study the fixed points of (2.8.7) on the $r = 0$ circle. If all these fixed points are hyperbolic the local phase portrait of $\dot{\mathbf{x}} = \mathbf{X}(\mathbf{x})$ at $\mathbf{x} = \mathbf{0}$ can be obtained. If not further blowing-up or normal form calculations may be necessary.

Example 2.8.1 Use polar blowing-up to find the topological type of the singularity at the origin of the system

$$\dot{x} = x^2 - 2xy, \qquad \dot{y} = y^2 - 2xy. \tag{2.8.8}$$

Solution. In polar coordinates, (2.8.8) becomes

$$\begin{aligned} \dot{r} &= r^2(\cos^3\theta - 2\cos^2\theta\sin\theta - 2\cos\theta\sin^2\theta + \sin^3\theta) = r^2R(r, \theta), \\ \dot{\theta} &= 3r\cos\theta\sin\theta(\sin\theta - \cos\theta) = r\Theta(r, \theta). \end{aligned} \tag{2.8.9}$$

In order to examine the $r = 0$ circle, observe that (2.8.9) is topologically equivalent to

$$\dot{r} = rR(r, \theta), \qquad \dot{\theta} = \Theta(r, \theta). \tag{2.8.10}$$

Setting $r = 0$ in (2.8.10) gives the flow on the $r = 0$ circle shown in Figure 2.14. Singularities occur at $\theta = 0$, π, $\pi/2$, $3\pi/2$, $\pi/4$ and $5\pi/4$ and Hartman's Theorem, applied at each of these points in turn, gives the topological types shown. For example, for $\theta = 0$ we have

$$\begin{pmatrix} \dot{r} \\ \dot{\theta} \end{pmatrix} = \begin{pmatrix} 1 & 0 \\ 0 & -3 \end{pmatrix} \begin{pmatrix} r \\ \theta \end{pmatrix}. \tag{2.8.11}$$

Thus $(r, \theta) = (0, 0)$ is a saddle point with unstable manifold tangent to the outward radial direction, and so on. In this case, these linearisations are easily done, but they can be rather tedious in some examples.

Finally, we can contract the $r = 0$ circle in Figure 2.14 onto the origin to obtain the local phase portrait shown in Figure 2.15. □

2.8.2 Directional blowing-up (Dumortier, 1978)

Consider the map $\mathbf{F}: \mathbb{R} \times (-\pi/2, \pi/2) \to \mathbb{R}^2$, defined by

$$\mathbf{F}(r, \theta) = (r \cos \theta, \tan \theta) = (u, v). \tag{2.8.12}$$

This map is a diffeomorphism; observe that

$$\mathbf{DF}(r, \theta) = \begin{pmatrix} \cos \theta & -r \sin \theta \\ 0 & \sec^2 \theta \end{pmatrix} \tag{2.8.13}$$

with determinant $\sec \theta \neq 0$ for $\theta \in (-\pi/2, \pi/2)$. Moreover, $\mathbf{F}$ takes the half circle $\{(r, \theta) | r = 0, -\pi/2 < \theta < \pi/2\}$ onto the v-axis with the point $\theta = 0$ at the origin of

Figure 2.14 The flow on, and near, the $r = 0$-circle for (2.8.10). The topological types of the fixed points are obtained by using Hartman's Theorem. Since $\dot{r}$ and $\dot{\theta}$ change sign when $\theta \mapsto \theta - \pi$, it is sufficient to consider only the singularities indicated.

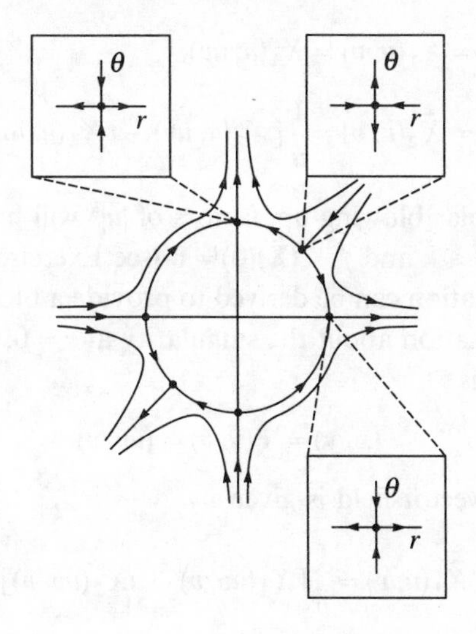

Figure 2.15 Local phase portrait for (2.8.8) at the origin obtained by allowing the radius of the $r = 0$-circle in Figure 2.14 to shrink to zero.

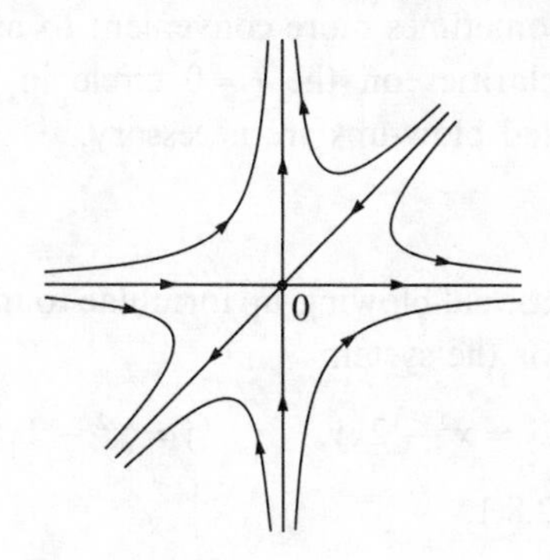

the uv-plane. It is important to note that $\mathbf{DF}(0, 0) = \mathbf{id}$, so that the linear part of the vector field $\mathbf{X}$ will be the *same* in both (r, θ) and (u, v) coordinates (see (1.5.12)). It follows that we can obtain the linearisation of $\mathbf{X}$ at the point $\theta = 0$ on the $r = 0$ circle by considering the linear part of $\mathbf{X}$ at the origin of the uv-plane.

The latter linearisation can be obtained directly from a Cartesian form of $\dot{\mathbf{x}} = \mathbf{X}(\mathbf{x})$. The change of coordinates $(u, v) \to (x, y)$ is given by the map $\mathbf{\Phi} \cdot \mathbf{F}^{-1} = \mathbf{\Psi}$, say, where

$$\mathbf{\Psi}(u, v) = (x, y) = (r \cos \theta, r \sin \theta) = (u, uv). \tag{2.8.14}$$

By restricting to $u > 0$, $\mathbf{\Psi}$ gives a diffeomorphism onto the half-space $x > 0$. The change of coordinates $(x, y) \to (u, v)$ given in (2.8.14) is called a *blow-up in the x-direction*, since it leads to information about the singularity on the $r = 0$ circle at $\theta = 0$, i.e. on the positive x-axis.

In (u, v)-coordinates, the Cartesian system $\dot{x} = X_1(x, y)$, $\dot{y} = X_2(x, y)$ is easily shown to become

$$\begin{aligned} \dot{u} &= \tilde{X}_1(u, v) = X_1(u, uv), \\ \dot{v} &= \tilde{X}_2(u, v) = \frac{1}{u}[X_2(u, uv) - vX_1(u, uv)]. \end{aligned} \tag{2.8.15}$$

Furthermore, as in polar blowing-up, factors of $|u|^k$ will have to be divided out if the jets, $j^l(\tilde{\mathbf{X}})(\mathbf{0}) \equiv \mathbf{0}$, $l \leqslant k$ and $j^{k+1}(\tilde{\mathbf{X}})(\mathbf{0}) \neq \mathbf{0}$ (see Exercise 2.8.4).

A similar transformation can be derived to provide a blow-up in the y-direction so as to obtain information about the singularity at $r = 0$, $\theta = \pi/2$. The change of coordinates required is

$$(x, y) = \bar{\mathbf{\Psi}}(u, v) = (uv, v) \tag{2.8.16}$$

and the transformed vector field is given by

$$\begin{aligned} \tilde{X}_1(u, v) &= \frac{1}{v}[X_1(uv, v) - uX_2(uv, v)], \\ \tilde{X}_2(u, v) &= X_2(uv, v). \end{aligned} \tag{2.8.17}$$

We have already noted that the linearisations involved in polar blowing-up can be rather lengthy. It is sometimes more convenient to use directional blowing-up to investigate the singularities on the $r = 0$ circle in the x- and y-directions, particularly when repeated blow-ups are necessary.

Example 2.8.2 Use directional blowing-up formulae to investigate the singularities at $r = 0$, $\theta = 0$ and $\pi/2$ for the system

$$\dot{x} = x^2 - 2xy, \qquad \dot{y} = y^2 - 2xy$$

considered in Example 2.8.1.

Solution. Blowing-up in the x-direction gives (see (2.8.15))

$$\tilde{\mathbf{X}}_{[0]}(u, v) = \left(u^2 - 2u^2v, \frac{1}{u}(u^2v^2 - 2u^2v - v(u^2 - 2u^2v)) \right). \tag{2.8.18}$$

Dividing by u we obtain

$$(u - 2uv, \; -3v + 3v^2). \tag{2.8.19}$$

Therefore, the singular point at $(r, \theta) = (0, 0)$ is of saddle-type. Notice also the linear part of (2.8.19) is precisely that given in (2.8.11).

For $\theta = \pi/2$ we blow-up in the y-direction to find

$$\frac{1}{v}\tilde{\mathbf{X}}_{[\pi/2]}(u, v) = (-3u + 3u^2, v - 2uv). \tag{2.8.20}$$

Again, (2.8.20) confirms that the singular point $(r, \theta) = (0, \pi/2)$ is a saddle. □

Figure 2.16 Illustration of the various stages of the calculation of the topological type of the cusp singularity by blowing-up techniques: (*a*) the first blow-up yields two non-hyperbolic fixed points; (*b*) the second blow-up gives four non-hyperbolic fixed points; (*c*) at the third blow-up singularities of the kind encountered in Example 2.8.1 occur; (*d*) the local phase portrait at the origin is as shown. All but one of the trajectories starting with $y_2 < 0$ flow through to $y_2 > 0$. Two trajectories approach the fixed point; one in forward, and the other in reverse, time. Their approach is tangent to the y_1-axis and in such senses as to form a curve with a cusp at the fixed point.

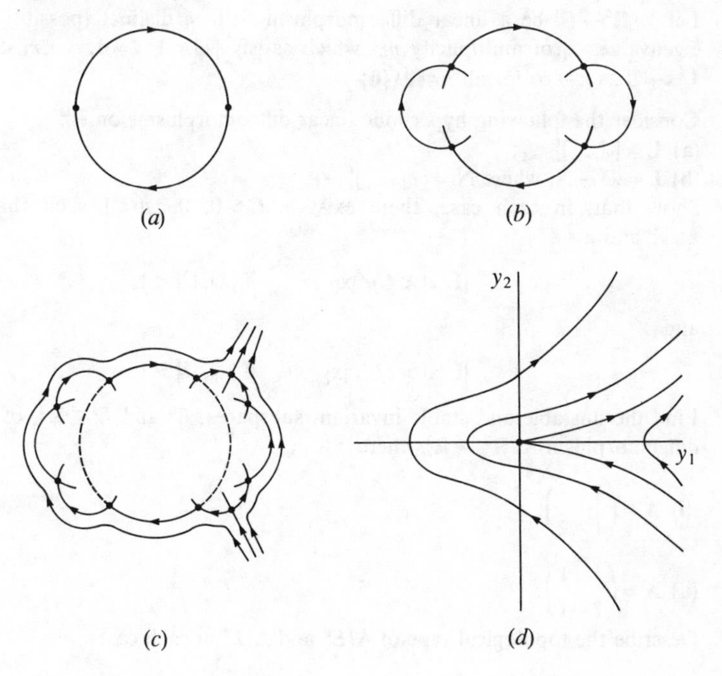

Finally, we return to the normal form given in Proposition 2.4.3. Recall that the degeneracy conditions $\mathrm{Det}\,\mathbf{DX}(\mathbf{x}^*) = \mathrm{Tr}\,\mathbf{DX}(\mathbf{x}^*) = 0$ were satisfied at this non-hyperbolic singularity. Given that the non-degeneracy conditions $\mathbf{DX}(\mathbf{x}^*) \neq \mathbf{0}$, $a_2 \neq 0$, $b_2 \neq 0$ are satisfied, a lengthy calculation (Takens, 1974a) involving three blow-ups (see Figure 2.16(*a*,*b*,*c*)) yields the local phase portrait shown in Figure 2.16(*d*) for $b_2 > 0$. When $b_2 < 0$ the orientation of the trajectories is reversed. In view of the cuspoidal nature of the separatrices in Figure 2.16(*d*), this is known as a *cusp singularity*.

We have now dealt with the topological types of the three principal non-hyperbolic fixed points for planar vector fields. It is perhaps worth noting that we frequently find ourselves in the situation where the *degeneracy* conditions appropriate to a particular singularity are satisfied but we are faced with very long calculations to verify that the corresponding *non-degeneracy* conditions hold. In this case, we can note that, since the non-degeneracy conditions are inequalities, it would be exceptional for them to fail to be satisfied. In these circumstances we say that the particular singularity will occur *generically*. For example, given that $\mathrm{Tr}\,\mathbf{DX}(\mathbf{x}^*) = \mathrm{Det}\,\mathbf{DX}(\mathbf{x}^*) = 0$ we can say that $\mathbf{x}^*$ is generically a cusp singularity because in the generic case $\mathbf{DX}(\mathbf{x}^*) \neq \mathbf{0}$, $a_2 \neq 0$ and $b_2 \neq 0$.

Exercises

2.1 Hyperbolic linear diffeomorphisms and flows

2.1.1 Let $\mathbf{L}\colon \mathbb{R}^l \to \mathbb{R}^l$ be a linear diffeomorphism with m distinct (possibly complex) eigenvalues λ_i of multiplicity m_i, which satisfy $|\lambda_i| < 1$, $i = 1, \ldots, m$. Show that $\mathbf{L}^k\mathbf{x} \to \mathbf{0}$ as $k \to \infty$ for all $\mathbf{x} \in \mathbb{R}^l \backslash \{\mathbf{0}\}$.

2.1.2 Consider the following hyperbolic linear diffeomorphisms on $\mathbb{R}^l$:
(a) $\mathbf{L} = [\lambda_i \delta_{ij}]_{i,j=1}^l$;
(b) $\mathbf{L} = \lambda\mathbf{I} + \mathbf{N}$, where $\mathbf{N} = [\delta_{i,j-1}]_{i,j=1}^l$.
Show that, in each case, there exist c, $C > 0$, $0 < \mu < 1$ such that, for all $\mathbf{x} \in \mathbb{R}^l$ and $n \in \mathbb{Z}^+$,

$$|\mathbf{L}^n\mathbf{x}| < C\mu^n|\mathbf{x}| \qquad \text{if } |\lambda_i|, |\lambda| < 1,$$

and

$$|\mathbf{L}^n\mathbf{x}| > c\mu^{-n}|\mathbf{x}| \qquad \text{if } |\lambda_i|, |\lambda| > 1.$$

2.1.3 Find the unstable and stable invariant subspaces, E^{u} and $E^{\mathrm{s}} \subseteq \mathbb{R}^2$, of the linear diffeomorphism $\mathbf{A}\colon \mathbb{R}^2 \to \mathbb{R}^2$ where

(i) $\mathbf{A} = \begin{pmatrix} 1 & 1 \\ 1 & 2 \end{pmatrix}$,

(ii) $\mathbf{A} = \begin{pmatrix} 1 & 1 \\ 2 & 1 \end{pmatrix}$.

Describe the topological type of $\mathbf{A}|E^{\mathrm{u}}$ and $\mathbf{A}|E^{\mathrm{s}}$ in each case.

2.1.4 Find the stable subspace $E^s \subseteq \mathbb{R}^3$ of the linear diffeomorphism

$$\mathbf{A} = \begin{pmatrix} 3/4 & 1/2 & -5/4 \\ -1/4 & -1/2 & -1/4 \\ -5/4 & 1/2 & 3/4 \end{pmatrix}$$

and describe the behaviour of $\mathbf{A}|E^s$.

2.1.5 Sketch typical orbits of linear diffeomorphisms in $\mathbb{R}^3$ with distinct, positive eigenvalues $\lambda_1, \lambda_2, \lambda_3$ such that:
(a) $\lambda_1, \lambda_2, \lambda_3 < 1$;
(b) $\lambda_1, \lambda_2 < 1, \lambda_3 > 1$.
What are the dimensions of E^s and E^u for cases (a) and (b) respectively? What is the effect on the orbits if the eigenvalue λ_3 in (b) is negative?

2.1.6 Verify Theorem 2.1.2 for the system $\dot{\mathbf{x}} = \mathbf{A}\mathbf{x}$ where

$$\mathbf{A} = \begin{pmatrix} 3 & 2 & 5 \\ -1 & -2 & -5 \\ 1 & -4 & -1 \end{pmatrix}.$$

2.1.7 Sketch the four topologically distinct types of phase portrait that can occur at a hyperbolic fixed point of a flow in $\mathbb{R}^3$.

2.1.8 Show that the number of distinct topological types of hyperbolic diffeomorphisms on $\mathbb{R}$ and $\mathbb{R}^2$ is 4 and 8 respectively. Generalise your arguments to show that there are $4n$ distinct topological types on $\mathbb{R}^n$.

2.2 Hyperbolic non-linear fixed points
2.2.1 Diffeomorphisms

2.2.1 Find the linearisations of the following local diffeomorphisms of the plane at the origin:

(a) $\mathbf{f}(x, y) = (2\sin(x) + xy, x - \frac{1}{2}y\cos(y))$;
(b) $\mathbf{f}(x, y) = (y\exp(y), \exp(2x) - 1)$.
Give the topological type of these linearisations. How is this behaviour related to that of the diffeomorphism $\mathbf{f}$?

2.2.2 Let $\mathbf{f}$ be the time-one flow map of the system $\dot{x} = y, \dot{y} = x - x^2$. Show that $\mathbf{f}$ has a hyperbolic fixed point of saddle-type at $(x, y) = (0, 0)$ and that $W^s_{\mathbf{f}}(\mathbf{0}) \cap W^u_{\mathbf{f}}(\mathbf{0})$ is a closed curve. Contrast this behaviour with that of the diffeomorphism in Figure 2.2.

2.2.3 Let $\mathbf{f}: \mathbb{R}^n \to \mathbb{R}^n$ be a diffeomorphism with a hyperbolic q-cycle. Prove that if $\mathbf{x}$ and $\mathbf{y}$ are periodic points of the same q-cycle, then there exist neighbourhoods U of $\mathbf{x}$ and V of $\mathbf{y}$, such that $\mathbf{f}^q|U$ is differentiably conjugate to $\mathbf{f}^q|V$.

2.2.4 Let $\mathbf{x}^*$ be a periodic point of the diffeomorphism $\mathbf{f}: \mathbb{R}^n \to \mathbb{R}^n$ with periodic orbit $\Lambda = \{\mathbf{x}^*, \mathbf{f}(\mathbf{x}^*), \ldots, \mathbf{f}^{q-1}(\mathbf{x}^*)\}$. The stable manifold $W^s(\Lambda)$ is defined to be $\bigcup_{i=0}^{q-1} W^s(\mathbf{f}^i(\mathbf{x}^*))$, where $W^s(\mathbf{f}^i(\mathbf{x}^*))$ is the stable manifold of the fixed point $\mathbf{f}^i(\mathbf{x}^*)$ of $\mathbf{f}^q$. Prove that if $\mathbf{y} \in W^s(\Lambda)$, then Λ is a subset of the ω-limit set of $\mathbf{y}$. Is the converse true?

2.2.2 Flows

2.2.5 Consider the flow

$$\dot{x} = x(1 - x^2), \qquad \dot{y} = y(1 - y^2) \tag{E2.1}$$

and its time-one map φ_1. Define $\mathbf{f}: \mathbb{R}^2 \to \mathbb{R}^2$ by

$$\mathbf{f}(x, y) = \varphi_1(y, -x). \tag{E2.2}$$

Show that $\mathbf{f}$ has two hyperbolic periodic orbits of period 4 and describe their global stable and unstable manifolds.

2.2.6 Let γ be a closed orbit in the flow φ of the vector field $\mathbf{X}$ and suppose that S_0 and S_1 are local sections at points $\mathbf{x}_0$ and $\mathbf{x}_1$ ($\neq \mathbf{x}_0$) of γ. Show that the local Poincaré maps $\mathbf{P}_0: S_0 \to S_0$ and $\mathbf{P}_1: S_1 \to S_1$ are C^1-conjugate.

2.2.7 Prove that the system

$$\begin{aligned} \dot{x}_1 &= -x_2 + x_1(x_1^2 + x_2^2 - 1), \\ \dot{x}_2 &= x_1 + x_2(x_1^2 + x_2^2 - 1), \\ \dot{x}_3 &= x_3, \end{aligned} \tag{E2.3}$$

has a hyperbolic periodic orbit.

2.3 Normal forms for vector fields

2.3.1 Prove that the transformation $\mathbf{x} = \mathbf{y} + \mathbf{h}_r(\mathbf{y})$, $\mathbf{x}, \mathbf{y} \in \mathbb{R}^n$, with $\mathbf{h}_r$ a vector of rth degree homogeneous polynomials in $\mathbf{y}$, has an inverse at $\mathbf{0} \in \mathbb{R}^n$ of the form

$$\mathbf{y} = \mathbf{x} - \mathbf{h}_r(\mathbf{x}) + O(|\mathbf{x}|^{r+1}). \tag{E2.4}$$

Check this result by showing that
(i) $x = y + y^2$ implies $y = x - x^2 + O(|x|^3)$,
and
(ii) $\begin{pmatrix} x_1 \\ x_2 \end{pmatrix} = \begin{pmatrix} y_1 \\ y_2 \end{pmatrix} + \begin{pmatrix} y_1^2 \\ y_1 y_2 \end{pmatrix}$ implies $\begin{pmatrix} y_1 \\ y_2 \end{pmatrix} = \begin{pmatrix} x_1 \\ x_2 \end{pmatrix} - \begin{pmatrix} x_1^2 \\ x_1 x_2 \end{pmatrix} + O(|\mathbf{x}|^3)$.

2.3.2 Consider the following transformations of the form

$$\mathbf{x} = \mathbf{y} + \mathbf{h}_2(\mathbf{y}) = \mathbf{H}(\mathbf{y}). \tag{E2.5}$$

In each case find $\mathbf{H}^{-1}$ by solving for y_1 and y_2 correct to terms $O(|\mathbf{x}|^3)$.

(a) $\begin{pmatrix} x_1 \\ x_2 \end{pmatrix} = \begin{pmatrix} y_1 \\ y_2 \end{pmatrix} + \begin{pmatrix} b y_1 y_2 + c y_2^2 \\ 0 \end{pmatrix}$;

(b) $\begin{pmatrix} x_1 \\ x_2 \end{pmatrix} = \begin{pmatrix} y_1 \\ y_2 \end{pmatrix} + \begin{pmatrix} a y_1^2 \\ 0 \end{pmatrix}$;

(c) $\begin{pmatrix} x_1 \\ x_2 \end{pmatrix} = \begin{pmatrix} y_1 \\ y_2 \end{pmatrix} + \begin{pmatrix} b y_1 y_2 + c y_2^2 \\ a y_2^2 \end{pmatrix}$.

2.3.3 Let $L_{\mathbf{A}}(\mathbf{h}_r(\mathbf{x}))$ be the Lie bracket of the vector fields $\mathbf{A}\mathbf{x}$ and $\mathbf{h}_r(\mathbf{x}) \in H^r$ on $\mathbb{R}^n$. Show that

(i) $L_{(\mathbf{A}+\mathbf{B})}(\mathbf{h}_r(\mathbf{x})) = L_{\mathbf{A}}(\mathbf{h}_r(\mathbf{x})) + L_{\mathbf{B}}(\mathbf{h}_r(\mathbf{x}))$, $\quad \mathbf{A}, \mathbf{B} \in L(\mathbb{R}^n)$,
(ii) $L_{\mathbf{A}}(\mathbf{h}_r(\mathbf{x}) + \mathbf{h}_r'(\mathbf{x})) = L_{\mathbf{A}}(\mathbf{h}_r(\mathbf{x})) + L_{\mathbf{A}}(\mathbf{h}_r'(\mathbf{x}))$, $\quad \mathbf{h}_r(\mathbf{x}), \mathbf{h}_r'(\mathbf{x}) \in H^r$.

2.3.4 Given that $\mathbf{A}=\begin{pmatrix}0 & 3\\ 2 & 0\end{pmatrix}$, calculate $L_{\mathbf{A}}\begin{pmatrix}\mathbf{x}^{\mathbf{m}}\\ 0\end{pmatrix}$ and $L_{\mathbf{A}}\begin{pmatrix}0\\ \mathbf{x}^{\mathbf{m}}\end{pmatrix}$, where $\mathbf{x}^{\mathbf{m}}=x_1^{m_1}x_2^{m_2}$. Hence obtain expressions for $L_{\mathbf{A}}$ acting on all possible monomials with $m_1+m_2=2$. Find the change of coordinates of the form $\mathbf{x}=\mathbf{y}+\mathbf{h}_2(\mathbf{y})$ which transforms the differential equation

$$\begin{pmatrix}\dot{x}_1\\ \dot{x}_2\end{pmatrix}=\begin{pmatrix}3x_2-x_1^2+7x_1x_2+3x_2^2\\ 2x_1+4x_1x_2+x_2^2\end{pmatrix} \tag{E2.6}$$

into the form

$$\begin{pmatrix}\dot{y}_1\\ \dot{y}_2\end{pmatrix}=\begin{pmatrix}3y_2\\ 2y_1\end{pmatrix}+O(|\mathbf{y}|^3). \tag{E2.7}$$

2.3.5 Let $\mathbf{X}(\mathbf{x})$ be a smooth vector field on $\mathbb{R}^2$ with $\mathbf{X}(\mathbf{0})=\mathbf{0}$ and $\mathbf{X}_1=\begin{pmatrix}3x_1\\ x_2\end{pmatrix}$. Show that the normal form of $\mathbf{X}$ can be written as

$$\begin{pmatrix}3y_1+Ky_2^3\\ y_2\end{pmatrix}+O(|\mathbf{y}|^{N+1}) \tag{E2.8}$$

for some real constant K and any integer $N\geqslant 3$.

2.3.6 Let $\mathbf{X}(\mathbf{x})$ be a smooth vector field with $\mathbf{X}(\mathbf{0})=\mathbf{0}$ and $\mathbf{X}_1=\begin{pmatrix}p\lambda x_1\\ -q\lambda x_2\end{pmatrix}$, where $p,q\in\mathbb{Z}^+$ and $\lambda>0$. Write down the resonant terms of order $p+q+1$.

2.3.7 Consider a planar vector field $\mathbf{X}(\mathbf{x})$ with linear part $(\lambda_1x_1,\lambda_2x_2)^{\mathrm{T}}$ such that $\lambda_1=2\lambda_2$. Prove that the normal form up to any order $N\geqslant 3$ is $\begin{pmatrix}\lambda_1x_1+cx_2^2\\ \lambda_2x_2\end{pmatrix}$. Find the corresponding normal form for $\lambda_1=m\lambda_2$, where m is a positive integer greater than 2.

2.3.8 Given that $\mathbf{A}=\begin{pmatrix}\lambda & 1\\ 0 & \lambda\end{pmatrix}$, calculate $L_{\mathbf{A}}\begin{pmatrix}\mathbf{x}^{\mathbf{m}}\\ 0\end{pmatrix}$ and $L_{\mathbf{A}}\begin{pmatrix}0\\ \mathbf{x}^{\mathbf{m}}\end{pmatrix}$ where $\mathbf{x}^{\mathbf{m}}=x_1^{m_1}x_2^{m_2}$. Hence obtain the matrix representation of $L_{\mathbf{A}}$ with respect to the basis

$$\left\{\begin{pmatrix}0\\ x_1^2\end{pmatrix},\begin{pmatrix}0\\ x_1x_2\end{pmatrix},\begin{pmatrix}0\\ x_2^2\end{pmatrix},\begin{pmatrix}x_1^2\\ 0\end{pmatrix},\begin{pmatrix}x_1x_2\\ 0\end{pmatrix},\begin{pmatrix}x_2^2\\ 0\end{pmatrix}\right\}.$$

Show that the eigenvalues of $L_{\mathbf{A}}$ are given by $m_1\lambda_1+m_2\lambda_2-\lambda_i$, $i=1,2$, with $\lambda_1=\lambda_2=\lambda$ and $m_1+m_2=r=2$.

2.3.9 Generalise Exercise 2.3.8 to find the matrix representative of $L_{\mathbf{A}}\colon H^r\to H^r$, $r\geqslant 2$ when $\mathbf{A}=\begin{pmatrix}\lambda & 1\\ 0 & \lambda\end{pmatrix}$. Show that the eigenvalues of $L_{\mathbf{A}}$ are also repeated with value $\lambda(r-1)$. Show that this result is consistent with $\Lambda_{\mathbf{m},i}$ given by (2.3.12) and deduce that $L_{\mathbf{A}}^{-1}$ exists if and only if $\lambda\neq 0$.

2.3.10 Prove that the vector field $\mathbf{X}(\mathbf{x})=(3x_1+3x_1^2,x_2)^{\mathrm{T}}$ can be transformed to $\tilde{\mathbf{X}}(\mathbf{y})=(3y_1+6y_1^3+O(y_1^4),y_2)^{\mathrm{T}}$ by the transformation $x_1=y_1+y_1^2$, $x_2=y_2$ (see Example 2.3.1(b)).

2.4 Non-hyperbolic singular points of vector fields

2.4.1 In order to assign a codimension to subsets of the set of all 2×2 real matrices, associate the matrix $\mathbf{A} = \begin{pmatrix} a & b \\ c & d \end{pmatrix}$ with the element (a, b, c, d) of $\mathbb{R}^4$. If a given subset, S, of matrices corresponds to a subset of points in a subspace of $\mathbb{R}^4$ with dimension n, then the codimension of S is defined to be $4 - n$.

Give minimal conditions which a, b, c, d must satisfy if $\mathbf{A}$ is to have the following properties:
(1) both eigenvalues non-zero;
(2) exactly one zero eigenvalue;
(3) two zero eigenvalues but $\mathbf{A}$ is not the null-matrix.
For $j = 1, 2, 3$ obtain the codimension of the subset, S_j, of 2×2 real matrices satisfying condition (j) above. What does this imply about the degeneracy level of the (linear) vector fields for which (2.4.1) and (2.4.3) are valid?

2.4.2 Sketch phase portraits for the following non-linear perturbations of the non-hyperbolic systems shown in Figure 2.5:

(a) $\dot{x} = x$, $\dot{y} = y^2$;
(b) $\dot{x} = y + x(x^2 + y^2)$, $\dot{y} = -x + y(x^2 + y^2)$;
(c) $\dot{x} = y$, $\dot{y} = x^2$.

2.4.3 Show that the vector field

$$\mathbf{X}(\mathbf{x}) = \begin{pmatrix} x_2 + ax_1^2 + bx_1x_2 + cx_2^2 \\ dx_1^2 + ex_1x_2 + fx_2^2 \end{pmatrix}, \tag{E2.9}$$

is transformed to

$$\tilde{\mathbf{X}}(\mathbf{x}) = \begin{pmatrix} y_2 + (a + D)y_1^2 + (b + E - 2A)y_1y_2 - (c + F - B)y_2^2 \\ dy_1^2 + (e - 2D)y_1y_2 + (f - E)y_2^2 \end{pmatrix} \tag{E2.10}$$

by

$$\mathbf{x} = \mathbf{y} + \begin{pmatrix} Ay_1^2 + By_1y_2 + Cy_2^2 \\ Dy_1^2 + Ey_1y_2 + Fy_2^2 \end{pmatrix}. \tag{E2.11}$$

Choose $A, B, \ldots, F$ so that:

(i) $\tilde{\mathbf{X}}(\mathbf{y}) = \begin{pmatrix} y_2 + \alpha y_1^2 \\ \beta y_1^2 \end{pmatrix} + O(|\mathbf{y}|^3)$;

(ii) $\tilde{\mathbf{X}}(\mathbf{y}) = \begin{pmatrix} y_2 \\ \gamma y_1^2 + \delta y_1y_2 \end{pmatrix} + O(|\mathbf{y}|^3)$.

Give expressions for $\alpha, \beta, \gamma, \delta$ in terms of $a, b, \ldots, f$.

2.4.4 Show that the systems

$$\dot{x}_1 = x_2 + ax_1^2, \qquad \dot{x}_2 = bx_1^2,$$

and

$$\dot{y}_1 = y_2, \qquad \dot{y}_2 = cy_1^2 + dy_1y_2, \tag{E2.12}$$

are smoothly conjugate by the change of variables $x_1 = y_1$, $x_2 = y_2 - ay_1^2$. What are the relationships between the constants a, b and c, d?

2.4.5 Let the vector field $\mathbf{X}$ on $\mathbb{R}^n$ have a non-hyperbolic fixed point at $\mathbf{x}=\mathbf{0}$. Use the real Jordan form of $\mathbf{DX}(\mathbf{0})$ to show that the normal form of $\mathbf{X}$ contains resonant terms of order r for all $r \geqslant 2$.

2.5 Normal forms for diffeomorphisms

2.5.1 Show that the operator $M_{\mathbf{A}}: H^r \to H^r$ defined by equation (2.5.6) is a linear map. Prove that if $\mathbf{A}=[\lambda_i \delta_{ij}]_{i,j=1}^n$, then the eigenvalues of $M_{\mathbf{A}}$ are $\boldsymbol{\lambda}^{\mathbf{m}} - \lambda_i$, where $\boldsymbol{\lambda}^{\mathbf{m}} = \lambda_1^{m_1}\lambda_2^{m_2}\cdots\lambda_n^{m_n}$ and $m_1 + \cdots + m_n = r$, with corresponding eigenvectors $\mathbf{x}^{\mathbf{m}}\mathbf{e}_i$.

2.5.2 Let the map $f: \mathbb{R} \to \mathbb{R}$ have a fixed point at the origin with the eigenvalue -1. Show that f can be written in the form

$$f(x) = -x + cx^3 + O(|x|^5). \tag{E2.13}$$

Prove that the origin is a stable fixed point for $c > 0$. Show that if f is originally in the form

$$f(x) = -x + c_2x^2 + c_3x^3 + O(|x|^4) \tag{E2.14}$$

then $c = c_3 + c_2^2$.

2.5.3 Consider a map $f: \mathbb{C} \to \mathbb{C}$ of the form $\sum_{i=1}^{\infty} a_i z^i$ with $f(0)=0$ and eigenvalue $\lambda = \exp(2\pi i p/q)$. Show that there exists a change of coordinates such that f can be written in the form

$$f(z) = \lambda z + cz^{q+1} + O(|z|^{2q+1}). \tag{E2.15}$$

2.5.4 Find a transformation of the form $z = w + a\bar{w}^3$ such that

$$f(z) = \exp(i\alpha)z + \bar{z}^3 + z^2\bar{z}, \tag{E2.16}$$

where $\alpha \neq 2\pi p/q$, $q = 1, 2, 3, 4$, becomes

$$\tilde{f}(w) = \exp(i\alpha)w + w^2\bar{w} + O(|w|^4). \tag{E2.17}$$

2.5.5 Let $B^r = M_{\mathbf{A}}(H^r)$ be the image subspace of H^r when $\mathbf{A}$ is:

(a) $\begin{pmatrix} 1 & 0 \\ 0 & \lambda \end{pmatrix}$, $\lambda \neq 1$;

(b) $\begin{pmatrix} 1 & 1 \\ 0 & 1 \end{pmatrix}$.

In each case, find a complementary subspace G^r of B^r in H^r. Show that the normal form of a diffeomorphism $\mathbf{f}: \mathbb{R}^2 \to \mathbb{R}^2$, with $\mathbf{f}(\mathbf{0}) = \mathbf{0}$ and $\mathbf{Df}(\mathbf{0}) = \mathbf{A}$, can be written as:

(a) $\mathbf{f}(\mathbf{x}) = \left(x_1 + \sum_{r \geqslant 2} a_r x_1^r, \quad \lambda x_2 + \sum_{r \geqslant 2} b_r x_1^{r-1} x_2 \right)^{\mathrm{T}}$;

(b) $\mathbf{f}(\mathbf{x}) = \left(x_1 + x_2 + \sum_{r \geqslant 2} a_r x_1^r, \quad x_2 + \sum_{r \geqslant 2} b_r x_1^r \right)^{\mathrm{T}}$.

2.6 Time-dependent normal forms

2.6.1 Let $\mathbf{M}$ be a non-singular, real, $n \times n$ matrix with Jordan form $\mathbf{J} = [\lambda_i \delta_{ij}]_{i,j=1}^n$. Show that $\ln \mathbf{M}$ is similar to the matrix $[\ln \lambda_i \delta_{ij}]_{i,j=1}^n$. Hence deduce that if $\lambda_i \in \mathbb{R}^+$,

$i=1,\ldots,n$, then $\ln \mathbf{M}$ is real. Give a counterexample to show that the converse is not true.

2.6.2 Consider the non-singular, $n \times n$ Jordan block matrix $\mathbf{J} = \lambda\mathbf{I} + \mathbf{N}$, with $\lambda \in \mathbb{C}$ and $\mathbf{N} = [N_{ij}]_{i,j=1}^{n}$, where $N_{ij} = 1$ if $j = i+1$ and zero otherwise. Assume that $\ln \mathbf{J}$ is upper triangular and show that it is given by:
(a) $\ln \lambda\mathbf{I} + (\mathbf{N}/\lambda)$ for $n=2$;
(b) $\ln \lambda\mathbf{I} + (\mathbf{N}/\lambda) - (\mathbf{N}^2/2\lambda^2)$ for $n=3$.
Verify that these results are consistent with the general form

$$\ln \mathbf{J} = \ln \lambda\mathbf{I} + \mathbf{R}, \tag{E2.18}$$

where $\mathbf{R}$ is the matrix power series obtained from the Maclaurin expansion of $\ln(1+x)$ by replacing x by $\mathbf{N}/\lambda$.

2.6.3 Let $\mathbf{M}$ be a non-singular, real, $n \times n$ matrix. Use the results of Exercises 2.6.1 and 2.6.2 to show that there is a (possibly complex) matrix $\mathbf{L}$ such that $\mathbf{M} = \exp(\mathbf{L})$.
An alternative form of Floquet's Theorem (see Hale, 1969) states:
Every fundamental matrix $\mathbf{Q}(t)$ *for* $\dot{\mathbf{x}} = \mathbf{A}(t)\mathbf{x}$, $\mathbf{A}(t+2\pi) = \mathbf{A}(t)$, $t \in \mathbb{R}$, *has the form* $\mathbf{Q}(t) = \mathbf{U}(t)\exp(\mathbf{C}t)$, *with* $\mathbf{U}(t)$ *and* $\mathbf{C}$ $n \times n$ *matrices, where* $\mathbf{C}$ *is constant and* $\mathbf{U}(t)$ *is* 2π*-periodic in* t.
Show that this statement is equivalent to that given in Theorem 2.6.1 with $\mathbf{C} = \mathbf{\Lambda}$ and $\mathbf{U}(t) = \mathbf{B}(t) = \boldsymbol{\varphi}(t,0)\exp(-\mathbf{\Lambda}t)$.

2.6.4 Prove that, for any *real*, non-singular, $n \times n$ matrix $\mathbf{M}$, there is a *real* matrix $\mathbf{L}$ such that $\mathbf{M}^2 = \exp(\mathbf{L})$.
Consider the real linear system $\dot{\mathbf{x}} = \mathbf{A}(t)\mathbf{x}$, $\mathbf{A}(t+2\pi) = \mathbf{A}(t)$, $t \in \mathbb{R}$. Show that there is a *real* change of variable $\mathbf{x} = \mathbf{B}(t)\mathbf{y}$, with $\mathbf{B}(t)$ 4π*-periodic* in t, such that $\dot{\mathbf{y}} = \mathbf{\Lambda}\mathbf{y}$ where $\mathbf{\Lambda}$ is *real* and independent of t.

2.6.5 Show that the system

$$\begin{pmatrix} \dot{x}_1 \\ \dot{x}_2 \end{pmatrix} = \begin{pmatrix} -\sin 2t & \cos 2t - 1 \\ \cos 2t + 1 & \sin 2t \end{pmatrix} \begin{pmatrix} x_1 \\ x_2 \end{pmatrix} \tag{E2.19}$$

has two linearly independent solutions

$$\mathbf{x}_{\pm} = \begin{pmatrix} \exp(\pm t)(\ \cos t \mp \sin t) \\ \exp(\pm t)(\pm\cos t + \sin t) \end{pmatrix}. \tag{E2.20}$$

Obtain the matrix $\boldsymbol{\varphi}(t,0)$ and find $\mathbf{\Lambda}$ such that $\boldsymbol{\varphi}(2\pi,0) = \exp(2\pi\mathbf{\Lambda})$. By constructing $\mathbf{B}(t) = \boldsymbol{\varphi}(t,0)\exp(-\mathbf{\Lambda}t)$ check that $\mathbf{x} = \mathbf{B}(t)\mathbf{y}$ is a change of variables such that $\dot{\mathbf{y}} = \mathbf{\Lambda}\mathbf{y}$.

2.6.6 Show that the autonomous system $\dot{\mathbf{x}} = \mathbf{\Lambda}\mathbf{x}$ is transformed into

$$\dot{\mathbf{y}} = \mathbf{\Lambda}\mathbf{y} - \left(L_{\mathbf{\Lambda}} + \frac{\partial}{\partial t}\right)\mathbf{h}_r(\mathbf{y},t) + O(|\mathbf{y}|^{r+1}, t), \tag{E2.21}$$

by the change of variables $\mathbf{x} = \mathbf{y} + \mathbf{h}_r(\mathbf{y},t)$. Here the components of $\mathbf{h}_r$ are homogeneous polynomials of degree r in $\mathbf{y}$ and $L_{\mathbf{\Lambda}}$ is the Lie bracket operator.
Let H^r be the vector space of vector fields with basis $\{x_1^{m_1}x_2^{m_2}\exp(i\nu t)\mathbf{e}_j \mid m_1+m_2=r, j=1,2\}$ for some fixed integer ν. Consider the operator $(L_{\mathbf{\Lambda}} + \partial/\partial t)\colon H^r \to H^r$ with $\mathbf{\Lambda} = \begin{pmatrix} \lambda & 1 \\ 0 & \lambda \end{pmatrix}$. Show that the eigenvalues of $L_{\mathbf{\Lambda}} + \partial/\partial t$ are of the form $i\nu + m_1\lambda_1 + m_2\lambda_2 - \lambda_i$ with $\lambda_1 = \lambda_2 = \lambda$.

2.6.7 Show that, if the real system

$$\dot{\mathbf{x}} = \begin{pmatrix} \alpha & -\beta \\ \beta & \alpha \end{pmatrix} \mathbf{x} \tag{E2.22}$$

is diagonalised over the complex numbers $\mathbb{C}$, such that the transformed coordinates (z_1, z_2) are taken relative to complex conjugate eigendirections, then $z_1 = \bar{z}_2 = z$. Hence, deduce that the real system is effectively replaced by a single complex differential equation

$$\dot{z} = (\alpha + i\beta)z. \tag{E2.23}$$

Find the time-2π flow maps $\mathbf{P}_1 : \mathbb{R}^2 \to \mathbb{R}^2$ and $P_2 : \mathbb{C} \to \mathbb{C}$ of the systems (E2.22) and (E2.23) respectively. Use $z = x + iy$ to deduce $\mathbf{P}_1$ from P_2.

2.6.8 Explain why the resonance condition (2.6.14)

$$i\nu + \mathbf{m}\cdot\boldsymbol{\lambda} - \lambda_j = 0, \qquad \sum_i m_i = r$$

determines, for a given r, only *finitely* many Fourier terms.

Find the resonant terms, up to order 5 in $|z|$, for a time-dependent differential equation

$$\dot{z} = i\omega z + O(|z|^2, t) \tag{E2.24}$$

when (a) ω is irrational; and (b) $\omega = \frac{2}{5}$.

2.6.9 Suppose $\mathbf{\Lambda} = \mathbf{0}$ in (2.6.6). Use (2.6.14) to show there are no time-dependent resonant terms.

Find a change of variable, 2π-periodic in time, which reduces the equation

$$\dot{x} = x^2 \cos^2 t - cx^3 \tag{E2.25}$$

to the form

$$\dot{x} = ax^2 + bx^3 + O(|x|^4; t). \tag{E2.26}$$

Find the constants a, b and describe the behaviour of solutions close to $x = 0$.

2.6.10 Consider the system

$$\begin{pmatrix} \dot{z} \\ \dot{\bar{z}} \end{pmatrix} = \begin{pmatrix} i\omega & 0 \\ 0 & -i\omega \end{pmatrix} \begin{pmatrix} z \\ \bar{z} \end{pmatrix} + \mathbf{f}(z, \bar{z}, t) \tag{E2.27}$$

discussed in Example 2.6.1, where $z, \bar{z} \in \mathbb{C}$, $\omega \in \mathbb{R}$ and $\mathbf{f}$ is 2π-periodic in t. Use (2.6.14) to confirm that if $\begin{pmatrix} z^{m_1}\bar{z}^{m_2}\exp(i\nu t) \\ 0 \end{pmatrix}$ is a resonant term of $\mathbf{f}(z, \bar{z}, t)$ then so is $\begin{pmatrix} 0 \\ z^{m_2}\bar{z}^{m_1}\exp(-i\nu t) \end{pmatrix}$.

2.7 Centre manifolds

2.7.1 Let $\exp(\mathbf{A}t) : \mathbb{R}^n \to \mathbb{R}^n$ be a non-hyperbolic linear flow with a two-dimensional centre eigenspace $E^c \subseteq \mathbb{R}^n$. Describe the three topological types of flow that can occur on E^c. Which of these flows give rise to an unbounded motion on E^c?

2.7.2 Let $\exp(\mathbf{A}t) : \mathbb{R}^n \to \mathbb{R}^n$ be a non-hyperbolic flow with centre eigenspace E^c that is a proper subset of $\mathbb{R}^n$. Let $\mathbf{x} \in \mathbb{R}^n \backslash E^c$. Show that $\mathbf{x}$ is a wandering point.

2.7.3 Show that the system

$$\dot{x} = -x^5, \qquad \dot{y} = y \tag{E2.28}$$

has a centre manifold of the form $y = h(x)$, where

$$h(x) = \begin{cases} c_1 \exp(-\frac{1}{4}x^{-4}), & x > 0 \\ 0, & x = 0, \\ c_2 \exp(-\frac{1}{4}x^{-4}), & x < 0 \end{cases} \tag{E2.29}$$

for any choice of the constants $c_1, c_2 \in \mathbb{R}$. What is the differentiability class of these manifolds? What is the topological type of the flow on each manifold?

2.7.4 Prove that the curve

$$y = \begin{cases} C_1 |x|^{b/a}, & x < 0, \\ C_2 |x|^{b/a}, & x \geqslant 0, \end{cases} \tag{E2.30}$$

$C_1, C_2 \in \mathbb{R}$ with $a, b > 0$, is invariant for the system

$$\dot{x} = ax, \qquad \dot{y} = by. \tag{E2.31}$$

Is this invariant set a centre manifold? What is its maximum differentiability class in general?

2.7.5 Approximate the equation of the centre manifold of the system

$$\begin{aligned} \dot{x} &= \mu x - x^3, \\ \dot{y} &= y + x^4, \\ \dot{\mu} &= 0, \end{aligned} \tag{E2.32}$$

by substituting $y = \sum_{i=2, j=0}^{\infty} a_{ij} x^i \mu^j$ into $(\mu x - x^3)\, \mathrm{d}y/\mathrm{d}x = y + x^4$ and comparing coefficients of x^i for $i = 2, 3, \ldots, 6$. Hence show that y is C^4 in x if $\mu < \frac{1}{4}$ and C^6 if $\mu < \frac{1}{6}$.

2.7.6 Show that the system $\dot{x} = x^3$, $\dot{y} = 1 + 2y$ has a non-hyperbolic fixed point with centre manifold $y = -\frac{1}{2} + C\exp(-1/x^2)$ for any choice of the constant C.

If the centre manifold is assumed to be analytic of the form $y = \sum_{i=0}^{\infty} a_i x^i$, and the a_i are found by substitution of the series in the system, show that only the case $C = 0$ is found. Explain this apparent discrepancy.

2.7.7 Show that the differential equation

$$\dot{x} = x^3, \qquad \dot{y} = 2y - 2x^2 \tag{E2.33}$$

does not have an analytic centre manifold.

2.7.8 Consider the system

$$\begin{aligned} \dot{x} &= ax^3 + x^3 y, \\ \dot{y} &= -y + y^2 + xy + x^2 - xy^2. \end{aligned} \tag{E2.34}$$

Show that there is a centre manifold of the form $y = x^2 + O(x^3)$. Hence show that the origin is a weak stable node for $a < 0$ and weak saddle for $a > 0$.

2.7.9 Show that the stability of the system

$$\dot{x} = x + ay^2, \qquad \dot{y} = -xy \tag{E2.35}$$

on its centre manifold is determined by the sign of a. Sketch the local phase portrait at $\mathbf{x} = \mathbf{0}$.

2.7.10 Consider the system

$$\dot{x} = ax^r + by^s, \qquad \dot{y} = -y, \tag{E2.36}$$

where $a \in \mathbb{R} \backslash \{0\}$ and $r, s \in \mathbb{Z}^+$ with $r < 2s$. Show that the origin is an asymptotically stable fixed point *only* if $a < 0$ and r is odd.

2.7.11 Consider the differential equation

$$\dot{u} = v, \qquad \dot{v} = -v + \alpha u^2 + \beta vu. \tag{E2.37}$$

Make a linear transformation, $\mathbf{u} = \mathbf{Mx}$, $\mathbf{u} = (u, v)^{\mathrm{T}}$, $\mathbf{x} = (x, y)^{\mathrm{T}}$, to put the linear part of (E2.37) into Jordan form. Find the form of the centre manifold in the xy-plane correct to terms of order 3. Hence sketch the local phase portrait at $\mathbf{x} = \mathbf{0}$ for $\alpha > \beta > 0$.

2.8 Blowing-up techniques on $\mathbb{R}^2$

2.8.1 Polar blowing-up

2.8.1 Use polar blowing-up to determine the topological type of the singularities at the origin of the following system:

$$\dot{x} = x^2 + ay^2, \qquad \dot{y} = bxy, \tag{E2.38}$$

for (i) $a > 0$, $b < 1$; and (ii) $a < 0$, $b > 1$.

2.8.2 Find the topological types of the following systems at $(x, y) = \mathbf{0}$ by using polar blowing-up:
(a) $\dot{x} = x(2y + x)$, $\dot{y} = y(y + 2x)$;
(b) $\dot{x} = x^2$, $\dot{y} = y(2x - y)$.

2.8.2 Directional blowing-up

2.8.3 Show that directional blowing-up along the y-axis does not resolve the nature of the singularity at the origin of the system

$$\dot{x} = x(\lambda + ay), \qquad \dot{y} = by^2, \qquad a, b, \lambda > 0. \tag{E2.39}$$

2.8.4 Observe that the formulae for directional blowing-up can also be used along the negative x- and y-axis. Why must any dividing out of the transformed vector field be by $|u|^k$ or $|v|^k$ and not u^k or v^k in this case? Find the directional blow-ups of

$$\dot{x} = x^2 - 2xy, \qquad \dot{y} = y^2 - 2xy, \tag{E2.40}$$

on the negative x- and y-axes to confirm the results of Example 2.8.1.

2.8.5 Use directional blowing-up to find the topological type of the singularity at $\mathbf{0}$ of the system

$$\dot{x} = x^2 - y^2, \qquad \dot{y} = 2xy. \tag{E2.41}$$

2.8.6 Use blowing-up techniques to determine the topological types of the singularities at the origin in the following systems:

$$\text{(a)}\ \dot{x} = y + x^3, \qquad \dot{y} = x^3; \tag{E2.42}$$
$$\text{(b)}\ \dot{x} = x^2 - y^2, \qquad \dot{y} = -2xy. \tag{E2.43}$$

Explain why the technique fails to determine the topological type of the singularity at $(x, y) = \mathbf{0}$ in the system $\dot{x} = y + x^3$, $\dot{y} = -x^3$.

3

Structural stability, hyperbolicity, and homoclinic points

In applications we require our mathematical models to be robust. By this we mean that their qualitative properties should not change significantly when the model is subjected to small, allowable perturbations. If we wish to make these ideas more precise, then we must have some class of perturbations in mind and some way of deciding when they are small. From a theoretical stand point, this means that we regard our model as a member of some chosen space, $\mathscr{S}$, of dynamical systems to which we attach an appropriate metric. It is then possible to give meaning to the idea that a perturbation is 'close' to the original model. A dynamical system whose topological properties are shared (in a sense that must be properly defined) by all sufficiently close neighbouring systems is said to be *structurally stable*.

Structural stability, like hyperbolicity (see §§ 2.1 and 2.2), is a property of individual dynamical systems and we can ask if this property is, in some sense, typical of the elements of the space $\mathscr{S}$. The subset of all structurally stable systems is *open*. This follows directly from the definition of structural stability itself. Clearly, every structurally stable system lies in an open set, each element of which is also structurally stable. Thus the subset of structurally stable systems is a union of open sets and is therefore open itself. It follows that a structurally stable system cannot be approximated arbitrarily closely by structurally unstable systems. However, in some cases the subset of structurally stable systems can be shown to be *dense* in the space $\mathscr{S}$. This means that every structurally unstable system can be approximated arbitrarily closely by structurally stable systems. In such cases, we say that structural stability is a *generic property* of $\mathscr{S}$. In general, a property is said to be generic if it is shared by a *residual subset* of the space of systems involved. Such a subset is a countable intersection of open dense sets. We will, however, encounter only open dense sets in this chapter.

In § 3.1, we show that a linear flow is structurally stable if and only if it is hyperbolic and that hyperbolicity of flow is a generic property of linear transformations on $\mathbb{R}^n$. Thus the structurally stable linear flows are characterised by a single, hyperbolic fixed point at the origin and structural stability of flow is a generic property of linear transformations.

Studies of flows on two-dimensional compact manifolds (see § 3.3) show that the structurally stable systems have non-wandering sets which are characterised by hyperbolic fixed points and closed orbits, together with the global requirement that no saddle connections occur. Structural stability is, once again, a generic property of these flows.

The above findings led to two conjectures about flows on manifolds of dimension $n \geqslant 2$:

(i) structural stability is a generic property of such flows;
(ii) the structurally stable flows are characterised in the same way as flows on two-manifolds.

Neither of these conjectures is correct. A counterexample to the first was given by Smale (1966). We do not discuss this example here; the interested reader can consult Arnold & Avez, 1968, pp. 196–200. In §§ 3.4 and 3.5 we describe two types of diffeomorphism on two-manifolds that are structurally stable but cannot be characterised in the manner described in (ii). Remember (see § 1.7) these diffeomorphisms correspond to flows in three dimensions by suspension. The *Anosov automorphisms* of the two-torus (see § 3.4) and the *horseshoe diffeomorphism* of the sphere (see § 3.5) have complicated non-wandering sets and to characterise them we must extend our notion of hyperbolic sets beyond fixed points and closed orbits (see § 3.6). The systems referred to in (ii) are now known as *Morse–Smale systems* (see Nitecki, 1971).

The horseshoe diffeomorphism also plays a central role in our understanding of the complex dynamics described in § 1.9. In particular, it can be shown that it has orbits that behave in a random or *chaotic* way (see §§ 3.5.2 and 3). In §§ 3.6 and 3.7 we explain how this behaviour is related to the 'two-dimensional, chaotic orbits' associated with hyperbolic fixed or periodic points of some planar maps (see Figures 1.39–42). The occurrence of *homoclinic* (or *heteroclinic*) *points* is the central feature of this discussion. They arise when the stable and unstable manifolds of a hyperbolic point (or points) intersect transversely and it can be shown that the map must then contain embedded horseshoes.

3.1 Structural stability of linear systems

Let $L(\mathbb{R}^n)$ be the set of real linear transformations of $\mathbb{R}^n$ to itself. Define the *norm* of an $n \times n$ matrix $\mathbf{A} = [a_{ij}]$ to be $\|\mathbf{A}\| = \sum_{i,j=1}^{n} |a_{ij}|$. An *ε-neighbourhood* of $\mathbf{A}$ is given by $N_\varepsilon(\mathbf{A}) = \{\mathbf{B} \in L(\mathbb{R}^n) \mid \|\mathbf{B} - \mathbf{A}\| < \varepsilon\}$. Each $\mathbf{B} \in N_\varepsilon(\mathbf{A})$ is said to be ε-close to $\mathbf{A}$. We are now able to give a formal definition of structural stability for linear flows and diffeomorphisms on $\mathbb{R}^n$.

Definition 3.1.1 *A linear flow*, $\exp(\mathbf{A}t)$: $\mathbb{R}^n \to \mathbb{R}^n$, (*or diffeomorphism*, **A**) *is said to be* structurally stable *in* $L(\mathbb{R}^n)$ *if there is an* ε*-neighbourhood of* **A**, $N_\varepsilon(\mathbf{A}) \subseteq L(\mathbb{R}^n)$, *such that, for every* $\mathbf{B} \in N_\varepsilon(\mathbf{A})$, $\exp(\mathbf{B}t)$ (*or* **B**) *is topologically equivalent* (*conjugate*) *to* $\exp(\mathbf{A}t)$ (*or* **A**).

The following result shows that, in this linear case, the structurally stable systems can be completely characterised.

Proposition 3.1.1 *A linear flow or diffeomorphism on* $\mathbb{R}^n$ *is structurally stable in* $L(\mathbb{R}^n)$ *if and only if it is hyperbolic.*

Proof. A linear flow $\exp(\mathbf{A}t)$ is hyperbolic if *all* the eigenvalues of the matrix **A** have non-zero real parts (see Definition 2.1.2). The eigenvalues of any ε-close matrix **B** differ from those of **A** by terms $O(\varepsilon)$ (see Exercise 3.1.1). Thus, by making ε sufficiently small we can ensure that the eigenvalues of **B** are near enough to those of **A** for their real parts to be non-zero. Moreover, **A** and **B** will then have the same number, $n_s(n_u)$, of eigenvalues with negative (positive) real parts. Theorem 2.1.2 then implies that $\exp(\mathbf{A}t)$ and $\exp(\mathbf{B}t)$ are both equivalent to the flow of $\dot{\mathbf{x}} = -\mathbf{x}$, $\dot{\mathbf{y}} = \mathbf{y}$, where $\mathbf{x} \in \mathbb{R}^{n_s}$ and $\mathbf{y} \in \mathbb{R}^{n_u}$. Thus **A** is structurally stable.

Conversely, suppose the flow $\exp(\mathbf{A}t)$ is *not* hyperbolic. Then **A** has at least one eigenvalue with zero real part. However, $\mathbf{B} = \mathbf{A} + \varepsilon\mathbf{I}$ is hyperbolic for almost all $\varepsilon \neq 0$ and can be made arbitrarily close to **A** by taking ε sufficiently small. Thus, the non-hyperbolic flow $\exp(\mathbf{A}t)$ is *not* structurally stable. Hence, if a linear flow is structurally stable it must be hyperbolic (see Figure 3.1).

The proof of Proposition 3.1.1 for diffeomorphisms follows similar lines; it is considered in Exercise 3.1.2. □

It is important to note that Definition 3.1.1 specifies a space of systems ($L(\mathbb{R}^n)$) to which the perturbations must belong. Whether or not a given system is structurally stable depends on the choice of this space. For example, let $CL(\mathbb{R}^2) \subseteq L(\mathbb{R}^2)$ be the subspace of linear transformations with pure imaginary, non-zero eigenvalues. If $\mathbf{A} \in CL(\mathbb{R}^2)$, it is structurally stable in $CL(\mathbb{R}^2)$ but structurally unstable in $L(\mathbb{R}^2)$. Clearly, if $\mathbf{B} \in CL(\mathbb{R}^2)$ and is ε-close to **A**, then **B** has pure imaginary eigenvalues that are numerically close to those of **A**. Thus, the flows $\exp(\mathbf{A}t)$ and $\exp(\mathbf{B}t)$ are both of centre type and therefore topologically equivalent (see Exercise 1.6.1). Hence **A** is structurally stable in $CL(\mathbb{R}^2)$. Of course, $\mathbf{A} \in CL(\mathbb{R}^2)$ is not structurally stable in $L(\mathbb{R}^2)$ (see Figure 3.2). This follows from Proposition 3.1.1 because **A** is not hyperbolic.

The relationship between hyperbolic and structurally stable flows in $L(\mathbb{R}^n)$ allows us to show that 'structural stability of flow' is a generic property of linear transformations. Let $SF(\mathbb{R}^n) \subseteq L(\mathbb{R}^n)$ denote the set of linear transformations which give rise to structurally stable flows on $\mathbb{R}^n$.

Proposition 3.1.2 *The set $SF(\mathbb{R}^n)$ is open and dense in $L(\mathbb{R}^n)$.*

Proof. As we have already noted in the opening remarks to this chapter, $SF(\mathbb{R}^n)$ must be open. More precisely, by Definition 3.1.1, each $\mathbf{A} \in SF(\mathbb{R}^n)$ lies in an ε-neighbourhood, $N_\varepsilon(\mathbf{A})$, of topologically equivalent flows. Since $N_\varepsilon(\mathbf{A})$ is open, each of its elements must also be structurally stable, i.e. $N_\varepsilon(\mathbf{A}) \subseteq SF(\mathbb{R}^n)$. Hence $SF(\mathbb{R}^n)$ is open because it is a union of open sets.

It follows from Proposition 3.1.1 that $SF(\mathbb{R}^n) = HF(\mathbb{R}^n) (\subseteq L(\mathbb{R}^n))$ the set of linear transformations which give rise to hyperbolic flows on $\mathbb{R}^n$. However, $HF(\mathbb{R}^n)$ is a dense subset of $L(\mathbb{R}^n)$. To show this, suppose $\mathbf{A} \notin HF(\mathbb{R}^n)$. Then there exist linear transformations, $\mathbf{B}$, arbitrarily close to $\mathbf{A}$ for which $\exp(\mathbf{B}t)$ is hyperbolic (e.g.

Figure 3.1 Some examples of phase portraits of structurally stable ((*a*), (*b*)) and structurally unstable ((*c*), (*d*)) linear flows on $\mathbb{R}^3$: (*a*) $\lambda_1, \lambda_2 = \alpha \pm i\beta$, $\alpha > 0$, $\lambda_3 < 0$; (*b*) $\lambda_1, \lambda_2, \lambda_3 < 0$; (*c*) $\lambda_1 = 0$, $\lambda_2, \lambda_3 < 0$; (*d*) $\lambda_1 > 0$, $\lambda_2, \lambda_3 = \pm i\beta$.

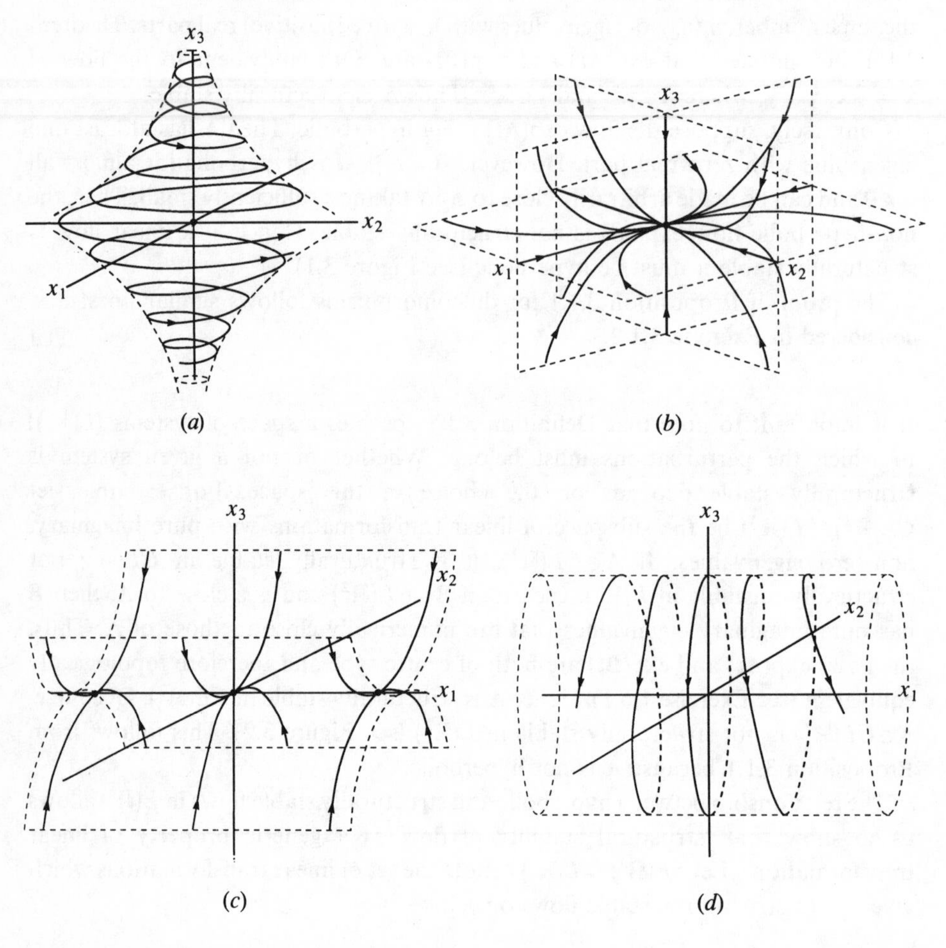

$\mathbf{B} = \mathbf{A} + \varepsilon\mathbf{I}$, with ε small). Thus, every element of $L(\mathbb{R}^n)$ is arbitrarily close to points of $HF(\mathbb{R}^n)$ and $HF(\mathbb{R}^n) = SF(\mathbb{R}^n)$ is a dense subset of $L(\mathbb{R}^n)$. Hence, structural stability of flow is a generic property of $L(\mathbb{R}^n)$. □

The analogous result for hyperbolic, linear diffeomorphisms is dealt with in Exercise 3.1.2.

3.2 Local structural stability (Hirsch & Smale, 1974, Chapter 16)

An important outcome of the discussion in § 3.1 is that hyperbolic fixed points of linear flows and diffeomorphisms persist under sufficiently small linear perturbations. This result can be extended to the local flows or diffeomorphisms defined in the neighbourhood of hyperbolic fixed points of non-linear systems. Of course, we must choose an appropriate class of perturbations and specify when they are small.

Let U be an open subset of $\mathbb{R}^n$ and $\mathrm{Vec}^1(U)$ be the set of all C^1-vector fields defined on U. The magnitude of each vector field $\mathbf{X} \in \mathrm{Vec}^1(U)$ is taken to be its *C^1-norm*, $\|\mathbf{X}\|_1$, where

$$\|\mathbf{X}\|_1 = \sup_{\mathbf{x} \in U} \left\{ \sum_{i=1}^{n} |X^i(\mathbf{x})| + \sum_{i,j=1}^{n} \left| \frac{\partial X^i(\mathbf{x})}{\partial x_j} \right| \right\}. \tag{3.2.1}$$

Thus, if $\mathbf{X}(\mathbf{x}) = (X^1(\mathbf{x}), \ldots, X^n(\mathbf{x}))^{\mathrm{T}}$ then $\|\mathbf{X}\|_1$ is 'small' when $X^i(\mathbf{x})$ *and* $\partial X^i(\mathbf{x})/\partial x_j$, $i, j = 1, \ldots, n$, are 'small' for all $\mathbf{x} \in U$. We now define

$$N_\varepsilon(\mathbf{X}) = \{\mathbf{Y} \in \mathrm{Vec}^1(U) \mid \|\mathbf{X} - \mathbf{Y}\|_1 < \varepsilon\}$$

to be an *ε-neighbourhood* of $\mathbf{X}$ in $\mathrm{Vec}^1(U)$. The use of the C^1-norm is often emphasised by saying that $\mathbf{Y} \in N_\varepsilon(\mathbf{X})$ is *ε–C^1-close* to (or is an *ε–C^1-perturbation* of) $\mathbf{X}$. This highlights the fact that the components of $\mathbf{X}$ and $\mathbf{Y}$ *and* their first derivatives are close throughout U.

We can now make our opening remarks more precise.

Figure 3.2 The flow in (*a*) is an allowed perturbation of the centre, (*b*), in $L(\mathbb{R}^2)$ but it is not an element of $CL(\mathbb{R}^2)$. For a sufficiently weak spiral, (*a*) can be made arbitrarily close to (*b*) in $L(\mathbb{R}^2)$.

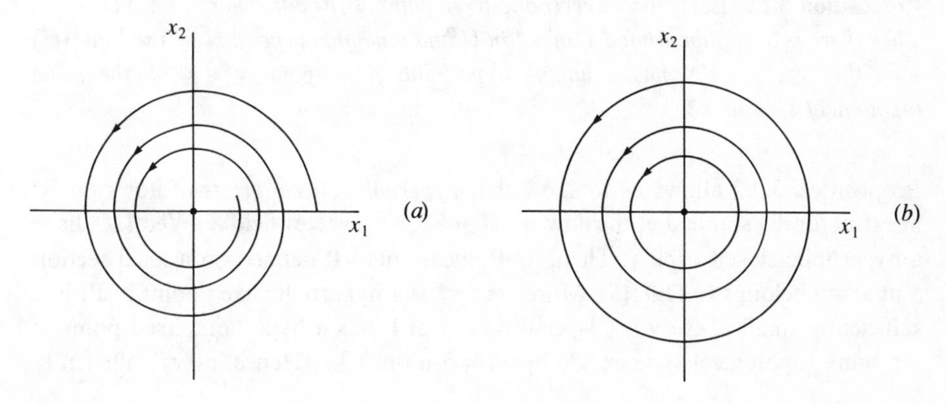

Proposition 3.2.1 *Let* $\mathbf{X} \in \mathrm{Vec}^1(U)$ *have a hyperbolic singularity at* $\mathbf{x} = \mathbf{x}^*$. *Then there exists a neighbourhood* V *of* $\mathbf{x}^*$ *in* U *and a neighbourhood* N *of* $\mathbf{X}$ *in* $\mathrm{Vec}^1(U)$ *such that each* $\mathbf{Y} \in N$ *has a unique hyperbolic singular point* $\mathbf{y}^* \in V$. *Moreover, the linearised flow,* $\exp(D\mathbf{Y}(\mathbf{y}^*)t)$ *has stable and unstable eigenspaces of the same dimension as* $\exp(D\mathbf{X}(\mathbf{x}^*)t)$.

Notice the perturbed fixed point $\mathbf{y}^*$ does not, in general, coincide with $\mathbf{x}^*$. However, given any $\delta > 0$, N can be chosen such that $|\mathbf{y}^* - \mathbf{x}^*| < \delta$ for all $\mathbf{Y} \in N$. Furthermore, by using Proposition 3.2.1 in conjunction with Hartman's Theorem (Theorem 2.2.3) we can deduce that there are neighbourhoods of $\mathbf{x}^*$ and $\mathbf{y}^*$ on which $\mathbf{X}$ and $\mathbf{Y}$ are topologically equivalent. Proposition 3.2.1 gives equality of dimension of the stable eigenspaces of the linearised flows $\exp(D\mathbf{Y}(\mathbf{y}^*)t)$ and $\exp(D\mathbf{X}(\mathbf{x}^*)t)$. It follows that these linear flows are topologically equivalent (see Theorem 2.1.2). The Hartman Theorem states that there is a neighbourhood of $\mathbf{x}^*$ ($\mathbf{y}^*$) on which the flow of $\dot{\mathbf{x}} = \mathbf{X}(\mathbf{x})$ ($\dot{\mathbf{y}} = \mathbf{Y}(\mathbf{y})$) is topologically conjugate to $\exp(D\mathbf{X}(\mathbf{x}^*)t)$ ($\exp(D\mathbf{Y}(\mathbf{y}^*)t)$). Thus the local, C^0-equivalence of the flows of $\mathbf{X}$ and $\mathbf{Y}$ follows from

$$\varphi_t | U^{\mathrm{H}}_{\mathbf{x}^*} \sim \exp(D\mathbf{X}(\mathbf{x}^*)t) \sim \exp(D\mathbf{Y}(\mathbf{y}^*)t) \sim \psi_t | U^{\mathrm{H}}_{\mathbf{y}^*}, \tag{3.2.2}$$

where φ_t and ψ_t are the flows of $\mathbf{X}$ and $\mathbf{Y}$ respectively. $U^{\mathrm{H}}_{\mathbf{x}^*}$ and $U^{\mathrm{H}}_{\mathbf{y}^*} \subseteq U$ denote the neighbourhoods on which Hartman's Theorem is valid. There is, therefore, a sense in which $\varphi_t \colon U \to \mathbb{R}^n$ is structurally stable in a neighbourhood of $\mathbf{x}^*$. Namely that for every pair $(\mathbf{Y}, \mathbf{y}^*)$ with $\mathbf{Y} \in N$ there is a neighbourhood $U^{\mathrm{H}}_{\mathbf{y}^*} \subseteq U$ of $\mathbf{y}^*$ such that $\psi_t | U^{\mathrm{H}}_{\mathbf{y}^*}$ is C^0-equivalent to $\varphi_t | U^{\mathrm{H}}_{\mathbf{x}^*}$. Thus, one might say that φ_t is *locally structurally stable* at $\mathbf{x}^*$. Alternatively we can observe that (3.2.2) means that the *topological type* of the fixed point (see opening paragraph of Chapter 2) is preserved under all sufficiently small C^1-perturbations and say that the *type of fixed point* is structurally stable.

Hyperbolic fixed points of diffeomorphisms also persist under sufficiently small C^1-perturbations. Let $\mathrm{Diff}^1(U)$ be the set of C^1-diffeomorphisms $\mathbf{f} \colon U(\subseteq \mathbb{R}^n) \to \mathbb{R}^n$ with the C^1-norm. Then the following result parallels Proposition 3.2.1 for flows.

Proposition 3.2.2 *Let* $\mathbf{x}^*$ *be a hyperbolic fixed point of the diffeomorphism* $\mathbf{f} \colon U \to \mathbb{R}^n$. *Then there is a neighbourhood* V *of* $\mathbf{x}^*$ *in* U *and a neighbourhood* N *of* $\mathbf{f}$ *in* $\mathrm{Diff}^1(U)$ *such that every* $\mathbf{g} \in N$ *has a unique hyperbolic fixed point* $\mathbf{y}^* \in V$ *of the same topological type as* $\mathbf{x}^*$.

Proposition 3.2.2 allows us to show that *hyperbolic closed orbits* of flows on $\mathbb{R}^n$ are structurally stable. Let the flow $\varphi_t \colon U \to \mathbb{R}^n$, with vector field $\mathbf{X} \in \mathrm{Vec}^1(U)$, have a hyperbolic closed orbit γ. Then, its Poincaré map $\mathbf{P}$ defined on a local section S at $\mathbf{x}^* \in \gamma$ belongs to $\mathrm{Diff}^1(S)$. Moreover, $\mathbf{x}^*$ is a hyperbolic fixed point of $\mathbf{P}$. For sufficiently small ε, every ε–C^1-perturbation of $\mathbf{P}$ has a hyperbolic fixed point of the same topological type as $\mathbf{x}^*$, by Proposition 3.2.2. Hence, every sufficiently small C^1-perturbation of $\mathbf{X}$ gives rise to a flow with a hyperbolic closed orbit of the same topological type as γ.

3.3 Flows on two-dimensional manifolds

The characterisation of structurally stable flows on two-manifolds provides a good illustration of the technical complications that can arise when we try to extend the local discussion of § 3.2 to include global phenomena. Recall that the results obtained in § 3.2 depended on the existence of sufficiently small neighbourhoods of the unperturbed and perturbed fixed points on which topological equivalence could be established. These neighbourhoods were sufficiently small subsets of the open set U on which $\mathbf{X}$ was defined and therefore we were not involved with the behaviour of $\mathbf{X}$ on the boundary, ∂U, of U. Indeed, as far as the local discussion was concerned we had not considered whether or not $\mathbf{X}$ had a natural extension to ∂U. Moreover, since we were involved with perturbations differing from $\mathbf{X}$ only on sufficiently small neighbourhoods of the singular point $\mathbf{x}^*$, the question of whether $\|\mathbf{X}-\mathbf{Y}\|_1$ was finite for *all* $\mathbf{X}, \mathbf{Y}$ in $\text{Vec}^1(U)$ was not relevant. Complications of this kind cannot be ignored if precise statements are to be made about global structural stability and they give rise to a number of technical conditions in the resulting theorems.

Let us begin by considering vector fields on $\mathbb{R}^2$. We can ensure that $\|\mathbf{X}-\mathbf{Y}\|_1$ is defined for all $\mathbf{X}, \mathbf{Y}$ by restricting the discussion to compact subsets of $\mathbb{R}^2$. Therefore, let $D^2 = \{\mathbf{x} \in \mathbb{R}^2 \,|\, |\mathbf{x}| \leqslant 1\}$ and let ∂D^2 denote its boundary. To ensure that the vector fields involved are well defined on ∂D^2, we will assume they are defined on an open set U containing D^2 and then take their restriction to the unit disc. Let $\text{Vec}^1(D^2)$ be the set of all C^1-vector fields, $\mathbf{X}$, defined in this way equipped with the C^1-norm

$$\|\mathbf{X}\|_1 = \max_{\mathbf{x} \in D^2} \left\{ \sum_{i=1}^{2} |X^i| + \sum_{i,j=1}^{2} \left| \frac{\partial X^i}{\partial x_j} \right| \right\}. \tag{3.3.1}$$

Definition 3.3.1 *A vector field* $\mathbf{X}$ *in* $\text{Vec}^1(D^2)$ *is said to be* structurally stable *if there exists a neighbourhood,* N, *of* $\mathbf{X}$ *in* $\text{Vec}^1(D^2)$ *such that the flow of every* $\mathbf{Y}$ *in* N *is topologically equivalent to that of* $\mathbf{X}$ *on* D^2.

The above precautions are not enough to focus attention on the structural instabilities occurring in the *interior of* D^2. Unfortunately, instabilities associated with the behaviour of the vector fields at the boundary of the disc can still occur. In the absence of further constraints, vector fields that are tangent to ∂D^2 are still present in $\text{Vec}^1(D^2)$. Topologically distinct C^1-perturbations of such vector fields are illustrated in Figure 3.3. These unwanted structural instabilities can be excluded by confining the discussion to those vector fields that are *transverse to* ∂D^2. Clearly, a vector field that satisfies this requirement is a member of one of two disjoint subsets of $\mathrm{Vec}^1(D^2)$: it either points *into* or *out of* D^2 at every point of ∂D^2. Let $\mathrm{Vec}^1_{\mathrm{in}}(D^2)$ be the set of vector fields defined on the disc D^2 in the manner described above and such that $\mathbf{X}(\mathbf{x})$ points into D^2 for every $\mathbf{x}$ in ∂D^2. Then the following theorems generalise Propositions 3.1.1 and 3.1.2 for linear vector fields.

Theorem 3.3.1 (Peixoto) *Let* $\mathbf{X}$ *belong to* $\mathrm{Vec}^1_{\mathrm{in}}(D^2)$. *Then* $\mathbf{X}$ *is structurally stable if and only if its flow satisfies*:

(i) *all fixed points are hyperbolic*;
(ii) *all closed orbits are hyperbolic*;
(iii) *there are no orbits connecting saddle points.*

Notice that items (i) and (ii) in Theorem 3.3.1 simply ensure local structural stability of the fixed points and closed orbits in the flow of $\mathbf{X}$. It is really only item (iii) that involves a global property of the flow.

Theorem 3.3.2 *The subset of vector fields in* $\mathrm{Vec}^1_{\mathrm{in}}(D^2)$ *that are structurally stable is open and dense in* $\mathrm{Vec}^1_{\mathrm{in}}(D^2)$.

Obviously, parallel results could be stated for those vector fields that point out of D^2 at every point of ∂D^2. Indeed, the flows of vector fields in $\mathrm{Vec}^1_{\mathrm{in}}(D^2)$ and $\mathrm{Vec}^1_{\mathrm{out}}(D^2)$ are in 1:1 correspondence by time reversal.

By stereographic projection (see Figure 3.4(*a*)), the unit disc D^2 is homeomorphic to a closed cap, C^2, based on the south pole of the sphere S^2. The boundary condition on ∂D^2 can then be replaced by considering flows on S^2 with a single repelling fixed point in $S^2\backslash C^2$. This is conveniently placed at the north pole (see Figure 3.4(*b*)). Provided that this additional fixed point is hyperbolic, a structurally stable vector field on D^2 is also structurally stable on S^2. This follows from the work of Peixoto (1962) who extended Theorems 3.3.1 and 3.3.2 to vector fields on two-dimensional, compact manifolds. Let M be a two-dimensional, compact manifold without boundary and let $\mathrm{Vec}^1(M)$ be the set of C^1-vector fields on M

Figure 3.3 The vector field in $\mathrm{Vec}^1(D^2)$ with a tangency at $\mathbf{x}_0 \in \partial D^2$ in (*a*) is not structurally stable as the perturbation (*b*) shows. In (*b*), an orbit leaves D^2 with increasing time.

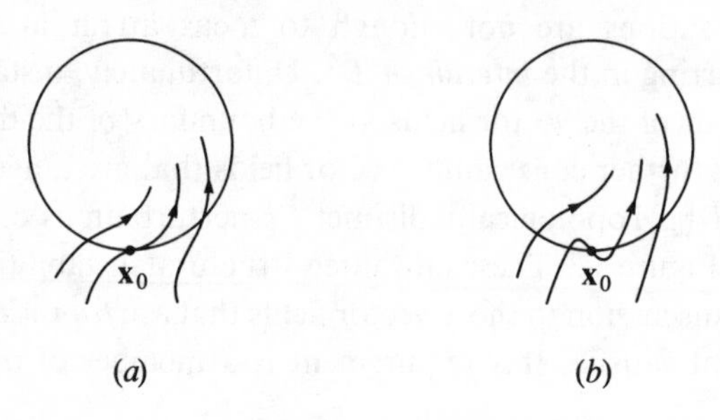

with the C^1-norm. This norm is defined by imposing the C^1-norm on each of the charts of a finite atlas for M. Then, Peixoto's result may be stated as follows.

Theorem 3.3.3 (Peixoto) *A vector field in* $\mathrm{Vec}^1(M)$ *is structurally stable if and only if its flow satisfies*:

(i) *all fixed points are hyperbolic*;
(ii) *all closed orbits are hyperbolic*;
(iii) *there are no orbits connecting saddle points*;
(iv) *the non-wandering set consists only of fixed points and periodic orbits.*

Moreover, if M is orientable the set of structurally stable C^1-vector fields forms an open dense subset of $\mathrm{Vec}^1(M)$.

Here orientable simply means that two distinct sides of M can be recognised. The sphere, torus, pretzel, etc. are all examples of orientable manifolds.

The statement of Theorem 3.3.3 contains an additional condition (iv) that does not appear in Theorem 3.3.1. Its role can be illustrated as follows. Consider the irrational flow on the torus T^2. This flow satisfies items (i)–(iii) vacuously. There are no fixed points, closed orbits or saddle connections. However, it fails to satisfy item (iv) because the non-wandering set is the whole of T^2. Therefore, the irrational flow is not structurally stable on T^2. Clearly, there exist ε–C^1-close rational flows for which every orbit is closed. Some examples of structurally stable flows on S^2 and T^2 are shown in Figure 3.5.

It should perhaps be noted that, since M is compact, flows on it can only have finitely many fixed and periodic points if they are all hyperbolic. By the Hartman

Figure 3.4 (*a*) The stereographic projection from the north pole N of the sphere. The circle $\mathscr{C}$ projects onto the boundary of D^2. (*b*) A vector field in a neighbourhood of N which cuts $\mathscr{C}$ transversely.

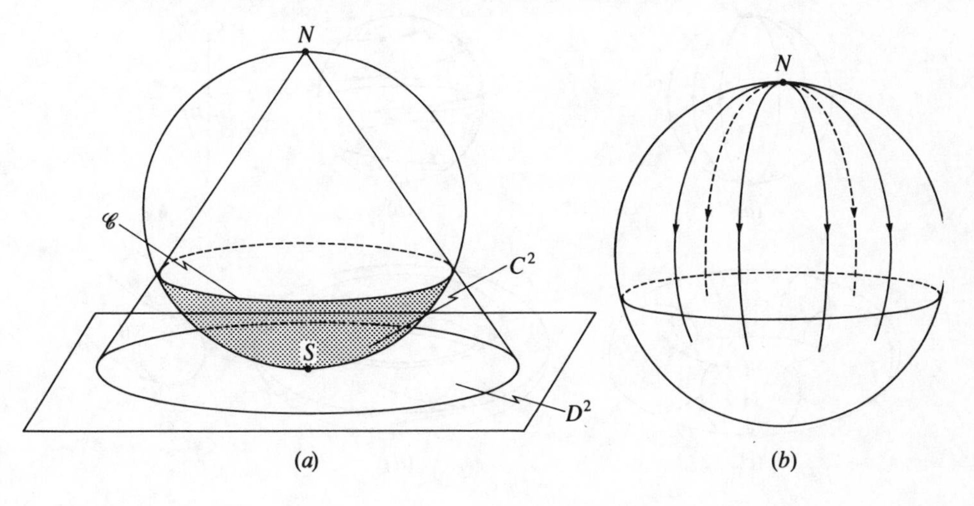

Theorem, the flow in the vicinity of a hyperbolic fixed point is topologically conjugate to that of its linearisation. The latter has an isolated fixed point at the origin, hence hyperbolic points must be isolated. This means that fixed points cannot accumulate at a hyperbolic fixed point. Consequently, only a finite number of fixed points can occur on a compact manifold if they are *all* hyperbolic.

We might try to extend the discussion of structural stability to non-compact sets such as the whole plane. Of course, a finite C^1-norm can no longer be guaranteed but an obvious approach is to consider restrictions to compact subsets. Clearly, if a vector field is structurally unstable on any compact subset of the plane then it can be deemed structurally unstable on the whole plane. This approach does have practical merit since in applications we are rarely involved with variables passing to infinity, rather they become very large but finite. The following example illustrates this idea.

Example 3.3.1 Show that the vector field, **X**, of the differential equation

$$\dot{x} = 2x - x^2, \qquad \dot{y} = -y + xy \tag{3.3.2}$$

is *not* structurally stable on any compact subset of the plane with the line segment joining the singular points of **X** in its interior.

Solution. The system (3.3.2) has saddle points at $\mathbf{x}^* = (0, 0)$ and $\mathbf{y}^* = (2, 0)$. On the x-axis, $\dot{y} = 0$ and so there is an orbit connecting these hyperbolic fixed points. The phase portrait is shown in Figure 3.6(a). Let D be any compact set containing the common separatrix of $\mathbf{x}^*$ and $\mathbf{y}^*$. Notice that the stable manifold of $\mathbf{x}^*$ is the line $x = 0$ while the line $x = 2$ is the unstable manifold for $\mathbf{y}^*$. This means that

Figure 3.5 Examples of structurally stable phase portraits on the sphere and torus. (a) $\dot{\theta} = \sin\theta$, $\dot{\varphi} = 0$; (b) $\dot{\theta} = \sin 2\theta$, $\dot{\varphi} = 1$; (c) $\dot{\theta} = 1$, $\dot{\varphi} = \sin\varphi$; ($d$) $\dot{\theta} = \sin 2\theta$, $\dot{\varphi} = 1$.

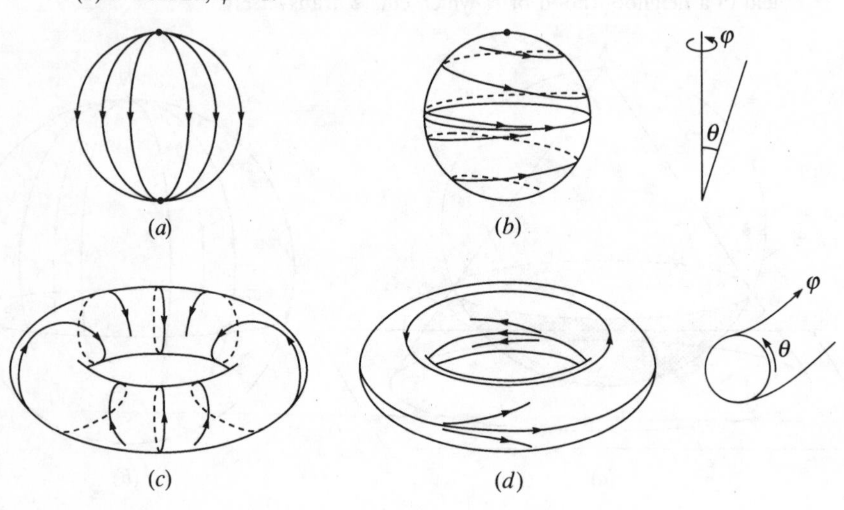

there are points $\mathbf{x}$ in the boundary of D for which $\mathbf{X}(\mathbf{x})$ points both into and out of D. Therefore, we are (technically) *not* able to apply Theorem 3.3.1. However, consider the one-parameter family of systems

$$\dot{x} = 2x - x^2, \qquad \dot{y} = -y + xy + \mu(2x - x^2). \tag{3.3.3}$$

The vector field of (3.3.3) can be made ε–C^1-close to that of (3.3.2) on *any* compact subset D of the type described above by taking μ sufficiently small. The phase portrait for (3.3.3) with $\mu > 0$ is shown in Figure 3.6(b). The points (0, 0) and (2, 0) are saddle points for all real μ. However, for non-zero μ, $\dot{y} \neq 0$ on the x-axis between these points. Moreover, the stable separatrix at (2, 0) is tangent to $y = \frac{2}{3}\mu x$. Therefore, for $\mu > 0$ there is no saddle connection between the fixed points. Hence, the flows for $\mu = 0$ and $\mu > 0$ are topologically distinct. The vector field in (3.3.2) is consequently structurally unstable on every compact subset of the plane containing the saddle connection. □

It is perhaps worth noting that, in view of the role played by the boundary condition on ∂D^2 in Theorem 3.3.1, we can say that a vector field that fails to satisfy the conditions (i)–(iii) will certainly be structurally unstable independently of the boundary condition on ∂D^2. Of course, the converse is not true unless the vector field is transverse to the boundary.

Example 3.3.1 illustrates a useful way of giving meaning to structural instability on the plane. Some other examples are shown in Figure 3.7. However, we must not be misled into believing that if a vector field is structurally stable on arbitrarily large compact subsets of the plane then it is structurally stable on the whole plane. The following is a counterexample to this erroneous conjecture.

Example 3.3.2 Show that there are arbitrarily large compact subsets of the plane on which the system

$$\dot{x} = -x, \qquad \dot{y} = \sin(\pi y)\exp(-y^2) \tag{3.3.4}$$

Figure 3.6 Phase portraits for (a) (3.3.2), (b) (3.3.3) with $\mu > 0$.

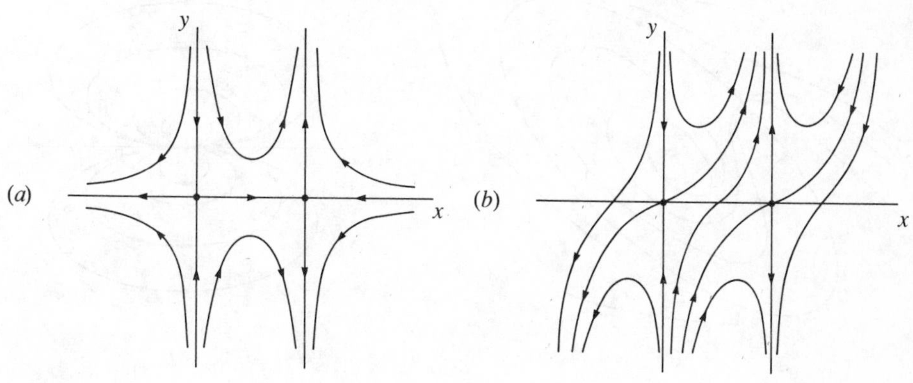

is structurally stable. Verify that the topological type of (3.3.4) on the whole plane is changed by the addition of the perturbation $(0, \mu)$ to $(\dot{x}, \dot{y})$, however small the value of $\mu \neq 0$.

Solution. Figure 3.8(*a*) shows $\dot{y}$ as a function of y. It follows that (3.3.4) has fixed points at $(\dot{y}, y) = (0, p)$, $p \in \mathbb{Z}$. These fixed points are alternately stable nodes and saddle points as shown in Figure 3.8(*b*). Suppose D is a compact subset of $\mathbb{R}^2$ whose boundary intersects the y-axis at $(0, y_l)$, $(0, y_u)$ with $y_l < y_u$ and $y_l, y_u \neq p$ for any integer p. If D is such that $\dot{y}(y_l) > 0$ and $\dot{y}(y_u) < 0$ then (3.3.4) is structurally stable on D by Theorem 3.3.1. Clearly, arbitrarily large D can be constructed in this way.

Consider the family of vector fields defined by $\dot{\mathbf{x}} = \mathbf{X}_\mu(\mathbf{x})$, where

$$\dot{x} = -x, \qquad \dot{y} = \sin(\pi y)\exp(-y^2) + \mu. \tag{3.3.5}$$

Figure 3.7 Some examples of structurally unstable phase portraits on $\mathbb{R}^2$. In each case the structural instability is apparent in the restriction of the flow to a compact subset of the plane.

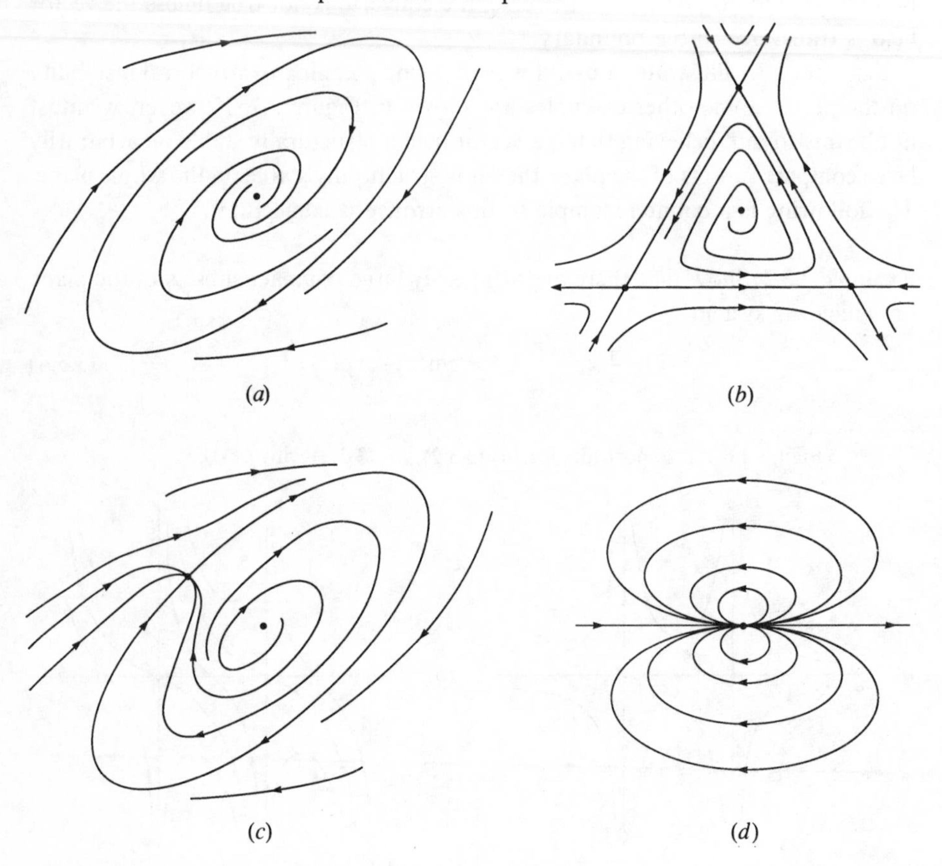

The vector field of (3.3.4) is $\mathbf{X}_0$ and

$$\|\mathbf{X}_0 - \mathbf{X}_\mu\|_1 = \sup_{\mathbb{R}^2}\left(\sum_{i=1}^{2} |X_0^i - X_\mu^i| + \sum_{i,j=1}^{2} \left| \frac{\partial X_0^i}{\partial x_j} - \frac{\partial X_\mu^i}{\partial x_j} \right| \right) = \mu. \tag{3.3.6}$$

Thus $\mathbf{X}_\mu$ can be made ε–C^1-close to $\mathbf{X}_0$ on the whole plane. However, $\mathbf{X}_\mu$, $\mu \neq 0$, has only finitely many fixed points (see the broken curve on Figure 3.8(*a*)). Hence, the flow of (3.3.5) is topologically distinct from that of (3.3.4) for any non-zero value of μ. □

Notice that the topological type of the flow of $\mathbf{X}_0$ in Example 3.3.2 changes when μ departs from zero, even though it has only hyperbolic fixed points and no saddle connections. Since the plane is not compact, infinitely many hyperbolic fixed points can occur. If we wish to maintain contact with Theorem 3.3.1, we must specify a boundary condition at infinity. However, if we require that all vector fields point inward on the boundaries of all sufficiently large discs, we are led back to Theorem 3.3.3 via stereographic projection. In other words, we can obtain structurally stable flows on $\mathbb{R}^2$ from structurally stable flows on S^2, however, they will have only finitely many hyperbolic fixed points and their behaviour at infinity will correspond to having a hyperbolic fixed point at the north pole of the sphere (see Figure 3.9).

Figure 3.8 Phase portrait of (3.3.4) in relation to plot of $\dot{y}$ as a function of y. All zeroes of $\dot{y}$ are simple. The fixed point (x^*, y^*) is a saddle if $\mathrm{d}\dot{y}/\mathrm{d}y|_{y^*} > 0$ and a node if $\mathrm{d}\dot{y}/\mathrm{d}y|_{y^*} < 0$. The broken curve in (*a*) is a plot of $\dot{y}$ versus y for (3.3.5). The decay of the amplitude of the oscillations in $\dot{y}$ means that there are no fixed points for $\exp(-y^2) < |\mu|$, i.e. for $y^2 > -\ln(|\mu|)$.

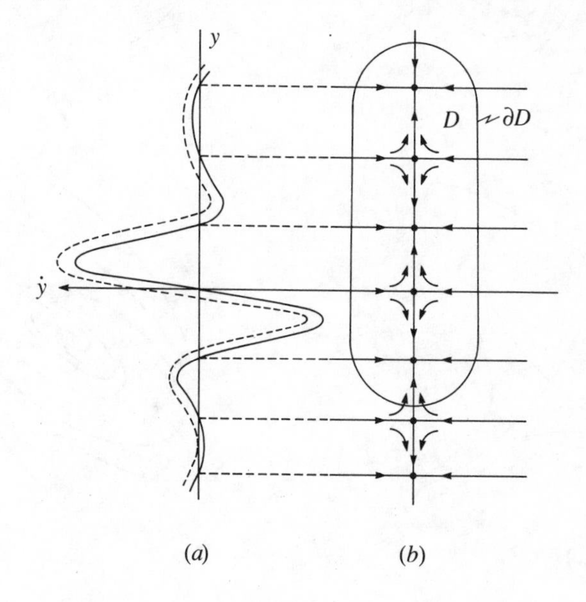

3.4 Anosov diffeomorphisms

The relationship between flows on n-manifolds with global sections and diffeomorphisms on manifolds of dimension $n-1$ (via Poincaré maps) was discussed in § 1.7. It is not surprising, therefore, that there is an analogous result to Theorem 3.3.3 for diffeomorphisms on a compact 1-manifold without boundary. Topologically, there is only one such connected manifold; namely the circle, S^1.

Let $\mathrm{Diff}^1(S^1)$ be the space of orientation-preserving C^1-diffeomorphisms on S^1 with the C^1-norm. Then Peixoto's theorem for diffeomorphisms on the circle can be stated as follows.

Theorem 3.4.1 (Peixoto) *A diffeomorphism $f \in \mathrm{Diff}^1(S^1)$ is structurally stable if and only if its non-wandering set consists of finitely many fixed points or periodic orbits*

Figure 3.9 Some examples of structurally stable phase portraits on $\mathbb{R}^2$ derived from flows on S^2. All fixed points and periodic orbits are assumed hyperbolic. To ensure stability at infinity the corresponding vector fields on S^2 have an unstable hyperbolic fixed point at the north pole (cf. Figure 3.4).

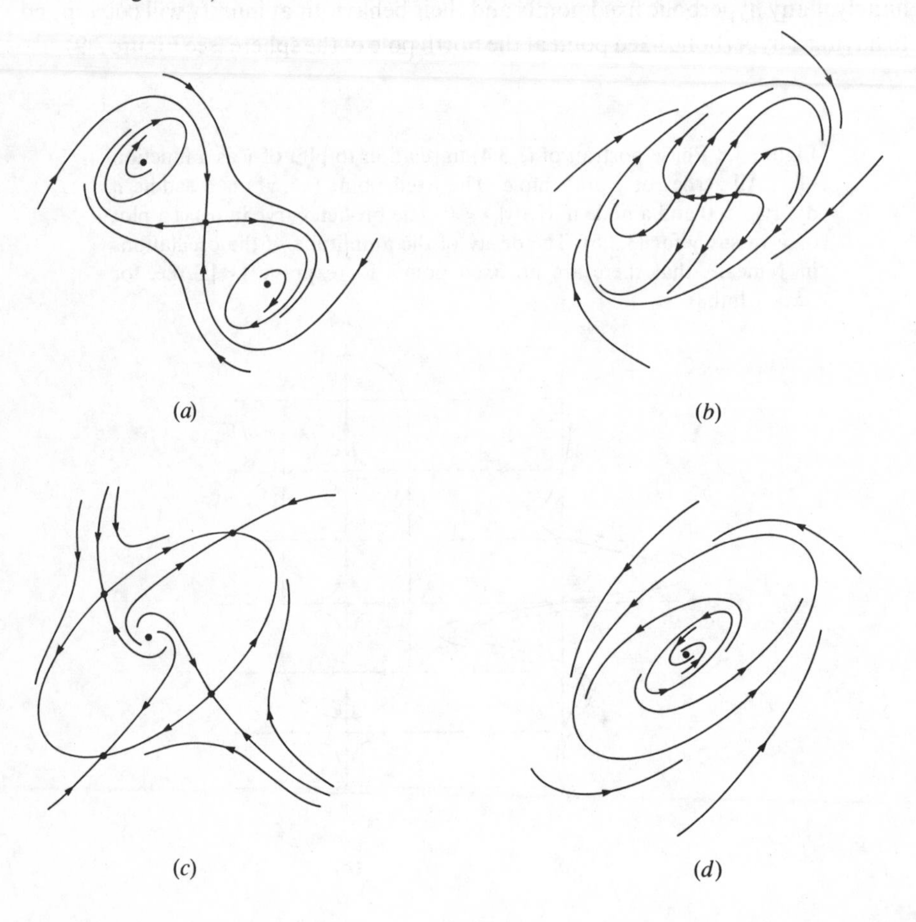

all of which are hyperbolic. Moreover, the structurally stable diffeomorphisms form an open dense subset of $\mathrm{Diff}^1(S^1)$.

Recall (see Proposition 1.5.1) that if f has periodic points then its rotation number $\rho(f)$ is rational. Therefore, the structurally stable diffeomorphisms on S^1 have rational rotation number (the converse of this statement is not true (see Exercise 3.4.1)). If f is structurally stable with rotation number $\rho(f) = p/q$, in lowest terms, then its dynamics are very simple. It has an even number of period-q cycles with stable and unstable periodic points alternating around the circle (see Figure 1.23).

It was hoped that generalisations of the behaviour described in Theorems 3.3.3 and 3.4.1 would not only characterise structurally stable systems in higher dimensions but would also prove to be generic. To this end, 'Morse–Smale' vector fields and diffeomorphisms were defined (Chillingworth, 1976, p. 231 and Nitecki, 1971, p. 88). Unfortunately, it was found that, while such systems were structurally stable, their properties did not characterise structurally stable systems in higher dimensions. In particular, it was shown that there were structurally stable diffeomorphisms on manifolds of dimension $n \geqslant 2$ (corresponding to vector fields on manifolds of dimension $n+1 \geqslant 3$, by suspension) whose non-wandering sets contained infinitely many periodic points. The *Anosov diffeomorphisms of the torus*, T^n, are a subset of $\mathrm{Diff}^1(T^n)$ that exhibit this behaviour. We can describe a diffeomorphism $\mathbf{f}$ on T^n in terms of a 'lift' in much the same way as we did for diffeomorphisms of the circle (see § 1.2.2). In this case, the lift $\bar{\mathbf{f}}$ is a diffeomorphism on $\mathbb{R}^n$ which satisfies

$$\mathbf{f}(\boldsymbol{\pi}(\mathbf{x})) = \boldsymbol{\pi}(\bar{\mathbf{f}}(\mathbf{x})), \tag{3.4.1}$$

for each $\mathbf{x} \in \mathbb{R}^n$, where $\boldsymbol{\pi}\colon \mathbb{R}^n \to T^n$ is given by

$$\begin{aligned} \boldsymbol{\pi}(\mathbf{x}) = \boldsymbol{\pi}((x_1, \ldots, x_n)^{\mathrm{T}}) &= (x_1 \bmod 1, \ldots, x_n \bmod 1)^{\mathrm{T}} \\ &= (\theta_1, \ldots, \theta_n)^{\mathrm{T}} = \boldsymbol{\theta} \end{aligned} \tag{3.4.2}$$

(see Figure 3.10). If $\mathbf{k} \in \mathbb{Z}^n$, then (3.4.1) implies

$$\boldsymbol{\pi}(\bar{\mathbf{f}}(\mathbf{x}+\mathbf{k})) = \mathbf{f}(\boldsymbol{\pi}(\mathbf{x}+\mathbf{k})) = \mathbf{f}(\boldsymbol{\pi}(\mathbf{x})) = \boldsymbol{\pi}(\bar{\mathbf{f}}(\mathbf{x})), \tag{3.4.3}$$

for all $\mathbf{x} \in \mathbb{R}^n$. Continuity of $\bar{\mathbf{f}}$ then gives,

$$\bar{\mathbf{f}}(\mathbf{x}+\mathbf{k}) = \bar{\mathbf{f}}(\mathbf{x}) + \mathbf{l}(\mathbf{k}), \tag{3.4.4}$$

where $\mathbf{l}(\mathbf{k}) \in \mathbb{Z}^n$ (see (1.2.10) *et seque*). All lifts of diffeomorphisms on T^n must satisfy (3.4.4).

We will begin by describing a special subset of the Anosov diffeomorphisms that are known as the *Anosov automorphisms*. The lift of an Anosov automorphism $\mathbf{f}$ is a hyperbolic, linear diffeomorphism, $\mathbf{A}\colon \mathbb{R}^n \to \mathbb{R}^n$, which is such that:

$$\mathbf{A}\colon \mathbb{Z}^n \to \mathbb{Z}^n; \tag{3.4.5a}$$

$$\operatorname{Det} \mathbf{A} = \pm 1. \tag{3.4.5b}$$

Together (3.4.5a and b) ensure that $\mathbf{A}^{-1}\colon \mathbb{Z}^n \to \mathbb{Z}^n$ and it follows that both $\mathbf{A}$ and $\mathbf{A}^{-1}$ satisfy (3.4.4). Moreover, given that $\mathbf{f}(\pi(\mathbf{x})) = \pi(\mathbf{A}\mathbf{x})$ and $\mathbf{g}(\pi(\mathbf{x})) = \pi(\mathbf{A}^{-1}\mathbf{x})$, then

$$\mathbf{f}(\mathbf{g}(\pi(\mathbf{x}))) = \mathbf{f}(\pi(\mathbf{A}^{-1}\mathbf{x})) = \pi(\mathbf{A}\mathbf{A}^{-1}\mathbf{x}) = \pi(\mathbf{x}) \tag{3.4.6}$$

and

$$\mathbf{g}(\mathbf{f}(\pi(\mathbf{x}))) = \mathbf{g}(\pi(\mathbf{A}\mathbf{x})) = \pi(\mathbf{A}^{-1}\mathbf{A}\mathbf{x}) = \pi(\mathbf{x}). \tag{3.4.7}$$

Thus, $\mathbf{f}^{-1}\colon T^n \to T^n$ exists and has lift $\mathbf{A}^{-1}$. Finally, observe that $\mathbf{f}(\mathbf{x})$ and $\mathbf{f}^{-1}(\mathbf{x})$ are differentiable because $\mathbf{A}\mathbf{x}$ and $\mathbf{A}^{-1}\mathbf{x}$ are obviously so and π is a local diffeomorphism. Hence $\mathbf{f}$ is a diffeomorphism on T^n.

Hyperbolic, linear diffeomorphisms on $\mathbb{R}^n$ are often referred to as *automorphisms* because they are isomorphisms of the group $\mathbb{R}^n$ with itself. The diffeomorphism $\mathbf{f}$ defined above is called an automorphism to distinguish it from other Anosov diffeomorphisms whose lift is not linear. A general definition of Anosov diffeomorphisms can be found in Arnold (1973, p. 126) or Nitecki (1971, p. 103) but, thanks to the following result due to Manning (1974), it will be sufficient for us to consider only the automorphisms here.

Theorem 3.4.2 *Every Anosov diffeomorphism* $\mathbf{f}$ *of* T^n, *such that* $\Omega(\mathbf{f}) = T^n$, *is topologically conjugate to some Anosov automorphism of* T^n.

Coupled with Theorem 3.4.2, the following result shows that such Anosov diffeomorphisms have complicated non-wandering sets.

Proposition 3.4.1 *A point* $\boldsymbol{\theta} \in T^n$ *is a periodic point of the Anosov automorphism* $\mathbf{f}\colon T^n \to T^n$ *if and only if* $\boldsymbol{\theta} = \pi(\mathbf{x})$, *where* $\mathbf{x} \in \mathbb{R}^n$ *has rational coordinates.*

Figure 3.10 (*a*) Angular coordinates on T^2 measured in units of 2π. (*b*) The map π identifies each point $(x_1, x_2) \in \mathbb{R}^2$ with a point $(\theta_1, \theta_2) \in \{(x_1, x_1) | \; 0 \leqslant x_1 < 1, 0 \leqslant x_2 < 1\}$ which in turn defines a unique point on T^2.

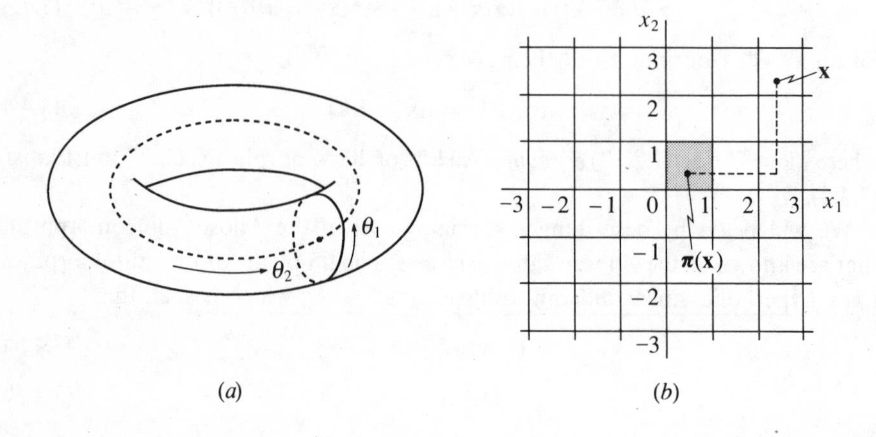

Proof. Let $\boldsymbol{\theta}$ be a periodic point of $\mathbf{f}$, then $\mathbf{f}^q(\boldsymbol{\theta}) = \boldsymbol{\theta}$, for some positive integer q. Suppose $\mathbf{x} \in \mathbb{R}^n$ satisfies $\pi(\mathbf{x}) = \boldsymbol{\theta}$, then

$$\pi(\mathbf{x}) = \boldsymbol{\theta} = \mathbf{f}^q(\boldsymbol{\theta}) = \mathbf{f}^q(\pi(\mathbf{x})) = \pi(\mathbf{A}^q\mathbf{x}). \tag{3.4.8}$$

This means that

$$(\mathbf{A}^q - \mathbf{I})\mathbf{x} = \mathbf{m}, \tag{3.4.9}$$

where $\mathbf{x} = (x_1, \ldots, x_n)^{\mathrm{T}}$ and $\mathbf{m} = (m_1, \ldots, m_n)^{\mathrm{T}}$, $m_i \in \mathbb{Z}$, $i = 1, \ldots, n$. Since $\mathbf{A}$ is a hyperbolic matrix, $(\mathbf{A}^q - \mathbf{I})^{-1}$ exists and (3.4.9) has solution

$$\mathbf{x} = (\mathbf{A}^q - \mathbf{I})^{-1}\mathbf{m}. \tag{3.4.10}$$

Now, $\mathbf{A}^q - \mathbf{I}$ is an integer matrix and therefore $(\mathbf{A}^q - \mathbf{I})^{-1}$ has rational elements. Hence $\mathbf{x}$ has rational coordinates.

Conversely, if $\boldsymbol{\theta} \in T^n$ has representative $\mathbf{x} = ((p_1^{(0)}/r), \ldots, (p_n^{(0)}/r))^{\mathrm{T}}$, where $p_i^{(0)}$, $r \in \mathbb{Z}$ with $r \neq 0$, then, for any $k \in \mathbb{Z}$,

$$\mathbf{A}^k\mathbf{x} = \left(\frac{p_1^{(k)}}{r}, \ldots, \frac{p_n^{(k)}}{r}\right)^{\mathrm{T}}, \tag{3.4.11}$$

for some integers $p_1^{(k)}, \ldots, p_n^{(k)}$. However, there are at most r^n points on T^n that can be represented in this way and, therefore, there is a $q > 0$ such that $\pi(\mathbf{A}^q\mathbf{x}) = \pi(\mathbf{x})$. □

Proposition 3.4.1 not only implies that $\mathbf{f}$ has infinitely many periodic points; it also shows that the periodic points are dense in the torus. All these points lie in the non-wandering set Ω of $\mathbf{f}$ and, since Ω is closed (see Exercise 1.4.2), we conclude that $\Omega = T^n$.

The final piece of the argument against Morse–Smale systems was provided by Mather (1967).

Theorem 3.4.3 (Mather) *The Anosov diffeomorphisms on T^n are structurally stable in* $\mathrm{Diff}^1(T^n)$.

Thus, the Anosov diffeomorphisms on T^n, $n \geqslant 2$, are examples of structurally stable diffeomorphisms on a compact manifold whose non-wandering set contains infinitely many points. What is more, the dynamics on Ω is very complicated involving infinitely many periodic orbits densely distributed over the torus. In fact, every periodic point is hyperbolic (see Exercise 3.4.2). Since T^n is compact, there can only be finitely many such points with a given period q on the torus. However, periodic points of infinitely many periods can be shown to occur (see Exercise 3.4.3) making up the infinite set predicted by Proposition 3.4.1. To gain some insight into how all this complexity arises let us consider the following well quoted example (see Arnold, 1983; Arnold & Avez, 1968).

Let $\mathbf{A}: \mathbb{R}^2 \to \mathbb{R}^2$ be given by

$$\mathbf{A} = \begin{pmatrix} 1 & 1 \\ 1 & 2 \end{pmatrix}. \tag{3.4.12}$$

It is easily verified that $\mathbf{A}$ satisfies (3.4.5). The behaviour of $\mathbf{A}$ on $\mathbb{R}^2$ is simple: it has a saddle point at $\mathbf{x} = \mathbf{0}$ with stable and unstable eigenspaces given by the straight lines

$$y = \left(\frac{1 - 5^{1/2}}{2}\right)x \quad \text{and} \quad y = \left(\frac{1 + 5^{1/2}}{2}\right)x, \tag{3.4.13}$$

respectively. The complexity arises when it is mapped down onto T^2. Forward iterations of $\mathbf{A}$ on $\mathbb{R}^2$ have the effect of contracting and expanding along the two perpendicular directions in (3.4.13) as shown in Figure 3.11(*a*). The unit square $B_1 = \{(x, y) | 0 \leqslant x < 1, 0 \leqslant y < 1\}$ is mapped onto thinner and thinner parallelograms (Figure 3.11(*b*) and (*c*)). The slopes of the longer diagonal of these parallelograms are rational but they approach the irrational $\frac{1}{2}(1 + 5^{1/2})$ for large numbers of iterations. When the points in these images of B_1 under $\mathbf{A}$ are identified with points on T^2 (see Figure 3.11(*b*) and (*c*)), it can be seen that repeated application of $\mathbf{f}$ has the effect of distributing any subset of T^2 more and more evenly over the whole torus. An alternative way of seeing this is to recognise that, for any $\mathbf{x} \in \mathbb{R}^2$, $\mathbf{A}^N\mathbf{x}$ can be made arbitrarily close to the line $y = \frac{1}{2}(1 + 5^{1/2})x$ by taking N to be sufficiently large. Since $\frac{1}{2}(1 + 5^{1/2})$ is irrational, this straight line represents a curve, W^{u}, that winds densely around the torus (see Figure 3.12).

The stable eigenspace of the saddle point in Figure 3.11(*a*) also corresponds to a densely wound curve, W^{s}, on T^2. The key to the complexity of the dynamics of $\mathbf{f}$ lies in the fact that these stable and unstable manifolds intersect in a dense set of transverse homoclinic points (see Figure 3.12). A *homoclinic point* is one that lies in both the stable and the unstable manifolds of a fixed or periodic point. Such points are said to be *transverse* if they arise from a transverse, rather than a tangential, intersection of the manifolds. Observe that if $\boldsymbol{\theta}^\dagger$ is a homoclinic point, i.e. $\boldsymbol{\theta}^\dagger \in W^{\mathrm{s}} \cap W^{\mathrm{u}}$, then $\mathbf{f}(\boldsymbol{\theta}^\dagger) \in W^{\mathrm{s}} \cap W^{\mathrm{u}}$ because $\boldsymbol{\theta}^\dagger \in W^{\mathrm{s,u}}$. Hence $\mathbf{f}(\boldsymbol{\theta}^\dagger)$ is a homoclinic point. Thus the dynamics of these homoclinic points is confined to the dense set of intersection points of W^{s} and W^{u}.

The reader must not confuse these homoclinic points with the periodic points of $\mathbf{f}$. Recall, by Proposition 3.4.1, periodic points of $\mathbf{f}$ have representatives $\mathbf{x} \in \mathbb{R}^2$ with rational coordinates. However, the stable and unstable eigenspaces of $\mathbf{A}$ have irrational slope (see (3.4.13)) and, apart from $\mathbf{x} = \mathbf{0}$, no point on them has rational coordinates. We shall see later (see §3.7) that the occurrence of transverse homoclinic points is indicative of complicated dynamical behaviour.

It is also worth noting that $\mathbf{A}^q$ is a lift of $\mathbf{f}^q$ for any q. Thus if $\mathbf{f}$ has a periodic point $\boldsymbol{\theta}^*$ of period q, then its stable and unstable manifolds are densely wound curves on T^2 parallel to W^{s} and W^{u}, respectively. This is because, for each $\mathbf{x}^*$

Figure 3.11 Illustration of the toral automorphism given by (3.4.12): (*a*) linear saddle point of $\mathbf{A} = \begin{pmatrix} 1 & 1 \\ 1 & 2 \end{pmatrix}$; (*b*) image $\mathbf{A}(B_1)$ of the unit square B_1 under $\mathbf{A}$; (*c*) image $\mathbf{A}^2(B_1)$. The shading in (*b*) and (*c*) indicates how $\pi(\mathbf{A}(B_1))$ defines $\mathbf{f}\colon T^2 \to T^2$ and $\pi(\mathbf{A}^2(B_1))$ gives $\mathbf{f}^2\colon T^2 \to T^2$.

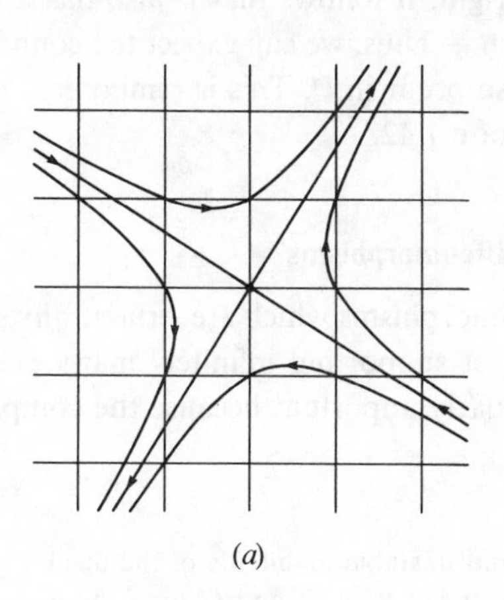

(*a*)

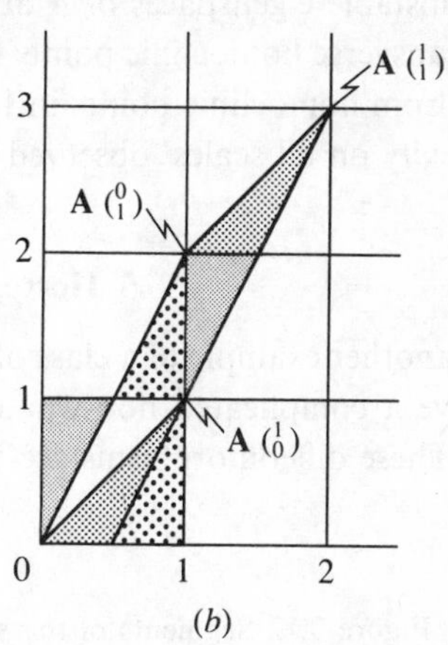

(*b*)

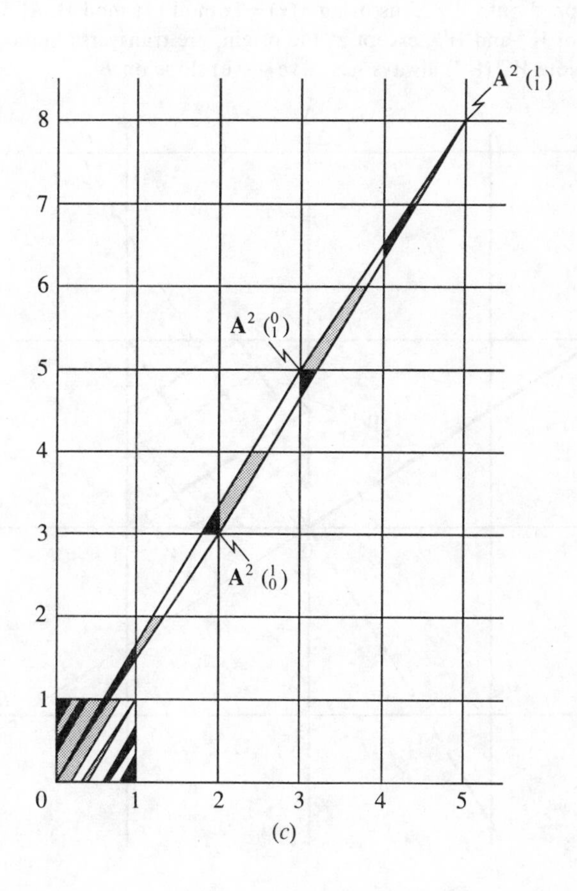

(*c*)

satisfying $\boldsymbol{\theta}^* = \pi(\mathbf{x}^*)$, we can write

$$\mathbf{A}^q(\mathbf{x}) = \mathbf{A}^q(\mathbf{x}^*) + \mathbf{A}^q(\mathbf{x} - \mathbf{x}^*). \tag{3.4.14}$$

Thus, the stable/unstable eigendirections at $\mathbf{x}^*$ are given by *translates* of the stable/unstable eigenspaces of $\mathbf{A}$ at the origin. It follows that $\mathbf{f}^q$ also has a dense set of transverse homoclinic points for each q. Thus, we can expect the complexity arising from homoclinic points in $\mathbf{f}$, to also occur in $\mathbf{f}^q$. This is reminiscent of the 'complexity on all scales' observed in Figure 1.42.

3.5 Horseshoe diffeomorphisms

This is another example of a class of diffeomorphisms which are structurally stable and have a complicated non-wandering set supporting infinitely many periodic orbits. These diffeomorphisms are particularly important because the complexity

Figure 3.12 Segments of the stable and unstable manifolds of the fixed point, $\pi(\mathbf{0})$, of $\mathbf{f}$ on T^2. Segments AB and $A'B'$ of E^u and E^s for $\mathbf{A}$ have been mapped onto the torus using $\pi(\mathbf{x}) = (x \bmod 1, y \bmod 1)$. All intersections of W^s and W^u, except at the origin, are transverse homoclinic points. Note W^u (W^s) always has +ve (−ve) slope on B_1.

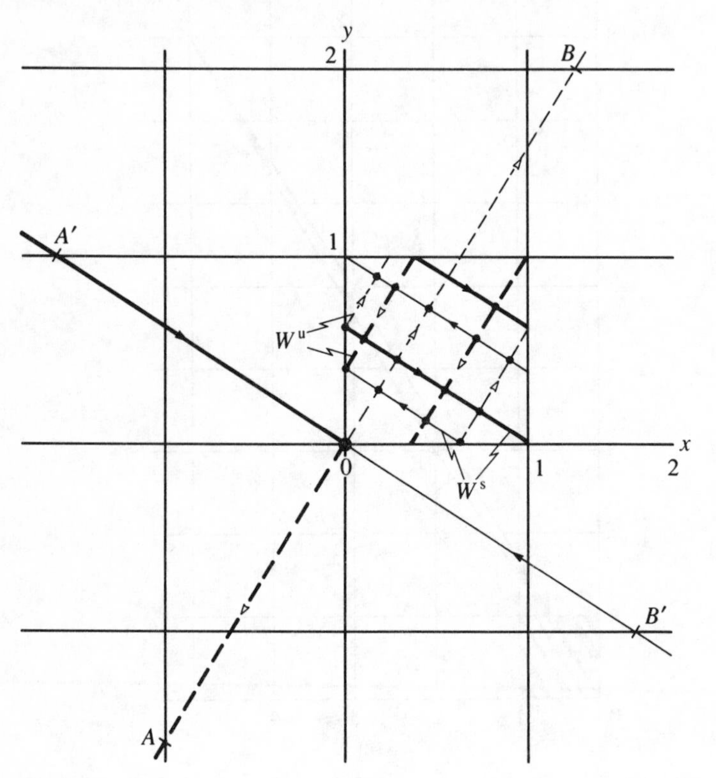

exhibited by them can be shown to occur in any map that has transverse homoclinic points (see § 3.7).

3.5.1 The canonical example

Consider a diffeomorphism $\mathbf{f}: Q \to \mathbb{R}^2$, where $Q = \{(x, y) \mid |x|, |y| \leqslant 1\}$, that is constructed in the following way. Each point $(x, y) \in Q$ is first mapped to $(5x, y/5)$ and Q is mapped onto the rectangular region $R = \{(x, y) \mid |x| \leqslant 5, |y| \leqslant 1/5\}$. This region can be divided into fifths by the lines $|x| = 1$ and 3. The map $\mathbf{f}$ is completed by bending the central fifth of the rectangle and placing the resulting horseshoe-shaped region on the plane in such a way that its second and fourth fifths intersect with Q in Q_0 and Q_1 as shown in Figure 3.13(*a*). Observe that, if P_0, P_1 denote the pre-images of Q_0, Q_1, respectively (i.e. $\mathbf{f}(P_i) = Q_i$, $i = 0, 1$; see Figure 3.13(*b*)), then $\mathbf{f} \mid P_i$ is linear for $i = 0, 1$ (see Exercise 3.5.1).

We can show that $\mathbf{f}$ has a complicated invariant set by considering the sequence of subsets of Q defined inductively by

$$Q^{(n+1)} = \mathbf{f}(Q^{(n)}) \cap Q, \tag{3.5.1}$$

$n \in \mathbb{Z}^+$, with $Q^{(1)} = Q_0 \cup Q_1$. It is not difficult to see that $Q^{(2)} = \mathbf{f}(Q_0 \cup Q_1) \cap Q$ consists of four horizontal strips lying inside $Q_0 \cup Q_1$ (see Figure 3.14(*a*)). Moreover, it is apparent that $Q^{(1)} \supset Q^{(2)} \supset \cdots \supset Q^{(n)} \supset \cdots$ and $Q^{(n)}$ consists of 2^n horizontal strips (see Figure 3.14(*b*)). If we consider the intersections of the sets $Q^{(n)}$ with the y-axis then the relationship between the resulting subintervals for successive values of n is easily recognised as a prescription for the construction of a Cantor set. It follows that the intersection $\bigcap_{n \in \mathbb{Z}^+} Q^{(n)}$ is a Cartesian product of an interval in x with a Cantor set of y-values. In a similar way, iterations of the inverse of $\mathbf{f}$ can be used to obtain an analogous set of vertical strips. Some care is needed here because $\mathbf{f}^{-1} \mid Q$ is only defined on the subset $Q^{(1)} = Q_0 \cup Q_1$ of Q. To avoid this difficulty, we take

$$Q^{(0)} = P_0 \cup P_1 = \mathbf{f}^{-1}(Q^{(1)}) \tag{3.5.2}$$

and define

$$Q^{(-n)} = \mathbf{f}^{-1}(Q^{(-(n-1))} \cap Q^{(1)}) = \mathbf{f}^{-1}(Q^{(-(n-1))} \cap \mathbf{f}(Q)), \tag{3.5.3}$$

for $n \in \mathbb{Z}^+$. The latter equality in (3.5.3) follows because $Q^{(-(n-1))}$ is a subset of the two vertical strips $Q^{(0)} = P_0 \cup P_1$ for all $n \in \mathbb{Z}^+$. Thus, for each $n \in \mathbb{Z}^+$, the intersection of $Q^{(-(n-1))}$ with $Q^{(1)} = Q_0 \cup Q_1$ is the same as its intersection with the whole horseshoe $\mathbf{f}(Q)$. The map $\mathbf{f}^{-1}$ stretches $Q^{(-(n-1))} \cap Q^{(1)}$ linearly by a factor of five in the y-direction, contracts it by five in the x-direction and replaces it on the square as shown in Figure 3.15(*a,b*) for $n = 1, 2$, respectively. The sets $Q^{(0)}, Q^{(-1)}, Q^{(-2)}, \ldots$ then have 2, 4, 8, ... vertical strips, respectively (see Figure 3.15(*c*)) and $\bigcap_{n \in \mathbb{N}} Q^{(-n)}$ is the Cartesian product of an interval in y with a Cantor

Figure 3.13 (*a*) Construction of **f** on $Q = ABCD$, showing rectangular region R divided into fifths labelled from left to right. (*b*) Pre-images P_0, P_1 of Q_0, Q_1, respectively, are vertical strips consisting of the second and fourth fifths of $Q = ABCD$.

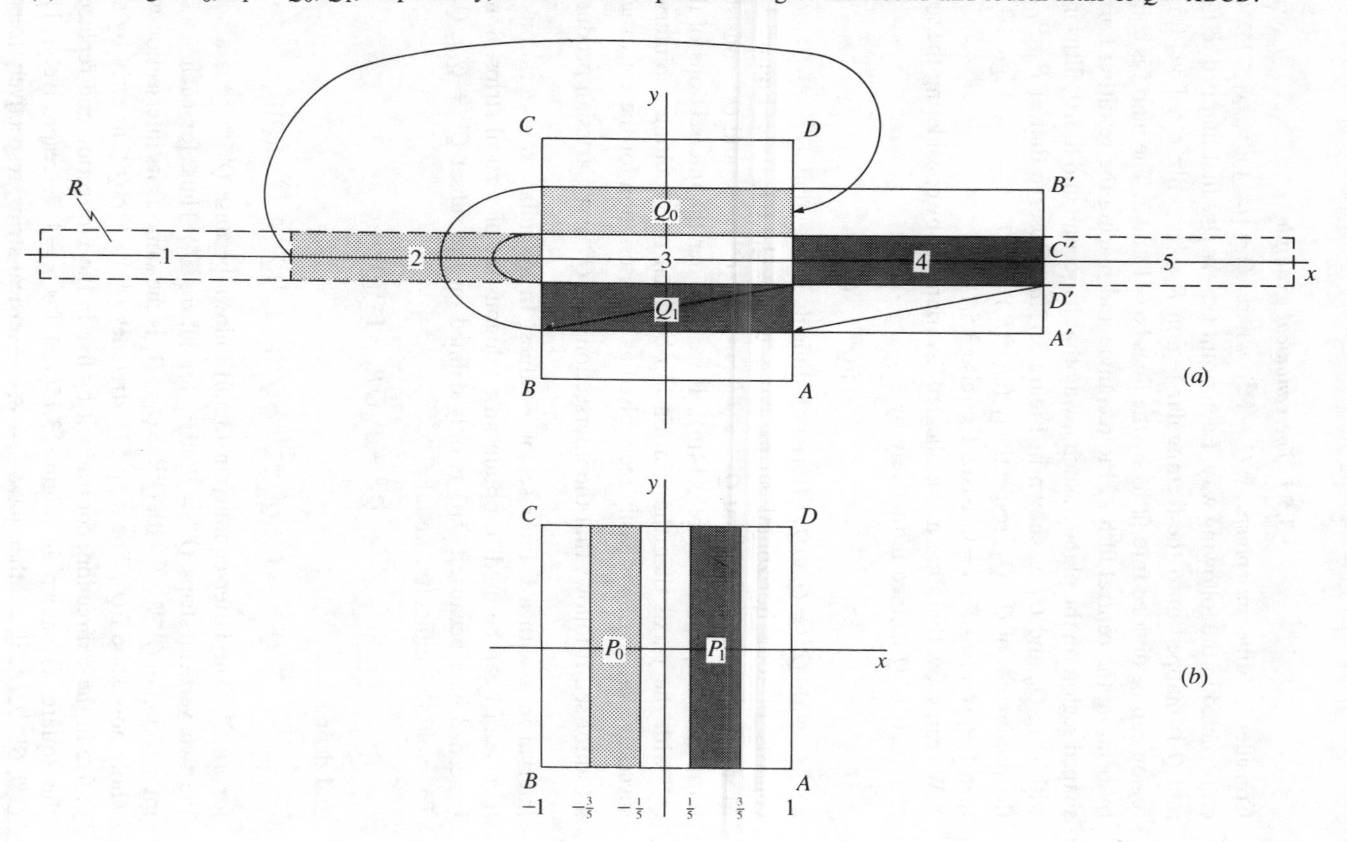

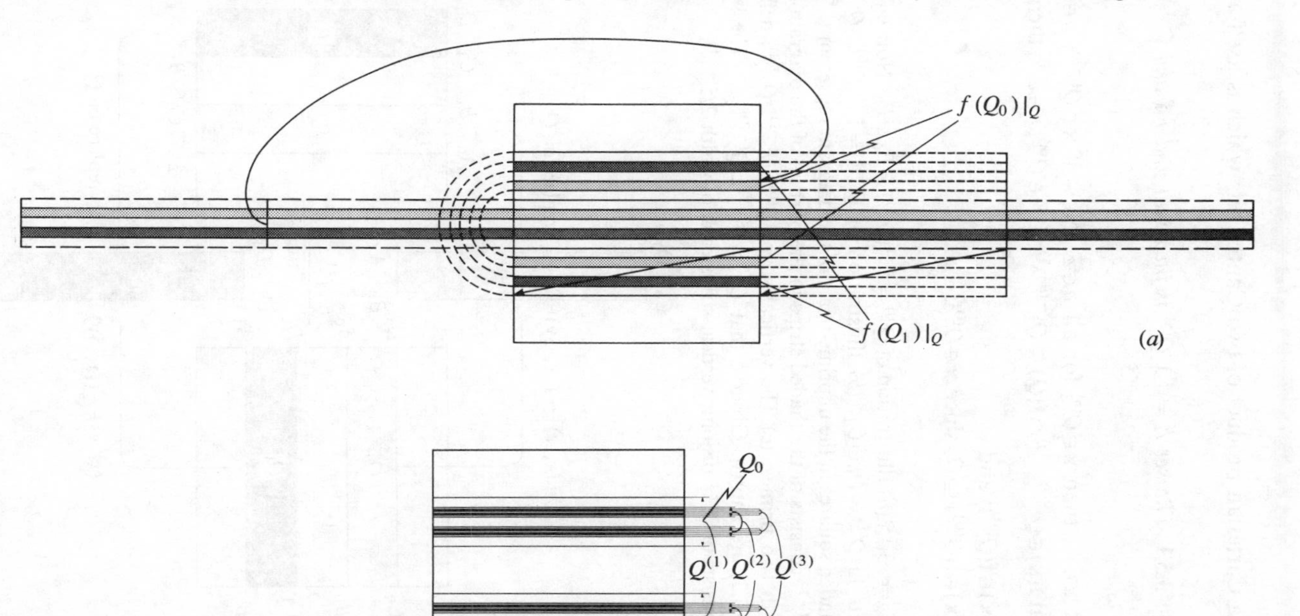

Figure 3.14 (*a*) Construction of $\mathbf{f}(Q_0 \cup Q_1)$: stretching/contraction yields two strips within the rectangle R; folding yields four horizontal strips for $Q^{(2)} = \mathbf{f}(Q^{(1)}) \cap Q$. Images of Q_0 (Q_1) at each stage are shown. (*b*) Illustration of $Q^{(n)}$ for $n = 1, 2, 3$. $Q^{(n)}$ consists of 2^n disjoint horizontal strips whose width, $2/5^n$, rapidly decreases with increasing n; indeed, the eight strips of $Q^{(3)}$ (shown in black) are barely resolved in this diagram.

set of x-values. If we now define

$$\Lambda = \bigcap_{n \in \mathbb{Z}} Q^{(n)} \tag{3.5.4}$$

then Λ is a Cartesian product of two Cantor sets which is itself a Cantor set.

Proposition 3.5.1 *The set* $\Lambda = \bigcap_{n \in \mathbb{Z}} Q^{(n)}$ *is invariant under* $\mathbf{f}$ *and* $\mathbf{f}^{-1}$.

Proof. Let $\mathbf{x} \in \Lambda$, then $\mathbf{x} \in Q^{(n)}$ for all $n \in \mathbb{Z}$. Now if $\mathbf{x} \in Q^{(-n)}$, $n \in \mathbb{N}$, then (3.5.3) implies that $\mathbf{f}(\mathbf{x}) \in Q^{(-(n-1))} \cap \mathbf{f}(Q) \subseteq Q^{(-(n-1))}$. If $\mathbf{x} \in Q^{(n)}$, $n \in \mathbb{Z}^+$, then observe that:

(i) $\mathbf{f}(\mathbf{x}) \in \mathbf{f}(Q^{(n)})$; and
(ii) $\mathbf{f}(\mathbf{x}) \in \mathbf{f}(Q^{(0)}) \subseteq Q$, since $\mathbf{x} \in Q^{(0)}$.

Figure 3.15 (*a*) Illustration of $Q^{(0)} = \mathbf{f}^{-1}(Q^{(1)}) = P_0 \cup P_1$. Notice $\mathbf{f}^{-1}$ only acts in $Q^{(1)} = Q_0 \cup Q_1$. (*b*) Illustration of $Q^{(-1)} = \mathbf{f}^{-1}(Q^{(0)} \cap Q^{(1)})$. The shaded squares in the unprimed part of the diagram represent $Q^{(0)} \cap Q^{(1)}$. $Q^{(-1)}$ consists of the shaded strips in the portion of the diagram labelled with two primes. (*c*) The vertical strips $Q^{(0)}$, $Q^{(-1)}$, $Q^{(-2)}$, defined by (3.5.3) are shown. Observe that $Q^{(0)} \supset Q^{(-1)} \supset Q^{(-2)}, \ldots$. The set $Q^{(-n)}$ consists of 2^{n+1} disjoint vertical strips each of width $2/5^{(n+1)}$.

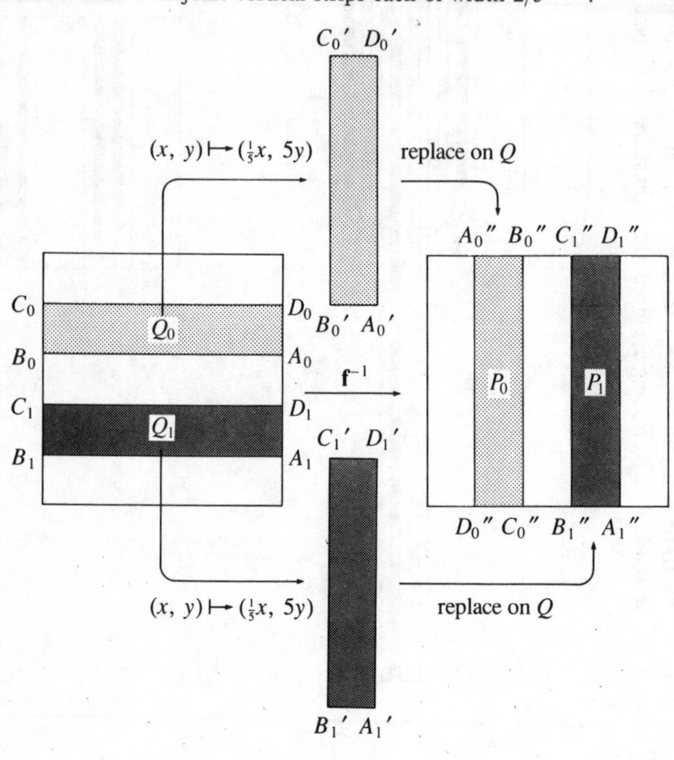

(*a*)

Together (i) and (ii) imply $\mathbf{f}(\mathbf{x}) \in \mathbf{f}(Q^{(n)}) \cap Q = Q^{(n+1)}$. Thus, if $\mathbf{x} \in Q^{(n)}$, for all $n \in \mathbb{Z}$, then $\mathbf{f}(\mathbf{x}) \in Q^{(n+1)}$ for all $n \in \mathbb{Z}$. Hence $\mathbf{f}(\mathbf{x}) \in \Lambda$.

Similar arguments (see Exercise 3.5.2), with the roles of (3.5.1) and (3.5.3) reversed, show that Λ is invariant under $\mathbf{f}^{-1}$ and, therefore, $\mathbf{f}(\Lambda) = \Lambda$. □

The map **f**, as we have defined it up to now, is *not* a diffeomorphism of the square

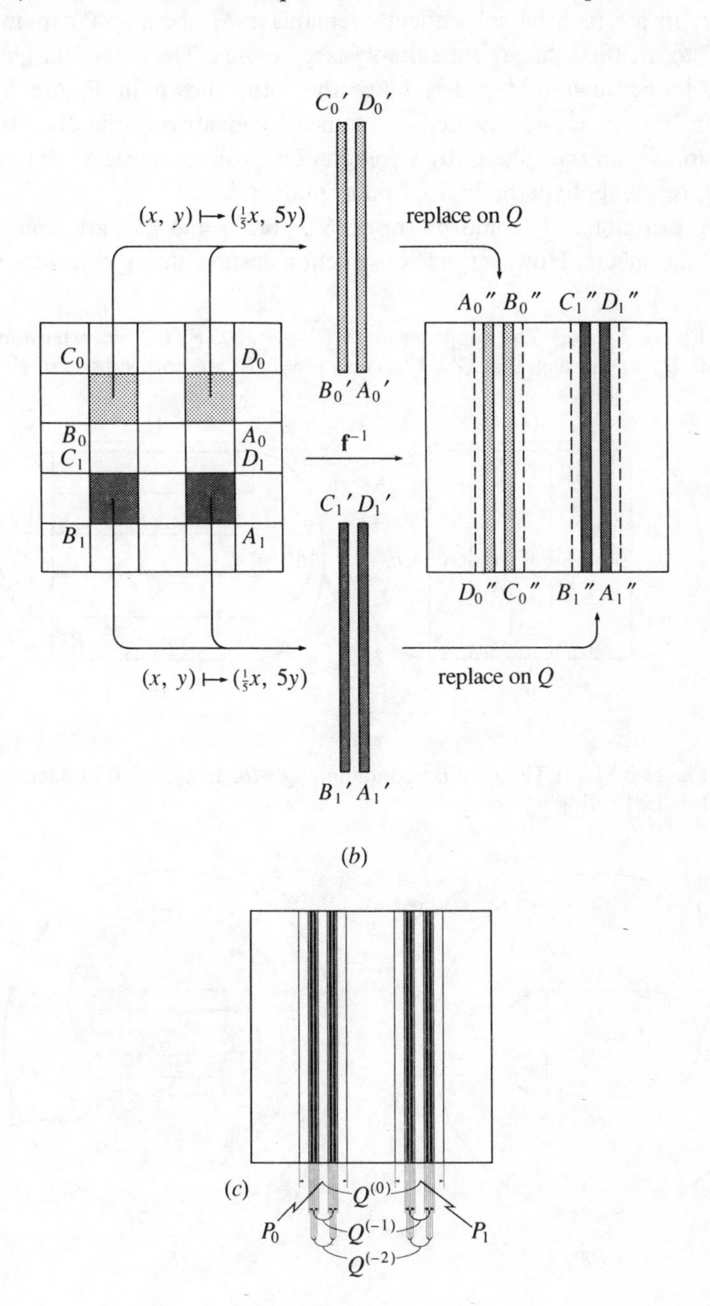

Q ($\mathbf{f}(Q) \nsubseteq Q$). Moreover, it does not have an obvious connection with diffeomorphisms on compact manifolds without boundary. However, a diffeomorphism, $\mathbf{g}: S^2 \to S^2$ can be constructed such that $\mathbf{f}$ is the restriction of $\mathbf{g}$ to a subset of the sphere. The first step in the construction is to extend the map $\mathbf{f}$ to a capped square Q' as indicated in Figure 3.16. The extension $\mathbf{f}'$ is constructed in such a way that $\mathbf{f}'|F$ has a unique, attracting, hyperbolic fixed point. This means that once a point is mapped into F its orbit subsequently remains in F. The map $\mathbf{f}'$ can, in turn, be extended to a closed disc D^2 of suitably large radius. The extension $\mathbf{g}': D^2 \to D^2$ is taken to be such that $\mathbf{g}'(D^2)$ takes the form shown in Figure 3.17. The diffeomorphism $\mathbf{g}: S^2 \to S^2$ is finally obtained by identifying the disc, D^2, on $\mathbb{R}^2$ with a cap, C^2, on the sphere (by stereographic projection, see § 3.3) and adding a unique, repelling, hyperbolic fixed point in $S^2 \backslash C^2$.

Since $\mathbf{g}$ is a global diffeomorphism on S^2, both $\mathbf{g}$ and $\mathbf{g}^{-1}$ are defined for all points of the sphere. However, its construction ensures that $\mathbf{g}$ coincides with $\mathbf{f}$ or

Figure 3.16 (*a*) The capped square $Q' = G \cup Q \cup F$; (*b*) the extension $\mathbf{f}': Q' \to Q'$ is such that $G' = \mathbf{f}'(G)$ and $F' = \mathbf{f}'(F)$ are both subsets of F.

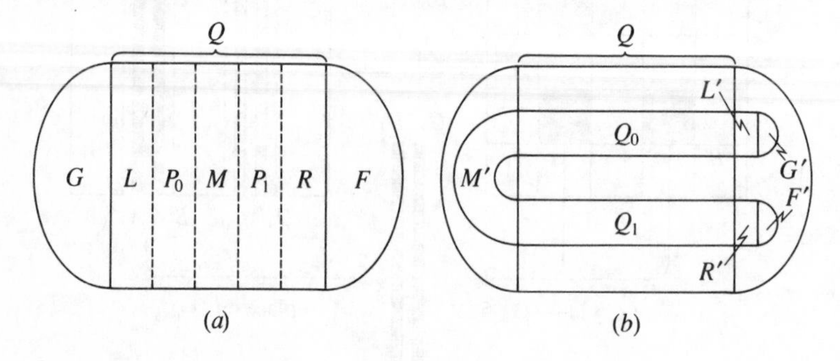

Figure 3.17 (*a*) The disc D^2 containing Q'; (*b*) image of D^2 under $\mathbf{g}'$ (shaded) with $\mathbf{g}'|Q' = \mathbf{f}'$.

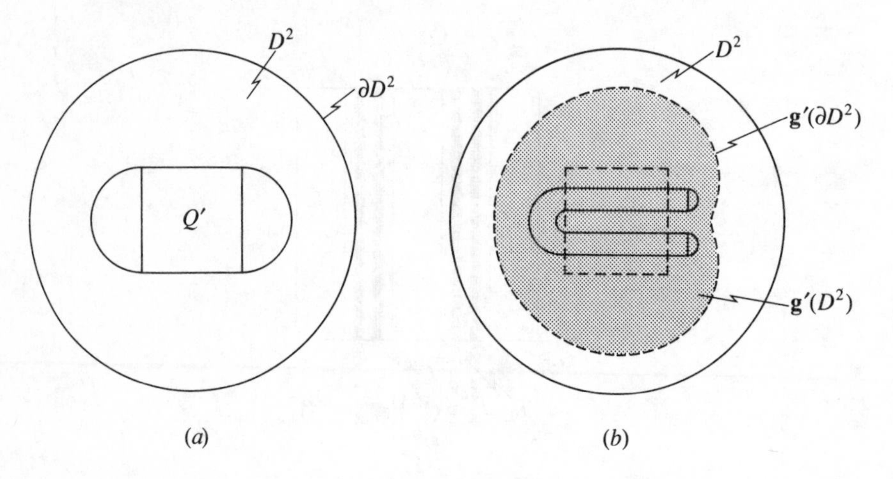

$\mathbf{f}'$ when restricted appropriately. Here we have taken the notational liberty of not distinguishing between these restrictions on the sphere and their representatives on $\mathbb{R}^2$ via stereographic projection. This distinction does not play a significant role in our discussion and, once noted, should not lead to any confusion. The unstable fixed point in S^2/C^2 means that the ordinary points of this set move towards C^2 under $\mathbf{g}$. As we have shown in Figure 3.17(*b*), $\mathbf{g}|(D^2\backslash Q')$ is again a contraction and, therefore, $\mathbf{g}$ essentially delivers points to Q'. On Q', $\mathbf{g}$ behaves in the same way as $\mathbf{f}'$. We already know that $\mathbf{f}'$ has an invariant Cantor set Λ arising from its restriction, $\mathbf{f}$, to Q, but what of the points $S^2\backslash\Lambda=\Lambda^c$? The following proposition provides part of the answer to this question. It states that those points of Q' that do not lie on the infinite set of vertical line segments, $\bigcap_{n\in\mathbb{N}} Q^{(-n)}$, are eventually swept into F.

Proposition 3.5.2 *The orbits under* $\mathbf{g}$ *of points in* $Q\backslash(\bigcap_{n\in\mathbb{N}} Q^{(-n)})$ *ultimately approach the stable fixed point of* $\mathbf{g}$ *in* F.

Proof. Figure 3.16 shows what happens to the various parts of Q under a single application of $\mathbf{g}|Q'=\mathbf{f}'$. The left ($L$) and right ($R$) fifths of Q are mapped, together with G and F, into F. Since F contains a unique, attracting fixed point, the orbit of any point in F approaches this point asymptotically. Points in the mid-fifth, M, of Q suffer the same fate after one more iteration. Such points are mapped into G by $\mathbf{f}'$ and into F by $\mathbf{f}'^2$. Only points in $P_0\cup P_1$ remain in Q (in fact in $Q_0\cup Q_1$) after one application of $\mathbf{f}'$. In other words, points in $Q\backslash Q^{(0)}$ enter F after at most two iterations of $\mathbf{g}$.

Let us focus attention on the partition of the square Q provided by $Q^{(-1)}$, rather than $Q^{(0)}$. Observe that points in $Q^{(0)}\backslash(Q^{(0)}\cap Q^{(-1)})$ are mapped into L, M, R after one iteration of $\mathbf{g}|Q=\mathbf{f}$ (see Figure 3.15(*b*)) and thence into F after two or three iterations. Thus we conclude that all points in $(Q\backslash Q^{(0)})\cup(Q^{(0)}\backslash(Q^{(0)}\cap Q^{(-1)}))=Q\backslash(Q^{(0)}\cap Q^{(-1)})$ enter F after at most three applications of $\mathbf{g}$. Similarly, if we consider the partition of Q provided by $Q^{(-2)}$, we conclude that points in $Q^{(-1)}\backslash Q^{(-2)}=(Q^{(0)}\cap Q^{(-1)})\backslash(Q^{(0)}\cap Q^{(-1)}\cap Q^{(-2)})$ have images under $\mathbf{g}$ in $Q^{(0)}\backslash(Q^{(0)}\cap Q^{(-1)})$. Therefore, if $\mathbf{x}\in Q\backslash\bigcap_{n=0}^{2} Q^{(-n)}$ then $\mathbf{g}^k(\mathbf{x})\in F$ for $k>4$. Thus, we conclude inductively that all points in $Q\backslash\bigcap_{n\in\mathbb{N}} Q^{(-n)}$ ultimately enter F. □

In view of the construction of Λ, it is clear that $\Lambda\subseteq\bigcap_{n\in\mathbb{N}} Q^{(-n)}$. Moreover, it is not difficult to show that $\bigcap_{n\in\mathbb{N}} Q^{(-n)}$ is invariant under $\mathbf{g}$ (see Exercise 3.5.3). Bearing in mind that $\bigcap_{n\in\mathbb{N}} Q^{(-n)}$ is a set of straight line segments parallel to the y-axis and that $\mathbf{g}$ involves a contraction along that direction, it is not surprising that points in $(\bigcap_{n\in\mathbb{N}} Q^{(-n)})\backslash\Lambda$ have orbits that approach Λ asymptotically (see Exercise 3.5.3). It must be emphasised that these orbits are *not* confined to a single vertical line subset of $\bigcap_{n\in\mathbb{N}} Q^{(-n)}$.

Let us now turn to the dynamics of points in $S^2\backslash Q'$. Unlike $\mathbf{f}^{-1}$, $\mathbf{g}^{-1}$ is defined for all $\mathbf{x}\in Q$ (see Figure 3.18). Of course, $\mathbf{g}^{-1}|Q^{(1)}\equiv\mathbf{f}^{-1}$ is as illustrated in Figure 3.15 but $\mathbf{g}^{-1}(Q\backslash Q^{(1)})\subseteq S^2\backslash Q'=Q'^{c}$. This means that points in $Q\backslash Q^{(1)}$ are the images under $\mathbf{g}$ of points lying *outside* Q'. We have already discussed the fate of such images under forward iterations of $\mathbf{g}$. Since $\Lambda\cap(Q\backslash Q^{(1)})=\varnothing$, points in $(Q\backslash Q^{(1)})\cap(\bigcap_{n\in\mathbb{N}} Q^{(-n)})$ have orbits approaching Λ asymptotically; while those in $(Q\backslash Q^{(1)})\backslash\bigcap_{n\in\mathbb{N}} Q^{(-n)}$ have orbits entering F. However, under reverse iterations of $\mathbf{g}$

Figure 3.18 Illustration of $\mathbf{g}^{-1}$ showing that points starting in $Q\backslash Q^{(1)}=T\cup MH\cup B$ are mapped out of Q' by $\mathbf{g}^{-1}$. Observe that $S^2\backslash Q'$ contains a unique, *stable*, hyperbolic fixed point of $\mathbf{g}^{-1}$ so that the orbits of these points under $\mathbf{g}^{-1}$ do not return to Q'.

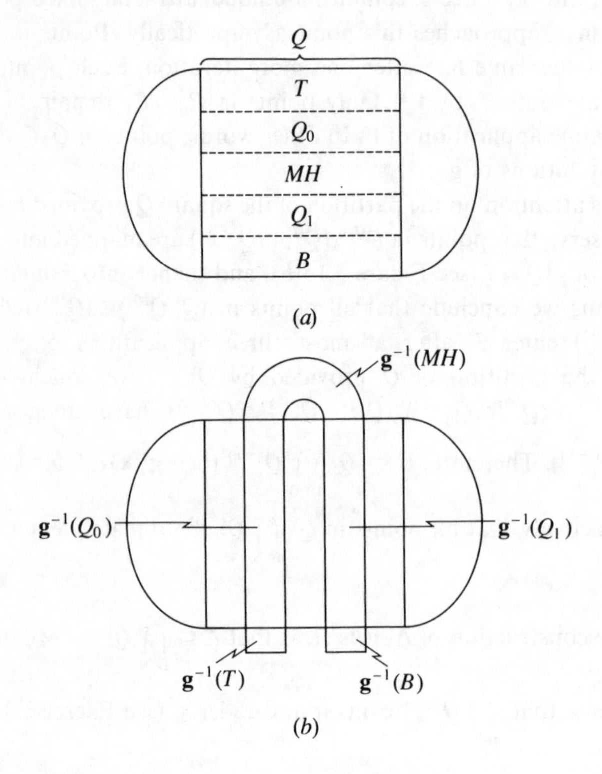

we are able to extend the invariant set $\bigcap_{n\in\mathbb{N}} Q^{(-n)}$ onto the whole sphere. The resulting set of points is called the *inset*, in(Λ) *of* Λ, i.e.

$$\text{in}(\Lambda) = \{\mathbf{x} \in S^2 \,|\, \mathbf{g}^n(\mathbf{x}) \to \Lambda \text{ as } n \to \infty\}. \tag{3.5.5}$$

Similar arguments to those presented above, with $\mathbf{g}^{-1}$ replacing $\mathbf{g}$, lead to analogous conclusions about the infinite set of horizontal lines $\bigcap_{n\in\mathbb{Z}^+} Q^{(n)}$ (see Exercise 3.5.3). It follows that there is a set of points, out(Λ) $\subseteq G \cup Q \cup F$, the outset of Λ, whose orbits approach Λ under reverse iterations of $\mathbf{g}$, i.e.

$$\text{out}(\Lambda) = \{\mathbf{x} \in S^2 \,|\, \mathbf{g}^{-n}(\mathbf{x}) \to \Lambda \text{ as } n \to \infty\}. \tag{3.5.6}$$

The role of the set Λ in the dynamics of $\mathbf{g}$ is clearly analogous to that of a saddle point in simpler diffeomorphisms. The inset and outset of Λ generalise the stable and unstable manifolds of the saddle. We will return to sets possessing this more general hyperbolic structure in § 3.6 but now we must consider the dynamics of $\mathbf{g}$ on Λ.

3.5.2 Dynamics on symbol sequences

Let Σ be the set of all bi-infinite sequences of the binary symbols $\{0, 1\}$, i.e. $\Sigma = \{\sigma \,|\, \sigma: \mathbb{Z} \to \{0, 1\}\}$. The elements, σ, of Σ are called *symbol sequences* and they are defined by specifying $\sigma(n) = \sigma_n \in \{0, 1\}$ for each $n \in \mathbb{Z}$. We will write $\sigma = \{\sigma_n\}_{n=-\infty}^{\infty} = \{\ldots \sigma_{-2}\sigma_{-1}\sigma_0 \cdot \sigma_1\sigma_2 \ldots\}$. Our aim is to study the dynamics of the map $\alpha: \Sigma \to \Sigma$ defined by

$$\alpha(\sigma)_n = \sigma_{n-1}, \tag{3.5.7}$$

$n \in \mathbb{Z}$. This is known as a *left-shift* on Σ because it corresponds to moving the binary point one symbol to the left.

Proposition 3.5.3 *The left shift $\alpha: \Sigma \to \Sigma$ has periodic orbits of all periods as well as aperiodic orbits.*

A point $\sigma^* \in \Sigma$ is periodic if

$$\alpha^q(\sigma^*) = \sigma^*, \tag{3.5.8}$$

$q \in \mathbb{Z}^+$. If q is the least, positive integer for which (3.5.8) is satisfied then σ^* is said to be of period q. It is not difficult to see that (3.5.8) will be satisfied if and only if $\sigma_n^* = \sigma_{n+q}^*$, for all $n \in \mathbb{Z}$. It is then easy to find periodic points of α with any given period, q. The required sequence, σ^*, is generated by repetition of a block of symbols of length q that is itself not composed of repetitions of any of its sub-blocks. For example, the point

$$\sigma^* = \{\ldots 11010\overline{11110111011010} \cdot 11110 \ldots\} \tag{3.5.9}$$

has period-14, while

$$\sigma^* = \{\ldots 10\overline{1011010}\,\overline{1011010} \cdot 101 \ldots\} \tag{3.5.10}$$

satisfies $\alpha^{14}(\sigma^*) = \sigma^*$ but has period-7 because $\alpha^7(\sigma^*) = \sigma^*$, also. It is equally straightforward to show that α has aperiodic orbits. For instance,

$$\sigma = \{\ldots \underbrace{1 \ldots 110}_{n} \ldots \underbrace{11110}_{4}\underbrace{1110}_{3}\underbrace{110}_{2}\underbrace{10}_{1} \cdot 1011011101111 \ldots\}, \tag{3.5.11}$$

which contains symbol blocks of the type shown for all $n \in \mathbb{Z}^+$, is such that there is no $q \in \mathbb{Z}^+$ such that $\alpha^q(\sigma) = \sigma$.

Proposition 3.5.4 *There is a topology in which the periodic points of α are dense in Σ.*

There is a natural way of defining how close two symbol sequences are to one another. Given two sequences in Σ, we can obtain the length of the largest symbol block, centred on the binary point, on which they agree. The larger the size of this block the closer the two sequences are deemed to be. We are then able to define the limit of a sequence of elements in Σ. A sequence $\{\sigma^{(m)}\}_{m=0}^{\infty} \subseteq \Sigma$ is said to tend to $\sigma \in \Sigma$ as $m \to \infty$, if, given $N \in \mathbb{Z}^+$, there exists $M \in \mathbb{Z}^+$ such that $\sigma_n^{(m)} = \sigma_n$ for $-(N-1) \leqslant n \leqslant N$, when $m > M$. Clearly, if $\sigma^{(m)} \to \sigma$ as $m \to \infty$ then $\sigma^{(m)}$ and σ agree on increasingly large central blocks. For example, the sequence $\sigma^{(m)}$ defined by

$$\sigma_n^{(m)} = \begin{cases} 1 & -(m-1) \leqslant n \leqslant m, m \in \mathbb{Z}^+ \\ 0 & \text{otherwise}, \end{cases} \tag{3.5.12}$$

converges, as $m \to \infty$, to the sequence σ with $\sigma_n = 1$ for all $n \in \mathbb{Z}$.

With the above definition of convergence, periodic points of α are dense in Σ. This follows because, given any $\sigma \in \Sigma$, there is a sequence of periodic sequences $\{\sigma^{(m)}\}_{m=0}^{\infty}$ which tends to σ as $m \to \infty$. Each sequence $\sigma^{(m)}$ is simply taken to be periodic with period $2m$ and such that $\sigma_n^{(m)} = \sigma_n$ for $-(m-1) \leqslant n \leqslant m$. As an example, let σ be the aperiodic sequence (3.5.11), for which

$$\begin{aligned} \sigma^{(1)} &= \ldots \overline{01}\,\overline{01}\,\overline{0 \cdot 1}\,\overline{01}\,\overline{01} \ldots, \\ \sigma^{(2)} &= \ldots \overline{1010}\,\overline{10 \cdot 10}\,\overline{1010} \ldots, \\ \sigma^{(3)} &= \ldots \overline{010101}\,\overline{010 \cdot 101}\,\overline{010101} \ldots, \\ \sigma^{(4)} &= \ldots \overline{10101011}\,\overline{1010 \cdot 1011}\,\overline{10101011} \ldots \end{aligned} \tag{3.5.13}$$

and so on.

Proposition 3.5.5 *The left shift $\alpha: \Sigma \to \Sigma$ has a dense orbit on Σ.*

To justify Proposition 3.5.5 we must show that α has an orbit on Σ that approaches every point of Σ arbitrarily closely. Let $\sigma \in \Sigma$ be such that σ_{-n} for $n \in \mathbb{N}$ is given by the following ordered lists of symbol blocks:

(i) all blocks of length 1, i.e. $\{0\}, \{1\}$;
(ii) all blocks of length 2, i.e. $\{0, 0\}, \{0, 1\}, \{1, 0\}, \{1, 1\}$;
(ii) all blocks of length 3; and so on.

All possible blocks of all lengths are included in $\{\sigma_{-n}\}_{n=0}^{\infty}$; σ_n, $n \in \mathbb{Z}^+$, can be chosen arbitrarily. The orbit of σ under α contains $\{\alpha^m(\sigma) | m \in \mathbb{N}\}$. Now, by construction σ contains any given symbol block of length N in its left hand half. After sufficiently many applications of α this block will be centrally placed about the binary point. Since N is arbitrary, any element of Σ can be approximated arbitrarily closely by some point on the orbit of σ under α.

In view of the rather special construction used above to obtain a sequence σ whose orbit under α is dense in Σ, the reader may feel that such sequences are in some sense rare or atypical. This is *not* the case. In fact, most binary bi-infinite sequences contain any prescribed block of symbols (see Hardy, 1979) and therefore have a dense orbit under α. The particular example chosen above is carefully ordered purely to make the argument more convincing.

Having established some properties of the left shift $\alpha: \Sigma \to \Sigma$, we must reveal our motive for examining the dynamics of this map: namely to obtain a symbolic description of the dynamics of the horseshoe diffeomorphism on Λ. Before doing this, it is worth noting that the validity of Propositions 3.5.3–5 does not depend upon the *binary* nature of the sequences in Σ. Similar results can be derived for sequences of m-symbols, $\{0, 1, \ldots, m-1\}$ say (see Exercise 3.5.5). Binary symbol sequences allow us to deal with the horseshoe map of § 3.5.1. However, there are more sophisticated maps of this type (see Exercise 3.6.5) whose 'symbolic dynamics' involve sequences of m symbols with $m > 2$.

3.5.3 Symbolic dynamics for the horseshoe diffeomorphism

In this section we show that the restriction of the horseshoe diffeomorphism to the invariant set Λ is topologically conjugate to the left shift α on Σ. The key idea is that the points of Λ can be 'coded' as bi-infinite sequences of $\{0, 1\}$.

Recall that $\Lambda = \bigcap_{n \in \mathbb{Z}} Q^{(n)}$, where $Q^{(n)}$, $n \in \mathbb{Z}^+$, is the disjoint union of 2^n-horizontal strips on the square Q, while $Q^{(-n)}$, $n \in \mathbb{N}$, is the union of 2^{n+1} similar vertical strips. As Figures 3.14(b) and 3.15(c) illustrate these sets of strips are 'nested', i.e. $Q^{(1)} \supset Q^{(2)} \supset \cdots \supset Q^{(n)} \supset \cdots$ and $Q^{(0)} \supset Q^{(-1)} \supset \cdots \supset Q^{(-n)} \supset \cdots$. Thus, $\Lambda^{(N)} = \bigcap_{n=-(N-1)}^{N} Q^{(n)} = Q^{(-(N-1))} \cap Q^{(N)}$ is the disjoint union of 2^{2N} squares of side $2/5^N$ (see Figure 3.19). Clearly, as $N \to \infty$, the size of the squares tends to zero, their number becomes infinite and $\Lambda^{(N)} \to \Lambda$. The coding of the points of Λ follows from the fact that each square of $\Lambda^{(N)}$ can be uniquely represented by a symbol block, $\sigma^{(N)} = \{\sigma_{-(N-1)} \cdots \sigma_0 \cdot \sigma_1 \ldots \sigma_N\}$, $\sigma_n \in \{0, 1\}$, of length $2N$.

Any given strip in $Q^{(n)}$ can be allocated either 0 or 1 in the following way.

Consider the vertical strips P_0 and P_1. Observe that

$$Q^{(1)} = \mathbf{g}(P_0) \cup \mathbf{g}(P_1) = Q_0 \cup Q_1 \tag{3.5.14}$$

where $Q_0 \cap Q_1 = \varnothing$ (see Figure 3.13). Furthermore,

$$Q^{(2)} \subseteq \mathbf{g}^2(P_0) \cup \mathbf{g}^2(P_1) \tag{3.5.15}$$

with $\mathbf{g}^2(P_0) \cap \mathbf{g}^2(P_1) = \varnothing$ (see Figure 3.14(*a*)). In general, for $n \in \mathbb{Z}^+$,

$$Q^{(n)} \subseteq \mathbf{g}^n(P_0) \cup \mathbf{g}^n(P_1) \tag{3.5.16}$$

and $\mathbf{g}^n(P_0) \cap \mathbf{g}^n(P_1)$ is always empty because $P_0 \cap P_1 = \varnothing$ and $\mathbf{g}$ is a diffeomorphism. Thus a horizontal strip of $Q^{(n)}$ lies either in $\mathbf{g}^n(P_0)$ or $\mathbf{g}^n(P_1)$. We allocate the symbol 0 to a strip of $Q^{(n)}$ if it is a subset of $\mathbf{g}^n(P_0)$ and the symbol 1 if it lies in $\mathbf{g}^n(P_1)$ (see Figure 3.20). Obviously, these symbols alone do not provide a unique description of each horizontal strip in $Q^{(n)}$ for $n \geqslant 2$, however, they can be used to obtain one. For example, two strips of $Q^{(2)}$ have been allocated the symbol 0 but they are distinguished by the fact that one lies in $\mathbf{g}(P_0)$ (i.e. strip 0 of $Q^{(1)}$) and the other lies in $\mathbf{g}(P_1)$ (i.e. strip 1 of $Q^{(1)}$). Hence the strips in $Q^{(2)}$ can be

Figure 3.19 Illustration of $\bigcap_{n=-(N-1)}^{N} Q^{(n)}$ for (*a*) $N = 1$; (*b*) $N = 2$. The square regions defined by (3.5.17) with $\sigma^{(2)}$ given by $\{11 \cdot 01\}$ and $\{10 \cdot 11\}$ are indicated.

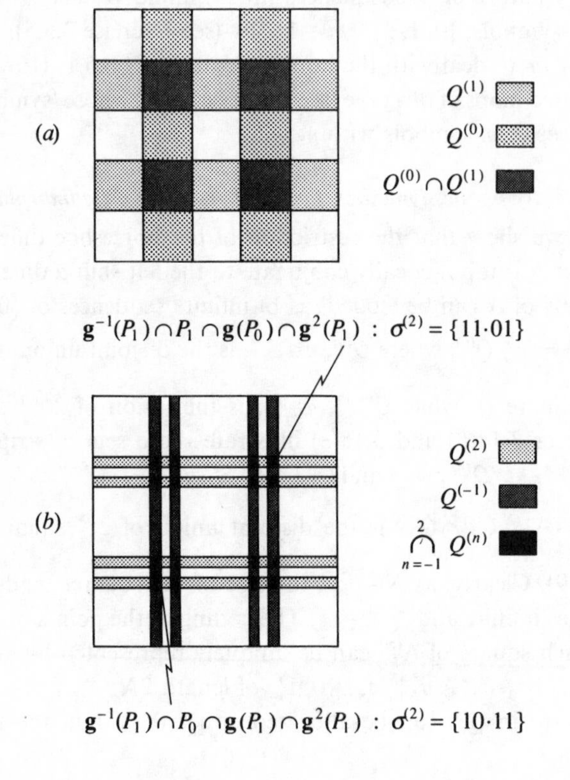

uniquely labelled by giving two symbols: the first specifying a strip in $Q^{(1)}$ so that the second uniquely determines a strip in $Q^{(2)}$ (see Figure 3.20). Similarly, the strips of $Q^{(3)}$ can be uniquely labelled by starting from the unique labelling of the strips in $Q^{(2)}$ and appending the symbols allocated to $Q^{(3)}$. It follows that the strips of $Q^{(n)}$ are uniquely specified by a set of n of the symbols $\{0, 1\}$. Similar arguments can be carried through for $Q^{(-n)}$ by considering the images of P_0 and P_1 under powers of $\mathbf{g}^{-1}$ (see Exercise 3.5.8). A vertical strip of $Q^{(-n)}$, $n \in \mathbb{N}$, is allocated the symbol i if it is a subset of $\mathbf{g}^{-n}(P_i)$, $i = 0, 1$. For $n \in \mathbb{Z}^+$, unique labels for the strips of $Q^{(-n)}$ are obtained by appending these allocated symbols to those of the strips in $Q^{(-(n-1))}$ (see Figure 3.21). Notice we have appended symbols on the left so that the order in the strip label matches that of the negative integers. Finally, we

Figure 3.20 Coding of strips in $Q^{(n)}$ for (*a*) $n = 1$; (*b*) $n = 2$; (*c*) $n = 3$. The symbol allocated to each strip is shown on the left and the unique code for the strip is given on the right.

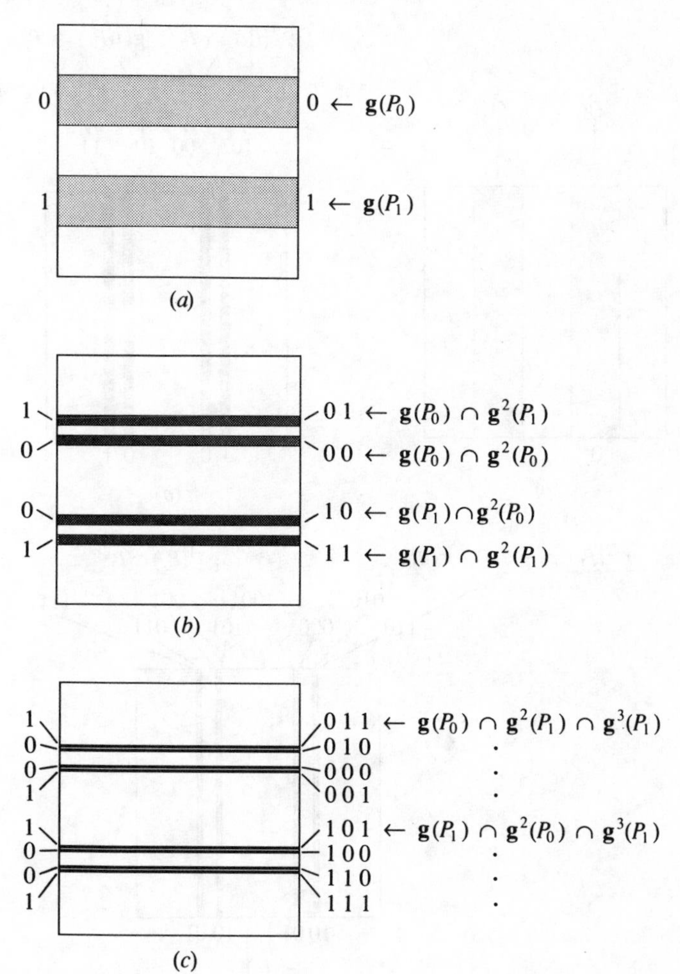

can construct the symbol blocks representing the squares occurring in $\Lambda^{(N)}$. Each such square is the intersection of one of the vertical strips of $Q^{(-(N-1))}$ with one of the horizontal strips of $Q^{(N)}$. If the vertical strip has label $\sigma_{-(N-1)}, \ldots, \sigma_{-1}, \sigma_0$ and the horizontal strip has label $\sigma_1, \ldots, \sigma_N$, the symbol block representing the square is taken to be $\sigma^{(N)} = \{\sigma_{-(N-1)}, \ldots, \sigma_{-1}, \sigma_0 \cdot \sigma_1, \ldots, \sigma_N\}$. Thus, for example $\{11.01\}$ and $\{10.11\}$, respectively, represent the top right hand and bottom left hand squares in the illustration of $\Lambda^{(2)}$ given in Figure 3.19(*b*). It is not difficult to show (see Exercise 3.5.11) that the square represented by the symbol block $\sigma^{(N)}$

Figure 3.21 Coding for strips of $Q^{(-n)}$ for $n = 0, 1, 2$: (*a*) $Q^{(0)}$; (*b*) $Q^{(-1)}$; (*c*) $Q^{(-2)}$. Unique labels for the strips are given above and allocated symbols below. Notice that, to match the negative integers, symbols are appended to the left rather than to the right.

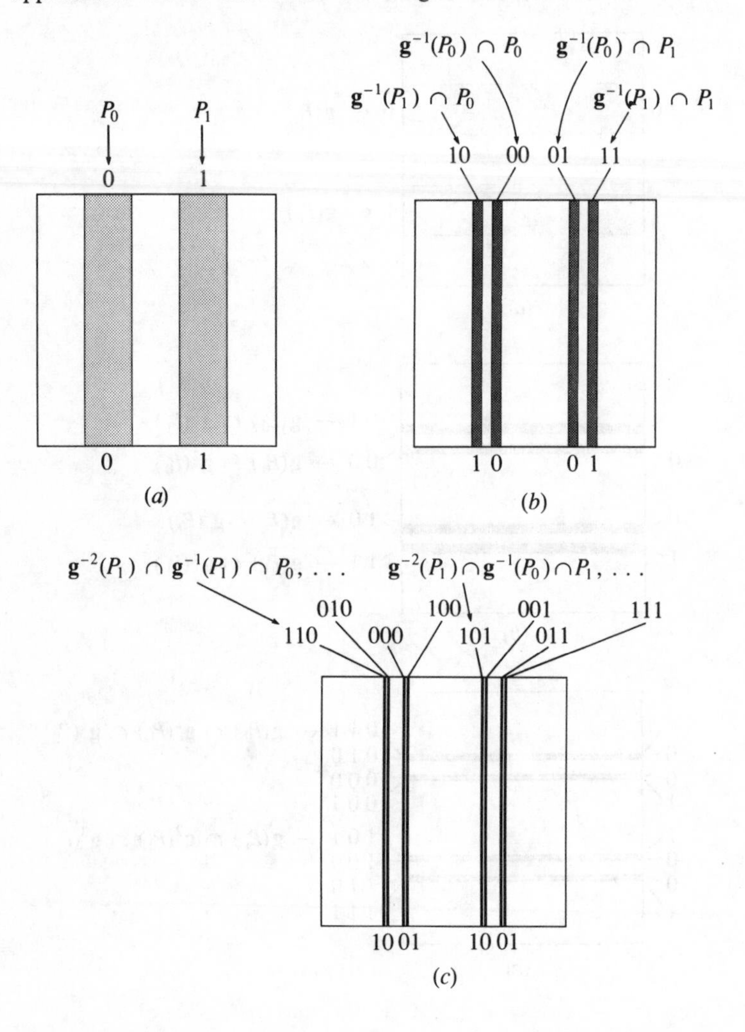

is given by

$$\left(\bigcap_{n=-(N-1)}^{N} \mathbf{g}^n(P_{\sigma_n}) \right) \cap Q. \tag{3.5.17}$$

In the limit $N \to \infty$, the above construction assigns a unique, bi-infinite binary sequence with each point of Λ. Moreover, (3.5.17) allows any such sequence to be converted to a unique point of Λ. We have therefore constructed a bijection $\mathbf{h}: \Sigma \to \Lambda$.

Proposition 3.5.6 *The bijection* $\mathbf{h}: \Sigma \to \Lambda$ *defined above is a homeomorphism that exhibits the topological conjugacy of* $\mathbf{g}: \Lambda \to \Lambda$ *and* $\alpha: \Sigma \to \Sigma$.

Proof. The nested nature of vertical and horizontal strips defining Λ means that sequences that are close, in the sense that they agree over large central blocks, map under $\mathbf{h}$ to points of Λ that are geometrically close together. Similarly if two points of Λ are geometrically close, the symbol sequences agree over a large central block because it is only for N sufficiently large that such points are distinguished in $\Lambda^{(N)}$. Thus $\mathbf{h}$ is a homeomorphism.

Let $\sigma \in \Sigma$ and

$$\mathbf{x} = \mathbf{h}(\sigma) = \left[\bigcap_{n \in \mathbb{Z}} \mathbf{g}^n(P_{\sigma_n}) \right] \cap Q. \tag{3.5.18}$$

Then

$$\begin{aligned} \mathbf{g}(\mathbf{h}(\sigma)) = \mathbf{g}(\mathbf{x}) &= \mathbf{g}\left(\left[\bigcap_{n \in \mathbb{Z}} \mathbf{g}^n(P_{\sigma_n}) \right] \cap Q \right), \\ &= \left[\bigcap_{n \in \mathbb{Z}} \mathbf{g}^{n+1}(P_{\sigma_n}) \right] \cap Q, \\ &= \left[\bigcap_{n \in \mathbb{Z}} \mathbf{g}^n(P_{\sigma_{n-1}}) \right] \cap Q, \\ &= \mathbf{h}(\alpha(\sigma)). \end{aligned} \tag{3.5.19}$$

Therefore $\mathbf{h}$ exhibits the conjugacy of $\mathbf{g}$ and α. □

Proposition 3.5.6 implies that the complexity exhibited by the orbits of points in Σ under α (see § 3.5.2) also occurs in the orbits of points of Λ under $\mathbf{g}$. Thus $\mathbf{g}|\Lambda$ has infinitely many periodic points, its periodic points are dense in Λ and it has orbits that are themselves dense subsets of Λ.

Another feature of the dynamics of $\alpha: \Sigma \to \Sigma$, that has important repercussions for $\mathbf{g}|\Lambda$, is that there are points in Σ whose orbit under α is aperiodic. Since $\alpha(\sigma)$ and σ are not, in general, close in Σ, these orbits wander throughout Σ in an apparently disorganised way. Similarly, their counterpart in the orbits of $\mathbf{g}|\Lambda$ move around Λ by hopping from point to point in a random or *chaotic* way. Indeed, invariant sets like Λ are often referred to as *chaotic sets* (see § 3.6) because of the presence of such orbits.

While the dynamics of $\mathbf{g}|\Lambda$ is very complicated, we must not forget that the dynamics of $\mathbf{g}|\Lambda^c$ suggests that Λ is, in some sense, hyperbolic. In the following section, we consider how such sets fit into a general theoretical framework.

3.6 Hyperbolic structure and basic sets

Let us return to the hyperbolic nature of the invariant set Λ of the horseshoe diffeomorphism $\mathbf{g}: S^2 \to S^2$. In fact, Λ is said to have a *hyperbolic structure* or to be a *hyperbolic set* for $\mathbf{g}$. Our aim in this section is to explain this statement and to introduce an important theorem about diffeomorphisms whose non-wandering set, Ω, has a hyperbolic structure.

It is helpful to review our previous encounters with hyperbolicity (see § 2.1 and 2.2). The striking feature is that, thus far, we have only had to consider hyperbolic fixed points. Non-trivial hyperbolic sets such as a hyperbolic periodic orbit, or a normally hyperbolic invariant circle, are defined in terms of a hyperbolic fixed point of a related map ($\mathbf{f}^q$ or $\mathscr{F}$ in § 2.2). We are then involved with the local behaviour of a map at a fixed point in a Euclidean or Banach space. In such cases, the hyperbolic nature of the fixed point is given in terms of the eigenvalues of the derivative map ($\mathrm{D}\mathbf{f}^q$ or $\mathrm{D}\mathscr{F}$). It is not possible to use this approach to characterise the hyperbolicity of the invariant set, Λ, of the horseshoe diffeomorphism. However, it is useful to consider why such an approach fails. There are two problem areas.

(i) The horseshoe diffeomorphism is defined on a manifold (the sphere) and not a Euclidean space. This means that the generalisation of the derivative map to this situation must be considered.

(ii) The complexity of Λ is such that it is not possible to formulate the problem in terms of a fixed point of some related map. For example, Λ contains aperiodic orbits which do not correspond to a fixed point of $\mathbf{g}^q$, for any $q \in \mathbb{Z}^+$. Thus, having introduced the appropriate generalisation of the derivative map, our definition of hyperbolicity must allow for the fact that $\mathbf{x}$ and $\mathbf{g}(\mathbf{x})$ are different points in Λ.

Let us begin by considering how the results of § 2.2 can be applied to a diffeomorphism $\mathbf{f}: M \to M$ when M is an n-dimensional, differentiable manifold that is not a subset of $\mathbb{R}^n$ (see Figure 3.22). The derivative map, $\mathrm{D}\mathbf{f}(\mathbf{x}^*): \mathbb{R}^n \to \mathbb{R}^n$, used to discuss the hyperbolic fixed point, $\mathbf{x}^*$, of $\mathbf{f}: \mathbb{R}^n \to \mathbb{R}^n$ in § 2.2, is replaced by the tangent map $T\mathbf{f}_{\mathbf{x}^*}: TM_{\mathbf{x}^*} \to TM_{\mathbf{x}^*}$, where $TM_{\mathbf{x}^*}$ is the tangent space to M at $\mathbf{x}^*$. Recall (see § 1.3) that $TM_{\mathbf{x}}$ can be defined, for any $\mathbf{x} \in M$, in terms of equivalence classes of curves on M with the same tangent vector at $\mathbf{x}$. To see the connection with the behaviour of $\mathbf{f}$ near $\mathbf{x}^*$, let $\boldsymbol{\eta}(t)$, with $t \in I \subseteq \mathbb{R}$, $0 \in I$ and $\boldsymbol{\eta}(0) = \mathbf{x}^*$, be a parametrised curve on M passing through $\mathbf{x}^*$. To find the tangent vector at $\mathbf{x}^*$, we need to differentiate $\boldsymbol{\eta}(t)$ with respect to t and (see § 1.1) this can only be done by using a local chart, $(U_\alpha, \mathbf{h}_\alpha)$ say, containing $\mathbf{x}^*$. The local representatives, $\tilde{\mathbf{f}}, \tilde{\boldsymbol{\eta}}$ and $\widetilde{\mathbf{f}\cdot\boldsymbol{\eta}}$, of $\mathbf{f}$, $\boldsymbol{\eta}$ and $\mathbf{f}\cdot\boldsymbol{\eta}$, respectively, in $(U_\alpha, \mathbf{h}_\alpha)$ (or, more concisely, α-representatives) are given by

$$\tilde{\mathbf{f}}_\alpha = \mathbf{h}_\alpha \cdot \mathbf{f} \cdot \mathbf{h}_\alpha^{-1}, \qquad \tilde{\boldsymbol{\eta}}_\alpha = \mathbf{h}_\alpha \cdot \boldsymbol{\eta}, \qquad \widetilde{(\mathbf{f} \cdot \boldsymbol{\eta})}_\alpha = \mathbf{h}_\alpha \cdot (\mathbf{f} \cdot \boldsymbol{\eta}). \tag{3.6.1}$$

They satisfy the equation

$$\widetilde{(\mathbf{f} \cdot \boldsymbol{\eta})}_\alpha(t) = \tilde{\mathbf{f}}_\alpha(\tilde{\boldsymbol{\eta}}_\alpha(t)) \tag{3.6.2}$$

which, provided M is a C^1-manifold, can be differentiated to give

$$\dot{\widetilde{(\mathbf{f} \cdot \boldsymbol{\eta})}}_\alpha(0) = \mathrm{D}\tilde{\mathbf{f}}_\alpha(\tilde{\boldsymbol{\eta}}_\alpha(0))\dot{\tilde{\boldsymbol{\eta}}}_\alpha(0) \tag{3.6.3}$$

at $t = 0$. The vectors $(\widetilde{\mathbf{f}_\alpha \cdot \boldsymbol{\eta}_\alpha})(0)$ and $\dot{\tilde{\boldsymbol{\eta}}}_\alpha(0)$ are α-representatives of elements of $TM_{\mathbf{x}^*}$ in $(U_\alpha, \mathbf{h}_\alpha)$. Strictly speaking, they lie in the tangent space to U_α at $\tilde{\mathbf{x}}_\alpha^* = \tilde{\boldsymbol{\eta}}_\alpha(0)$ but, as $TU_{\tilde{\mathbf{x}}_\alpha^*}$ is a replica of $\mathbb{R}^n$, this distinction is not always apparent. The derivative map $\mathrm{D}\tilde{\mathbf{f}}_\alpha(\tilde{\mathbf{x}}_\alpha^*)$ is the local representative of the tangent map $T\mathbf{f}_{\mathbf{x}^*}$. As the opening remarks to §2.2 suggest, $\mathbf{x}^* \in M$ is said to be a hyperbolic fixed point of $\mathbf{f}: M \to M$ if $\tilde{\mathbf{x}}_\alpha^*$ is a hyperbolic fixed point of $\tilde{\mathbf{f}}_\alpha$ in the sense of Definition 2.2.1, i.e. if $\mathrm{D}\tilde{\mathbf{f}}_\alpha(\tilde{\mathbf{x}}_\alpha^*)$ has no eigenvalue with unit modulus. What is more, if we assign a metric to $TU_{\tilde{\mathbf{x}}_\alpha^*}$ then (see Exercise 2.1.2) hyperbolicity of $\tilde{\mathbf{x}}_\alpha^*$ corresponds to imposing bounds on $|\mathrm{D}\tilde{\mathbf{f}}_\alpha(\tilde{\mathbf{x}}_\alpha^*)^n\mathbf{v}|$ for all $\mathbf{v}$ in the stable and unstable eigenspaces of $\mathrm{D}\tilde{\mathbf{f}}_\alpha(\tilde{\mathbf{x}}_\alpha^*)$. Of course, it is only by making such an assignment that we can define a metric on $TM_{\mathbf{x}^*}$.

Figure 3.22 For a hyperbolic periodic orbit, the Invariant Manifold Theorem gives the existence of stable and unstable manifolds in each chart for $\mathbf{f}^q$. The charting homeomorphisms allow them to be transferred to M (shown here as S^2). $E^{\mathrm{s,u}}_{\mathbf{x}_l}$ are tangent to the images of $W^{\mathrm{s,u}}_{\mathbf{x}_l}$ on any local chart.

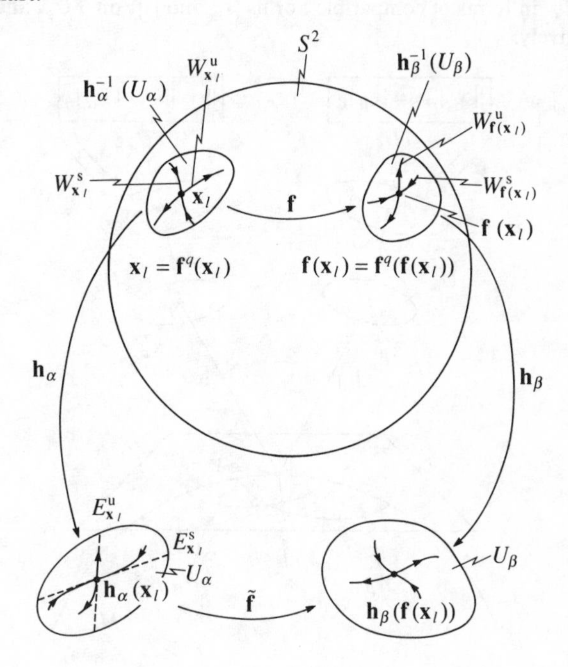

Indeed, if $\mathbf{x}^*$ lies in the overlap of two charts $(U_\alpha, \mathbf{h}_\alpha)$ and $(U_\beta, \mathbf{h}_\beta)$, then we are only allowed to choose metrics on $TU_{\tilde{\mathbf{x}}_\alpha^*}$ and $TU_{\tilde{\mathbf{x}}_\beta^*}$ that, for all $\mathbf{v} \in TM_{\mathbf{x}^*}$, satisfy $|\tilde{\mathbf{v}}_\alpha|_\alpha = |\tilde{\mathbf{v}}_\beta|_\beta$, where $\tilde{\mathbf{v}}_\alpha$ ($\tilde{\mathbf{v}}_\beta$) is the α- (β-)-representative of $\mathbf{v}$ (see Figure 3.23). The common value defines $\|\mathbf{v}\|_{\mathbf{x}^*}$ for any $\mathbf{v}$ in $TM_{\mathbf{x}^*}$. Compatible metrics, such that $\|\mathbf{v}\|_{\mathbf{x}}$ is positive definite, defined at all points $\mathbf{x}$ of all overlaps of an atlas provides a Riemannian structure for M (see Exercise 3.6.2). If M is equipped with a Riemannian structure then we can express the hyperbolic nature of $\mathbf{x}^*$ in a coordinate-free way by requiring that:

(i) $TM_{\mathbf{x}^*} = E^s_{\mathbf{x}^*} \oplus E^u_{\mathbf{x}^*}$, where $E^{s(u)}_{\mathbf{x}^*}$ is the stable (unstable) eigenspace of $T\mathbf{f}_{\mathbf{x}^*}$;

(ii) there exist c, $C > 0$ and $0 < \mu < 1$ such that, for every $n \in \mathbb{Z}^+$,

$$\|(T\mathbf{f}_{\mathbf{x}^*})^n(\mathbf{v})\|_{\mathbf{x}^*} < C\mu^n \|\mathbf{v}\|_{\mathbf{x}^*} \text{ for all } \mathbf{v} \in E^s_{\mathbf{x}^*}, \tag{3.6.4}$$

$$\|(T\mathbf{f}_{\mathbf{x}^*})^n(\mathbf{v})\|_{\mathbf{x}^*} > c\mu^{-n} \|\mathbf{v}\|_{\mathbf{x}^*} \text{ for all } \mathbf{v} \in E^u_{\mathbf{x}^*}. \tag{3.6.5}$$

With this in mind, let us consider an alternative definition of a hyperbolic periodic orbit.

Let $\mathbf{f}: M \to M$ have a q-periodic orbit, $\Lambda^{(q)} = \{\mathbf{x}_0, \mathbf{x}_1, \ldots, \mathbf{x}_{q-1}\}$. Of course, each point $\mathbf{x}_l = \mathbf{f}^l(\mathbf{x}_0)$ is a fixed point of $\mathbf{f}^q$, but we will resist the temptation to use this to test for the hyperbolicity of $\Lambda^{(q)}$. Instead, let us use the approach discussed above. The new feature in this case is that $T\mathbf{f}_{\mathbf{x}_l}$ maps $TM_{\mathbf{x}_l}$ to $TM_{\mathbf{f}(\mathbf{x}_l)}$, where $f(\mathbf{x}_l) \neq \mathbf{x}_l$. However, the Riemann structure on M allows us to deal with this change because it provides a norm, $\|\cdot\|_{\mathbf{x}}$, for every $TM_{\mathbf{x}}$, $\mathbf{x} \in M$. We must therefore

Figure 3.23 Schematic representation of the definition of a norm, $\|\cdot\|_{\mathbf{x}^*}$, on $TM_{\mathbf{x}^*}$ in terms of compatible norms $|\cdot|_\alpha$ and $|\cdot|_\beta$ on $TU_{\tilde{\mathbf{x}}_\alpha^*}$ and $TU_{\tilde{\mathbf{x}}_\beta^*}$, respectively.

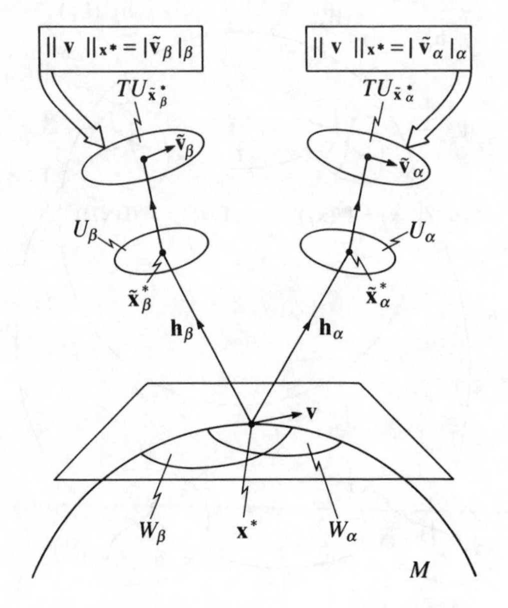

generalise (3.6.1)–(3.6.5) as follows. Let $\mathbf{x}_l$ and $\mathbf{f}(\mathbf{x}_l)$ belong to charts $(U_\alpha, \mathbf{h}_\alpha)$ and $(U_\beta, \mathbf{h}_\beta)$, respectively, then

$$\tilde{\mathbf{f}}_{\alpha\beta} = \mathbf{h}_\beta \cdot \mathbf{f} \cdot \mathbf{h}_\alpha^{-1}, \qquad \tilde{\boldsymbol{\eta}}_\alpha = \mathbf{h}_\alpha \cdot \boldsymbol{\eta}, \qquad \widetilde{(\mathbf{f} \cdot \boldsymbol{\eta})} = \mathbf{h}_\beta \cdot (\mathbf{f} \cdot \boldsymbol{\eta}) \tag{3.6.6}$$

with

$$\widetilde{(\mathbf{f} \cdot \boldsymbol{\eta})}(t) = \tilde{\mathbf{f}}_{\alpha\beta}(\tilde{\boldsymbol{\eta}}_\alpha(t)). \tag{3.6.7}$$

Differentiating with respect to t and setting $t = 0$ gives

$$\tilde{\mathbf{v}}_\beta = \dot{\widetilde{(\mathbf{f} \cdot \boldsymbol{\eta})}}(0) = \mathrm{D}\tilde{\mathbf{f}}_{\alpha\beta}(\tilde{\boldsymbol{\eta}}_\alpha(0))\dot{\tilde{\boldsymbol{\eta}}}_\alpha(0) = \mathrm{D}\tilde{\mathbf{f}}_{\alpha\beta}(\tilde{\mathbf{x}}_l)\tilde{\mathbf{v}}_\alpha. \tag{3.6.8}$$

Now, $\tilde{\mathbf{v}}_\alpha = \dot{\tilde{\boldsymbol{\eta}}}_\alpha(0)$ is a α-representative of $\mathbf{v} \in TM_{\mathbf{x}_l}$ whilst $\tilde{\mathbf{v}}_\beta = \dot{\widetilde{\mathbf{f} \cdot \boldsymbol{\eta}}}(0)$ is a β-representative of $T\mathbf{f}_{\mathbf{x}_l}(\mathbf{v}) \in TM_{\mathbf{f}(\mathbf{x}_l)}$. Thus, the tangent map $T\mathbf{f}_{\mathbf{x}_l}: TM_{\mathbf{x}_l} \to TM_{\mathbf{f}(\mathbf{x}_l)}$ and the familiar eigenspace decomposition of $TM_{\mathbf{x}_l}$ can no longer be used. Instead we require that, for each $\mathbf{x}_l$, there exist subspaces $E^{\mathrm{s}}_{\mathbf{x}_l}$ and $E^{\mathrm{u}}_{\mathbf{x}_l}$ such that $TM_{\mathbf{x}_l} = E^{\mathrm{s}}_{\mathbf{x}_l} \oplus E^{\mathrm{u}}_{\mathbf{x}_l}$ and $T\mathbf{f}_{\mathbf{x}_l}(E^{\mathrm{s,u}}_{\mathbf{x}_l}) = E^{\mathrm{s,u}}_{\mathbf{f}(\mathbf{x}_l)}$. Of course, the existence of such a decomposition is assured if $\mathbf{x}_l$ is a hyperbolic fixed point of $\mathbf{f}^q$. Finally, it is important to remember that the appropriate norms must be used in the generalisations of (3.6.4) and (3.6.5), i.e.

$$\|(T\mathbf{f}_{\mathbf{x}_l})^n \mathbf{v}\|_{\mathbf{f}^n(\mathbf{x}_l)} < C\mu^n \|\mathbf{v}\|_{\mathbf{x}_l} \text{ for all } \mathbf{v} \in E^{\mathrm{s}}_{\mathbf{x}_l}, \tag{3.6.9}$$

$$\|(T\mathbf{f}_{\mathbf{x}_l})^n \mathbf{v}\|_{\mathbf{f}^n(\mathbf{x}_l)} > c\mu^{-n} \|\mathbf{v}\|_{\mathbf{x}_l} \text{ for all } \mathbf{v} \in E^{\mathrm{u}}_{\mathbf{x}_l}. \tag{3.6.10}$$

The above discussion motivates the following definition of hyperbolic structure for more general invariant sets.

Definition 3.6.1 *An invariant set* Λ *is said to be* hyperbolic *for* $\mathbf{f}$ (*or to have a* hyperbolic structure) *if for each* $\mathbf{x} \in \Lambda$ *the tangent space* $TM_{\mathbf{x}}$ *splits into two linear subspaces* $E^{\mathrm{s}}_{\mathbf{x}}$, $E^{\mathrm{u}}_{\mathbf{x}}$ *such that*:

(i) $T\mathbf{f}_{\mathbf{x}}(E^{\mathrm{s,u}}_{\mathbf{x}}) = E^{\mathrm{s,u}}_{\mathbf{f}(\mathbf{x})}$;
(ii) (3.6.9) *and* (3.6.10), *with* $\mathbf{x}_l \mapsto \mathbf{x}$, *are satisfied for all positive integers* n;
(iii) *the subspaces* $E^{\mathrm{s}}_{\mathbf{x}}$, $E^{\mathrm{u}}_{\mathbf{x}}$ *depend continuously on* $\mathbf{x} \in \Lambda$.

Item (iii) is trivially satisfied if Λ is a periodic orbit, since the points $\mathbf{x} \in \Lambda$ are isolated. However, it is an important technical restriction for invariant sets containing a dense orbit or a dense subset of periodic orbits.

It is not difficult to accept that the invariant set $\Lambda = \bigcap_{n \in \mathbb{Z}} Q^{(n)}$ is a hyperbolic set for the horseshoe diffeomorphism $\mathbf{g}$. Observe (see § 3.5.1) that the set of vertical line segments, $\bigcap_{n \in \mathbb{N}} Q^{(-n)}$, on the square Q give rise to curves on S^2 analogous to the stable manifold of a periodic orbit. Similarly, the horizontal line segments, $\bigcap_{n \in \mathbb{Z}^+} Q^{(n)}$, lead to the analogue of the unstable manifold of the periodic orbit. At each point $\mathbf{x} \in \Lambda$ we can identify tangents to these curves to obtain $E^{\mathrm{s}}_{\mathbf{x}}$ and $E^{\mathrm{u}}_{\mathbf{x}}$. Moreover, this splitting into $E^{\mathrm{s}}_{\mathbf{x}}$ and $E^{\mathrm{u}}_{\mathbf{x}}$ depends continuously on $\mathbf{x}$, because, for any $\mathbf{x}, \mathbf{x}' \in \Lambda$, $E^s_{\mathbf{x}}$ and $E^s_{\mathbf{x}'}$, (or $E^u_{\mathbf{x}}$ and $E^u_{\mathbf{x}'}$) are tangent to diffeomorphic images of parallel line segments on the square Q. Finally, the contraction on $E^s_{\mathbf{x}}$ and expansion on $E^u_{\mathbf{x}}$ satisfy (3.6.9) and (3.6.10) with $\mu > \frac{1}{5}$ and $c = C = 1$, so that Λ has a hyperbolic structure.

In § 1.4, we noted that fixed points and periodic orbits are invariant sets that frequently appear to attract or repel the orbits of points not contained in them. What is more, they are rather special in so far as they have no proper subsets that are themselves invariant. The following theorem for diffeomorphisms whose non-wandering set, Ω, has a hyperbolic structure, provides the theoretical basis for these observations.

Theorem 3.6.1 *Let* $\mathbf{f}: M \to M$ *be a diffeomorphism on a compact manifold without boundary with a hyperbolic non-wandering set* Ω. *If the periodic points of* $\mathbf{f}$ *are dense in* Ω, *then* Ω *can be written as a disjoint union of finitely many* basic sets Ω_i, *i.e.*

$$\Omega = \Omega_1 \cup \Omega_2 \cup \cdots \cup \Omega_k. \tag{3.6.11}$$

Each Ω_i *is closed, invariant and contains a dense orbit of* $\mathbf{f}$. *Moreover, the splitting of* Ω *into basic sets is unique and* M *can be decomposed as a disjoint union*

$$M = \bigcup_{i=1}^{k} \mathrm{in}(\Omega_i), \tag{3.6.12}$$

where

$$\mathrm{in}(\Omega_i) = \{\mathbf{x} \in M \mid \mathbf{f}^m(\mathbf{x}) \to \Omega_i,\ m \to \infty\} \tag{3.6.13}$$

is the inset of Ω_i.

Diffeomorphisms with hyperbolic non-wandering set, Ω, and periodic orbits dense in Ω are usually referred to as *axiom-A* diffeomorphisms (see Chillingworth, 1976, p. 240; Nitecki, 1971, p. 189). Clearly, any diffeomorphism whose non-wandering set consists of a finite number of fixed points or periodic orbits is axiom-A. Moreover, fixed points and periodic orbits are closed, invariant sets that trivially contain a dense orbit, i.e. they are basic sets.

Theorem 3.6.1 does not merely give a decomposition of Ω. Equation (3.6.12) states that every $\mathbf{x} \in M$ belongs to the inset (or equivalently, the outset) of one and only one basic set. This means that the wandering points move between the basic sets approaching those that are attracting asymptotically. Some simple examples are illustrated in Figure 3.24.

The horseshoe diffeomorphism, $\mathbf{g}$, on the sphere is a more substantial example. The non-wandering set, Ω, of this diffeomorphism consists of the invariant set $\Lambda = \bigcap_{n \in \mathbb{Z}} Q^{(n)}$ and the two fixed points; one stable and one unstable. Ω has a hyperbolic structure and Proposition 3.5.4 shows that the periodic points of $\mathbf{g}$ are dense in Ω, so Theorem 3.6.1 applies. Obviously, each fixed point is a basic set

(Ω_1 unstable; Ω_2 stable, say) but can Λ be decomposed into a number of basic sets? Propositions 3.5.5 and 3.5.6 show that Λ contains a dense orbit of $\mathbf{g}$. This means that further decomposition of Λ is out of the question and, since it is also closed (it is a Cantor set) and invariant, the only remaining basic set (Ω_3) is Λ itself. Basic sets of this type are referred to as *chaotic sets* (see § 3.5.3 and Exercise 3.5.11).

The Anosov automorphisms provide another illustration of Theorem 3.6.1 involving a chaotic basic set. Recall that the periodic points of these maps are dense in T^n and the non-wandering set is the whole torus. Moreover, in the two-dimensional example with $\mathbf{A} = \begin{pmatrix} 1 & 1 \\ 1 & 2 \end{pmatrix}$, discussed in § 3.4, it is clear that $E^s_{\mathbf{x}_l}$ and $E^u_{\mathbf{x}_l}$ are given by $(1, (1 - 5^{1/2})/2)$ and $(1, (1 + 5^{1/2})/2)$, respectively, at every periodic point $\mathbf{x}_l$ (see Exercise 3.4.2). Continuity requires that this be so for each $\mathbf{x} \in \Omega$ because the periodic points are dense. The splitting of the tangent space is

Figure 3.24 Illustrations of Theorem 3.6.1 where the basic sets are fixed points and periodic orbits. (*a*) $\mathbf{f}: S^2 \to S^2$ has non-wandering set, Ω, consisting of two basic sets, Ω_1 and Ω_2 – both fixed points. All wandering points have α-limit set Ω_1 and ω-limit set Ω_2. (*b*) The basic sets Ω_1 and Ω_2 are unstable fixed points, Ω_3 is a saddle-like 4-cycle and Ω_4 is a stable 4-cycle. The dynamics of the wandering points are shown schematically on the right. (*c*) The basic sets are all fixed points in this case: Ω_1 is unstable; Ω_2 is stable and Ω_3, Ω_4 are saddle-like. Once again a schematic representation of the dynamics of the wandering points is given.

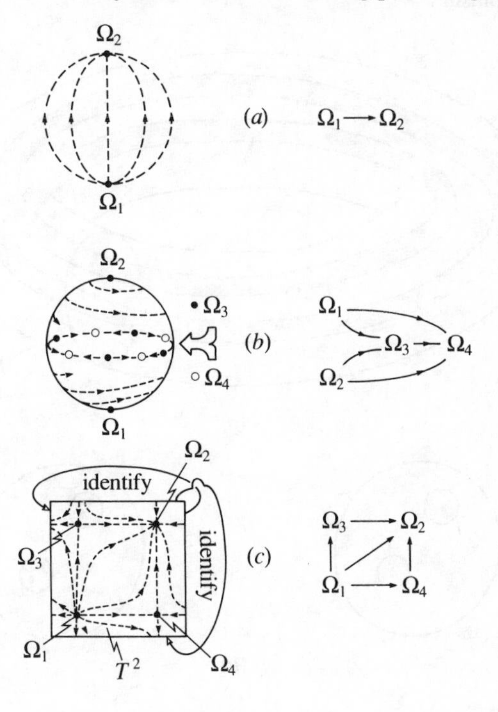

therefore trivially continuous, being the same at every point of Ω. Hyperbolic rates of contraction and expansion follow from the hyperbolicity of $\mathbf{A}$ (see Exercise 3.6.4). Hence Ω has a hyperbolic structure. In this case, there is only a single basic set $\Omega_1 = \Omega = T^2$ and it follows from Theorem 3.6.1 that the toral automorphism must have an orbit which is dense in the torus.

A further example is the transformation, $\mathbf{f}$, of the solid torus, $T = S^1 \times D^2$, shown in Figure 3.25. The torus is treated rather like a solid rubber ring. It is stretched (with consequent loss of cross-sectional area), twisted and folded to fit inside itself. Repeated application of this transformation results in longer and longer tori, wrapped around T increasingly many times. If the disc D^2 is a cross-section of T, then $D^2 \cap \left(\bigcap_{n=1}^{\infty} \mathbf{f}^n(T) \right)$ is a Cantor set. Thus, the ω-limit set of $\mathbf{f}$ is locally the product of a Cantor set and a one-manifold. This example has the important property that the chaotic basic set is an *attractor*.

The set $\Lambda = \bigcap_{n \in \mathbb{Z}} Q^{(n)}$ in the horseshoe diffeomorphism has only a one-dimensional inset. This means that most orbits are not asymptotic to Λ and this makes Λ

Figure 3.25 (*a*) Illustration of a transformation $\mathbf{f}$ of the solid torus T which has an attracting chaotic set. The image, $\mathbf{f}(T)$, of T under $\mathbf{f}$ is shown shaded. (*b*) Intersections of successive images of T under $\mathbf{f}$ with a cross section D^2 of the torus. Notice that $\mathbf{f}^{-1}$ is not defined on the whole torus, however, only forward iterations are required to observe the attracting set. The mapping $\mathbf{f}$ is sometimes called the 'spinning diffeomorphism'.

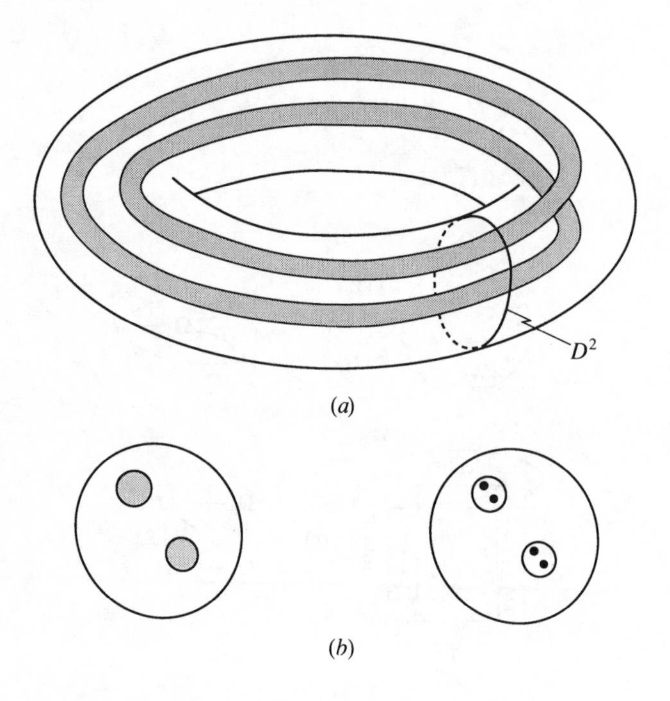

difficult to observe in numerical experiments. In principle, we can find points whose orbits under **g** remain in a given neighbourhood of Λ for an arbitrary number of iterations. However, in practice most plotted orbits spend at most a few iterations near Λ before vanishing into the sink in F. This is because, with finite computer arithmetic, we are unable to approximate $\mathrm{in}(\Lambda)$ closely enough to sustain a presence near Λ in the face of repeated five-fold expansion (see Exercise 3.5.4). Thus naive computer experiments involving the orbits of wandering points do not reveal much about the position of Λ let alone the dynamics on it. Some feeling for the latter aspect of **g** can be obtained by using symbolic dynamics (see Exercise 3.5.11).

In view of these practical difficulties, it is not surprising that chaotic basic sets are much easier to observe in numerical experiments if they are attracting. Attracting chaotic sets – often referred to as *strange attractors* – have been observed in a wide variety of computer experiments (see Figures 3.26–3.30). Detailed documentation of this area can be found in Gumowski & Mira, 1980; Helleman, 1980; Lichtenberg & Lieberman, 1982; Sparrow, 1982). Such attracting sets are not fully understood and may not be basic sets in the sense of Theorem 3.6.1. However, they do appear to have the common property that points in them occur on finer and finer scales. For example, in Figure 3.27, the Hénon attractor appears to be one-dimensional and to consist of a number of segments. Closer examination reveals that each 'segment' consists of several closely spaced curves of similar shape

Figure 3.26 *The Duffing attractor* (see Guckenheimer & Holmes, 1983, pp. 82–91 & 191–3). The Duffing equation can be written in the form

$$\dot{x} = y, \qquad \dot{y} = x - x^3 + \varepsilon(a\cos\theta - by), \qquad \dot{\theta} = 1.$$

This system is periodic in θ and the phase space can be taken as $M = \mathbb{R}^2 \times S^1$. Every surface $\theta = \text{constant}$ is a global Poincaré section so that the system behaviour is completely described by the Poincaré map $\mathbf{P}_{\varepsilon,\theta}$. Numerical approximations to $\mathbf{P}_{\varepsilon,\theta}$ appear to have a chaotic attracting set – the Euler approximation is shown in this diagram for $\varepsilon a = 0.4$ and $\varepsilon b = 0.25$. The structure of $\mathbf{P}_{\varepsilon,\theta}$ is discussed in greater detail in § 3.8.

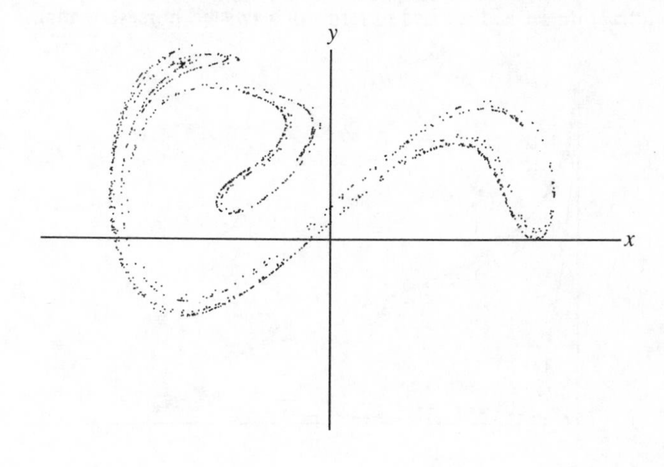

Figure 3.27 *The Hénon attractor* (see Hénon, 1976). The map **f** producing this attractor is defined by

$$(x, y) \overset{\mathbf{f}}{\mapsto} (y - ax^2 + 1, bx),$$

where $a, b \in \mathbb{R}$. The attracting set for $a = 1.4$ and $b = 0.3$ is shown. It arises from the repeated folding and stretching brought about by the action of **f**. When magnified the attractor is found to consist of many curves, of similar shape to those resolved above, occurring very close together. This 'braided' nature of the attractor appears to be repeated on all scales.

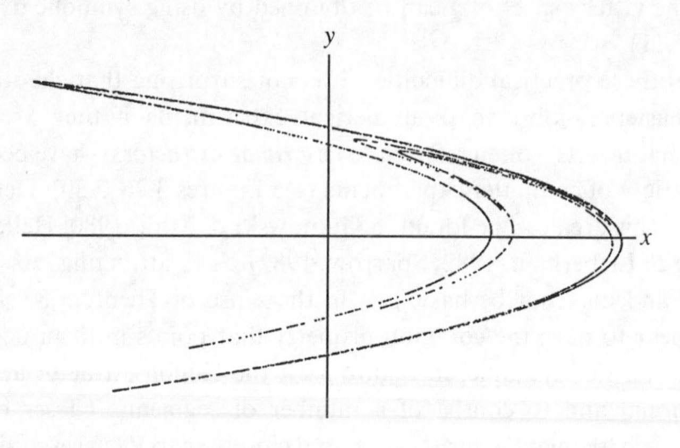

Figure 3.28 Guckenheimer *et al.* (1977) discussed a Leslie model of a density dependent population with two age classes of size x and y. It is a discrete-time model and the dynamics of the two classes are represented by the map

$$(x, y) \mapsto (r[x + y] \exp(-[x + y]/10), x),$$

where r is a real parameter. The map appears to exhibit chaotic behaviour for $r \geqslant 17$ and a typical orbit for $r = 20$ is shown here. More numerical details can be found in Guckenheimer *et al.* (1977), where the origin of the attracting set is discussed in terms of a twisted horseshoe map.

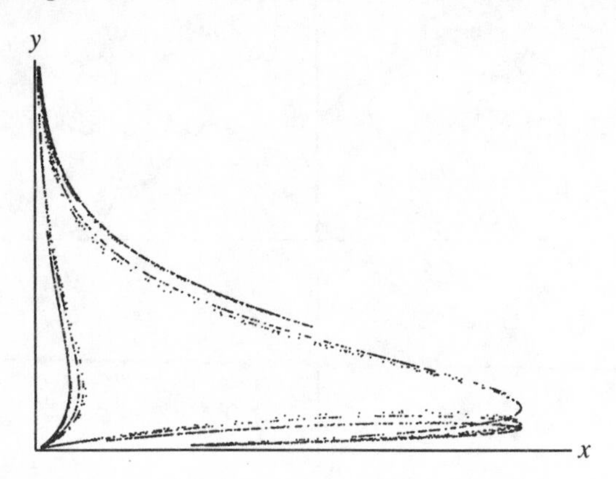

Figure 3.29 *The Lorenz attractor* (Lorenz, 1963). It must be pointed out that there is a theorem corresponding to Theorem 3.6.1 for flows so that strange attracting sets can also arise in flows that are not the suspension of a diffeomorphism. The Lorentz equations

$$\dot{x} = 10(y - x), \qquad \dot{y} = x(28 - z) - y, \qquad \dot{z} = xy - (8/3)z,$$

have fixed points at $(\pm 6(2^{1/2}), \pm 6(2^{1/2}), 27)$. The system does not have a global section so the projection onto the xz-plane is shown. The orbit generated by using the Euler method with step length of 0.005 and initial point $(x, y, z) = (0.1, 0, 0)$ is plotted. The projected orbit switches between revolving about $(x, z) = (+6(2^{1/2}), 27)$ and $(x, z) = (-6(2^{1/2}), 27)$ in an apparently random way.

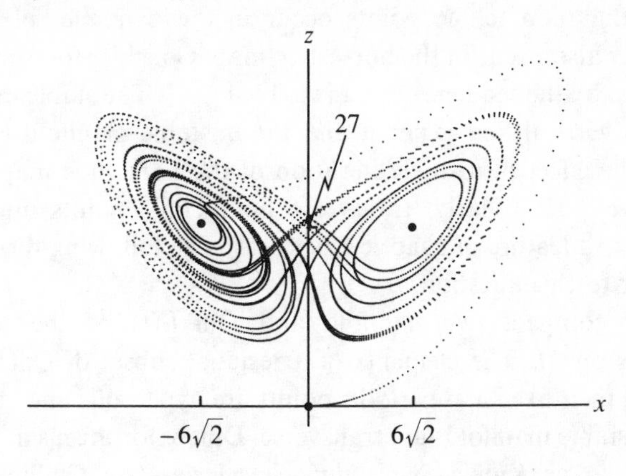

Figure 3.30 *The Rössler attractor* (Rössler, 1979). This is another three-dimensional flow exhibiting an attracting set with complex dynamics. The system equations are

$$\dot{x} = -(y + z), \qquad \dot{y} = x + ey, \qquad \dot{z} = f + xz - \mu z.$$

A perspective view of an orbit near to the attractor is shown for $e = 0.17$, $f = 0.4$ and $\mu = 8.5$. It is obtained by using the Euler method with step length 0.005 to approximate the trajectory through $(x, y, z) = (1, 0, 0)$ and plotting $u = x + y$, $v = y + z$.

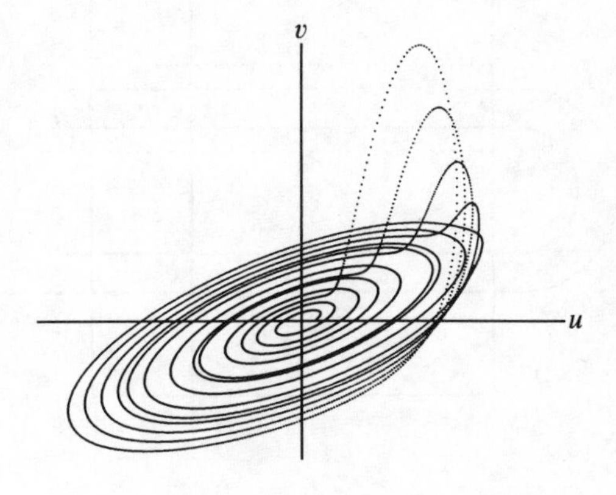

that are not resolved on the scale of Figure 3.27. Further magnification shows that each of the latter 'curves' has a similar structure, and so on. The attractor is said to have a 'braided' nature which is repeated on all scales. The reader will recall that the chaotic basic set Λ of the horseshoe diffeomorphism has this property (see Exercise 3.5.9). In fact, a theoretical connection with the basic set of the horseshoe diffeomorphism can be made in some cases: namely when homoclinic points occur.

3.7 Homoclinic points

We have seen that homoclinic points occur in the dynamics of Anosov automorphisms. They also occur in the horseshoe map. Consider for example the fixed point represented by the sequence $\{\ldots 1111 \cdot 1111 \ldots\}$. The stable manifold of this point on Q is a vertical line segment and the unstable manifold is a horizontal line segment. The effect of a single iteration of the horseshoe map $\mathbf{f}$ of § 3.5.1 is shown in Figure 3.31. Clearly, transverse homoclinic points must occur. Are homoclinic points a feature of chaotic basic sets? The following theorem provides a partial answer to this question.

Let M be a compact two-manifold and $\mathrm{Diff}^1(M)$ be the set of all C^1-diffeomorphisms on M. The elements of a residual subset of $\mathrm{Diff}^1(M)$ have the property that all their fixed and periodic points are hyperbolic and all intersections of stable and unstable manifolds are transverse. Diffeomorphisms in this subset are usually referred to as *Kupka–Smale diffeomorphisms* (see Chillingworth, 1976; p. 227; Nitecki, 1971, p. 83).

Figure 3.31 A transverse homoclinic point $\mathbf{x}^\dagger$ of the fixed point $\mathbf{x}^* = \{\ldots 11 \cdot 11 \ldots\}$ of the horseshoe map.

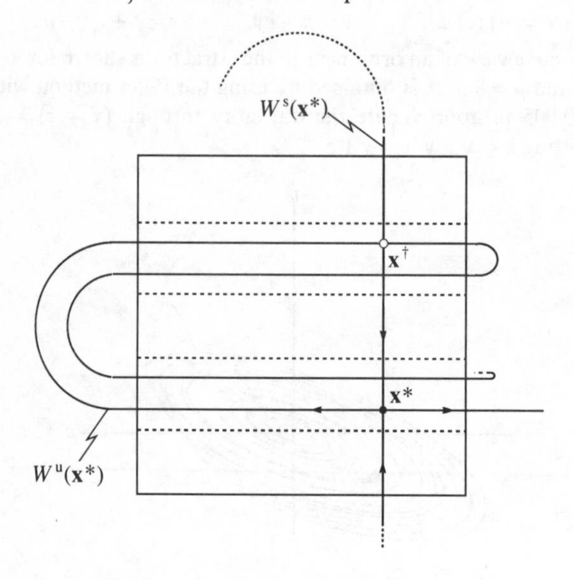

Theorem 3.7.1 (Smale–Birkhoff) *Let* $\mathbf{f} \in \mathrm{Diff}^1(M)$ *be Kupka–Smale and* $\mathbf{x}^\dagger$ *be a transverse homoclinic point of a periodic point* $\mathbf{x}^*$ *of* $\mathbf{f}$. *Then there is a closed subset* Λ *of* $\Omega(\mathbf{f})$, *containing* $\mathbf{x}^\dagger$, *such that:*

(i) Λ *is a Cantor set;*

(ii) $\mathbf{f}^p(\Lambda) = \Lambda$ *for some* $p \in \mathbb{Z}^+$;

(iii) $\mathbf{f}^p$ *restricted to* Λ *is topologically conjugate to a shift on two symbols.*

A point $\mathbf{x}^\dagger$ is a homoclinic point of a periodic point $\mathbf{x}^*$ of period q if it lies at an intersection ($\neq \mathbf{x}^*$) of the stable and unstable manifolds of the fixed point of $\mathbf{f}^q$ at $\mathbf{x}^*$.

The idea behind the proof of Theorem 3.7.1 is illustrated in Figure 3.32. If the stable and unstable manifolds of the hyperbolic saddle point $\mathbf{x}^*$ intersect at some point $\mathbf{x}_1^\dagger$, then they must intersect infinitely many times. Recall (see § 3.4) if $\mathbf{x}_1^\dagger \in W^s \cap W^u$ then $\mathbf{f}^m(\mathbf{x}_1^\dagger) \in W^s \cap W^u$ for every $m \in \mathbb{Z}$. Figure 3.32 illustrates the effect this constraint has on the two manifolds if we attempt to return them directly to $\mathbf{x}^*$ itself. As the unstable manifold approaches the saddle point the loops between adjacent homoclinic points are stretched parallel to W^u_{loc} and squeezed parallel to W^s_{loc}. The manifold therefore undergoes oscillations of increasing amplitude and decreasing period. The fate of the stable manifold is similar under reverse iterations resulting in the *homoclinic tangle* shown in Figure 3.32.

The connection with shifts on two-symbol sequences is apparent if we consider the images of a small 'parallelogram' R, containing $\mathbf{x}_1^\dagger$ and with sides parallel to W^s and W^u, under forward and reverse iterations. For $m > 0$, the mth iteration of $\mathbf{f}$ stretches $\mathbf{f}^{(m-1)}(R)$ along W^u and contracts it along W^s. Remember, $\mathbf{f}^{(m-1)}(\mathbf{x}_1^\dagger)$ is a homoclinic point and belongs to $\mathbf{f}^{(m-1)}(R)$ for every m. Eventually, for some $N \in \mathbb{Z}^+, \mathbf{f}^N(R)$, takes the horseshoe shape R_1 (see Figure 3.32). For reverse iterations

Figure 3.32 Illustration of the homoclinic tangle occurring at a hyperbolic saddle point. The parallelogram R has images $R_1 = \mathbf{f}^N(R)$ and $R_0 = f^{-N'}(R)$ intersecting in a horseshoe configuration.

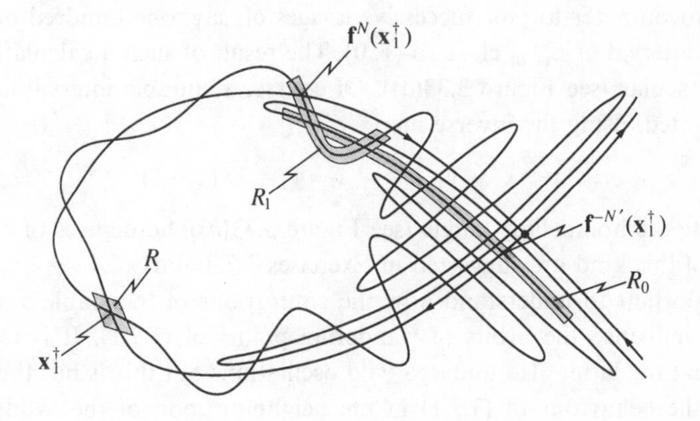

the roles of W^s and W^u are reversed and, for some $N' \in \mathbb{Z}^+$, $\mathbf{f}^{(-N')}(R) = R_0$, where R_1 and R_0 intersect as shown in Figure 3.32. Clearly, if $p = N + N'$, $\mathbf{f}^p(R_0) = R_1$ and we would expect $\mathbf{f}^p$ to exhibit horseshoe-like behaviour, i.e. be conjugate to a left shift on two-symbols. The homoclinic point referred to in Theorem 3.7.1 would in this case be $\mathbf{x}^\dagger = \mathbf{f}^{-N'}(\mathbf{x}_1^\dagger)$.

Theorem 3.7.1 means that $\mathbf{f}$ exhibits all the complexity of the left shift $\alpha: \Sigma \to \Sigma$ discussed in §3.5.2. In particular, in every neighbourhood of a transverse homoclinic point of $\mathbf{f}$, there is a periodic point. By Theorem 3.7.1, the transverse homoclinic point $\mathbf{x}^\dagger \in \Lambda$ and $\mathbf{f}^p|\Lambda$ is topologically conjugate to the left shift $\alpha: \Sigma \to \Sigma$. However, by Proposition 3.5.4 the periodic points of α are dense in Σ. Hence, periodic points of $\mathbf{f}^p|\Lambda$ are dense in Λ and, therefore, there is a periodic point of $\mathbf{f}$ arbitrarily close to $\mathbf{x}^\dagger$. Thus there are infinitely many periodic points in any neighbourhood of $\mathbf{x}^\dagger$.

It is important to realise that Theorem 3.7.1 employs sufficient conditions to ensure the existence of Λ. As Smale has pointed out (see Smale, 1963), we might expect a similar result to hold with weaker constraints on $\mathbf{f}$. Figure 3.32 suggests that the key requirement is that the stable and unstable manifolds of a hyperbolic fixed or periodic point intersect transversely. With this in mind, the following example shows that the remarkable phenomena described above really do occur. Let us examine the planar map

$$x_1 = x + y_1, \qquad y_1 = y + kx(x-1), \tag{3.7.1}$$

numerically, for $0 < k < 4$. This map has fixed points at $(x, y) = (0, 0)$ and $(1, 0)$ for all values of k. The fixed point at $(0, 0)$ is non-hyperbolic. The linear approximation to (3.7.1) at $(0, 0)$ is conjugate to an anticlockwise rotation through angle θ, where

$$2 \sin\theta = [k(4-k)]^{1/2}, \qquad 2\cos\theta = (2-k). \tag{3.7.2}$$

Linearisation at $(1, 0)$ shows that this fixed point is a hyperbolic saddle point with $E^s_{(1,0)}$ and $E^u_{(1,0)}$ given by

$$v = u\{-k - [k(4+k)]^{1/2}\}/2 \quad \text{and} \quad v = u\{-k + [k(4+k)]^{1/2}\}/2, \tag{3.7.3}$$

respectively, where (u, v) are local coordinates at $(1, 0)$. It is not difficult to then use a microcomputer to plot successive images of, say, one hundred points lying in a small interval of $E^u_{(1,0)}$ close to $(1, 0)$. The result of such a calculation can be quite spectacular (see Figure 3.33(a)). Of course, a suitable interval along $E^s_{(1,0)}$ can be iterated, using the inverse map

$$x = x_1 - y_1, \qquad y = y_1 - kx(x-1), \tag{3.7.4}$$

to complete the homoclinic tangle (see Figure 3.33(b)). Some uses of a computer program of this kind are suggested in Exercises 3.7.3 and 3.7.4.

It is important to understand how the contortions of the stable and unstable manifolds influence the orbits of wandering points of (3.7.1). It is tempting to imagine that the latter also undergo wild oscillations but this is not the case. For example, the behaviour of (3.7.1) in the neighbourhood of the saddle point is determined by Hartman's Theorem. Thus, since the eigenvalues of the linearisation at $(1, 0)$ are both positive (see Exercise 3.7.3), the orbits of individual points pass the saddle point as shown in Figure 2.1(e).

It is easily shown that the derivative map of (3.7.1) has positive determinant for all $(x, y) \in \mathbb{R}^2$ (see Exercise 3.7.5). A diffeomorphism, $\mathbf{f}: \mathbb{R}^2 \to \mathbb{R}^2$, with this property is said to be orientation-preserving (see Chillingworth, 1976, p. 139). A planar closed curve γ can be oriented in two ways depending on whether an observer walking along the oriented curve finds the region enclosed by γ on his right- or left-hand side. When $\mathrm{Det}(\mathbf{Df}(\mathbf{x})) > 0$ for all $\mathbf{x} \in \mathbb{R}^2$, it can be proved (see Exercise 3.7.6) that the orientation of the image of γ under $\mathbf{f}$ must be the same as that of γ. Now consider the closed region S_0, with boundary γ_0, shown in Figure 3.34(a) and let γ_0 be oriented according to the sense of description of the unstable manifold. It follows that the image of S_0 under (3.7.1) must be one of the regions

Figure 3.33 (a) Approximation to the unstable manifold of a hyperbolic saddle point of the planar map (3.7.1) at $(1, 0)$ for $k = 1.5$. (b) Homoclinic tangle for (3.7.1) obtained by adding to (a) an approximation to the stable manifold at $(1, 0)$. The latter is obtained by reverse iteration of a small interval of $E^s_{(1,0)}$ close to $(1, 0)$ (see Exercise 3.7.3).

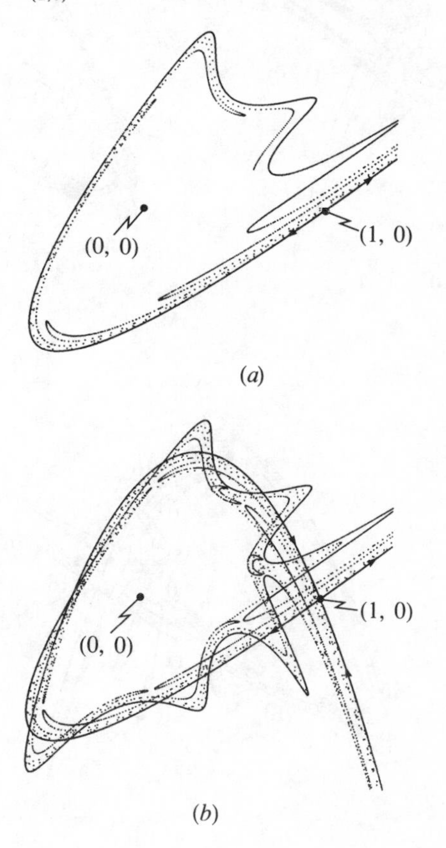

Figure 3.34 (*a*) Plot of the stable and unstable manifolds of the saddle point at (1, 0) of (3.7.1). Since the map is orientation-preserving, the image of the manifold loop S_0 must be one of loops S_i, $i = 1, 2, \ldots$, with the same orientation as S_0. It is not difficult to see that the orientation of the loops $\bar{S}_i$ is opposite to that of S_0. In fact, for (3.7.1), the image of S_i is S_{i+1} (see Figure 3.34(*c*)) but this is not the case in general. For example, the image of S_i under the square of (3.7.1) is S_{i+2}. (*b*) The images of S_0 under iterates of the inverse of (3.7.1) are the regions S_{-i} which wrap further around the fixed point at (0, 0) as i increases. (*c*) Numerical plot of the orbit of the point $P = (0.64, -0.094)$ under (3.7.1). It sweeps around (0, 0) twice, passing near to the saddle point on each occasion, before arriving in S_0 at the fifteenth iteration. Subsequent iterates are carried away to infinity under the influence of the saddle point. Note that, since the manifold loops become extremely narrow and close together, the number of revolutions of the orbit about (0, 0) before expulsion to infinity can depend sensitively on the choice of initial point.

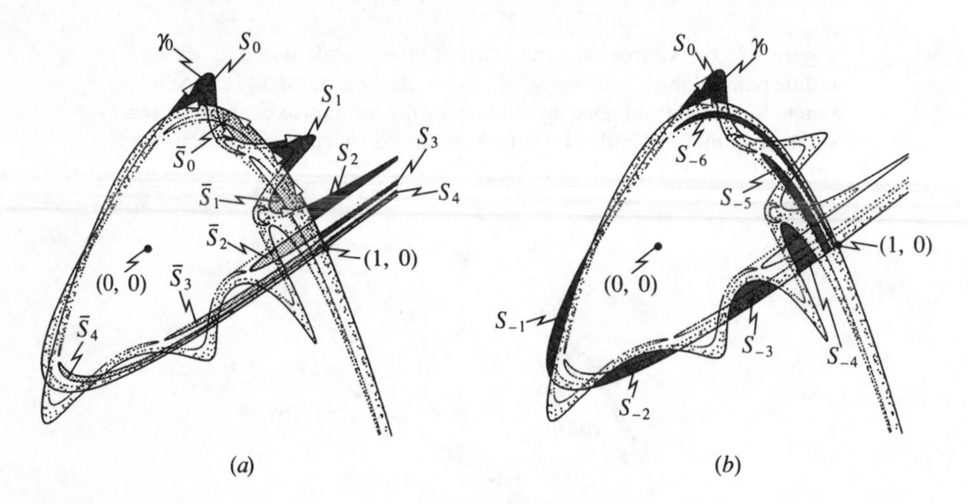

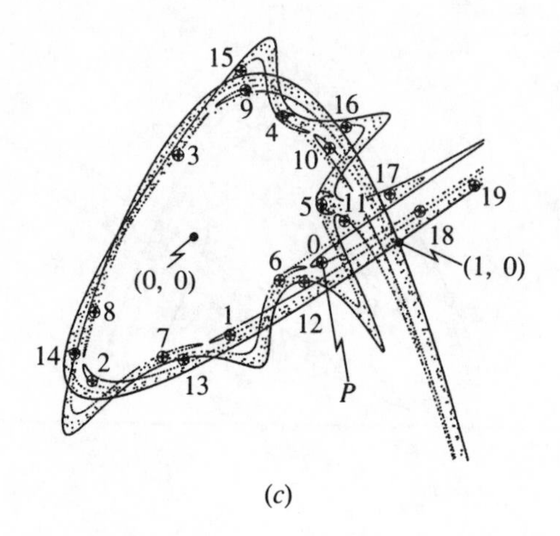

S_i, $i = 1, 2, \ldots$, with the same orientation as S_0, and not one of $\bar{S}_i$, $i = 0, 1, 2, \ldots$, for which the orientation is reversed. Thus points in S_0 are ultimately swept off to infinity under the influence of the saddle point at $(1, 0)$. Similarly, points in $\bar{S}_0$ are swept around the fixed point at $(0, 0)$ and fed back into the vicinity of the saddle point once again. The role of this movement about $(0, 0)$ in the dynamics of (3.7.1) is best understood by considering images of S_0 under powers of the inverse map. The pre-images of S_0 are a subset of the regions S_{-i}, $i = 1, 2, \ldots$, shown in Figure 3.34(*b*). Observe that, as i increases, these regions stretch further around $(0, 0)$. Indeed, for each $N \in \mathbb{Z}^+$, there is an $i(N)$ such that $S_{-i(N)}$ wraps around $(0, 0)$ N times. It follows that there are points in $S_{-i(N)}$ whose orbit makes N trips around $(0, 0)$ before it appears in S_0 and subsequently sweeps out to infinity. It is not difficult to confirm these ideas numerically. An orbit exhibiting this behaviour is shown in Figure 3.34(*c*).

Similar orbits were shown in Figure 1.39 and 1.40 for the Hénon area-preserving map. This is no coincidence. Hénon has shown (see Hénon, 1969) that every quadratic, area-preserving, planar map, with rotational linear part at the origin, is conjugate to the form (1.9.40). It is easily verified that the derivative of the map (3.7.1) has unit determinant for all $(x, y) \in \mathbb{R}^2$ (see Exercise 3.7.5)). Thus (3.7.1) and (1.9.40) must exhibit the same dynamics. For our present purpose, (3.7.1) has the advantage that the saddle point remains at $(1, 0)$ for all k, so that $E^{\mathrm{u}}_{(1,0)}$ and $E^{\mathrm{s}}_{(1,0)}$ are easily calculated.

In the above discussion, we have assumed that the stable and unstable manifolds that intersect one another come from a single fixed point $\mathbf{x}^*$. Recall that Theorem 3.7.1 includes the case where the stable and unstable manifolds involved are associated with a fixed point of $\mathbf{f}^q$. Similarly, if $\mathbf{x}^*$ is a periodic point of period greater than one, then, for example, the unstable manifold of $\mathbf{x}^*$ may intersect transversely with the stable manifold of $\mathbf{f}(\mathbf{x}^*)$ (see Figure 3.35(*a*)). Once again, the manifolds oscillate wildly because images of homoclinic points are homoclinic points. Given that the unstable manifold of $\mathbf{f}(\mathbf{x}^*)$ also intersects the stable manifold of $\mathbf{x}^*$ transversely, then consideration of the images under $\mathbf{f}$ of a suitable parallelogram, R, again indicates that some power of $\mathbf{f}$ behaves like a horseshoe map (see Figure 3.35(*b*)).

This construction is also relevant to quadratic, area-preserving maps of the plane. Suppose $\mathbf{x}^*$ has period-q and homoclinic points arise in the manner described above at each point of the periodic orbit, i.e. in the above argument $\mathbf{x}^* \mapsto \mathbf{f}^{(m-1)}(\mathbf{x}^*)$ and $\mathbf{f}(\mathbf{x}^*) \mapsto \mathbf{f}^{m'}(\mathbf{x}^*)$, $m' = m \bmod q$, for $m = 1, \ldots, q$. Then we obtain a chain of homoclinic tangles as shown in Figure 3.36. In this case, the orbit of a point such as P in this figure could sweep around the whole periodic orbit before being fed back into the vicinity of $\mathbf{x}^*$ at a different point, P'. Because of the massive stretching along the unstable manifold at each periodic point, the position of P' depends sensitively on that of P.

There is evidence of this kind of behaviour in the maps (1.9.40) and (3.7.1). The 'two-dimensional' orbits shown in Figures 1.41 and 1.42 are associated with a hyperbolic periodic orbit, they are generated by iterating a single point and their extent is similar to that of the expected homoclinic tangles (see Gumowski & Mira, 1980, p. 303). In this situation, there is a good reason (see Figure 6.17) why orbits of this kind do not escape from the influence of the periodic orbit. Therefore, the plotted iterates of a single point appear to fill out the two-dimensional region in an apparently random way.

3.8 The Melnikov function

In this section we describe a method for proving that transverse homoclinic points occur in the Poincaré maps of certain types of flow in three dimensions. This

Figure 3.35 (*a*) Illustration of the unstable manifold of the periodic point $\mathbf{x}^*$ intersecting the stable manifold of $\mathbf{f}(\mathbf{x}^*)$ transversely at $\mathbf{x}_1^\dagger$ and hence at infinitely many other homoclinic points. (*b*) The parallelogram R is iterated forward to R_1 and in reverse to R_2. The map from R_2 to R_1 is horseshoe-like.

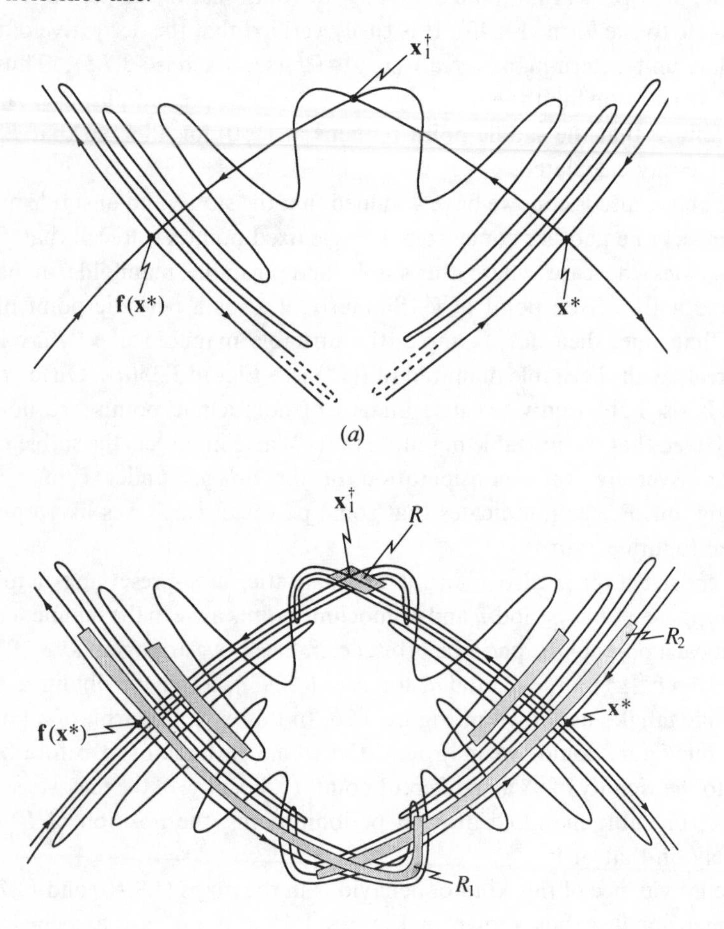

method is particularly interesting here because it can be applied to the Duffing equation which appears, numerically, to have a chaotic, attracting set (see Figure 3.26).

Consider the planar differential equation

$$\dot{\mathbf{x}} = \mathbf{f}_0(\mathbf{x}) \tag{3.8.1}$$

which has a hyperbolic saddle point at $\mathbf{x} = \mathbf{0}$ and assume there is a homoclinic saddle connection, Γ, as shown in Figure 3.37. Now consider the product flow in $\mathbb{R}^2 \times S^1$ defined by

$$\dot{\mathbf{x}} = \mathbf{f}_0(\mathbf{x}), \qquad \dot{\theta} = 1. \tag{3.8.2}$$

The saddle point of (3.8.1) at $\mathbf{x} = \mathbf{0} \in \mathbb{R}^2$ becomes a periodic orbit $\gamma_0 = \{(\mathbf{x}, \theta) \in \mathbb{R}^2 \times S^1 | \mathbf{x} = \mathbf{0}, \theta \in S^1\}$ of saddle type. Moreover, the unstable manifold of γ_0, $W^u(\gamma_0)$, intersects the stable manifold, $W^s(\gamma_0)$, in the cylindrical surface

Figure 3.36 Chain of homoclinic tangles that can arise on a hyperbolic periodic orbit.

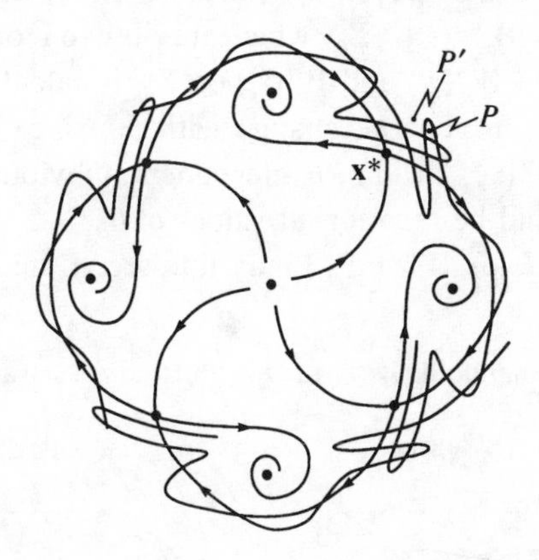

Figure 3.37 Phase portrait for $\dot{\mathbf{x}} = \mathbf{f}_0(\mathbf{x})$. The origin is a hyperbolic saddle point and Γ is a homoclinic saddle connection.

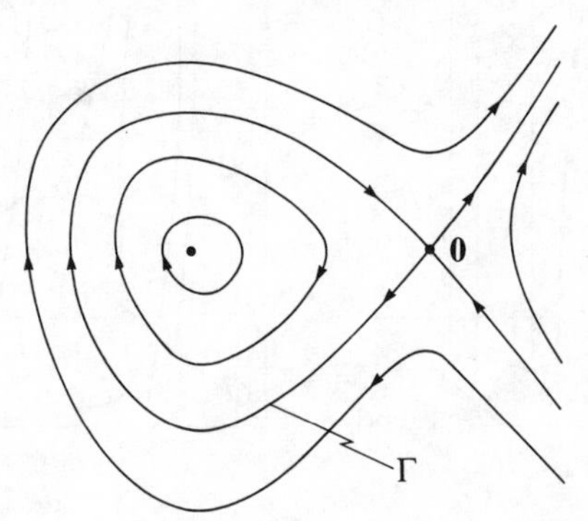

$\Gamma \times S^1 \subseteq \mathbb{R}^2 \times S^1$. This behaviour is non-generic. In particular, the stable and unstable manifolds of the corresponding fixed point of the Poincaré map, $\mathbf{P}_0$, of (3.8.2) do not intersect transversely. The Melnikov method applies to small perturbations of (3.8.2) of the form

$$\dot{\mathbf{x}} = \mathbf{f}_0(\mathbf{x}) + \varepsilon \mathbf{f}_1(\mathbf{x}, \theta); \qquad \dot{\theta} = 1 \tag{3.8.3}$$

with $\varepsilon \in \mathbb{R}^+$ and $\mathbf{f}_1(\mathbf{x}, \theta) = \mathbf{f}_1(\mathbf{x}, \theta + 2\pi)$. For sufficiently small ε, it follows from Proposition 3.2.2 that (3.8.3) also has a hyperbolic periodic orbit, γ_ε, close to γ_0. However, the invariant manifolds, $W^{\mathrm{u}}(\gamma_\varepsilon)$ and $W^{\mathrm{s}}(\gamma_\varepsilon)$, need not intersect to form a cylinder (see Figure 3.38). The Melnikov function is related to the 'distance' between these two manifolds.

Let $\mathbf{x}_0 \in \mathbb{R}^2$ be a point of the saddle connection Γ in the unperturbed system (3.8.1). Take a perpendicular section L to the saddle connection at $\mathbf{x}_0$. We use the point $\mathbf{x}_0$ and the section L in the $\theta = \theta_0$-plane, Σ_{θ_0}, as follows. Consider the perturbed system and the intersections of γ_ε, $W^{\mathrm{u}}(\gamma_\varepsilon)$ and $W^{\mathrm{s}}(\gamma_\varepsilon)$ with Σ_{θ_0}. This is equivalent to studying the Poincaré map $\mathbf{P}_{\varepsilon,\theta_0}\colon \Sigma_{\theta_0} \to \Sigma_{\theta_0}$ of the flow (3.8.3). $\mathbf{P}_{\varepsilon,\theta_0}$ will have a hyperbolic saddle point, $\mathbf{x}^*_{\varepsilon,\theta_0}$, near to $\mathbf{x} = \mathbf{0}$, with stable and unstable manifolds, $W^{\mathrm{u,s}}(\mathbf{x}^*_{\varepsilon,\theta_0}) = W^{\mathrm{u,s}}(\gamma_\varepsilon) \cap \Sigma_{\theta_0}$, which are close to Γ on Σ_{θ_0} (see Figure 3.39).

The distance between $W^{\mathrm{u}}(\gamma_\varepsilon)$ and $W^{\mathrm{s}}(\gamma_\varepsilon)$ on Σ_{θ_0} is calculated along L. Observe that this distance will, in general, change with θ_0, since $\varepsilon > 0$ implies that the curves $W^{\mathrm{u}}(\mathbf{x}^*_{\varepsilon,\theta_0})$ and $W^{\mathrm{s}}(\mathbf{x}^*_{\varepsilon,\theta_0})$ will be θ_0-dependent. Obviously, for the special case $\varepsilon = 0$, the distance would be zero for all values of θ_0.

Of course, the manifolds $W^{\mathrm{u,s}}(\mathbf{x}^*_{\varepsilon,\theta_0})$ may intersect L many times, however, on

Figure 3.38 The manifolds $W^{\mathrm{u}}(\gamma_\varepsilon)$ *and* $W^{\mathrm{s}}(\gamma_\varepsilon)$ for (*a*) $\varepsilon = 0$ and (*b*) $\varepsilon > 0$.

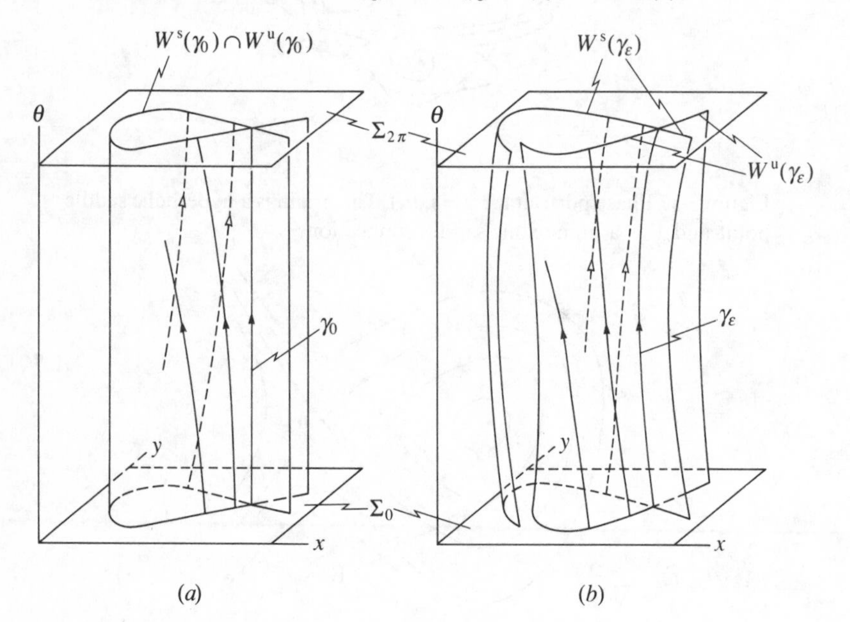

each curve there will be a unique point of intersection $A^{u,s}$, closest to $\mathbf{x}_0$ (see Figure 3.39). Let $(\mathbf{x}^{u,s}(t; \theta_0, \varepsilon), t)$, $t \in \mathbb{R}$, be the unique trajectory of (3.8.3) passing through $A^{u,s}$ at $t = \theta_0$, i.e. $A^{u,s}$ is the point $\mathbf{x}^{u,s}(\theta_0; \theta_0, \varepsilon) \in \Sigma_{\theta_0}$. We then define the time-dependent distance function,

$$\Delta_\varepsilon(t, \theta_0) = \mathbf{f}_0(\mathbf{x}_0(t - \theta_0)) \wedge [\mathbf{x}^u(t; \theta_0, \varepsilon) - \mathbf{x}^s(t; \theta_0, \varepsilon)], \tag{3.8.4}$$

where $\mathbf{x}_0(t)$ is the homoclinic trajectory of (3.8.1) with $\mathbf{x}_0(0) = \mathbf{x}_0$. In (3.8.4) the wedge product is defined by $\mathbf{a} \wedge \mathbf{b} = a_1 b_2 - a_2 b_1$ where $\mathbf{a}, \mathbf{b} \in \mathbb{R}^2$ have Cartesian coordinates (a_1, a_2) and (b_1, b_2), respectively. It follows that $\Delta_\varepsilon(t, \theta_0)$ is $|\mathbf{f}_0(\mathbf{x}_0(t - \theta_0))|$ times the component of the vector $[\mathbf{x}^u(t; \theta_0, \varepsilon) - \mathbf{x}^s(t; \theta_0, \varepsilon)]$ perpendicular to $\mathbf{f}_0(\mathbf{x}_0(t - \theta_0))$. The latter vector is, of course, tangent to Γ at $\mathbf{x}_0(t - \theta_0)$. Thus, $\Delta_\varepsilon(\theta_0, \theta_0)/|\mathbf{f}_0(\mathbf{x}_0)|$ is the distance between $W^u(\gamma_\varepsilon)$ and $W^s(\gamma_\varepsilon)$ measured along L on Σ_{θ_0}.

We can obtain a useful form for $\Delta_\varepsilon(\theta_0, \theta_0)$ by studying (3.8.4) more closely. Let

$$\mathbf{x}^u(t; \theta_0, \varepsilon) = \mathbf{x}_0(t - \theta_0) + \varepsilon \mathbf{x}_1^u(t, \theta_0) + O(\varepsilon^2) \tag{3.8.5}$$

and

$$\mathbf{x}^s(t; \theta_0, \varepsilon) = \mathbf{x}_0(t - \theta_0) + \varepsilon \mathbf{x}_1^s(t, \theta_0) + O(\varepsilon^2), \tag{3.8.6}$$

where $\mathbf{x}_1^u, \mathbf{x}_1^s$ are first variations with respect to ε. Thus, (see Exercise 3.8.1)

$$\dot{\mathbf{x}}_1^{u,s}(t, \theta_0) = \mathrm{D}\mathbf{f}_0(\mathbf{x}_0(t - \theta_0))\mathbf{x}_1^{u,s}(t, \theta_0) + \mathbf{f}_1(\mathbf{x}_0(t - \theta_0), t). \tag{3.8.7}$$

Now define

$$\Delta_\varepsilon^{u,s}(t, \theta_0) = \mathbf{f}_0(\mathbf{x}_0(t - \theta_0)) \wedge \varepsilon \mathbf{x}_1^{u,s}(t, \theta_0), \tag{3.8.8}$$

so that $\Delta_\varepsilon(t, \theta_0)$ in (3.8.4) can be written in the form

$$\Delta_\varepsilon(t, \theta_0) = \Delta_\varepsilon^u(t, \theta_0) - \Delta_\varepsilon^s(t, \theta_0) + O(\varepsilon^2). \tag{3.8.9}$$

Figure 3.39 The intersections of γ_ε, $W^u(\gamma_\varepsilon)$ and $W^s(\gamma_\varepsilon)$ with Σ_{θ_0} for $\varepsilon = 0$ and $\varepsilon > 0$.

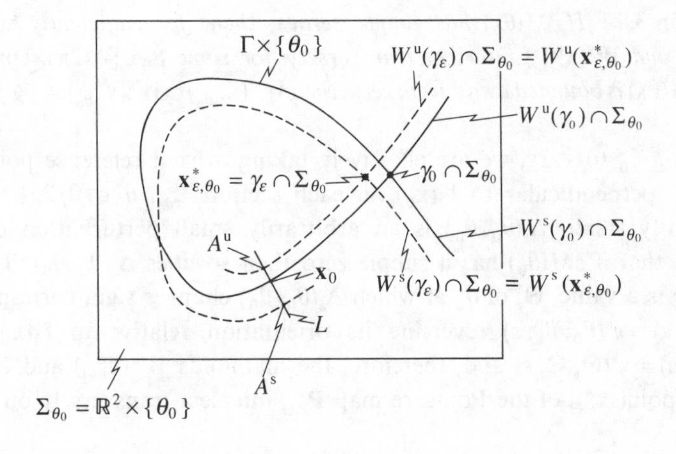

We can obtain differential equations for Δ_ε^u and Δ_ε^s. It can be shown that, since $\dot{\mathbf{x}}_0(t-\theta_0)=\mathbf{f}_0(\mathbf{x}_0(t-\theta_0))$,

$$\dot{\Delta}_\varepsilon^u(t,\theta_0)=\varepsilon[\mathrm{Tr}(\mathbf{Df}_0(\mathbf{x}_0(t-\theta_0)))\mathbf{f}_0(\mathbf{x}_0(t-\theta_0))\wedge \mathbf{x}_1^u(t,\theta_0) + \mathbf{f}_0(\mathbf{x}_0(t-\theta_0))\wedge \mathbf{f}_1(\mathbf{x}_0(t-\theta_0),t)]. \quad (3.8.10)$$

The expression (3.8.10) is greatly simplified if $\mathbf{f}_0$ is a Hamiltonian vector field, as it is for the Duffing equation, for then $\mathrm{Tr}(\mathbf{Df}_0(\mathbf{x}))\equiv 0$ (see (1.9.21 and 24)) and

$$\dot{\Delta}_\varepsilon^u(t,\theta_0)=\varepsilon\mathbf{f}_0(\mathbf{x}_0(t-\theta_0))\wedge \mathbf{f}_1(\mathbf{x}_0(t-\theta_0),t). \quad (3.8.11)$$

Integration of (3.8.11) from $t=-\infty$ to $t=\theta_0$ gives

$$\Delta_\varepsilon^u(\theta_0,\theta_0)=\varepsilon\int_{-\infty}^{\theta_0}\mathbf{f}_0(\mathbf{x}_0(t-\theta_0))\wedge \mathbf{f}_1(\mathbf{x}_0(t-\theta_0),t)\,\mathrm{d}t. \quad (3.8.12)$$

Here we have noted that $\Delta_\varepsilon^u(-\infty,\theta_0)=0$ because $\mathbf{x}_0(-\infty)=\mathbf{0}=\mathbf{f}_0(\mathbf{0})$. A similar calculation leads to

$$\Delta_\varepsilon^s(\theta_0,\theta_0)=-\varepsilon\int_{\theta_0}^{\infty}\mathbf{f}_0(\mathbf{x}_0(t-\theta_0))\wedge \mathbf{f}_1(\mathbf{x}_0(t-\theta_0),t)\,\mathrm{d}t. \quad (3.8.13)$$

and therefore,

$$\Delta_\varepsilon(\theta_0,\theta_0)=\varepsilon\int_{-\infty}^{\infty}\mathbf{f}_0(\mathbf{x}_0(t-\theta_0))\wedge \mathbf{f}_1(\mathbf{x}_0(t-\theta_0),t)\,\mathrm{d}t+O(\varepsilon^2). \quad (3.8.14)$$

Finally, we define the *Melnikov function*, $M(\theta_0)$, by

$$M(\theta_0)=\int_{-\infty}^{\infty}\mathbf{f}_0(\mathbf{x}_0(t-\theta_0))\wedge \mathbf{f}_1(\mathbf{x}_0(t-\theta_0),t)\,\mathrm{d}t, \quad (3.8.15)$$

so that

$$\Delta_\varepsilon(\theta_0,\theta_0)=\varepsilon M(\theta_0)+O(\varepsilon^2). \quad (3.8.16)$$

Proposition 3.8.1 *If $M(\theta_0)$ has simple zeroes, then, for sufficiently small $\varepsilon>0$, $W^u(\mathbf{x}_{\varepsilon,\theta_0}^*)$ and $W^s(\mathbf{x}_{\varepsilon,\theta_0}^*)$ intersect transversely for some $\theta_0\in[0,2\pi)$. On the other hand, if $M(\theta_0)$ is bounded away from zero, then $W^u(\mathbf{x}_{\varepsilon,\theta_0}^*)\cap W^s(\mathbf{x}_{\varepsilon,\theta_0}^*)=\varnothing$ for all θ_0.*

In allowing θ_0 to vary, we are effectively taking a fixed reference point $\mathbf{x}_0$ and section L, perpendicular to $\mathbf{f}_0(\mathbf{x}_0)$, in each section Σ_{θ_0}, $\theta_0\in[0,2\pi)$. By taking ε sufficiently small, $\Delta_\varepsilon(\theta_0,\theta_0)$ is an arbitrarily small perturbation of $\varepsilon M(\theta_0)$. It follows that if $\varepsilon M(\theta_0)$ has a simple zero then so does $\Delta_\varepsilon(\theta_0,\theta_0)$. This means that there is a value, Θ, of θ_0 at which $\Delta_\varepsilon(\theta_0,\theta_0)$ changes sign, corresponding to $\mathbf{x}^u(\theta_0;\theta_0,\varepsilon)-\mathbf{x}^s(\theta_0;\theta_0,\varepsilon)$ reversing its orientation relative to $\mathbf{f}_0(\mathbf{x}_0)$. Clearly $\mathbf{x}^u(\Theta;\Theta,\varepsilon)=\mathbf{x}^s(\Theta;\Theta,\varepsilon)$ and, therefore, the manifolds $W^u(\mathbf{x}_{\varepsilon,\Theta}^*)$ and $W^s(\mathbf{x}_{\varepsilon,\Theta}^*)$ of the fixed point $\mathbf{x}_{\varepsilon,\Theta}^*$ of the Poincaré map $\mathbf{P}_{\varepsilon,\Theta}$ intersect transversely on L near to

$\mathbf{x}_0$. Of course, all the Poincaré maps $\mathbf{P}_{\varepsilon,\theta_0}$, $\theta_0 \in [0, 2\pi)$, are topologically conjugate (see Exercise (1.7.3)) and, consequently, $W^{\mathrm{u}}(\mathbf{x}^*_{\varepsilon,\theta_0})$ and $W^{\mathrm{s}}(\mathbf{x}^*_{\varepsilon,\theta_0})$ must intersect transversely for all $\theta_0 \in [0, 2\pi)$ (although, obviously, not always near to $\mathbf{x}_0$ (see Figure 3.40). Equally, if $M(\theta_0)$ is bounded away from zero, then, for sufficiently

Figure 3.40 The manifolds $W^{\mathrm{u}}(\gamma_\varepsilon)$ and $W^{\mathrm{s}}(\gamma_\varepsilon)$ intersect in a homoclinic trajectory that ultimately approaches γ_ε as $t \to \infty$. When $\Delta_\varepsilon(\theta_0, \theta_0)$ has simple zeroes, this trajectory passes through the section $L \times [0, 2\pi)$ at least twice. An impression of the nature of the homoclinic trajectory can be gained by recalling that $\theta = 2\pi$ is to be identified with $\theta = 0$. Thus the segment B_0B_1 continues as B_1B_2 and C_0C_1 as C_1C_2. The trajectory itself is $\bigcup_{n\in\mathbb{Z}} (B_{n-1}B_n) \cup (C_{n-1}C_n)$. It follows that corresponding pairs of zeroes occur for any choice of $\mathbf{x}_0 \in \Gamma$. Moreover, if $M(\theta_0)$ is bounded away from zero on $[0, 2\pi)$, then it is so, independently of the choice of $\mathbf{x}_0$, and no homoclinic points occur. For given θ_0, the stable and unstable manifolds of the fixed point $\mathbf{x}^*_{\varepsilon,\theta_0}$ of the Poincaré map $\mathbf{P}_{\varepsilon,\theta_0}$ are obtained by taking the corresponding section in this figure.

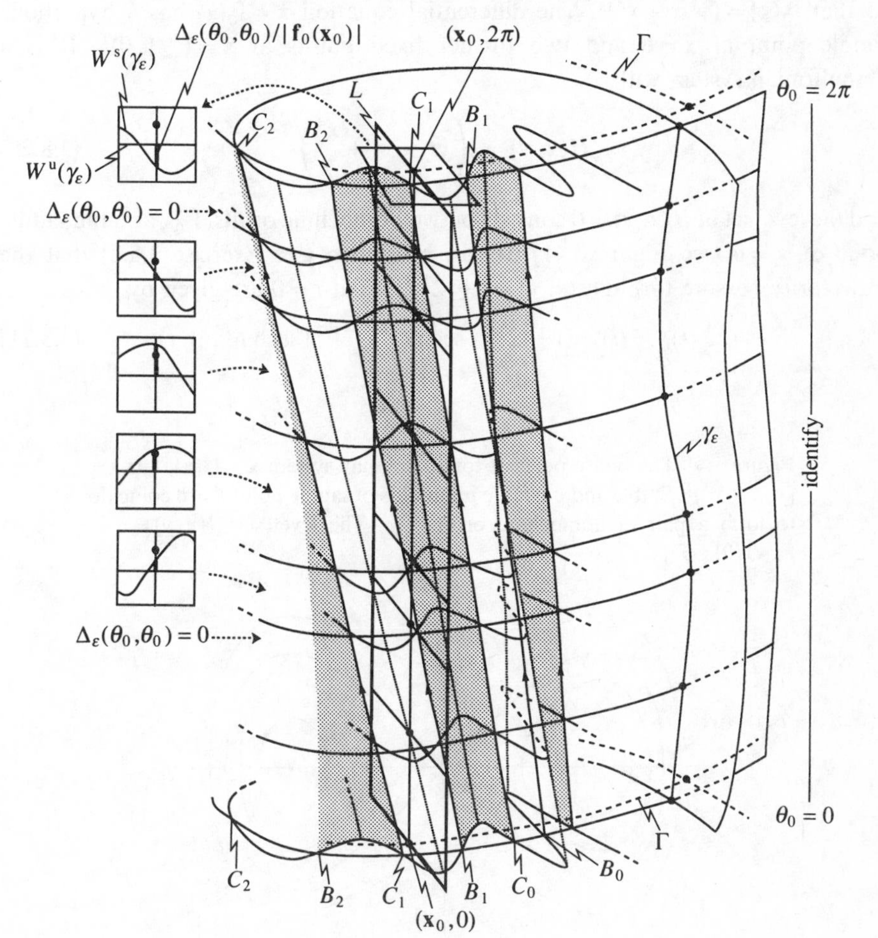

small ε, so is $\Delta_\varepsilon(\theta_0, \theta_0)$. This, in turn, means that transverse homoclinic points do not occur on L for any $\theta_0 \in [0, 2\pi)$. As Figure 3.40 shows, this conclusion does not depend on the choice of $\mathbf{x}_0 \in \Gamma$ through which L passes. Hence, there are no homoclinic points.

Example 3.8.1 Show that the Poincaré map of the Duffing equation

$$\dot{x} = y, \qquad \dot{y} = x - x^3 + \varepsilon(a \cos \theta - by), \qquad \dot{\theta} = 1, \tag{3.8.17}$$

$a, b > 0$, has transverse homoclinic points, for sufficiently small values of ε, provided

$$\frac{a}{b} > \frac{4 \cosh(\pi/2)}{3(2^{1/2})\pi}. \tag{3.8.18}$$

Solution. When $\varepsilon = 0$, (3.8.17) becomes

$$\dot{x} = y, \qquad \dot{y} = x - x^3, \qquad \dot{\theta} = 1, \tag{3.8.19}$$

so that $\mathbf{f}_0(\mathbf{x}) = (y, x - x^3)^{\mathrm{T}}$. The differential equation $\dot{\mathbf{x}} = \mathbf{f}_0(\mathbf{x})$ has a hyperbolic saddle point at $\mathbf{x} = \mathbf{0}$ and two further fixed points at $\mathbf{x} = (\pm 1, 0)^{\mathrm{T}}$. It is a Hamiltonian system with

$$H(x, y) = \tfrac{1}{2}\left(y^2 - x^2 + \frac{x^4}{2}\right) \tag{3.8.20}$$

and the level set of $H(x, y) = 0$ consists of two homoclinic orbits, $\Gamma_0^{\pm}$, and the saddle point at $\mathbf{x} = \mathbf{0}$ (see Figure 3.41). It can be shown (see Exercise 3.8.2) that the trajectories passing through $(x, y) = (\pm(2^{1/2}), 0)$ at $t = 0$ are given by

$$(x^{\pm}(t), y^{\pm}(t)) = (\pm(2^{1/2}) \operatorname{sech} t, \mp(2^{1/2}) \operatorname{sech} t \tanh t). \tag{3.8.21}$$

Figure 3.41 The phase portrait for the planar system $\dot{\mathbf{x}} = \mathbf{f}_0(\mathbf{x})$, $\mathbf{f}_0(\mathbf{x}) = (y, x - x^3)^{\mathrm{T}}$. Stable and unstable manifolds of saddle point $\mathbf{x} = \mathbf{0}$ coincide to form a pair of homoclinic orbits $\Gamma_0^{\pm}$. The level set $H(x, y) = 0$ is $\Gamma_0^- \cup \{\mathbf{0}\} \cup \Gamma_0^+$.

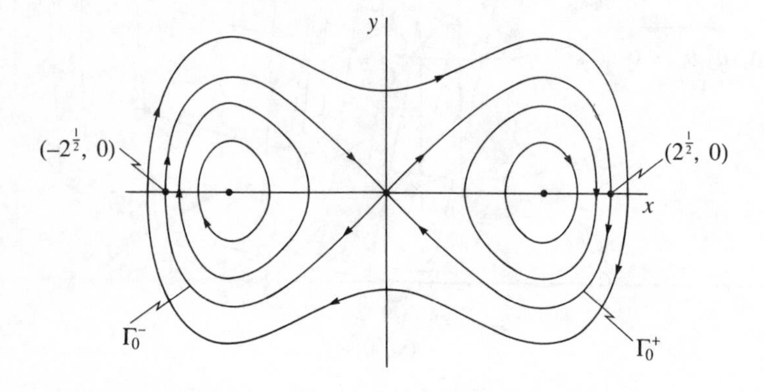

Comparison of (3.8.17) with (3.8.3) gives

$$\mathbf{f}_1(\mathbf{x}, \theta) = (0, a\cos\theta - by)^{\mathrm{T}}, \tag{3.8.22}$$

which satisfies $\mathbf{f}_1(\mathbf{x}, \theta) = \mathbf{f}_1(\mathbf{x}, \theta + 2\pi)$. It follows (from (3.8.15)) that the Melnikov function for the homoclinic orbit Γ_0^+ is

$$M(\theta_0) = -2^{1/2}\int_{-\infty}^{\infty} \operatorname{sech}(t-\theta_0)\tanh(t-\theta_0)[a\cos(t) + 2^{1/2}b\operatorname{sech}(t-\theta_0)\tanh(t-\theta_0)]\,\mathrm{d}t. \tag{3.8.23}$$

The change of variable of integration $t \mapsto t - \theta_0$ gives

$$M(\theta_0) = -2^{1/2}a\int_{-\infty}^{\infty} \operatorname{sech}((t)\tanh(t)\cos(t+\theta_0)\,\mathrm{d}t - 2b\int_{-\infty}^{\infty} \operatorname{sech}^2(t)\tanh^2(t)\,\mathrm{d}t. \tag{3.8.24}$$

The latter integral is easily evaluated, while the former can be simplified by writing $\cos(t+\theta_0) = \cos(t)\cos(\theta_0) - \sin(t)\sin(\theta_0)$ and noting that

$$\int_{-\infty}^{\infty} \operatorname{sech}(t)\tanh(t)\cos(t)\,\mathrm{d}t = 0 \tag{3.8.25}$$

because the integrand is an odd function of t. Thus,

$$M(\theta_0) = 2^{1/2}a\sin(\theta_0)\int_{-\infty}^{\infty} \operatorname{sech}(t)\tanh(t)\sin(t)\,\mathrm{d}t - \frac{4b}{3}. \tag{3.8.26}$$

The integral occurring in (3.8.26) can be evaluated using the method of residues (see Exercise 3.8.4)) and we finally obtain

$$M(\theta_0) = -\frac{4b}{3} + 2^{1/2}\pi a\operatorname{sech}\left(\frac{\pi}{2}\right)\sin(\theta_0). \tag{3.8.27}$$

Clearly, if (3.8.18) is satisfied $M(\theta_0)$ has simple zeroes and, by Proposition 3.8.1, transverse homoclinic points must occur. On the other hand, if the reverse inequality is satisfied, $M(\theta_0)$ is bounded away from zero and Proposition 3.8.1 implies that there are no homoclinic points. □

There is one remaining possibility for the system (3.8.17): namely that

$$\frac{a}{b} = \frac{4\cosh(\pi/2)}{3(2^{1/2})\pi}. \tag{3.8.28}$$

In this case, $M(\theta_0)$ has a double zero at $\theta_0 = 3\pi/2$. This corresponds to $W^{\mathrm{u}}(\mathbf{x}^*_{\varepsilon,3\pi/2})$ and $W^{\mathrm{s}}(\mathbf{x}^*_{\varepsilon,3\pi/2})$ meeting tangentially rather than transversely. As before, the orbit of such a homoclinic point under $\mathbf{P}_{\varepsilon,3\pi/2}$ consists entirely of tangential intersections

Figure 3.42 (After Ueda, in Guckenheimer & Holmes, 1983, p. 192.) Stable and unstable manifolds for the Poincaré map of the Duffing equation (3.8.17) with $\varepsilon b = 0.25$ and (*a*) $\varepsilon a = 0.11$; (*b*) $\varepsilon a = 0.19$; (*c*) $\varepsilon a = 0.30$. Observe the tangency of the stable and unstable manifolds in (*b*).

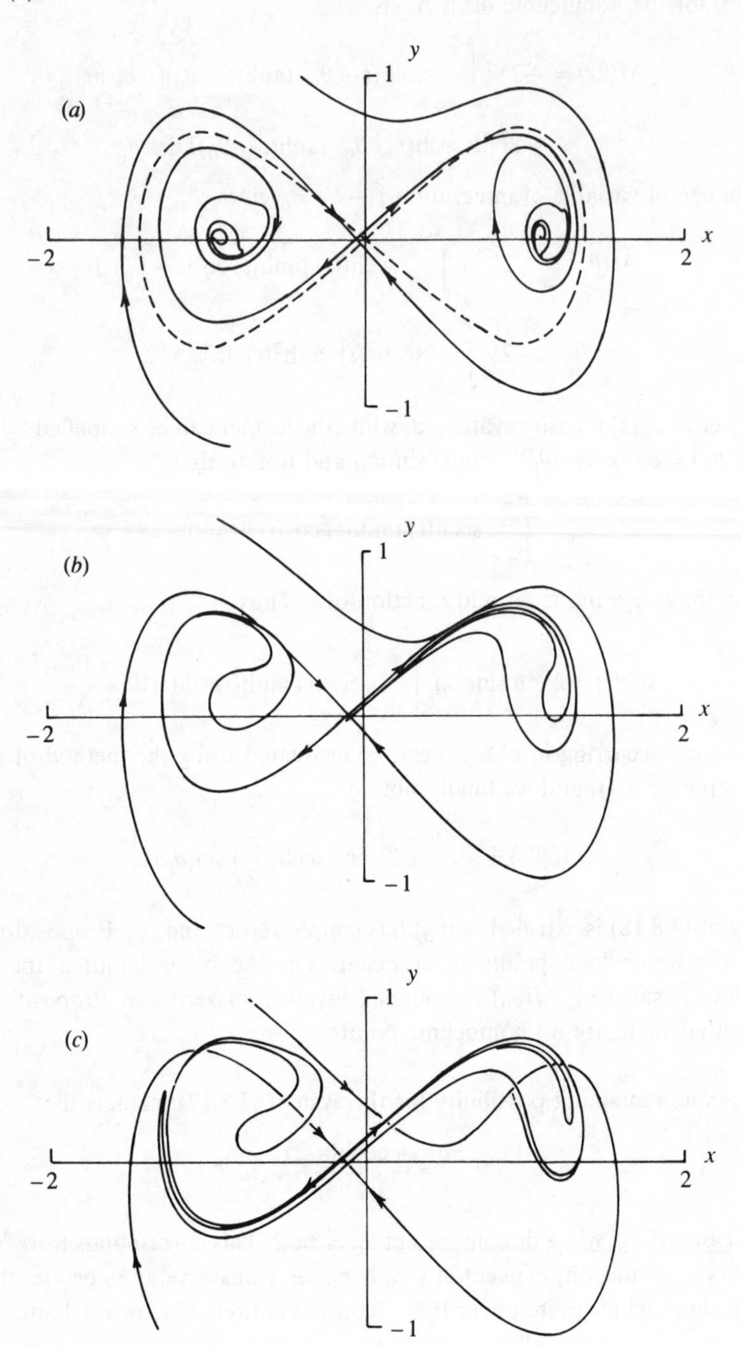

Figure 3.43 (After Ueda, in Guckenheimer & Holmes, 1983, p. 90.) Comparison of the shape of the unstable manifold of the saddle point and the attracting set for the Poincaré map of the Duffing equation (3.8.17): when (*a*) $\varepsilon a = 0.40$, $\varepsilon b = 0.25$; (*b*) $\varepsilon a = 0.30$, $\varepsilon b = 0.20$.

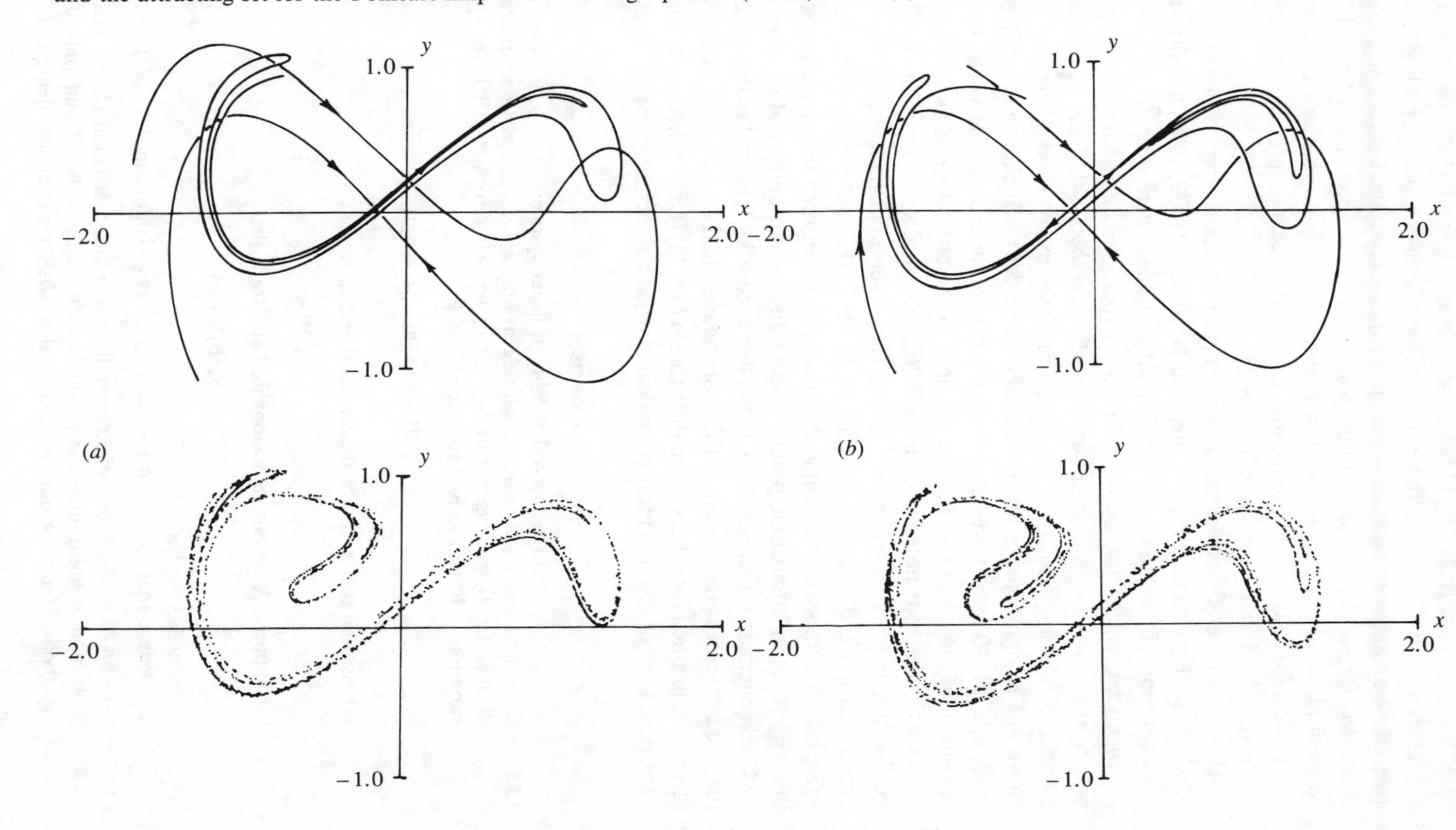

of $W^u(\mathbf{x}^*_{\varepsilon,3\pi/2})$ and $W^s(\mathbf{x}^*_{\varepsilon,3\pi/2})$. Moreover, since $\mathbf{P}_{\varepsilon,\theta_0}$ and $\mathbf{P}_{\varepsilon,\theta_0'}$ are topologically conjugate for all θ_0 and θ_0', these homoclinic tangencies occur in all $\mathbf{P}_{\varepsilon,\theta_0}$. Ueda (see Guckenheimer & Holmes, 1983, p. 192) has computed stable and unstable manifolds for the hyperbolic saddle point of the Poincaré map of (3.8.17) and some of his results are reproduced in Figure 3.42. It is not difficult to verify that the value of a/b at which homoclinic tangencies occur numerically (see Figure 3.42(*b*)), is in close agreement with (3.8.28).

The occurrence of homoclinic tangencies has important repercussions the details of which are beyond the scope of this text. Newhouse (1979, 1980) has shown that if such a tangency occurs at $\mathbf{x}_1$ for $\mathbf{f} \in \mathrm{Diff}^r(\mathbb{R}^2)$, then there is an $\tilde{\mathbf{f}}$ ε–C^r-close to $\mathbf{f}$ for which tangencies also occur stably in a hyperbolic invariant set. This set lies near to the orbit of $\mathbf{x}_1$ and is known as a *wild hyperbolic set*. $\tilde{\mathbf{f}}$ also has an infinite number of stable periodic orbits – or 'infinitely many sinks' – as the title of Newhouse's original paper had it. We refer the reader to Guckenheimer & Holmes, 1983, pp. 331–40 for a more detailed description of these ideas. However, this kind of behaviour may occur in $\mathbf{P}_{\varepsilon,\theta}$ for a/b near to the critical value (3.8.28).

As we have already noted (see Figure 3.26), numerical approximations to (3.8.17) exhibit a complicated attracting set. Such a set appears even in the Euler approximation and it is then not difficult to verify that a/b must exceed a critical value before it appears. This suggests a connection between the attracting set and the occurrence of homoclinic points. Indeed, the careful numerical work of Ueda (in Guckenheimer & Holmes, 1983, p. 90) (see Figure 3.43) has led to the conjecture that the attractor is the closure of the unstable manifold of the saddle point. While this can be justified for a/b less than the value in (3.8.28) (Guckenheimer & Holmes, 1983, p. 91), the situation is more complicated when homoclinic points are present.

Exercises

3.1 Structural stability of linear systems

3.1.1 Consider a real, $n \times n$ matrix, $\mathbf{A}$, with eigenvalues $\lambda_1, \ldots, \lambda_n$ that are not necessarily distinct. Let $\mathbf{B}$, with eigenvalues $\mu_1, \ldots, \mu_n$, be ε-close to $\mathbf{A}$ in $L(\mathbb{R}^n)$. The spectral variation of $\mathbf{B}$ with respect to $\mathbf{A}$ is defined by

$$S_{\mathbf{A}}(\mathbf{B}) = \max_j [\min_i (|\lambda_i - \mu_j|)]. \tag{E3.1}$$

(a) Assume that $\mathbf{A}$ can be diagonalised and show that

$$S_{\mathbf{A}}(\mathbf{B}) < \varepsilon.$$

(b) Suppose $\mathbf{A}$ cannot be diagonalised and show that

$$S_{\mathbf{A}}(\mathbf{B}) < (n\varepsilon)^{1/n}$$

provided $\varepsilon < 1/n$.

(c) Deduce that $\{\mu_1, \ldots, \mu_n\} \to \{\lambda_1, \ldots, \lambda_n\}$ as $\varepsilon \to 0$ for any $\mathbf{A} \in L(\mathbb{R}^n)$.

3.1.2 Let $SD(\mathbb{R}^n)$ be the subset of structurally stable linear diffeomorphisms in $L(\mathbb{R}^n)$. Show that a linear diffeomorphism is structurally stable if and only if it is hyperbolic. Hence, or otherwise, show that $SD(\mathbb{R}^n)$ is open and dense in $L(\mathbb{R}^n)$.

3.1.3 Let S be the subspace of $L(\mathbb{R}^n)$ defined by $\left\{\begin{pmatrix}1 & 0\\ 0 & \lambda\end{pmatrix}\middle|\, \lambda \neq 0 \text{ or } 1\right\}$. Show that every linear diffeomorphism in S is structurally stable within S but not within $L(\mathbb{R}^2)$.

3.1.4 Consider the subspace, $O(\mathbb{R}^2)$, of $L(\mathbb{R}^2)$ defined by $\{\mathbf{A}\,|\,\mathbf{A}^{\mathrm{T}}\mathbf{A} = \mathbf{I}, \mathbf{A} \in L(\mathbb{R}^2)\}$. Show that no element of $O(\mathbb{R}^2)$ is structurally stable in $L(\mathbb{R}^2)$.

3.2 Local structural stability

3.2.1 Let the vector field $\mathbf{X}(\mathbf{x}) \in \mathrm{Vec}^1(U)$, $U \subseteq \mathbb{R}^n$ and open, have a hyperbolic fixed point at $\mathbf{x}^* = \mathbf{0} \in U$ and suppose that $\tilde{\mathbf{X}}(\mathbf{x})$ is an ε–C^1-perturbation of $\mathbf{X}$. Verify Proposition 3.2.1 for the special case when $\tilde{\mathbf{X}} - \mathbf{X}$ is (a) constant; (b) linear; (c) $O(|\mathbf{x}|^k)$, $k \geqslant 2$.

3.2.2 Find $\eta = \eta(\varepsilon)$ such that each of the following vector fields is ε–C^1-close, on $U = \{(\theta, r)|r < 2\}$, to $\dot{r} = r(1-r)$, $\dot{\theta} = 1$;
(a) $\dot{r} = r(1 + \eta - r)$, $\dot{\theta} = 1$;
(b) $\dot{r} = r(1 - r + \eta r^2)$, $\dot{\theta} = 1$;
(c) $\dot{r} = (1 + \eta)r(1 - r)$, $\dot{\theta} = 1$.
Verify that the flows (a)–(c) all have a hyperbolic periodic orbit near to $r = 1$ for sufficiently small values of ε.

3.2.3 Show that the non-trivial fixed point $\mathbf{x}^* = (c/f, a/b)^{\mathrm{T}}$ of the Volterra–Lotka vector field

$$\mathbf{X}(\mathbf{x}) = ((a - by)x, -(c - fx)y)^{\mathrm{T}}, \tag{E3.2}$$

$a, b, c, f > 0$, is non-hyperbolic. Find a first integral for the system $\dot{\mathbf{x}} = \mathbf{X}(\mathbf{x})$ and determine the topological type of $\mathbf{x}^*$.

Consider vector fields of the form $\mathbf{X} + \mathbf{X}_\delta$ on a disc of radius $R > |\mathbf{x}^*|$, where:
(a) $\mathbf{X}_\delta = (-\delta x, -\delta y)^{\mathrm{T}}$;
(b) $\mathbf{X}_\delta = (-\delta x^2, 0)^{\mathrm{T}}$.
Choose δ in each case such that $\|\mathbf{X}_\delta\|_1 < \varepsilon$. If ε is sufficiently small, show that $\mathbf{X} + \mathbf{X}_\delta$ has a fixed point $\mathbf{y}^*$ near to $\mathbf{x}^*$ for both perturbations but that the topological type of $\mathbf{y}^*$ is the same as $\mathbf{x}^*$ for (a), while it is different for (b). Explain why this result is consistent with Proposition 3.2.1?

3.3 Flows on two-dimensional manifolds

3.3.1 All of the following vector fields are structurally unstable on $\mathbb{R}^2$. To which of these examples does Theorem 3.3.1 apply? Use the theorem, where applicable, to explain the nature of the instability. For the remaining examples construct ε–C^1-close systems to exhibit their structural instability in $\mathrm{Vec}^1(\mathscr{D})$, where $\mathscr{D}$ is the closed disc of radius 2 centred on the origin.
(a) $\dot{r} = -r(r-1)^2$, $\dot{\theta} = 1$;
(b) $\dot{r} = r(1-r)$, $\dot{\theta} = \sin^2(\theta)$;
(c) $\dot{x} = -2y(1-x^2) + xB(x)$, $\dot{y} = 2x(1-y^2) + yB(y)$;
where $B(x) = \exp\{-x^2/(1-x^2)\}$ for $|x| < 1$ and $= 0$ for $|x| \geqslant 1$.

3.3.2 The flows $\boldsymbol{\varphi}_t$: $\mathbb{R}^2 \to \mathbb{R}^2$ of the following systems give rise to flows on the torus $T^2 = \{(\theta_1, \theta_2)|0 \leqslant \theta_1, \theta_2 < 1\}$ by taking mod 1 in both components of $\boldsymbol{\varphi}_t$. Use Peixoto's Theorem to show that these toral flows are structurally unstable.
(a) $\dot{x} = \sin(2\pi x)$, $\dot{y} = 0$;
(b) $\dot{x} = 1$, $\dot{y} = 2$.
Illustrate these instabilities by giving topologically distinct systems which are ε–C^1-close for arbitrarily small ε.

3.3.3 Consider the system on the cylinder $C = \{(\theta, r) | 0 \leqslant \theta < 2\pi, r \in \mathbb{R}\}$, given by

$$\dot{r} = \frac{r}{(1 + r^2)} \cos(2\pi r), \; \dot{\theta} = 1. \tag{E3.3}$$

Show that it is structurally stable on every set $S_n = \{(\theta, r) | -n \leqslant r \leqslant n\}$, $n \in \mathbb{Z}^+$, but that it is not structurally stable on C.

3.4 Anosov diffeomorphisms

3.4.1 Use Peixoto's Theorem to show that none of the following diffeomorphisms $f \in \mathrm{Diff}^1(S^1)$ are structurally stable:
(a) $f(\theta) = (\theta + \alpha) \bmod 1$, $\alpha \in \mathbb{Q}$;
(b) $f(\theta) = (\theta + \alpha) \bmod 1$, $\alpha \in \mathbb{R} \backslash \mathbb{Q}$;
(c) $f(\theta) = (\theta + \sin^2(2\pi\theta)) \bmod 1$;
(d) $f(\theta) = (\theta + \frac{1}{2} + 0.1 \sin^2(2\pi\theta)) \bmod 1$.

3.4.2 Show that all of the periodic points of an Anosov automorphism $\mathbf{f}: T^n \to T^n$ are hyperbolic. Let $\boldsymbol{\theta}^*$ be a periodic point of $\mathbf{f}$ of period $q \in \mathbb{Z}^+$. Give expressions for the stable and unstable manifolds of the periodic orbit containing $\boldsymbol{\theta}^*$.

3.4.3 Let the Anosov automorphism $\mathbf{f}: T^2 \to T^2$ have lift $\mathbf{A}: \mathbb{R}^2 \to \mathbb{R}^2$ given by (3.4.12). Prove that $\mathbf{f}$ has periodic points of every *prime* period-q by showing that there exists $\mathbf{x} \in \mathbb{R}^2$ such that $\mathbf{A}^q\mathbf{x} - \mathbf{x} \in \mathbb{Z}^n$, where $\mathbf{x}$ does not represent a fixed point.

3.4.4 Let $\mathbf{f}$ and $\mathbf{g}$ be Anosov automorphisms of T^n with lifts given by automorphisms $\mathbf{A}, \mathbf{B}: \mathbb{R}^n \to \mathbb{R}^n$. Prove that $\mathbf{A}$ and $\mathbf{B}$ are similar matrices if $\mathbf{f}$ and $\mathbf{g}$ are differentiably conjugate. Conversely, show that, if $\mathbf{A}$, $\mathbf{B}$ are lifts of Anosov automorphisms $\mathbf{f}$, $\mathbf{g}$ and they are similar by a matrix $\mathbf{C}$ with integer entries and determinant ± 1, then $\mathbf{f}$ and $\mathbf{g}$ are differentiably conjugate.

3.4.5 Show that the Anosov automorphism $\mathbf{f}$ induced by

$$\mathbf{A} = \begin{pmatrix} 1 & 3 \\ 1 & 2 \end{pmatrix}: \mathbb{R}^2 \to \mathbb{R}^2 \tag{E3.4}$$

has a saddle point at $\boldsymbol{\pi}(\mathbf{0})$, where $\boldsymbol{\pi}: \mathbb{R}^2 \to T^2$ is the map given in (3.4.2). Find the equation of the separatrices of $\mathbf{A}$ at $\mathbf{0}$ and show that they have irrational slope. What are the implications of this on the torus? Show that the point on the torus given by $\boldsymbol{\pi}(\mathbf{x}^\dagger)$, where $\mathbf{x}^\dagger =$ Anosov auto$(13^{1/2} - 1)/2(13^{1/2}), 1/13^{1/2})$, is a transverse homoclinic point. How does the Anosov automorphism considered in this question differ from that given by (3.4.12)?

3.5 Horseshoe diffeomorphisms

3.5.1 The canonical example

3.5.1 Obtain explicit equations for the horseshoe map $\mathbf{f}: Q \to \mathbb{R}^2$ on its restriction to $P_0 \cup P_1 \subset Q$ (see Figure 3.13). Verify that $(x, y) \overset{\mathbf{f}}{\mapsto} (f_1(x), f_2(x, y))$.

3.5.2 Let Λ be the Cantor set of the horseshoe diffeomorphism $\mathbf{f}$ as defined in (3.5.4). Complete the proof of Proposition 3.5.1 by showing that Λ is invariant under $\mathbf{f}^{-1}$. Hence show that $\mathbf{f}(\Lambda) = \Lambda$.

3.5.3 Let $\mathbf{g}: S^2 \to S^2$ be the globally extended horseshoe diffeomorphism. Prove that:
(a) $\bigcap_{n \in \mathbb{N}} Q^{(-n)}$ is invariant under $\mathbf{g}$ and $\bigcap_{n \in \mathbb{Z}^+} Q^{(n)}$ is invariant under $\mathbf{g}^{-1}$;
(b) $\bigcap_{n \in \mathbb{N}} Q^{(-n)}$ ($\bigcap_{n \in \mathbb{Z}^+} Q^{(n)}$) is the inset (outset) of Λ for $\mathbf{g}$ on Q.

3.5.4 Suppose $\mathbf{g}: Q \to \mathbb{R}^2$ is the horseshoe map and take $\mathbf{x}_0$, with coordinates (x_0, y_0), to be a point of $Q \backslash \Lambda$. Let $d = d(\mathbf{x}_0, \Lambda) = \min_{(x,y) \in \Lambda} \{|x_0 - x|\}$ be the horizontal distance from $\mathbf{x}_0$ to Λ. Find the maximum value, $N(d)$, of n such that $\mathbf{f}^n(\mathbf{x}_0) \in Q$.

3.5.2 Dynamics on symbol sequences

3.5.5 Let Σ_S be the set of all bi-infinite sequences on the m symbols $S = \{0, 1, \ldots, m-1\}$ and $\alpha: \Sigma_S \to \Sigma_S$ be the left shift $\alpha(\sigma)_n = \sigma_{n-1}$ (see (3.5.7)). Prove that:
(a) α has periodic orbits of all periods as well as aperiodic orbits;
(b) there is a natural topology on Σ_S for which the periodic points of α are dense in Σ_S;
(c) there exist dense orbits of α on Σ_S.

3.5.6 Let α (β) be the left (right) shift on the symbol sequence space Σ. Show that:
(a) α is a homeomorphism and $\alpha^{-1} = \beta$;
(b) α is topologically conjugate to β.

3.5.7 Let $\alpha: \Sigma_S \to \Sigma_S$ be either a left or right shift on Σ_S, the space of symbol sequences with $S = \{0, 1, \ldots, m-1\}$. Show that there are m^q sequences σ such that $\alpha^q(\sigma) = \sigma$. Let m_k be the number of period-k points of α and $K = \{k \mid k \in \mathbb{Z}^+ \;\&\; k \mid q\}$. Prove that

$$\sum_{k \in K} m_k = m^q \tag{E3.5}$$

for $m, q \in \mathbb{Z}^+$. For $m = 2$, use this result to find the number of period-12 orbits.

3.5.3 Symbolic dynamics for the horseshoe diffeomorphism

3.5.8 Let $\mathbf{f}: Q \to \mathbb{R}^2$ be the horseshoe map. Use (3.5.3) to prove that each vertical strip of $Q^{(-n)}$, $n \in \mathbb{N}$, can be described by a binary sequence of n-symbols.

3.5.9 Let $(x, y) \in \Lambda$, the invariant Cantor set of the horseshoe map $\mathbf{f}: Q \to \mathbb{R}^2$. Define $\{a_i\}, \{b_i\}_{i=1}^{\infty}$ as follows.
(i) Consider the nested vertical strips given by $\bigcap_{n \in \mathbb{N}} Q^{(-n)}$. If $x \in \begin{cases} P_0 \\ P_1 \end{cases}$ then $a_1 = \begin{cases} -1 \\ +1 \end{cases}$.

For $i \geqslant 2$, $a_i = \begin{cases} -1 \\ +1 \end{cases}$ if x lies to the $\begin{cases} \text{left} \\ \text{right} \end{cases}$ of the previous strip in the nesting.

(ii) For the nested horizontal strips $\bigcap_{n \in \mathbb{Z}^+} Q^{(n)}$, define $b_1 = \begin{cases} +1 \\ -1 \end{cases}$ if $y \in \begin{cases} Q_0 \\ Q_1 \end{cases}$ and

$b_i = \begin{cases} +1 \\ -1 \end{cases}$, $i \geqslant 2$, if y lies to the $\begin{cases} \text{top} \\ \text{bottom} \end{cases}$ of the previous strip in the nesting.

Show that each point (x, y) of Λ can be written in the form

$$\left(2 \sum_{i=1}^{\infty} a_i/5^i,\ 2 \sum_{i=1}^{\infty} b_i/5^i\right),\quad a_i, b_i = \pm 1.$$

Hence show that the subset of Λ in the quadrant $x, y > 0$ is homeomorphic, by the five-fold magnification $(x, y) \mapsto (5x - 2, 5y - 2)$, to the whole of Λ. What does this imply about the structure of the set Λ?

3.5.10 Use the form for $\mathbf{x} \in \Lambda$ given in Exercise 3.5.9 to find the coordinates in Q of the fixed and period-2 points of the horseshoe map $\mathbf{f}: Q \to \mathbb{R}^2$.

3.5.11 Let $\mathbf{f}: Q \to \mathbb{R}^2$ be the horseshoe map. In the notation of §3.5.3, the set $\Lambda^{(N)} = \bigcap_{n=-(N-1)}^{N} Q^{(n)}$ consists of 2^{2N} connected components, each one being a square region of side $2/5^N$ lying within Q.

(a) Prove that (3.5.17) uniquely associates a symbol block $\sigma^{(N)} = \{\sigma_{-(N-1)}, \ldots, \sigma_N\}$ with each connected component, $\kappa(\sigma^{(N)})$ say, of $\Lambda^{(N)}$. Locate $\kappa(\sigma^{(N)})$ for the following symbol blocks $\sigma^{(N)}$:

$$\text{(i) } \{1\cdot 1\}; \qquad \text{(ii) } \{11\cdot 00\}; \qquad \text{(iii) } \{010\cdot 101\}.$$

(b) Let $\eta^{(N)} = \{\eta_{-(N-1)}, \ldots, \eta_N\}$ and $\nu^{(N)} = \{\nu_{-(N-1)}, \ldots, \nu_N\}$ be two symbol blocks of length $2N$. Explain how to choose a point $\mathbf{x} \in \Lambda$ in $\kappa(\eta^{(N)})$ such that $\mathbf{f}^{2N}(\mathbf{x}) \in \kappa(\nu^{(N)})$. Hence show that there exist points in $\kappa(\eta^{(N)})$ whose orbit under $\mathbf{f}^{2N}$ visits every connected component of $\Lambda^{(N)}$ in any desired order.

(c) What restriction must be imposed on the elements of the sequence σ if the orbit of the point $\mathbf{x} = \mathbf{h}(\sigma) \in \Lambda$ is to remain in a particular component, $\kappa(\sigma^{(N)})$ say, for k applications of $\mathbf{f}$? What is the maximum number of connected components that can be reached from $\kappa(\sigma^{(N)})$ in k iterations of $\mathbf{f}$?

(d) What aspect of the dynamics of $\mathbf{f}|\Lambda$ do the observations (a)–(c) reflect?

3.5.12 Recall that the horseshoe map $\mathbf{f}: Q \to \mathbb{R}^2$ satisfies $(x, y) \overset{\mathbf{f}}{\mapsto} (f_1(x), f_2(x, y))$. Verify this property for $\mathbf{f}|\Lambda$ by considering the left shift $\alpha: \Sigma \to \Sigma$ that is conjugate to $\mathbf{f}$. Show that $f_1: [-1, 1] \to \mathbb{R}$ has a repelling invariant Cantor set.

3.5.13 The *Baker's* transformation $\mathbf{B}: T^2 \to T^2$ is defined by

$$x_1 = 2x \bmod 1, \qquad y_1 = \tfrac{1}{2}(2x - x_1 + y) \bmod 1$$

for $(x \bmod 1, y \bmod 1) \in T^2$ (Arnold and Avez, 1968, Appendix 7).

Describe the effect of this transformation on the rectangles $P_0 = [0, \tfrac{1}{2}) \times [0, 1)$ and $P_1 = [\tfrac{1}{2}, 1) \times [0, 1)$. Show that every point $\mathbf{x} \in T^2$ can be written in the form

$$\mathbf{x} = \mathbf{h}(\sigma) = \bigcap_{n=-\infty}^{\infty} \mathbf{B}^n(P_{\sigma_n}),$$

where $\sigma = \{\sigma_n\}_{n=-\infty}^{\infty}$ is a bi-infinite sequence of $\{0, 1\}$. Use this result to show that

$$\mathbf{B}(\mathbf{x}) = \bigcap_{n=-\infty}^{\infty} \mathbf{B}^n(P_{\alpha(\sigma)_n}), \tag{E3.6}$$

where $\alpha: \Sigma \to \Sigma$ is the left shift.

Prove that $\mathbf{h}: \Sigma \to T^2$ is not one-to-one by finding $\mathbf{h}(\sigma_1)$ and $\mathbf{h}(\sigma_2)$, where $\sigma_1 = \{\dot{0}1\cdot 0\dot{}\}$ and $\sigma_2 = \{\dot{1}0\cdot 0\dot{}\}$ ($\dot{i}$ and $i\dot{}$ indicate indefinite recurrence of the symbol i to left and right, respectively). What is the general form of points $\mathbf{x} \in T^2$ for which $\mathbf{h}$ fails to be injective? Given that these problem points can be disregarded (see Arnold & Avez, 1968, p. 125), show that the non-wandering set of $\mathbf{B}$ is the whole of T^2.

3.6 Hyperbolic structure and basic sets

3.6.1 Let $\mathbf{g}: \mathbb{R}^2 \to \mathbb{R}^2$, given by $\mathbf{g}(\mathbf{x}) = (g_1(\mathbf{x}), g_2(\mathbf{x}))^{\mathrm{T}}$, $\mathbf{x} = (x_1, x_2)^{\mathrm{T}}$, be a smooth map and $\gamma: (-1, 1) \to \mathbb{R}^2$ be a smooth curve. Show that the tangents to the curves γ

and $\mathbf{g}\cdot\gamma$ at $t=0$ are related by the equation

$$\left.\frac{\mathrm{d}(\mathbf{g}\cdot\gamma)}{\mathrm{d}t}\right|_{t=0} = \mathrm{D}\mathbf{g}(\gamma(0))\left.\frac{\mathrm{d}\gamma}{\mathrm{d}t}\right|_{t=0} \tag{E3.7}$$

where $\mathrm{D}\mathbf{g}(\mathbf{x})$ is the matrix of partial derivatives $\left(\frac{\partial g_i}{\partial x_j}\right)_{i,j=1}^{2}$. Illustrate this result by finding the image under $\mathbf{g}(x_1, x_2) = (\exp(x_1 + x_2), x_1 x_2)^{\mathrm{T}}$ of the curves: (a) $x_1 = t$, $x_2 = t\cos t$; (b) $x_1 = t + t^2$, $x_2 = \tan t$.

Show that the curves (a) and (b) have a common tangent at $\mathbf{x} = \mathbf{0}$. Verify that the image curves under $\mathbf{g}$ also have a common tangent that is given by (E3.7).

3.6.2 Consider the representation of S^1 shown in Figure E3.1. Find the overlap map h_{12} between U_1 and U_2.

The norms $\|\cdot\|_{x_1}$ on TU_{1x_1} and $\|\cdot\|_{x_2}$ on TU_{2x_2} are compatible if

$$\|v\|_{x_1} = \|\mathrm{D}h_{12}(x_2)v\|_{x_2}. \tag{E3.8}$$

(a) Prove that the Euclidean norms on U_1 and U_2 are not compatible.

(b) Find the norm on $U_2 \backslash P_2$ which is compatible by h_{12} to the Euclidean norm on U_1. Show that this norm cannot be extended continuously to the whole of U_2.

(c) Verify that, at each point x_i of U_i

$$\langle u, v\rangle_i = 16uv/(4 + x_i^2)^2,$$

is a positive definite inner product on TU_{ix_i}. Show that the norms $\|v\|_i = \langle v, v\rangle_i$ are compatible.

3.6.3 Let M be a differentiable manifold and $\mathbf{x}^*$ be a fixed point of the diffeomorphism $\mathbf{f}\colon M \to M$. Let $\mathbf{h}\colon W \subset M \to U$ be a chart at $\mathbf{x}^*$ and $\mathbf{h}(\mathbf{x}^*) = \tilde{\mathbf{x}}^*$. Show that the eigenvalues of the derivative $\mathrm{D}(\mathbf{hfh}^{-1})(\tilde{\mathbf{x}}^*)$ are independent of the choice of the chart $(U, \mathbf{h})$ at $\mathbf{x}^*$. What is the significance of this result in relation to the problem of defining a hyperbolic fixed point on M?

3.6.4 Show that the Anosov automorphism $\mathbf{f}\colon T^2 \to T^2$ given by $\mathbf{A}\colon \mathbb{R}^2 \to \mathbb{R}^2$, where $\mathbf{A} = \begin{pmatrix} 1 & 1 \\ 1 & 2 \end{pmatrix}$, satisfies the hyperbolicity conditions (3.6.9 and 3.6.10) at each point of $\Omega = T^2$.

Figure E3.1 Stereographic projection from $(S^1 \backslash P_2) \to U_1$ and $(S^1 \backslash P_1) \to U_2$ provides an atlas for the unit circle $S^1 \subset \mathbb{R}^2$.

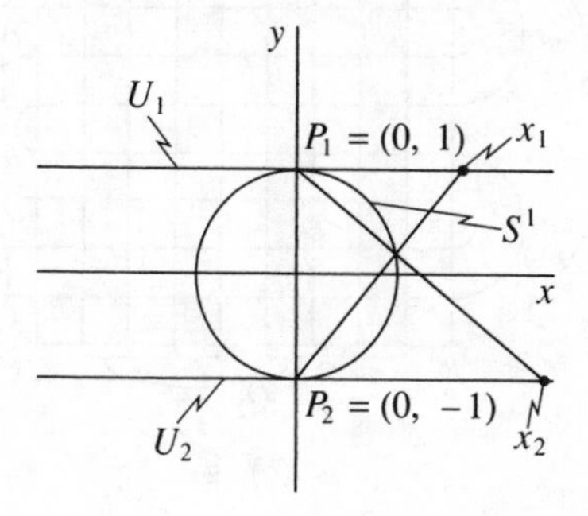

3.6.5 Consider the diffeomorphisms $\mathbf{g}_1, \mathbf{g}_2 \colon S^2 \to S^2$ defined in Figure E3.2. Outline arguments to show that $\mathbf{g}_i$, $i = 1, 2$, has an invariant Cantor set $\Lambda_i \subseteq Q$ such that $\mathbf{g}_i | \Lambda_i$ is conjugate to a shift on $m(i)$-symbols, where $m(1) = 3$ and $m(2) = 4$. What are the basic sets of $\mathbf{g}_1$ and $\mathbf{g}_2$? For both maps, draw schematic diagrams illustrating the dynamics of the wandering points. Is $\mathbf{g}_1 | \Lambda_1$ conjugate to $\mathbf{g}_2 | \Lambda_2$?

3.6.6 Find which of the following diffeomorphisms of the torus T^2 satisfy the hypotheses of Theorem 3.6.1 and describe their basic sets:

(a) $\mathbf{f}_1(\boldsymbol{\pi}(\mathbf{x})) = \boldsymbol{\pi}(\mathbf{A}\mathbf{x})$, $\mathbf{A} = \begin{pmatrix} 3 & 4 \\ 4 & 5 \end{pmatrix}$;

Figure E3.2 The restrictions of $\mathbf{g}_1$ and $\mathbf{g}_2$ to the capped square Q' are shown in (*a*) and (*b*), respectively. On each component of $Q_i^{(0)}$, $i = 1, 2$, $\mathbf{g}_i$ is assumed to be linear. The map $\mathbf{g}_1$ is a contracting diffeomorphism on both F_1 and G_1 with hyperbolic fixed points $P_1 \in F_1' = \mathbf{g}_1(F_1)$ and $P_2 \in G_1' = \mathbf{g}_1(G_1)$. Observe that $G_2' = \mathbf{g}_2(G_2) \subset F_2$ and $\mathbf{g}_2 | F_2$ is a contracting diffeomorphism with hyperbolic fixed point $P_3 \in F_2' = \mathbf{g}_2(F_2)$. On $S^2 \backslash Q'$, both $\mathbf{g}_1$ and $\mathbf{g}_2$ have a single repelling hyperbolic fixed point P_0. It may be assumed that Theorem 3.6.1 applies to both $\mathbf{g}_1$ and $\mathbf{g}_2$.

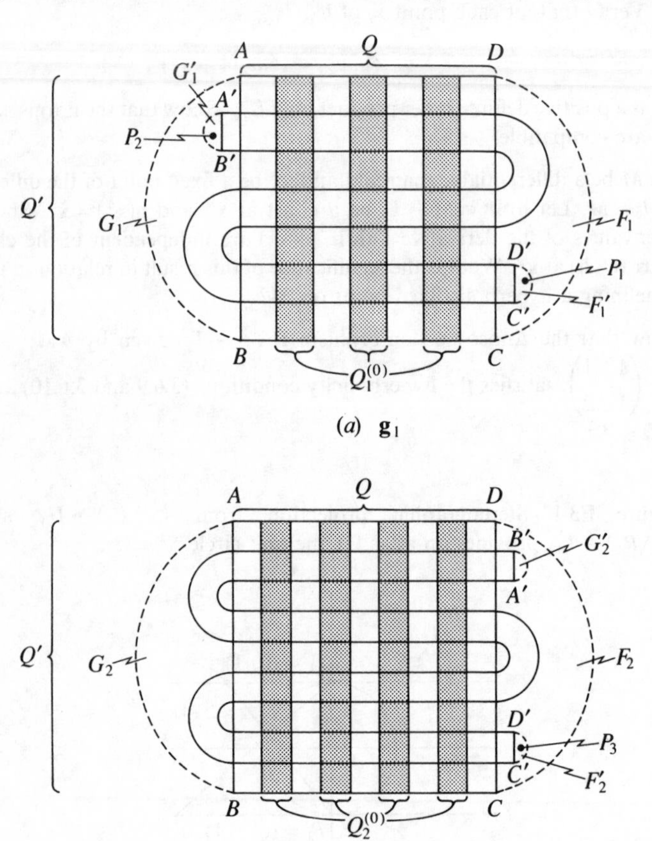

(*a*) $\mathbf{g}_1$

(*b*) $\mathbf{g}_2$

(b) $\mathbf{f}_2(\pi(\mathbf{x})) = \pi(\mathbf{x}+\mathbf{b}),\ \mathbf{b} = \begin{pmatrix} 1/2 \\ 3^{1/2} \end{pmatrix}$;

(c) $\mathbf{f}_3(\pi(\mathbf{x})) = \pi(\boldsymbol{\varphi}_1(\mathbf{x}))$, where $\boldsymbol{\varphi}_1$ is the time-one map of the system $\begin{pmatrix} \dot{x} \\ \dot{y} \end{pmatrix} = \begin{pmatrix} \sin(2\pi x) \\ \sin(2\pi y) \end{pmatrix}$.

3.6.7 Obtain the Hénon attractor using a microcomputer to plot the iterates of the map

$$x_1 = y - 1.4x^2 + 1, \qquad y_1 = 0.3x, \tag{E3.9}$$

with initial value $(1, 0)$. Choose x and y scales such that the square $Q = \{(x, y) \mid |x| < 1, |y| < 1\}$ fills a large portion of the screen. Observe the braided nature of the attractor by magnification and shifts of the origin.

3.7 Homoclinic points

3.7.1 Use the explicit form of the horseshoe map $\mathbf{f}: Q \to \mathbb{R}^2$ on $P_0 \cup P_1$ (see Exercise 3.5.1) to locate the fixed points of $\mathbf{f}$ and their stable and unstable manifolds. Show that there exist transverse homoclinic points at $(x, y) = (-\frac{1}{3}, -\frac{1}{3})$ and $(\frac{1}{2}, \frac{1}{2})$. Hence obtain Figure 3.31. Show that any homoclinic point $\mathbf{x}^\dagger$ of a periodic orbit on Λ is itself an element of Λ.

3.7.2 Let $\mathbf{f}: M \to M$ satisfy the requirements of Theorem 3.7.1, i.e. $\mathbf{f}$ is a Kupka–Smale diffeomorphism with a transverse homoclinic point associated with one of its periodic points. Show that $\tilde{\Lambda} = \bigcup_{i=0}^{p-1} \mathbf{f}^i(\Lambda)$ is a Cantor set such that $\mathbf{f}(\tilde{\Lambda}) = \tilde{\Lambda}$.

3.7.3 Find the eigenvalues of the linearisation of (3.7.1) at the fixed point $(x, y) = (1, 0)$ and verify that both are positive for $k > 0$. Show that the eigendirections are given by $y = \{-k \pm [k(k+4)]^{1/2}\}(x-1)/2$. For $k = 1.5$ take an interval of approximate length 0.0001 containing 100 points on the appropriate branch of the unstable manifold of the saddle fixed point. Plot 15 iterates of each point under (3.7.1) to obtain a numerical approximation to the unstable manifold at $(1, 0)$. Use the inverse map (3.7.4) to complete the homoclinic tangle shown in Figure 3.33. Modify the program to exhibit the image of each successive iteration of the interval separately. Observe the repeated stretching and folding around the origin.

3.7.4 Use the program developed in Exercise 3.7.3 to study how the extent of the homoclinic tangle depends on k. Plot the tangle for $k = 0.4$, 0.8, 1.2, 1.6 and 2.0. Comment on your results.

3.7.5 Show that the derivative map of (3.7.1) has unit determinant for all $(x, y) \in \mathbb{R}^2$. Given that there exists a linear conjugacy between (3.7.1) and (1.9.40), find a relation between k and α. Modify the program used in Exercise 1.9.9 to generate orbit plots for (3.7.1) corresponding to Figures 1.38 and 1.39.

3.7.6 Let $\gamma(t)$, $t \in I \subset \mathbb{R}$, define a closed curve, γ, in the plane oriented with increasing t. Suppose that $\boldsymbol{\Gamma}(s)$, $s \in J \subset \mathbb{R}$, defines a segment of a planar curve, Γ, that intersects γ transversely. Assume that the point of intersection is given by $\mathbf{x}_0 = \gamma(0) = \boldsymbol{\Gamma}(0)$ and that $\boldsymbol{\Gamma}(s)$ lies inside γ for $s > 0$. Verify that $\dot{\gamma}(0) \wedge \dot{\boldsymbol{\Gamma}}(0)$ determines the orientation of γ.

Let $\mathbf{f}: \mathbb{R}^2 \to \mathbb{R}^2$ be a diffeomorphism and show that

$$(\dot{\overline{\mathbf{f}\cdot\gamma}})(0) \wedge (\dot{\overline{\mathbf{f}\cdot\boldsymbol{\Gamma}}})(0) = \mathrm{Det}\, \mathbf{Df}(\mathbf{x}_0)[\dot{\gamma}(0) \wedge \dot{\boldsymbol{\Gamma}}(0)] \tag{E3.10}$$

gives the orientation of $\mathbf{f}\cdot\gamma$.

3.7.7 Let $\mathbf{f}: Q \to \mathbb{R}^2$ be the horseshoe map and $\mathbf{x}^*$ be a periodic point of period $q > 1$, on the invariant Cantor set Λ. Use symbolic dynamics to construct a point $\mathbf{x}^\dagger$ of Λ which is homoclinic to the periodic orbit containing $\mathbf{x}^*$ (cf. Figure 3.35). Can the transverse nature of the homoclinic point be detected by the symbolic dynamics?

3.8 The Melnikov function

3.8.1 Consider the solutions of

$$\dot{\mathbf{x}} = \mathbf{f}_0(\mathbf{x}) + \varepsilon \mathbf{f}_1(\mathbf{x}, t) \tag{E3.11}$$

given by (3.8.5) and (3.8.6). Obtain (3.8.7) by substitution and comparison of order ε terms. Hence deduce (3.8.10).

3.8.2 Show that $H(x, y) = \frac{1}{2}(y^2 - x^2 + \frac{1}{2}x^4)$ is a Hamiltonian for the system $\dot{x} = y$, $\dot{y} = x - x^3$ and verify that the level set $H = 0$ consists of a saddle point at $\mathbf{x} = \mathbf{0}$ and two homoclinic orbits Γ_0^+ and Γ_0^- given by

$$(x^\pm(t), y^\pm(t)) = (\pm 2^{1/2} \operatorname{sech}(t), \mp 2^{1/2} \operatorname{sech}(t) \tanh(t)).$$

3.8.3 If the system

$$\dot{\mathbf{x}} = \mathbf{f}_0(\mathbf{x}) + \varepsilon \mathbf{f}_1(\mathbf{x}, t), \tag{E3.12}$$

where $\mathbf{f}_1$ has period $2\pi/\omega$, has a homoclinic orbit $\mathbf{x}_0(t)$, for $\varepsilon = 0$, then the corresponding Melnikov function is

$$M(\theta_0) = \int_{-\infty}^{+\infty} \mathbf{f}_0(\mathbf{x}_0(t)) \wedge \mathbf{f}_1(\mathbf{x}_0(t), t + \theta_0)\, \mathrm{d}t, \tag{E3.13}$$

$\theta_0 \in [0, 2\pi/\omega)$. Use Γ_0^+ obtained in Exercise 3.8.2 to find a Melnikov function for the system

$$\dot{x} = y, \qquad \dot{y} = x - x^3 + \varepsilon \cos(\omega t). \tag{E3.14}$$

3.8.4 Prove that $\displaystyle\int_{-\infty}^{+\infty} \operatorname{sech}(t) \tanh(t) \sin(\omega t)\, \mathrm{d}t = \pi\omega \operatorname{sech}(\pi\omega/2)$ by using contour integration on a rectangle in $\mathbb{C}$ with vertices at $(\pm R, 0), (\pm R, i\pi)$ and letting $R \to \infty$.

3.8.5 The Melnikov function given in (3.8.15) can also be used to indicate the separation of stable and unstable manifolds of two different saddle points in a Hamiltonian system (the so-called heteroclinic case). Show that the saddle connections, $\Gamma_0^\pm$, between the fixed points $(-\pi, 0)$ and $(\pi, 0)$ of the system

$$\dot{x} = y, \qquad \dot{y} = -\sin(x) + \varepsilon(a - by) \tag{E3.15}$$

with $\varepsilon = 0$, are given by

$$(x_0(t), y_0(t)) = (\pm 2 \arctan(\sinh(t)), \pm 2 \operatorname{sech}(t)).$$

Calculate the Melnikov function for (E3.15) along these orbits and show that

$$M(\theta_0) = \pm 2a\pi - 8b, \tag{E3.16}$$

for Γ_0^+, Γ_0^-, respectively. Explain why $M(\theta_0)$ is constant.

3.8.6 The Sine–Gordon equation

$$\dot{x} = y, \qquad \dot{y} = -\sin(x) + \varepsilon(a\cos(\omega t) - by), \tag{E3.17}$$

$a, b > 0$, with $\varepsilon = 0$, has saddle connection orbits, $\Gamma_0^{\pm}$, between fixed points at $(\pm\pi, 0)$. Calculate the Melnikov function for (E3.17) along $\Gamma_0^{\pm}$ and show that it can be written in the form

$$M(\theta_0) = \frac{1}{\omega}\left\{\pm\frac{2a\pi\omega\cos(\omega\theta_0)}{\cosh(\pi\omega/2)} - 8b\right\}. \tag{E3.18}$$

Describe the regions of the (a, b)-plane for which transverse heteroclinic points occur.

4

Local bifurcations I: Planar vector fields and diffeomorphisms on $\mathbb{R}$

4.1 Introduction

Dynamical models frequently consist of a *family* of differential equations or diffeomorphisms. An important step in analysing such models is to recognise the topologically distinct types of behaviour that can occur in the family. This leads us to focus attention on those members of the family at which topological changes, or 'bifurcations', are possible. More precisely, let $\mathbf{X}: \mathbb{R}^m \times \mathbb{R}^n \to \mathbb{R}^n$ ($\mathbf{f}: \mathbb{R}^m \times \mathbb{R}^n \to \mathbb{R}^n$) be an m-parameter, C^r-family of vector fields (diffeomorphisms) on $\mathbb{R}^n$, i.e. $(\boldsymbol{\mu}, \mathbf{x}) \mapsto \mathbf{X}(\boldsymbol{\mu}, \mathbf{x})$ ($\mathbf{f}(\boldsymbol{\mu}, \mathbf{x})$), $\boldsymbol{\mu} \in \mathbb{R}^m$, $\mathbf{x} \in \mathbb{R}^n$. The family $\mathbf{X}$ ($\mathbf{f}$) is said to have a *bifurcation point* at $\boldsymbol{\mu} = \boldsymbol{\mu}^*$ if, in every neighbourhood of $\boldsymbol{\mu}^*$, there exist values of $\boldsymbol{\mu}$ such that the corresponding vector fields $\mathbf{X}(\boldsymbol{\mu}, \cdot) = \mathbf{X}_{\boldsymbol{\mu}}(\cdot)$ (diffeomorphisms $\mathbf{f}(\boldsymbol{\mu}, \cdot) = \mathbf{f}_{\boldsymbol{\mu}}(\cdot)$) exhibit topologically distinct behaviour.

Obviously, all structurally stable members of the family are excluded from such considerations, since, by Definition 3.3.1, their topological properties must be shared by all nearby family members. It was pointed out in Chapter 3 that *non-hyperbolic* fixed points of flows or diffeomorphisms are locally structurally unstable. Our aim in this chapter is to investigate the bifurcations associated with the simplest examples of non-hyperbolic fixed points on $\mathbb{R}^n$ for $n \leqslant 2$. The following example illustrates some of the possibilities for the one-dimensional vector field $X_0(x) = -x^2$ which has a non-hyperbolic singular point at $x = 0$.

Example 4.1.1 Consider the following one-parameter families of vector fields on $\mathbb{R}$:

(a) $$X(\mu, x) = \mu - x^2;$$

(b) $$X(\mu, x) = \mu x - x^2; \qquad (4.1.1)$$

(c) $$X(\mu, x) = -(1 + \mu^2)x^2.$$

Sketch diagrams in the μx-plane illustrating the local bifurcations, if any, that these families exhibit at $\mu = 0$.

Solution. This problem simply involves sketching the phase portraits for $\dot{x} = X_\mu(x)$, with (μ, x) near $(0, 0)$. It is in this sense that we are involved with *local* bifurcations. The resulting information is then collected on a diagram in the μx-plane.

(a) For $\mu < 0$, the differential equation $\dot{x} = \mu - x^2$ is such that $\dot{x} < 0$ for all $x \in \mathbb{R}$. When $\mu = 0$, there is a non-hyperbolic fixed point at $x = 0$ but $\dot{x} < 0$ for all $x \neq 0$. Finally for $\mu > 0$, there are two fixed points: one stable at $+\mu^{1/2}$ and a second, unstable, at $-\mu^{1/2}$. The required bifurcation diagram is shown in Figure 4.1(*a*).

(b) $X(\mu, x)$ has singular points at $x = 0$ and $x = \mu$ for $\mu \neq 0$. If $\mu > 0$, then $x = \mu$ is stable and $x = 0$ is unstable. The stabilities are reversed when $\mu < 0$. At $\mu = 0$, there is one singularity at $x = 0$ and $\dot{x} < 0$ for all $x \neq 0$. This leads to the bifurcation diagram shown in Figure 4.1(*b*).

(c) The differential equation $\dot{x} = X(\mu, x)$ has one fixed point at $x = 0$ for all μ and $\dot{x} < 0$ for all μ and $x \neq 0$. It follows that the same phase portrait is obtained for each $\mu \in \mathbb{R}$ and no bifurcation takes place. □

It should be emphasised that in the solution to Example 4.1.1 we are only involved with bifurcations occurring sufficiently close to $(\mu, x) = (0, 0)$ in the μx-plane. Typically we will consider the behaviour of smooth families of vector fields $\mathbf{X}$ (diffeomorphisms $\mathbf{f}$) defined in some sufficiently small neighbourhood of $(\boldsymbol{\mu}^*, \mathbf{x}^*)$ in $\mathbb{R}^m \times \mathbb{R}^n$, where $\mathbf{X}_{\boldsymbol{\mu}^*}$ $(\mathbf{f}_{\boldsymbol{\mu}^*})$ has a non-hyperbolic fixed point at $\mathbf{x} = \mathbf{x}^*$. In other words, we deal with *local families* of vector fields (diffeomorphisms) *at* $(\boldsymbol{\mu}^*, \mathbf{x}^*)$. This being the case, it is often convenient to introduce local coordinates in $\mathbb{R}^m \times \mathbb{R}^n$ and arrange that $(\boldsymbol{\mu}^*, \mathbf{x}^*) = (\mathbf{0}, \mathbf{0})$.

Example 4.1.1 also highlights the fact that structural instability of $\mathbf{x}^*$ is a *necessary*, but *not sufficient*, condition for $\boldsymbol{\mu}^*$ to be a bifurcation point of the family. Observe that family (*c*) has $X_0(x) = -x^2$ but no bifurcation takes place. We say that such a family does not *unfold* the singularity in X_0. More importantly, Example 4.1.1 shows that different families, containing a given singularity, may unfold it to different degrees. For example, three topological types of phase portrait occur in family (a), two in family (b) and only one in family (c). Moreover, the types occurring in (b) and (c) are subsets of those occurring in (a).

We can make these observations more precise by introducing the following definitions (see Arnold, 1983, p. 264).

Definition 4.1.1 *Any local family*, $\mathbf{X}(\boldsymbol{\mu}, \mathbf{x})$, *at* $(\mathbf{0}, \mathbf{0})$ *is said to be an* unfolding *of the vector field* $\mathbf{X}(\mathbf{0}, \mathbf{x}) = \mathbf{X}_0(\mathbf{x})$. *When* $\mathbf{X}_0(\mathbf{x})$ *has a singularity at* $\mathbf{x} = \mathbf{0}$, $\mathbf{X}(\boldsymbol{\mu}, \mathbf{x})$ *is referred to as an* unfolding of the singularity.

Definition 4.1.2 *Two local families* $\mathbf{X}$ *and* $\mathbf{Y}$ *are said to be* equivalent *if there is a continuous mapping* $\mathbf{h}: N \subseteq \mathbb{R}^m \times \mathbb{R}^n \to \mathbb{R}^n$ *at* $(\mathbf{0}, \mathbf{0})$, *satisfying* $\mathbf{h}(\mathbf{0}, \mathbf{0}) = \mathbf{0}$, *such that, for each* $\boldsymbol{\mu}$, $\mathbf{h}_{\boldsymbol{\mu}}(\cdot) = \mathbf{h}(\boldsymbol{\mu}, \cdot)$ *is a homeomorphism which exhibits the topological equivalence of the phase portrait of* $\mathbf{X}_{\boldsymbol{\mu}}$ *to that of* $\mathbf{Y}_{\boldsymbol{\mu}}$.

Definition 4.1.3 *A local family* $\mathbf{X}: \mathbb{R}^m \times \mathbb{R}^n \to \mathbb{R}^n$ *is induced by the family* $\mathbf{Y}: \mathbb{R}^l \times \mathbb{R}^n \to \mathbb{R}^n$ *by means of a continuous map* $\boldsymbol{\varphi}: \mathbb{R}^m \to \mathbb{R}^l$, $\boldsymbol{\varphi}(\mathbf{0}) = \mathbf{0}$, *if* $\mathbf{X}(\boldsymbol{\mu}, \mathbf{x}) = \mathbf{Y}(\boldsymbol{\varphi}(\boldsymbol{\mu}), \mathbf{x})$.

Figure 4.1 Phase portraits for flows with $\dot{x}$ given by (4.1.1(a)–(c)) are presented on the μx-plane.

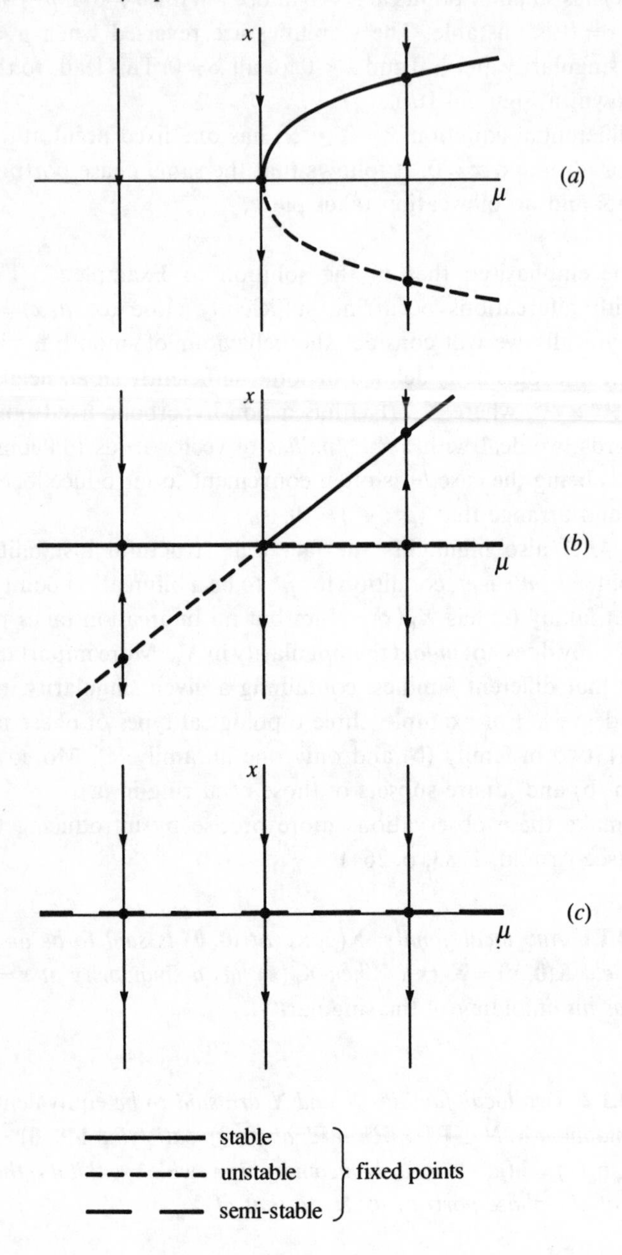

Similar definitions apply to families of diffeomorphisms, however, *topological equivalence* in Definition 4.1.2 is then replaced by *topological conjugacy*.

Example 4.1.2 Show that the following local families of vector fields at $(0, 0)$ are equivalent to families induced by $Y(v, y) = v - y^2$:

$$\begin{aligned}
&\text{(a)} && \dot{x} = \mu x - x^2; \\
&\text{(b)} && \dot{x} = -(1+\mu^2)x^2; \\
&\text{(c)} && \dot{x} = \mu_0 + \mu_1 x - x^2; \\
&\text{(d)} && \dot{x} = \mu_0 - x^2 - \mu_r x^r,\ r \geqslant 3;
\end{aligned} \tag{4.1.2}$$

where μ_0, μ_1, μ_r, μ and v are real parameters.

Solution. (a) Observe that $X(\mu, x) = \mu x - x^2 = \dfrac{\mu^2}{4} - \left(x - \dfrac{\mu}{2}\right)^2$. Let $y = h(\mu, x) = x - (\mu/2)$. Then, for each μ,

$$\dot{y} = \dot{x} = \frac{\mu^2}{4} - \left(x - \frac{\mu}{2}\right)^2 = \frac{\mu^2}{4} - y^2 = Z(\mu, y), \tag{4.1.3}$$

and the local families X and Z are equivalent by Definition 4.1.2. The bifurcation diagram for Z is shown in Figure 4.2. However,

$$Z(\mu, y) = \frac{\mu^2}{4} - y^2 = Y(\varphi(\mu), y), \tag{4.1.4}$$

where $\varphi\colon \mathbb{R} \to \mathbb{R}$ is defined by

$$\varphi(\mu) = \frac{\mu^2}{4}. \tag{4.1.5}$$

Thus, $Z(\mu, y)$ is induced by $Y(v, y)$.

Figure 4.2 Bifurcation diagram for the family $Z(\mu, y) = (\mu^2/4) - y^2$. Observe the similarity between this diagram and Figure 4.1(*b*).

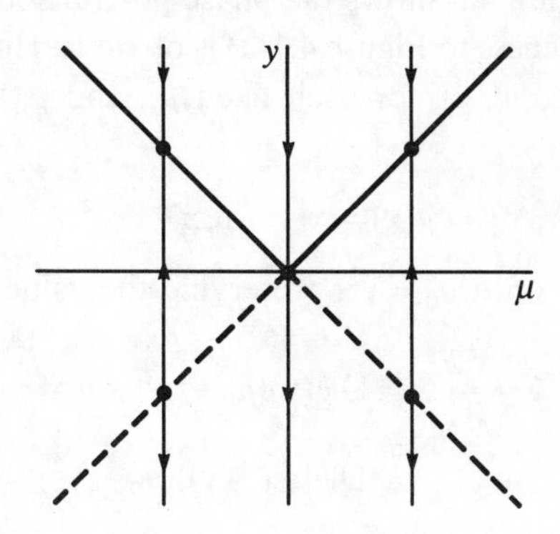

(b) Let $\dot{x} = -(1+\mu^2)x^2 = X(\mu, x)$. Introduce the change of variable $y = \alpha x$, so that

$$\dot{y} = \alpha\dot{x} = -\alpha(1+\mu^2)\frac{y^2}{\alpha^2}. \tag{4.1.6}$$

If we choose $\alpha = (1+\mu^2)$, (4.1.6) becomes

$$\dot{y} = -y^2 = Z(\mu, y). \tag{4.1.7}$$

Thus, for each μ, $y = h_\mu(x) = (1+\mu^2)x$ exhibits the topological equivalence of $X_\mu(x)$ and $Z_\mu(y)$. Now, the family $Z(\mu, y) = -y^2$, for all μ, is induced by $Y(v, y) = v - y^2$ by means of the continuous function $\varphi(\mu) \equiv 0$, since $Y(\varphi(\mu), y^2) = -y^2$.

(c) Write

$$\begin{aligned} X(\mu_0, \mu_1, x) &= \mu_0 + \mu_1 x - x^2, \\ &= \left(\mu_0 + \frac{\mu_1^2}{4}\right) - \left(x - \frac{\mu_1}{2}\right)^2, \end{aligned} \tag{4.1.8}$$

and let $y = x - (\mu_1/2)$ so that

$$\dot{y} = \dot{x} = \left(\mu_0 + \frac{\mu_1^2}{4}\right) - y^2 = Z(\mu_0, \mu_1, y). \tag{4.1.9}$$

Thus X and Z are equivalent. However, Z is induced by $Y(v, y) = v - y^2$ by $\varphi\colon \mathbb{R}^2 \to \mathbb{R}$, where

$$\varphi(\mu_0, \mu_1) = \mu_0 + \frac{\mu_1^2}{4}. \tag{4.1.10}$$

(d) For any given value of μ_r, the fixed points of

$$X(\mu_0, \mu_r, x) = \mu_0 - x^2 - \mu_r x^r, \qquad r \geqslant 3, \tag{4.1.11}$$

lie on the curve

$$\mu_0 = x^2 + \mu_r x^r \tag{4.1.12}$$

in the $\mu_0 x$-plane. For sufficiently small x, the form of this curve is determined by the quadratic term. Figure 4.3 shows the phase portraits for $\dot{x} = X(\mu_0, \mu_r, x)$ with μ_r held constant; the likeness to Figure 4.1(*a*) is obvious. However, to demonstrate the equivalence of $X(\mu_0, \mu_r, x)$, for each fixed μ_r, and $\mu_0 - y^2$, observe that the change of variable,

$$y = \psi(x) = x(1 + \mu_r x^{r-2})^{1/2}, \tag{4.1.13}$$

takes the curve (4.1.12) onto $\mu_0 = y^2$, preserving the value of μ_0. Moreover,

$$\begin{aligned} \dot{y} = \frac{d\psi}{dx}\dot{x} &= D\psi(x)(\mu_0 - x^2 - \mu_r x^r) \\ &= D\psi(\psi^{-1}(y))(\mu_0 - y^2). \end{aligned} \tag{4.1.14}$$

Now, (4.1.13) gives

$$D\psi(x) = 1 + \tfrac{1}{2}\mu_r(r-1)x^{r-2} + \cdots, \tag{4.1.15}$$

for small x, μ_r. Therefore, there exists a neighbourhood of $(\mu_r, x) = (0, 0)$ on which $D\psi(x) > 0$ and $\mathrm{sign}(\dot{y}) = \mathrm{sign}(D\psi(x)\dot{x}) = \mathrm{sign}(\dot{x})$. It follows that, for each sufficiently small μ_r, the vector field in (4.1.14) is topologically equivalent to $\mu_0 - y^2$. Finally, equivalence of the families $X(\mu_0, \mu_r, x)$ and $Z(\mu_0, \mu_r, y) = \mu_0 - y^2$ follows from Definition 4.1.2 with $h: \mathbb{R}^2 \times \mathbb{R} \to \mathbb{R}$ defined by

$$h(\mu_0, \mu_r, x) = x(1 + \mu_r x^{r-2})^{1/2}. \tag{4.1.16}$$

Clearly, $Y(v, y)$ induces $Z(\mu_0, \mu_r, y)$ by taking $v = \varphi(\mu_0, \mu_r) = \mu_0$. □

It can be shown that *every* smooth local family $X(\boldsymbol{\mu}, x)$, satisfying $X(\mathbf{0}, x) = X_{\mathbf{0}}(x) = -x^2$, is equivalent to one induced by $Y(v, y) = v - y^2$. As Exercise 4.1.1

Figure 4.3 Fixed points as a function of μ_0 for the family of (4.1.11) with $\mu_r > 0$ when: (*a*) r is even; (*b*) r is odd. Observe how they are displaced from the curve $\mu_0 = x^2$. For $\mu_r < 0$, the relative positions of $\mu_0 = x^2$ and $\mu_0 = x^2 + \mu_r x^r$ are reversed. Note that these diagrams are qualitatively the same as Figure 4.1(*a*)

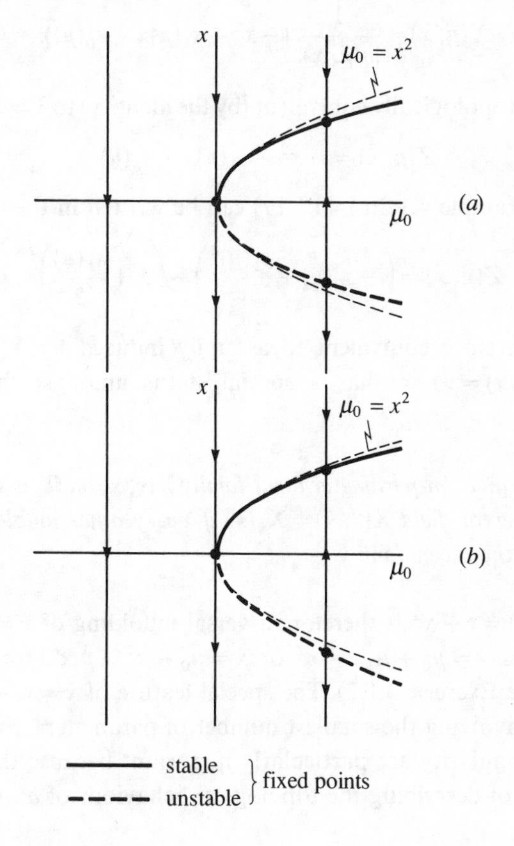

indicates, a proof of this result is not easily achieved along the lines suggested by Example 4.1.2. A more satisfactory approach makes use of the 'Malgrange Preparation Theorem' (see Chow & Hale, 1982; Golubitsky & Guillemin, 1973).

Theorem 4.1.1 (Malgrange) *Let $F(\boldsymbol{\mu}, x)$, $\boldsymbol{\mu} \in \mathbb{R}^m$, be a smooth, real-valued function defined on a neighbourhood of the origin in $\mathbb{R}^m \times \mathbb{R}$. Furthermore, suppose that $F(\mathbf{0}, x) = x^k g(x)$, where $g(x)$ is smooth in a neighbourhood of $x = 0$ and $g(0) \neq 0$. Then there exists a function $q(\boldsymbol{\mu}, x)$, smooth in a neighbourhood of $(\boldsymbol{\mu}, x) = (\mathbf{0}, 0)$, and functions $s_i(\boldsymbol{\mu})$, $i = 0, \ldots, k-1$, smooth in a neighbourhood of $\boldsymbol{\mu} = \mathbf{0}$, such that*

$$q(\boldsymbol{\mu}, x)F(\boldsymbol{\mu}, x) = x^k + \sum_{i=0}^{k-1} s_i(\boldsymbol{\mu})x^i. \tag{4.1.17}$$

Some elementary features of this subtle result are considered in Exercises 4.1.4 and 5. From the point of view of the present discussion, it is apparent that any smooth, m-parameter unfolding $X(\boldsymbol{\mu}, x)$ of the singular vector field $\dot{x} = -x^2$ satisfies the hypotheses of the Theorem 4.1.1 with $k = 2$ and $g(x) \equiv -1$. It is not difficult to show that $q(\mathbf{0}, 0) = -1$ so that $q(\boldsymbol{\mu}, x) < 0$ on a neighbourhood of $(\boldsymbol{\mu}, x) = (\mathbf{0}, 0)$. Thus (4.1.17) gives

$$X(\boldsymbol{\mu}, x) = \frac{1}{|q(\boldsymbol{\mu}, x)|}(-x^2 - s_1(\boldsymbol{\mu})x - s_0(\boldsymbol{\mu})), \tag{4.1.18}$$

and $\dot{x} = X(\boldsymbol{\mu}, x)$ is topologically equivalent (by the identity) to $\dot{x} = Z(\boldsymbol{\mu}, x)$, where

$$Z(\boldsymbol{\mu}, x) = -x^2 - s_1(\boldsymbol{\mu})x - s_0(\boldsymbol{\mu}). \tag{4.1.19}$$

However, as in Example 4.1.2(c), (4.1.19) can be written in the form

$$Z(\boldsymbol{\mu}, x) = \left(-s_0(\boldsymbol{\mu}) + \frac{s_1(\boldsymbol{\mu})^2}{4}\right) - \left(x + \frac{s_1(\boldsymbol{\mu})}{2}\right)^2 \tag{4.1.20}$$

and $Z(\boldsymbol{\mu}, x)$ is therefore equivalent to a family induced by $Y(\nu, x) = \nu - x^2$. It follows that $Y(\nu, x) = \nu - x^2$ has a special status amongst the unfoldings of $\dot{x} = -x^2$.

Definition 4.1.4 *A given m-parameter local family* $\mathbf{X}(\boldsymbol{\nu}, \mathbf{x})$ *on* $\mathbb{R}^n$ *is said to be a* versal unfolding *of the vector field* $\mathbf{X}(\mathbf{0}, \mathbf{x}) = \mathbf{X}_0(\mathbf{x})$ *if every other unfolding is equivalent to one induced by the given family.*

The family $X(\nu, x) = \nu - x^2$ is therefore a versal unfolding of $\dot{x} = -x^2$. However, so, for example, is $\dot{x} = \mu_0 + \mu_1 x - x^2$ or $\dot{x} = \mu_0 - x^2 - \mu_r x^r$, $r \geqslant 3$, because they induce $X(\nu, x)$ (see Exercise 4.1.2). The special feature of $\dot{x} = \nu - x^2$ is that it is a versal unfolding involving the smallest number of parameters. Such (mini) versal unfoldings of a singularity are particularly important because they represent the most concise way of describing the topological behaviour of *all* vector fields that can occur near $\mathbf{X}_0(\mathbf{x})$ in smoothly varying families. In other words, they exhibit, in the simplest manner, all the bifurcational behaviour that can be associated with the singularity in $\mathbf{X}_0(\mathbf{x})$.

The local family $X(\nu, x) = \nu - x^2$ is also representative of the unfoldings of all singularities of the form

$$\dot{x} = -ax^2, \qquad a > 0. \tag{4.1.21}$$

Obviously, it is not itself an unfolding of these singularities but it is easy to see that their versal unfoldings will be equivalent to it. For example, consider the unfolding

$$\dot{x} = \nu - ax^2, \qquad a > 0. \tag{4.1.22}$$

Let $y = \alpha x$, so that

$$\dot{y} = \alpha\dot{x} = \alpha\left(\nu - a\frac{y^2}{\alpha^2}\right), \tag{4.1.23}$$

and choose $\alpha = a^{1/2}$ to obtain

$$\dot{y} = a^{1/2}(\nu - y^2). \tag{4.1.24}$$

Finally, observe that, for each ν, the time rescaling $t \mapsto a^{1/2}t$ preserves orientation of trajectories and, therefore, (4.1.24) is topologically equivalent to

$$\dot{y} = (\nu - y^2), \tag{4.1.25}$$

by the identity. Unfoldings of singularities (4.1.21) with $a < 0$ can also be included if order-reversing homeomorphisms of $\mathbb{R}$ are allowed in the definition of topological equivalence (see Exercise 4.1.3).

Example 4.1.3 Show that the unfolding,

$$\dot{x} = \mu_0 + \mu_1 x + \mu_2 x^2 - x^3, \tag{4.1.26}$$

$\mu_i \in \mathbb{R}$, $i = 0, 1, 2$, of the singular vector field $\dot{x} = -x^3$ is equivalent to a local family induced by

$$\dot{x} = \nu_1 x + \nu_2 x^2 - x^3, \tag{4.1.27}$$

$\nu_i \in \mathbb{R}$, $i = 1, 2$. Sketch the bifurcation diagram for (4.1.27).

Solution. Given that $\mu_i \in \mathbb{R}$, $i = 0, 1, 2$, the cubic equation

$$\mu_0 + \mu_1 x + \mu_2 x^2 - x^3 = 0 \tag{4.1.28}$$

has roots λ_i, $i = 1, 2, 3$, that are either all real or one is real and the other two are complex conjugates of one another. Suppose that either $\lambda_i \in \mathbb{R}$, $i = 1, 2, 3$ and $\lambda_1 \leqslant \lambda_2 \leqslant \lambda_3$ or $\lambda_1 = \bar{\lambda}_3 \in \mathbb{C}$ and $\lambda_2 \in \mathbb{R}$. Since λ_i, $i = 1, 2, 3$, are the roots of (4.1.28),

(4.1.26) can be written as

$$\dot{x} = -(x-\lambda_1)(x-\lambda_2)(x-\lambda_3). \tag{4.1.29}$$

Let $y = h(\boldsymbol{\mu}, x) = (x-\lambda_2)$, then

$$\dot{y} = \dot{x} = -y(y+\alpha)(y+\beta), \tag{4.1.30}$$

with $\alpha = \lambda_2 - \lambda_1$ and $\beta = \lambda_2 - \lambda_3$. Now define the real parameters

$$\nu_1 = -\alpha\beta, \qquad \nu_2 = -(\alpha+\beta), \tag{4.1.31}$$

and write (4.1.30) as

$$\dot{y} = \nu_1 y + \nu_2 y^2 - y^3. \tag{4.1.32}$$

Since λ_2 depends continuously on μ_i, $i = 0, 1, 2$, it follows that (4.1.26) is equivalent to a family which is induced by (4.1.27).

The fixed points of (4.1.27) are given by

$$x = 0 \qquad \text{and} \qquad x = \tfrac{1}{2}[\nu_2 \pm (\nu_2^2 + 4\nu_1)^{1/2}]. \tag{4.1.33}$$

Therefore: (4.1.27) has:

(i) one fixed point at $x = 0$ for $\nu_2^2 < -4\nu_1$;
(ii) two fixed points, one at $x = 0$ and one (multiplicity 2) at $x = \nu_2/2$, for $\nu_2^2 = -4\nu_1$;
(iii) three fixed points for $\nu_2^2 > -4\nu_1 > 0$;
(iv) two fixed points, one at $x = 0$ (multiplicity 2) and one at $x = \nu_2$ for $\nu_1 = 0$;
(v) three fixed points for $\nu_1 > 0$.

The bifurcation diagram for (4.1.27) is shown in Figure 4.4. The orientation of the trajectories in the phase portraits shown is obtained by inspection of (4.1.27) and its linearisations. □

The Malgrange Preparation Theorem allows us to show that any smooth, m-parameter unfolding, $Y(\boldsymbol{\eta}, x)$, of $\dot{x} = -x^3$ is topologically equivalent to

$$\dot{x} = -x^3 - s_2(\boldsymbol{\eta})x^2 - s_1(\boldsymbol{\eta})x - s_0(\boldsymbol{\eta}). \tag{4.1.34}$$

Clearly (4.1.34) is induced by (4.1.26) with $\mu_i = \varphi_i(\boldsymbol{\eta}) = -s_i(\boldsymbol{\eta})$, $i = 0, 1, 2$. Thus (4.1.26) is a versal unfolding of the singularity $\dot{x} = -x^3$. We can then deduce, from Example 4.1.3, that (4.1.27), which involves only two parameters, is also versal. Now if $m = 1$ then the functions s_i in (4.1.34) depend on a single variable $\eta \in \mathbb{R}$. This, in turn, implies that the μ_i, and hence the ν_i, cannot be independent for one-parameter families. Thus, one-parameter unfoldings $Y(\eta, x)$ can only give rise to a subset of the bifurcations in Figure 4.4 and are consequently not versal. We therefore conclude that (4.1.27) is a (mini) versal unfolding of the singularity $\dot{x} = -x^3$. Observe that at least two parameters are required to completely unfold this singularity.

Examples 4.1.2 and 4.1.3 illustrate the relationship between the level of degeneracy of a singularity (see § 2.4) and the minimum number of parameters required to fully unfold it. Notice $\dot{x} = -x^2$ satisfies the single degeneracy condition $\mathrm{d}X/\mathrm{d}x|_{x=0} = 0$ and the non-degeneracy condition $\mathrm{d}^2X/\mathrm{d}x^2|_{x=0} \neq 0$. On the other hand, $\dot{x} = -x^3$ satisfies $\mathrm{d}X/\mathrm{d}x|_{x=0} = \mathrm{d}^2X/\mathrm{d}x^2|_{x=0} = 0$ and $\mathrm{d}^3X/\mathrm{d}x^3|_{x=0} \neq 0$. In general, the number of parameters required to completely unfold a singularity increases with its level of degeneracy. For example, a (mini) versal unfolding of $\dot{x} = -x^k$ is

$$\dot{x} = v_0 + v_1 x + \cdots + v_{k-2} x^{k-2} - x^k \tag{4.1.35}$$

(see Dumortier, 1978, and Exercise 4.1.6).

4.2 Saddle-node and Hopf bifurcations

In this section we consider the versal unfoldings of the non-hyperbolic singularities (2.4.1) and (2.4.2) in planar vector fields. These singularities are of similar status in the sense that a single degeneracy condition is satisfied in each case. They are examples of *co-dimension 1 singularities*.

4.2.1 Saddle-node bifurcation (Sotomayor, 1974)

Let $\mathbf{X}_0$ be a smooth vector field with a saddle-node singularity at the origin. Then $\mathrm{Det}\,\mathrm{D}\mathbf{X}_0(\mathbf{0}) = 0$ and $\mathbf{X}_0$ has normal form

$$\begin{pmatrix} x_1(\lambda + a_2 x_2) \\ b_2 x_2^2 \end{pmatrix} + O(|\mathbf{x}|^3), \tag{4.2.1}$$

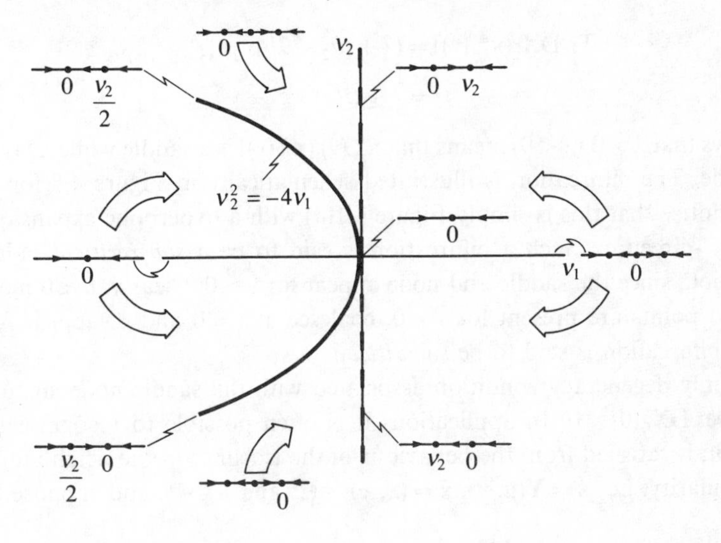

Figure 4.4 Fixed point bifurcation diagram for the local family (4.1.27). Bifurcations occur on the curve $v_2^2 = -4v_1$ and $v_1 = 0$.

where $\lambda = \mathrm{Tr}\, \mathbf{DX}_0(\mathbf{0})$ and b_2 are non-zero (see Proposition 2.4.1, p. 80, and Example 2.7.4, p. 101).

Proposition 4.2.1 *The local family*

$$\mathbf{X}(\nu, x) = \begin{pmatrix} x_1(\lambda + a_2 x_2) \\ \nu + b_2 x_2^2 \end{pmatrix} + O(|\mathbf{x}|^3), \tag{4.2.2}$$

is a versal unfolding of the saddle-node singularity (4.2.1).

The versality of (4.2.2) follows essentially from that of $\dot{x} = \nu + b_2 x_2^2$ (Exercises 4.1.3 and 6) and the Centre Manifold Theorem extended to families of vector fields (Palis & Takens, 1977).

Consider (4.2.2) when the $O(|\mathbf{x}|^3)$ terms are absent. Suppose $b_2 = -b$, $b > 0$, then there are no fixed points for $\nu < 0$ and two, given by $\mathbf{x} = \mathbf{x}^*_\pm(\nu)$, for $\nu > 0$. Here

$$\mathbf{x}^*_\pm(\nu) = (x^*_{1\pm}(\nu), x^*_{2\pm}(\nu))^{\mathrm{T}} = \left(0, \pm\left(\frac{\nu}{b}\right)^{1/2}\right)^{\mathrm{T}}. \tag{4.2.3}$$

Now,

$$\mathbf{DX}_\nu(\mathbf{x}^*_\pm(\nu)) = \begin{pmatrix} \lambda + a_2 x_2 & a_2 x_1 \\ 0 & -2b x_2 \end{pmatrix}\Bigg|_{\mathbf{x}^*_\pm(\nu)} \tag{4.2.4}$$

so that

$$\begin{aligned} \mathrm{Det}\, \mathbf{DX}_\nu(\mathbf{x}^*_\pm(\nu)) &= -2b x_2(\lambda + a_2 x_2)|_{x_2 = \pm(\nu/b)^{1/2}} \\ &= \mp 2b\lambda\left(\frac{\nu}{b}\right)^{1/2} + O(\nu) \end{aligned} \tag{4.2.5}$$

and

$$\begin{aligned} \mathrm{Tr}\, \mathbf{DX}_\nu(\mathbf{x}^*_\pm(\nu)) &= (\lambda + (a_2 - 2b)x_2)|_{x_2 = \pm(\nu/b)^{1/2}} \\ &= \lambda \pm O(\nu)^{1/2}. \end{aligned} \tag{4.2.6}$$

It follows that $\lambda > 0$ ($\lambda < 0$) means that $\mathbf{x}^*_+(\nu)$ ($\mathbf{x}^*_-(\nu)$) is a saddle while $\mathbf{x}^*_-(\nu)$ ($\mathbf{x}^*_+(\nu)$) is a node. The bifurcation is illustrated schematically in Figure 4.5 for the case $\lambda > 0$. Notice that this is simply Figure 4.1(*a*) with a hyperbolic expansion added in the x_1-direction. Such a bifurcation is said to be a *supercritical* saddle-node bifurcation, since the saddle and node appear for $\nu > 0$. Clearly, $b < 0$ means that the fixed points are present for $\nu < 0$, coalesce at $\nu = 0$ and disappear for $\nu > 0$. Such a bifurcation is said to be *subcritical*.

The only degeneracy condition associated with the saddle-node singularity in $\mathbf{X}_0$ is $\mathrm{Det}\, \mathbf{DX}_0(\mathbf{0}) = 0$. In applications it is often possible to recognise that this condition is satisfied from the behaviour of the isoclines in the neighbourhood of the singularity. Let $\dot{\mathbf{x}} = \mathbf{Y}(\boldsymbol{\mu}, \mathbf{x})$, $\mathbf{x} = (x, y)^{\mathrm{T}} \in \mathbb{R}^2$ and $\boldsymbol{\mu} \in \mathbb{R}^m$, and suppose that, for

$\mu = \mu^*$, the $\dot{x} = 0$ and $\dot{y} = 0$ isoclines are curves (see Exercises 4.2.1 and 2) that intersect *tangentially* rather than *transversely* (see Figure 4.6). For example, suppose $\mathbf{Y}_{\mu^*}(\mathbf{x}) = (f(x, y), g(x, y))^T$, where f, g are C^1-functions. Then, given that $D_y f(0,0)$ and $D_y g(0,0) \neq 0$ (where $D_x f = \partial f/\partial x$ etc.), the equations

$$f(x, y) = 0, \qquad g(x, y) = 0, \tag{4.2.7}$$

define curves $y = y_f(x)$ and $y = y_g(x)$, respectively, by the Implicit Function Theorem. Clearly, when tangential intersection occurs

$$\left.\frac{dy_f}{dx}\right|_{\mathbf{x}^*} = \left.\frac{dy_g}{dx}\right|_{\mathbf{x}^*}. \tag{4.2.8}$$

Figure 4.5 Supercritical saddle-node bifurcation in the truncation of (4.2.2) for $b_2 = -b$ with $b, \lambda > 0$. This figure is qualitatively unchanged when $O(|\mathbf{x}|^3)$ terms are present.

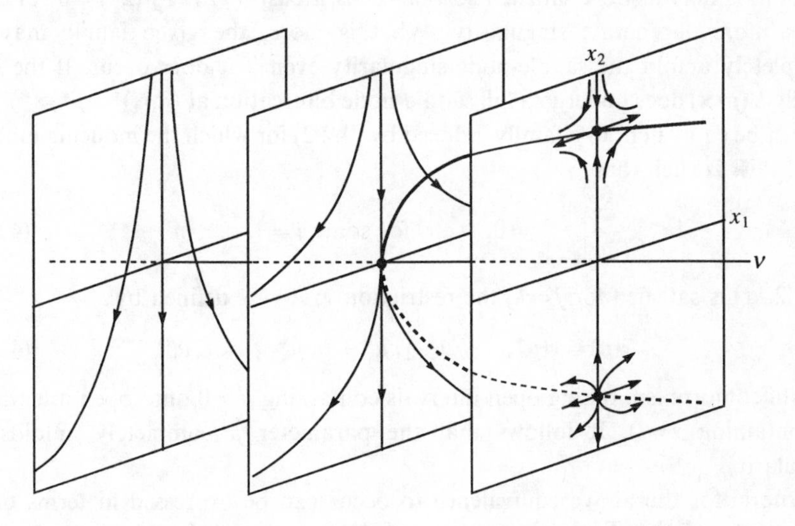

Figure 4.6 Intersection of the $\dot{x} = 0$ and $\dot{y} = 0$ isoclines: (*a*) tangential; (*b*) transverse. Tangency of isoclines at $\mathbf{x}^*$ implies that $y_f(x)$ and $y_g(x)$ have equal slope at $\mathbf{x}^*$. Generically (see (*a*)) $[d^2(y_f - y_g)/dx^2]_{\mathbf{x}^*} \neq 0$. Some non-generic configurations of isoclines for which the equal slope condition is satisfied are considered in Exercise 4.2.3.

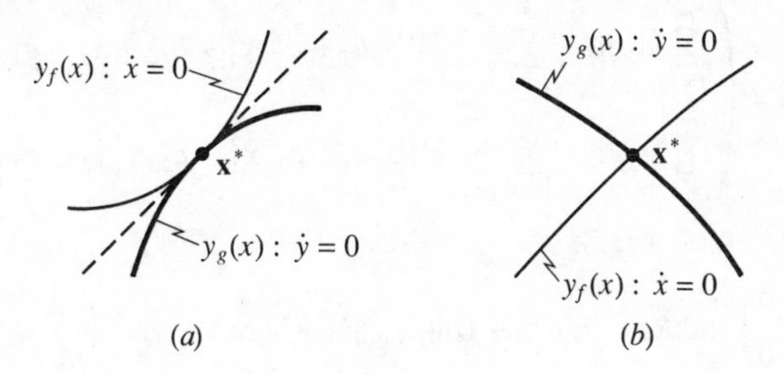

Differentiation of (4.2.7) with respect to x gives

$$-\frac{\mathrm{D}_x f(\mathbf{x}^*)}{\mathrm{D}_y f(\mathbf{x}^*)} = \left.\frac{\mathrm{d}y_f}{\mathrm{d}x}\right|_{\mathbf{x}^*} = \left.\frac{\mathrm{d}y_g}{\mathrm{d}x}\right|_{\mathbf{x}^*} = -\frac{\mathrm{D}_x g(\mathbf{x}^*)}{\mathrm{D}_y g(\mathbf{x}^*)}. \tag{4.2.9}$$

Thus

$$(\mathrm{D}_x f \mathrm{D}_y g - \mathrm{D}_y f \mathrm{D}_x g)|_{\mathbf{x}^*} = \mathrm{Det}\, \mathrm{D}\mathbf{Y}_{\boldsymbol{\mu}^*}(\mathbf{x}^*) = 0. \tag{4.2.10}$$

We therefore conclude that tangency of the $\dot{x}=0$ and $\dot{y}=0$ isoclines at $\mathbf{x}^*$, for $\boldsymbol{\mu} = \boldsymbol{\mu}^*$, implies $\mathrm{Det}\, \mathrm{D}\mathbf{Y}_{\boldsymbol{\mu}^*}(\mathbf{x}^*) = 0$. This phenomenon commonly occurs for models in which a fixed point is created (see Exercise 4.2.4 and 5). It is easy to see from Figure 4.6(*a*) that small perturbations of the isoclines can lead to situations with either zero or two fixed points.

While tangency of isoclines at $\mathbf{x}^*$ ensures that $\mathrm{Det}\, \mathrm{D}\mathbf{Y}_{\boldsymbol{\mu}^*}(\mathbf{x}^*) = 0$, it does *not* guarantee that $\mathbf{x}^*$ is a saddle-node singularity. For example, other degeneracy conditions may also be satisfied so that $\mathbf{x}^*$ is a cusp ($\mathrm{Tr}\, \mathrm{D}\mathbf{Y}_{\boldsymbol{\mu}^*}(\mathbf{x}^*) = 0$) or some other more degenerate singularity. What is more, the given family may not completely unfold the saddle-node singularity even if it does occur. If the local family $\mathbf{Y}(\boldsymbol{\mu}, \mathbf{x})$ does undergo a full saddle-node bifurcation at $(\boldsymbol{\mu}, \mathbf{x}) = (\boldsymbol{\mu}^*, \mathbf{x}^*)$, then it must be equivalent to a family induced by (4.2.2) for which the inducing function $v\colon \mathbb{R}^m \to \mathbb{R}$ is such that

$$\left.\frac{\partial v}{\partial \mu_j}\right|_{\boldsymbol{\mu}=\boldsymbol{\mu}^*} \neq 0, \qquad \text{for some } j = 1, \ldots, m. \tag{4.2.11}$$

If (4.2.11) is satisfied for $j = k$, the restriction $\bar{v}\colon \mathbb{R} \to \mathbb{R}$ defined by

$$\bar{v}(\mu) = v(\mu_1^*, \ldots, \mu_{k-1}^*, \mu_k^* + \mu, \mu_{k+1}^*, \ldots, \mu_m^*) \tag{4.2.12}$$

is a diffeomorphism taking open intervals containing $\mu = 0$ onto open intervals of v containing $v = 0$. It follows that the parameter μ completely unfolds the singularity.

Criteria for the above equivalence to occur can be expressed in terms of the coefficients of the Taylor expansion of $\mathbf{Y}(\boldsymbol{\mu}, \mathbf{x})$ about $(\boldsymbol{\mu}, \mathbf{x}) = (\boldsymbol{\mu}^*, \mathbf{x}^*)$. It is convenient to assume that an appropriate linear transformation has been made to reduce the linear part of $\mathbf{Y}_{\boldsymbol{\mu}^*}(\mathbf{x})$ to Jordan form. Moreover, if $(\boldsymbol{\mu}^*, \mathbf{x}^*) \neq (\mathbf{0}, \mathbf{0})$ then local coordinates should be taken in $\mathbb{R}^m \times \mathbb{R}^2$ before this reduction is carried out. Thus let the Taylor expansion of $\mathbf{Y}(\boldsymbol{\mu}, \mathbf{x})$ be given by

$$\begin{pmatrix} \dot{x}_1 \\ \dot{x}_2 \end{pmatrix} = \mathbf{J}\mathbf{x} + \begin{pmatrix} \sum_{j=1}^{m} a_{1j}\mu_j + \sum_{i=1}^{2}\sum_{j=1}^{m} b_{ij}^{(1)} x_i \mu_j + c_{11}x_1^2 + c_{12}x_1x_2 + c_{13}x_2^2 + R_1 \\ \sum_{j=1}^{m} a_{2j}\mu_j + \sum_{i=1}^{2}\sum_{j=1}^{m} b_{ij}^{(2)} x_i \mu_j + c_{21}x_1^2 + c_{22}x_1x_2 + c_{23}x_2^2 + R_2 \end{pmatrix}, \tag{4.2.13}$$

with $\mathbf{J} = \begin{pmatrix} \lambda & 0 \\ 0 & 0 \end{pmatrix}$ and $R_i = O(|\mu|^2) + O(|(\boldsymbol{\mu}, \mathbf{x})|^3)$, where $|(\boldsymbol{\mu}, \mathbf{x})| = \left(\sum_{j=1}^{m} \mu_j^2 + \sum_{i=1}^{2} x_i^2\right)^{1/2}$.

Proposition 4.2.2 *Let the coefficients in* (4.2.13) *satisfy*:

(i) $$c_{23} \neq 0; \tag{4.2.14}$$

(ii) $$a_{2j} \neq 0, \text{ for some } j = 1, \ldots, m. \tag{4.2.15}$$

Then there is a differentiable change of coordinates $\mathbf{x} \mapsto \mathbf{x}(\boldsymbol{\mu}, \mathbf{x})$, $\boldsymbol{\mu} \mapsto \nu(\boldsymbol{\mu})$ *at* $(\boldsymbol{\mu}, \mathbf{x}) = (\mathbf{0}, \mathbf{0}) \in \mathbb{R}^m \times \mathbb{R}^2$ *such that the 2-jet of* $\mathbf{Y}(\boldsymbol{\mu}, \mathbf{x})$ *is given by that of* (4.2.2).

A guide to the normal form calculations necessary to prove Proposition 4.2.2 is given in Exercise 4.2.6. The central idea is that the family $\dot{\mathbf{x}} = \mathbf{Y}(\boldsymbol{\mu}, \mathbf{x})$ can be thought of as the differential equation

$$\begin{pmatrix} \dot{\mathbf{x}} \\ \dot{\boldsymbol{\mu}} \end{pmatrix} = \begin{pmatrix} \mathbf{Y}(\boldsymbol{\mu}, \mathbf{x}) \\ \mathbf{0} \end{pmatrix} = \tilde{\mathbf{Y}}(\boldsymbol{\mu}, \mathbf{x}). \tag{4.2.16}$$

Normal form calculations can then be carried out on the extended vector field $\tilde{\mathbf{Y}}(\boldsymbol{\mu}, \mathbf{x})$ along lines similar to those presented in § 2.3. What emerges is that (4.2.14) is a non-degeneracy condition ensuring that the singularity is a saddle-node, while (4.2.15) is simply (4.2.11) for this problem. The use of Proposition 4.2.2 is illustrated in Exercise 4.2.7.

4.2.2 Hopf bifurcation (Hassard *et al*, 1981; Marsden & McCracken, 1976)

Let $\mathbf{X}_0$ be a smooth vector field with a Hopf singularity at the origin. Then $\operatorname{Tr} D\mathbf{X}_0(\mathbf{0}) = 0$ and the normal form of $\mathbf{X}_0$ can be written,

$$\begin{pmatrix} 0 & -\beta \\ \beta & 0 \end{pmatrix}\begin{pmatrix} x_1 \\ x_2 \end{pmatrix} + (x_1^2 + x_2^2)\left\{a_1\begin{pmatrix} x_1 \\ x_2 \end{pmatrix} + b_1\begin{pmatrix} -x_2 \\ x_1 \end{pmatrix}\right\} + O(|\mathbf{x}|^5), \tag{4.2.17}$$

where $\beta = (\operatorname{Det} D\mathbf{X}_0(\mathbf{0}))^{1/2} > 0$ and $a_1 \neq 0$ (see Proposition 2.4.2, p. 81, and the discussion following (2.7.25, 26) on p. 102.

Proposition 4.2.3 *The local family*

$$\mathbf{X}(\nu, \mathbf{x}) = \begin{pmatrix} \nu & -\beta \\ \beta & \nu \end{pmatrix}\begin{pmatrix} x_1 \\ x_2 \end{pmatrix} + (x_1^2 + x_2^2)\left\{a_1\begin{pmatrix} x_1 \\ x_2 \end{pmatrix} + b_1\begin{pmatrix} -x_2 \\ x_1 \end{pmatrix}\right\} + O(|\mathbf{x}|^5) \tag{4.2.18}$$

is a versal unfolding of the Hopf singularity (4.2.17).

In the absence of the terms of order $|\mathbf{x}|^5$, the local family (4.2.18) is more tractable in plane polar coordinates (r, θ). It is easily shown that

$$\dot{r} = r(\nu + a_1 r^2), \qquad \dot{\theta} = \beta + b_1 r^2. \tag{4.2.19}$$

Let us assume that $a_1 < 0$. The phase portrait of (4.2.19) for $\nu < 0$ consists of a hyperbolic, stable focus at the origin. When $\nu = 0$, $\dot{r} = a_1 r^3$ and the origin is still asymptotically stable, though it is no longer hyperbolic. For $\nu > 0$, $\dot{r} = 0$ for $r = (\nu/|a_1|)^{1/2}$ as well as for $r = 0$. It follows that for $\nu > 0$ there is a stable limit cycle, of radius proportional to $\nu^{1/2}$, surrounding a hyperbolic, unstable focus at the origin (see Figure 4.7(*a*)). This is called a *supercritical* Hopf bifurcation. If $a_1 > 0$, then the limit cycle occurs for $\nu < 0$: it is unstable and surrounds a stable fixed point. As ν increases, the radius of the cycle decreases to zero at $\nu = 0$, where the fixed point at the origin becomes a weakly unstable focus. For $\nu > 0$, $(x_1, x_2)^{\mathrm{T}} = \mathbf{0}$ is unstable and hyperbolic. This is known as a *subcritical* Hopf bifurcation (see Figure 4.7(*b*)). Let $\mathbf{Y}(\boldsymbol{\mu}, \mathbf{x})$, $\boldsymbol{\mu} \in \mathbb{R}^m$, $\mathbf{x} \in \mathbb{R}^2$ be a given family of vector fields. Generically, a Hopf bifurcation occurs at $(\boldsymbol{\mu}^*, \mathbf{x}^*)$ when $\mathrm{Tr}\, \mathrm{D}\mathbf{Y}_{\boldsymbol{\mu}}(\mathbf{x}^*)$ passes through zero at $\boldsymbol{\mu} = \boldsymbol{\mu}^*$. Thus, it is associated with a change of stability of the fixed point $\mathbf{x}^*$. The algebraic criteria that the family $\mathbf{Y}(\boldsymbol{\mu}, \mathbf{x})$ must satisfy, to ensure that a complete Hopf bifurcation takes place, are embodied in the following theorem, where μ is any one of μ_j, $j = 1, \ldots, m$ and $\mathbf{Y}(\mu, \mathbf{x}) = \mathbf{Y}(\boldsymbol{\mu}, \mathbf{x})|_{\mu_k = \mu_k^*,\, k \neq j}$.

Figure 4.7 Illustration of the Hopf bifurcation undergone by the local family (4.2.18): (*a*) supercritical ($a_1 < 0$); (*b*) subcritical ($a_1 > 0$). The qualitative features of these diagrams do not depend on whether or not the $O(|\mathbf{x}|^5)$-terms are present. Note that $\nu \mapsto -\nu$ in (4.2.18) induces families corresponding to the time-reversal of (a) and (b).

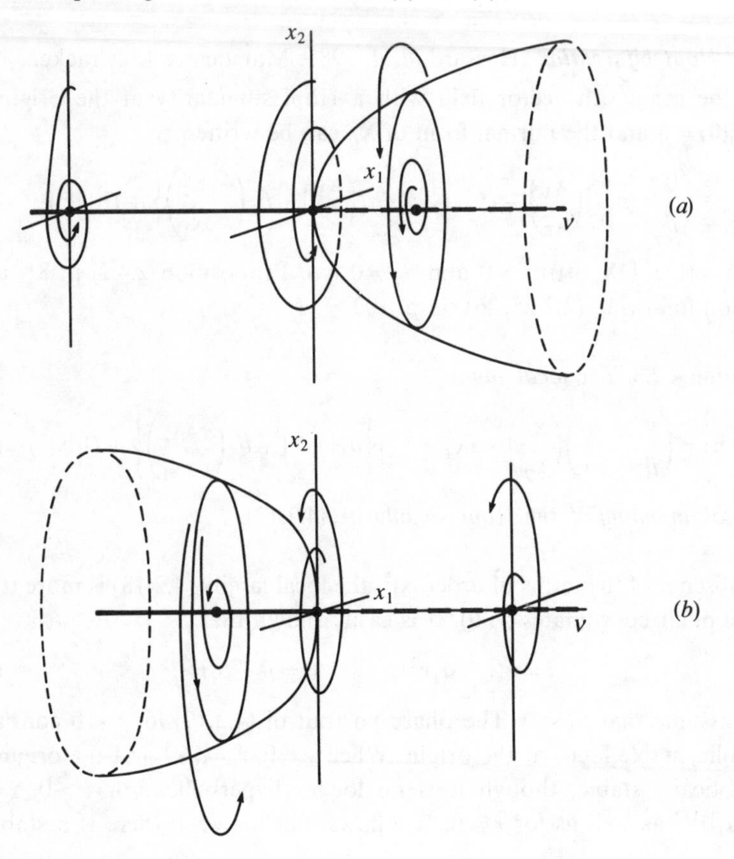

Theorem 4.2.1 (Hopf Bifurcation Theorem) *Suppose the parametrised system* $\dot{\mathbf{x}} = \mathbf{Y}(\mu, \mathbf{x})$, $\mathbf{x} \in \mathbb{R}^2$, $\mu \in \mathbb{R}$, *has a fixed point at the origin for all values of the real parameter* μ. *Furthermore, suppose the eigenvalues,* $\lambda_1(\mu)$ *and* $\lambda_2(\mu)$, *of* $\mathrm{D}\mathbf{Y}(\mu, \mathbf{0})$ *are pure imaginary for* $\mu = \mu^*$. *If the real part of the eigenvalues,* $\mathrm{Re}\,\lambda_1(\mu)$ $(= \mathrm{Re}\,\lambda_2(\mu))$, *satisfies*

$$\frac{\mathrm{d}}{\mathrm{d}\mu}\{\mathrm{Re}\,\lambda_1(\mu)\}|_{\mu=\mu^*} > 0 \tag{4.2.20}$$

and the origin is an asymptotically stable fixed point when $\mu = \mu^*$ *then:*

(*a*) $\mu = \mu^*$ *is a bifurcation point of the system;*

(*b*) *for* $\mu \in (\mu_1, \mu^*)$, *some* $\mu_1 < \mu^*$, *the origin is a stable focus;*

(*c*) *for* $\mu \in (\mu^*, \mu_2)$, *some* $\mu_2 > \mu^*$, *the origin is an unstable focus surrounded by a stable limit cycle, whose size increases with* μ.

Theorem 4.2.1 can be applied directly to a given family without any of the pre-conditoning required for Proposition 4.2.2. Items (a)–(c) describe what we have called a supercritical Hopf bifurcation and, indeed, the theorem can be trivially applied to (4.2.18) to show that the $O(|\mathbf{x}|^5)$ terms do not affect the bifurcational behaviour obtained from (4.2.19). For (4.2.18), $\lambda_1(v) = v + \mathrm{i}\beta$, which is pure imaginary for $v = 0$ and satisfies

$$\frac{\mathrm{d}}{\mathrm{d}v}(\mathrm{Re}\,\lambda_1(v))|_{v=0} = 1 > 0. \tag{4.2.21}$$

This condition in Theorem 4.2.1 ensures that $\mathrm{Re}\,\lambda_1(\mu)$ increases transversely through zero, or equivalently, $\lambda_1(\mu)$ crosses the imaginary axis of the complex λ-plane transversely from left to right. This ensures that $\mathbf{Y}(\mu, \mathbf{x})$ completely unfolds the Hopf singularity at $\mathbf{x} = \mathbf{0}$ given that the coefficient a_1 in the normal form of $\mathbf{Y}_{\mu^*}(\mathbf{x})$ is non-zero.

We have already noted that if $a_1 < 0$ then the origin is an asymptotically stable fixed point of (4.2.18) with $v = 0$. Theorem 4.2.1 highlights the fact that a limit cycle can still occur, provided the origin is asymptotically stable, even if $a_1 = 0$. For example, suppose $a_1, \ldots, a_{l-1} = 0$ and $a_l < 0$, then (4.2.19) becomes

$$\dot{r} = r(v + a_l r^{2l}), \qquad \dot{\theta} = \beta + b_1 r^2. \tag{4.2.22}$$

It follows that there is still a stable limit cycle for $v > 0$, whose radius, $r = (v/|a_l|)^{1/2l}$, increases with v. However, we shall see in §4.3.2 that (4.2.18) is no longer a versal unfolding of the corresponding singularity.

It is sometimes possible to establish that the origin is an asymptotically stable fixed point of $\mathbf{Y}_{\mu^*}$ by using a Liapunov function. Failing this, we can resort to using normal form calculations to obtain a_1. Assume $\mathbf{Y}_{\mu^*}$ has been pre-conditioned so that its linear part is in Jordan form, i.e.

$$\mathbf{Y}_{\mu^*}(\mathbf{x}) = \mathbf{J}\mathbf{x} + \begin{pmatrix} f(\mathbf{x}) \\ g(\mathbf{x}) \end{pmatrix}, \tag{4.2.23}$$

where $\mathbf{J} = \begin{pmatrix} 0 & -\beta \\ \beta & 0 \end{pmatrix}$ and $\mathbf{x} = (x, y)^{\mathrm{T}}$. Then, it can be shown that (see Guckenheimer & Holmes, 1983, pp. 152–6)

$$16a_1 = \left\{ (f_{xxx} + f_{xyy} + g_{xxy} + g_{yyy}) + \frac{1}{\beta}(f_{xy}[f_{xx} + f_{yy}] - g_{xy}[g_{xx} + g_{yy}] - f_{xx}g_{xx} + f_{yy}g_{yy}) \right\}\Bigg|_{\mathbf{x}=0} \tag{4.2.24}$$

The Hopf bifurcation has been widely studied and has enjoyed considerable success in applications (see Hassard *et al*, 1981; Marsden & McCracken, 1976).

4.3 Cusp and generalised Hopf bifurcations

The bifurcations considered in this section are associated with singularities in planar vector fields with co-dimension greater than unity. Such singularities are only completely unfolded by local families depending on more than one parameter.

4.3.1 Cusp bifurcation (Bogdanov, 1981a,b; Takens, 1974b)

Let $\mathbf{X}_0$ be a smooth vector field with a cusp singularity at the origin. Then $\mathbf{DX}_0(\mathbf{0}) \neq \mathbf{0}$ but $\mathrm{Tr}\,\mathbf{DX}_0(\mathbf{0}) = \mathrm{Det}\,\mathbf{DX}_0(\mathbf{0}) = 0$ and $\mathbf{X}_0$ has normal form

$$\begin{pmatrix} x_2 + a_2x_1^2 \\ b_2x_1^2 \end{pmatrix} + O(|\mathbf{x}|^3), \tag{4.3.1}$$

where $a_2 \neq 0$ and $b_2 \neq 0$ (see Proposition 2.4.3, p. 82, and the discussion on p. 108).

Proposition 4.3.1 *The local family*

$$\mathbf{X}(\nu, \mathbf{x}) = \begin{pmatrix} x_2 + \nu_2x_1 + a_2x_1^2 \\ \nu_1 + b_2x_1^2 \end{pmatrix} + O(|\mathbf{x}|^3) \tag{4.3.2}$$

is a versal unfolding of the cusp singularity (4.3.1).

We have already pointed out in §2.4 that the normal form (4.3.1) is not unique. Our knowledge of the properties of versal unfoldings of cusp singularities owes much to the work of Bogdanov (1981a,b) who used the alternative normal form (2.4.12). He showed that any two-parameter unfolding of a cusp singularity is equivalent to one induced by one of the families

$$\dot{x}_1 = x_2, \qquad \dot{x}_2 = \eta_1 + \eta_2 x_1 + x_1^2 \pm x_1 x_2. \tag{4.3.3}$$

Notice that the cusp singularity requires at least two parameters to completely unfold it. The bifurcation set consists of a collection of curves in the $\mathbf{v}$-plane. Let us consider (4.3.2) without the terms $O(|\mathbf{x}|^3)$ and with $a_2 < 0, b_2 > 0$, for definiteness. Observe that there are no fixed points for $v_1 > 0$. If $v_1 = 0$, with $v_2 \neq 0$, the origin of the phase plane is a fixed point with linearisation

$$\begin{pmatrix} v_2 & 1 \\ 0 & 0 \end{pmatrix} \begin{pmatrix} x_1 \\ x_2 \end{pmatrix}. \tag{4.3.4}$$

Hence, Det $\mathbf{DX}_{(0,v_2)}(\mathbf{0}) = 0$ and, since $b_2 > 0$, $\mathbf{x} = \mathbf{0}$ is a saddle-node. Hence (see Exercise 4.3.1) the v_2-axis, excluding the origin, is a line of saddle-node bifurcations (γ_0 in Figure 4.8).

For $v_1 < 0$, there are two fixed points

$$\begin{aligned} \mathbf{x}_1^* &= (\varepsilon, -v_2\varepsilon - a_2\varepsilon^2)^{\mathrm{T}}, \\ \mathbf{x}_2^* &= (-\varepsilon, v_2\varepsilon - a_2\varepsilon^2)^{\mathrm{T}}, \end{aligned} \tag{4.3.5}$$

where $\varepsilon = (-v_1/b_2)^{1/2} > 0$. Now

$$\mathbf{DX}_{\mathbf{v}}(\mathbf{x}) = \begin{pmatrix} v_2 + 2a_2x_1 & 1 \\ 2b_2x_1 & 0 \end{pmatrix}, \tag{4.3.6}$$

so that

$$\text{Det } \mathbf{DX}_{\mathbf{v}}(\mathbf{x}) = -2b_2x_1, \tag{4.3.7}$$

$$\text{Tr } \mathbf{DX}_{\mathbf{v}}(\mathbf{x}) = v_2 + 2a_2x_1. \tag{4.3.8}$$

Figure 4.8 Sketch showing the bifurcation curves for the versal unfolding (4.3.2) of the cusp singularity: γ_0, saddle-node; γ_1, Hopf; and γ_2, saddle connection. Note that both γ_1 and γ_2 are tangential to the v_2-axis at the origin.

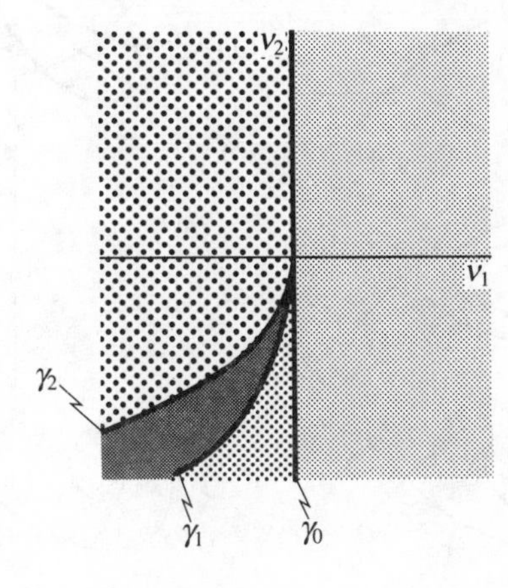

Given that $b_2 > 0$, (4.3.7) implies that $\mathbf{x}_1^*$ is a saddle. Moreover, for sufficiently small $|v_1|$ (and hence ε) (4.3.7) and (4.3.8) give $[\mathrm{Tr}\, \mathrm{D}\mathbf{X}_\mathbf{v}(\mathbf{x}_2^*)]^2 > 4 \,\mathrm{Det}\, \mathrm{D}\mathbf{X}_\mathbf{v}(\mathbf{x}_2^*) > 0$ so that $\mathbf{x}_2^*$ is a node: stable for $v_2 < 0$; unstable for $v_2 > 0$.

Let us focus attention on the stable node occurring at $\mathbf{x}_2^*$ for $v_2 < 0$. Here

$$\mathrm{Tr}\, \mathrm{D}\mathbf{X}_\mathbf{v}(\mathbf{x}_2^*) = v_2 - 2a_2\varepsilon. \tag{4.3.9}$$

As v_1 (<0) decreases, ε increases, and $\mathrm{Tr}\, \mathrm{D}\mathbf{X}_\mathbf{v}(\mathbf{x}_2^*)$ becomes zero for

$$v_2 = 2a_2\varepsilon = 2a_2\left(-\frac{v_1}{b_2}\right)^{1/2} < 0, \tag{4.3.10}$$

since $a_2 < 0$. Generically, this defines a line of Hopf bifurcations (γ_1 in Figure 4.8) tangent to the v_2-axis at the origin in the $\mathbf{v}$-plane (see Exercise 4.3.2).

The essentials of the phase portraits obtained as (v_1, v_2) travels along the path $\mathscr{C}$ in the $\mathbf{v}$-plane is shown schematically in Figure 4.9. As we move away from γ_1, the limit cycle created in the Hopf bifurcation grows in size and eventually meets the saddle point, $\mathbf{x}_1^*$, to form a *saddle connection*. Such a feature is not structurally stable (see § 3.3) and the connection breaks as shown in Figure 4.9(*e*). This means

Figure 4.9 Phase portraits occurring in the subfamily of the versal unfolding (4.3.2) corresponding to the path $\mathscr{C}$ in the $\mathbf{v}$-plane. (*a*) $\mathbf{v}$ lies between γ_0 and γ_1: the node arising on γ_0 becomes a focus as γ_1 is approached. (*b*) $\mathbf{v} \in \gamma_1$ and $\mathbf{x}_2^*$ is a weakly stable focus. (*c*) $\mathbf{v}$ lies between γ_1 and γ_2: the limit cycle expands with distance from γ_1. (*d*) $\mathbf{v} \in \gamma_2$: the limit cycle has reached $\mathbf{x}_1^*$ forming a saddle connection. (*e*) Beyond γ_2 the saddle connection breaks and all but one of the trajectories from the unstable focus escape to infinity.

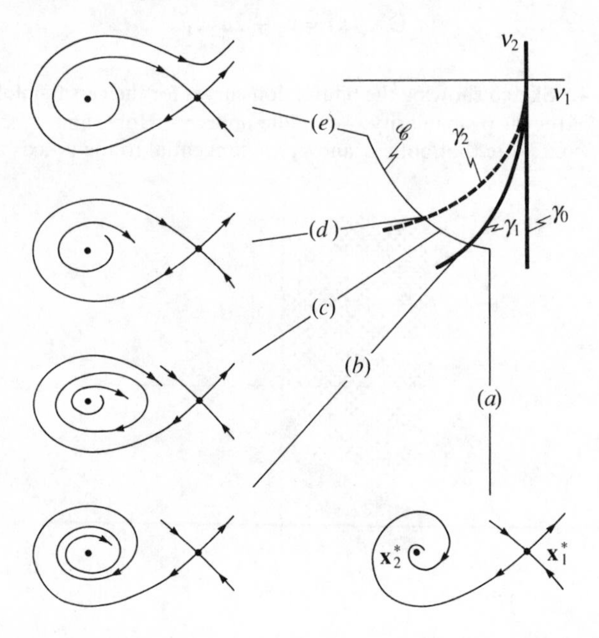

there is another curve (γ_2 in Figure 4.8) on which a *saddle connection bifurcation* occurs. In fact, γ_2 can be shown to be tangent to the ν_2-axis also (see Bogdanov 1981a,b; Takens, 1974b). The unstable focus at $\mathbf{x}_2^*$ in Figure 4.9(*e*) becomes an unstable node as $\nu_1 \to 0_-$ with $\nu_2 > 0$ (see comments following (4.3.8)) and the node and saddle coalesce for parameter points on the positive ν_2-axis.

The $O(|\mathbf{x}|^3)$ terms, neglected above, do not alter the bifurcation diagram qualitatively and the phase portraits for (4.3.2) in the complement of the bifurcation curves are shown in Figure 4.10. It must be emphasised that a proper derivation of the topological behaviour of this family and the proof of its versality is a major undertaking (see Bogdanov, 1981a,b; Takens, 1974b). At best, the above discussion indicates that the bifurcations described are plausible. For example, to prove that a saddle connection takes place is a non-trivial task. Saddle connections occur in simple Hamiltonian systems where their presence shows up in the level curves of the Hamiltonian. A subtle transformation (see Takens, 1974b and Exercise 4.3.3) is required to obtain a relationship between (4.3.2) and a family containing a Hamiltonian vector field with a homoclinic saddle connection. What is more, the detection of the saddle-node, Hopf and saddle-connection bifurcation curves is not sufficient to prove the versality of (4.3.2) (see Bogdanov, 1981b). It is necessary to establish that there is only one limit cycle which is created by the Hopf bifurcation

Figure 4.10 Structurally stable phase portraits occurring in the complement of the bifurcation curves for the family (4.3.2) with $a_2 < 0$, $b_2 > 0$. Similar diagrams (with γ_1 and γ_2 in different quadrants) arise for other choices of the signs of a_2 and b_2. The crucial feature is the stability of the limit cycle: stable for $a_2 b_2 < 0$; unstable for $a_2 b_2 > 0$.

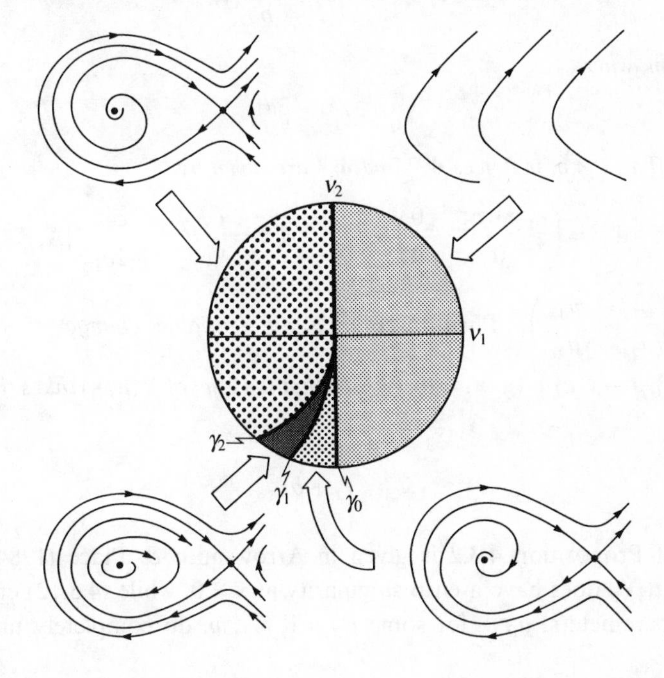

and destroyed by the saddle connection. This is to ensure that there are no further bifurcations involving the creation of pairs of limit cycles (see §4.3.2, and Takens, 1973b, 1974b). The proof of the uniqueness of the limit cycle is highly non-trivial (see Arnold, 1983, p. 282; Bogdanov, 1981a).

In analysing the cusp bifurcation we are involved with a local bifurcation arising from a singularity of degeneracy level two. It is worth noting that this *local* problem predicts the existence of a *global* bifurcation – the saddle connection. If we can show that a given family $\mathbf{Y}(\boldsymbol{\mu}, \mathbf{x})$ undergoes a cusp bifurcation at $(\boldsymbol{\mu}^*, \mathbf{x}^*)$ then the occurrence of a saddle connection for some $\boldsymbol{\mu}$ near $\boldsymbol{\mu}^*$ is assured. Of course, as in our previous discussions, $\mathrm{Tr}\, D\mathbf{Y}_{\boldsymbol{\mu}^*}(\mathbf{x}^*) = \mathrm{Det}\, D\mathbf{Y}_{\boldsymbol{\mu}^*}(\mathbf{x}^*) = 0$ are only necessary conditions for a cusp singularity at $(\boldsymbol{\mu}^*, \mathbf{x}^*)$. The following proposition gives algebraic criteria for a cusp bifurcation to occur.

Assume local coordinates at $(\boldsymbol{\mu}^*, \mathbf{x}^*)$ and that the linear part of $\mathbf{Y}_{\boldsymbol{\mu}^*}$ is in Jordan form. The Taylor expansion of the family $\mathbf{Y}(\boldsymbol{\mu}, \mathbf{x})$ then takes the form (4.2.13) with $\mathbf{J} = \begin{pmatrix} 0 & 1 \\ 0 & 0 \end{pmatrix}$.

Proposition 4.3.2 *Let the coefficients in* (4.2.13) *satisfy*:

(i) $$c_{11} + \tfrac{1}{2}c_{22} = a \neq 0, \qquad c_{21} = b \neq 0; \tag{4.3.11}$$

(ii) $$e_i a_{2j} - e_j a_{2i} \neq 0, \qquad \textit{for some } i, j = 1, \ldots, m. \tag{4.3.12}$$

In (4.3.12)

$$e_j = e_{1j}^{(1)} + e_{2j}^{(2)} - \frac{a}{b} e_{1j}^{(2)}, \tag{4.3.13}$$

$j = 1, \ldots, m$, *where*

$$e_{lj}^{(k)} = b_{lj}^{(k)} - d_{lj}^{(k)}, \tag{4.3.14}$$

$l, k = 1, 2$. *The* $2 \times m$ *matrices* $\mathbf{d}^{(1)}$ *and* $\mathbf{d}^{(2)}$ *are given by*

$$\mathbf{d}^{(1)} = \begin{pmatrix} c_{12} + c_{23} & 0 \\ 0 & 0 \end{pmatrix} \mathbf{A}, \qquad \mathbf{d}^{(2)} = \begin{pmatrix} c_{22} & c_{23} \\ c_{23} & -2c_{13} \end{pmatrix} \mathbf{A}, \tag{4.3.15}$$

with $\mathbf{A} = \begin{pmatrix} a_{11} \cdots a_{1m} \\ a_{21} \cdots a_{2m} \end{pmatrix}$. *Then there is a differentiable change of coordinates* $\mathbf{x} \mapsto \mathbf{x}(\boldsymbol{\mu}, \mathbf{x})$, $\boldsymbol{\mu} \mapsto \mathbf{v}(\boldsymbol{\mu})$ *at* $(\boldsymbol{\mu}, \mathbf{x}) = (\mathbf{0}, \mathbf{0})$ *such that the 2-jet of* $\mathbf{Y}(\boldsymbol{\mu}, \mathbf{x})$ *takes the form*

$$\begin{aligned} \dot{x}_1 &= x_2 + v_2 x_1 + a x_1^2, \\ \dot{x}_2 &= v_1 + b x_1^2. \end{aligned} \tag{4.3.16}$$

A proof of Proposition 4.3.2 is given in Arrowsmith & Place (1984): (4.3.11) ensures that $\mathbf{Y}_0$ does have a cusp singularity at $\mathbf{x} = \mathbf{0}$, while (4.3.12) ensures that a pair of parameters, μ_i, μ_j for some $i, j = 1, \ldots, m$, do completely unfold it. In

Arrowsmith & Place (1984), it is shown that $\mathbf{v}: \mathbb{R}^m \to \mathbb{R}^2$ takes the form

$$v_1(\boldsymbol{\mu}) = \sum_{j=1}^{m} a_{2j}\mu_j + O(|\boldsymbol{\mu}|^2), \tag{4.3.17}$$

$$v_2(\boldsymbol{\mu}) = \sum_{j=1}^{m} e_j\mu_j + O(|\boldsymbol{\mu}|)^2. \tag{4.3.18}$$

If (4.3.12) is satisfied for $i = i_0$ and $j = j_0$, then the restriction

$$\bar{\mathbf{v}}(\mu_{i_0}, \mu_{j_0}) = \mathbf{v}(0, \ldots, 0, \mu_{i_0}, 0, \ldots, 0, \mu_{j_0}, 0, \ldots, 0) \tag{4.3.19}$$

is a diffeomorphism of the $\mu_{i_0}\mu_{j_0}$-plane onto the $\mathbf{v}$-plane in the neighbourhood of their origins. It follows that the bifurcation curves are tangent to the direction $(-a_{2j_0}, a_{2i_0})$ (i.e. $\bar{v}_1(\mu_{i_0}, \mu_{j_0}) = 0$) in the $\mu_{i_0}\mu_{j_0}$-plane.

The application of the results presented above to a model for tumour growth is considered in Exercises 4.3.4–4.3.6. Other models exhibiting cusp bifurcations are discussed in Arrowsmith & Place (1984).

4.3.2 Generalised Hopf bifurcations
(Takens, 1973b, Guckenheimer & Holmes, 1983)

These bifurcations provide a good example of the relationship between the level of degeneracy of a singularity and the number of parameters appearing in its (mini) versal unfolding. The vector field (4.2.17) is obtained by assuming $a_1 \neq 0$ in the normal form

$$\begin{pmatrix} 0 & -\beta \\ \beta & 0 \end{pmatrix}\begin{pmatrix} x_1 \\ x_2 \end{pmatrix} + \sum_{k=1}^{\infty} (x_1^2 + x_2^2)^k \left[a_k \begin{pmatrix} x_1 \\ x_2 \end{pmatrix} + b_k \begin{pmatrix} -x_2 \\ x_1 \end{pmatrix} \right], \tag{4.3.20}$$

(see 2.4.6)). Further degeneracy can be introduced by taking

$$a_1 = a_2 = \cdots = a_{l-1} = 0 \qquad \text{and} \qquad a_l \neq 0 \tag{4.3.21}$$

as noted in §2.7. A vector field with normal form (4.3.20), for which (4.3.21) is satisfied, is said to have a *generalised Hopf singularity of type l* at $(x_1, x_2)^{\mathrm{T}} = \mathbf{0}$. Clearly, (4.2.17) is of type 1.

When a Hopf singularity of type l occurs, the following non-linear scaling reduces (4.3.20) to a simpler form. Let

$$f(x_1, x_2) = \left(\beta + \sum_{k=1}^{\infty} b_k (x_1^2 + x_2^2)^k \right) \tag{4.3.22}$$

and observe (see Exercise 4.3.7) that (4.3.20) can then be written as

$$f(x_1, x_2)\left[\begin{pmatrix} 0 & -1 \\ 1 & 0 \end{pmatrix}\begin{pmatrix} x_1 \\ x_2 \end{pmatrix} + \sum_{k=l}^{\infty} a_k'(x_1^2 + x_2^2)^k \begin{pmatrix} x_1 \\ x_2 \end{pmatrix} \right], \tag{4.3.23}$$

where $a_l' = a_l/\beta$. Since $\beta > 0$, $f(x_1, x_2)$ is greater than zero for all (x_1, x_2) sufficiently close to the origin. Thus, there is a neighbourhood of $(x_1, x_2)^{\mathrm{T}} = \mathbf{0}$ on which

(4.3.20) is topologically equivalent to the simpler vector field

$$\begin{pmatrix} 0 & -1 \\ 1 & 0 \end{pmatrix}\begin{pmatrix} x_1 \\ x_2 \end{pmatrix} + \sum_{k=l}^{\infty} a'_k (x_1^2 + x_2^2)^k \begin{pmatrix} x_1 \\ x_2 \end{pmatrix}. \tag{4.3.24}$$

A further radial scaling allows us to replace a'_l by $\mathrm{sgn}(a'_l) = \mathrm{sgn}(a_l)$ in (4.3.24) while maintaining topological equivalence. We distinguish the two possibilities by referring to a singularity of $(l, +)$ or $(l, -)$ type. The following theorem, given by Takens (1973b), provides a versal unfolding for (4.3.24).

Proposition 4.3.3 *The versal unfolding of a generalised Hopf singularity of type l given in* (4.3.24) *is*

$$\begin{pmatrix} 0 & -1 \\ 1 & 0 \end{pmatrix}\begin{pmatrix} x_1 \\ x_2 \end{pmatrix} + \sum_{k=l}^{\infty} a'_k (x_1^2 + x_2^2)^k \begin{pmatrix} x_1 \\ x_2 \end{pmatrix} + \sum_{i=0}^{l-1} v_i (x_1^2 + x_2^2)^i \begin{pmatrix} x_1 \\ x_2 \end{pmatrix}, \tag{4.3.25}$$

where $v_i \in \mathbb{R}$, $i = 0, 1, 2, \ldots, l-1$.

The unfoldings (4.3.25) are of two types depending on the sign of a'_l. Here we will only describe the $(l, +)$ case. It can be shown (see Exercise 4.3.7) that $(l, -)$ is simply related to a time reversal of $(l, +)$ so that the stabilities of fixed points and limit cycles are reversed but no new bifurcations occur.

Observe that the angular equation of the polar form of (4.3.25) is simply $\dot{\theta} = 1$. Thus, for the $(l, +)$ case, the bifurcational behaviour is given by the radial equation

$$\dot{r} = \sum_{i=0}^{l-1} v_i r^{2i+1} + r^{2l+1} + O(r^{2l+3}). \tag{4.3.26}$$

As in § 4.2.2, this behaviour is independent of the terms of order r^{2l+3} and, therefore, we will consider (4.3.26) when these terms are absent. Clearly, limit cycles are then predicted for the values of r given by the positive zeroes of the polynomial in $r^2 = \rho$ on the right hand side of (4.3.26). Thus, for $l = 1$ we examine the equation

$$\rho^2 + v_0 = 0 \tag{4.3.27}$$

and conclude (see § 4.2.2) that a limit cycle of radius $(-v_0)^{1/2}$ occurs only for $v_0 < 0$. The trivial solution $r = 0$ is present for all v_0 corresponding to a fixed point at the origin for each member of the one-parameter family (4.3.25).

For $l = 2$, the equation

$$\rho^2 + v_1 \rho + v_0 = 0 \tag{4.3.28}$$

is obtained and non-trivial zeroes are given by

$$\rho = [-v_1 \pm (v_1^2 - 4v_0)^{1/2}]/2. \tag{4.3.29}$$

It follows (see Exercise 4.3.8) that limit cycles occur as follows:

(i) two for $v_1 < 0$ and $v_1^2 - 4v_0 > 0$;

(ii) one for $\nu_0 < 0$;
(iii) none for $\nu_0 > 0$ and $\nu_1 > 0$ or $\nu_1^2 - 4\nu_0 < 0$.

The bifurcation diagram is shown in Figure 4.11. Note the *double limit cycle* bifurcation in which a limit cycle of non-zero radius, created at the bifurcation point, develops into a pair of limit cycles of opposite stability.

For $l = 3$ the situation is even more complicated. Here

$$\dot{r} = r^7 + \nu_2 r^5 + \nu_1 r^3 + \nu_0 r \tag{4.3.30}$$

and we must find where the three-parameter family of cubic equations

$$P_\nu(\rho) = \rho^3 + \nu_2 \rho^2 + \nu_1 \rho + \nu_0 = 0 \tag{4.3.31}$$

has positive solutions. There are no such solutions when $\nu_0, \nu_1, \nu_2 > 0$. However, positive solutions will appear (i.e. bifurcation points will occur) when (4.3.31) has multiple roots. It can be seen that the positions of the turning points of $P_\nu(\rho)$ are independent of ν_0. For example, if $\nu_2 = 0$ then $P_\nu(\rho)$ has a minimum at $\rho = (-\nu_1/3)^{1/2}$ and a maximum at $\rho = -(-\nu_1/3)^{1/2}$ (see Figure 4.12). For sufficiently large ν_0

Figure 4.11 The bifurcation diagram for the type-(2, +) Hopf bifurcation is shown along with the structurally stable topological types of phase portrait that occur in the complement of the bifurcation curves. A non-degenerate supercritical (subcritical) Hopf bifurcation takes place as ν_0 increases through zero with $\nu_1 < 0\,(>0)$ and there is a double limit cycle bifurcation on the semi-parabola Γ. As ν_0 decreases through Γ, a non-hyperbolic limit cycle of finite radius appears and subsequently splits into two hyperbolic cycles.

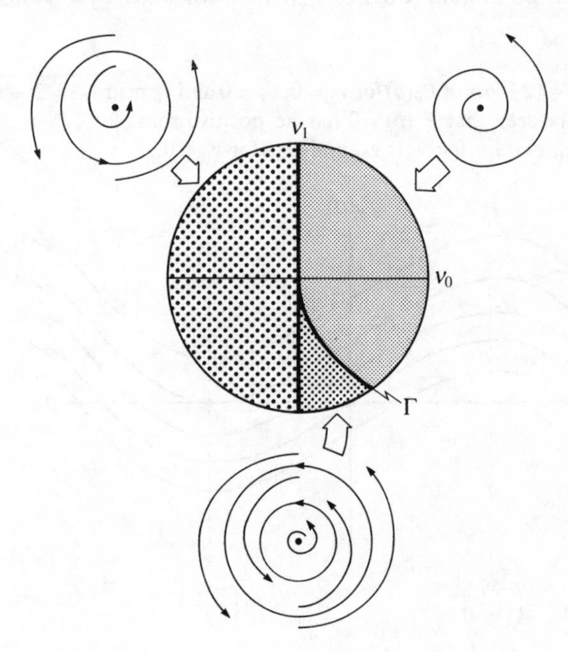

there are no positive roots of (4.3.31). As ν_0 decreases the graph of $P_\nu(\rho)$ simply shifts downward and when $\nu_0 = \nu_{0c} = 2(-\nu_1/3)^{3/2}$ a double root of (4.3.31) appears at $\rho = (-\nu_1/3)^{1/2}$. This root splits into two positive, simple zeroes of P_ν as ν_0 decreases below ν_{0c} and the smaller of these approaches $\rho = 0$ as $\nu_0 \to 0$. For $\nu_0 < 0$ there is only one positive root of (4.3.31). It follows that the vector field undergoes a double limit-cycle bifurcation at $\nu_0 = \nu_{0c}$. The sign of $\dot{r}$ is given by $\mathrm{sgn}(P_\nu(\rho))$ and therefore the limit cycle with the smaller radius is stable while the other is unstable. The radius of the smaller cycle shrinks to zero at $\nu_0 = 0$ and the stability of the fixed point at the origin changes as ν_0 passes through zero. In fact, the vector field undergoes a type-1 (non-degenerate) Hopf bifurcation at the origin when $\nu_0 = 0$. It is not difficult to show (see Exercise 4.3.9(b)) that an unstable limit cycle grows from the origin as ν_0 decreases through zero for $\nu_1 \geqslant 0$, leading to the bifurcation curves shown in Figure 4.13 for $\nu_2 = 0$.

For fixed $\nu_2 > 0$, the bifurcation curves in the $\nu_0\nu_1$-plane are similar to those $\nu_2 = 0$ (see Exercise 4.3.9(c)) but for $\nu_2 < 0$ sets of three limit cycles can occur. Equation (4.3.31) has a triple root when it is satisfied together with

$$3\rho^2 + 2\nu_2\rho + \nu_1 = 0 \tag{4.3.32}$$

and

$$6\rho + 2\nu_2 = 0. \tag{4.3.33}$$

Clearly, (4.3.33) has a positive solution $\rho = -\nu_2/3$ only for $\nu_2 < 0$. Elimination of ρ from (4.3.32) yields $\nu_1 = \nu_2^2/3$ and then (4.3.31) gives $\nu_0 = \nu_2^3/27$. Thus the curve $(\nu_0, \nu_1, \nu_2) = (\nu_2^3/27, \nu_2^2/3, \nu_2)$, with $\nu_2 < 0$, is a line of triple roots of (4.3.31). An investigation of the double roots of $P_\nu(\rho) = 0$ for fixed $\nu_2 < 0$ shows that double

Figure 4.12 Plots of $P_\nu(\rho)$ for $\nu_2 = 0$, $\nu_1 < 0$ and ν_0 near $\nu_{0c} = 2(-\nu_1/3)^{3/2}$. It can be seen that $P_\nu(\rho) = 0$ has no positive roots for $\nu_0 > \nu_{0c}$, one for $\nu_0 = \nu_{0c}$, two for $0 < \nu_0 < \nu_{0c}$ and one for $\nu_0 < 0$.

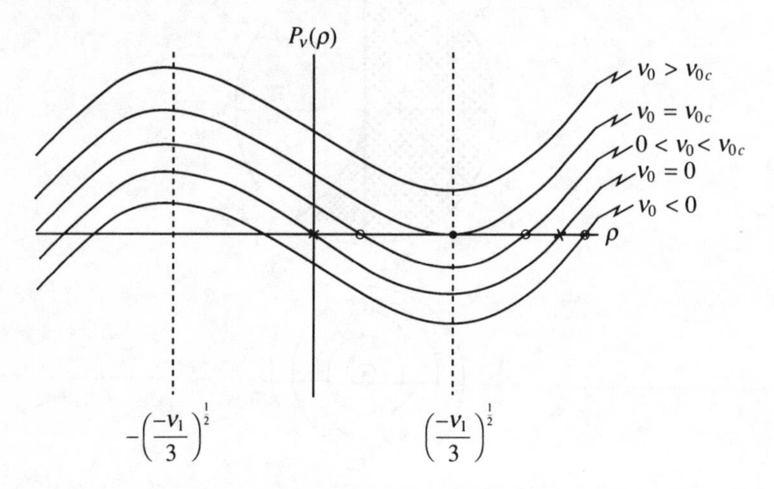

limit-cycle bifurcations occur on the cuspoidal curve, δ, shown in Figure 4.14. The v_1-axis remains a line of non-degenerate Hopf bifurcations and three limit cycles are simultaneously present for (v_0, v_1) in the region bounded by the v_1-axis and the cuspoidal curve (see Exercises 4.3.10 & 11). The complete bifurcation diagram for the $(3, +)$ Hopf bifurcation is shown in Figure 4.15.

For $l > 3$ the complexity of the bifurcation set continues to increase with l and the structurally stable regions become more difficult to locate. However, it is clear that there is, for any integer k between 0 and l, a region of parameter space with k limit cycles present. It follows that there will be at least $l + 1$ structurally stable regions in the parameter space.

4.4 Diffeomorphisms on ℝ (Arnold, 1983; Whitley, 1983)

In this section we consider the analogues, for diffeomorphisms, of the one-dimensional families discussed in §4.1. It is perhaps worth emphasising that we are dealing with *local* families throughout this chapter. In this section in particular we will frequently introduce families of maps that are *not* diffeomorphisms on the

Figure 4.13 Intersection of the type-$(3, +)$ Hopf bifurcation diagram with the plane $v_2 = 0$. Note that: (*a*) the negative v_1-axis is a line of non-denerate supercritical (i.e. type-$(1, -)$) Hopf bifurcations with increasing v_0; (*b*) the positive v_1-axis is a line of non-degenerate subcritical (i.e. type-$(1, +)$) Hopf bifurcations with increasing v_0; and (*c*) the curve $v_0 = 2(-v_1/3)^{3/2}$ is a line of double limit-cycle bifurcations: a pair of limit cycles of finite radius being created as v_0 decreases through this curve.

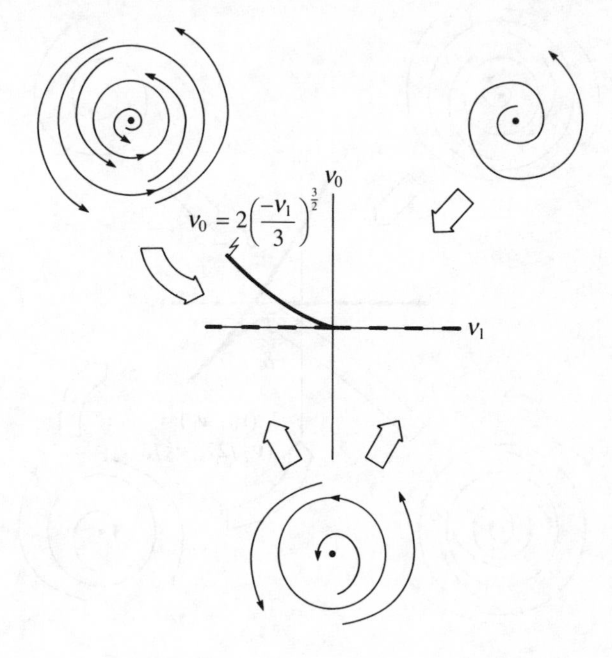

whole of ℝ but they *do* define *local* families of diffeomorphisms when suitably restricted. Only these restrictions appear in our discussion. To dwell on this point repeatedly throughout the section would complicate the notation and obscure the results presented. Therefore, we will content ourselves with periodic reminders to the reader of the local character of the families under consideration.

Let $f(\mu, x)$, $\mu, x \in \mathbb{R}$, be a smooth, one-parameter family of diffeomorphisms for which $f(0, x) = f_0(x)$ has a non-hyperbolic fixed point at the origin, i.e. $f_0(0) = 0$ and $|Df_0(0)| = |D_x f(\mathbf{0})| = 1$. In contrast to the vector field problem of § 4.1, there are *two* cases to consider: $D_x f(\mathbf{0}) = +1$ and $D_x f(\mathbf{0}) = -1$. The normal forms for f_0 under these conditions were considered in Example 2.5.1; here we must examine their unfoldings.

Taylor expansion of $f(\mu, x)$ about $(0, 0)$ gives

$$f_\mu(x) = D_x f(\mathbf{0})x + D_\mu f(\mathbf{0})\mu + \tfrac{1}{2}[x^2 D_{xx} f(\mathbf{0}) + 2\mu x D_{\mu x} f(\mathbf{0}) + \mu^2 D_{\mu\mu} f(\mathbf{0})] + O(|(\mu, x)|^3). \quad (4.4.1)$$

Figure 4.14 A section through the $(3, +)$ Hopf bifurcation diagram with $\nu_2 < 0$ and fixed. Double limit-cycle bifurcations take place on the cuspoidal curve, δ, and non-degenerate Hopf bifurcations occur on the ν_1-axis when $\nu_1 \neq 0$. The point $(\nu_0, \nu_1) = (\nu_2^3/27, \nu_2^2/3)$ is a triple limit-cycle bifurcation point. Three limit cycles are present in the phase portraits of systems corresponding to parameter values in the shaded region.

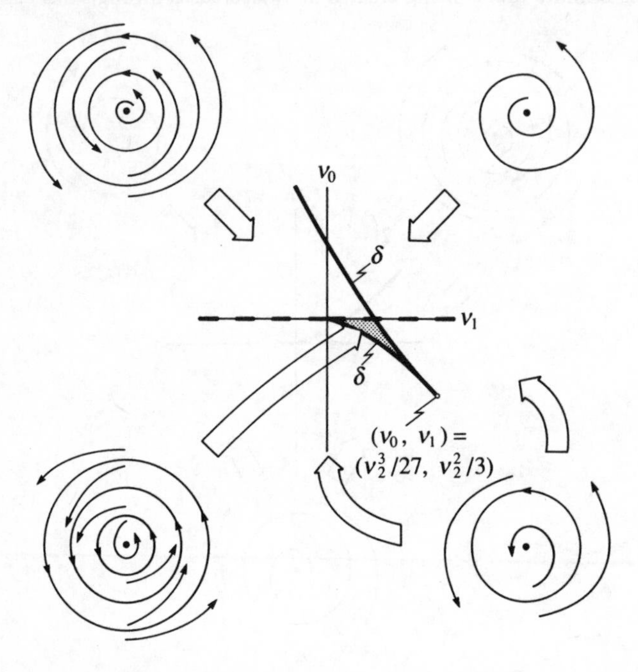

Gathering together terms in x^r, $r = 0, 1, 2, \ldots$, we obtain

$$f_\mu(x) = \sum_{r=0}^{N} a_r(\mu)x^r + O(x^{N+1}), \tag{4.4.2}$$

where $a_r(\mu)$ is a smooth function of μ for each r. Moreover,

$$a_0(\mu) = \mu \mathrm{D}_\mu f(\mathbf{0}) + \tfrac{1}{2}\mu^2 \mathrm{D}_{\mu\mu} f(\mathbf{0}) + \cdots, \tag{4.4.3a}$$

$$a_1(\mu) = \pm 1 + \mu \mathrm{D}_{\mu x} f(\mathbf{0}) + \cdots, \tag{4.4.3b}$$

$$a_2(\mu) = \tfrac{1}{2}\mathrm{D}_{xx} f(\mathbf{0}) + \cdots, \tag{4.4.3c}$$

$$\vdots$$

Figure 4.15 The (3, +) Hopf bifurcation surfaces divide the parameter space into the regions (a)–(d). Structurally stable phase portraits illustrating the behaviour of systems belonging to these regions are shown. The bifurcations that occur on the surfaces are generically as follows:

$$\left.\begin{matrix}(c)\to(a)\\(b)\to(d)\\(a)\to(b)\end{matrix}\right\}\text{ Hopf bifurcations; }\quad\left.\begin{matrix}(c)\to(b)\\(a)\to(d)\end{matrix}\right\}\text{ double limit cycle.}$$

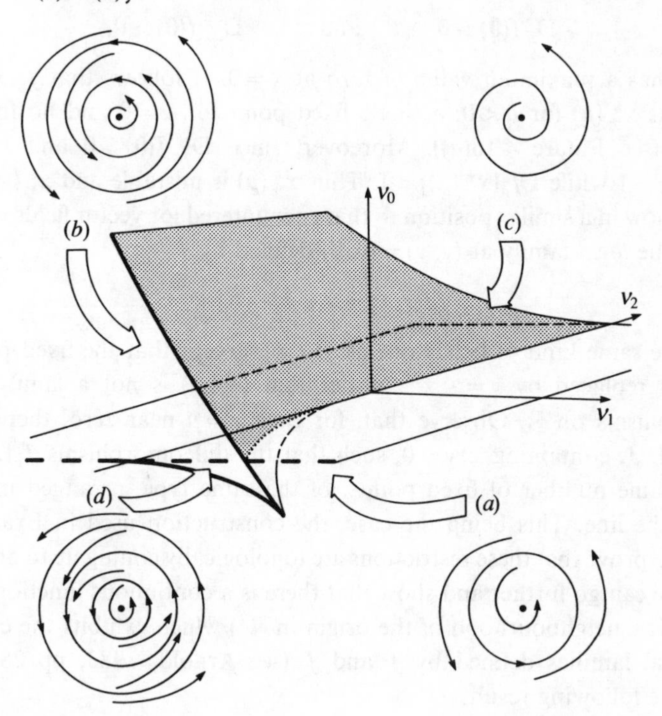

4.4.1 $D_x f(\mathbf{0}) = +1$: *the fold bifurcation*

The fixed points of $f_\mu(x)$ are given by

$$
\begin{aligned}
g(\mu, x) &= f(\mu, x) - x \\
&= a_0(\mu) + (a_1(\mu) - 1)x + a_2(\mu)x^2 + O(x^3) = 0,
\end{aligned} \tag{4.4.4}
$$

where $a_1(\mu) - 1 = \mu D_{\mu x} f(\mathbf{0}) + \cdots = O(\mu)$. By hypothesis $f(\mathbf{0}) = 0$ so that $(\mu, x) = (0, 0)$ satisfies (4.4.4). Given that $D_\mu f(\mathbf{0}) \neq 0$, the Implicit Function Theorem gives the existence of a smooth function $\mu(x)$, defined on an interval $(-\varepsilon, \varepsilon)$, $\varepsilon > 0$, such that $\mu(0) = 0$ and

$$
g(\mu(x), x) = 0, \tag{4.4.5}
$$

for each $x \in (-\varepsilon, \varepsilon)$.

Differentiation of (4.4.5) with respect to x gives

$$
D\mu(0) = -\frac{D_x g(\mathbf{0})}{D_\mu g(\mathbf{0})} = 0, \tag{4.4.6}
$$

since $D_x g(\mathbf{0}) = a_1(0) - 1 = 0$ and $D_\mu g(\mathbf{0}) = Da_0(0) = D_\mu f(\mathbf{0}) \neq 0$. A second differentiation then yields

$$
D^2\mu(0) = -\frac{D_{xx} g(\mathbf{0})}{D_\mu g(\mathbf{0})} = -\frac{D_{xx} f(\mathbf{0})}{D_\mu f(\mathbf{0})}. \tag{4.4.7}
$$

If we assume that

$$
D_\mu f(\mathbf{0}) > 0 \qquad \text{and} \qquad D_{xx} f(\mathbf{0}) > 0, \tag{4.4.8}
$$

then $\mu(x)$ has a maximum value of zero at $x = 0$. It follows that $f_\mu(x)$ has two fixed points $x^*_\pm(\mu)$ for $\mu < 0$, a single fixed point for $\mu = 0$ and no fixed points for $\mu > 0$ (see Figure 4.16(*a*)). Moreover, since $D_{xx} f(\mathbf{0}) > 0$ and $D_x f(\mathbf{0}) = 1$, $Df_\mu(x^*_+(\mu)) > 1$ while $Df_\mu(x^*_-(\mu)) < 1$. Thus $x^*_+(\mu)$ is unstable and $x^*_-(\mu)$ is stable.

We are now in a similar position to that encountered for vector fields in Example 4.1.2(d). The *local* family at $(v, y) = (0, 0)$ defined by

$$
\tilde{f}(v, y) = v + y + y^2 \tag{4.4.9}
$$

exhibits the same kind of behaviour as $f(\mu, x)$, except that the fixed point curve $\mu = \mu(x)$ is replaced by $v = -x^2$. Notice that (4.4.9) is not a family of *global* diffeomorphisms on $\mathbb{R}$. Observe that, for each $v = \mu$ near zero, there are open intervals, I, J, containing $x, y = 0$, such that the diffeomorphisms $f_\mu|I$ and $\tilde{f}_\mu|J$ have the same number of fixed points, of the same type, arranged in the same order on the line. This being the case, the construction used in Example 1.5.1 allows us to prove that these restrictions are topologically conjugate to one another. In fact, one can go further and show that there is a continuous function $h: U \to \mathbb{R}$ (where U is a neighbourhood of the origin in $\mathbb{R}^2$) which exhibits the equivalence of the local families defined by f and $\tilde{f}$ (see Arnold, 1983, pp. 285–6). This leads to the following result.

Proposition 4.4.1 (see Arnold, 1983, p. 285; Whitley, 1983, p. 181) *Let the smooth family of diffeomorphisms* $f(\mu, x)$ *satisfy* $f(\mathbf{0}) = 0$ *and* $\mathrm{D}_x f(\mathbf{0}) = +1$. *Then, given that* $\mathrm{D}_\mu f(\mathbf{0}) > 0$ *and* $\mathrm{D}_{xx} f(\mathbf{0}) > 0$, *there is a neighbourhood of* $(\mu, x) = (0, 0)$ *on which* $f(\mu, x)$ *is equivalent to the local family defined by* (4.4.9) *at* $(v, y) = (0, 0)$.

Notice that the condition $\mathrm{D}_\mu f(\mathbf{0}) > 0$ in Proposition 4.4.1 is sufficient to ensure that $f(\mu, x)$ completely unfolds the singularity at $(\mu, x) = (0, 0)$. This point is illustrated by the following example.

Example 4.4.1 Let the family $f(\mu, x)$ of diffeomorphisms on ℝ satisfy the following

Figure 4.16 Bifurcation diagrams for: (*a*) $f(\mu, x)$ (see 4.4.4)) and (*b*) $\tilde{f}(\mu, x)$ given by (4.4.9). These local families are said to undergo a *fold bifurcation* when the parameter value is equal to zero.

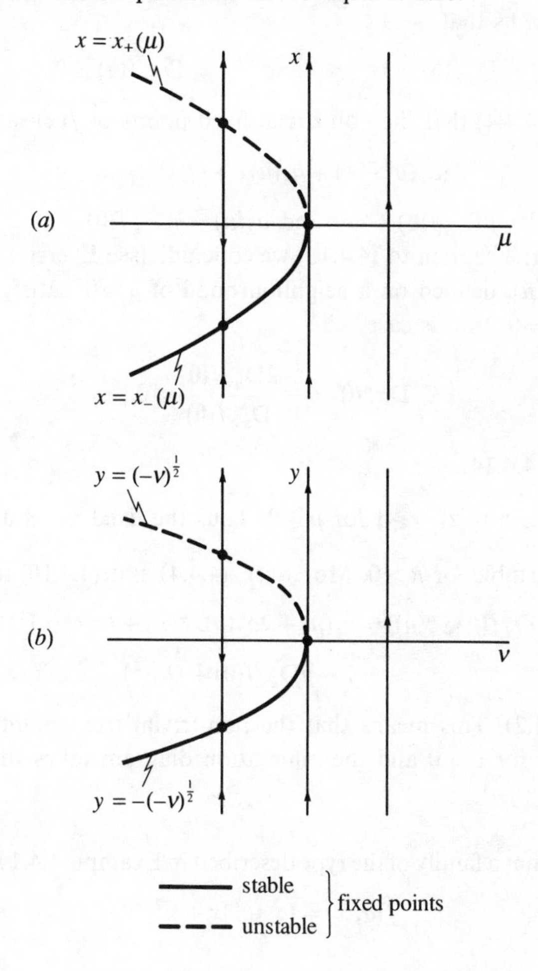

conditions:

(i) there is a fixed point at $x = 0$ for all μ, i.e.

$$f(\mu, 0) \equiv 0; \tag{4.4.10}$$

(ii) the eigenvalue of the linearisation of f_μ at $x = 0$, $\lambda(\mu) = \mathrm{D}_x f(\mu, 0)$, is such that

$$\lambda(0) = 1 \quad \text{and} \quad \left.\frac{\mathrm{d}\lambda(\mu)}{\mathrm{d}\mu}\right|_{\mu=0} > 0; \tag{4.4.11}$$

(iii)

$$\mathrm{D}_{xx} f(\mathbf{0}) > 0. \tag{4.4.12}$$

Sketch the bifurcation diagram for f in the μx-plane.

Solution. Equation (4.4.10) implies

$$a_0(\mu) \equiv 0, \tag{4.4.13}$$

while (4.4.11) means that

$$\mathrm{D}_x f(\mathbf{0}) = 1 \quad \text{and} \quad \mathrm{D}_{\mu x} f(\mathbf{0}) > 0. \tag{4.4.14}$$

It follows from (4.4.4) that the non-trivial fixed points of $f(x)$ satisfy

$$(a_1(\mu) - 1) + a_2(\mu)x + O(x^2) = 0, \tag{4.4.15}$$

where $(a_1(\mu) - 1) = \mu \mathrm{D}_{\mu x} f(\mathbf{0}) + \cdots$ and $a_2(\mu) = \frac{1}{2}\mathrm{D}_{xx} f(\mathbf{0}) + \cdots$. If we apply the Implicit Function Theorem to (4.4.15) we conclude (see Exercise 4.4.2) that there is a function $x^*(\mu)$, defined on a neighbourhood of $\mu = 0$, satisfying (4.4.15) and such that $x^*(0) = 0$. In this case

$$\mathrm{D}x^*(0) = -\frac{2\mathrm{D}_{\mu x} f(\mathbf{0})}{\mathrm{D}_{xx} f(\mathbf{0})} < 0 \tag{4.4.16}$$

by (4.4.12) and (4.4.14).

Since $\left.\frac{\mathrm{d}\lambda(\mu)}{\mathrm{d}\mu}\right|_{\mu=0} > 0$, $\lambda(\mu) \gtrless 1$ for $\mu \gtrless 0$. Thus the fixed point at $x = 0$ is stable for $\mu < 0$ and unstable for $\mu > 0$. Moreover, (4.4.4) and (4.4.16) imply

$$\begin{aligned} \mathrm{D}_x f(\mu, x^*(\mu)) &= a_1(\mu) + 2a_2(\mu)x^*(\mu) + O(x^*(\mu)^2) \\ &= 1 - \mu \mathrm{D}_{\mu x} f(\mathbf{0}) + O(\mu^2) \end{aligned} \tag{4.4.17}$$

(see Exercise 4.4.2). This means that the non-trivial fixed point is unstable for $\mu < 0$ and stable for $\mu > 0$ and the bifurcation diagram takes the form given in Figure 4.17. □

It can be shown that a family of the type described in Example 4.4.1 is equivalent to

$$\hat{f}(\sigma, x) = (\sigma + 1)x + x^2 \tag{4.4.18}$$

on a neighbourhood of $(\sigma, x) = (0, 0)$. The fixed points of $\hat{f}(x)$ are given by $\sigma x + x^2 = -\frac{1}{4}\sigma^2 + (\frac{1}{2}\sigma + x)^2 = 0$. The transformation $x = h_\sigma(y) = y - \frac{1}{2}\sigma$ leads to

$$h_\sigma^{-1}(\hat{f}_\sigma(h_\sigma(y))) = -\tfrac{1}{4}\sigma^2 + y + y^2 = \tilde{f}_{\nu(\sigma)}(y), \tag{4.4.19}$$

where $\nu(\sigma) = -\frac{1}{4}\sigma^2$. Since $h(\sigma, y) = y - \frac{1}{2}\sigma$ is continuous, (4.4.18) is equivalent to (4.4.19) which, in turn, is induced by (4.4.9). Since there is no value of σ for which (4.4.18 & 19) have no fixed points, the singularity at $(\sigma, x) = (0, 0)$ is not completely unfolded.

4.4.2 $\mathrm{D}_x f(\mathbf{0}) = -1$*: the flip bifurcation*

For μ near zero, f_μ is a decreasing diffeomorphism in a neighbourhood of $x = 0$ and it therefore has only one fixed point. More precisely, if we define $g(\mu, x)$ by

$$g(\mu, x) = f(\mu, x) - x, \tag{4.4.20}$$

then $g(\mathbf{0}) = 0$ and $\mathrm{D}_x g(\mathbf{0}) = -2$. The Implicit Function Theorem now gives the existence of a unique, smooth map $x = x^*(\mu)$, defined on an interval containing $\mu = 0$, such that

$$g(\mu, x^*(\mu)) = f(\mu, x^*(\mu)) - x^*(\mu) \equiv 0. \tag{4.4.21}$$

Of course, $x^*(0) = 0$ and differentiation of (4.4.21) with respect to μ yields

$$\mathrm{D}x^*(0) = -\frac{\mathrm{D}_\mu g(\mathbf{0})}{\mathrm{D}_x g(\mathbf{0})} = \tfrac{1}{2}\mathrm{D}_\mu f(\mathbf{0}). \tag{4.4.22}$$

Thus given that $\mathrm{D}_\mu f(\mathbf{0}) \neq 0$, we conclude that

$$x^*(\mu) = \frac{\mu}{2}\mathrm{D}_\mu f(\mathbf{0}) + O(\mu^2). \tag{4.4.23}$$

In order to reveal the bifurcational behaviour of the family, we transform it into normal form. Local coordinates are introduced so that the fixed point remains

Figure 4.17 Bifurcation diagram for $f(\mu, x)$ in Example 4.4.1. Notice that $x^*(\mu)$ has a negative slope at the origin (see 4.4.16)). This bifurcation is known as a *transcritical bifurcation* (Whitley, 1983, p. 185).

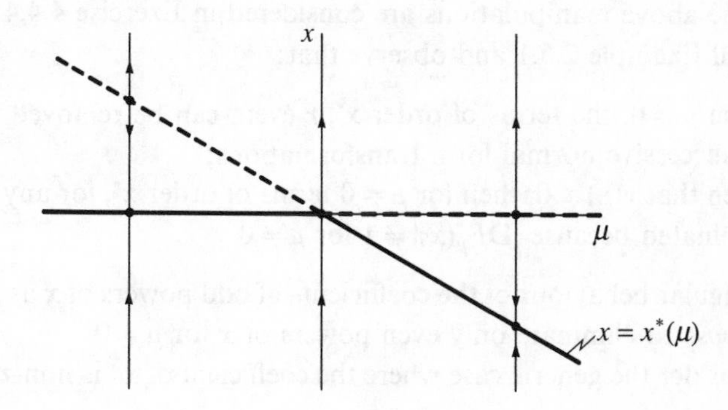

at the origin for all μ, i.e. $x \mapsto x - x^*(\mu)$, $\mu \mapsto \mu$. A Taylor expansion then gives

$$G(\mu, x) = g(\mu, x + x^*(\mu)) = \sum_{r=1}^{N} b_r(\mu)x^r + O(x^{N+1}), \tag{4.4.24}$$

where

$$b_r(\mu) = \frac{1}{r!} \mathrm{D}_x^r g(\mu, x^*(\mu)). \tag{4.4.25}$$

Equations (4.4.2) and (4.4.20) show that

$$b_1(\mu) = (a_1(\mu) - 1) + 2a_2(\mu)x^*(\mu) + 3a_3(\mu)x^*(\mu)^2 + \cdots,$$

and (4.4.26)

$$b_r(\mu) = a_r(\mu) + (r+1)a_{r+1}(\mu)x^*(\mu) + \cdots,$$

for $r \geqslant 2$. If we now define $F(\mu, x)$ by

$$G(\mu, x) = F(\mu, x) - x, \tag{4.4.27}$$

we obtain

$$F(\mu, x) = (b_1(\mu) + 1)x + \sum_{r=2}^{N} b_r(\mu)x^r + O(x^{N+1}). \tag{4.4.28}$$

However, by (4.4.26),

$$b_1(\mu) + 1 = a_1(\mu) + 2a_2(\mu)x^*(\mu) + 3a_3(\mu)x^*(\mu)^2 + \cdots, \tag{4.4.29}$$

and (4.4.3b and c), together with (4.4.23), then give

$$b_1(\mu) + 1 = \nu(\mu) - 1,$$

where

$$\nu(\mu) = \mu\{\mathrm{D}_{\mu x} f(\mathbf{0}) + \tfrac{1}{2}\mathrm{D}_{xx} f(\mathbf{0})\mathrm{D}_\mu f(\mathbf{0})\} + O(\mu^2). \tag{4.4.30}$$

Thus, (4.4.28) can be written

$$F(\mu, x) = (\nu(\mu) - 1)x + \sum_{r=2}^{N} b_r(\mu)x^r + O(x^{N+1}). \tag{4.4.31}$$

Details of the above manipulations are considered in Exercise 4.4.4.

Now recall Example 2.5.1 and observe that:

(i) when $\mu = 0$, the terms of order x^r, r even, can be removed from (4.4.31) by successive normal form transformations;

(ii) given that $\nu(\mu) \not\equiv 0$, then for $\mu \neq 0$ terms of order x^r, for any $r \geqslant 2$, can be eliminated because $|\mathrm{D}F_\mu(x)| \neq 1$ for $\mu \neq 0$.

To avoid singular behaviour of the coefficients of odd powers of x as μ approaches zero, we *choose* to eliminate only even powers of x for $\mu \neq 0$.

Let us consider the generic case where the coefficient of x^3 is non-zero for $\mu = 0$, i.e.

$$\tilde{F}(\mu, x) = (\nu(\mu) - 1)x + \tilde{b}_3(\mu)x^3 + O(x^5). \tag{4.4.32}$$

It must be emphasised that $\tilde{b}_3(\mu)$ occurring in (4.4.32) is not, in general, the same as $b_3(\mu)$ in (4.4.31) because the cubic terms in the latter may be altered when the quadratic terms are transformed away. The details of this normal form calculation are considered in Exercise (4.4.5) with the result that $\tilde{b}_3(\mu)$ in (4.4.32) is given by

$$\tilde{b}_3(\mu) = -\tfrac{1}{12}\mathrm{D}_{xxx}f^2(\mathbf{0}) + O(\mu). \tag{4.4.33}$$

The scaling $x \mapsto \alpha x$ with $\alpha^2 = |\tilde{b}_3(\mu)|$ reduces the magnitude of the coefficient of x^3 in (4.4.32) to unity.

Each of the above transformations assumes that μ and x are confined to some sufficiently small neighbourhood of $(\mu, x) = (0, 0)$ and their composition is a continuous function of μ and x. Thus, we conclude that, on a neighbourhood of $(\mu, x) = (0, 0)$, the family $f(\mu, x)$ is equivalent to the family

$$\alpha\tilde{F}(\mu, \alpha^{-1}x) = (\nu(\mu) - 1)x \pm x^3 + O(x^5), \tag{4.4.34}$$

where the sign is given by $-\mathrm{sgn}(\mathrm{D}_{xxx}f^2(\mathbf{0}))$. The family (4.4.34) is induced by the local family defined by

$$\bar{f}(\nu, x) = (\nu - 1)x \pm x^3 + O(x^5) \tag{4.4.35}$$

at $(\nu, x) = (0, 0)$. The inducing function $\nu = \nu(\mu)$ is a bijection on an interval containing $\mu = 0$ provided that the coefficient of μ in (4.4.30) is non-zero.

Example 4.4.2 Consider the local families defined by

$$\bar{f}(\nu, x) = (\nu - 1)x \pm x^3 \tag{4.4.36}$$

in the neighbourhood of the origin. Show that $x = 0$ is a fixed point of $\bar{f}_\nu$ for any ν and discuss its stability. Find the fixed points of $\bar{f}_\nu^2$ and determine their stability. Sketch a bifurcation diagram in the νx-plane illustrating both 1- and 2-cycles for both families.

Solution. Fixed points of $\bar{f}_\nu$ satisfy $x = \bar{f}_\nu(x)$ so $x = 0$ is clearly a fixed point for any ν. Moreover, $\mathrm{D}\bar{f}_\nu(0) = (\nu - 1)$ so, for ν near zero, this fixed point is stable (unstable) for $\nu > 0$ ($\nu < 0$) in both families.

Now,

$$\begin{aligned}\bar{f}_\nu^2(x) &= \bar{f}_\nu(\bar{f}_\nu(x))\\ &= (\nu - 1)^2 x \pm x^3(\nu - 1)[1 + (\nu - 1)^2] + O(x^4).\end{aligned} \tag{4.4.37}$$

Thus the fixed points of $\bar{f}_\nu^2$ satisfy

$$x[(\nu^2 - 2\nu) \pm (\nu - 1)(\nu^2 - 2\nu + 2)x^2 + O(x^3)] = 0 \tag{4.4.38}$$

and $\bar{f}_\nu^2$ has non-trivial fixed points given by

$$x^2 = \pm \frac{2\nu - \nu^2}{(\nu - 1)(\nu^2 - 2\nu + 2)} + O(\nu^{3/2})$$
$$= \mp \nu(1 + O(\nu^{1/2})). \tag{4.4.39}$$

It follows that when the + sign is operative in (4.4.36), (4.4.39) has:

(i) no real solutions for $\nu > 0$, and
(ii) two real solutions, $x = x_\pm^* = \pm(-\nu)^{1/2} + O(\nu)$, for $\nu < 0$.

The stability of the fixed points $x_\pm^*$ is determined by

$$D\bar{f}_\nu^2(x_\pm^*) = D\bar{f}_\nu(x_+^*)D\bar{f}_\nu(x_-^*), \tag{4.4.40}$$

(see Exercise 4.4.6). Now, $D\bar{f}_\nu(x) = (\nu - 1) + 3x^2$ and (4.4.40) gives

$$D\bar{f}_\nu^2(x_\pm^*) = [(\nu - 1) - 3\nu(1 + O(\nu^{1/2}))]^2$$
$$= [-1 - 2\nu + O(\nu^{3/2})]^2$$
$$= 1 + 4\nu + O(\nu^{3/2}). \tag{4.4.41}$$

Since ν is small and negative, both fixed points of $\bar{f}_\nu^2$ are stable and the bifurcation diagram is as shown in Figure 4.18(*a*).

When the − sign is operative in (4.4.36), $\bar{f}_\nu^2$ has two fixed points for $\nu > 0$ (i.e. $x_\pm^* = \pm\nu^{1/2} + O(\nu)$), $D\bar{f}_\nu(x) = (\nu - 1) - 3x^2$ and (4.4.41) is unchanged. However, since $\nu > 0$, $D\bar{f}_\nu^2(x_\pm^*) > 1$ and the fixed points of $\bar{f}_\nu^2$ are unstable. The bifurcation diagram is shown in Figure 4.18(*b*). □

It can be shown that the $O(x^5)$-terms in (4.4.35) do not change the bifurcational behaviour described in Example 4.4.2. Their main effect is to alter the quantitative, but not qualitative, details of the zero sets of $\bar{f}^2(\nu, x) - x$ (see Exercise 4.4.6). It follows that a bifurcation of the form shown in Figure 4.18 occurs generically when $f_{\mu^*}(x^*) = x^*$ and $Df_{\mu^*}(x^*) = -1$. More precisely, the following proposition can be proved. It is stated in terms of local coordinates at (μ^*, x^*).

Proposition 4.4.2 (see Arnold, 1983, p. 286; Whitley, 1983, p. 182) *Let the smooth family of diffeomorphisms $f(\mu, x)$ satisfy:*

(i) $$f(\mathbf{0}) = 0; \tag{4.4.42}$$

(ii) $$D_x f(\mathbf{0}) = -1; \tag{4.4.43}$$

(iii) $$D_{xxx} f^2(\mathbf{0}) < 0. \tag{4.4.44}$$

Then $f(\mu, x)$ is topologically equivalent to a local family induced by

$$\bar{f}(\nu, x) = (\nu - 1)x + x^3, \tag{4.4.45}$$

at $(\nu, x) = (0, 0)$.

If, in addition, the line of fixed points, $x^*(\mu)$, is such that

$$\frac{\mathrm{d}}{\mathrm{d}\mu}\{\mathrm{D}_x f(\mu, x^*(\mu))\}|_{\mu=0} > 0, \tag{4.4.46}$$

then $f(\mu, x)$ completely unfolds f_0.

The proof of Proposition 4.4.2 is essentially contained in the premable to Example 4.4.2. The inequality (4.4.44) is a non-degeneracy condition which ensures that the cubic term is non-zero in the normal form of f (see (4.4.33)). The condition (4.4.46) (see Exercise 4.4.4) ensures that μ completely unfolds the resulting singularity. The directions of the inequalities in (4.4.44) and (4.4.46) have only been chosen for definiteness. For example, $\mathrm{D}_{xxx} f^2(\mathbf{0}) > 0$ would simply change the sign of x^3 in (4.4.45); while the reverse inequality in (4.4.46) would have the effect of reversing the sign of ν in (4.4.45). Such *inequalities* are, of course, generically

Figure 4.18 Bifurcation diagram for the local family $\bar{f}(\nu, x)$ given by (4.4.36) when (*a*) + sign; (*b*) − sign; is operative. The ν-axis is a line of fixed points for $\bar{f}_\nu$: stable for $\nu > 0$ and unstable for $\nu < 0$. The fixed points of $\bar{f}_\nu^2$ lie on the quadratic curve defined by (4.4.39). These points correspond to a 2-cycle in $\bar{f}_\nu$: stable in (*a*) and unstable in (*b*). The local family $\bar{f}(\nu, x)$ is said to undergo a *flip bifurcation* at $\nu = 0$.

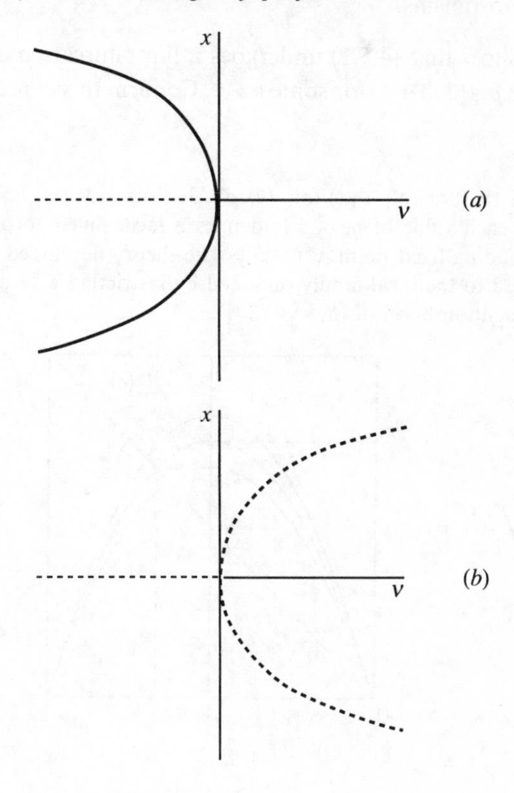

satisfied. We will refer to the bifurcation illustrated in Figure 4.18 as a *flip bifurcation* but is also called a *period doubling bifurcation*. It is clear that in this local bifurcation a 2-cycle is created at a change of stability of a fixed point (or 1-cycle). However, the significance of the latter terminology is only fully appreciated when the bifurcations in families of maps on an interval are considered.

4.5 The logistic map

Our aim in this section is to illustrate how the theory developed in §4.4 can be applied to the bifurcations in certain families of maps on an interval. A typical example is provided by the discrete logistic equation in population dynamics (see Guckenheimer *et al*, 1977; May, 1976; 1983), i.e.

$$F(\rho, x) = \rho x(1 - x). \tag{4.5.1}$$

For $0 < \rho \leqslant 4$, F_ρ: $[0, 1] \to [0, 1]$, but F_ρ is not a diffeomorphism on $[0, 1]$ for any ρ, because $\mathrm{D}F_\rho(\frac{1}{2}) = 0$. However, (4.5.1) has a non-trivial fixed point at $x^* = (\rho - 1)/\rho \in [0, 1]$ for $1 \leqslant \rho \leqslant 4$. Moreover,

$$\mathrm{D}F_\rho(x^*) = \rho(1 - 2x^*) = 2 - \rho, \tag{4.5.2}$$

so that this non-trivial fixed point is non-hyperbolic for $\rho = 3$. What is more, (as Figure 4.19 shows) for ρ near to 3 and x near to $x^* = \frac{2}{3}$, (4.5.1) defines a local family of diffeomorphisms.

Example 4.5.1 Show that (4.5.1) undergoes a flip bifurcation at $\rho = 3$, creating a stable 2-cycle for $\rho \in (3, 3 + \varepsilon)$ for some $\varepsilon > 0$. Confirm this conclusion by sketching F_ρ^2 for ρ near 3.

Figure 4.19 Plots of $F_\rho(x)$ for: (*a*) $\rho < 3$; (*b*) $\rho = 3$; (*c*) $\rho > 3$. Observe that for each value of ρ, (4.5.1) defines a local diffeomorphism at the corresponding fixed point $x^*(\rho) \simeq \frac{2}{3}$. The theory developed in §4.4 can be applied to the local family obtained by restricting F to a sufficiently small neighbourhood of $(\rho, x) = (3, \frac{2}{3})$.

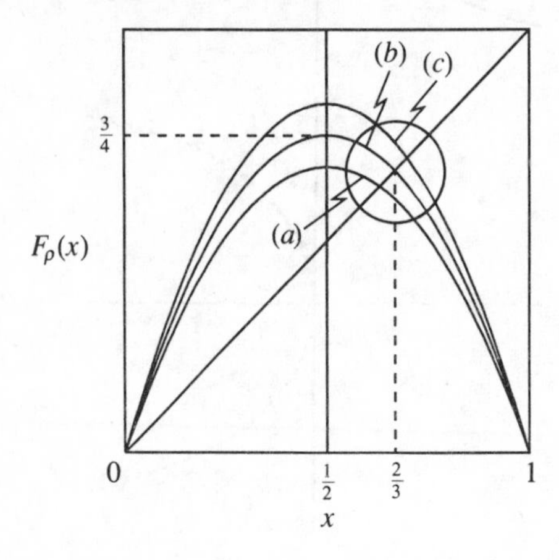

Solution. At $\rho = 3$, (4.5.2) implies $DF_3(x^*) = -1$. Now,

$$F_3^2(x) = F_3(F_3(x)) = 9x - 36x^2 + 54x^3 - 27x^4, \qquad (4.5.3)$$

and $\rho = 3$ means that $x^* = (\rho - 1)/\rho = \frac{2}{3}$. Therefore

$$D_{xxx}F^2(3, \tfrac{2}{3}) = 54(6 - 12x^*) = -108 < 0, \qquad (4.5.4)$$

as required by Proposition 4.4.2. However,

$$\frac{d}{d\rho}[D_x F(\rho, x^*(\rho))]_{\rho=3} = \frac{d}{d\rho}(2-\rho)|_{\rho=3} = -1 < 0. \qquad (4.5.5)$$

It follows that (see remarks following Proposition 4.4.2) the local family, defined by (4.5.1) in the neighbourhood of $(3, \frac{2}{3})$, is topologically equivalent to (4.4.45) with $\nu \mapsto -\nu$. We therefore conclude that its bifurcation diagram is given by Figure 4.18(*a*) with the sign of ν reversed. Thus, there is an $\varepsilon > 0$ such that, for $\rho \in (3, 3 + \varepsilon)$, F_ρ has a stable 2-cycle.

Sketches of $F_\rho^2(x)$ for σ near 3 are shown in Figure 4.20. They are obtained as indicated in Figure 4.21. □

As Figure 4.21 suggests, the effect of increasing ρ, in the neighbourhood of $\rho = 3$, is to increase the maximum values and reduce the minimum value of F_ρ^2. This results in $DF_\rho^2(x_i^*)$, $i = 1, 2$, (see Figure 4.20(*c*)) decreasing as ρ increases (see Exercise 4.5.2). Eventually, $DF_\rho^2(x_i^*) = DF_\rho(x_1^*)DF_\rho(x_2^*)$ becomes equal to -1, for $i = 1, 2$, and the fixed points x_i^* of F_ρ^2 both undergo flip bifurcations creating 2-cycles of F_ρ^2, i.e. 4-cycles of F_ρ. It can be shown that these 4-cycles, in turn, undergo flip bifurcations giving 8-cycles and so on (see Exercise 4.5.3). Thus cycles of length $2, 4, 8, \ldots, 2^k, \ldots$ appear in succession as ρ increases above 3. Numerical experiments (Collet & Eckmann, 1980; May, 1976) show that the range of ρ for which the 2^k-cycles remain stable, decreases as k increases and the stable 2^k-cycles accumulate at $\rho = \rho_c = 3.5700\cdots$ as $k \to \infty$.

Example 4.5.2 Sketch the graph of $F_\rho^3(x)$ for $\rho_0 < \rho < 4$, where ρ_0 is the largest root of

$$\frac{\rho^2}{4}\left(1 - \frac{\rho}{4}\right) = \frac{1}{2} \qquad (4.5.6)$$

Hence, show that F_ρ develops 3-cycles, both stable and unstable, for some ρ in this interval.

Solution. The form of $F_\rho^3(x)$ may be obtained as indicated in Figure 4.22. $F_\rho^3(x)$ first reaches a maximum at $x = x_m^{(2)}$, where $F_\rho^2(x_m^{(2)}) = \frac{1}{2}$; it then falls to a minimum at $x = x_m^{(1)}$ where $F_\rho^2(x)$ reaches its maximum of $\rho/4$. As x increases from $x_m^{(1)}$ to $\frac{1}{2}$,

$F_\rho^2(x)$ decreases from $\rho/4$ to $F_\rho^2(\frac{1}{2})$. Now,

$$F_\rho^2(\tfrac{1}{2}) = \rho[\rho x(1-x)][1-\rho x(1-x)]|_{x=\frac{1}{2}} = \frac{\rho^2}{4}\left(1-\frac{\rho}{4}\right), \tag{4.5.7}$$

so that $F_\rho^2(\frac{1}{2}) = 0$ for $\rho = 0, 4$ and has a unique maximum on $(0, 4)$. Given that $\rho > \rho_0$, then $F_\rho^2(\frac{1}{2}) < \frac{1}{2}$ and $F_\rho(F_\rho^2(x))$ increases to a maximum of $\rho/4$ at $x = x_{\mathrm{m}}^{(3)}$ (see Figure 4.22). Finally, $F_\rho^3(x)$ falls to a minimum at $x = \frac{1}{2}$ as $F_\rho^2(x)$ decreases from $\frac{1}{2}$ to $\dfrac{\rho^2}{4}\left(1-\dfrac{\rho}{4}\right)$. We conclude that $F_\rho^3(x)$ has four maxima and three minima as shown in Figure 4.22.

Figure 4.20 Sketches of $F_\rho^2(x)$ for: (*a*) $\rho < 3$; (*b*) $\rho = 3$; (*c*) $\rho > 3$; illustrating the flip bifurcation at $\rho = 3$ for the logistic map (4.5.1). Observe that as ρ increases through three the fixed point at $x^*(\rho)$ becomes unstable and the additional fixed points $x_1^*(\rho)$, $x_2^*(\rho)$, satisfying $x_1^*(\rho) < x^*(\rho) < x_2^*(\rho)$, are created. These fixed points of F_ρ^2 correspond to a stable 2-cycle in F_ρ.

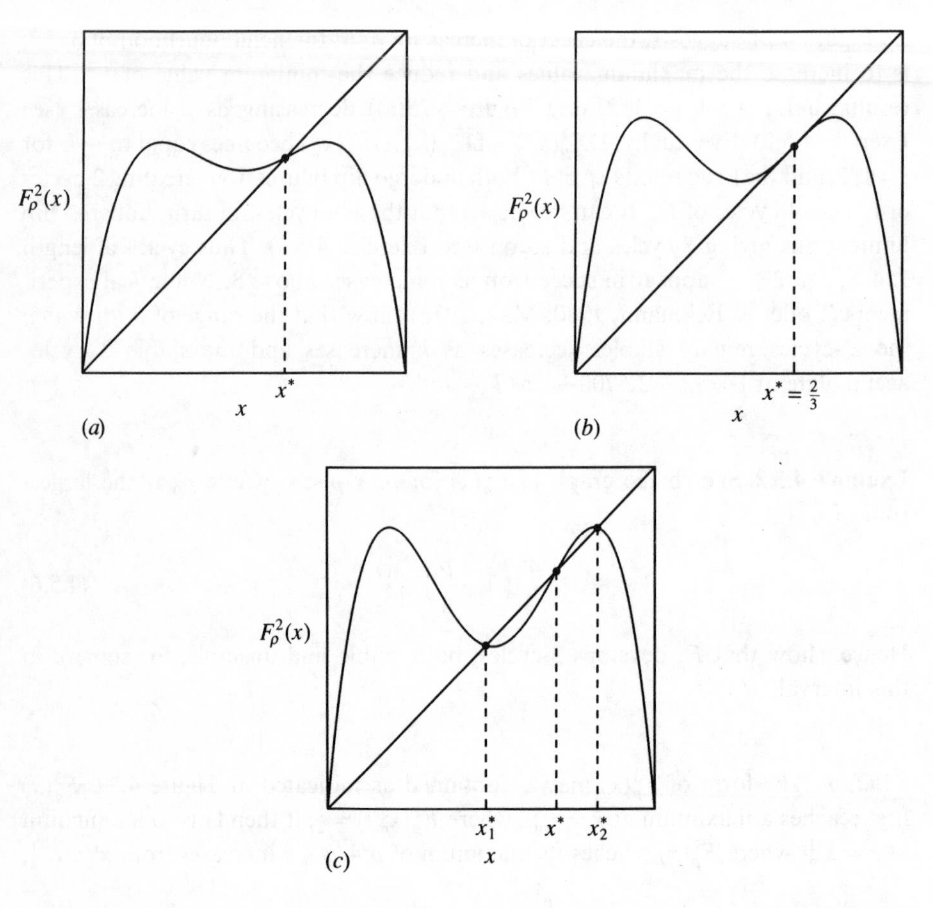

Figure 4.21 Illustration of how the single, symmetric maximum in $F_\rho(x)$, $\rho \simeq 3$, results in symmetrically placed twin maxima in $F_\rho^2(x)$. For ρ near 3, $\max_{0 \leqslant x \leqslant 1}(F_\rho(x)) = \rho/4 > \frac{1}{2}$. Hence F_ρ^2 passes through a maximum at $x_m^{(1)} = \{1 - [1 - (2/\rho)]^{1/2}\}/2$, where $F_\rho(x) = \frac{1}{2}$, and falls to a minimum at $x = \frac{1}{2}$, where $F_\rho(x) = \rho/4 > \frac{1}{2}$. Since $F_\rho(x) = F_\rho(1-x)$, the remainder of F_ρ^2 is given by reflection in $x = \frac{1}{2}$, in particular $x_m^{(1)'} = 1 - x_m^{(1)}$. Clearly, the larger ρ, the closer $\rho/4$ is to unity and the smaller the value of $F_\rho^2(\frac{1}{2})$.

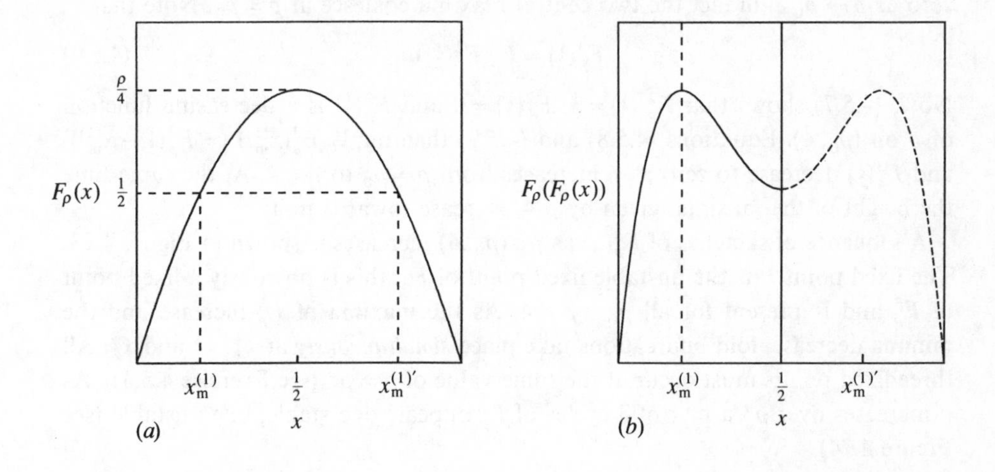

Figure 4.22 Sketches illustrating how the graph of $F_\rho^3(x)$ can be obtained from that of $F_\rho^3(x)$ for $0 < x \leqslant \frac{1}{2}$. $F_\rho^3(x) = F_\rho^3(1-x)$ then gives the graph of $F_\rho^3(x)$ for $\frac{1}{2} < x < 1$ (shown broken). Notice (cf. Figure 4.21) that the number of maxima has doubled. These maxima will, in turn, generate even more maxima in F_ρ^4 and so on. The root of this increasingly complicated behaviour is the single maximum of F_ρ at $x = \frac{1}{2}$ with $F_\rho(\frac{1}{2}) = \rho/4 > \frac{1}{2}$.

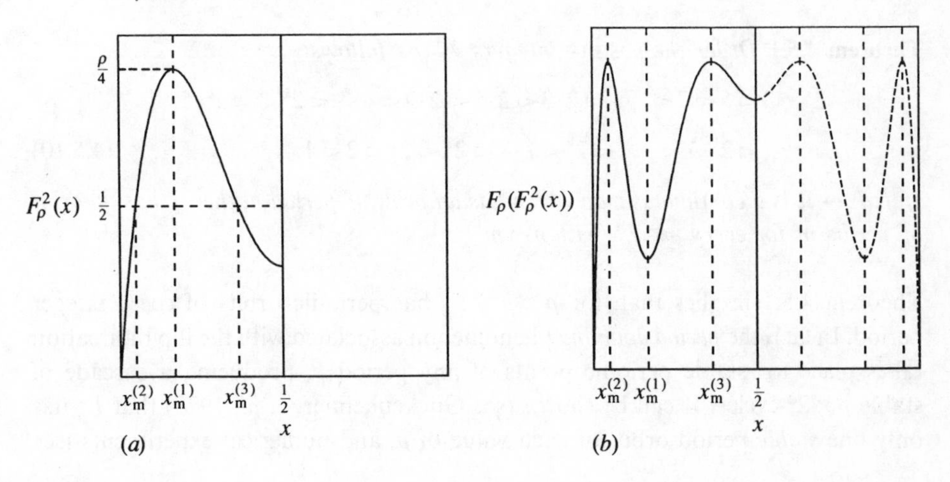

The above argument shows that the value of F_ρ^3 at the minimum where $x = x_{\mathrm{m}}^{(1)}$ is the same as that of F_ρ^2 at $x = \frac{1}{2}$. Observe,

$$F_\rho^3(x_{\mathrm{m}}^{(1)}) = F_\rho(F_\rho^2(x_{\mathrm{m}}^{(1)})) = F_\rho\left(\frac{\rho}{4}\right) = F_\rho^2(\tfrac{1}{2}) \tag{4.5.8}$$

by (4.5.7). What is more, the depth of the local minimum in F_ρ^3 at $x = \frac{1}{2}$ tends to zero as $\rho \to \rho_{0+}$; in fact the two central maxima coalesce at $\rho = \rho_0$. Note that

$$F_\rho^3(\tfrac{1}{2}) = F_\rho(F_\rho^2(\tfrac{1}{2})). \tag{4.5.9}$$

Now, (4.5.7) shows that $F_{\rho_0}^2(\frac{1}{2}) = \frac{1}{2}$, $F_4^2(\frac{1}{2}) = 0$ and $F_\rho^2(\frac{1}{2})$ is a decreasing function of ρ on $(\rho_0, 4)$. Equations (4.5.8) and (4.5.9) then imply $F_\rho^3(x_{\mathrm{m}}^{(1)})$ $(= F_\rho^3(1 - x_{\mathrm{m}}^{(1)}))$ and $F_\rho^3(\frac{1}{2})$ decrease to zero as ρ increases from $\rho = \rho_0$ to $\rho = 4$. At the same time the height of the maxima, given by $\rho/4$, increase towards unity.

A sequence of sketches of $F_\rho^3(x)$ as $\rho \in (\rho_0, 4)$ increases is shown in Figure 4.23. The fixed point P is the unstable fixed point of F_ρ: this is obviously a fixed point of F_ρ^3 and is present for all $\rho_0 < \rho < 4$. As the maxima of F_ρ^3 increase and the minima decrease, fold bifurcations take place *simultaneously* at x_1^*, x_2^* and x_3^*. All three fixed points must occur at the same value of $\rho = \rho^*$ (see Exercise 4.5.4)). As ρ increases over ρ^* a pair of 3-cycles of F_ρ appear: one stable, one unstable (see Figure 4.24). □

It is clear that a stable 3-cycle of F_ρ exists for $\rho \in (\rho^*, \rho^* + \varepsilon)$ for some $\varepsilon > 0$, but (see Figure 4.25) as ρ continues to increase, the slope of F_ρ^3 at the stable period-3 points decreases and eventually reaches -1. Flip bifurcations then occur creating stable cycles of period 6. As with the stable 2^k-cycles, there follows a cascade of stable 3×2^k-cycles. Numerical experiments estimate $\rho^* = 3.8284\cdots$ (see May, 1976) and show that the stable 3.2^k-cycles accumulate at $\rho = 3.8495\cdots$.

In fact, the dynamics of F_ρ for $\rho \geqslant \rho^*$ is even more complicated than the above remarks suggest, as the following result (due to Sarkovskii, 1964) implies.

Theorem 4.5.1 *Order the positive integers $\mathbb{Z}^+$ as follows:*

$$\begin{aligned} &3 \lhd 5 \lhd 7 \lhd \cdots \lhd 2\cdot 3 \lhd 2\cdot 5 \lhd 2\cdot 7 \lhd \cdots \lhd 2^k\cdot 3 \lhd 2^k\cdot 5 \\ &\lhd 2^k\cdot 7 \lhd \cdots \lhd 2^k \lhd \cdots \lhd 2^3 \lhd 2^2 \lhd 2 \lhd 1. \end{aligned} \tag{4.5.10}$$

If $f\colon \mathbb{R} \to \mathbb{R}$ is a continuous map which has an orbit of period n, then f has an orbit of period m for every $m \in \mathbb{Z}^+$ with $n \lhd m$.

Theorem 4.5.1 implies that, for $\rho \geqslant \rho^*$, F_ρ has periodic orbits of *every* integer period. In fact, the *period doubling* phenomenon associated with the flip bifurcation takes place for stable periodic points of any period p, producing a cascade of stable $p \times 2^k$-cycles. It can be shown (see Guckenheimer, *et al*, 1977) that F_ρ has only one *stable* period orbit for each value of ρ, and numerical experiments (see

May, 1976; Cvitanovic, 1984) show that each cascade of stable $p \times 2^k$-cycles occurs in a small 'window' of ρ values, accumulating at the right hand end of this interval. The fact that the case $p = 3$ is the last to appear was first observed by Li & Yorke (1975). They also showed that for $\rho > \rho^*$ infinitely many asymptotically aperiodic orbits exist as well. Although this aspect of the dynamics of the logistic map is not an application of local bifurcation theory, we will take this opportunity to consider it in a little more detail. Not only will this allow us to emphasise that it is the 'folded' (non-diffeomorphic) character of F_ρ: $[0, 1] \to [0, 1]$ that is responsible for the complexity of the dynamics, but also we can highlight the similarity between the dynamics of F_ρ and that of the horseshoe map considered in § 3.5.

Consider a value of $\rho > \rho^*$ for which there is a stable 3-cycle, $\Lambda^{(3)}$, with periodic

Figure 4.23 Plots of $F_\rho^3(x)$ for: (*a*) $\rho_0 < \rho < \rho^*$; (*b*) $\rho = \rho^*$; (*c*) $\rho^* < \rho < 4$, where ρ_0 is given by (4.5.6). Observe that for $\rho = \rho^*$ non-hyperbolic fixed points appear at x_i^*, $i = 1, 2, 3$, and each develops into a pair of hyperbolic fixed points as ρ increases above ρ^*. This behaviour is typical of the fold bifurcation discussed in § 4.4.

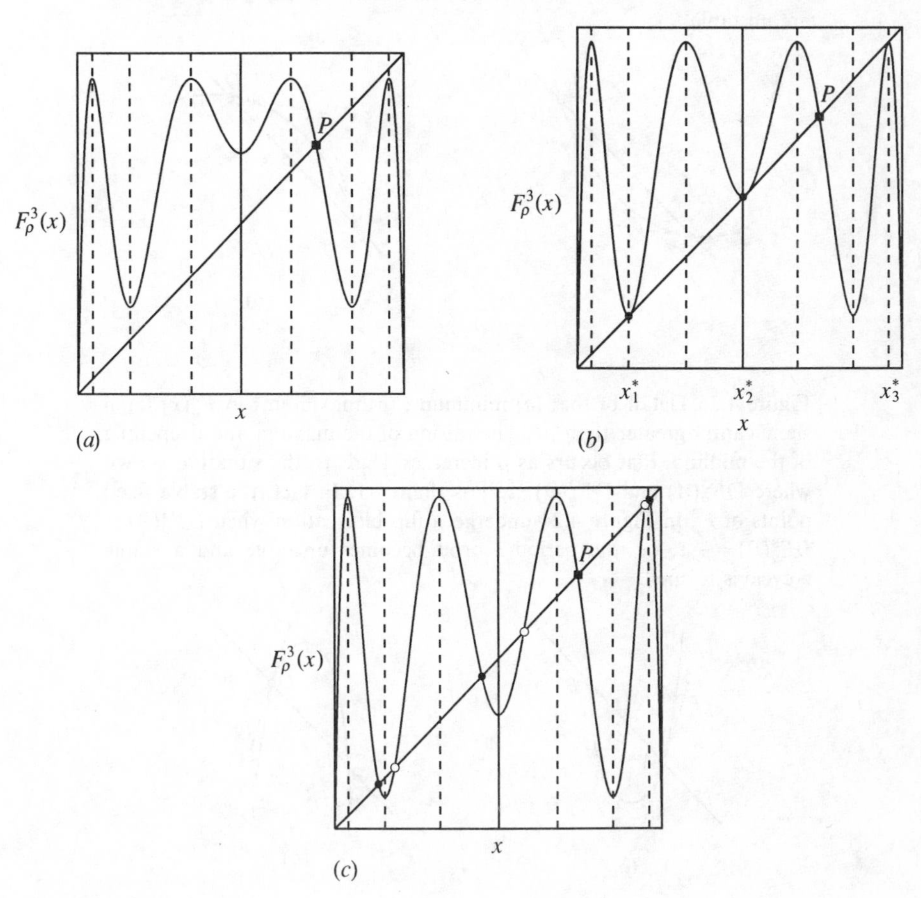

points x_1^*, x_2^*, x_3^*. For example (see Guckenheimer, *et al*, 1977), since

$$\begin{aligned} DF_\rho^3(x_i^*) &= DF_\rho(F_\rho^2(x))DF_\rho(F_\rho(x))DF_\rho(x)|_{x_i^*} \\ &= \prod_{j=1}^{3} DF_\rho(x_j^*), \end{aligned} \tag{4.5.11}$$

for each $i = 1, 2, 3$, we can ensure stability by choosing $DF_\rho(x_2^*) = 0$, i.e. $x_2^* = \frac{1}{2}$. Thus if $x_2^* = \frac{1}{2}$ is to be a fixed point of F_ρ^3, ρ must be a solution of

$$F_\rho^3(\tfrac{1}{2}) = \frac{\rho^3}{4}\left(1 - \frac{\rho}{4}\right)\left(1 - \frac{\rho^2}{4}\left(1 - \frac{\rho}{4}\right)\right) = \tfrac{1}{2} \tag{4.5.12}$$

which is less than 4. Numerical solution of (4.5.12) gives $\rho \simeq 3.832\cdots$. The resulting 3-cycle is shown in Figure 4.26.

Figure 4.24 Detail of the: (*a*) minimum; (b) maximum in $F_\rho^3(x)$ for ρ greater than, but sufficiently close to, ρ^*. It is clear that in (*a*) $|DF_\rho^3(A)| < 1$, $DF_\rho^3(B) > 1$ while in (*b*) $|DF_\rho^3(D)| < 1$, $DF_\rho^3(C) > 1$. It follows that the fixed points of F_ρ^3 at A, D are stable and those at B, C are unstable.

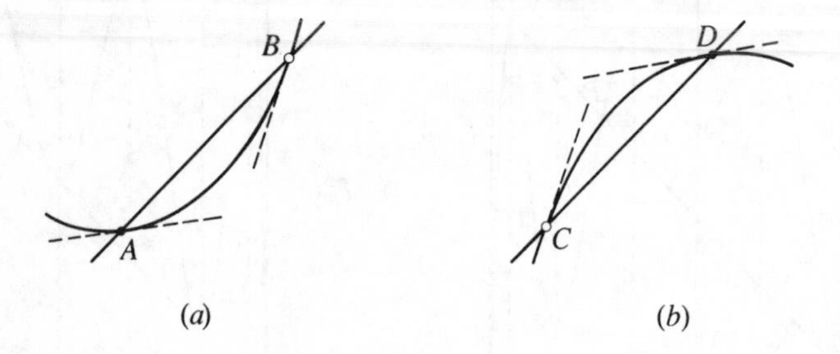

Figure 4.25 Detail of the: (*a*) minimum; (*b*) maximum; in $F_\rho^3(x)$ for ρ significantly greater than ρ^*. The raising of the maxima and deepening of the minima, that occurs as ρ increases, leads to the situation shown where $DF_\rho^3(A)$ and $DF_\rho^3(D)$ are less than -1. In fact, the stable fixed points of F_ρ^3 in Figure 4.24 undergo a flip bifurcation when $DF_\rho^3(A) = DF_\rho^3(D) = -1$, i.e. the period-3 orbit becomes unstable and a stable 6-cycle is formed.

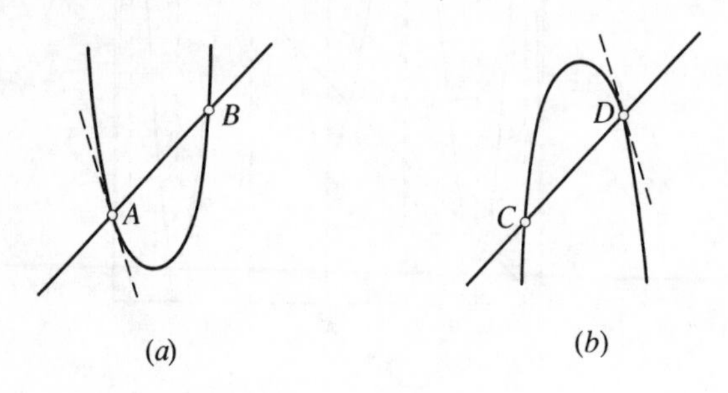

The periodic points x_i^*, $i = 1, 2, 3$ divide the interval [0, 1] into four intervals α, β, γ, δ as shown in Figure 4.26(*b*). The images of these intervals under F_ρ are easily recognised. The interval [0, 1] is stretched (non-uniformly) to a length of $2 \times \rho/4$, folded on itself and replaced as shown in Figure 4.26(*b*). This stretching and folding resembles a variant of the horseshoe map in which the vertical coordinate has been allowed to shrink to zero.

Using $F_\rho(\alpha)$ to denote the image of α under F_ρ, etc, we can describe the action of F_ρ by

$$F_\rho(\alpha) = \alpha \cup \beta, \qquad F_\rho(\beta) = \gamma,$$
$$F_\rho(\gamma) = \gamma \cup \beta, \qquad F_\rho(\delta) = \alpha. \tag{4.5.13}$$

In a similar manner to that used for the horseshoe map, we can associate the orbit of any point $x_0 \in [0, 1]$ with an infinite sequence $s(x_0) = \{a_k\}_0^\infty$ involving the four symbols α, β, γ, δ. We define

$$a_k = j \qquad \text{if } F_\rho^k(x_0) \in j \backslash \Lambda^{(3)}, \qquad j = \alpha, \beta, \gamma, \delta. \tag{4.5.14}$$

In the event that $F_\rho^k(x_0) \in \Lambda^{(3)}$ then

$$a_k = \begin{cases} \alpha \\ \beta \\ \gamma \end{cases} \quad \text{if} \quad F_\rho^k(x_0) = \begin{cases} x_1^* \\ x_2^*. \\ x_3^* \end{cases} \tag{4.5.15}$$

Figure 4.26 (*a*) Illustration of the 3-cycle in the iteration $x_{n+1} = F_\rho(x_n)$ for $\rho = 3.832\ldots$. Observe that $x_0 = x_1^*$ gives $x_1 = F_\rho(x_1^*) = x_2^*$, $x_2 = F_\rho(x_1) = F_\rho(x_2^*) = x_3^*$ and $x_3 = F_\rho(x_2) = F_\rho(x_3^*) = x_1^*$. (*b*) Geometrical interpretation of the effect of F_ρ on the interval [0, 1]. Let $\alpha = [0, x_1^*]$, $\beta = [x_1^*, x_2^*]$, $\gamma = [x_2^*, x_3^*]$, $\delta = [x_3^*, 1]$, then F_ρ stretches and folds these intervals as shown.

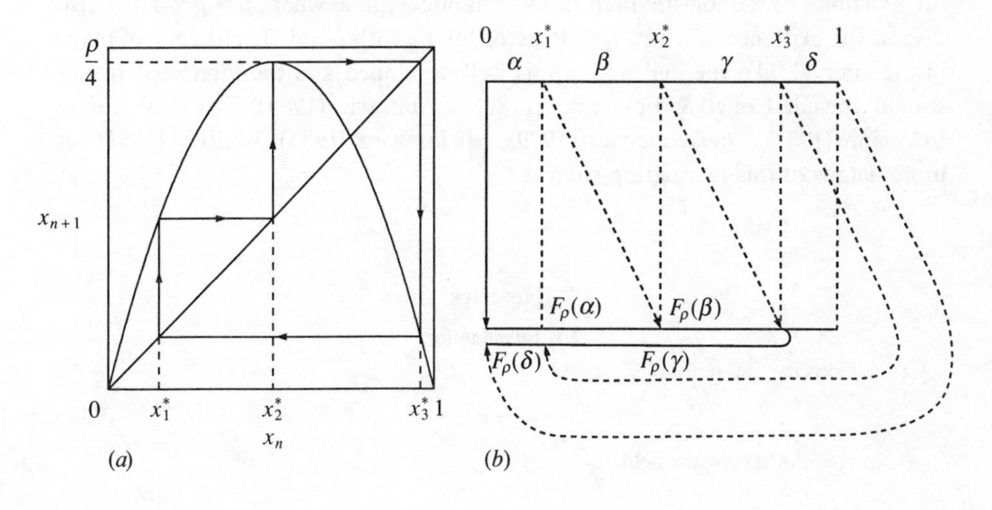

Thus, for example:

$$\begin{aligned}
x_0 &= 0 \quad \text{gives } s(0) = \{\alpha, \alpha, \ldots, \alpha, \ldots\}; \\
x_0 &= 1 \quad \text{gives } s(1) = \{\delta, \alpha, \ldots, \alpha, \ldots\}; \\
x_0 &= x_1^* \text{ gives } s(x_1^*) = \{\overline{\alpha, \beta, \gamma}, \overline{\alpha, \beta, \gamma}, \ldots\}; \\
x_0 &= x_2^* \text{ gives } s(x_2^*) = \{\overline{\beta, \gamma, \alpha}, \overline{\beta, \gamma, \alpha}, \ldots\}; \\
x_0 &= x_3^* \text{ gives } s(x_3^*) = \{\overline{\gamma, \alpha, \beta}, \overline{\gamma, \alpha, \beta}, \ldots\}.
\end{aligned} \tag{4.5.16}$$

Any two distinct points of [0, 1], however close together they may be, give rise to different sequences. Of course, if x and y are both interior points of the same subinterval α, β, γ or δ and the distance between them is very small, then given that $s(x) = \{a_k\}_0^\infty$ and $s(y) = \{b_k\}_0^\infty$ we can have

$$a_k = b_k \qquad \text{for } k \leqslant N, \tag{4.5.17}$$

where N is a large positive integer. However, the repeated stretching of the interval $[x, y]$ at each application of F_ρ will ultimately lead to $F_\rho^k(x)$ and $F_\rho^k(y)$ occurring in distinct intervals for some k with the result that $s(x) \neq s(y)$.

Not every sequence of the symbols α, β, γ, δ corresponds to an orbit of F_ρ. The rules for constructing allowed sequences are given by (4.5.13). For example, $F_\rho(\beta) = \gamma$ means that sequences having two consecutive symbols equal to β do not correspond to orbits of F_ρ. Similarly, $F_\rho(\delta) = \alpha$ implies any occurrence of δ must be followed by α. Moreover, since (4.5.13) shows that no subintervals map into δ, this symbol can only occur at the beginning of a sequence. It is also apparent from (4.5.13) that the orbits of points in α and δ ultimately end up in $\beta \cup \gamma$. Thus periodic orbits of F_ρ will only be found in the union of these subintervals. The existence of allowed sequences that do not tend to any periodic sequence, e.g. $\beta\gamma\beta\gamma\gamma\beta\gamma\gamma\gamma\beta\gamma\gamma\gamma\gamma\beta\cdots$, demonstrates that F_ρ has aperiodic orbits.

Guckenheimer *et al* (1977) use the symbolic dynamics defined above to study the dynamics of the logistic map in the 'chaotic' regime where $\rho > \rho^*$. They also discuss the existence of a strange attractor for F_ρ with $\rho = 4$. The theory of maps of the interval, like the logistic map, is well developed and the interested reader should consult Collet & Eckmann (1980), Cvitanovic (1984), Devaney (1986), Eckmann (1983), Guckenheimer (1979), Misiurewicz (1983), Whitley (1983) for more details of this fascinating subject.

Exercises

4.1 Introduction

4.1.1 Consider the unfolding

$$\dot{x} = \mu_0 + \mu_1 x - x^2 - \mu_r x^r, \tag{E4.1}$$

$r \geqslant 3$, of the vector field $\dot{x} = -x^2$.

(a) Show that the change of variable (4.1.13) leads to the topological equivalence of (E4.1) and

$$\dot{y} = \mu_0 + \mu_1 y - y^2 - \frac{\mu_1 \mu_r}{2} y^{r-1} + O(y^r), \qquad \text{(E4.2)}$$

$r \geqslant 3$. Explain why this result presents an obstacle to the generalisation of the arguments presented in Example 4.1.2.

(b) Let $y = x - (\mu_1/2)$ and show that (E4.1) implies

$$\dot{y} = \left(\mu_0 + \frac{\mu_1^2}{4}\right) - r\mu_r\left(\frac{\mu_1}{2}\right)^{r-1} y - \left[1 + \frac{r(r-1)}{2}\mu_r\left(\frac{\mu_1}{2}\right)^{r-2}\right] y^2 + O(y^3). \qquad \text{(E4.3)}$$

Compare (E4.3) with (4.1.9). Why is this 'completion of the square' not immediately helpful in this case?

4.1.2 (a) Let $\dot{x} = X(\boldsymbol{\mu}, x)$, $(\boldsymbol{\mu}, x) \in \mathbb{R}^{m_1} \times \mathbb{R}$ and $\dot{y} = Y(\mathbf{v}, y)$, $(\mathbf{v}, y) \in \mathbb{R}^{m_2} \times \mathbb{R}$ be two smooth families. Define $X \sim Y$ if the system $\dot{x} = X(\boldsymbol{\mu}, x)$ is equivalent to a family induced by $\dot{y} = Y(\mathbf{v}, y)$. Prove that $\sim$ is reflexive and transitive. Give an example to show that, in general, '$\sim$' is not symmetric.

(b) Find families, $\mathbf{h}$, of equivalence homeomorphisms and inducing functions $\boldsymbol{\varphi}$ to show that $X \sim Y$ and $Y \sim X$ for:

(i) $X(\mu_0, \mu_1, x) = \mu_0 + \mu_1 x - x^2$, $Y(v_0, y) = v_0 - y^2$; (E4.4)

(ii) $X(\mu_0, \mu_1, x) = \mu_0 + \mu_1 x - x^3$, $Y(v_0, v_1, y) = v_0 y + v_1 y^2 - y^3$. (E4.5)

4.1.3 Assume that $\dot{x} = x^2$ has versal unfolding $\dot{x} = v + x^2$. Deduce that $\dot{y} = ay^2$, $a \neq 0$, has versal unfolding $\dot{y} = \mu + ay^2$.

4.1.4 Given $F(\mu, x) = x^2 + \mu$, verify that the following functions q, s_0, s_1 satisfy the conclusion, (4.1.17), of the Malgrange Preparation Theorem:

(a) $q(\mu, x) \equiv 1$, $s_0(\mu) = \mu$, $s_1(\mu) \equiv 0$; (E4.6)

(b) $q(\mu, x) = \begin{cases} 1 + \dfrac{\exp(-1/\mu^2)}{F(\mu, x)}, & \mu > 0 \\ 1, & \mu \leqslant 0 \end{cases}$,

$$s_0(\mu) = \begin{cases} \exp(-1/\mu^2) + \mu, & \mu > 0 \\ \mu, & \mu \leqslant 0 \end{cases}, \qquad \text{(E4.7)}$$

$s_1(\mu) \equiv 0$.

4.1.5 Consider the smooth function

$$F(\mu, x) = \mu x^3 + x^2 + \mu^2 x + \mu. \qquad \text{(E4.8)}$$

(a) Show that $F(\mu, x)$ does *not* satisfy

$$q(\mu, x) F(\mu, x) = x^2 + \mu^2 x + \mu, \qquad \text{(E4.9)}$$

where $q(\mu, x)$ is a smooth function.

(b) Use

$$F(\mu, x) = (b_0(\mu) + b_1(\mu)x)(x^2 + s_1(\mu)x + s_0(\mu)), \qquad \text{(E4.10)}$$

with b_0, b_1 smooth, to obtain $q(\mu, x) = (1 + \mu x)^{-1}$, $s_1(\mu) \equiv 0$, $s_0(\mu) = \mu$ in (4.1.17).

(c) Use the functions given in (E4.7) (see Exercise 4.1.4) to show that there exist non-analytic but smooth functions $\hat{q}$ and $\hat{s}_0$ such that

$$\hat{q}(\mu, x)F(\mu, x) = x^2 + \hat{s}_0(\mu). \tag{E4.11}$$

4.1.6 Let $U \subseteq \mathbb{R}^m \times \mathbb{R}$ be a neighbourhood of the origin $(\boldsymbol{\mu}, x) = (\mathbf{0}, 0)$, $\boldsymbol{\mu} \in \mathbb{R}^m$, $x \in \mathbb{R}$ and $C^\infty(U, \mathbb{R})$ denote the set of all smooth functions from U to $\mathbb{R}$. Assume $F \in C^\infty(U, \mathbb{R})$ satisfies $F(\mathbf{0}, x) = x^k g(x)$, with $g(0) \neq 0$ and g smooth in a neighbourhood of $x = 0$. Then the Mather Division Theorem (Golubitsky & Guillemin, 1973, Chow & Hale, 1982) can be stated as follows:

Theorem (Mather) *Let F be as defined above and G be any smooth, real-valued function defined on a neighbourhood of the origin in $\mathbb{R}^m \times \mathbb{R}$. Then there are functions q, $r \in C^\infty(V, \mathbb{R})$, V a neighbourhood of the origin in $\mathbb{R}^m \times \mathbb{R}$, such that*

$$G(\boldsymbol{\mu}, x) = q(\boldsymbol{\mu}, x)F(\boldsymbol{\mu}, x) + r(\boldsymbol{\mu}, x), \tag{E4.12}$$

where

$$r(\boldsymbol{\mu}, x) = \sum_{i=0}^{k-1} r_i(\boldsymbol{\mu})x^i. \tag{E4.13}$$

(a) Show that the Malgrange Preparation Theorem (Theorem 4.1.1) is a consequence of this result and derive the alternative form

$$F(\boldsymbol{\mu}, x) = Q(\boldsymbol{\mu}, x)\left(x^k + \sum_{i=0}^{k-1} s_i(\boldsymbol{\mu})x^i\right), \tag{E4.14}$$

where $Q \in C^\infty(W, \mathbb{R})$, for some neighbourhood $W \subseteq \mathbb{R}^m \times \mathbb{R}$ containing the origin, and $Q(\mathbf{0}, 0) = g(0)$.

(b) Use (E4.14) to obtain the versal unfolding

$$\dot{x} = -x^k + \sum_{i=0}^{k-1} v_i x^i \tag{E4.15}$$

of $\dot{x} = -x^k$. Deduce that:

(i) $\dot{x} = -x^k + \sum_{i=0}^{k-2} \xi_i x^i$ (E4.16)

is also versal;

(ii) $\dot{x} = -x^k + \sum_{i=1}^{k-1} \eta_i x^i$ (E4.17)

is versal for k odd but not for k even.

4.2 Saddle-node and Hopf bifurcations

4.2.1 Saddle-node bifurcation

4.2.1 Consider the systems, $\dot{\mathbf{x}} = \mathbf{X}(\mathbf{x}) = (f(\mathbf{x}), g(\mathbf{x}))^{\mathrm{T}}$, $\mathbf{x} = (x, y)^{\mathrm{T}}$, where $\mathbf{X}(\mathbf{x})$ is given by:

(i) $(x^2 - y^2, 2xy)^{\mathrm{T}}$;

(ii) $(y - x^2, x^2)^{\mathrm{T}}$;

(iii) $(x^2, 2x-y)^T$; (E4.18)

(iv) $(y-\sinh x, y-x)^T$;

(v) $(1+y-x-\cosh x, y-x)^T$.

(a) For each system, verify that the $\dot{x}=0$ and $\dot{y}=0$ isoclines are curves in the plane that intersect at $\mathbf{x}^*=\mathbf{0}$. Which of the systems satisfies the genericity condition $f_x(\mathbf{0})$, $f_y(\mathbf{0})$, $g_x(\mathbf{0})$, $g_y(\mathbf{0}) \neq 0$? What property of the isoclines distinguishes this generic case?

(b) Confirm that Det $D\mathbf{X}(\mathbf{0})=0$ for every system. For which systems do the $\dot{x}=0$ and $\dot{y}=0$ isoclines intersect tangentially? Which of these tangential intersections is generic in the sense that the separation of the isoclines depends quadratically on x or y?

4.2.2 Consider the system $\dot{\mathbf{x}}=\mathbf{X}(\mathbf{x})$, $\mathbf{x}=(x, y)^T$, where

$$\mathbf{X}(\mathbf{x})=\begin{pmatrix} -\frac{1}{2}y+x(1-x^2-y^2) \\ \frac{1}{2}x+y(1-x^2-y^2)-\frac{1}{2} \end{pmatrix}. \tag{E4.19}$$

Show that $\{(x, y)|\dot{y}=0\}$ is the disjoint union of the point $(x, y)=(\frac{1}{2}, \frac{1}{2})$ and a curve lying in the half-plane $y \leqslant y_1$, where $0<y_1<(5^{1/2}-1)/4$. Verify that $\mathbf{x}^*=(\frac{1}{2}, \frac{1}{2})$ is a fixed point of (E4.19) and calculate Det $D\mathbf{X}(\mathbf{x}^*)$ and Tr $D\mathbf{X}(\mathbf{x}^*)$.

4.2.3 Sketch the $\dot{x}=0$ and $\dot{y}=0$ isoclines of the system $\dot{\mathbf{x}}=\mathbf{X}_r(\mathbf{x})=(y-x^r, y)^T$ for $r=2, 3, 4$. In each case confirm that these isoclines are curves which intersect tangentially at $\mathbf{x}=\mathbf{0}$ and verify that Det $D\mathbf{X}(\mathbf{0})=0$. Show that:

(a) there is no unfolding of $\mathbf{X}_2$ for which more than two fixed points bifurcate from the origin;

(b) there is an unfolding of $\mathbf{X}_3$ for which three fixed points bifurcate from the origin;

(c) there are unfoldings of $\mathbf{X}_4$ for which two and four fixed points bifurcate from the origin but none for which exactly three hyperbolic fixed points occur.

4.2.4 A special case of Resigno and De Lisi's model for the growth of a spherical tumour (see Arrowsmith & Place, 1984) has dynamical equations

$$\dot{x}=[-\lambda_1+y^{2/3}(1-x)/(1+x)]x,$$
$$\dot{y}=[\lambda_2 y^{1/3}-x/(1+x)]y^{2/3}, \tag{E4.20}$$

$0<x \leqslant 1$, $y>0$, $\lambda_1, \lambda_2>0$. Show that the x-coordinates of the non-trivial fixed points (i.e. $x^*, y^*>0$), of (E4.20) satisfy

$$\psi(x)=\frac{x^2(1-x)}{(1+x)^3}=\lambda_1\lambda_2^2. \tag{E4.21}$$

Deduce that (E4.20) has two, one or no fixed points according to whether $\lambda_1\lambda_2^2$ is $<, =, >$ (1/27). Confirm, numerically, that the $\dot{x}=0$ and $\dot{y}=0$ isoclines of (E4.20) intersect tangentially for $\lambda_1\lambda_2^2=1/27$.

4.2.5 (a) Consider the following specialisation of a prey-predator model (see Arrowsmith & Place, 1984) due to Tanner,

$$\dot{H}=\frac{11}{5}H\left(1-\frac{H}{11}\right)-\frac{10HP}{(2+3H)},$$
$$\dot{P}=P\left(1-\frac{P(\beta+H)}{\gamma H}\right), \tag{E4.22}$$

where H (P) is the herbivore (predator) population and β, γ are positive constants. Sketch configurations of $\dot{H}=0$ and $\dot{P}=0$ isoclines to indicate how β and γ may be chosen so that (E4.22) has: (i) zero; (ii) one; (iii) two non-trivial fixed points.

(b) A model for the interaction between a plant population (p) and its animal pollinator population (a), given by Soberon *et al.* (see Arrowsmith & Place, 1984), takes the form

$$\dot{a} = a(K-a) + \frac{ap}{1+p},$$
$$\dot{p} = -\gamma p + \frac{ap}{1+p}, \tag{E4.23}$$

when all but two of its parameters are set to unity. Show that for $\gamma = \frac{1}{2}$ the non-trivial $\dot{a}=0$ and $\dot{p}=0$ isoclines intersect tangentially for $K = K^* = 2^{1/2}-1$ at $(a, p) = (a^*, p^*) = (1/2^{1/2}, 2^{1/2}-1)$.

4.2.6 Consider the m-parameter family of vector fields defined by (4.2.13) when the terms $R_{1,2}$ are absent. Examine the $(m+2)$-dimensional system formed by the addition of $\dot{\boldsymbol{\mu}} = \mathbf{0}$ as in (4.2.16).

(a) Show that normal form transformations on this extended system enable all parameter-free terms to be removed, except for those of type x_1x_2 in the first component and x_2^2 in the second, at the expense of modifying some of the $x_i\mu_j$-terms.

(b) Show that $x_1\mu_k$-terms, $k = 1, \ldots, m$, can be eliminated from $\dot{x}_2$ and $x_2\mu_k$-terms can be removed from $\dot{x}_1$ at the expense of changing terms $O(|\boldsymbol{\mu}|^2)$.

(c) Disregarding terms of order greater than or equal to 3 in $|(\boldsymbol{\mu}, \mathbf{x})|$, use a centre manifold argument to show that (4.2.13) is equivalent to a family induced by (4.2.2) with $a_2 = 0$.

4.2.2 Hopf bifurcation

4.2.7 Consider the system (E4.23), discussed in Exercise 4.2.5(b), with $\gamma = \frac{1}{2}$. Define $a = a^* + x_1$, $p = p^* + x_2$, $K = K^* + \kappa$ and show that

$$\dot{\mathbf{x}} = \begin{pmatrix}\dot{x}_1\\ \dot{x}_2\end{pmatrix} = \begin{pmatrix} \dfrac{\kappa}{2^{1/2}} - \dfrac{x_1}{2^{1/2}} + \dfrac{x_2}{2(2^{1/2})} + \kappa x_1 - x_1^2 + \frac{1}{2}x_1x_2 - \frac{1}{4}x_2^2 \\ \left(1 - \dfrac{1}{2^{1/2}}\right)x_1 - \frac{1}{2}\left(1 - \dfrac{1}{2^{1/2}}\right)x_2 + \frac{1}{2}x_1x_2 - \frac{1}{4}x_2^2 \end{pmatrix} + O(|\mathbf{x}|^3). \tag{E4.24}$$

Verify that the transformation $\mathbf{x} = \mathbf{M}\mathbf{y}$, $\mathbf{y} = (y_1, y_2)^{\mathrm{T}}$, with

$$\mathbf{M} = \begin{pmatrix} 1 & 1 \\ 1-2^{1/2} & 2 \end{pmatrix}, \tag{E4.25}$$

reduces the linear part of (E4.24) to Jordan form and hence use Proposition 4.2.2 to confirm that (E4.23) undergoes a saddle-node bifurcation as K increases through K^*.

4.2.8 Examine how the hypotheses of the Hopf Bifurcation Theorem (Theorem 4.2.1) fail for the following systems:

$$\begin{aligned}
&\text{(a) } \dot{r} = r(\mu^2 - r^2),\ \dot{\theta} = 1;\\
&\text{(b) } \dot{r} = \mu r(r - \mu)(r - 2\mu),\ \dot{\theta} = 1;\\
&\text{(c) } \dot{r} = \mu r(r - \mu)(r - 2\mu) - r^3,\ \dot{\theta} = 1;\\
&\text{(d) } \dot{r} = \mu r,\ \dot{\theta} = 1.
\end{aligned} \tag{E4.26}$$

In each case exhibit the bifurcations that do occur.

4.2.9 Consider the family of differential equations

$$\begin{pmatrix}\dot{x}_1\\ \dot{x}_2\end{pmatrix} = \begin{pmatrix}\mu & -1\\ 1 & \mu\end{pmatrix}\begin{pmatrix}x_1\\ x_2\end{pmatrix} + ar^2\begin{pmatrix}x_1\\ x_2\end{pmatrix} + br^2\begin{pmatrix}-x_2\\ x_1\end{pmatrix}, \tag{E4.27}$$

where $r^2 = (x_1^2 + x_2^2)$, $b, \mu \in \mathbb{R}$ and $a < 0$. Use the polar form of (E4.27) to show that it undergoes a supercritical Hopf bifurcation with increasing μ at $\mu = 0$.

Prove, without using Theorem 4.2.1, that the family of differential equations

$$\begin{pmatrix}\dot{x}_1\\ \dot{x}_2\end{pmatrix} = \begin{pmatrix}\mu+2 & -5\\ 1 & \mu-2\end{pmatrix}\begin{pmatrix}x_1\\ x_2\end{pmatrix} + (x_1^2 - 4x_1x_2 + 5x_2^2)\begin{pmatrix}x_1\\ x_2\end{pmatrix} \tag{E4.28}$$

undergoes a Hopf bifurcation at $\mu = 0$. How does the bifurcation occurring in (E4.28) differ from that shown by (E4.27)?

4.2.10 Use Theorem 4.2.1 to show that the one-parameter family of differential equations

$$\begin{aligned}\dot{x} &= (1+\mu)x - y + x^2 - xy,\\ \dot{y} &= 2x - y + x^2,\end{aligned} \tag{E4.29}$$

undergoes a supercritical Hopf bifurcation with increasing μ at $\mu = 0$.

4.2.11 Prove that, both the Rayleigh equation

$$\ddot{x} + \dot{x}^3 - \mu\dot{x} + x = 0, \tag{E4.30}$$

and the Van der Pol equation

$$\ddot{x} + \dot{x}(\mu - x^2) + x = 0, \tag{E4.31}$$

undergo Hopf bifurcations at $\mu = 0$. In both cases, sketch phase portraits for $\mu <, =, > 0$ and state what type of Hopf bifurcation takes place.

4.2.12 The kinetic equations for the interaction of two chemicals with concentrations x and y are

$$\dot{x} = a - (b+1)x + x^2y, \qquad \dot{y} = bx - x^2y, \tag{E4.32}$$

where a, b are positive constants. For fixed a, show that a supercritical Hopf bifurcation takes place, with increasing b, at $b = b^* = a^2 + 1$.

4.3 Cusp and generalised Hopf bifurcations

4.3.1 Cusp bifurcation

4.3.1 Consider the system

$$\dot{x}_1 = x_2 + v_2x_1 + a_2x_1^2, \qquad \dot{x}_2 = v_1 + b_2x_1^2, \tag{E4.33}$$

with $b_2 > 0$ and v_2 held constant at a non-zero value. Let $\mathbf{x} = \mathbf{M}\mathbf{y}$, with

$\mathbf{x} = (x_1, x_2)^{\mathrm{T}}, \mathbf{y} = (y_1, y_2)^{\mathrm{T}}, \mathbf{M} = \begin{pmatrix} 1 & -1 \\ 0 & \nu_2 \end{pmatrix}$, and obtain $\dot{\mathbf{y}}$. Hence show that (E4.33) undergoes a saddle-node bifurcation with increasing ν_1 at $\nu_1 = 0$. Sketch phase portraits for the normal form of (E4.33) with $\nu_1 <, =, > 0$ when : (a) $\nu_2 > 0$; (b) $\nu_2 < 0$. Why is the bifurcation a supercritical one in both (a) and (b)?

4.3.2 Show that the system

$$\dot{x}_1 = x_2 + \nu_2 x_1 + a_2 x_1^2, \qquad \dot{x}_2 = \nu_1 + b_2 x_1^2, \tag{E4.34}$$

where $a_2 < 0$, $b_2 > 0$, undergoes a supercritical Hopf bifurcation:
(a) with increasing ν_2 at $\nu_2 = \nu_2^* = 2a_2(-\nu_1/b_2)^{1/2}$ for fixed $\nu_1 < 0$;
(b) with decreasing ν_1 at $\nu_1 = \nu_1^* = -b_2(\nu_2/2a_2)^2$ for fixed $\nu_2 < 0$.

4.3.3 Show that the singular transformation

$$x_1 = \tau^2 \bar{u}, \qquad x_2 = \tau^3 \bar{v}, \qquad \nu_1 = \tau^4 \bar{\nu}_1, \qquad \nu_2 = \tau \bar{\nu}_2, \qquad \mathbf{X} = \tau \bar{\mathbf{X}} \tag{E4.35}$$

on the vector field

$$\mathbf{X} = (x_2 + \nu_2 x_1 + a x_1^2 + O(|\mathbf{x}|^3), \nu_1 + b x_1^2 + O(|\mathbf{x}|^3))^{\mathrm{T}}, \tag{E4.36}$$

$a < 0$, $b > 0$, gives

$$\bar{\mathbf{X}}_{(\bar{\nu}_1, \bar{\nu}_2, \tau)} = (\bar{v} + \bar{\nu}_2 \bar{u} + a\tau \bar{u}^2 + O(\tau^2), \bar{\nu}_1 + b\bar{u}^2 + O(\tau^2))^{\mathrm{T}}. \tag{E4.37}$$

Given $\bar{\nu}_1 = -1$, demonstrate that the half-plane $\nu_1 < 0$ of $\nu_1\nu_2$-space is in 1–1 correspondence with the half-plane $\tau > 0$ of $(\bar{\nu}_2, \tau)$-space. Show that $\bar{\mathbf{X}}_{(-1,0,0)}$ is a Hamiltonian vector field and obtain its phase portrait. Show that, for $\bar{\nu}_1 = -1$ and $\bar{\nu}_2 < 0$, a supercritical Hopf bifurcation occurs with increasing τ at $\tau^* = \bar{\nu}_2 b^{1/2}/2a > 0$.

4.3.4 Let $\dot{\mathbf{x}} = \mathbf{X}(\mathbf{x})$, $\mathbf{x} = (x, y)^{\mathrm{T}}$, be defined by the dynamical equations (E4.20) in Exercise 4.2.4. Show that, at any non-trivial fixed point $\mathbf{x}^*$,

$$\mathrm{Tr}\, D\mathbf{X}(\mathbf{x}^*) = \frac{\lambda_2}{3}\left[1 - 6\left(\frac{\lambda_1}{\lambda_2}\right)\frac{x}{(1-x^2)}\right]_{\mathbf{x}^*}. \tag{E4.38}$$

Use (E4.38), together with the result obtained in Exercise 4.2.4, to deduce that (E4.20), with $\lambda_1 = \lambda_1^* = 4^{-2/3}/3$ and $\lambda_2 = \lambda_2^* = 4^{1/3}/3$, has a fixed point $\mathbf{x}^* = (\frac{1}{2}, \frac{1}{4})^{\mathrm{T}}$ for which $\mathrm{Det}\, D\mathbf{X}(\mathbf{x}^*) = \mathrm{Tr}\, D\mathbf{X}(\mathbf{x}^*) = 0$.

4.3.5 The 2-jet of (E4.20) in the point $(\lambda_1, \lambda_2, x, y) = (\lambda_1^*, \lambda_2^*, \frac{1}{2}, \frac{1}{4})$ (see Exercise 4.3.4), can be shown to be

$$\begin{pmatrix} -\frac{1}{2}\mu_1 - \varepsilon x_1 + \varepsilon x_2 - \frac{4}{3}\varepsilon x_1^2 - \frac{2}{3}\varepsilon x_1 x_2 - \frac{2}{3}\varepsilon x_2^2 - \mu_1 x_1 \\ \frac{1}{4}\mu_2 - \varepsilon x_1 + \varepsilon x_2 + \frac{2}{3}\varepsilon x_1^2 - \frac{8}{3}\varepsilon x_1 x_2 + \frac{4}{3}\varepsilon x_2^2 + \mu_2 x_2 \end{pmatrix}, \tag{E4.39}$$

where

$$x_1 = x - \tfrac{1}{2}, \qquad x_2 = y - \tfrac{1}{4}, \qquad \mu_1 = \lambda_1 - \lambda_1^*, \qquad \mu_2 = \lambda_2 - \lambda_2^*, \qquad \varepsilon = 4^{1/3}/9.$$

Verify that the transformation $\mathbf{x} = (x_1, x_2)^{\mathrm{T}} = \mathbf{M}\mathbf{y}$, $\mathbf{y} = (y_1, y_2)^{\mathrm{T}}$, with $\mathbf{M} = \begin{pmatrix} \varepsilon & 0 \\ \varepsilon & 1 \end{pmatrix}$ reduces the linear part of (E4.39) to real Jordan form. Hence, use Proposition 4.3.2 to show that:
(a) the singularity in Exercise 4.3.4 is a cusp singularity;
(b) the parameters μ_1, μ_2 completely unfold it.
What is the stability of the limit cycles predicted by this calculation?

4.3.6 Let $\dot{\mathbf{x}} = \mathbf{X}(\mathbf{x})$, $\mathbf{x} = (x, y)^{\mathrm{T}}$, be defined by (E4.20) in Exercise 4.2.4. Show that at any non-trivial fixed point, $\mathbf{x}^*$,

$$\operatorname{Det} \mathrm{D}\mathbf{X}(\mathbf{x}^*) = \left[2\lambda_2 x y^{2/3} \left(\frac{1}{x} - 2 \right) \Big/ (3(1+x)^2) \right]_{\mathbf{x}^*}. \tag{E4.40}$$

Given that (E4.20) undergoes a cusp bifurcation at $(x, y) = (\frac{1}{2}, \frac{1}{4})$ for $(\lambda_1, \lambda_2) = (4^{-2/3}/3, 4^{1/3}/3)$, show that:

(a) the saddle-node bifurcation curve is given by $\lambda_2^2 = 1/(27\lambda_1)$;

(b) the Hopf bifurcation curve takes the form

$$(\lambda_1(s), \lambda_2(s)) = \left(\frac{1-s}{[36(1+s)]^{1/3}}, \frac{6s\lambda_1(s)}{(1-s^2)} \right) \tag{E4.41}$$

for $0 < s < \frac{1}{2}$.

4.3.2 Generalised Hopf bifurcations

4.3.7 Show that the vector field

$$\begin{pmatrix} 0 & -\beta \\ \beta & 0 \end{pmatrix} \begin{pmatrix} x_1 \\ x_2 \end{pmatrix} + \sum_{k=1}^{\infty} (x_1^2 + x_2^2)^k \left[a_k \begin{pmatrix} x_1 \\ x_2 \end{pmatrix} + b_k \begin{pmatrix} -x_2 \\ x_1 \end{pmatrix} \right], \tag{E4.42}$$

with $\beta > 0$, and $a_1 = a_2 = \cdots = a_{l-1} = 0$, $a_l \neq 0$, can be written as

$$f(x_1, x_2) \left[\begin{pmatrix} 0 & -1 \\ 1 & 0 \end{pmatrix} \begin{pmatrix} x_1 \\ x_2 \end{pmatrix} + \sum_{k=l}^{\infty} a_k' (x_1^2 + x_2^2)^k \begin{pmatrix} x_1 \\ x_2 \end{pmatrix} \right], \tag{E4.43}$$

where $f(x_1, x_2) = \beta + \sum_{k=1}^{\infty} b_k (x_1^2 + x_2^2)^k$ and a_k' is defined inductively for $k \geqslant l$ by $a_l' = a_l/\beta$ and

$$a_k' = (a_k - a_{k-1}' b_1 - a_{k-2}' b_2 - \cdots - a_l' b_{k-l})/\beta, \qquad k > l. \tag{E4.44}$$

Find a radial scaling such that a_l' can be replaced by ± 1. Show that a vector field of type $(l, -)$ is topologically equivalent to the time reversal of one of type $(l, +)$.

4.3.8 Show that the bifurcation diagram for the system

$$\dot{r} = v_0 r + v_1 r^3 + r^5, \; \dot{\theta} = 1, \tag{E4.45}$$

is that shown in Figure 4.11 and obtain the phase portraits of the structurally stable systems occurring in the complement of the bifurcation curves. Investigate the transitions between these systems by sketching phase portraits on the bifurcation curves themselves.

4.3.9 Consider the polynomial

$$P_{\mathbf{v}}(\rho) = \rho^3 + v_2 \rho^2 + v_1 \rho + v_0, \tag{E4.46}$$

where $\mathbf{v} = (v_0, v_1, v_2) \in \mathbb{R}^3$, $\rho \in \mathbb{R}$.

(a) Assume $v_2 = 0$ and show that, for $v_1 < 0$, $P_{\mathbf{v}}(\rho)$ has a minimum at $\rho = (-v_1/3)^{1/2}$ and a maximum at $\rho = -(-v_1/3)^{1/2}$. Sketch $P_{\mathbf{v}}(\rho)$ for $\mathbf{v} = (0, v_1, 0)$ with $v_1 <, =, > 0$.

(b) Use sketches of $P_{\mathbf{v}}(\rho)$, with $\mathbf{v} = (v_0, v_1, 0)$, to show that an unstable limit cycle grows from the origin of the system (4.3.30) as v_0 decreases through zero with $v_2 = 0$ and $v_1 \geqslant 0$.

(c) Explain how the diagrams drawn in (a) and (b) must be modified if $\nu_2 > 0$. Given $C > 0$, in what respect does the $\nu_2 = C$-section of the type (3, +) Hopf bifurcation diagram differ from Figure 4.13.

4.3.10 Consider the polynomial $P_\nu(\rho)$ defined in (E4.46).

(a) Explain how the diagrams drawn in (a) and (b) of Exercise 4.3.9 have to be modified for $\nu_2 < 0$.

(b) Verify that these modified diagrams are consistent with the section of the (3, +) Hopf bifurcation diagram shown in Figure 4.14 for $\nu_2 < 0$.

4.3.11 Consider the $\nu_2 = C$-section ($C < 0$) of the (3, +) Hopf bifurcation diagram shown in Figure 4.14.

(a) Show that the double limit-cycle bifurcation curves are given by

$$\nu_0^{\pm} = \frac{\nu_2^3}{27} + \frac{1}{3}\nu_2\tilde{\nu} \mp \tilde{\nu}\left(\frac{-\tilde{\nu}}{3}\right)^{1/2}, \tag{E4.47}$$

where $\tilde{\nu} = \nu_1 - \nu_2^2/3$.

(b) Confirm that ν_0^+ and ν_0^- meet at the point $P = \left(\frac{\nu_2^3}{27}, \frac{\nu_2^2}{3}, \nu_2\right)$ and that

$$\left.\frac{d\nu_0^+}{d\nu_1}\right|_P = \left.\frac{d\nu_0^-}{d\nu_1}\right|_P.$$

(c) Show that ν_0^+ has zero slope at $\nu_1 = 0$ for $\nu_2 > 0$ (see Figure 4.13) and ν_0^- has zero slope at $\nu_1 = 0$ for $\nu_2 < 0$ (see Figure 4.14).

4.4 Diffeomorphisms on $\mathbb{R}$

4.4.1 $\mathrm{D}_x f(\mathbf{0}) = +1$*: the fold bifurcation*

4.4.1 Deduce (4.4.6) and (4.4.7) for the curve of fixed points $\mu = \mu(x)$ given by (4.4.5). Sketch bifurcation diagrams corresponding to Figure 4.16(*a*) when $\mathrm{D}_\mu f(\mathbf{0})$ and $\mathrm{D}_{xx} f(\mathbf{0})$ are respectively

(a) positive, negative,
(b) negative, positive,
(c) negative, negative.

4.4.2 Use the Implicit Function Theorem to show that (4.4.5) has solution $x = x^*(\mu)$ for $\mu \in (-\varepsilon, \varepsilon)$, $\varepsilon > 0$, and obtain

$$\mathrm{D}x^*(0) = \frac{-2\mathrm{D}_{\mu x} f(\mathbf{0})}{\mathrm{D}_{xx} f(\mathbf{0})} < 0. \tag{E4.48}$$

Use this result to show that

$$\mathrm{D}_x f(\mu, x^*(\mu)) = 1 - \mu \mathrm{D}_{\mu x} f(\mathbf{0}) + O(\mu^2). \tag{E4.49}$$

Hence determine the stability of the fixed point at $x = x^*(\mu)$ for $\mu < 0$, > 0.

4.4.3 (The pitchfork bifurcation (see Whitley, 1983, p. 185)) Let $f(\mu, x)$ be a smooth one-parameter family of diffeomorphisms satisfying

(i) $f(\mu, -x) = -f(\mu, x)$; (E4.50)

(ii) $\mathrm{D}_x f(\mu, 0) = \lambda(\mu)$, $\lambda(0) = 1$ and $\left.\frac{d\lambda}{d\mu}\right|_{\mu=0} > 0$; (E4.51)

(iii) $D_{xxx}f(\mathbf{0}) < 0$. (E4.52)

Show that, for (μ, x) sufficiently close to $(0, 0)$,

$$f(\mu, x) = \lambda(\mu)x + a_3(\mu)x^3 + O(x^5), \tag{E4.53}$$

where $a_3(0) < 0$.

Deduce that, on some neighbourhood of $(\mu, x) = (0, 0)$,

(a) f_μ has a stable fixed point at the origin for $\mu < 0$;

(b) for $\mu > 0$, f_μ has an unstable fixed point at the origin and stable fixed points at

$$x^*_\pm(\mu) = \pm\left[6\mu \left.\frac{d\lambda}{d\mu}\right|_{\mu=0} \Big/ |D_{xxx}f(\mathbf{0})|\right]^{1/2}(1 + O(\mu)).$$

Sketch the bifurcation diagram for f in the μx-plane.

4.4.2 $D_x f(\mathbf{0}) = -1$: *the flip bifurcation*

4.4.4 Consider the function $G(\mu, x)$ defined in (4.4.24). Use (4.4.2), (4.4.3) and (4.4.20) to show that

$$\begin{aligned} F(\mu, x) &= G(\mu, x) + x \\ &= (\nu(\mu) - 1)x + \sum_{r=2}^{N} b_r(\mu)x^r + O(x^{N+1}), \end{aligned} \tag{E4.54}$$

where

$$b_r(\mu) = \sum_{i=r}^{N} \binom{i}{r} a_i(\mu)x^*(\mu)^{i-r} \tag{E4.55}$$

and

$$\nu(\mu) = \mu \frac{d}{d\mu}\{D_x f(\mu, x^*(\mu))\}|_{\mu=0} + O(\mu^2). \tag{E4.56}$$

4.4.5 Show that the family of diffeomorphisms

$$F(\mu, x) = (\nu(\mu) - 1)x + \sum_{r=2}^{3} b_r(\mu)x^r + O(x^4), \tag{E4.57}$$

defined in (4.4.31), is equivalent to

$$\tilde{F}(\mu, x) = (\nu(\mu) - 1)x + \tilde{b}_3(\mu)x^3 + O(x^4), \tag{E4.58}$$

where

$$\tilde{b}_3(\mu) = b_3(\mu) + \frac{2b_2(\mu)^2}{(\nu(\mu) - 1)(\nu(\mu) - 2)}. \tag{E4.59}$$

Verify that

$$\tilde{b}_3(\mu) = -\tfrac{1}{12}D_{xxx}f^2(\mathbf{0}) + O(\mu). \tag{E4.60}$$

4.4.6 Consider $\bar{f}^2_\nu$ defined by (4.4.37) with the negative sign operative. Show that $\bar{f}^2_\nu$ has fixed points, $x^*_\pm$, satisfying (4.4.39) and verify that $x^*_\pm$ are not fixed points of $\bar{f}_\nu$. Hence show that

$$D\bar{f}^2_\nu(x^*_+) = D\bar{f}_\nu(x^*_+)D\bar{f}_\nu(x^*_-) \tag{E4.61}$$

and confirm that the 2-cycle is unstable in this case. How are the above calculations affected if $\bar{f}_\nu(x)$ contains $O(x^5)$-terms, as in (4.4.35)?

4.4.7 Check that each of the following families of maps undergoes the bifurcation indicated
(a) $x \mapsto \mu - x^2$, fold, $\mu = -\frac{1}{4}$,
(b) $x \mapsto \mu - x^2$, flip, $\mu = \frac{3}{4}$,
(c) $x \mapsto \mu x(1-x)$, transcritical, $\mu = 1$,
(d) $x \mapsto \mu x - x^3$, pitchfork, $\mu = 1$.
Sketch the local bifurcation diagrams in the μx-plane for each case.

4.5 The logistic map

4.5.1 Consider the logistic map, $F_\rho(x) = \rho x(1-x)$, $\rho \in (0, 4]$ and $x \in [0, 1]$. Use a microcomputer to plot the graphs of F_ρ. F_ρ^2, F_ρ^4, F_ρ^8 for $\rho = 3.6$ and the line $y = x$. Label the periodic points of F_ρ of period 1, 2, 4 and 8.

4.5.2 Plot graphs of F_ρ^2, where F_ρ is the logistic map, for $\rho = 3.4$, 3.5, 3.6 and 3.7. Observe that the slope of F_ρ^2 at two of its non-trivial fixed points decreases through -1 with increasing ρ. Check that a flip bifurcation occurs by showing that period-4 points are created.

4.5.3 For the logistic map $F_\rho(x) = \rho x(1-x)$, plot the graphs of F_ρ^4 and F_ρ^8 for $\rho = 3.0$, 3.2, 3.4, 3.6, 3.8, 4.0. Between which two consecutive pairs of ρ-values are: (a) period-4; (b) period-8; points first created? Make a closer investigation for $\rho \in [3.9, 4.0]$ to show that further period-4 orbits arise at fold bifurcations in F_ρ^4.

4.5.4 Show that two period-3 periodic orbits are created between $\rho = 3.8$ and $\rho = 3.9$ by finding the fixed points of F_ρ^3 graphically. Indicate which of the periodic orbits is initially stable. Why is the graph of F_ρ^3 simultaneously tangent at three distinct points of the line $y = x$ as ρ increases?

4.5.5 (a) Let $\bar{f}: \mathbb{R} \to \mathbb{R}$ be given by $\bar{f}(x) = 2x$. Use the standard projection $\pi(x) = x \bmod 1$ to verify that $\bar{f}$ is the lift of a well-defined, non-injective map $f: S^1 \to S^1$. Consider the subclass Σ' of bi-infinite 2-symbol sequences generated by the binary forms of the real numbers. Show that $\bar{f}$ is topologically conjugate to a right shift on Σ'. Deduce that:
(i) the set of all periodic points of f is dense in S^1;
(ii) there is an orbit of f that is dense in S^1.
(b) A map $F: I \to I \subseteq \mathbb{R}$ has the property that its orbits depend sensitively on initial conditions if there exists an $\varepsilon > 0$ such that for every $x \in I$ and every open set N containing x, there is a point $y \in N$ for which $|F^n(x) - F^n(y)| > \varepsilon$, for some $n \in \mathbb{Z}^+$.
Show that the logistic map $F_\rho: [0, 1] \to [0, 1]$, $\rho = 4$, and $f: S^1 \to S^1$ defined in (a) satisfy $\Phi F_4 = f\Phi$, where $\Phi: S^1 \to I$ is given by $\Phi(\theta) = \sin^2 \theta$. Hence deduce that the logistic map F_4 has the following 'chaotic' features:
(i) it has a set of periodic orbits that is dense in [0, 1];
(ii) it has an orbit that is dense in [0, 1],
(iii) it satisfies the requirements for sensitive dependence on initial conditions with $\varepsilon = \frac{1}{2}$.

5

Local bifurcations II: Diffeomorphisms on $\mathbb{R}^2$

5.1 Introduction

A planar diffeomorphism $\mathbf{f}_0$ has a non-hyperbolic fixed point at $\mathbf{x} = \mathbf{0}$ if at least one of the eigenvalues of $\mathbf{Df}_0(\mathbf{0})$ has modulus equal to unity. Let $\mathbf{Df}_0(\mathbf{0})$ have eigenvalues λ_1 and λ_2, $|\lambda_1| \geqslant |\lambda_2|$, then $\mathbf{x} = \mathbf{0}$ is non-hyperbolic if

$$|\lambda_1| = 1, \; |\lambda_2| < 1, \tag{5.1.1}$$

$$|\lambda_1| > 1, \; |\lambda_2| = 1, \tag{5.1.2}$$

$$|\lambda_1| = |\lambda_2| = 1. \tag{5.1.3}$$

The local bifurcational behaviour of generic one-parameter families, $\mathbf{f}(\mu, \mathbf{x})$, for which $\mathbf{f}_0(\cdot) = \mathbf{f}(0, \cdot)$ satisfies (5.1.1) or (5.1.2) is determined by the corresponding one-dimensional family discussed in §4.4 (see Arnold, 1983, p. 286). For example, if (5.1.1) holds with $\lambda_1 = 1$, then, for $(\mu, \mathbf{x})$ near $(0, \mathbf{0})$, $\mathbf{f}(\mu, \mathbf{x})$ is equivalent to the local family at $(0, \mathbf{0})$ defined by

$$\tilde{\mathbf{f}}(\mu, \mathbf{x}) = (\mu + x + x^2, \pm(y/2))^{\mathrm{T}}, \tag{5.1.4}$$

where $\mathbf{x} = (x, y)^{\mathrm{T}}$ and the sign in the second component is determined by that of λ_2. On the other hand, if $\lambda_1 = -1$, then $\mathbf{f}(\mu, \mathbf{x})$ is locally equivalent to

$$\tilde{\mathbf{f}}(\mu, \mathbf{x}) = ((\mu - 1)x \pm x^3, \pm(y/2))^{\mathrm{T}}. \tag{5.1.5}$$

The sign in the first component of (5.1.5) is determined by that of the cubic term in the normal form of $\mathbf{f}_0$. The local families in (5.1.4) and (5.1.5) are examples of what are called *suspensions* of the one-dimensional families appearing in the first component. The bifurcation diagrams for (5.1.4) and (5.1.5) are illustrated in Figure 5.1. When (5.1.2) is satisfied, similar suspensions can be constructed in which the hyperbolic eigenvalue gives rise to an expansion rather than a contraction (see Exercise 5.1.1).

When both eigenvalues of $\mathbf{Df}_0(\mathbf{0})$ have modulus unity, as in (5.1.3), there are a number of possibilities for both $\mathbf{Df}_0(\mathbf{0})$ and (λ_1, λ_2):

(i) $$\lambda_1 = \bar{\lambda}_2 = \exp(2\pi i\beta), \tag{5.1.6}$$

with $\beta \neq 0, \frac{1}{2}$ (here $^-$ denotes complex conjugate);

(ii)
$$\lambda_1 = \lambda_2 = 1 \text{ or } \lambda_1 = \lambda_2 = -1, \tag{5.1.7}$$

but $\mathbf{Df}_0(\mathbf{0})$ cannot be diagonalised;

(iii)
$$\lambda_1 = 1 \text{ and } \lambda_2 = -1; \tag{5.1.8}$$

(iv)
$$\lambda_1 = \lambda_2 = 1 \text{ or } \lambda_1 = \lambda_2 = -1, \tag{5.1.9}$$

and $\mathbf{Df}_0(\mathbf{0})$ can be diagonalised (see $\beta = 0, \frac{1}{2}$ in (5.1.6)).

Generically, (5.1.3) is satisfied by λ_1 and λ_2 of the type described in (i). However, if $\beta = 0$ or $\frac{1}{2}$ the generic situation is given by (ii) and not (iv). Cases (i) and (ii) present some new and rather subtle problems which will occupy us for the remainder of this chapter. It is tempting to think that (iii) is simpler to deal with than (ii) because the eigenvalues are distinct. Unfortunately, this is not the case. Exercises 5.1.3 and 4 make it clear that it is not sufficient to consider the product of the 'fold' and 'flip' families mentioned above, What is more, while both (ii) and

Figure 5.1 Illustration of the bifurcation diagrams for the families $\tilde{\mathbf{f}}(\mu, \mathbf{x})$ appearing in (5.1.4) and (5.1.5) assuming the positive sign in the second component. (*a*) The broken lines indicate the behaviour of the orbits of $\tilde{\mathbf{f}}_\mu$ defined by (5.1.4) for: (i) $\mu < 0$; (ii) $\mu = 0$; (iii) $\mu > 0$. This local family is said to undergo a *saddle-node bifurcation* at $\mu = 0$. (*b*) Here $\tilde{\mathbf{f}}_\mu$ is given by (5.1.5). For μ near zero, the sign of the first component of $\tilde{\mathbf{f}}_\mu$ alternates. Therefore, in this case, the broken lines indicate the behaviour of $\tilde{\mathbf{f}}_\mu^2$ and the non-trivial fixed points that occur for $\mu > 0$ correspond to a 2-cycle in $\tilde{\mathbf{f}}_\mu$.

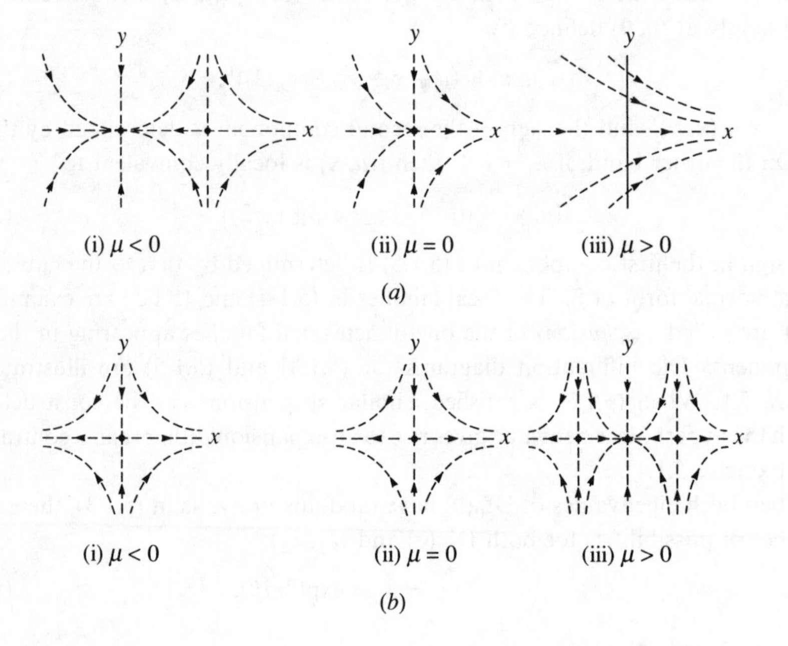

(iii) have linear co-dimension two (see Exercises 5.1.2 and 5), the latter is not amenable to the vector field approximation methods, used to analyse (i) and (ii), because one of its eigenvalues is negative. The negative eigenvalue problem can be overcome by considering the square of the map (see § 5.5.3) but for (iii) $\mathbf{Df}_0^2(\mathbf{0})$ is the identity which means that $\mathbf{f}_0^2$ is of the type described in (iv). Case (iv) is of linear co-dimension four (see Exercise 5.1.5) and is therefore of a higher order of difficulty than (i) and (ii). We therefore focus attention on unfoldings of $\mathbf{f}_0$ satisfying (i) and (ii).

Let $\mathbf{f}\colon \mathbb{R} \times \mathbb{R}^2 \to \mathbb{R}^2$ be a smooth, one-parameter family of maps such that $\mathbf{f}(\mu, \mathbf{x}) = \mathbf{f}_\mu(\mathbf{x})$ has a fixed point at $\mathbf{x} = \mathbf{0}$ for all $\mu \in \mathbb{R}$ (see Exercise 5.1.6). Suppose further that, at $\mu = 0$, the eigenvalues of $\mathbf{Df}_0(\mathbf{0})$ satisfy (5.1.6) with $\beta \neq 0, \frac{1}{2}$. It follows that the linearisation of $\mathbf{f}_0$ at $\mathbf{x} = \mathbf{0}$ is an anticlockwise rotation by an angle $2\pi\beta$ and there is a basis for $\mathbb{R}^2$ such that

$$\mathbf{Df}_0(\mathbf{0}) = \begin{pmatrix} \cos 2\pi\beta & -\sin 2\pi\beta \\ \sin 2\pi\beta & \cos 2\pi\beta \end{pmatrix}. \tag{5.1.10}$$

What is the topological behaviour of such a family for $(\mu, \mathbf{x}) \in U$, some sufficiently small neighbourhood of $(0, \mathbf{0}) \in \mathbb{R} \times \mathbb{R}^2$? The restriction of $\mathbf{f}$ to U, $\mathbf{f}|U$, defines a one-parameter, local family of diffeomorphisms. Given that $\mathbf{Df}_\mu(\mathbf{0})$ has eigenvalues $\lambda(\mu)$, $\bar{\lambda}(\mu)$ then, generically, $[\mathrm{d}|\lambda(\mu)|/\mathrm{d}\mu]_{\mu=0} \neq 0$ and the stability of the fixed point at $\mathbf{x} = \mathbf{0}$ changes as μ passes through zero. This is the analogue for diffeomorphisms of the situation encountered in the Hopf bifurcation for vector fields (see § 4.2.2).

The typical case of this kind has β irrational, since the rationals have zero measure in $[0, 1)$. It is then not difficult to show that an invariant circle does occur in the local family $\mathbf{f}|U$ in a similar manner to the vector field problem (see § 5.3). However, while for vector fields the dynamic on the invariant circle is topologically trivial, it is not so for the family of maps. Generically, as the invariant circle grows, the rotation number of the diffeomorphism on it passes through both rational and irrational values near β; each rational value occurring for an interval of parameter values. The length of these intervals decreases with the size of the invariant circle so that the creation of the invariant circle itself (i.e. $\mu = 0$) represents an accumulation of bifurcations (from above if the invariant circle occurs for $\mu > 0$). Such complicated behaviour as this has no counterpart in the vector field case and to understand how it arises we must consider *two-parameter* unfoldings of $\mathbf{f}_0$. The relevant feature of two-parameter families of diffeomorphisms on the circle is simply illustrated by the example (see Arnold, 1983, p. 108)

$$f(\alpha, \varepsilon; \theta) = \theta + \alpha + \varepsilon \sin \theta, \tag{5.1.11}$$

$\theta \in [0, 2\pi]$, $\varepsilon \in [0, 1)$, which we examine in detail in § 5.2.

The additional freedom afforded by two-parameter unfoldings also allows us to consider the situation where β is rational, equal to p/q in lowest terms, where $q \geqslant 3$. Normal form calculations (see §§ 5.3 and 5.4) show that resonant terms of the form $\bar{z}^{q-1}$ occur *in addition* to those present when β is irrational. If $q \geqslant 5$ (weak resonance) it can be shown that these additional resonances play a similar, but not determining, role for all q. However, this is no longer true for $q \leqslant 4$ (strong resonance) where the additional resonances lead to bifurcations that are characterised by the value of q.

Following Arnold & Takens (see Arnold, 1983, pp. 292—313; Takens, 1974b) we describe a systematic approach to these resonance phenomena by approximating $\mathbf{f}_\mu^q$, $\mu \in \mathbb{R}$, by the time-2π map of an autonomous vector field. The construction of the approximation is described in § 5.5. This approach has the advantage that it can be extended to include the generic case when $q = 1$ and $q = 2$ (see (ii) above). For these values of q, $\mathbf{Df}_0(\mathbf{0})$ has real eigenvalues and is not a rotation. Bifurcation diagrams for the resulting families of vector fields are presented in § 5.6 and their relation to the corresponding local family $\mathbf{f}|U$ is discussed in § 5.7.

5.2 Arnold's circle map

Consider the following two-parameter family of diffeomorphisms $f_{(\alpha,\varepsilon)}: S^1 \to S^1$,

$$f_{(\alpha,\varepsilon)}(\theta) = \theta + \alpha + \varepsilon \sin \theta, \tag{5.2.1}$$

$\varepsilon \in [0, 1)$, θ, $\alpha \in [0, 2\pi]$, with 0 and 2π identified. For what values of (α, ε) does $f_{(\alpha,\varepsilon)}$ have rotation number p/q in lowest terms, where $p \in \{0, 1, \ldots, q-1\}$, $q \in \mathbb{Z}^+$? It is convenient to measure θ in units of 2π, i.e. $2\pi\theta' = \theta$, so that the iteration $\theta_{n+1} = f_{(\alpha,\varepsilon)}(\theta_n)$ becomes

$$\begin{aligned} \theta'_{n+1} &= \frac{1}{2\pi} f_{(\alpha,\varepsilon)}(2\pi\theta'_n) = f'_{(\alpha',\varepsilon')}(\theta'_n) \\ &= \theta'_n + \alpha' + \varepsilon' \sin 2\pi\theta', \end{aligned} \tag{5.2.2}$$

where $2\pi\alpha' = \alpha$ and $2\pi\varepsilon' = \varepsilon$. Dropping primes, we can write

$$\theta_{n+1} = f_{(\alpha,\varepsilon)}(\theta_n) = \theta_n + \alpha + \varepsilon \sin 2\pi\theta_n, \tag{5.2.3}$$

with α, $\theta \in [0, 1]$ and $\varepsilon \in [0, 1/2\pi)$. Finally, we obtain the lift, $\bar{f}_{(\alpha,\varepsilon)}$, of $f_{(\alpha,\varepsilon)}$ as

$$\bar{f}_{(\alpha,\varepsilon)}(x) = x + \alpha + \varepsilon \sin 2\pi x, \tag{5.2.4}$$

$x \in \mathbb{R}$. This clearly satisfies $\bar{f}_{(\alpha,\varepsilon)}(x + 1) = \bar{f}_{(\alpha,\varepsilon)}(x) + 1$, as required for an orientation-preserving diffeomorphism on S^1 (see § 1.2).

Proposition 5.2.1 *The rotation number $\rho(f_{(\alpha,\varepsilon)}) = p/q$ if and only if*

$$\bar{f}^q_{(\alpha,\varepsilon)}(x) - (x + p) = 0 \tag{5.2.5}$$

for some $x \in \mathbb{R}$.

Proof. Clearly, if (5.2.5) is satisfied for some $x = x_0$ then

$$\bar{f}^{q}_{(\alpha,\varepsilon)}(x_0) = x_0 + p, \tag{5.2.6}$$

and therefore

$$\bar{f}^{nq}_{(\alpha,\varepsilon)}(x_0) = x_0 + np. \tag{5.2.7}$$

Thus

$$\rho(f_{(\alpha,\varepsilon)}) = \operatorname*{Lim}_{n\to\infty} \frac{\bar{f}^{n}_{(\alpha,\varepsilon)}(x_0) - x_0}{n} \bmod 1$$

$$= \operatorname*{Lim}_{n\to\infty} \frac{\bar{f}^{nq}_{(\alpha,\varepsilon)}(x_0) - x_0}{nq} \bmod 1 = p/q. \tag{5.2.8}$$

Conversely, observe that (5.2.4) implies

$$\bar{f}^{q}_{(\alpha,\varepsilon)}(x) = x + q\alpha + F(\alpha, \varepsilon, x). \tag{5.2.9}$$

Let $\alpha = (p/q) + \beta$ to obtain

$$\bar{f}^{q}_{(\alpha,\varepsilon)}(x) = x + p + G_{p/q}(\beta, \varepsilon, x), \tag{5.2.10}$$

where $G_{p/q}(\beta, \varepsilon, x) = q\beta + F((p/q) + \beta, \varepsilon, x)$. Thus if (5.2.5) is *not* satisfied for some $x \in \mathbb{R}$, then

$$G_{p/q}(\beta, \varepsilon, x) \neq 0 \tag{5.2.11}$$

for *all* $x \in \mathbb{R}$. Since $G_{p/q}$ is periodic in x (see Exercise 5.2.1), this means that $G_{p/q}$ is bounded away from zero. It follows that

$$|\rho(f_{(\alpha,\varepsilon)}) - p/q| \geqslant \min_{x\in\mathbb{R}} |q^{-1} G_{p/q}(\beta, \varepsilon, x)| > 0. \tag{5.2.12}$$

Hence, $\rho(f_{(\alpha,\varepsilon)}) = p/q$ if and only if (5.2.5) is satisfied for some $x \in \mathbb{R}$. □

It is not difficult to show that

$$F(\alpha, \varepsilon, x) = \varepsilon \sum_{k=0}^{q-1} \sin[2\pi \bar{f}^{k}_{(\alpha,\varepsilon)}(x)], \tag{5.2.13}$$

and

$$\sin[2\pi \bar{f}^{k}_{(\alpha,\varepsilon)}(x+1)] = \sin[2\pi \bar{f}^{k}_{(\alpha,\varepsilon)}(x)], \tag{5.2.14}$$

$k = 0, 1, \ldots, q-1$, so that G is bounded and attains its maximum and minimum value on $[0, 1]$. Thus, for each ε, there is an interval of β on which $G_{p/q}(\beta, \varepsilon, x) = 0$ for some $x \in [0, 1]$ (see Figure 5.2).

How do the end-points of this interval of β depend on ε? For $q = 1$, $p = 0$ and

$$G_{0/1}(\beta, \varepsilon, x) = \beta + \varepsilon \sin 2\pi x = 0 \tag{5.2.15}$$

has solutions for some $x \in [0, 1]$ provided $\beta \leqslant \pm\varepsilon$. Since 0 and 1 are identified, we conclude that $\rho(f_{(\alpha,\varepsilon)})=0$ for (α, ε) in the linear wedge-shaped regions shown in Figure 5.3(*a*). Given that $q \geqslant 2$, we can approximate the boundary of the region in which $\rho(f_{(\alpha,\varepsilon)})=p/q$ for $\varepsilon \ll 1$. First observe that (5.2.13) implies

$$|F(\alpha, \varepsilon, x)| \leqslant q\varepsilon \tag{5.2.16}$$

for any α, x. Therefore, $|\beta| \leqslant \varepsilon$ is a necessary condition for $G_{p/q}(\beta, \varepsilon, x)$ to be zero for some x. Thus, if ε is small so is β and we can consider the Taylor expansion for $F((p/q)+\beta, \varepsilon, x)$ about $(\beta, \varepsilon)=(0, 0)$. To this end we next observe that, for $q \geqslant 2$,

$$\bar{f}^k_{(\alpha,\varepsilon)}(x)=\begin{cases} x+k\alpha+\varepsilon \sum_{l=0}^{k-1} \sin(2\pi \bar{f}^l_{(\alpha,\varepsilon)}(x)), & k=1,\ldots,q-1; \\ x, & k=0. \end{cases} \tag{5.2.17}$$

Hence, from (5.2.13),

$$\begin{aligned} F(\alpha, \varepsilon, x) &= \varepsilon\left\{\sin 2\pi x+\sum_{k=1}^{q-1} \sin\left[2\pi\left(x+\frac{kp}{q}+k\beta+\varepsilon \sum_{l=0}^{k-1} \sin[2\pi \bar{f}^l_{(\alpha,\varepsilon)}(x)]\right)\right]\right\} \\ &= \varepsilon\left\{\sin 2\pi x+\sum_{k=1}^{q-1} \sin\left[2\pi\left(x+\frac{kp}{q}\right)\right]\right\}+O(\varepsilon^{r+1}\beta^s\colon r+s=1), \end{aligned} \tag{5.2.18}$$

where $\alpha=(p/q)+\beta$. It is easily shown (see Exercise 5.2.2) that

$$\sum_{k=1}^{q-1} \sin\left[2\pi\left(x+\frac{kp}{q}\right)\right]=-\sin 2\pi x; \tag{5.2.19}$$

and we conclude that, for $q \geqslant 2$, $G_{p/q}(\beta, \varepsilon, x)$ takes the form

$$G_{p/q}(\beta, \varepsilon, x)=q\beta+g_0(x)\beta\varepsilon+g_1(x)\varepsilon^2+O(\varepsilon^{r+1}\beta^s\colon r+s=2). \tag{5.2.20}$$

Figure 5.2 Plots of $G_{1/2}(\beta, \varepsilon, x)$ (see (5.2.27)) for $\varepsilon=0.025$ and (*a*) $\beta=\beta_{\mathrm{m}}(\varepsilon)\simeq 0.001$; (*b*) $\beta=0$; (*c*) $\beta=-\beta_{\mathrm{m}}(\varepsilon)$. Observe that $G_{1/2}(\beta, \varepsilon, x)=0$ for some $x \in [0, 1]$ provided $\beta \in [-\beta_{\mathrm{m}}(\varepsilon), \beta_{\mathrm{m}}(\varepsilon)]$.

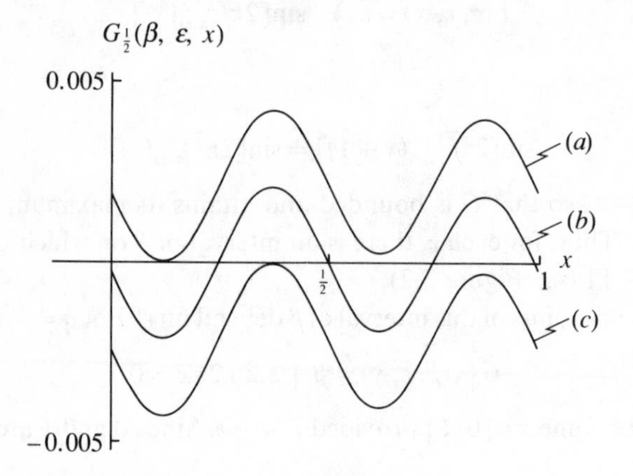

Let $G_{p/q}(\beta, \varepsilon, x) = 0$ define β as a function of ε and x in the form

$$\beta(\varepsilon, x) = \beta_0(x) + \varepsilon\beta_1(x) + \varepsilon^2\beta_2(x) + O(\varepsilon^3). \tag{5.2.21}$$

Substitute (5.2.21) into (5.2.20) and compare coefficients of powers of ε to obtain

$$\beta_0(x) = \beta_1(x) \equiv 0, \qquad \beta_2(x) = -\frac{g_1(x)}{q}. \tag{5.2.22}$$

It follows that $\beta(\varepsilon, x)$ is at least quadratic in ε for all $q \geqslant 2$.

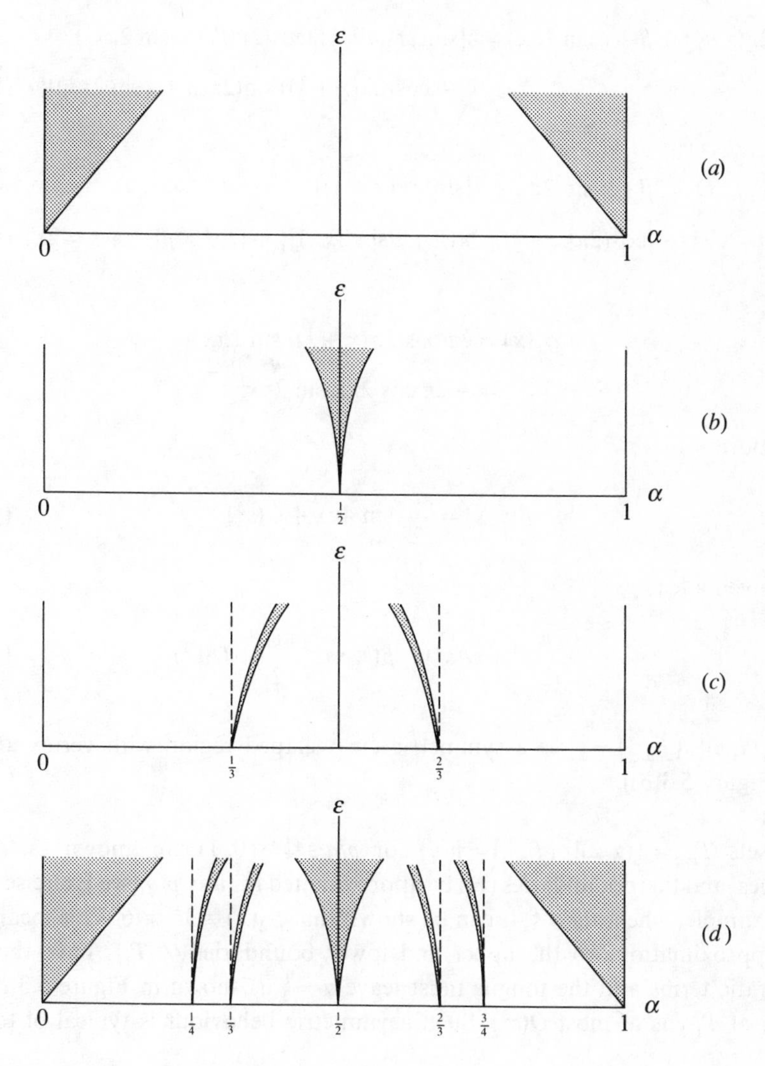

Figure 5.3 Arnold tongues for the circle map (5.2.3): (*a*) $T_{0/1}$; (*b*) $T_{1/2}$; (*c*) $T_{1/3}$ and $T_{2/3}$; (*d*) schematic illustration of $\{T_{p/q} | 1 \leqslant q \leqslant 4\}$. There is a tongue $T_{p/q} = \{(\alpha, \varepsilon) | \rho(f_{(\alpha,\varepsilon)}) = p/q\}$ for every rational number $p/q \in [0, 1)$ but its width decreases rapidly with increasing q. The system of tongues is symmetric about $\alpha = \frac{1}{2}$.

Example 5.2.1 Given that $p/q = \frac{1}{2}$, find $g_1(x)$. Describe the set $\{(\alpha, \varepsilon) | \rho(f_{(\alpha,\varepsilon)}) = \frac{1}{2}\}$ and draw a diagram illustrating this set in the α,ε-plane.

Solution. Observe that

$$\bar{f}^2_{(\alpha,\varepsilon)}(x) = x + 2\alpha + \varepsilon \sin 2\pi x + \varepsilon \sin 2\pi(x + \alpha + \varepsilon \sin 2\pi x). \tag{5.2.23}$$

Let $\alpha = \frac{1}{2} + \beta$ to obtain

$$\bar{f}^2_{(\alpha,\varepsilon)}(x) = x + 1 + 2\beta + \varepsilon \sin 2\pi x + \varepsilon \sin 2\pi(x + \tfrac{1}{2} + \beta + \varepsilon \sin 2\pi x). \tag{5.2.24}$$

Therefore

$$G_{1/2}(\beta, \varepsilon, x) = 2\beta + \varepsilon \sin 2\pi x + \varepsilon \sin 2\pi(x + \tfrac{1}{2} + \beta + \varepsilon \sin 2\pi x). \tag{5.2.25}$$

Now,

$$\begin{aligned}\varepsilon \sin 2\pi(x + \tfrac{1}{2} + \beta + \varepsilon \sin 2\pi x) = {} & \varepsilon\{\sin(2\pi(x + \tfrac{1}{2})) \cos(2\pi(\beta + \varepsilon \sin 2\pi x)) \\ & + \cos(2\pi(x + \tfrac{1}{2})) \sin(2\pi(\beta + \varepsilon \sin 2\pi x))\}\end{aligned} \tag{5.2.26}$$

and

$$\begin{aligned}G_{1/2}(\beta, \varepsilon, x) = {} & 2\beta + \varepsilon \sin 2\pi x + \varepsilon\{\sin(2\pi(x + \tfrac{1}{2})) \\ & + \cos(2\pi(x + \tfrac{1}{2}))[2\pi(\beta + \varepsilon \sin 2\pi x)]\} + O(\varepsilon^{r+1}\beta^s : r + s = 2).\end{aligned} \tag{5.2.27}$$

Thus

$$\begin{aligned}g_1(x) &= 2\pi \cos(2\pi(x + \tfrac{1}{2})) \sin 2\pi x, \\ &= -2\pi \cos 2\pi x \sin 2\pi x.\end{aligned} \tag{5.2.28}$$

Therefore

$$\beta(\varepsilon, x) = \frac{\pi\varepsilon^2}{2} \sin 4\pi x + O(\varepsilon^3). \tag{5.2.29}$$

For given $\varepsilon \ll 1$,

$$-\frac{\pi}{2}\varepsilon^2 + O(\varepsilon^3) \leqslant \beta(\varepsilon, x) \leqslant \frac{\pi\varepsilon^2}{2} + O(\varepsilon^3) \tag{5.2.30}$$

and $\{(\alpha, \varepsilon) | \rho(f_{(\alpha,\varepsilon)}) = \frac{1}{2}\}$ is a symmetric cusp-shaped region with vertex at $\alpha = \frac{1}{2}$ (see Figure 5.3(*b*)). □

The sets $T_{p/q} = \{(\alpha, \varepsilon) | \rho(f_{(\alpha,\varepsilon)}) = p/q\}$ for $p/q \in \mathbb{Q} \cap [0, 1)$ are known as 'Arnold tongues' and their boundaries can be approximated for any p/q (see Exercise 5.2.3). For example, when $p/q = \frac{1}{3}$, it can be shown that $\beta_2(x) \equiv 3^{1/2}\pi/6$. This means that the approximations to the upper and lower boundaries of $T_{1/3}$ have the same quadratic terms and the tongue must leave $\alpha = \frac{1}{3}$ as shown in Figure 5.3(*c*). The width of $T_{1/3}$ is at most $O(\varepsilon^3)$. Such asymmetric behaviour is typical of tongues with $q \geqslant 3$ and the symmetry exhibited by $T_{0/1}$ and $T_{1/2}$ is exceptional. It is also useful to note that the system of tongues is symmetric about $\alpha = \frac{1}{2}$. More precisely, $T_{(1-p/q)}$ is the image of $T_{p/q}$ under reflection in $\alpha = \frac{1}{2}$. This means that $T_{2/3}$ takes the form shown in Figure 5.3(*c*). Similar considerations for $q = 4$ yield the schematic representation of $\{T_{p/q} | 1 \leqslant q \leqslant 4\}$ given in Figure 5.3*d*.

There is a separate tongue for every rational in [0, 1). The greater the value of q the thinner is the tongue, but each one still has a positive width for $\varepsilon > 0$. It follows that the dependence of $\rho(f_{(\alpha,\varepsilon)})$ on α, for fixed $\varepsilon > 0$, is rather subtle. For each rational number, $p/q \in [0, 1)$, there is an interval of values of α for which $\rho(f_{(\alpha,\varepsilon)}) = p/q$. However, the length of the (p/q)-interval diminishes rapidly with increasing q and, in fact, the measure of the totality of tongues for $0 < \varepsilon < \varepsilon_0 \ll (2\pi)^{-1}$, $0 \leqslant \alpha \leqslant 1$ is small compared with ε_0. This means that, on selecting a member of the family at random, it will have an irrational rotation number with probability near one as $\varepsilon \to 0$. This is in sharp contrast to the behaviour of circle diffeomorphisms for which (see Theorem 3.4.1) a rational rotation number is a generic property.

Analogous results to those described above can be shown to hold for any analytic or sufficiently smooth unfolding of a rotation. In particular, this is true for families of the form

$$f_{(\alpha,\varepsilon)}(\theta) = \theta + \alpha + \varepsilon a(\theta), \tag{5.2.31}$$

where $a(\theta)$ is an arbitrary analytic function on S^1.

5.3 Irrational rotations

When $\mathbf{Df}_0(\mathbf{0})$ is an irrational rotation, a straightforward normal form calculation shows that an invariant circle occurs. We have already considered this calculation for $\mu = 0$ in Examples 2.5.2 and 3, where it was shown that $\mathbf{f}_0$ could be transformed into the complex form

$$\tilde{f}_0(z) = \lambda(0)z + a_3 z|z|^2 + O(|z|^5), \tag{5.3.1}$$

where $a_3 = \tilde{a}_{21} \in \mathbb{C}$ in (2.5.31) and $\lambda(0) = \exp(2\pi i\beta)$. Let $\lambda(\mu), \bar{\lambda}(\mu)$ be the eigenvalues of $\mathbf{Df}_\mu(\mathbf{0})$. Then, given that $[(\mathrm{d}|\lambda(\mu)|/\mathrm{d}\mu)]_{\mu=0} \neq 0$, $|\lambda(\mu)| \neq 1$ for $\mu \neq 0$ and the $z|z|^2$-term in (5.3.1) is no longer resonant. Therefore, it could be removed by a suitable transformation. However, this would destroy the continuity of $\tilde{f}(\mu, z)$ in μ. We therefore choose not to remove this term and conclude that $f(\mu, z)$ is equivalent to

$$\tilde{f}_\mu(z) = \lambda(\mu)z\{1 + a(\mu)|z|^2 + R(\mu, z)\}, \tag{5.3.2}$$

where $\lambda(\mu) = |\lambda(\mu)|\exp(2\pi i\beta(\mu))$ and $a(\mu) = a_3(\mu)/\lambda(\mu)$ depends smoothly on μ. In (5.3.2), $R(\mu, z)$ is $O(|z|^4)$.

Let $a(\mu) = c(\mu) + id(\mu)$, then, in the absence of the remainder, $R(\mu, z)$,

$$|\tilde{f}_\mu(z)| = |\lambda(\mu)|\,|z|\,|(1 + a(\mu)|z|^2)|. \tag{5.3.3}$$

Clearly, $\tilde{f}_\mu$ maps the set of circles centred on $z=0$ onto itself, because $|\tilde{f}_\mu(z)|$ is independent of $\arg(z)$. Invariant circles arise if and only if $|\tilde{f}_\mu(z)|=|z|$, i.e. if and only if

$$|\lambda(\mu)|\,|(1+a(\mu)|z|^2)|=1; \tag{5.3.4}$$

or, equivalently, when

$$|a(\mu)|^2|z|^4+2c(\mu)|z|^2+1-\frac{1}{|\lambda(\mu)|^2}=0. \tag{5.3.5}$$

For generic families, $c(0)=c\neq 0$ and $[(\mathrm{d}|\lambda(\mu)|/\mathrm{d}\mu)]_{\mu=0}=b\neq 0$, so that

$$|\lambda(\mu)|=1+b\mu+O(\mu^2), \tag{5.3.6}$$

$$c(\mu)=c+O(\mu). \tag{5.3.7}$$

It can then be shown (see Exercise 5.3.1) that (5.3.5) has a solution of the form

$$|z|^2=r_0(\mu)^2=-\frac{b\mu}{c}+O(\mu^2), \tag{5.3.8}$$

where $b=\frac{1}{2}[\mathrm{d}(\mathrm{Det}\,\mathbf{Df}_\mu(\mathbf{0})/\mathrm{d}\mu]_{\mu=0}$ and $c=\mathrm{Re}(a(0))=\mathrm{Re}(a_3(0)\exp(-2\pi i\beta))$. Furthermore, when $|z|=r_0(\mu)$,

$$\arg(\tilde{f}_\mu(z))-\arg(z)=2\pi\beta(\mu)+\arg(1+a(\mu)r_0(\mu)^2),$$

$$=2\pi\beta+O(\mu), \tag{5.3.9}$$

where $\beta(0)=\beta\in\mathbb{R}\backslash\mathbb{Q}$. Therefore, given that $(b/c)<0$ (>0), (5.3.8) implies the existence of an invariant circle for $\mu>0$ (<0). The radius of this circle is $O(\mu^{1/2})$ and it is stable, surrounding an unstable fixed point at $z=0$ (see Exercise 5.3.2). Equation (5.3.9) shows that on the invariant circle the mapping reduces to a rotation, because $\arg(\tilde{f}_\mu(z))-\arg(z)$ is independent of $\arg(z)$.

The phenomena described above essentially parallel those occurring at a supercritical Hopf bifurcation in a family of vector fields (see §4.2.2). However, this parallel cannot be maintained when $R(\mu, z)$ is present in (5.3.2). Although an invariant circle still exists and its stability is preserved (see Whitley, 1983), the dynamic on it need no longer be a pure rotation. To understand how this difference arises, let us return to the normal form calculations. It can be shown (see Exercise 5.3.3) that $\mathbf{f}(\mu, \mathbf{x})$ is equivalent to

$$\tilde{f}(\mu, z)=\lambda(\mu)z\{1+A_N(\mu,|z|^2)+R_N(\mu, z)\}, \tag{5.3.10}$$

where $A_N(\mu,|z|^2)=\sum_{m=1}^{N}(a_{2m+1}(\mu)/\lambda(\mu))|z|^{2m}$ and $R_N(\mu, z)$ is $O(|z|^{2N+2})$, $N\in\mathbb{Z}^+$. In the absence of $R_N(\mu, z)$,

$$|\tilde{f}_\mu(z)|=|\lambda(\mu)|\,|z|\,|(1+A_N(\mu,|z|^2))| \tag{5.3.11}$$

and

$$\arg(\tilde{f}_\mu(z)) - \arg(z) = 2\pi\beta(\mu) + \arg(1 + A_N(\mu, |z|^2)). \qquad (5.3.12)$$

Once again (5.3.11) implies the existence of an invariant circle of radius $r_0(\mu)$ (see Exercise 5.3.3), while (5.3.12), with $|z|^2 = r_0(\mu)^2$, shows that $\tilde{f}_\mu$ restricted to the invariant circle again reduces to a rotation. Thus the difference we are looking for cannot be attributed to the inevitable resonance terms of any finite order. Even if $\lim_{N\to\infty} A_N(\mu, |z|^2)$ exists, there may be a part of $\tilde{f}_\mu(z)$ which cannot be described by a power series. This '$O(\infty)$-part', which is omitted from the normal form calculations, will, in general, depend on z and not $|z|^2$.

When $R_N(\mu, z)$ is present, the task of demonstrating the existence of an invariant circle is more difficult. An outline of a proof of existence is given in Whitley (1983). There are two important points for us to note here. First, the invariant circle is 'topological' rather than 'geometrical' (although it is obtained as a perturbation of the geometrical circle $|z| = r_0(\mu)$). Second, it may no longer be differentiable. Let C denote the invariant circle and $r_0(\mu, 2\pi\theta)$ be the radial distance from $z = 0$ to C in the direction defined by θ. Then $C = \{z \mid |z| = r_0(\mu, 2\pi\theta), 0 \leqslant \theta < 1\}$. Now take arguments on both sides of (5.3.10) and define $f^C_\mu(\theta)$ as follows:

$$\arg(\tilde{f}_\mu(r_0(\mu, 2\pi\theta)\exp(2\pi i\theta)) = 2\pi f^C_\mu(\theta) = 2\pi\theta + 2\pi\beta(\mu)$$
$$+ \arg\{1 + A_N(\mu, r_0(\mu, 2\pi\theta)^2) + R_N(\mu, r_0(\mu, 2\pi\theta)\exp(2\pi i\theta))\}. \qquad (5.3.13)$$

Thus

$$f^C_\mu(\theta) = \theta + \beta + \Phi(\mu, \theta), \qquad (5.3.14)$$

where $\beta(0) = \beta \in \mathbb{R}\backslash\mathbb{Q}$ and $\theta \in [0, 1)$. We conclude that the restriction, f^C_μ, of $\tilde{f}_\mu$ to C is a diffeomorphism of the circle that is no longer necessarily reducible to a rotation.

We have already discussed the typical behaviour of diffeomorphisms on S^1 in §§ 1.2 and 3.4. Recall that if the rotation number, $\rho(f^C_\mu)$, is rational then f^C_μ will, in general, have an even number of periodic points, alternately stable and unstable, arranged around the circle (see Figure 1.23). These points will be alternately node- and saddle-type periodic points of $\tilde{\mathbf{f}}_\mu: \mathbb{R}^2 \to \mathbb{R}^2$ (see Figure 5.4). If $\rho(f^C_\mu)$ is irrational then (see Proposition 1.5.1) f^C_μ has no periodic points. Moreover, if f^C_μ is twice differentiable then Denjoy's Theorem implies that it is topologically conjugate to an irrational rotation by $\rho(f^C_\mu)$.

In § 5.1, we noted that, for the generic case, $\rho(f^C_\mu)$ passes through both rational and irrational values as μ varies. To understand the nature of the dependence of $\rho(f^C_\mu)$ on μ, we must consider two-parameter unfoldings of $\mathbf{f}_0$. Let $\mathbf{f}(\boldsymbol{\mu}, \mathbf{x})$, $\boldsymbol{\mu} = (\mu_1, \mu_2) \in \mathbb{R}^2$, be a smooth two-parameter family with a fixed point at $\mathbf{x} = \mathbf{0}$ for all $\boldsymbol{\mu}$ near $\mathbf{0}$ and such that $\mathbf{Df_0(0)}$ is given by (5.1.10). Since the parameters play a passive role in the normal form calculations, (5.3.10) generalises to

$$\tilde{f}(\boldsymbol{\mu}, z) = \lambda(\boldsymbol{\mu})z\left\{1 + \sum_{m=1}^{N} \tilde{a}_{2m+1}(\boldsymbol{\mu})|z|^{2m} + R_N(\boldsymbol{\mu}, z)\right\} \tag{5.3.15}$$

where $\tilde{a}_{2m+1}(\boldsymbol{\mu}) = a_{2m+1}(\boldsymbol{\mu})/\lambda(\boldsymbol{\mu})$. With the notation defined in Figure 5.5, we

Figure 5.4 Local behaviour of a planar diffeomorphism in the neighbourhood of an attracting invariant circle with rational rotation number. Circle diffeomorphisms with rational rotation number generically have an even number of alternately stable and unstable periodic points. These points become node/foci and saddle points, respectively, in the planar map. Note that if the invariant circle is repelling, the saddle points form at the stable, rather than the unstable, points of the restriction to the circle.

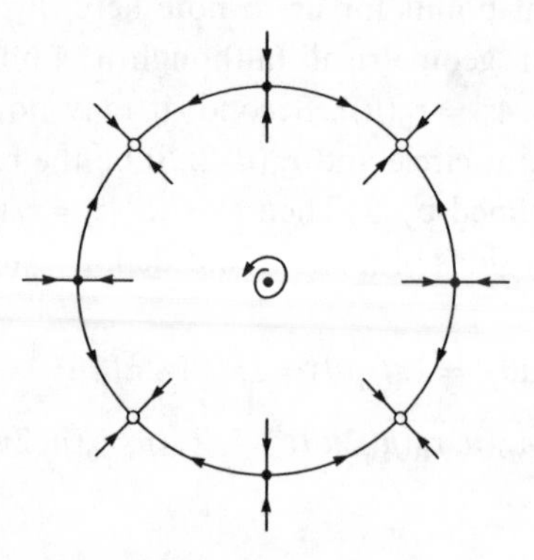

Figure 5.5 Diagram illustrating the reparametrisation that transforms (5.3.15) into (5.3.17). Conditions for the map $\lambda: U \to V$ to be a local diffeomorphism are considered in Exercise 5.3.4.

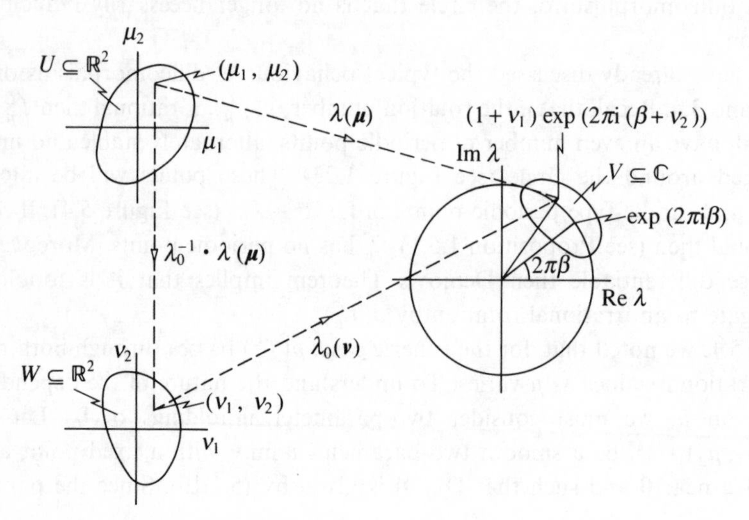

observe that the map taking $(\mu_1, \mu_2) \in U \subseteq \mathbb{R}^2$ to $\lambda(\mu_1, \mu_2) \in V \subseteq \mathbb{C}$ is generically a local diffeomorphism. Convenient coordinates for $\lambda \in V$ are (v_1, v_2) defined by

$$v_1 = |\lambda| - 1, \; 2\pi v_2 = \arg(\lambda) - 2\pi\beta. \tag{5.3.16}$$

It follows that there is a non-singular change of parameters $(\mu_1, \mu_2) \mapsto (v_1, v_2)$ such that

$$\tilde{f}_{\mathbf{v}}(z) = (1 + v_1)\exp[2\pi i(\beta + v_2)]z\left\{1 + \sum_{m=1}^{N} \tilde{a}_{2m+1}(\mathbf{v})|z|^{2m} + R_N(\mathbf{v}, z)\right\}. \tag{5.3.17}$$

This parametrisation is particularly useful. For example, when $N = 1$ and $R_1(\mathbf{v}, z) = O(|z|^4)$ is neglected, (5.3.17) implies the existence of an invariant circle C given by

$$r_0(\mathbf{v})^2 = -\frac{v_1}{\operatorname{Re}\tilde{a}_3(\mathbf{0})} + O(|\mathbf{v}|^2). \tag{5.3.18}$$

Moreover, in the same approximation,

$$\begin{aligned} f_{\mathbf{v}}^C(\theta) &= \frac{1}{2\pi}\arg(\tilde{f}_{\mathbf{v}}(r_0(\mathbf{v})\exp(2\pi i\theta))) \\ &= \theta + \beta + v_2 - \frac{v_1}{2\pi}\frac{\operatorname{Im}\tilde{a}_3(\mathbf{0})}{\operatorname{Re}\tilde{a}_3(\mathbf{0})} + O(|\mathbf{v}|^2). \end{aligned} \tag{5.3.19}$$

Thus, v_1 primarily determines the size of the invariant circle, while v_2, for given v_1, controls the rotation number of the restriction $f_{\mathbf{v}}^C$. When $R_1(\mathbf{v}, z)$ is included in (5.3.17), an invariant circle still exists for $v_1 > 0$ (<0) given that $\operatorname{Re}\tilde{a}_3(\mathbf{0}) < 0$ (>0) (see Whitley, 1983, p. 194). As in the one-parameter case (see (5.3.14)), $r_0(\mathbf{v})$ is replaced by $r_0(\mathbf{v}, 2\pi\theta)$ and (5.3.19) is replaced by

$$f_{\mathbf{v}}^C(\theta) = \theta + \beta + v_2 + \Phi(\mathbf{v}, \theta). \tag{5.3.20}$$

Observe that when the invariant circle is small

$$\Phi(\mathbf{v}, \theta) = \frac{1}{2\pi}\operatorname{Im}\tilde{a}_3(\mathbf{0})r_0(\mathbf{v}, 2\pi\theta)^2 + O(r_0(\mathbf{v}, 2\pi\theta)^4), \tag{5.3.21}$$

and $r_0(\mathbf{v}, 2\pi\theta) \simeq r_0(\mathbf{v})$ for all θ. Thus (5.3.18) implies $\Phi(\mathbf{v}, \theta)$ is small if v_1 is small. With this property of $\Phi(\mathbf{v}, \theta)$ in mind let us compare (5.3.20) with Arnold's circle map (5.2.3), identifying $\beta + v_2$ with α and v_1 with ε.

If we assume that $f_{\mathbf{v}}^C$ exhibits analogous behaviour to $f_{(\alpha,\varepsilon)}$, then we expect to find a tongue, $T_{p/q}$, associated with each rational p/q near β. If $\mathbf{v} \in T_{p/q}$ then $\rho(f_{\mathbf{v}}^C) = p/q$. Figure 5.6 illustrates this state of affairs in the complete λ-plane. Of course, there are infinitely many such tongues emanating from any arc on the unit circle containing $\exp(2\pi i\beta)$. To explain the behaviour of $\rho(f_\mu^C)$ as a function of μ for our original one-parameter family $\mathbf{f}(\mu, \mathbf{x})$, it only remains to observe that the family f_μ^C, $\mu \in \mathbb{R}$, is generically represented by a curve, $\mathscr{C}$, in the λ-plane passing transversely through $\lambda = \exp(2\pi i\beta)$ when $\mu = 0$ (see Figure 5.6). For the supercritical case, the curve $\mathscr{C}$ lies inside the unit circle for $\mu < 0$ and outside it for $\mu > 0$. Thus, for $\mu > 0$, $\rho(f_\mu^C)$ takes the rational value p/q near β on an interval of μ corresponding to the intersection of $\mathscr{C}$ with the tongue $T_{p/q}$. The cusp-like shape of the tongues means that the length of this interval in μ decreases as μ approaches zero from above. Furthermore, there are infinitely many irrational numbers between any two rational numbers, so as $\mathscr{C}$ passes from one tongue to another $\rho(f_\mu^C)$ takes irrational values also. It is now clear that $\mu = 0$ marks the onset of a multitude of bifurcations in the family $\mathbf{f}(\mu, \mathbf{x})$.

The above explanation of the dependence of $\rho(f_\mu^C)$ on μ is based on the assumption that f_ν^C behaves in a similar manner to $f_{(\alpha,\varepsilon)}$. To prove this we must investigate two-parameter unfoldings of rational rotations and show that there is a tongue $T_{p/q}$ emanating from $\lambda = \exp(2\pi i p/q)$.

5.4 Rational rotations and weak resonance

If $\beta \in [0, 1)$ is rational, equal to p/q in lowest terms, then $q \in \mathbb{Z}^+$ and $p \in \{0, 1, \ldots, q-1\}$. Since $\beta = 0$ and $\beta = \frac{1}{2}$ are to be excluded (see (5.1.6)), we will confine our discussion to $q \geqslant 3$ and $p > 0$. When $\mathbf{Df}_0(\mathbf{0})$ is a rational rotation satisfying these requirements, additional resonance terms appear in the normal form calculations. Recall from Example 2.5.3 that the resonance condition for terms of order r is

$$\lambda^{m_1}\bar{\lambda}^{m_2} - \lambda = 0, \tag{5.4.1}$$

where $\lambda = \lambda(0) = \exp(2\pi i\beta)$ and $m_1 + m_2 = r$. The inevitable resonances (which

Figure 5.6 The conjecture that f_ν^C behaves in the same way as $f_{(\alpha,\varepsilon)}$ leads to Arnold tongues at each point of the unit circle in the complex λ-plane for which $\beta + \nu_2$ is rational. Note that, generically, the curve $\mathscr{C}$ crosses the tongues near β transversely. This accounts for the dynamics on the invariant circle arising in the one-parameter family of planar diffeomorphisms represented by the curve $\mathscr{C}$.

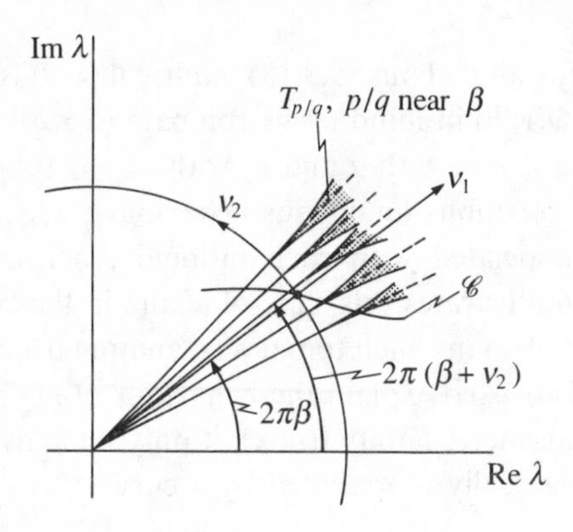

occur whether or not β is rational) correspond to $m_1 = m + 1$, $m_2 = m$ with $m = (r-1)/2$, $r \geqslant 3$ and odd. However, if $\lambda = \exp(2\pi i p/q)$ then $\lambda^q = \bar{\lambda}^q = 1$ and (5.4.1) is satisfied if

$$m_1 - m_2 - 1 = lq, \tag{5.4.2}$$

$l \in \mathbb{Z}$. We see that additional resonances occur for:

$$m_1 = m + lq + 1, \qquad m_2 = m, \qquad l > 0; \tag{5.4.3}$$

$$m_1 = m, \qquad m_2 = m - lq - 1, \qquad l < 0; \tag{5.4.4}$$

with $m \geqslant 0$ and $m_1 + m_2 = r$. It can be shown that the lowest order additional resonance term takes the form $\bar{z}^{q-1}$. Moreover, it follows (see Exercise 5.4.1) that when $\beta = p/q \in \mathbb{Q} \cap (0, 1)$, (5.3.17) can be replaced by

$$\tilde{f}(\mathbf{v}, z) = \lambda(\mathbf{v})z + a_3(\mathbf{v})z^2\bar{z} + b(\mathbf{v})\bar{z}^2 + O(|z|^4), \tag{5.4.5}$$

for $q = 3$, and

$$\tilde{f}(\mathbf{v}, z) = \lambda(\mathbf{v})z + \sum_{m=1}^{[\frac{1}{2}(q-2)]} a_{2m+1}(\mathbf{v})z^{m+1}\bar{z}^m + b(\mathbf{v})\bar{z}^{q-1} + O(|z|^q) \tag{5.4.6}$$

for $q \geqslant 4$. In these equations $\lambda(\mathbf{v}) = (1 + v_1)\exp\{2\pi i((p/q) + v_2)\}$ and $[\cdot]$ denotes the integer part of $\cdot$.

The lowest order inevitable resonance term is given by $z|z|^2 = O(|z|^3)$ and therefore we might hope that the new resonant terms do not play a significant role for $q \geqslant 5$. That this is the case is the content of the following theorem, the statement of which involves the polar form of (5.4.6). Let $z = r\exp(2\pi i\theta)$ and $\tilde{f}_{\mathbf{v}}(z) = R\exp(2\pi i\Theta)$, then it can be shown (see Exercise 5.4.2) that

$$R = (1 + v_1)r + \sum_{m=1}^{[\frac{1}{2}(q-2)]} c_{2m+1}(\mathbf{v})r^{2m+1} + \tilde{c}_{q-1}(\mathbf{v}, \theta)r^{q-1} + O(r^q), \tag{5.4.7}$$

$$\Theta = \theta + \frac{p}{q} + v_2 + \sum_{m=1}^{[\frac{1}{2}(q-2)]} d_{2m}(\mathbf{v})r^{2m} + \tilde{d}_{q-2}(\mathbf{v}, \theta)r^{q-2} + O(r^{q-1}). \tag{5.4.8}$$

Here c_{2m+1}, d_{2m}, $\tilde{c}_{q-1}$ and $\tilde{d}_{q-2}$ are smooth functions of v_1 and v_2 near $\mathbf{v} = \mathbf{0}$. Moreover, $\tilde{c}_{q-1}$, $\tilde{d}_{q-2}$ take the form

$$A(\mathbf{v})\cos(2\pi q\theta) + B(\mathbf{v})\sin(2\pi q\theta). \tag{5.4.9}$$

Theorem 5.4.1 *Consider the family (5.4.6) for $q \geqslant 5$. For all sufficiently small $\mathbf{v}$ with $|\lambda(\mathbf{v})| > 1$ $(|\lambda(\mathbf{v})| < 1)$, $\tilde{f}_{\mathbf{v}}$ has an attracting (repelling) invariant circle if $c_3 = c_3(0) < 0$ $(c_3 > 0)$.*

Furthermore, if $b = b(\mathbf{0}) \neq 0$ and $d_2 = d_2(\mathbf{0}) \neq 0$, then the map has two orbits of period-q on the circle, one stable and the other unstable, for values of $\mathbf{v}$ lying within *a tongue in the parameter plane with boundaries given by*

$$\nu_2 \simeq \frac{d_2}{c_3}\nu_1 \pm \frac{|b|}{2\pi|c_3|^{\frac{1}{2}(q-2)}}\nu_1^{\frac{1}{2}(q-2)}. \tag{5.4.10}$$

The situation considered in Theorem 5.4.1 (i.e. where $q \geqslant 5$ and the inevitable resonances dominate) is known as *weak resonance*. Indeed, Theorem 5.4.1 is often referred to as the 'Weak Resonance Theorem'. For *strong resonance*, when $q \leqslant 4$, it can be shown that bifurcations characterised by the value of q occur. We will return to strong resonance in §5.7 when we have developed techniques which allow us to justify this statement.

Details of the proof of Theorem 5.4.1 would not be appropriate here. However, given the existence of an invariant circle, C, the possibility of a tongue of $\boldsymbol{\nu}$-values for which $\rho(f_{\boldsymbol{\nu}}^C) = p/q$ is apparent. For example, if it is assumed that C is a geometrical circle of radius $(-\nu_1/c_3)^{1/2}$ and the $O(r^{q-1})$-terms in (5.4.8) are neglected, then the resulting approximation to $f_{\boldsymbol{\nu}}^C$ can be shown to have rotation number p/q if and only if $\boldsymbol{\nu}$ belongs to a tongue of the form given in (5.4.10) (see Exercise 5.4.3(a)). However, a proper treatment of the problem requires a more sophisticated approach and a number of carefully chosen changes of variable. The interested reader should consult Whitley (1983, pp. 203–4).

The content of Theorem 5.4.1 is represented diagrammatically in Figure 5.7. Weak resonance differs from the case where β is irrational in the appearance of the resonance tongue (region (*c*) in Figure 5.7) where the rotation number of the restriction to the invariant circle is known to be p/q. As q increases, the tongue narrows (cf Exercise 5.4.3(b)) and for β irrational can be thought of as degenerating to a line.

Figure 5.7 needs careful interpretation. In particular, it would be incorrect to assume that all members of the family $\mathbf{f}(\boldsymbol{\nu}, \mathbf{x})$ with $\boldsymbol{\nu}$ belonging to a given region (*a*), (*b*), (*c*) or (*d*) are topologically conjugate to one another. Indeed, as Figure 5.6 suggests, the rotation number of $f_{\boldsymbol{\nu}}^C$ for $\boldsymbol{\nu} \in (b)$ or (*d*) will be determined by the distribution of tongues for rationals near p/q. Equally, within the region (*c*), the existence of the stable and unstable q-cycles is not sufficient to ensure that all maps in (*c*) are topologically conjugate to one another. Thus Figure 5.7 presents only a small subset of bifurcations occurring in the family. This is in contrast to the bifurcation diagrams for vector fields in Chapter 4, where family members in regions like (*a*)–(*d*) were all topologically equivalent. This state of affairs reflects the fact that, for these vector fields, the normal form determines the type up to topological equivalence. However, the normal forms of the above diffeomorphisms do not determine their type up to topological conjugacy. This is even true of their restrictions to the invariant circle.

Theorem 5.4.1 confirms the conjectures made at the end of §5.3 for generic, one-parameter families. If we focus attention on the existence of the invariant circle, then we can state a Hopf bifurcation theorem for maps on $\mathbb{R}^2$.

Theorem 5.4.2 *Let $\mathbf{f}(\mu, \mathbf{x})$ be a one-parameter family of maps of the plane satisfying:*

(a) $\mathbf{f}_\mu(\mathbf{0}) = \mathbf{0}$ *for μ near 0;*
(b) $\mathrm{D}\mathbf{f}_\mu(\mathbf{0})$ *has two non-real eigenvalues $\lambda(\mu)$ and $\bar{\lambda}(\mu)$ for μ near 0 with $|\lambda(0)| = 1$;*
(c) $[\mathrm{d}|\lambda(\mu)|/\mathrm{d}\mu]_{\mu=0} > 0$;
(d) $\lambda = \lambda(0)$ *is not a qth root of unity for $q = 1, 2, 3$ or 4.*

Then there is a smooth μ-dependent change of coordinates such that

$$\mathbf{f}_\mu(\mathbf{x}) = \mathbf{g}_\mu(\mathbf{x}) + O(|\mathbf{x}|^5), \tag{5.4.11}$$

where, in polar coordinates,

$$\mathbf{g}_\mu(r, \theta) = (|\lambda(\mu)|r + c_3(\mu)r^3, \theta + \beta(\mu) + d_2(\mu)r^2) \tag{5.4.12}$$

and c_3, d_2, β are smooth functions of μ.

If $c_3(0) < 0$ ($c_3(0) > 0$) then, for $\mu < 0$ ($\mu > 0$), the origin is stable (unstable) and, for $\mu > 0$ ($\mu < 0$), the origin is unstable (stable) and surrounded by an attracting (repelling) invariant circle. When $c_3(0) < 0$ ($c_3(0) > 0$) the bifurcation at $\mu = 0$ is said to be supercritical (subcritical).

Figure 5.7 Diagrammatic representation of the results given in Theorem 5.4.1. While an invariant circle exists for all $\nu_1 > 0$, the restriction of (5.4.7, 8) to it is only shown to have q-cycles ($q \geqslant 5$) in the resonance tongue, (c), defined by (5.4.10). That the rotation number of the restriction is then p/q, follows from (5.4.8) (see Exercise 5.4.3). For $\nu_1 < 0$, (5.4.7) implies that the origin is a stable fixed point.

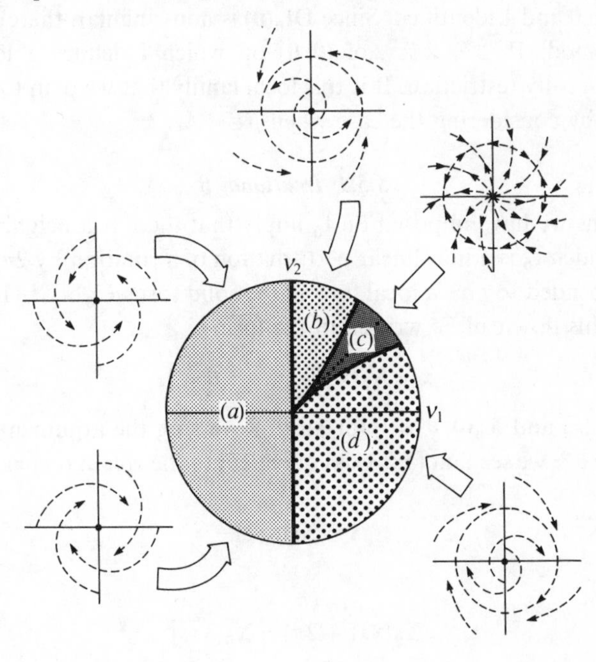

Theorem 5.4.2 gives a practical procedure for checking for the occurrence of an invariant circle in a family of planar maps, just as Theorem 4.2.1 did for vector fields. This parallel can be extended to include an expression for $c_3(0)$ in terms of the Taylor expansion of $\mathbf{f}_0$. By keeping track of the changes incurred in the removal of quadratic terms in Example 2.5.3, it can be shown (see Iooss, 1979, p. 30) that, if

$$f_0(z) = \lambda z + \sum_{r=2,3} \sum_{m_1+m_2=r} a_{m_1m_2} z^{m_1} \bar{z}^{m_2} + O(|z|^4), \tag{5.4.13}$$

then

$$c_3(0) = \operatorname{Re}(\bar{\lambda} a_{21}) - |a_{02}|^2 - \tfrac{1}{2}|a_{11}|^2 - \operatorname{Re}\left[\frac{(1-2\lambda)\bar{\lambda}^2}{(1-\lambda)} a_{11} a_{20}\right]. \tag{5.4.14}$$

This expression for $c_3(0)$ is the analogue of (4.2.24) in the vector field problem.

Theorem 5.4.2 and Theorem 4.2.1 exemplify a connection between the bifurcations given by normal form calculations for diffeomorphisms and corresponding vector field phenomena. This is not coincidental. In fact, any smooth family of maps containing $\mathbf{f}_0$ can be approximated locally by a family involving the time-2π maps of certain autonomous planar vector fields. The origin of these 'vector field approximations' is explained in the next section.

5.5 Vector field approximations

Let $\mathbf{f}\colon \mathbb{R}^l \times \mathbb{R}^2 \to \mathbb{R}^2$, $(\boldsymbol{\mu}, \mathbf{x}) \to \mathbf{f}(\boldsymbol{\mu}, \mathbf{x})$, be a smooth l-parameter family of maps such that $\mathbf{f}(\mathbf{0}, \mathbf{0}) = \mathbf{f}_0(\mathbf{0}) = \mathbf{0}$ and $D_{\mathbf{x}}\mathbf{f}(\mathbf{0}, \mathbf{0}) = D\mathbf{f}_0(\mathbf{0})$ has eigenvalues given by $\exp(\pm 2\pi i \beta)$, $\beta \in [0, 1]$ with 0 and 1 identified. Since $D\mathbf{f}_0(\mathbf{0})$ is non-singular, there is, generically, a neighbourhood, $W \subseteq \mathbb{R}^l \times \mathbb{R}^2$, of $(\mathbf{0}, \mathbf{0})$ on which $\mathbf{f}$ defines a local family of diffeomorphisms by restriction. It is this local family that we plan to approximate. Let us begin by considering the case when $\beta \in \mathbb{R} \backslash \mathbb{Q}$.

5.5.1 *Irrational* β

The conditions we have imposed on $\mathbf{f}_0$ imply that there is a neighbourhood, U_0, of $\mathbf{x} = \mathbf{0}$ on which $\mathbf{f}_0$ is a non-linear perturbation of a rotation by $2\pi\beta$. This means $\mathbf{f}_0$ can be suspended to give a local flow in the solid torus $U_0 \times S^1$. The differential equation of this flow can be written in the form

$$\dot{\mathbf{x}} = \mathbf{X}_0(\mathbf{x}, \theta), \qquad \dot{\theta} = 1, \tag{5.5.1}$$

where $\theta \in [0, 2\pi)$ and $\mathbf{X}_0(\mathbf{0}, \theta) = 0$ for all θ. Reversing the arguments given at the beginning of § 1.8, we see that (5.5.1) is equivalent to the *non-autonomous* differential equation

$$\dot{\mathbf{x}} = \mathbf{X}_0(\mathbf{x}, t), \tag{5.5.2}$$

where

$$\mathbf{X}_0(\mathbf{x}, t + 2\pi) = \mathbf{X}_0(\mathbf{x}, t) \tag{5.5.3}$$

and $\mathbf{X_0}(\mathbf{0}, t) = \mathbf{0}$ for all t. The period advance map for the periodic, non-autonomous system (5.5.2) is the Poincaré map of the flow (5.5.1) and this is $\mathbf{f_0}$ by construction.

We have already discussed normal form calculations for systems of the form (5.5.2) subject to (5.5.3) in Example 2.6.1. Recall that for irrational β, (5.5.2) can be written in the complex form

$$\dot{z} = \mathrm{i}\beta z + \sum_{s=1}^{k-1} c_s z|z|^{2s} + G(z, t), \tag{5.5.4}$$

where $G(z, t) = O(|z|^{2k+1})$ is 2π-periodic in t. Of course, (5.5.4) is equivalent to the autonomous system

$$\dot{z} = Z_0(z, \theta), \qquad \dot{\theta} = 1, \tag{5.5.5}$$

where $Z_0(z, \theta)$ is given by the right hand side of (5.5.4) with t replaced by θ. All the changes of variable used to obtain (5.5.5) from (5.5.2) can be interpreted as diffeomorphisms in the phase space of (5.5.1) (see Exercise 5.5.1). It follows that (5.5.1) and (5.5.5) are differentiably conjugate and consequently that $\mathbf{f_0}$ is differentiably conjugate to the Poincaré map, $\mathbf{P_0}$, obtained from (5.5.5) (see Exercise 2.6.7).

Now consider the truncated system

$$\dot{z} = \tilde{Z}_0(z), \qquad \dot{\theta} = 1, \tag{5.5.6}$$

where

$$\tilde{Z}_0(z) = \mathrm{i}\beta z + \sum_{s=1}^{k-1} c_s z|z|^{2s}. \tag{5.5.7}$$

Since z and θ are decoupled in (5.5.6), the Poincaré map, $\tilde{P}_0$, of this system, defined on the section $\theta = 0$, is given by the time-2π map, $\varphi_{2\pi}$, of the autonomous complex flow $\dot{z} = \tilde{Z}_0(z)$. Of course, $\tilde{P}_0$ differs from P_0 only in terms $O(|z|^{2k+1})$, where k is a positive integer. Thus P_0 can be approximated, up to terms of arbitrarily large but finite degree, by the time-2π map of the planar vector field $\tilde{Z}_0(z)$.

If we now consider $\boldsymbol{\mu} \neq \mathbf{0}$ and repeat the normal form calculations of Example 2.6.1 eliminating only those terms that are non-resonant for $\boldsymbol{\mu} = \mathbf{0}$ (see § 5.3); we obtain a smooth family of systems analogous to (5.5.5). For example, given that $\mathbf{f}_{\boldsymbol{\mu}}(\mathbf{0}) = \mathbf{0}$ for all $\boldsymbol{\mu}$ near $\mathbf{0}$, we obtain

$$\begin{aligned} \dot{z} &= Z_{\boldsymbol{\mu}}(z, \theta) = \omega(\boldsymbol{\mu}) z + \sum_{s=1}^{k-1} c_s(\boldsymbol{\mu}) z|z|^{2s} + G(\boldsymbol{\mu}, z, \theta), \\ \dot{\theta} &= 1, \end{aligned} \tag{5.5.8}$$

where $\omega(\mathbf{0}) = \mathrm{i}\beta$ and $G(\boldsymbol{\mu}, z, \theta)$ is $O(|z|^{2k+1})$. It follows that $\mathbf{f}_{\boldsymbol{\mu}}$ is differentiably conjugate to the Poincaré map, $\mathbf{P}_{\boldsymbol{\mu}}$, deduced from (5.5.8), the complex form of which is approximated by $\tilde{P}_{\boldsymbol{\mu}} = \varphi_{2\pi}^{\boldsymbol{\mu}}$, the time-$2\pi$ map of the complex vector field

$$\tilde{Z}_{\boldsymbol{\mu}}(z) = \omega(\boldsymbol{\mu}) z + \sum_{s=1}^{k-1} c_s(\boldsymbol{\mu}) z|z|^{2s}. \tag{5.5.9}$$

To summarise, this procedure provides a smooth, local family $\mathbf{P}: \mathbb{R}^l \times \mathbb{R}^2 \to \mathbb{R}^2$ equivalent to $\mathbf{f}$ which is approximated by the family $\tilde{\mathbf{P}}: \mathbb{R}^l \times \mathbb{R}^2 \to \mathbb{R}^2$ of time-2π maps given by (5.5.9). The extent to which $\tilde{\mathbf{P}}$ reflects the bifurcational behaviour of $\mathbf{P}$ (and hence $\mathbf{f}$) will be discussed in § 5.7.

5.5.2 *Rational* $\beta = p/q$, $q \geqslant 3$

For β of this type, $\mathbf{Df_0(0)}$ is still a non-trivial rotation and $\mathbf{f_0}$ once again suspends to give a flow on a solid torus. The construction of $\tilde{Z}_0(z)$ proceeds in the same way as in the β-irrational case, the only difference being in the complex normal form that is obtained. Instead of (5.5.4),

$$\dot{z} = \mathrm{i}\beta z + \sum_{s=1}^{S} c_s z|z|^{2s} + d\bar{z}^{(q-1)} \exp(\mathrm{i}pt) + G(z, t), \tag{5.5.10}$$

where $S = \max\{1, [(q-2)/2]\}$ and

$$G(z, t) = \begin{cases} O(|z|^4), & q = 3, \quad (5.5.11) \\ O(|z|^q), & q \geqslant 4, \quad (5.5.12) \end{cases}$$

is 2π-periodic in t (see Example 2.6.1). Unfortunately, the additional resonance term is time-dependent and a further change of variable is required in order to obtain an autonomous vector field. Let

$$z = \exp(\mathrm{i}\beta t)\zeta, \tag{5.5.13}$$

then

$$\begin{aligned} \dot{\zeta} &= -\mathrm{i}\beta\zeta + \exp(-\mathrm{i}\beta t)\dot{z}, \\ &= \sum_{s=1}^{S} c_s\zeta|\zeta|^{2s} + d\bar{\zeta}^{q-1} + G(\exp(\mathrm{i}\beta t)\zeta, t), \end{aligned} \tag{5.5.14}$$

since $\exp(-\mathrm{i}\beta t)\bar{z}^{q-1} = \bar{\zeta}^{q-1} \exp(-\mathrm{i}q\beta t) = \bar{\zeta}^{q-1} \exp(-\mathrm{i}pt)$. Thus (5.5.13) removes the time dependence of the additional resonance term but it is not 2π-periodic in t as were the normal form transformations. In fact, it is $2\pi q$-periodic in t and consequently so is (5.5.14). Notice,

$$G(\exp[\mathrm{i}\beta(t + 2\pi q)]\zeta, t + 2\pi q) = G(\exp(\mathrm{i}\beta t)\zeta, t) \tag{5.5.15}$$

since $\beta = p/q$ and $G(z, t)$ is 2π-periodic in t. Therefore $G(\exp(\mathrm{i}\beta t)\zeta, t)$ is $2\pi q$-periodic in t.

The transformation (5.5.13) also eliminates the linear z-dependence from (5.5.10). This is not surprising because (5.5.13) is the solution of $\dot{z} = \mathrm{i}\beta z$ passing through ζ at $t = 0$. This observation provides a useful geometrical interpretation of (5.5.13). The flow lines for $\dot{z} = \mathrm{i}\beta z$, $\dot{\theta} = 1$, form what is called Seifert's foliation (see Arnold, 1983), p. 170) of the solid torus $\mathbb{R}^2 \times S^1$ by circles. In the Seifert foliation of type (p, q) a twist of $2\pi p/q$ is imparted for each revolution around the torus (see Figure 5.8(a)). The same integral curves provide a simpler foliation of the q-sheeted covering of $\mathbb{R}^2 \times S^1$ (see Figure 5.8(*b*)). We can regard ζ as the base coordinate in Figure 5.8(*b*) and $t(\mathrm{mod}\ 2\pi q)$ as the coordinate on a fibre. In these coordinates, the t-dependence of ζ represents a deviation from the spiralling flow of $\dot{z} = \mathrm{i}\beta z$.

It is clear that the price we have paid for an autonomous vector field up to terms $O(|\zeta|^{q-1})$ is 2π-periodicity. What is more, the period advance map, $\mathbf{P}_0$, obtained from (5.5.14) is differentiably conjugate to $\mathbf{f}_0^q$ and not to $\mathbf{f}_0$ itself (see Exercise 5.5.2). A map approximating $\mathbf{f}_0$ can still be obtained from the vector field

$$\tilde{Z}_0(\zeta) = \sum_{s=1}^{S} c_s \zeta |\zeta|^{2s} + d\bar{\zeta}^{q-1} \tag{5.5.16}$$

by observing that $\tilde{Z}_0(\zeta)$ has $2\pi/q$-symmetry. This property of the normal form can be extended to arbitrary order. An interesting way of proving such a result, which involves the equivalence of the normal form calculations to averaging along the Seifert foliation, is considered in Exercise 5.5.3. However, for (5.5.16), it is easy to see that $\tilde{Z}_0(\zeta)$ is invariant under the transformation $\zeta \to \exp(2\pi \mathrm{i}/q)\zeta$.

Figure 5.8 (*a*) Seifert's foliation of type-(1, 3) for the solid torus $\mathbb{R}^2 \times S^1$, represented here as a cylinder whose ends are to be identified. Typical flow lines for $\dot{z} = \mathrm{i}z/3$, $\dot{\theta} = 1$ are shown. As θ increases from 0 to 2π, the flow-lines rotate through $2\pi/3$ about the axis of the cylinder. (*b*) Seifert's foliation of type-(1, 3) on the 3-sheeted covering of the solid torus, represented here by a cylinder of length 6π with ends identified. Here the 6π-periodicity of the flow-lines of $\dot{z} = \mathrm{i}z/3$, $\dot{\theta} = 1$ is easily recognised.

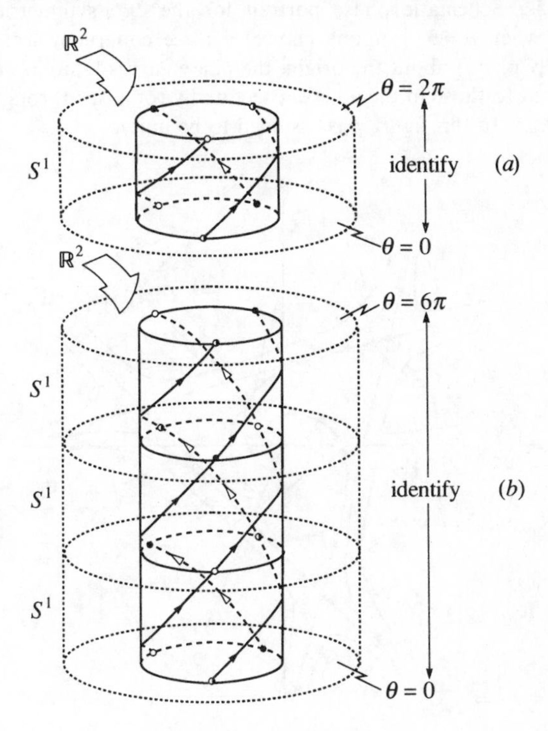

It follows that, if any disc centred on $\zeta = 0$ is divided into q congruent sectors, then the phase portrait of (5.5.16) is the same in each sector (see Figure 5.9). Now $\mathbf{f}_0^q$ is differentiably conjugate to $\mathbf{P}_0$ and, given that (5.5.16) has flow φ_t^0, the complex form of the latter is

$$P_0(\zeta) = \varphi_{2\pi q}^0(\zeta) + O(|\zeta|^q). \qquad (5.5.17)$$

Thus, from a topological point of view, $\varphi_{2\pi q}^0 : \mathbb{R}^2 \to \mathbb{R}^2$ approximates $\mathbf{f}_0^q$. In order to obtain a map which approximates $\mathbf{f}_0$ in the same sense, we require the appropriate qth root of $\varphi_{2\pi q}^0$. The following result (see Moser, 1968) shows that this qth root is determined uniquely.

Proposition 5.5.1 *Let* $\mathbf{M}: \mathbb{R}^2 \to \mathbb{R}^2$ *be such that* $\mathbf{M}(\mathbf{0}) = \mathbf{0}$ *and* $\mathrm{D}\mathbf{M}(\mathbf{0}) = \mathbf{R}_\beta$ *where* $\mathbf{R}_\beta$ *is given by* (5.1.10) *and* $\beta = p/q$ *is a* known *rational lying in* $[0, 1)$. *Define* Σ *to be the space spanned by the monomials that are resonant of order* $r \geqslant 2$ *for* $\mathbf{R}_\beta$. *Then the equation*

$$\mathbf{M}^q(\mathbf{x}) = \mathbf{x} + \mathbf{N}(\mathbf{x}) \qquad (5.5.18)$$

determines $\mathbf{M}$ *uniquely provided* $\mathbf{N}$ *lies in* Σ.

The statement of Proposition 5.5.1 involves the space, Σ, spanned by monomials that are resonant of order $r \geqslant 2$ for the real Jordan form $\mathbf{R}_\beta$ given in (5.1.10). The

Figure 5.9 Schematic phase portrait for the $2\pi/q$-symmetric system (5.5.16) when $q = 3$. For any choice of three congruent sectors symmetrically placed about the origin, the phase curves lying in one sector are related to those in the adjacent sectors by rotations through $+2\pi/3$ and $-2\pi/3$. In this figure p is assumed to be unity.

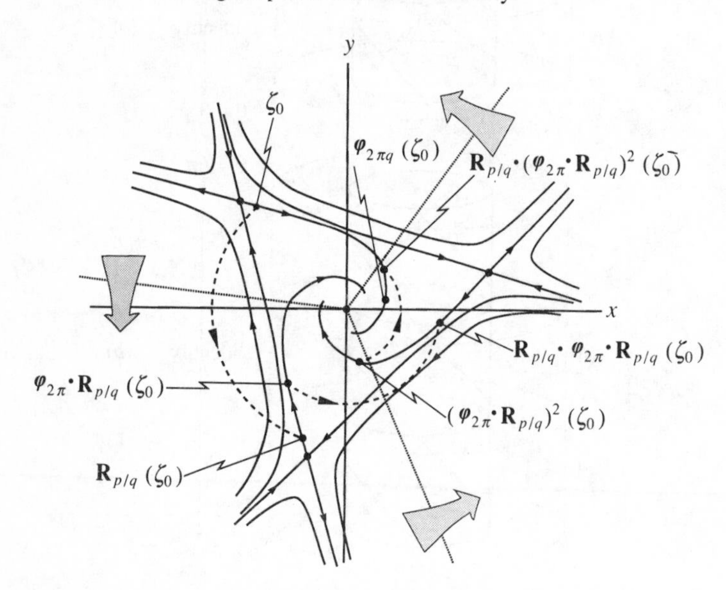

complex transformation $z = x + \mathrm{i}y$, $\bar{z} = x - \mathrm{i}y$ relates such monomials to

$$\left\{ \begin{pmatrix} z^{m_1}\bar{z}^{m_2} \\ 0 \end{pmatrix}, \begin{pmatrix} 0 \\ z^{m_2}\bar{z}^{m_1} \end{pmatrix} \middle| m_1 + m_2 = r, m_1 - m_2 - 1 = lq, l \in \mathbb{Z} \right\}, \tag{5.5.19}$$

(see (5.4.2)) which are the resonant terms for the (complex) Jordan form of $\mathbf{R}_\beta$. The reader will recall (see Example 2.5.2) that the resonance condition for the latter problem has the property that $z^{m_1}\bar{z}^{m_2}$ is resonant in the first component if and only if $z^{m_2}\bar{z}^{m_1}$ is resonant in the second. Therefore, only the first component need be retained in the so-called complex normal form. Clearly, if a function of (x, y) has complex form involving only those monomials in (5.5.19), then the function itself lies in Σ.

Since $\mathbf{DM}(\mathbf{0}) = \mathbf{R}_\beta$, the complex form of $\mathbf{M}$ can be written as

$$M(z) = \lambda z + \sum_{r=2}^{\infty} M_r(z, \bar{z}), \tag{5.5.20}$$

where $\lambda = \exp(2\pi \mathrm{i}\beta)$ and

$$M_r(z, \bar{z}) = \sum_{\substack{\mathbf{m} \\ m_1 + m_2 = r}} c_{\mathbf{m}}^{(r)} z^{m_1}\bar{z}^{m_2}. \tag{5.5.21}$$

The proof of Proposition 5.5.1 involves the comparison of corresponding terms in the complex forms $M^q(z)$ and $N(z)$. Such a comparison (see Exercise 5.5.4) shows that the coefficients $c_{\mathbf{m}}^{(r)}$ must satisfy

$$\sum_{\substack{\mathbf{m} \\ m_1 + m_2 = r}} c_{\mathbf{m}}^{(r)} z^{m_1}\bar{z}^{m_2}\lambda^{q-1}[1 + \lambda^{m_1 - m_2 - 1} + \lambda^{2(m_1 - m_2 - 1)} + \cdots$$
$$+ \lambda^{(q-1)(m_1 - m_2 - 1)}] = g_r(z, \bar{z}), \tag{5.5.22}$$

where g_r is determined by $N(z)$ and $M_k(z, \bar{z})$ with $k < r$. Moreover (see Exercise 5.5.5) if $\mathbf{N}$ lies in Σ, $g_r(z, \bar{z})$ only involves the first components of the monomials appearing in (5.5.19). Now,

$$\sum_{i=0}^{q-1} \lambda^{i(m_1 - m_2 - 1)} = \begin{cases} \dfrac{1 - \lambda^{q(m_1 - m_2 - 1)}}{1 - \lambda^{m_1 - m_2 - 1}} = 0, & \text{for } m_1 - m_2 - 1 \neq lq, \\ q \neq 0, & \text{for } m_1 - m_2 - 1 = lq, \end{cases} \tag{5.5.23}$$

since $\lambda^q = 1$. Hence $c_{\mathbf{m}}^{(r)}$ is uniquely determined and the non-linear part of $\mathbf{M}(\mathbf{x})$ must itself lie in Σ.

It is interesting to note that (5.5.23) holds if λ is *any* qth root of unity and it is important therefore that $\mathbf{DM}(\mathbf{0})$ is known. This is clearly the case for our approximation to $\mathbf{f}_0$ since we know $\mathbf{Df}_0(\mathbf{0}) = \mathbf{R}_{p/q}$. Furthermore, (5.5.16) can be used (see Exercise 5.5.6) to show that $\boldsymbol{\varphi}_t^0(\mathbf{x}) - \mathbf{x}$ lies in Σ for all t. Therefore Proposition 5.5.1 gives the existence of a unique qth root of $\boldsymbol{\varphi}_{2\pi q}^0$ with linear part equal to $\mathbf{R}_{p/q}$. For example, $\boldsymbol{\varphi}_{2\pi}^0$ satisfies $(\boldsymbol{\varphi}_{2\pi}^0)^q = \boldsymbol{\varphi}_{2\pi q}^0$ but $\mathrm{D}\boldsymbol{\varphi}_{2\pi}^0(\mathbf{0}) = \mathbf{I}$ and not $\mathbf{R}_{p/q}$. On the other hand, $(\mathbf{R}_{p/q} \cdot \boldsymbol{\varphi}_{2\pi}^0)$ has the correct linear behaviour but is it a

qth root of $\boldsymbol{\varphi}^{\mathbf{0}}_{2\pi q}$? Obviously,

$$(\mathbf{R}_{p/q}\cdot\boldsymbol{\varphi}^{\mathbf{0}}_{2\pi})^q = (\boldsymbol{\varphi}^{\mathbf{0}}_{2\pi}\cdot\mathbf{R}_{p/q})^q = (\mathbf{R}_{p/q})^q(\boldsymbol{\varphi}^{\mathbf{0}}_{2\pi})^q = \boldsymbol{\varphi}^{\mathbf{0}}_{2\pi q}. \tag{5.5.24}$$

provided $\mathbf{R}_{p/q}\cdot\boldsymbol{\varphi}^{\mathbf{0}}_{2\pi} = \boldsymbol{\varphi}^{\mathbf{0}}_{2\pi}\cdot\mathbf{R}_{p/q}$. Now, recall from Exercise 5.5.6 that $\boldsymbol{\varphi}^{\mathbf{0}}_{2\pi}(\mathbf{x}) - \mathbf{x} \in \Sigma$ and therefore $\mathbf{R}_{p/q}(\boldsymbol{\varphi}^{\mathbf{0}}_{2\pi} - \mathbf{I})(\mathbf{x}) = (\boldsymbol{\varphi}^{\mathbf{0}}_{2\pi} - \mathbf{I})\mathbf{R}_{p/q}(\mathbf{x})$ by Proposition 2.5.1. Hence $\boldsymbol{\varphi}^{\mathbf{0}}_{2\pi}$ commutes with $\mathbf{R}_{p/q}$ and

$$\mathbf{R}_{p/q}\cdot\boldsymbol{\varphi}^{\mathbf{0}}_{2\pi} = \boldsymbol{\varphi}^{\mathbf{0}}_{2\pi}\cdot\mathbf{R}_{p/q} \tag{5.5.25}$$

is the desired qth root of $\boldsymbol{\varphi}^{\mathbf{0}}_{2\pi q}$. It is this map that approximates the qualitative behaviour of $\mathbf{f_0}$.

Once again the calculations outlined above can be parallelled with $\mu \neq 0$. For example, if $\mathbf{f}_{\boldsymbol{\mu}}(\mathbf{0}) = \mathbf{0}$ near $\boldsymbol{\mu} = \mathbf{0}$, then equation (5.5.10) is replaced by

$$\dot{z} = \omega(\boldsymbol{\mu})z + \sum_{s=1}^{S} c_s(\boldsymbol{\mu})z|z|^{2s} + d(\boldsymbol{\mu})\bar{z}^{(q-1)}\exp(ipt) + G(\boldsymbol{\mu}, z, t) \tag{5.5.26}$$

and (5.5.14) by

$$\dot{\zeta} = [\omega(\boldsymbol{\mu}) - i\beta]\zeta + \sum_{s=1}^{S} c_s(\boldsymbol{\mu})\zeta|\zeta|^{2s} + d(\boldsymbol{\mu})\bar{\zeta}^{(q-1)} + G(\boldsymbol{\mu}, \exp(i\beta t)\zeta, t), \tag{5.5.27}$$

where $S = \max\{1, [(q-2)/2]\}$. Notice that the vector field in (5.5.27) is invariant under rotations by $2\pi/q$ for each value of $\boldsymbol{\mu}$. Finally, the behaviour of the family $\mathbf{f}\colon \mathbb{R}^l \times \mathbb{R}^2 \to \mathbb{R}^2$ is approximated locally by

$$(\boldsymbol{\mu}, \mathbf{x}) \mapsto (\boldsymbol{\mu}, \mathbf{R}_{p/q}\cdot\boldsymbol{\varphi}^{\boldsymbol{\mu}}_{2\pi}(\mathbf{x})) = (\boldsymbol{\mu}, \boldsymbol{\varphi}^{\boldsymbol{\mu}}_{2\pi}\cdot\mathbf{R}_{p/q}(\mathbf{x})). \tag{5.5.28}$$

5.5.3 *Rational* $\beta = p/q$, $q = 1, 2$

We have already noted that $\mathbf{Df_0}(\mathbf{0})$ has real equal eigenvalues $\lambda = \bar{\lambda} = 1$ for $q = 1$ and $\lambda = \bar{\lambda} = -1$ for $q = 2$. In the generic case, the Jordan form of $\mathbf{Df_0}(\mathbf{0})$ is:

$$\begin{pmatrix} 1 & 1 \\ 0 & 1 \end{pmatrix} \quad \text{for } q = 1; \tag{5.5.29}$$

and

$$\begin{pmatrix} -1 & 1 \\ 0 & -1 \end{pmatrix} \quad \text{for } q = 2. \tag{5.5.30}$$

Neither (5.5.29) nor (5.5.30) is a rotation and, since they are not diagonal, the normal form analysis presented in § 2.6 is not directly applicable. We must therefore re-examine the main steps of the calculation to verify that an approximating vector field can still be constructed. In the following, it will be convenient to assume that the reduction to Jordan form has been carried out so that $\mathbf{Df_0}(\mathbf{0})$ is equal to either (5.5.29) or (5.5.30).

(a) $q = 1$

In this case $\mathbf{Df_0(0)}$ is a linear shear map (see Figure 5.10) and therefore the suspension of $\mathbf{f_0}$ is still a flow on the solid torus. The period advance map of the linear part (see (2.6.3)) of the corresponding periodic non-autonomous differential equation is $\mathbf{Df_0(0)}$ itself. Moreover,

$$\mathbf{Df_0(0)} = \begin{pmatrix} 1 & 1 \\ 0 & 1 \end{pmatrix} = \exp(2\pi\mathbf{\Lambda}) \tag{5.5.31}$$

with $\mathbf{\Lambda} = \begin{pmatrix} 0 & (2\pi)^{-1} \\ 0 & 0 \end{pmatrix}$. Numerical inconvenience can be avoided by noting that $\mathbf{f_0}$ is linearly conjugate to a map with linear part $\begin{pmatrix} 1 & 2\pi \\ 0 & 1 \end{pmatrix}$. Therefore, without loss of generality, we can take

$$\mathbf{\Lambda} = \begin{pmatrix} 0 & 1 \\ 0 & 0 \end{pmatrix} \tag{5.5.32}$$

in (2.6.5). In the subsequent normal form calculation, we must solve the homological equation

$$\mathrm{L}_{\mathbf{\Lambda}}\mathbf{h}_r + \frac{\partial}{\partial t}\mathbf{h}_r = \mathbf{X}_r(\mathbf{x}, t) \tag{5.5.33}$$

(see (2.6.8)). Unfortunately, $\{\exp(i\nu t)\mathbf{x}^{\mathbf{m}}\mathbf{e}_j \mid m_1 + m_2 = r, \nu \in \mathbb{Z}, j = 1, 2\}$ are no longer eigenvectors of $\mathrm{L}_{\mathbf{\Lambda}} + \partial/\partial t$, because $\mathbf{\Lambda}$ is not diagonal. However, (see Exercise 2.6.6)

Figure 5.10 Illustration of the orbits of $\mathbf{Df_0(0)}$ given in (5.5.29). The lines $y =$ constant are invariant and orbits move to the right for $y > 0$ and to the left for $y < 0$. The x-axis is a line of fixed points. Such a transformation is referred to as a *linear shear* map. The square of the linear form (5.5.30) is also of this type, although the sense of description of the orbits is reversed. However, the orbit of (x_0, y_0) under the map (5.5.30) itself alternates between the lines $y = y_0$ and $y = -y_0$.

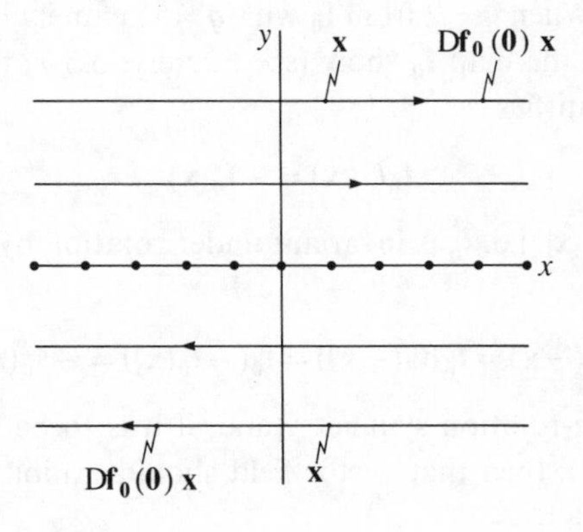

we can still show that $(\mathrm{L}_{\mathbf{\Lambda}} + \partial/\partial t)$ does not exist if there is an $(\mathbf{m}, j)$ such that

$$i\nu + \mathbf{m}\cdot\boldsymbol{\lambda} - \lambda_j = 0, \tag{5.5.34}$$

where $\mathbf{m}\cdot\boldsymbol{\lambda} = m_1\lambda_1 + m_2\lambda_2$ and $m_1 + m_2 = r$. Now (5.5.32) implies $\lambda_1 = \lambda_2 = 0$, so that (5.5.34) can only be satisfied if $\nu = 0$, i.e. if there are no time-dependent resonant terms. Moreover, given that $\nu = 0$, (5.5.34) reduces to the homological equation for a cusp singularity. Thus the time-independent resonant terms can be obtained just as in the proof of Proposition 2.4.3. We therefore conclude that the non-autonomous differential equation (5.5.2,3) can be reduced to the form

$$\begin{aligned}\begin{pmatrix}\dot{x}\\ \dot{y}\end{pmatrix} &= \begin{pmatrix}0 & 1\\ 0 & 0\end{pmatrix}\begin{pmatrix}x\\ y\end{pmatrix} + \sum_{s=2}^{k}\begin{pmatrix}a_s x^s\\ b_s x^s\end{pmatrix} + O(|\mathbf{x}|^{k+1}; t)\\ &= \tilde{\mathbf{X}}_0(\mathbf{x}) + O(|\mathbf{x}|^{k+1}; t),\end{aligned} \tag{5.5.35}$$

where $O(|\mathbf{x}|^{k+1}; t)$ is 2π-periodic in t. Given that $a_2 \neq 0$, $b_2 \neq 0$, we recognise that $\tilde{\mathbf{X}}_0(\mathbf{x})$ has a cusp singularity at $\mathbf{x} = \mathbf{0}$. If $\mathbf{P}_0$ is the period advance map of (5.5.35) and $\tilde{\mathbf{P}}_0$ is the time-2π map of $\tilde{\mathbf{X}}_0(\mathbf{x})$, then $\mathbf{f}_0$ is differentiably conjugate to $\mathbf{P}_0 = \tilde{\mathbf{P}}_0 + O(|\mathbf{x}|^{k+1})$, just as we had for $q \geqslant 3$.

(b) $q = 2$

In this case, although $\mathbf{f}_0$ can be suspended to obtain a flow on the solid torus, the period advance map of (2.6.3) for the resulting non-autonomous system is $\begin{pmatrix}-1 & 1\\ 0 & -1\end{pmatrix}$, which does not have a real logarithm. As we noted after Theorem 2.6.1, this means that a 4π-periodic change of coordinates is required to eliminate the time dependence in (2.6.3). Such a change of coordinates, followed by Lie bracket transformations that are 4π-periodic in t, yield a non-autonomous system of the form (5.5.35), where $O(|\mathbf{x}|^{k+1}; t)$ is 4π-periodic in t. Of course, the period advance map, $\mathbf{P}_0$, of this system is differentiably conjugate to $\mathbf{f}_0^2$ and not $\mathbf{f}_0$.

An odd feature of this calculation is that the resonant terms appear to be the same as those for $q = 1$. Apart from 4π-, rather than 2π-, periodicity in t, we appear to arrive at (5.5.35) for $q = 2$ also. However, we have omitted a vital feature of $\mathbf{f}_0$ that distinguishes $\mathbf{f}_0^2$ when $q = 2$ from $\mathbf{f}_0$ with $q = 1$: namely its symmetry. Normal form calculations on the map $\mathbf{f}_0$ show (see Exercise 5.5.7) that if $D\mathbf{f}_0(\mathbf{0})$ is given by (5.5.30) then $\mathbf{f}_0$ satisfies

$$\mathbf{f}_0(-\mathbf{x}) = -\mathbf{f}_0(\mathbf{x}) \tag{5.5.36}$$

to all finite orders in $|\mathbf{x}|$, i.e. $\mathbf{f}_0$ is invariant under rotation by $2\pi/q = \pi$ *when* $q = 2$. Now observe that

$$\mathbf{f}_0^2(-\mathbf{x}) = \mathbf{f}_0(\mathbf{f}_0(-\mathbf{x})) = \mathbf{f}_0(-\mathbf{f}_0(\mathbf{x})) = -\mathbf{f}_0^2(\mathbf{x}). \tag{5.5.37}$$

Clearly, $\mathbf{f}_0^2$ also has π-rotation symmetry and, if it is to be approximated by the time-2π map of $\tilde{\mathbf{X}}_0(\mathbf{x})$, then that vector field should exhibit the same symmetry.

This means that $a_s = b_s = 0$, for s even, in (5.5.35). It is perhaps worth noting that the topological type of such a vector field is determined if $a_3 \neq 0$ and $b_3 \neq 0$ (see Takens, 1974b).

Once the π-rotation symmetry of $\tilde{\mathbf{X}}_0(\mathbf{x})$ has been recognised, then we can construct an approximation to $\mathbf{f}_0$ itself along the lines described in § 5.5.2 for $q \geqslant 3$ (see Exercise 5.5.8). We obtain

$$\mathbf{f}_0 \sim \boldsymbol{\varphi}^0_{2\pi} \cdot \mathbf{R}_{1/2} + O(|\mathbf{x}|^{k+1}), \tag{5.5.38}$$

where $\boldsymbol{\varphi}^0$ is the flow of $\tilde{\mathbf{X}}_0(\mathbf{x})$, in agreement with (5.5.25). Moreover, as for $q \geqslant 3$ (see (5.5.26,7)), the normal form calculations described above can be used to generate smooth unfoldings of the singular vector field $\tilde{\mathbf{X}}_0(\mathbf{x})$ for q equal to both 1 and 2.

To sum up, for each value of $q \in \mathbb{Z}^+$, the time-dependent normal form calculations yield unfoldings of $\tilde{\mathbf{X}}_0$ or $\tilde{Z}_0$ that are invariant under rotations by $2\pi/q$. These families of vector fields are referred to as *equivariant unfoldings*. Clearly, if we wish to obtain as much information as possible about the behaviour of the families of approximating time-2π maps, then we require the *equivariant versal unfoldings* of $\tilde{\mathbf{X}}_0$ and $\tilde{Z}_0$. An unfolding is said to be *equivariantly versal* if every other equivariant unfolding is equivalent to an unfolding that is induced by the versal one. The bifurcations occurring in these very special families of vector fields are the subject of the next section.

5.6 Equivariant versal unfoldings for vector field approximations

Equivariant versal unfoldings are known for all the autonomous vector fields obtained in § 5.5, *except* for the case when $q = 4$. While these unfoldings reveal the local topological nature of all nearby equivariant vector fields, their time-2π maps provide only a subset of the maps that are near to $\mathbf{f}_0$. However, the bifurcations occurring in the families of time-2π maps must have some counterpart in the broader bifurcational picture associated with $\mathbf{f}_0$. In the present section, we aim to describe the bifurcation diagrams for the vector fields alone. In § 5.7, we will turn to the problem of how this information should be interpreted in terms of bifurcations in the unfoldings of $\mathbf{f}_0$.

When β is irrational the vector field approximation is given by (5.5.7) and we have already dealt with its versal unfolding. It is easily shown that (5.5.7) is the complex form of (4.2.17) with $c_1 = a_1 + ib_1$. Since there is no symmetry requirement for irrational β, its versal unfolding is given by the complex form of (4.2.18), i.e.

$$\dot{z} = \omega(v)z + c_1 z|z|^2 + O(|z|^5), \tag{5.6.1}$$

where $\omega(v) = v + i\beta$. Given that $\operatorname{Re} c_1 < 0$ (>0), there is a supercritical (subcritical) Hopf bifurcation at $v = 0$. Similarly, when $\beta = 0$ (i.e. $p/q \in \mathbb{Q} \cap [0, 1)$, with $q = 1$) we recognise that, generically, $\tilde{\mathbf{X}}_0(\mathbf{x})$ in (5.5.35) has a cusp singularity at $\mathbf{x} = \mathbf{0}$. Again, $2\pi/q$-symmetry imposes no restriction. Thus, given that $a_2 \neq 0$ and $b_2 \neq 0$, the (equivariant) versal unfolding of $\tilde{\mathbf{X}}_0(\mathbf{x})$ is given by (4.3.2) and its bifurcation diagram is illustrated in Figure 4.10.

We have not discussed the singularities occurring for $\beta = p/q \in \mathbb{Q} \cap (0, 1)$, with $q \geqslant 2$, earlier in the text. The linear parts of the vector fields involved all have zero trace and determinant, indeed, for $q \geqslant 3$ (see (5.5.16)) there are no linear terms at all. The problem of determining their topological types can be approached by using blowing-up techniques (see Exercise 2.8.6 and Dumortier, 1977). Equivariant versal unfoldings are available for $q \neq 4$ and our aim is to describe their bifurcation diagrams and make them plausible. Proofs of versality are major mathematical tasks beyond the scope of this text (Arnold, 1983; Bogdanov, 1981a,b; Carr, 1981; Takens, 1974b).

5.6.1 $q = 2$

The generic case is given by $\tilde{\mathbf{X}}_0(\mathbf{x})$ in (5.5.35) with $a_2 = b_2 = 0$, $a_3 \neq 0$ and $b_3 \neq 0$. It can be shown that, under these circumstances, (5.5.35) is topologically equivalent to

$$\dot{x} = y + x^3, \qquad \dot{y} = \delta x^3, \tag{5.6.2}$$

where $\delta = \pm 1$, or its time reversal. The equivariant versal unfolding, $\tilde{\mathbf{X}}^{\delta}_{\mathbf{v}}(\mathbf{x})$, of the singular vector field, $\tilde{\mathbf{X}}_0(\mathbf{x})$, defined by (5.6.2) is given by

$$\dot{x} = v_1 x + (1 + v_2)y + x^3, \qquad \dot{y} = -v_2 x + v_1 y + \delta x^3 \tag{5.6.3}$$

(see Takens, 1974b, p. 22). Notice that $\tilde{\mathbf{X}}^{\delta}_{\mathbf{v}}(\mathbf{x})$ is invariant under rotation by π for any choice of δ, $\mathbf{v}$. However, the bifurcation diagram for the family of vector fields defined by (5.6.3) does depend on the value of δ.

(a) $\delta = +1$

The bifurcation diagram is shown in Figure 5.11. Observe that (5.6.3) has a fixed point at the origin for all $\mathbf{v}$, with linearisation

$$\mathrm{D}\tilde{\mathbf{X}}^{+}_{\mathbf{v}}(\mathbf{0})\mathbf{x} = \begin{pmatrix} v_1 & 1 + v_2 \\ -v_2 & v_1 \end{pmatrix}\begin{pmatrix} x \\ y \end{pmatrix}. \tag{5.6.4}$$

Clearly,

$$\mathrm{Det}\,\mathrm{D}\tilde{\mathbf{X}}^{+}_{\mathbf{v}}(\mathbf{0}) = v_1^2 + v_2^2 + v_2, \tag{5.6.5}$$

$$\mathrm{Tr}\,\mathrm{D}\tilde{\mathbf{X}}^{+}_{\mathbf{v}}(\mathbf{0}) = 2v_1. \tag{5.6.6}$$

Thus, the origin is a saddle-point for $\mathbf{v}$ lying below the parabola, γ_1, given by

$$v_1^2 + v_2^2 + v_2 = 0 \tag{5.6.7}$$

and is a node/focus above this curve. The bifurcation taking place on γ_1 is a *symmetric saddle-node* bifurcation (see Exercise 5.6.1). This bifurcation takes a degenerate form at $v_1 = 0$. Let $\mathbf{v}$ lie on the v_2-axis, then (5.6.3) implies $\dot{y} = 0$ if

$x = 0$ or $\pm(\nu_2)^{1/2} = x^*_\pm$, $\nu_2 > 0$. Elimination of x^3 between $\dot{x} = 0$ and $\dot{y} = 0$ gives

$$y = -\frac{(\nu_1 + \nu_2)}{(1 + \nu_2 - \nu_1)} x, \tag{5.6.8}$$

so that $y = 0$ or $y^*_\pm = \mp \nu_2^{3/2}/(1 + \nu_2)$. Moreover,

$$\mathrm{Det}\, \mathrm{D}\tilde{\mathbf{X}}^+_\nu(\mathbf{x}^*_\pm) = -2(\nu_2 + \nu_2^2) < 0 \tag{5.6.9}$$

and the non-trivial fixed points are saddle points. Thus, as ν_2 increases through zero, with $\nu_1 = 0$, the saddle point at $\mathbf{x} = \mathbf{0}$ passes through a non-hyperbolic saddle (see Exercise 2.8.6), becomes a weak node/focus and throws off two saddle points. These saddle points are symmetrically placed about the origin.

The stability of the origin changes as ν_1 passes through zero for $\nu_2 > 0$ and therefore the positive ν_2-axis is (generically) a line of Hopf bifurcations. Calculations of a_1 in (4.2.24) shows that these bifurcations are subcritical. As ν_1 decreases, for fixed $\nu_2 > 0$, the saddle points remain stationary to first order in $|\boldsymbol{\nu}|$ and the limit cycle, created at the subcritical Hopf bifurcation, expands until a symmetric saddle connection takes place on the curve γ_3 in Figure 5.11 (see Exercise 5.6.2).

Figure 5.11 Bifurcation diagram for the equivariant versal unfolding $\tilde{\mathbf{X}}^+(\boldsymbol{\nu}, \mathbf{x})$ defined by (5.6.3) with $\delta = +1$. Phase portraits for $\tilde{\mathbf{X}}^+_\nu$ when $\boldsymbol{\nu}$ lies in the complement of the bifurcation curves are shown. Observe that all are invariant under rotation by π.

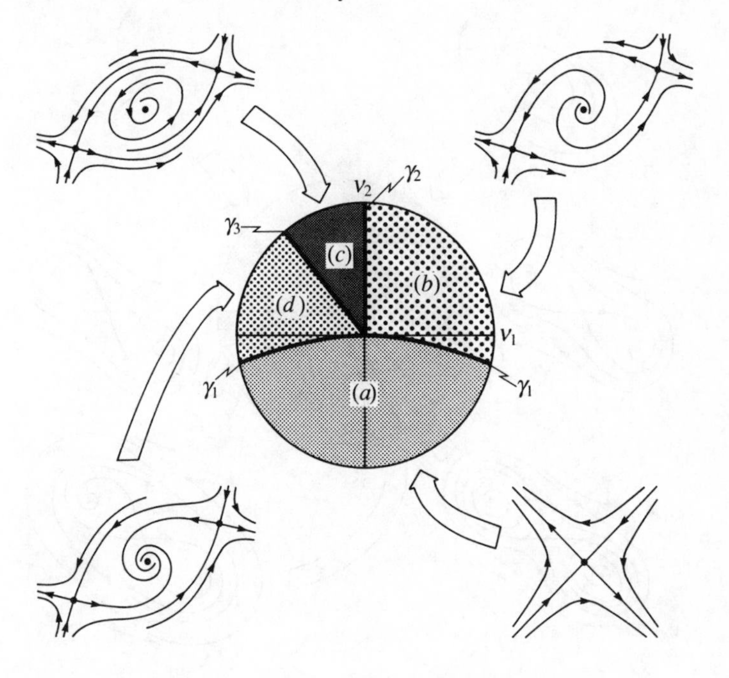

(b) $\delta = -1$

The bifurcation diagram for the family $\tilde{\mathbf{X}}^-(\mathbf{v}, \mathbf{x})$ is shown in Figure 5.12. The sense of the symmetric saddle-node bifurcation is reversed. For example, when $v_1 = 0$, $\dot{y} = 0$ for $x = 0$ or $\pm(-v_2)^{1/2}$, $v_2 < 0$; i.e. there are three fixed points *below* γ_1 in Figure 5.12. Moreover, it turns out (see Exercise 5.6.1) that Det $\mathrm{D}\tilde{\mathbf{X}}_{\mathbf{v}}^-(\mathbf{x}_\pm^*)$ is still given by (5.6.9) but, since $v_2 < 0$, Det $\mathrm{D}\tilde{\mathbf{X}}_{\mathbf{v}}^-(\mathbf{x}_\pm^*) > 0$ and the non-trivial fixed points are not saddles. Of course, the origin is still a saddle point below γ_1 and we conclude that two nodes and a saddle are created from the saddle-node. Recall that for $\delta = +1$ there were two saddles and one node. We shall see that this difference between $\delta = +1$ and $\delta = -1$ leads to a more complicated bifurcation diagram for the latter.

Let us start with $v_1, v_2 > 0$ and follow an anticlockwise, closed path in the $\mathbf{v}$-plane surrounding $\mathbf{v} = 0$. Close enough to the positive v_2-axis the origin is an unstable focus, a subcritical Hopf bifurcation takes place at $v_1 = 0$ and an unstable limit cycle is created surrounding a stable focus. As the curve γ_1 is approached with $v_1 < 0$, this limit cycle is still present and has finite size. On passing through γ_1, a pair of stable fixed points appear symmetrically from the origin (see Exercise

Figure 5.12 Bifurcation diagram for the family $\tilde{\mathbf{X}}^-(\mathbf{v}, \mathbf{x})$ given by (5.6.3) with $\delta = -1$. Observe that the phase portraits given are all π-rotation invariant. Note the symmetric Hopf and saddle-connection bifurcations. The double limit-cycle bifurcation is also of particular interest.

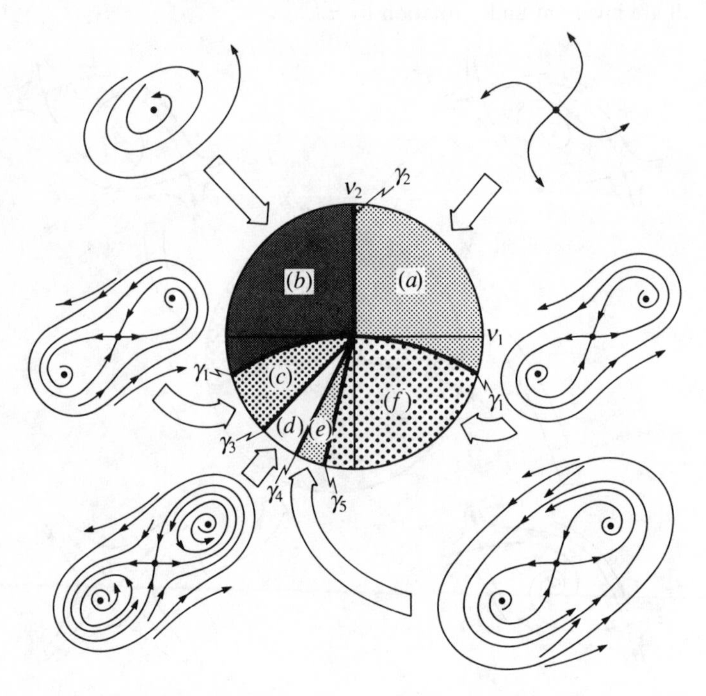

5.6.1) with

$$x^2 = x_{\pm}^{*2} = -\left\{v_2 + \frac{v_1(v_1 - v_2)}{(1 + v_2 + v_1)}\right\}. \quad (5.6.10)$$

For given v_1, $x_{\pm}^{*2}$ increases as v_2 decreases and, since

$$\mathrm{Tr}\, \mathrm{D}\tilde{\mathbf{X}}_{\mathbf{v}}^{-}(\mathbf{x}_{\pm}^{*}) = 3x_{\pm}^{*2} + 2v_1, \quad (5.6.11)$$

the non-trivial fixed points simultaneously undergo Hopf bifurcations on a curve γ_3 in Figure 5.12 defined by $3x_{\pm}^{*2} + 2v_1 = 0$ (see Exercise 5.6.3). If we continue to reduce v_2, the limit cycles surrounding the non-trivial fixed points increase in size until a symmetric connection occurs on γ_4 in Figure 5.12. This saddle connection breaks to give a limit cycle surrounding all three fixed points. Now the inner cycle expands as v_2 decreases and the outer one shrinks as v_1 increases. Eventually, the two limit cycles coalesce and disappear in what is a saddle-node bifurcation in the local Poincaré map. This double limit-cycle bifurcation takes place on γ_5 in Figure 5.12. Finally, as γ_1 is approached from below with $v_1 > 0$, the three fixed points move together and coalesce into a single node on crossing γ_1.

5.6.2 $q = 3$

The equivariant versal unfolding of the vector field (5.5.16) with $q = 3$ is the complex form

$$\dot{\zeta} = \varepsilon\zeta + c\zeta|\zeta|^2 + d\bar{\zeta}^2, \quad (5.6.12)$$

where ε, c, $d \in \mathbb{C}$. It is easily shown (see Exercise 5.6.4) that a scaling of ζ allows us to take $d = 1$. Furthermore, if we assume $d = 1$ and write $\varepsilon = v_1 + \mathrm{i}v_2$, $c = a + \mathrm{i}b$ and $\zeta = x + \mathrm{i}y$, then taking real and imaginary parts in (5.6.12) gives

$$\begin{aligned} \dot{x} &= v_1 x - v_2 y + ax(x^2 + y^2) - by(x^2 + y^2) + x^2 - y^2, \\ \dot{y} &= v_2 x + v_1 y + ay(x^2 + y^2) + bx(x^2 + y^2) - 2xy. \end{aligned} \quad (5.6.13)$$

In plane polar coordinates, (5.6.13) becomes

$$\begin{aligned} \dot{r} &= v_1 r + ar^3 + r^2 \cos 3\theta, \\ \dot{\theta} &= v_2 + br^2 - r \sin 3\theta. \end{aligned} \quad (5.6.14)$$

Eliminating θ from the fixed point equations of (5.6.14) gives the following equation for the radial coordinate of the non-trivial fixed points

$$(a^2 + b^2)r^4 - [1 - 2(v_1 a + v_2 b)]r^2 + (v_1^2 + v_2^2) = 0. \quad (5.6.15)$$

This equation has solutions

$$r_{\mathrm{p}}^2 = \frac{1 - 2(v_1 a + v_2 b)}{2(a^2 + b^2)}\left\{1 \pm \left(1 - \frac{4(a^2 + b^2)(v_1^2 + v_2^2)}{[1 - 2(v_1 a + v_2 b)]^2}\right)^{1/2}\right\}. \quad (5.6.16)$$

For small $|\mathbf{v}|$, (5.6.16) gives

$$r_p^2 = v_1^2 + v_2^2 + O(|\mathbf{v}|^3) \tag{5.6.17}$$

for non-trivial fixed points in the neighbourhood of the origin. Substitution of (5.6.17) into (5.6.14) with $\dot{r} = \dot{\theta} = 0$ provides three values of θ, separated by $2\pi/3$, corresponding to three non-trivial fixed points symmetrically placed on a circle of radius r_p.

Of course, $\mathbf{x} = \mathbf{0}$ is a fixed point of (5.6.13) for all values of $\mathbf{v}$. Moreover, the linearisation of (5.6.13) at $\mathbf{x} = \mathbf{0}$ has eigenvalues $\lambda(v_1, v_2) = v_1 + iv_2$ and these are pure imaginary for $v_1 = 0$ and $v_2 \neq 0$. If we consider $v_2 \neq 0$ and fixed, then it is straightforward to use Theorem 4.2.1 to show that (5.6.13) undergoes a Hopf bifurcation at $v_1 = 0$. Equation (4.2.24) allows us to show that this bifurcation is supercritical (subcritical) if $a < 0$ (>0) (see Exercise 5.6.5). Furthermore, the resulting limit cycle has approximate radius $r_0 = (v_1/|a|)^{1/2}$.

Suppose $a < 0$ and consider the one-parameter subfamily of vector fields corresponding to a circular arc in the $\mathbf{v}$-plane crossing the negative v_2-axis at $v_2 = -v$ travelling in an anticlockwise sense. For such a subfamily r_p is essentially fixed at v, while r_0 increases like $(v_1/|a|)^{1/2}$. Clearly, there will be a value of v_1 for which the invariant circle meets the non-trivial fixed points. It can be shown (see Exercise 5.6.6) that all of the non-trivial fixed points are saddles for any $\mathbf{v}$. When the limit cycle reaches these points a heteroclinic saddle connection forms (see Figure 5.13). Close enough to $\mathbf{v} = \mathbf{0}$, we can obtain an approximation to the saddle-connection bifurcation curve by equating r_p^2 and r_0^2, i.e. we require

$$-\frac{v_1}{a} + O(v_1^{3/2}) = v_1^2 + v_2^2 + O(|\mathbf{v}|^3) \tag{5.6.18}$$

or, since $v_1 \ll 1$,

$$v_1 = -av_2^2 + O(v_2^3). \tag{5.6.19}$$

This is consistent with the saddle-connection bifurcation curve given in Figure 5.13. Notice that, as in the cusp bifurcation, the limit cycle is destroyed in the saddle-connection bifurcation.

5.6.3 $q = 4$

In this section we study some of the bifurcations occurring in the unfolding

$$\dot{\zeta} = \varepsilon\zeta + c\zeta|\zeta|^2 + \bar{\zeta}^3 \tag{5.6.20}$$

of (5.5.16) with $q = 4$. Notice we have once again scaled ζ so that $d = 1$. The local family (5.6.20) is obtained from (5.5.16) by the addition of $\varepsilon\zeta$. This procedure is the same as that used to construct the equivariant versal unfoldings for $q \neq 4$. Unfortunately, (5.6.20) has not yet been proved to be equivariantly versal (Arnold,

1983, pp. 288–9). The polar form of (5.6.20) is

$$\begin{aligned} \dot{r} &= \nu_1 r + ar^3 + r^3 \cos 4\theta, \\ \dot{\theta} &= \nu_2 + br^2 - r^2 \sin 4\theta, \end{aligned} \tag{5.6.21}$$

where $\varepsilon = \nu_1 + i\nu_2$ and $c = a + ib$.

The non-trivial fixed points of (5.6.21) satisfy

$$(|c|^2 - 1)r^4 + 2(\nu_1 a + \nu_2 b)r^2 + (\nu_1^2 + \nu_2^2) = 0, \tag{5.6.22}$$

(see 5.6.15) which has solutions

$$r_p^2 = \frac{\xi_2 \pm \{\xi_2^2 + (1 - |c|^2)\xi_1^2\}^{1/2}}{(1 - |c|^2)}, \qquad |c|^2 \neq 1, \tag{5.6.23}$$

$$r_p^2 = -\xi_1^2/2\xi_2, \qquad |c|^2 = 1, \tag{5.6.24}$$

where $\xi_1^2 = \nu_1^2 + \nu_2^2$ and $\xi_2 = \nu_1 a + \nu_2 b$. There are three cases to consider.

Figure 5.13 Bifurcation diagram for the family of vector fields defined by the complex form (5.6.12) when $a = \mathrm{Re}(c) < 0$. The ν_2-axis is a line of supercritical Hopf (H) bifurcations with increasing ν_1 and the expansion of the resulting invariant circle leads to a heteroclinic saddle connection (s.c.) on the quadratic curve (5.6.19). Note that, for small r, the sense of rotation of the flow is determined by the sign of ν_2. If $a > 0$, then subcritical Hopf bifurcations take place on the ν_2-axis and the saddle-connection bifurcations occur at negative values of ν_1.

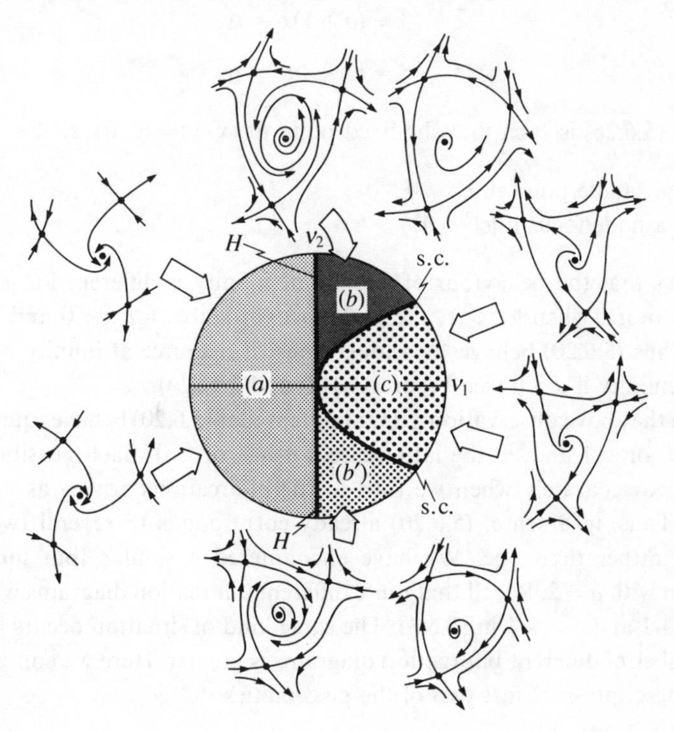

(a) $|c|^2 < 1$
Equation (5.6.23) gives only one positive value for r_p^2 for each $\xi_2 \in \mathbb{R}$. This leads to four non-trivial fixed points on solving (5.6.21) for the corresponding values of θ.

(b) $|c|^2 = 1$
It follows from (5.6.24) that (5.6.21) has four non-trivial fixed points provided $\xi_2 < 0$. These fixed points move out of any given neighbourhood of the origin as $\xi_2 \to 0_-$. There are no non-trivial fixed points for $\xi_2 > 0$.

(c) $|c|^2 > 1$
In this case, 8, 4, or 0 non-trivial fixed points can occur as follows. Suppose $\xi_2 < 0$, then (5.6.23) gives *two* positive values for r_p^2 if $\Xi = \xi_2^2 + (1 - |c|^2)\xi_1^2 > 0$. As Ξ approaches zero, these two positive solutions tend to the same value and coalesce when $\Xi = 0$. If $\Xi < 0$, then (5.6.23) has no real solutions. Thus, (5.6.21) has 8, 4, 0 non-trivial fixed points for $\Xi >, =, < 0$, respectively. When $\xi_2 \geqslant 0$, (5.6.22) has no positive solutions so there are no non-trivial fixed points.

For sufficiently large $|\zeta|$, (5.6.20) is approximated by

$$\dot{\zeta} = c\zeta|\zeta|^2 + \bar{\zeta}^3, \tag{5.6.25}$$

for any ε near zero. It can be shown (see Exercise 5.6.7) that (5.6.25) is topologically equivalent to the linear system

$$\begin{aligned} \dot{x} &= (a+1)x - by, \\ \dot{y} &= bx + (a-1)y. \end{aligned} \tag{5.6.26}$$

Clearly, (5.6.26) is such that the fixed point at $(x, y) = (0, 0)$ is:

(i) a saddle point if $|c|^2 < 1$;
(ii) a node/focus if $|c|^2 > 1$.

It follows that the behaviour of (5.6.20) at infinity is different for $|c|^2 <$ or > 1. What is more, when $|c|^2 > 1$, the node/focus is stable for $a < 0$ and unstable for $a > 0$. Thus (5.6.20) behaves as though there is a source at infinity if $a < 0$ and a sink at infinity if $a > 0$ (see Figures 1.1(*d*) and 1.12(*b*)).

From the above observations, it is apparent that (5.6.20) behaves quite differently for $|c|^2 <$ or > 1 and, in the latter case, for $a <$ or > 0. Each possibility must be considered separately when we turn to the bifurcations arising as ν_1 and ν_2 are varied. Thus, in a sense, (5.6.20) already corresponds to several two-parameter families rather than one. We have encountered a similar (but much simpler) situation with $q = 2$. Recall that quite different bifurcation diagrams were obtained for $\delta = +1$ and $\delta = -1$ in (5.6.3). The same kind of situation occurs for $q = 4$ but the number of different bifurcation diagrams is greater. Here we content ourselves with a description of just two of the possibilities.

Example 5.6.1 Illustrate the similarity between the bifurcations that occur in the family (5.6.20) for $|c|^2 < 1$ and those discussed in § 5.6.2.

Solution. We have already noted that when $|c|^2 < 1$ there are four non-trivial fixed points for every $\mathbf{v} \neq \mathbf{0}$. It can be shown (see Exercise 5.6.8) that the determinant of the linearisation of (5.6.20) at any non-trivial fixed point is given by

$$-4(v_1^2 + v_2^2) - 4(1 - |c|^2)r_p^4. \tag{5.6.27}$$

Thus the four non-trivial fixed points in question are saddles for each $\mathbf{v} \neq \mathbf{0}$. The origin ($\zeta = 0$) is a fixed point for all $\mathbf{v}$ and the trace of the linearisation of (5.6.20) at $\zeta = 0$ is $2v_1$ (cf. (5.6.13, 5.6.14 and 5.6.21)). The fixed point at $\zeta = 0$ therefore changes stability as v_1 passes through zero. It is not difficult to confirm (using Theorem 4.2.1 and (4.2.24)) that, for $v_2 \neq 0$, a supercritical (subcritical) Hopf bifurcation takes place if $a = \mathrm{Re}(c) < 0$ (>0). These observations parallel the discussion for $q = 3$ in § 5.6.2 and, indeed the bifurcation diagram for this case is similar to Figure 5.13. If $a < 0$, a sequence of phase portraits typical of those found on moving anticlockwise around a closed curve surrounding $\mathbf{v} = \mathbf{0}$ is shown in Figure 5.14. The limit cycle, created in the Hopf bifurcation at $v_1 = 0$, $v_2 < 0$, expands and is destroyed at a heteroclinic saddle connection. Notice that the sense of rotation of the flow about the origin is reversed on crossing the v_1-axis (see (5.6.21)). Finally, a limit cycle, created at the upper branch of the saddle-connection bifurcation curve, contracts and vanishes at a subcritical Hopf bifurcation on the positive v_2-axis. □

Figure 5.14 Illustration of the bifurcations occurring in the $2\pi/4$-symmetric (i.e. $q = 4$) family (5.6.20) when $|c^2| < 1$ and $a < 0$ (see Example 5.6.1). Note the similarity to the bifurcations taking place in Figure 5.13.

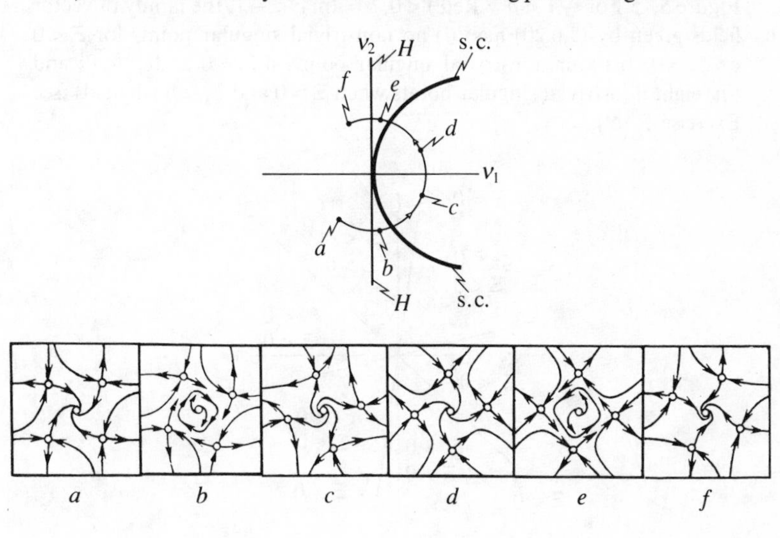

Example 5.6.2 Discuss the bifurcations occurring in the family (5.6.20) when $|c|^2 > 1$ and $a < 0$. Choose b so as to ensure a clockwise rotation of the flow for large r and restrict a so that Hopf bifurcations are possible at non-trivial fixed points.

Solution. Since $a < 0$, the trajectories of the flow of (5.6.20) spiral in from infinity. For large enough r, (5.6.21) shows that the sense of their rotation about the origin is determined by the sign of b provided $|b| > 1$. To ensure a clockwise rotation we must take $b < -1$.

Let us begin by considering the non-trivial fixed points of (5.6.20). We have already noted that (5.6.23) shows that such points only occur for $\xi_2 = av_1 + bv_2 < 0$ and $\Xi = \xi_2^2 + (1 - |c|^2)\xi_1^2 \geqslant 0$. It can be shown (see Exercise 5.6.9) that $\Xi = 0$ defines a pair of straight lines passing through $\mathbf{v} = \mathbf{0}$ and symmetrically placed relative to $v_2 = (-a/b)v_1$ on which $\xi_2 = 0$. It follows from our earlier discussion that four non-trivial fixed points occur when $\mathbf{v}$ lies on these $\Xi = 0$ lines *and* $\xi_2 < 0$ (see Figure 5.15). Observe that at such points $r_p^2 = \xi_2/(1 - |c|^2)$ and (5.6.27) takes the form $-4\Xi/(1 - |c|^2) = 0$. Generically, therefore these fixed points are saddle-nodes. When $\Xi > 0$, (5.6.22) has two distinct roots

$$r_{p>}^2 = \frac{\xi_2 - \Xi^{1/2}}{1 - |c|^2} \tag{5.6.28}$$

and

$$r_{p<}^2 = \frac{\xi_2 + \Xi^{1/2}}{1 - |c|^2}. \tag{5.6.29}$$

We expect one of these radii to correspond to the saddle points and the other to

Figure 5.15 For $-1 < a = \mathrm{Re}(c) < 0$, $b = \mathrm{Im}\, c < -1$, the family of vector fields given by (5.6.20) has: (i) no non-trivial singular points for $\Xi < 0$ or $\xi_2 > 0$; (ii) four non-trivial singular points if $\Xi = 0$ and $\xi_2 < 0$; and (iii) eight non-trivial singular points when $\Xi > 0$ and $\xi_2 < 0$ (shaded) (see Exercise 5.6.9).

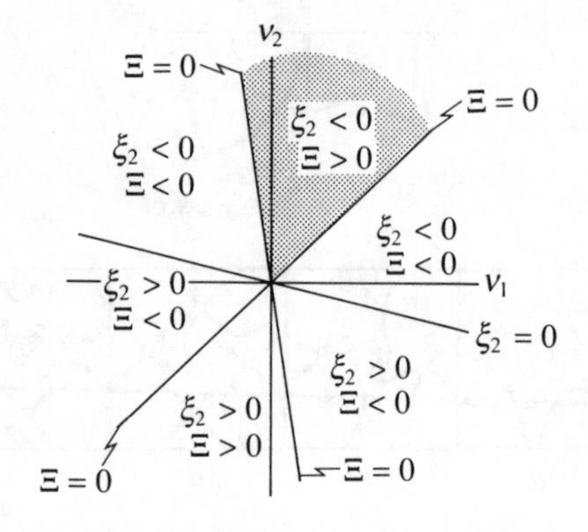

the node/foci that develop from the saddle-nodes on $\Xi = 0$. Indeed, it can be shown (see Exercise 5.6.8) that (5.6.27) can be recast in the form

$$-8(\Xi)^{1/2}\frac{(\Xi^{1/2} \pm \xi_2)}{(1-|c|^2)} = \begin{cases} -8(\Xi)^{1/2} r_{p<}^2 & \text{for +ve sign} \\ 8(\Xi)^{1/2} r_{p>}^2 & \text{for −ve sign} \end{cases} \tag{5.6.30}$$

so that the non-trivial fixed points nearer to the origin are saddle points.

The stability of the node/foci is determined by the trace of their linearisation. This is given by

$$2v_1 + 4ar_{p>}^2. \tag{5.6.31}$$

On equating (5.6.31) to zero, the stability of the node/foci can be shown to change on the straight line

$$v_2 = \frac{b}{2a}\left\{1 - \left(\frac{1-a^2}{b^2}\right)^{1/2}\right\} v_1, \tag{5.6.32}$$

provided $-1 < a < 0$ and $|b|$ is sufficiently large. Details of this calculation are outlined in Exercise 5.6.10. The bifurcations that occur when $a < -1$ are examined in Exercise 5.6.19.

We will focus attention on $a \in (-1, 0)$ and $b \ll -1$, when, generically, (5.6.32) defines a line of 'remote' Hopf bifurcations (i.e. Hopf bifurcations at the four non-trivial node/foci and therefore 'remote' from the origin). The nature of each of these Hopf bifurcations is given by the coefficient, a_1, of r^3 in the normal form at the non-trivial fixed point. This is only given by (4.2.24) after:

(i) local coordinates have been introduced at the fixed point;
(ii) the linear part of the vector field at the bifurcation point has been reduced to real Jordan form.

For given a and b, these calculations can be carried out numerically for any point on the line (5.6.32) in the neighbourhood of $\mathbf{v} = \mathbf{0}$. For example, for $a = -\frac{1}{2}$, $b = -5$, we find $a_1 > 0$ for $0 < v_1 < 1$ (see Exercise 5.6.12) and therefore conclude that the fixed points in question are weak repellors on the bifurcation curve (5.6.32). For these values of a and b, the slope of (5.6.32) is positive and the sign of (5.6.31) is as shown in Figure 5.16. It follows that a subcritical Hopf bifurcation takes place as the line (5.6.32) is crossed with v_1 increasing. This means that an unstable limit cycle exists for those parameter values for which (5.6.31) is sufficiently small and negative. Moreover, as (5.6.31) decreases from zero, the radius of the limit cycle increases and we might expect a remote, homoclinic saddle connection to occur (cf. §4.3.1).

It must not be forgotten that a supercritical Hopf bifurcation takes place at $\mathbf{x} = \mathbf{0}$ as v_1 increases across the v_2-axis. If $v_2 > 0$, then $\xi_2 < 0$, $\Xi > 0$ and the non-trivial fixed points are already present in the phase portrait when the limit cycle is created around the origin. Its radius grows like $(-v_1/a)^{1/2}$ (see Exercise 5.6.2), while the radial coordinate of the saddle points is given by (5.6.29). For sufficiently large $|b|$, it can be shown that $r^2_{p<}$ is essentially independent of v_1 (see Exercise 5.6.13) so that as v_1 increases the limit cycle expands to meet the saddle points in a heteroclinic saddle connection. If the v_2-axis is crossed with $v_2 < 0$ and v_1 increasing, there are no non-trivial fixed points in the phase portrait at the creation of the limit cycle. If we follow a circular arc in the $\mathbf{v}$-plane around $\mathbf{v} = \mathbf{0}$, the limit cycle is already present when the saddle-nodes appear. What is more, these fixed points can be created inside the limit cycle (see Exercise 5.6.13(c)). It is then possible for the remote Hopf and saddle-connection bifurcations to take place within the limit cycle surrounding the origin.

The complicated sequence of bifurcations arising on a closed path surrounding the origin of the $\mathbf{v}$-plane has been beautifully illustrated in Arnold, (1983, p. 310) and reproduced here in Figure 5.17. The reader should pay particular attention to how the limit cycle apparently 'passes through' the non-trivial fixed points by becoming a non-differentiable invariant circle that contains them. Note also the changes in basins of attraction of the node/foci that occur during this transitional phase. It is this transition that links the approaches from $v_2 > 0$ and $v_2 < 0$ that we have described above. □

5.6.4 $q \geqslant 5$

In this case, the versal unfolding of (5.5.16) subject to the requirement of $2\pi/q$-symmetry is

$$\dot{\zeta} = \varepsilon\zeta + \sum_{s=1}^{[\frac{1}{2}(q-2)]} c_s\zeta|\zeta|^{2s} + d\bar{\zeta}^{q-1}, \tag{5.6.33}$$

Figure 5.16 Saddle-node and remote Hopf bifurcation curves for the family (5.6.20) discussed in Example 5.6.2. The remote Hopf bifurcations take place on the straight line (5.6.32), where $\mathrm{Tr} = 2v_1 + 4ar^2_{p>} = 0$, and are subcritical with increasing v_1. Note also that the v_2-axis is a line of supercritical Hopf bifurcations at the origin and that eight non-trivial fixed points exist in the shaded region where $\Xi > 0$ and $\xi_2 < 0$.

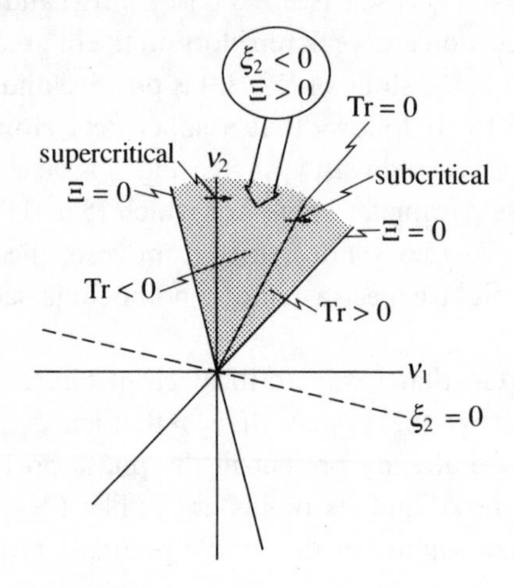

Figure 5.17 Illustration of the bifurcations occurring in the $2\pi/4$-symmetric family of vector fields defined by (5.6.20) when $a = \text{Re}(c) = -\frac{1}{2}$ and $b = \text{Im}(c) = -5$. Bifurcations occurring on two segments of the arc shown are discussed in the text: (i) (a)–(f) where the limit cycle is created when the non-trivial fixed points already exist; (ii) (p)–(j) on which the non-trivial fixed points are created inside the limit cycle and undergo remote Hopf and saddle-connection bifurcations. The diagrams (e)–(i) show how the limit cycle moves through the non-trivial fixed points as an invariant circle. Note that in (g) there is a non-differentiable invariant circle containing all eight non-trivial fixed points.

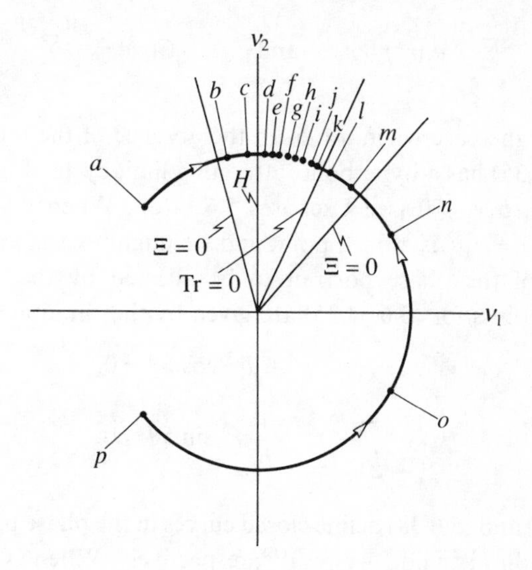

with $\varepsilon = \nu_1 + i\nu_2$, $c_s = a_s + ib_s$ and $d \in \mathbb{C}$. Given that $a_1 < 0$ (if not time reversal will make it so), it can be shown (see Exercise 5.6.14) that there is a complex scaling of ζ such that $a_1 = -1$ and $d \in \mathbb{R}$. It is then easily shown that the polar form of (5.6.33) can be written as

$$\dot{r} = r(\nu_1 - r^2 + F(r^2) + dr^{q-2}\cos q\theta), \tag{5.6.34}$$

$$\dot{\theta} = \nu_2 + b_1 r^2 + G(r^2) - dr^{q-2}\sin q\theta. \tag{5.6.35}$$

For $q = 5$, $F = G = 0$, while $q \geqslant 6$

$$F(r^2) = \sum_{s=2}^{[\frac{1}{2}(q-2)]} a_s(r^2)^s \qquad \text{and} \qquad G(r^2) = \sum_{s=2}^{[\frac{1}{2}(q-2)]} b_s(r^2)^s. \tag{5.6.36}$$

Let us consider the case when $q = 5$. In the absence of the θ-dependent terms, the system (5.6.34,35) has a hyperbolic, attracting limit cycle of radius $\nu_1^{1/2}$, $\nu_1 > 0$, provided that $\nu_2 + b_1\nu_1 \neq 0$ (see Exercises 5.6.14(c)). When $\nu_2 + b_1\nu_1 = 0$, every point of the circle $r = \nu_1^{1/2}$ is a fixed point and we might expect such a structurally unstable feature of the phase portrait to be affected by the $O(r^3)$-terms. The non-trivial fixed points for (5.6.34,35) are given by the simultaneous solutions of

$$\nu_1 - r^2 + dr^3\cos 5\theta = 0, \tag{5.6.37}$$

$$-\frac{\nu_2}{b_1} - r^2 + \frac{d}{b_1}r^3\sin 5\theta = 0. \tag{5.6.38}$$

Equations (5.6.37) and (5.6.38) define closed curves in the phase plane that oscillate about circles of radii $\nu_1^{1/2}$ and $(-\nu_2/b_1)^{1/2}$, respectively. When $\nu_2 + b_1\nu_1 = 0$, these isoclines intersect as shown in Figure 5.18(*c*) generating ten fixed points: five lying within $r = \nu_1^{1/2}$ and five lying outside this circle. It can be shown (see Exercise 5.6.16) that the fixed points with radial coordinate less than $\nu_1^{1/2}$ are saddle points, while the others are stable node/foci.

As $|\nu_2 + b_1\nu_1|$ increases from zero, the $\dot{r} = 0$ and $\dot{\theta} = 0$ isoclines pull apart as shown in Figures 5.18(*a*,*b*) and (*d*,*e*). The saddles and the node/foci move together in pairs and vanish in saddle-node bifurcations. The resulting saddle-node bifurcation curves (which mark the boundary of the region in the $\boldsymbol{\nu}$-plane for which non-trivial fixed points exist) can be approximated as follows. Let $r = \nu_1^{1/2} + r_1(\nu_1, \theta)$, where $r_1(\nu_1, \theta) = O(\nu_1)$, and use (5.6.37) to show that $\dot{r} = 0$ for

$$r = \nu_1^{1/2} + \tfrac{1}{2}d\nu_1\cos 5\theta + O(\nu_1^{3/2}). \tag{5.6.39}$$

Substitute (5.6.39) into (5.6.38) to obtain

$$\nu_2 + b_1\nu_1 = d\nu_1^{3/2}(\sin 5\theta - b_1\cos 5\theta) + O(\nu_1^2). \tag{5.6.40}$$

Now, (5.6.40) can only be satisfied if

$$|\nu_2 + b_1\nu_1| + O(\nu_1^2) < d\nu_1^{3/2}(1 + b_1^2)^{1/2}, \tag{5.6.41}$$

and (5.6.41) defines a region in the $\mathbf{v}$-plane with boundary

$$v_2 = -b_1 v_1 \pm d v_1^{3/2}(1 + b_1^2)^{1/2} + O(v_1^2). \tag{5.6.42}$$

Thus, non-trivial fixed points exist for (5.6.34,35) only in a cuspoidal region, or 'tongue' in the $\mathbf{v}$-plane which, to leading order, is symmetrically placed about the line $v_2 + b_1 v_1 = 0$.

For a general point within the tongue defined by (5.6.41), the $\dot{r} = 0$ and $\dot{\theta} = 0$ isoclines intersect transversely as shown in Figures 5.18(*b*)–(*d*). Examination of how the vector field crosses these isoclines shows that the unstable separatrices of the saddle points emanate into the loops formed by the interlacing of the $\dot{r} = 0$ and $\dot{\theta} = 0$ curves. With the aid of the five-fold rotational symmetry of the flow, it is then possible to argue that, generically, an unstable separatrix must be attracted into the nearest adjacent node/focus (see Exercise 5.6.17). Such a conclusion implies the existence of an invariant circle consisting of the union of the unstable

Figure 5.18 Relative positions of the $\dot{r} = 0$ and $\dot{\theta} = 0$ isoclines for the system (5.6.34, 35) as $|v_2 + b_1 v_1|$ increases from zero. The plots shown are obtained when $d = 1$, $b_1 = -1$, $v_1 = 0.05$ and v_2 is varied. Observe that the $\dot{\theta} = 0$ isocline expands through the $\dot{r} = 0$ isocline as $v_2 + b_1 v_1$ increases through zero. The values of v_2 for the tangencies shown in (*a*) and (*e*) to occur are approximated by (5.6.42).

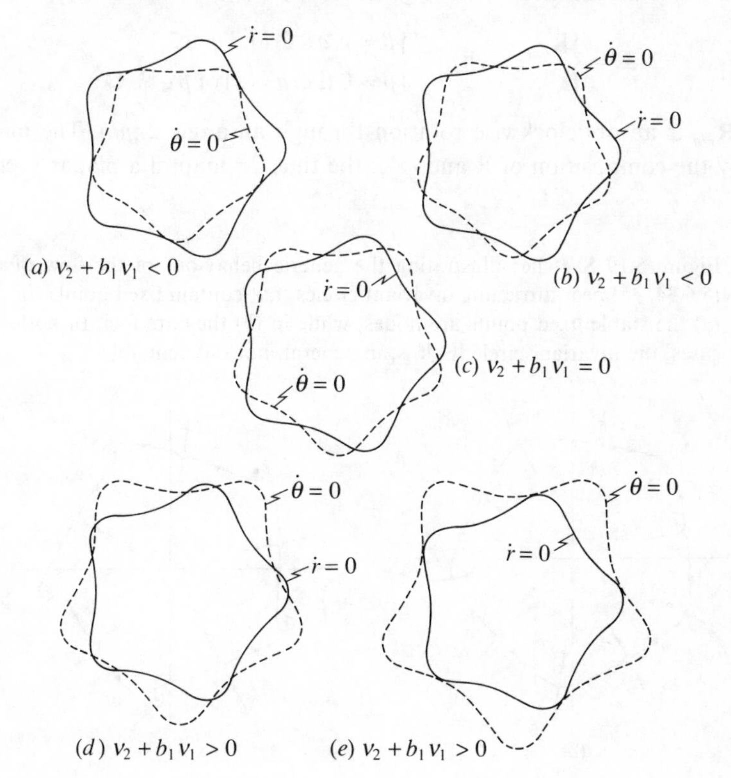

separatrices and the fixed points themselves. In general, such an invariant circle will be non-differentiable (see Figure 5.19). Finally, we conclude that the bifurcation diagram for (5.6.34,35) with $q = 5$ is as shown in Figure 5.20.

When $q > 5$, F and G are no longer zero, however, they depend only on r. Consequently, their role is to alter quantitative, rather than qualitative, features of Figure 5.20. The crucial change in (5.6.40) is that $\nu_1^{3/2} \mapsto \nu_1^{(q-2)/2}$, so that the index of the cusp is given by $(q-2)/2$, in agreement with (5.4.10).

5.7 Unfoldings of rotations and shears

We must not forget our purpose in considering the families of vector fields discussed in § 5.6. Recall our aim is to study the bifurcations occurring in local families of diffeomorphisms $\mathbf{f}: \mathbb{R}^l \times \mathbb{R}^2 \to \mathbb{R}^2$ that are unfoldings of $\mathbf{f_0}$, where $\mathbf{Df_0}(\mathbf{0})$ is either a non-trivial rotation or a shear. In § 5.5 we showed that, for each $\boldsymbol{\mu} \in \mathbb{R}^l$, $\mathbf{f}_{\boldsymbol{\mu}}$ is differentiably conjugate to a map $\mathbf{P}_{\boldsymbol{\mu}}$ of the form

$$\mathbf{P}_{\boldsymbol{\mu}} = \tilde{\mathbf{P}}_{\boldsymbol{\mu}} + O(|\mathbf{x}|^N), \tag{5.7.1}$$

where N is a finite, positive integer and

$$\tilde{\mathbf{P}}_{\boldsymbol{\mu}} = \varphi_{2\pi}^{\boldsymbol{\mu}} \cdot \mathbf{R}. \tag{5.7.2}$$

In (5.7.2), $\mathbf{R}$ is determined by the eigenvalues, $\exp(\pm 2\pi i\beta)$, of $\mathbf{Df_0}(\mathbf{0})$:

$$\mathbf{R} = \begin{cases} \mathbf{R}_{p/q} \\ \mathbf{I} \end{cases} \quad \text{if} \quad \begin{cases} \beta = p/q \in \mathbb{Q} \text{ and } q \geqslant 2 \\ \beta = 0 \text{ (i.e. } q = 1) \text{ or } \beta \in \mathbb{R}\backslash\mathbb{Q} \end{cases}, \tag{5.7.3}$$

where $\mathbf{R}_{p/q}$ is an anticlockwise rotation through an angle $2\pi p/q$. The map $\tilde{\mathbf{P}}_{\boldsymbol{\mu}}$ is given by the composition of $\mathbf{R}$ and $\varphi_{2\pi}^{\boldsymbol{\mu}}$: the time-$2\pi$ map of a planar vector field

Figure 5.19 Sketches illustrating the generic behaviour of the flow of (5.6.34, 35) near attracting invariant circles that contain fixed points. In (*a*) the stable fixed points are nodes, while in (*b*) they are foci. In both cases, the invariant circle itself is, in general, not differentiable.

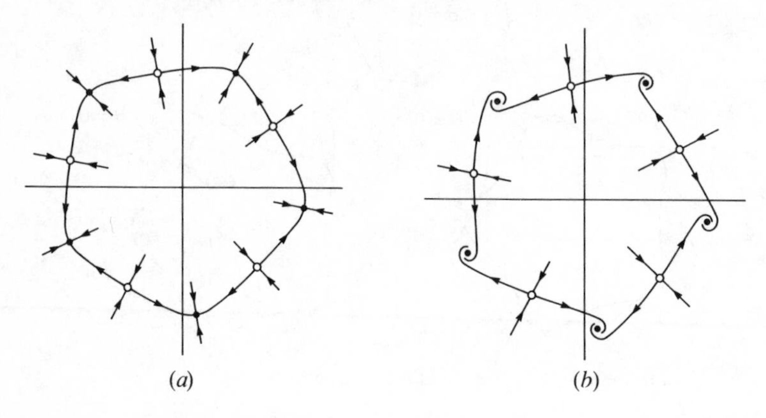

$\tilde{\mathbf{X}}_{\boldsymbol{\mu}}(\mathbf{x})$ that has $2\pi/q$-rotation symmetry. The family, $\tilde{\mathbf{X}}: \mathbb{R}^l \times \mathbb{R}^2 \to \mathbb{R}^2$, of $2\pi/q$-symmetric vector fields $\tilde{\mathbf{X}}_{\boldsymbol{\mu}}(\mathbf{x})$, $\boldsymbol{\mu} \in \mathbb{R}^l$, is an equivariant unfolding of the singular vector field $\tilde{\mathbf{X}}_{\mathbf{0}}$ involved in the approximation of $\mathbf{f}_{\mathbf{0}}$. It is therefore equivalent to a family induced by the equivariant versal unfolding, $\tilde{\mathbf{X}}(\boldsymbol{\nu}, \mathbf{x})$, of $\tilde{\mathbf{X}}_{\mathbf{0}}$.

We make use of the relationship between the unfoldings of $\mathbf{f}_{\mathbf{0}}$ and the unfoldings of its approximating vector field $\tilde{\mathbf{X}}_{\mathbf{0}}$ in an indirect way. We do not view § 5.5 as providing a means of constructing $\tilde{\mathbf{X}}(\boldsymbol{\mu}, \mathbf{x})$ for a given $\mathbf{f}(\boldsymbol{\mu}, \mathbf{x})$. Instead, we use the equivariant versal unfoldings of $\tilde{\mathbf{X}}_{\mathbf{0}}$ to provide insight into the bifurcations that could occur in unfoldings of $\mathbf{f}_{\mathbf{0}}$. For example, we observe that β determines which of the families in § 5.6 is relevant to $\mathbf{f}_{\mathbf{0}}$. If β is irrational then (5.6.1) (or 4.2.18) is required. If β rational and equal to p/q in lowest terms, then q determines which family from § 5.6 should be considered. In this way, β essentially fixes the bifurcation diagram for the vector field approximation that completely unfolds the singularity in $\tilde{\mathbf{X}}_{\mathbf{0}}$. For $q = 2$, there are two possibilities, and there are rather more for $q = 4$, but for other values of q we are led to a unique bifurcation diagram. Note that this diagram takes the same form for $q \geqslant 5$ (weak resonance), while the possible diagrams vary with q for $q \leqslant 4$ (strong resonance).

Figure 5.20 Bifurcation diagram for the family of $2\pi/5$-symmetric vector fields defined by (5.6.33) with $q = 5$. Non-trivial fixed points exist in the flows corresponding to values of $\boldsymbol{\nu}$ in the tongue approximated by (5.6.42). This diagram should be compared with Figure 5.7.

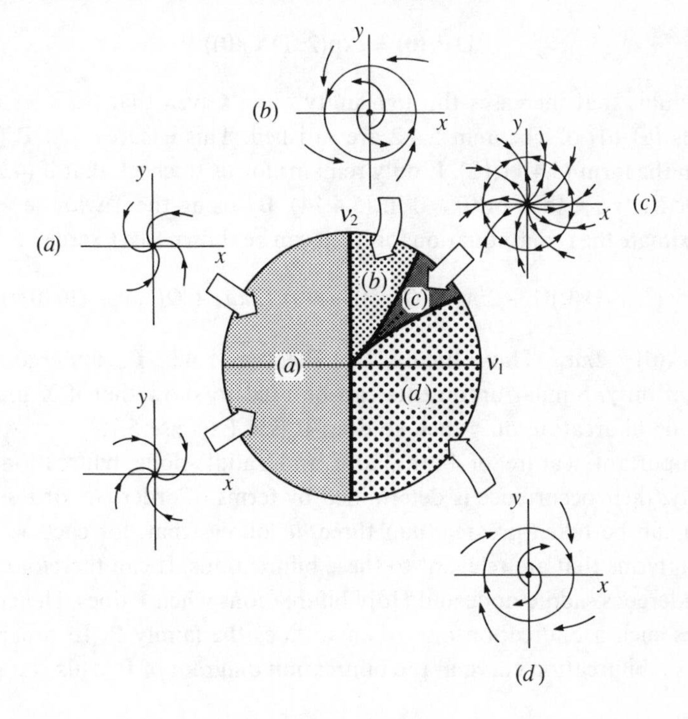

How do we interpret the bifurcation diagram(s) associated with a given value of β? The equivariant versal unfolding $\tilde{\mathbf{X}}(\mathbf{v}, \mathbf{x})$ provides a time-2π map, $\boldsymbol{\varphi}_{2\pi}^{\mathbf{v}}$, for (5.7.2). The composition with the rotation $\mathbf{R}$ has the effect of taking non-trivial fixed points of $\tilde{\mathbf{X}}_{\mathbf{v}}$ onto q-periodic points of $\tilde{\mathbf{P}}_{\mathbf{v}}$. For example, for $q = 2$, $\mathbf{x} = \mathbf{0}$ is a fixed point for both $\tilde{\mathbf{X}}_{\mathbf{v}}$ and $\tilde{\mathbf{P}}_{\mathbf{v}}$, while the non-trivial fixed points of $\tilde{\mathbf{X}}_{\mathbf{v}}$ form a 2-cycle for $\tilde{\mathbf{P}}_{\mathbf{v}}$, because $\mathbf{R} = \mathbf{R}_{1/2}$ in (5.7.2). It is perhaps worth noting that when $q = 1$ the non-trivial saddle and node/focus, that occur for some $\tilde{\mathbf{X}}_{\mathbf{v}}$ (see Figure 4.10), are fixed points for $\tilde{\mathbf{P}}_{\mathbf{v}}$ also, since $\mathbf{R} = \mathbf{I}$. Closed orbits in the phase portrait of $\tilde{\mathbf{X}}_{\mathbf{v}}$ result in closed invariant curves for $\tilde{\mathbf{P}}_{\mathbf{v}}$. For example, when $q = 1$ the limit cycle arising from the Hopf bifurcation (see Figure 4.10) corresponds to an invariant curve for $\tilde{\mathbf{P}}_{\mathbf{v}}$ simply because $\tilde{\mathbf{P}}_{\mathbf{v}} = \boldsymbol{\varphi}_{2\pi}^{\mathbf{v}}$. Alternatively, when $q = 4$ the four limit cycles surrounding the non-trivial foci (see Figure 5.17(k)) are distinct closed orbits for $\tilde{\mathbf{X}}_{\mathbf{v}}$ but all four closed invariant curves are sampled by a single orbit of $\tilde{\mathbf{P}}_{\mathbf{v}}$, with the orbit only returning to the original closed invariant curve after four iterations (cf. Figure 5.9). Of course, the nature of the dynamics for $\tilde{\mathbf{P}}_{\mathbf{v}}$ on a closed invariant curve will vary with $\mathbf{v}$ since both rational and irrational rotation numbers can be encountered.

If $\tilde{\mathbf{X}}$ undergoes a Hopf bifurcation when $\mathbf{v}$ crosses some curve γ in the $\mathbf{v}$-plane, then $\tilde{\mathbf{P}}_{\mathbf{v}}$ also undergoes a Hopf bifurcation. To verify this we need to check that the conditions given in Theorem 5.4.2 are satisfied by $\tilde{\mathbf{P}}_{\mathbf{v}}$. We will illustrate the calculation for $\beta \in \mathbb{R}\backslash\mathbb{Q}$, when the Hopf bifurcation takes place at $\mathbf{x} = \mathbf{0}$. The condition that the real part of the eigenvalues of $\mathrm{D}\tilde{\mathbf{X}}_{\mathbf{v}}(\mathbf{0})$ increases through zero on γ means that those of

$$\mathrm{D}\tilde{\mathbf{P}}_{\mathbf{v}}(\mathbf{0}) = \exp(2\pi \mathrm{D}\tilde{\mathbf{X}}_{\mathbf{v}}(\mathbf{0})) \tag{5.7.4}$$

have modulus that increases through unity on γ. Given that $\beta \in \mathbb{R}\backslash\mathbb{Q}$, it follows that items (a)–(d) of Theorem 5.4.2 are satisfied. This ensures that $\tilde{\mathbf{P}}_{\mathbf{v}}(\mathbf{x})$ can be written in the form (5.4.11,12). It only remains for us to check that if (4.2.24) gives $a_1 \neq 0$ for $\tilde{\mathbf{X}}_{\mathbf{v}}$, $\mathbf{v} \in \gamma$, then $c_3(0) \neq 0$ in (5.4.14). By using the Taylor series method to approximate the radial equation for $\tilde{\mathbf{X}}_{\mathbf{v}}$, it can be shown (see Exercise 5.7.1) that

$$r(2\pi) = r(0)(1 + 2\pi v_1 + O(v_1^2)) + r(0)^3(2\pi a_1 + O(v_1)) + O(r(0)^5) \tag{5.7.5}$$

so that $c_3(0) = 2\pi a_1$. Thus, finally, by Theorem 5.4.2, $\tilde{\mathbf{P}}_{\mathbf{v}}$ undergoes a Hopf bifurcation on γ. Similar arguments can be used to show that if $\tilde{\mathbf{X}}$ undergoes a saddle-node bifurcation on γ then so does $\tilde{\mathbf{P}}$ (see Exercise 5.7.2).

The important feature of both Hopf and saddle-node bifurcations is that, generically, their occurrence is determined by terms of order $|\mathbf{x}|^3$ or less. Since N in (5.7.1) can be taken greater than three, it follows that, for each $\mathbf{v}$, $\mathbf{P}_{\mathbf{v}}$ agrees with $\tilde{\mathbf{P}}_{\mathbf{v}}$ in terms that are relevant to these bifurcations. It can therefore be shown that $\mathbf{P}$ undergoes saddle-node and Hopf bifurcations when $\tilde{\mathbf{P}}$ does. Hence if $\tilde{\mathbf{X}}(\mathbf{v}, \mathbf{x})$ undergoes such a bifurcation on γ then so does the family $\mathbf{P}$. In other words, γ persists as a bifurcation curve in the bifurcation diagram of $\mathbf{P}$. This is not the case for the bifurcation curves involving saddle connections in the vector field approximation.

Uniqueness of solutions implies that if two trajectories of a flow intersect in a point, they must completely coincide. Thus, for flows, the unstable separatrix of a saddle meets the stable separatrix of a saddle in a single trajectory – a saddle connection (homoclinic or heteroclinic). This is in complete contrast to diffeomorphisms. We have already noted (see § 3.7) that, generically, stable and unstable manifolds of diffeomorphisms intersect transversely, resulting in homoclinic or heteroclinic tangles. This difference has important repercussions. Consider $q = 1$ for example. A homoclinic saddle connection occurs in the phase portrait of $\tilde{\mathbf{X}}_\nu$ for $\boldsymbol{\nu}$ on a curve γ_2 in the $\boldsymbol{\nu}$-plane (see Figure 4.10). Here $\tilde{\mathbf{P}}_\nu = \varphi^\nu_{2\pi}$, the time-$2\pi$ map of $\tilde{\mathbf{X}}_\nu$, so the trajectories of $\tilde{\mathbf{X}}_\nu$ are invariant curves for $\tilde{\mathbf{P}}_\nu$ and the family $\tilde{\mathbf{P}}$ undergoes a saddle connection bifurcation of the same kind on γ_2. However, for diffeomorphisms this represents highly non-generic behaviour. So much so that, no matter how large N is taken to be in (5.7.1), higher order terms can perturb $\tilde{\mathbf{P}}_\nu$, $\boldsymbol{\nu} \in \gamma_2$, so that the stable and unstable manifolds of the saddle for $\mathbf{P}_\nu$ do not

Figure 5.21 Homoclinic saddle connections in vector field approximations are replaced by homoclinic tangles in generic planar diffeomorphisms. This schematic diagram indicates how such tangles might behave as the parameters are varied in a family of diffeomorphisms. The two ways in which the manifolds move apart are illustrated by the extreme cases, (*a*) and (*c*), where only tangency remains.

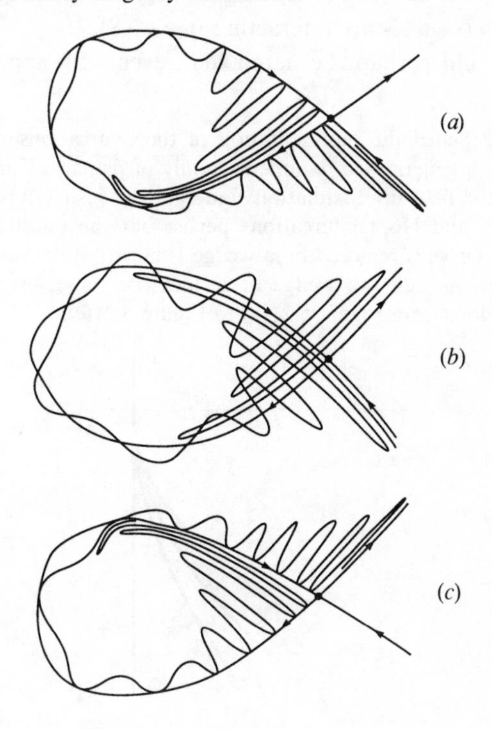

coincide but intersect transversely in a homoclinic tangle (see Figure 5.21(*b*)). It follows that the 'saddle connection' no longer takes place on a single curve. The curve is replaced by a narrow wedge of parameter values (see Figure 5.22). Between the boundaries, γ_2' and γ_2'', of the wedge, the stable and unstable manifolds move through one another as indicated in the sequence $(a) \to (b) \to (c)$ in Figure 5.21. As we move from γ_2' to γ_2'', infinitely many bifurcations occur corresponding to tangencies between stable and unstable manifolds forming and breaking, as they are drawn through one another. The dynamics of $\mathbf{P}_\mathbf{v}$, for $\mathbf{v}$ in the wedge, is extremely complicated. On a microcomputer, the orbit of a single point under $\mathbf{P}_\mathbf{v}$ appears to move within the region containing the tangle in an essentially random manner. Theoretically, the tangle can be shown to contain horseshoe maps and infinitely many sinks (see §3.8).

Similar phenomena occur at heteroclinic saddle connections and, indeed, at double limit-cycle bifurcations. In the latter case, the limit cycles in the approximating vector field move together, coincide and disappear. This happens for the corresponding invariant circles in $\tilde{\mathbf{P}}$ also. However, this degenerate behaviour cannot occur generically for $\mathbf{P}$. The invariant circles must intersect transversely and move through one another in some way over a range of parameter values. Let us assume that transverse intersections occur. Since the set of points of intersection of the circles must be invariant under $\mathbf{P}_\mathbf{v}$, we might not be surprised to find both stable (unstable) and saddle-type periodic points within this set. However, the stable and unstable manifolds of such points may intersect in homoclinic or heteroclinic tangles leading to very complicated dynamics in the region where the two circles are interacting (see §6.8).

In closing, it should perhaps be noted that, even with appropriate wedges of

Figure 5.22 Schematic representation of the bifurcations that can be expected in a generic two-parameter family of planar diffeomorphisms whose vector field approximation undergoes a cusp bifurcation. The saddle-node and Hopf bifurcations persist but the saddle-connection bifurcation curve is replaced by a wedge (shaded) of parameter values. As $\mathbf{v}$ moves through this wedge in a clockwise sense, the homoclinic tangle develops from Figure 5.21(*a*) to Figure 5.21(*c*).

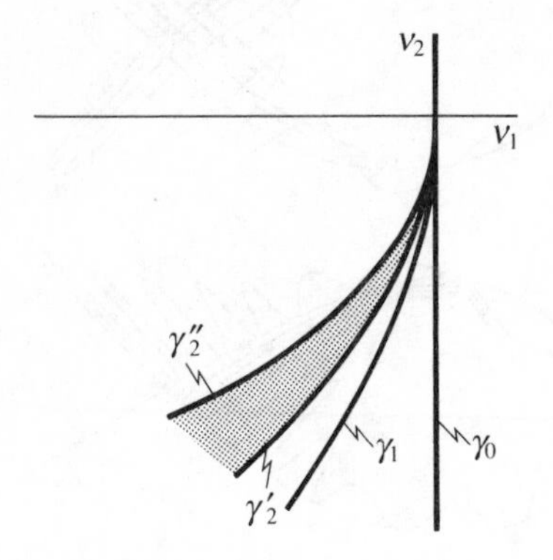

bifurcations added, the vector field approximation bifurcation diagrams present comparatively incomplete information relative to topological conjugacy. In a sense, such diagrams use a weaker equivalence relation (namely topological equivalence of vector field approximation) to focus attention on some gross features in the bifurcation diagrams of the corresponding families of planar maps.

Exercises

5.1 Introduction

5.1.1 Let $\mathbf{f}(\mu, \mathbf{x})$ be a smooth, generic, one-parameter family of planar diffeomorphisms for which the eigenvalues λ_1, λ_2 of $\mathbf{Df}_0(\mathbf{0})$ satisfy (5.1.2). Write down suspensions of the families (4.4.9) and (4.4.36) corresponding to (5.1.4) and (5.1.5) making clear how the signs involved are obtained from $\mathbf{f}$. Draw diagrams to illustrate the bifurcations taking place in each suspended family.

5.1.2 The co-dimension of an m-dimensional manifold in $\mathbb{R}^n$ is $n-m$. Let $\mathbf{A} = \begin{pmatrix} a & b \\ c & d \end{pmatrix}$.

Show that the set of points $(a, b, c, d) \in \mathbb{R}^4$ which satisfy $\operatorname{Tr} \mathbf{A} = 0$, $\operatorname{Det} \mathbf{A} = -1$ form a manifold of co-dimension two in $\mathbb{R}^4$.

5.1.3 Show that the normal form of a smooth family of diffeomorphisms $\mathbf{f}\colon \mathbb{R}^m \times \mathbb{R}^2 \to \mathbb{R}^2$ such that $\mathbf{f_0}(\mathbf{0}) = \mathbf{0}$, $\mathbf{Df_0}(\mathbf{0}) = \begin{pmatrix} 1 & 0 \\ 0 & -1 \end{pmatrix}$ has infinity jet

$$\begin{pmatrix} x + \sum_{i=0}^{\infty} a_{2i}(\boldsymbol{\mu}, x) y^{2i} \\ -y + \sum_{i=0}^{\infty} b_{2i+1}(\boldsymbol{\mu}, x) y^{2i+1} \end{pmatrix}, \tag{E5.1}$$

where $a_0(\mathbf{0}, x) = O(x^2)$, $b_1(\mathbf{0}, x) = O(x)$ and a_{2i} and b_{2i+1} are smooth functions for all $i \in \mathbb{N}^+$.

Verify that the local family, at $(\boldsymbol{\mu}, \mathbf{x}) = (\mathbf{0}, \mathbf{0})$, of product diffeomorphisms given by

$$\mathbf{f}(\mu_1, \mu_2, x, y) = \begin{pmatrix} \mu_1 + x + x^2 \\ (\mu_2 - 1)y - y^3 \end{pmatrix} \tag{E5.2}$$

is in normal form and describe its bifurcational behaviour in a neighbourhood of the origin in $\mathbb{R}^2 \times \mathbb{R}^2$.

5.1.4 Consider the one-parameter local family

$$\mathbf{f}(\mu, \mathbf{x}) = \begin{pmatrix} x + \mu x y \\ -y + \mu(x^2 - y^2) \end{pmatrix} \tag{E5.3}$$

at $(\mu, \mathbf{x}) = (0, \mathbf{0})$.

(a) Plot orbits of $\mathbf{f}_1$ to demonstrate that, for $\mu = 1$, there exist points $\mathbf{x}_0$ such that $\lim_{n \to \infty} \mathbf{f}_1^n(\mathbf{x}_0) = \lim_{n \to -\infty} \mathbf{f}_1^n(\mathbf{x}_0) = \mathbf{0}$.

(b) Prove that $\mathbf{f}_1$ and $\mathbf{f}_\mu$, $\mu \neq 0$, are differentiably conjugate.

(c) Use (a) and (b) to show that (E5.3) is *not* equivalent to a family induced by (E5.2).

5.1.5 Let $\mathbf{A} = \begin{pmatrix} a & b \\ c & d \end{pmatrix}$. Prove that Tr $\mathbf{A} = 2\delta$, $\delta = \pm 1$, and Det $\mathbf{A} = +1$ if and only if $a = 2\delta - d$ and

$$\tfrac{1}{4}(b-c)^2 = \tfrac{1}{4}(b+c)^2 + (d-\delta)^2. \tag{E5.4}$$

Show that the set of points $(a, b, c, d) \in \mathbb{R}^4$ which satisfy (E5.4) form a double cone in (b, c, d)-space with common vertex at $(0, 0, \delta)$. Hence deduce that the linear types in (5.1.7) form a manifold of co-dimension two in $\mathbb{R}^4$, while those in (5.1.9) correspond to one of co-dimension four.

5.1.6 Consider a smooth, one-parameter unfolding of $\mathbf{f}_0$, where $\mathbf{f}_0(\mathbf{0}) = \mathbf{0}$ and $\mathrm{D}\mathbf{f}_0(\mathbf{0})$ has both eigenvalues of modulus unity. For which of the possibilities (5.1.6)–(5.1.9) can it be assumed that this unfolding is equivalent to a local family with $\mathbf{f}_\mu(\mathbf{0}) = \mathbf{0}$, for μ in an interval containing 0.

5.2 Arnold's circle map

5.2.1 Consider the circle map $f_{(\alpha,\varepsilon)}$ defined in (5.2.3) with lift given by (5.2.4). For $k \in \mathbb{Z}^+$, prove that:

(a)

$$\bar{f}^k_{(\alpha,\varepsilon)}(x) = x + k\alpha + F_k(\alpha, \varepsilon, x), \tag{E5.5}$$

where

$$F_k(\alpha, \varepsilon, x) = \varepsilon \sum_{l=0}^{k-1} \sin[2\pi \bar{f}^l_{(\alpha,\varepsilon)}(x)]; \tag{E5.6}$$

(b)

$$\sin[2\pi \bar{f}^k_{(\alpha,\varepsilon)}(x+1)] = \sin[2\pi \bar{f}^k_{(\alpha,\varepsilon)}(x)]. \tag{E5.7}$$

Hence deduce that, for $\alpha = (p/q) + \beta$, $\bar{f}^q_{(\alpha,\varepsilon)}(x)$ is given by (5.2.10), where

$$G_{p/q}(\beta, \varepsilon, x+1) = G_{p/q}(\beta, \varepsilon, x). \tag{E5.8}$$

Explain why (E5.8) is important to the problem of finding $\{(\alpha, \varepsilon) \mid \rho(f_{(\alpha,\varepsilon)}) = p/q\}$.

5.2.2 The following results are useful in evaluations of $G_{p/q}(\beta, \varepsilon, x)$. Let $\eta = \exp(2\pi i p/q)$, $\bar{\eta} = \exp(-2\pi i p/q)$, $p/q \in [0, 1)$. Prove that:

$$\sum_{k=1}^{q-1} \eta^k = -1; \tag{E5.9}$$

$$\sum_{k=1}^{q-1} \eta^k(\eta^k - 1) = 0, \qquad q \geqslant 3; \tag{E5.10}$$

$$\sum_{k=1}^{q-1} \bar{\eta}^k(\eta^k - 1) = q. \tag{E5.11}$$

Use (E5.9) to deduce (5.2.19).

5.2.3 Extend (5.2.18) to show that, for small β, ε,

$$\begin{aligned} G_{p/q}(\beta, \varepsilon, x) = q\beta &+ \beta\varepsilon\left(2\pi \sum_{k=1}^{q-1} \cos\left[2\pi\left(x + \frac{kp}{q}\right)\right]\right) \\ &+ \varepsilon^2\left(2\pi \sum_{k=1}^{q-1} \cos\left[2\pi\left(x + \frac{kp}{q}\right)\right] \sum_{l=0}^{k-1} \sin\left[2\pi\left(x + \frac{lp}{q}\right)\right]\right) \\ &+ O(\varepsilon^{r+1}\beta^s : r+s=2). \end{aligned} \tag{E5.12}$$

Given that $\beta = O(\varepsilon^2)$, deduce that, for $q \geqslant 3$, $\beta_2(x)$ in (5.2.22) is given by

$$\beta_2(x) = \frac{\pi \sin(2\pi p/q)}{2(1-\cos(2\pi p/q))}. \qquad \text{(E5.13)}$$

Verify that $\{(\alpha, \varepsilon) | \rho(f_{(\alpha,\varepsilon)}) = p/q\}$ has boundaries given by

$$\alpha = \begin{cases} \dfrac{1}{4} + \dfrac{\pi}{2}\varepsilon^2 + O(\varepsilon^3) \\ \dfrac{1}{3} + \dfrac{(3)^{1/2}\pi}{6}\varepsilon^2 + O(\varepsilon^3) \\ \dfrac{2}{3} - \dfrac{(3)^{1/2}\pi}{6}\varepsilon^2 + O(\varepsilon^3) \\ \dfrac{3}{4} - \dfrac{\pi}{2}\varepsilon^2 + O(\varepsilon^3) \end{cases} \quad \text{for} \quad \begin{cases} p/q = \frac{1}{4} \\ p/q = \frac{1}{3} \\ p/q = \frac{2}{3} \\ p/q = \frac{3}{4}. \end{cases} \qquad \text{(E5.14)}$$

5.2.4 Given $f_{(\alpha,\varepsilon)}: S^1 \to S^1$ with lift $\bar{f}_{(\alpha,\varepsilon)}(x) = x + \alpha + \varepsilon \sin(2\pi x)$, show that the resonance tongues $T_{p/q} = \{(\alpha, \varepsilon) | \rho(f_{(\alpha,\varepsilon)}) = p/q\}$, $p/q \in \mathbb{Q} \cap [0, 1)$, are such that $(\alpha, \varepsilon) \in T_{p/q}$ if and only if $(1-\alpha, \varepsilon) \in T_{1-(p/q)}$.

5.3 Irrational rotations

5.3.1 Let $\mathbf{f}(\mu, \mathbf{x})$, $\mu \in \mathbb{R}$, $\mathbf{x} \in \mathbb{R}^2$, be the family of planar diffeomorphisms defined by the complex form

$$f(\mu, z) = \lambda(\mu) z (1 + a(\mu)|z|^2), \qquad \text{(E5.15)}$$

where $\lambda(0) = \exp(2\pi i \beta)$, $\beta \in \mathbb{R} \backslash \mathbb{Q}$ (see (5.3.2) with $R(\mu, z) \equiv 0$). Assume that $[\mathrm{d}|\lambda(\mu)|/\mathrm{d}\mu]_{\mu=0} = b > 0$ and that $a(\mu) = c(\mu) + id(\mu)$, where $c(0) = c < 0$ and $d(0) = d \neq 0$. Prove that (E5.15) has two invariant circles for μ small and positive. Show that one of these circles, C, has radius $r_0(\mu)$ given in (5.3.8) and explain why the other is not relevant to the local bifurcation in $\mathbf{f}(\mu, \mathbf{x})$ at $(\mu, \mathbf{x}) = (0, \mathbf{0})$.

5.3.2 Consider the family $f(\mu, z)$ defined by (E5.15) in Exercise 5.3.1.

(a) Show that the invariant circle, C, of radius $r_0(\mu)$ is stable for μ small and positive.

(b) Obtain $f_\mu | C$ and confirm that this is a rotation through an angle

$$2\pi\beta(\mu) - \frac{bd}{c}\mu + O(\mu^2), \qquad \text{(E5.16)}$$

where $\lambda(\mu) = |\lambda(\mu)| \exp(2\pi i \beta(\mu))$.

(c) Does (E5.15) have any invariant circles for μ small and negative? If so, do they figure in the local bifurcation at $(\mu, z) = (0, 0)$?

(d) Describe the bifurcation occurring in the local family defined by (E5.15) at $(\mu, z) = (0, 0)$.

5.3.3 Let $\mathbf{f}(\mu, \mathbf{x})$, $\mu \in \mathbb{R}$, $\mathbf{x} \in \mathbb{R}^2$, be a smooth, generic, one-parameter family of planar diffeomorphisms such that $\mathbf{f}(\mu, \cdot) = \mathbf{f}_\mu(\cdot)$ satisfies $\mathbf{f}_\mu(\mathbf{0}) = \mathbf{0}$ for $\mu \in (-\mu_0, \mu_0)$, $\mu_0 > 0$, and

$$\mathrm{D}\mathbf{f}_0(\mathbf{0}) = \begin{pmatrix} \cos 2\pi\beta & -\sin 2\pi\beta \\ \sin 2\pi\beta & \cos 2\pi\beta \end{pmatrix}, \qquad \beta \in \mathbb{R} \backslash \mathbb{Q}. \qquad \text{(E5.17)}$$

(a) Find the complex normal form of $\mathbf{f}_\mu$ for: (i) $\mu = 0$; (ii) $\mu \neq 0$. Explain why the complex normal form, $\tilde{f}$, of the family $\mathbf{f}: \mathbb{R} \times \mathbb{R}^2 \to \mathbb{R}^2$ is taken to be (5.3.10).
(b) Consider the truncation of $\tilde{f}_\mu(\cdot) = \tilde{f}(\mu, \cdot)$ obtained by dropping $O(|z|^r)$-terms, $r \geq 2N+3$, from (5.3.10). Assume that $\mathrm{Re}(a_3(0)/\lambda(0)) = c < 0$ and that $\mu > 0$ is small. Show that, for any $N \in \mathbb{Z}^+$, the truncated form has a unique invariant circle, C, whose radius tends to zero as $\mu \to 0$. Given that $[\mathrm{d}|\lambda(\mu)|/\mathrm{d}\mu]_{\mu=0} = b > 0$, show that the radius, $r_0(\mu)$, of C takes the form (5.3.8) independently of N.

5.3.4 Consider a two-parameter family of diffeomorphisms $\mathbf{f}_{\boldsymbol{\mu}}: \mathbb{R}^2 \to \mathbb{R}^2$, where $\mathbf{f}_{\boldsymbol{\mu}}(\mathbf{0}) = \mathbf{0}$ and $\mathrm{D}\mathbf{f}_{\boldsymbol{\mu}}(\mathbf{0})$ has complex eigenvalues $\lambda(\boldsymbol{\mu})$, $\bar{\lambda}(\boldsymbol{\mu})$ with $\lambda(\mathbf{0}) = \exp(2\pi i\beta)$. Show that the map $\lambda: U(\subseteq \mathbb{R}^2) \ni \mathbf{0} \to V(\subseteq \mathbb{C}) \ni \exp(2\pi i\beta)$ is a local diffeomorphism if and only if

$$\left[\mathrm{Im}\left(\frac{\partial\lambda}{\partial\mu_1}\cdot\frac{\partial\bar{\lambda}}{\partial\mu_2}\right)\right]_{\boldsymbol{\mu}=(\mu_1,\mu_2)=\mathbf{0}} \neq 0. \qquad \text{(E5.18)}$$

Explain how this condition is related to the elements of $\mathrm{D}\mathbf{f}_{\boldsymbol{\mu}}(\mathbf{0})$. Verify that the map $\lambda_0(\mathbf{v})$ defined in Figure 5.5 satisfies (E5.18) and hence deduce that $\lambda_0^{-1}\cdot\lambda: U \to W \subseteq \mathbb{R}^2 \ni \mathbf{0}$ is generically a local diffeomorphism.

Use the re-parametrised form (5.3.17) to obtain:
(a) the estimate (5.3.18) of the radius of the invariant circle C;
(b) the approximations (5.3.19) and (5.3.20,21) to $f_{\mathbf{v}}^{C}$.

5.4 Rational rotations and weak resonance

5.4.1 Let $\mathbf{f}(\boldsymbol{\mu}, \mathbf{x})$ be a smooth, two-parameter family of planar diffeomorphisms satisfying: (i) $\mathbf{f}_{\boldsymbol{\mu}}(\mathbf{0}) = \mathbf{0}$; (ii) $\mathrm{D}\mathbf{f}_{\mathbf{0}}(\mathbf{0})$ is a rotation through angle $2\pi p/q$, $q \geq 3$, $p/q \in \mathbb{Q} \cap (0, 1)$. Use (5.4.1) to show that the lowest-order additional resonance term in the complex normal form of $\mathbf{f}_{\mathbf{0}}$ is proportional to $\bar{z}^{q-1}$. Explain how the complex normal forms for the family $\mathbf{f}$ given in (5.4.5) and (5.4.6) are obtained.

5.4.2 Show that the polar form of (5.4.6), with $\lambda(\mathbf{v}) = (1 + v_1)\exp(2\pi i((p/q) + v_2))$, is given by (5.4.7) and (5.4.8). In particular, prove that

$$\tilde{c}_{q-1}(\mathbf{v}, \theta) = |b(\mathbf{v})| \cos(2\pi(q\theta - \varphi)), \qquad \text{(E5.19)}$$

$$\tilde{d}_{q-2}(\mathbf{v}, \theta) = -\frac{|b(\mathbf{v})|}{2\pi(1+v_1)} \sin(2\pi(q\theta - \varphi)), \qquad \text{(E5.20)}$$

where $2\pi\varphi = \arg(b(\mathbf{v}))$.

5.4.3 (a) Assume that $r^2 = -v_1/c_3$, $c_3 = c_3(0) < 0$, $v_1 > 0$, in (5.4.8) and ignore $O(r^{q-1})$-terms. Use the form of $\tilde{d}_{q-2}(\mathbf{v}, \theta)$ given in (E5.20) to show that the resulting diffeomorphism of the circle has rotation number p/q for values of $\mathbf{v} = (v_1, v_2)$ lying within a tongue in the $\mathbf{v}$-plane whose boundaries are approximated by

$$v_2 \simeq \frac{d_2}{c_3} v_1 \pm \frac{|b| v_1^{\frac{1}{2}(q-2)}}{2\pi|c_3|^{\frac{1}{2}(q-2)}}. \qquad \text{(E5.21)}$$

The notation used in (E5.21) is that defined in Theorem 5.4.1.

(b) Consider the functions $v_2^{\pm}(v_1)$ defined by

$$v_2^{\pm}(v_1) = \gamma_1 v_1 \pm \gamma_2 v_1^{\frac{1}{2}(q-2)}, \qquad \text{(E5.22)}$$

where $\gamma_1, \gamma_2 \in \mathbb{R}^+$. Show that $\mathrm{D}^n v_2^+(0) = \mathrm{D}^n v_2^-(0)$ for $n = 1, \ldots, [(q-4)/2]$. What does this result imply about the shape of the curves $v_2 = v_2^\pm(v_1)$, near $v_1 = 0$, as q increases through positive integer values? Illustrate your answer by plotting $v_2^\pm(v_1)$ for $0 \leqslant v_1 \leqslant \frac{1}{2}$, when $\gamma_1 = \gamma_2 = 1$ and q is equal to: (i) 5; (ii) 7; (iii) 9.

5.4.4 Show that the family

$$\mathbf{f}(\mu, \mathbf{x}) = \begin{pmatrix} (1+\mu)x + y + x^2 - 2y^2 \\ -x + (1+\mu)y + x^2 - x^3 \end{pmatrix} \tag{E5.23}$$

undergoes a Hopf bifurcation at $\mu = 0$. Are the invariant circles so formed attracting or repelling?

5.5 Vector field approximations

5.5.1 Irrational β

5.5.1 List the steps required to transform (5.5.2) into (5.5.5). Verify that each step involves a transformation that can be interpreted as a diffeomorphism in the phase space of (5.5.1). Deduce that (5.5.5) is the complex form of a system that is differentiably conjugate to (5.5.1).

5.5.2 Rational $\beta = p/q$, $q \geqslant 3$

5.5.2 Apply the transformation $z = \exp(\mathrm{i}\beta t)\zeta$ to (5.5.10) and obtain

$$\dot{\zeta} = \sum_{s=1}^{S} c_s \zeta |\zeta|^{2s} + d\bar{\zeta}^{q-1} + G(\exp(\mathrm{i}\beta t)\zeta, t). \tag{E5.24}$$

Verify that $G(\exp(\mathrm{i}\beta t)\zeta, t)$ is a $2\pi q$-periodic function of t. Let $\hat{P}_\mathbf{0}$, $P_\mathbf{0}$ be the period advance maps of (5.5.10) and (E5.24), respectively. Show that $P_\mathbf{0}$ is differentiably conjugate to $\hat{P}_\mathbf{0}^q$.

5.5.3 (Averaging in the Seifert foliation.) Consider the time-dependent complex form

$$\dot{z} = \mathrm{i}\beta z + \sum_{r=2}^{N} Z_r(z, t) + G(z, t), \tag{E5.25}$$

where

$$Z_r(z, t) = \sum_{\substack{\mathbf{m} \\ m_1 + m_2 = r}} a_\mathbf{m}(t) z^{m_1} \bar{z}^{m_2}, \tag{E5.26}$$

$$G(z, t) = O(|z|^{N+1}), \tag{E5.27}$$

and G, Z_r, $r = 2, \ldots, N$, are 2π-periodic in t. Let $z = \exp(\mathrm{i}\beta t)\zeta$ and verify that $\dot{\zeta}$:
(a) contains no terms linear in ζ;
(b) is $2\pi q$-periodic in t if $\beta = p/q$, p, q relatively prime.

The average over one period of the term of order $|\zeta|^r$ in $\dot{\zeta}$ is

$$\tilde{Z}_r(\zeta) = \frac{1}{2\pi q} \int_0^{2\pi q} \exp(-\mathrm{i}\beta t) Z_r(\exp(\mathrm{i}\beta t)\zeta, t)\,\mathrm{d}t. \tag{E5.28}$$

Prove that $\tilde{Z}_r(\zeta)$ only contains monomials $\zeta^{m_1}\bar{\zeta}^{m_2}$ of order r that satisfy the resonance condition (2.6.24). Show that, for any $r \geqslant 2$, $\tilde{Z}_r(\zeta)$ is invariant under rotation by $2\pi/q$.

5.5.4 Let M take the form in (5.5.20) and define $z_j = M^j(z)$, $j = 0, \ldots, q$. Show that:

(a) $$z_q = \lambda^q z + \sum_{r=2}^{\infty} [\lambda^{q-1} M_r(z, \bar{z}) + \lambda^{q-2} M_r(z_1, \bar{z}_1) + \cdots + M_r(z_{q-1}, \bar{z}_{q-1})], \quad \text{(E5.29)}$$

where $\lambda = \exp(2\pi i\beta)$;

(b) $$M_r(z_j, \bar{z}_j) = M_r(\lambda^j z, \bar{\lambda}^j \bar{z}) + O(|z|^{r+1}), \quad \text{(E5.30)}$$

$j = 1, \ldots, q-1$.
Deduce (5.5.22) and verify that $g_r(z, \bar{z})$ depends only on N and the coefficients $c_{\mathbf{m}}^{(k)}$ for $k < r$.

5.5.5 Consider z_q given in (E5.29) and let $\Sigma_{\mathbb{C}}$ denote the space spanned by the monomials $z^{m_1}\bar{z}^{m_2}$, where $m_1 + m_2 = r \geqslant 2$ and $m_1 - m_2 - 1 = lq$, $l \in \mathbb{Z}$ (see (5.5.19)). Given that $M_r \in \Sigma_{\mathbb{C}}$, for $r \geqslant 2$, prove that (5.5.22) determines M_r uniquely provided that the non-linear terms $N(z) \in \Sigma_{\mathbb{C}}$. What is the status of this calculation if $N(z) \notin \Sigma_{\mathbb{C}}$?

5.5.6 Consider the complex form

$$\dot{z} = \tilde{Z}_0(z) \quad \text{(E5.31)}$$

where $\tilde{Z}_0(z)$ lies in the space $\Sigma_{\mathbb{C}}$ defined in Exercise 5.5.5 (see (5.5.16)). Show that the linear part of the flow of (E5.31) is the identity and that its non-linear terms lie in $\Sigma_{\mathbb{C}}$.

5.5.3 Rational $\beta = p/q$, $q = 1, 2$

5.5.7 Let $\mathbf{f}: \mathbb{R}^2 \to \mathbb{R}^2$ be a local diffeomorphism at $\mathbf{0} \in \mathbb{R}^2$ such that $\mathbf{f}(\mathbf{0}) = \mathbf{0}$ and $\mathbf{Df}(\mathbf{0}) = \begin{pmatrix} -1 & 1 \\ 0 & -1 \end{pmatrix}$. Show that the k-jet of the normal form of $\mathbf{f}$ can be written as

$$j^k(\tilde{\mathbf{f}})(\mathbf{0})(x, y) = \begin{pmatrix} -x + y + \sum_{i=1}^{[(k-1)/2]} a_i x^{2i+1} \\ -y + \sum_{i=1}^{[(k-1)/2]} b_i x^{2i+1} \end{pmatrix}. \quad \text{(E5.32)}$$

Hence conclude that $j^k(\tilde{\mathbf{f}})(\mathbf{0})$ has π-rotation symmetry for any $k \in \mathbb{Z}^+$.

5.5.8 (a) Let $\boldsymbol{\varphi}_t^0: \mathbb{R}^2 \to \mathbb{R}^2$ be the flow of the vector field (5.5.35) for the $q = 2$ case with $k = \infty$. Show that the infinity jet of $\boldsymbol{\varphi}_t^0$ can be written in the form

$$\begin{pmatrix} x - (t/2\pi)y + \sum_{i=1}^{\infty} a_{2i+1}(t) x^{2i+1} \\ y + \sum_{i=1}^{\infty} b_{2i+1}(t) x^{2i+1} \end{pmatrix}, \quad \text{(E5.33)}$$

when $t \neq 0$.

(b) Given that $\mathbf{D\hat{P}_0} = \begin{pmatrix} -1 & 1 \\ 0 & -1 \end{pmatrix}$, prove that $\mathbf{\hat{P}}_0^2 = \boldsymbol{\varphi}_{4\pi}^0$ determines $\mathbf{\hat{P}_0}$ uniquely as $\boldsymbol{\varphi}_{2\pi}^0 \cdot \mathbf{R}_{1/2}$.

5.6 Equivariant versal unfoldings for vector field approximations

5.6.1 $q = 2$

5.6.1 Show that the non-trivial fixed points of the vector field $\tilde{\mathbf{X}}_\nu^\delta(\mathbf{x})$, $\delta = \pm 1$, defined in (5.6.3) can be written in the form $\mathbf{x}_\pm^* = (x_\pm^*, y_\pm^*)$, where

$$x_\pm^* = \left[\delta\left(\nu_2 + \frac{\nu_1(\nu_1 + \delta\nu_2)}{(1 + \nu_2 - \delta\nu_1)}\right)\right]^{1/2}, \tag{E5.34}$$

$$y_\pm^* = \frac{-(\nu_1 + \delta\nu_2)}{(1 + \nu_2 - \delta\nu_1)} x_\pm^*. \tag{E5.35}$$

Let γ_1 be the curve in the $\boldsymbol{\nu}$-plane defined by $\nu_2 + \nu_1^2 + \nu_2^2 = 0$. Show that:
(a) $\mathbf{x}_\pm^* = \mathbf{0}$ on γ_1 for both values of δ;
(b) the fixed points, $\mathbf{x}_\pm^*$, exist above γ_1 for $\delta = +1$ and below γ_1 for $\delta = -1$;
(c) $\operatorname{Det} D\tilde{\mathbf{X}}_\nu^\delta(\mathbf{x}_\pm^*) = -2(\nu_2 + \nu_1^2 + \nu_2^2)$.

5.6.2 Consider the two families of vector fields $\tilde{\mathbf{X}}^\delta(\boldsymbol{\nu}, \mathbf{x})$, $\delta = \pm 1$, defined by (5.6.3). Show that, for fixed $\nu_2 > 0$, both families undergo a subcritical Hopf bifurcation, with increasing ν_1, at $\nu_1 = 0$. Verify that:
(a) the radius, r_0, of the limit cycle involved in the Hopf bifurcation takes the form

$$r_0^2 = -\frac{\nu_1}{a_1} + O(\nu_1^2), \tag{E5.36}$$

where $a_1 > 0$;
(b) when $\delta = +1$, the radial distance, r_S, of the two saddle points from $\mathbf{x} = \mathbf{0}$ satisfies

$$r_S^2 = \nu_2 + O(|\boldsymbol{\nu}|)^2. \tag{E5.37}$$

What do (E5.36) and (E5.37) suggest about the bifurcation diagram for $\tilde{\mathbf{X}}^+(\boldsymbol{\nu}, \mathbf{x})$?

5.6.3 Consider the family $\tilde{\mathbf{X}}^-(\boldsymbol{\nu}, \mathbf{x})$ defined by (5.6.3) with $\delta = -1$. Show that $\operatorname{Tr} D\tilde{\mathbf{X}}_\nu^-(\mathbf{x}_\pm^*) = 0$ on a curve, γ_3, in the ν_1, ν_2-plane satisfying

$$\nu_2 = \tfrac{2}{3}\nu_1 + O(\nu_1^2), \tag{E5.38}$$

with $\nu_2 < 0$. Prove that, for fixed $\nu_2 < 0$, $\tilde{\mathbf{X}}^-(\boldsymbol{\nu}, \mathbf{x})$ undergoes a supercritical Hopf bifurcation at $\mathbf{x} = \mathbf{x}_\pm^*$ as ν_1 increases through γ_3. What does this imply about the limit cycles appearing in region (*d*) of Figure 5.12?

5.6.2 $q = 3$

5.6.4 Consider the complex differential equation

$$\dot{\zeta} = \varepsilon\zeta + c\zeta|\zeta|^2 + d\bar{\zeta}^{q-1}, \tag{E5.39}$$

$q \geqslant 3$. Find the scaling of the variable ζ that reduces the coefficient, d, of $\bar{\zeta}^{q-1}$ to unity. Show that, when $d = 1$, the polar form of (E5.39) is

$$\dot{r} = \nu_1 r + ar^3 + r^{q-1}\cos q\theta, \qquad \dot{\theta} = \nu_2 + br^2 - r^{q-2}\sin q\theta, \tag{E5.40}$$

where $\zeta = r\exp(i\theta)$, $\varepsilon = \nu_1 + i\nu_2$ and $c = a + ib$. Hence verify (5.6.14) and (5.6.21).

5.6.5 Use Theorem 4.2.1 to show that, for any $q \geqslant 3$, the complex differential equation

$$\dot{\zeta} = \varepsilon\zeta + c\zeta|\zeta|^2 + \bar{\zeta}^{q-1}, \tag{E5.41}$$

$\zeta = x + iy$, $\varepsilon = \nu_1 + i\nu_2$, $\nu_2 \neq 0$, $c = a + ib$, undergoes a Hopf bifurcation, with increasing v_1, at $(v_1, x, y) = (0, 0, 0)$. Evaluate a_1 defined by (4.2.24) to prove that this bifurcation is supercritical (subcritical) if $a = \text{Re}(c) < 0\ (>0)$.

5.6.6 Let $\mathbf{X}(\mathbf{x})$ be a planar vector field. Prove that, when expressed in terms of plane polar coordinates (r, θ), Det $\mathbf{DX}(\mathbf{x})$ takes the form

$$\text{Det}\,\mathbf{DX}(\mathbf{x}) = \frac{1}{r}\left\{\left(X_r + \frac{\partial X_\theta}{\partial \theta}\right)\frac{\partial X_r}{\partial r} + \left(X_\theta - \frac{\partial X_r}{\partial \theta}\right)\frac{\partial X_\theta}{\partial r}\right\}, \tag{E5.42}$$

where $X_r = \dot{r}$ and $X_\theta = r\dot{\theta}$.

Take $\mathbf{X}$ to be defined by (5.6.14) and show that, at any non-trivial singular point $\mathbf{x}^*$,

$$\text{Det}\,\mathbf{DX}(\mathbf{x}^*) = -3(v_1^2 + v_2^2) + 3(a^2 + b^2)|\mathbf{x}^*|^4. \tag{E5.43}$$

Use (5.6.17) to show that all the non-trivial singular points of (5.6.14) are of saddle type.

5.6.3 $q = 4$

5.6.7 Show that the complex differential equation $\dot{\zeta} = c\zeta|\zeta|^2 + \bar{\zeta}^3$ (see (5.6.25)) is topologically equivalent to the linear system

$$\begin{pmatrix}\dot{x}\\ \dot{y}\end{pmatrix} = \begin{pmatrix} a+1 & -b \\ b & a-1\end{pmatrix}\begin{pmatrix}x\\ y\end{pmatrix}, \tag{E5.44}$$

where $c = a + ib$. Verify that the singular point at the origin in (E5.44) is: (a) a saddle for $|c|^2 < 1$; (b) a stable node/focus for $|c|^2 > 1$ and $a < 0$; (c) an unstable node/focus for $|c|^2 > 1$ and $a > 0$.

5.6.8 Let $\mathbf{X}(\mathbf{x})$ be the planar vector field with polar form

$$\dot{r} = v_1 r + ar^3 + r^3\cos 4\theta, \qquad \dot{\theta} = v_2 + br^2 - r^2 \sin 4\theta, \tag{E5.45}$$

(see (5.6.21)). Use (E5.42) in Exercise 5.6.6 to show that, at any non-trivial singular point, $\mathbf{x}^*$, of $\mathbf{X}$,

$$\text{Det}\,\mathbf{DX}(\mathbf{x}^*) = -4(v_1^2 + v_2^2) - 4(1 - |c|^2)|\mathbf{x}^*|^4. \tag{E5.46}$$

(a) Suppose (see Example 5.6.1) that $|c|^2 = a^2 + b^2 < 1$ and deduce that any non-trivial singular point of $\mathbf{X}$ is of saddle-type.

(b) Assume (see Example 5.6.2) that:
   (i) $b < -1$, $a < 0$;
   (ii) $\Xi = \xi_2^2 - (|c|^2 - 1)\xi_1^2 > 0$, where $\xi_1^2 = v_1^2 + v_2^2$, $\xi_2 = av_1 + bv_2 < 0$;
   (iii) $r_{p>}^2$ and $r_{p<}^2$ are defined by (5.6.28) and (5.6.29), respectively.

Show that

$$\text{Det}\,\mathbf{DX}(\mathbf{x}^*) = \begin{cases} -8(\Xi)^{1/2} r_{p<}^2 & \text{if} \quad |\mathbf{x}^*| = r_{p<} \\ 8(\Xi)^{1/2} r_{p>}^2 & \text{if} \quad |\mathbf{x}^*| = r_{p>} \end{cases}. \tag{E5.47}$$

5.6.9 Consider the function $\Xi = \xi_2^2 - (|c|^2 - 1)\xi_1^2$, where $\xi_1^2 = v_1^2 + v_2^2$, $\xi_2 = av_1 + bv_2$ and $|c|^2 = a^2 + b^2 > 1$. Show that the level curves of Ξ in the v_1, v_2-plane are a family of hyperbolae with asymptotes ($\Xi = 0$) that are symmetrically placed with respect to the line $\xi_2 = 0$. Find the slopes of the asymptotes and investigate their behaviour as b varies in $(-\infty, -1]$ with a fixed in $(-1, 0)$. Compute slopes for the $\Xi = 0$ lines when $a = -\frac{1}{2}$ and $b = -5$ and draw a diagram showing these and the $\xi_2 = 0$

line. Indicate where the system with these parameter values (see Example 5.6.2) has 0, 4 and 8 non-trivial fixed points.

5.6.10 Let **X** be a planar vector field with polar components X_r and X_θ (see Exercise 5.6.6). Prove that

$$\mathrm{Tr}\,\mathbf{DX}(\mathbf{x}) = \frac{1}{r}\frac{\partial}{\partial r}(rX_r) + \frac{\partial X_\theta}{\partial \theta}. \tag{E5.48}$$

Take **X** to be defined by (5.6.21) and let $\mathbf{x}^*$ be any one of the four non-trivial node/foci encountered in Example 5.6.2. Obtain $\mathrm{Tr}\,\mathbf{DX}(\mathbf{x}^*)$ and show that it changes sign on the line

$$v_2 = \frac{b}{2a}\left\{1 - \left(\frac{1-a^2}{b^2}\right)^{1/2}\right\} v_1 \tag{E5.49}$$

in the v_1, v_2-plane, provided $a^2 < 1$ and $|b|$ is sufficiently large. Verify that $\mathrm{Tr}\,\mathbf{DX}(\mathbf{x}^*) = 0$ on the line (E5.49) for $a = -\frac{1}{2}$, $b = -5$.

5.6.11 Compute the slope of the $\mathrm{Tr}\,\mathbf{DX}(\mathbf{x}^*) = 0$ line, given in (E5.49), for $a = -\frac{1}{2}$ and $b = -5$ and add this line to the diagram produced in Exercise 5.6.9. Observe that $\mathrm{Tr}\,\mathbf{DX}(\mathbf{x}^*) = 0$ in the region where $\Xi > 0$. Prove that this is the case for any (a, b) satisfying $-1 < a < 0$ and $b < -1$.

5.6.12 The nature of the remote Hopf bifurcation occurring on the line (E5.49) in Exercise 5.6.10 can be determined by using (4.2.24). Unfortunately, this requires the vector field at the bifurcation value, $\boldsymbol{v}^*$, to be expressed in terms of local coordinates at the singularity $\mathbf{x}^*$. Moreover, the linear part of the vector field must be in real Jordan form (see (4.2.23)). Write a computer program to carry out these transformations for given v_1 and to calculate the coefficient a_1 in (4.2.24). Show that when $a = -\frac{1}{2}$ and $b = -5$ subcritical remote Hopf bifurcations occur as v_1 increases through the $\mathrm{Tr}\,\mathbf{DX}(\mathbf{x}^*) = 0$ line.

5.6.13 Assume that v_2 is a positive constant and consider the expression for $r_{\mathrm{p}<}^2$, given in (5.6.29), when $-1 < a < 0$, $b \ll -1$ and $v_1 \geqslant 0$.

(a) Show that the non-trivial saddle points discussed in Example 5.6.2 do not exist for $v_1 > v_{1m}$, where

$$v_{1m} = \frac{v_2(a-1)}{b}(1 + O(b^{-2})). \tag{E5.50}$$

(b) Examine the dependence of Ξ on $v_1 \in [0, v_{1m}]$, for fixed v_2, and prove that

$$r_{\mathrm{p}<}^2 < \frac{bv_2}{1-a^2-b^2} + O\left(\frac{v_2}{|b|^3}\right). \tag{E5.51}$$

(c) Explain why you would not expect $r_{\mathrm{p}<}^2$ to vary strongly on $[0, v_{1m}/3]$ when $a = -\frac{1}{2}$. Carry out a graphical comparison of $r_{\mathrm{p}<}^2(v_1)$ and its $v_1 = 0$ value for $a = -\frac{1}{2}$, $b = -5$ and $v_2 = 0.5$. Plot the estimate, $(-v_1/a)$, of the square of the radius of the Hopf limit cycle on the same diagram. What does the result suggest about the bifurcation diagram studied in Example 5.6.2 of the text?

5.6.4 $q \geqslant 5$

5.6.14 Consider the complex differential equation

$$\dot{\zeta} = \varepsilon\zeta + \sum_{s=1}^{[\frac{1}{2}(q-2)]} c_s \zeta |\zeta|^{2s} + d\bar{\zeta}^{q-1}, \tag{E5.52}$$

where $\zeta = x + iy$, $\varepsilon = v_1 + iv_2$ and $c_s = a_s + ib_s$, with $a_1 < 0$.

(a) Find the complex scaling of the variable ζ such that $a_1 \to -1$ and $d \to d' \in \mathbb{R}$.

(b) Assume $a_1 = -1$, $d \in \mathbb{R}$ and $q \geqslant 6$. Show that the polar form of (E5.52) is given by (5.6.34 and 35) with F and G defined by (5.6.36). Explain why $F = G = 0$ for $q = 5$.

(c) Suppose $a_1 = -1$, $d = 0$ and $q = 5$ in (E5.52). Show that the circle of radius $v_1^{1/2}$ centred on the origin is a hyperbolic invariant circle for any v_2 but a hyperbolic limit cycle only if $v_2 \neq -b_1 v_1$. Given these values of a, d and q, does (E5.52) have any hyperbolic non-trivial fixed points when $v_2 = -b_1 v_1$?

5.6.15 Consider the planar differential equation with polar form given in (5.6.34 and 35) with $q = 5$. Assume that v_1, v_2 are small and satisfy $v_2 + b_1 v_1 = 0$ with $v_1 > 0$ and $b_1 < 0$. Show that the flow of the differential equation has ten fixed points, with angular coordinates spaced at intervals of $\pi/5$. Verify that five of these fixed points lie inside the circle $r = v_1^{1/2}$ and five of them lie outside it.

5.6.16 Let $\mathbf{X}$ be the planar vector field with polar form (5.6.34 and 35) with $q = 5$. Use (E5.42) in Exercise 5.6.6 to show that at any non-trivial singular point, $\mathbf{x}^*$, of $\mathbf{X}$

$$\text{Det } D\mathbf{X}(\mathbf{x}^*) = \{-5[(v_1 - r^2)(3v_1 - r^2) + (v_2 + b_1 r^2)(3v_2 + b_1 r^2)]\}|_{\mathbf{x}^*}. \tag{E5.53}$$

Deduce that the non-trivial fixed points lying inside (outside) the circle $r = v_1^{1/2}$ in Exercise 5.6.15 are saddle points (stable node/foci).

5.6.17 Consider the system defined in (5.6.34 and 35) when $q = 5$ and $v_2 + b_1 v_1 = 0$. Plot the $\dot{r} = 0$ and $\dot{\theta} = 0$ isoclines on the same diagram for the particular case where $d = 1$, $b_1 = -1$, $v_1 = 0.05$. Confirm the existence of ten fixed points as predicted in Exercise 5.6.15. Indicate:

(i) the sense in which the vector field crosses the isoclines;

(ii) the sign of $\dot{r}$ and $\dot{\theta}$ in the regions of the plane adjoining the isoclines.

Use this information to determine which of the fixed points are node/foci. Show that the unstable manifold of any saddle point emanates into the region between the isoclines and not into either the region inside, or the region outside, both isoclines. Hence deduce that, in this case, all ten fixed points must lie on a single (in general non-differentiable) invariant circle.

5.6.18 Repeat the diagrammatic analysis of the system discussed in Exercise 5.6.17 when $v_1 = 0.05$ and $v_2 + b_1 v_1 = 0.015$. Sketch trajectories to confirm that this information alone does not preclude the existence of an invariant circle containing only the five saddle points. Show that such a circle cannot exist if the point $(r, \theta) = (r_m, \pi/5)$, where r_m satisfies $dr_m^3 + r_m^2 - v_1 = 0$, lies in the basin of attraction of the nearest node/focus. Hence verify numerically that all ten fixed points are contained in the same invariant circle in this case also.

5.6.19 Consider the equivariant unfolding (5.6.21), discussed in § 5.6.3, when *both* a and $b < -1$.

(a) Plot $\dot{r} = 0$ and $\dot{\theta} = 0$ isoclines for: (i) $a = -5$, $b = -5$, and $v_1 = v_2 = 0.05$; (ii) $a = -1.25$, $b = -5$, $v_1 = 0.05$ and $v_2 = 0.85$; and compare them with the circle of radius $v_1^{1/2}$. Confirm the existence of eight fixed points in both (i) and (ii).

(b) Note the sign of $\dot{r}$ and $\dot{\theta}$ on the isoclines and in the regions adjoining them. Deduce that (i) is only consistent with the existence of an invariant circle of radius $\simeq \nu_1^{1/2}$ containing all eight fixed points. Sketch trajectories to show that (ii) is consistent with an invariant circle of radius $\simeq \nu_1^{1/2}$ with all eight fixed points remote from it.
(c) Plot the saddle-node bifurcation lines in the $\boldsymbol{\nu}$-plane for (i) and (ii) and locate the parameter points for which isoclines have been plotted.
(d) What do the results of (a)–(c) suggest about the bifurcations that are possible for the $q = 4$ case when $a, b < -1$? How do these possibilities differ from the $q = 5$ case and what is the main reason for these differences?

5.7 Unfoldings of rotations and shears

5.7.1 Let the family, $\tilde{\mathbf{X}}(\boldsymbol{\nu}, \mathbf{x})$, defined by the polar form

$$\dot{r} = \nu_1 r + a_1 r^3 + O(r^5), \qquad \dot{\theta} = \nu_2 + \beta + b_1 r^2 + O(r^4), \tag{E5.54}$$

where $\beta \in \mathbb{R}\backslash\mathbb{Q} \cap (0, 1)$ and $a_1 < 0$, be a vector field approximation for an unfolding of $\mathbf{f}_0$. Show that the family $\tilde{\mathbf{P}}(\boldsymbol{\nu}, \mathbf{x})$, defined by (5.7.2), can be written in the form

$$\begin{aligned} r(2\pi) &= r(0)(1 + 2\pi\nu_1 + O(\nu_1^2)) + r(0)^3(2\pi a_1 + O(\nu_1) + O(r(0)^5), \\ \theta(2\pi) &= \theta(0) + \beta + \nu_2 + r(0)^2(2\pi b_1 + 4\pi^2 b_1 \nu_1 + O(\nu_1^2)) + O(r(0)^4). \end{aligned} \tag{E5.55}$$

Verify that both $\tilde{\mathbf{X}}$ and $\tilde{\mathbf{P}}$ undergo a supercritical Hopf bifurcation as ν_1 increases through zero with $\nu_2 \neq 0$. What happens if $a_1 > 0$?

5.7.2 Let $\tilde{\mathbf{X}}(\boldsymbol{\nu}, \mathbf{x})$ be one of the families discussed in § 5.6 with a saddle-node bifurcation curve, γ say. A one-parameter subfamily, $\mathbf{Y}$, corresponding to a curve in the $\boldsymbol{\nu}$-plane that intersects γ transversely at $\boldsymbol{\nu}^*$, undergoes a saddle-node bifurcation in which a fixed point is created at $\mathbf{x}^*$. Assume that, when expressed in terms of local coordinates, $(\mu, \mathbf{y})$, at $(\boldsymbol{\nu}^*, \mathbf{x}^*)$, $\mathbf{Y}$ takes the form

$$\mathbf{Y}(\mu, \mathbf{y}) = \begin{pmatrix} y_1(\lambda + F(\mu, \mathbf{y})) \\ G(\mu, \mathbf{y}) \end{pmatrix}, \tag{E5.56}$$

where $\mathbf{y} = (y_1, y_2)^{\mathrm{T}}$ and F, G are smooth functions with $F(0, \mathbf{0}) = G(0, \mathbf{0}) = 0$. Use Proposition 4.4.1 to show that the family of time-2π maps of (E5.56) also undergoes a saddle-node bifurcation at $(\mu, \mathbf{y}) = (0, \mathbf{0})$.

5.7.3 Sketch phase portraits of the vector field approximation $\tilde{\mathbf{X}}_{\boldsymbol{\nu}}$ for $\boldsymbol{\nu}$ lying on the saddle connection bifurcation curve in: (a) Figure 5.11; (b) Figure 5.12; (c) Figure 5.13. Why would you *not* expect such saddle connections to appear in the family given in (5.7.1)? In each of the cases (a)–(c), sketch the generic behaviour of the stable and unstable manifolds of the saddle points for the family $\mathbf{P}$. How is the saddle-connection bifurcation curve for the vector field approximation reflected in the bifurcation diagram of $\mathbf{P}$?

5.7.4 Consider the family (5.6.33) with $q = 5$. Examine the phase portraits occurring in the subfamily corresponding to a circular arc in the $\boldsymbol{\nu}$-plane which passes from a point in region (*b*) of Figure 5.20, through the resonance tongue, to a point in region (*d*). Sketch the behaviour of the vector field on the invariant circle for representative members of this subfamily. In the light of this information, would you expect the corresponding subfamily of $\mathbf{P}(\boldsymbol{\nu}, \mathbf{x})$ in (5.7.1) to behave differently from that derived from $\tilde{\mathbf{P}}(\boldsymbol{\nu}, \mathbf{x})$ given by (5.7.2)? Explain why similar expectations would not be justified in relation to the arc (*a*)–(*n*) in Figure 5.17 where $q = 4$.

6

Area-preserving maps and their perturbations

No text on dynamics would be complete without some theoretical discussion of area-preserving maps. We have already seen that numerical experiments on such maps yield extremely complicated phenomena and in §§6.2–6.5 we describe the major theorems that account for these observations. However, the study of area-preserving maps has also influenced current thinking in the non-area-preserving case. It seems appropriate, therefore, to end this introductory text with a selection of examples that illustrate some of the avenues of research that have arisen from this interaction (see §§6.6–6.8).

6.1 Introduction

Although our primary objective is to consider area-preserving maps, their connection with vector fields, via Hamiltonian flows, should not be overlooked. Indeed, vector field approximations play an important role throughout this chapter. The following considerations help us to appreciate the significance of area-preservation in the two cases.

It is important to realise that preserving areas in phase space represents a severe restriction of the vector fields and maps that we have studied thus far. For example, the flow of a vector field $\mathbf{X}\colon \mathbb{R}^2 \to \mathbb{R}^2$ is area-preserving if div $\mathbf{X}(\mathbf{x}) \equiv 0$ (see (1.9.23)). Let us impose this condition on the infinity jet $\mathbf{X}(\mathbf{x})$ of the vector field defined in (4.3.20), say. Observe that

$$\operatorname{div} \mathbf{X}(\mathbf{x}) = \sum_{l=1}^{\infty} (2l+1)a_l(x_1^2 + x_2^2)^l \tag{6.1.1}$$

and therefore the flow of $\mathbf{X}$ is area-preserving if and only if $a_l = 0$ for all $l \in \mathbb{Z}^+$. This means (see (2.7.25,26)) that the infinity jet of (4.3.20) must take the polar form

$$(\dot{\theta}, \dot{r}) = \left(\beta + \sum_{l=1}^{\infty} b_l r^{2l}, 0\right) \tag{6.1.2}$$

if its flow is to be area-preserving. Thus, by the criteria used in §4.3.2, area-preservation imposes an infinite level of degeneracy on this system. Of course, the area-preserving vector field is not simply a vector field with a Hopf singularity of infinite degeneracy since the order infinity terms of the latter will generically have non-zero divergence. The following theorem due to Birkhoff shows that (6.1.2) is, in fact, the normal form for a planar vector field with area-preserving flow in the neighbourhood of a stable fixed point (see Exercise 1.9.3).

For volume-preserving vector fields on $\mathbb{R}^{2n}$, the normal form is given in terms of the Hamiltonian function, $H(\mathbf{q}, \mathbf{p})$, whose quadratic part is assumed to have been reduced to the form

$$\frac{1}{2}\sum_{i=1}^{n} \omega_i(q_i^2 + p_i^2) \tag{6.1.3}$$

by a symplectic linear transformation (see Exercise 6.1.1). It should be noted that a symplectic transformation on $\mathbb{R}^2$ is area-preserving (see Exercise 6.1.2). Here $\{\omega_i\}_{i=1}^n$ are the characteristic frequencies of the n-F system (assumed distinct) and $(\mathbf{q}, \mathbf{p})$ are local coordinates at the fixed point. Following Arnold (1978), we make the following definitions.

Definition 6.1.1 *The frequencies, $\omega_1, \ldots, \omega_n$, are said to satisfy a* resonance condition *of order K if there are integers k_i, $i = 1, \ldots, n$, not all zero such that*

$$k_1\omega_1 + \cdots + k_n\omega_n = 0, \tag{6.1.4}$$

where $|k_1| + \cdots + |k_n| = K$.

Definition 6.1.2 *A* Birkhoff normal form *of degree s* for a Hamiltonian *is a polynomial of degree s in the canonical coordinates (Q_i, P_i), $i = 1, \ldots, n$, which is actually a polynomial of degree $[s/2]$ in the variables $\tau_i = (Q_i^2 + P_i^2)/2$.*

Theorem 6.1.1 *Assume that $\{\omega_i\}_{i=1}^n$ does not satisfy any resonance relation of order less than or equal to s. Then there is a canonical coordinate system in a neighbourhood of the fixed point such that the Hamiltonian is reduced to a Birkhoff normal form of degree s up to terms of order $s + 1$, i.e.*

$$H(\mathbf{q}, \mathbf{p}) = H_s(\mathbf{Q}, \mathbf{P}) + R, \qquad R = O(|(\mathbf{Q}, \mathbf{P})|^{s+1}). \tag{6.1.5}$$

For vector fields on the plane, the full power of Theorem 6.1.1 is not required. There is only one characteristic frequency so that resonance cannot occur at any order. Thus,

$$H(q, p) = \sum_{l=1}^{m} \alpha_l \left(\frac{Q^2 + P^2}{2}\right)^l + O(|(\mathbf{Q}, \mathbf{P})|^{2m+2}) \tag{6.1.6}$$

for any $m \in \mathbb{Z}^+$. The polar form of the infinity jet of Hamilton's equations with this Hamiltonian is just (6.1.2) with $\beta = -\alpha_1$ and $b_l = -(l+1)\alpha_{l+1}/2^l$, $l = 1, 2 \ldots$ (see Exercise 6.1.3).

The resonance condition given in Definition 6.1.1 is slightly different to that encountered in Chapter 2, in that the k_i may be negative (cf. Definition 2.3.1). This difference arises from the symplectic nature of the transformations used to produce this normal form and the fact that the Hamiltonian, rather than the vector field, is transformed. Consider, for example, a system with only one degree of freedom. The generating function

$$F_2(q, P) = qP + S_N(q, P), \tag{6.1.7}$$

(see Percival & Richards, 1982, p. 90) where $S_N(q, P)$ is a homogeneous polynomial of degree $N > 2$, gives rise to the symplectic transformation

$$p = P + \mathrm{D}_q S_N(q, P) = P + \mathrm{D}_q S_N(Q, P) + O(|Q, P|^N), \tag{6.1.8}$$

$$q = Q - \mathrm{D}_P S_N(q, P) = Q - \mathrm{D}_P S_N(Q, P) + O(|Q, P|^N), \tag{6.1.9}$$

(see Exercise 6.1.4) where $\mathrm{D}_q S_N(Q, P) = [\partial S_N/\partial q]|_{(Q,P)}$, etc. To understand how the change in the resonance condition comes about, it is sufficient to consider the effect of this transformation on the quadratic terms in the Hamiltonian. This can easily be done in terms of the complex variables

$$z = q + \mathrm{i}p, \qquad \bar{z} = q - \mathrm{i}p, \tag{6.1.10}$$

when the quadratic part of the Hamiltonian takes the form $-\mathrm{i}\omega z\bar{z}$ (see Exercise 6.1.4). Substitution of

$$\mathscr{S}_N(z, \bar{z}) = S_N(q(z, \bar{z}), P(z, \bar{z})) = \sum_{\substack{m,\bar{m} \\ m+\bar{m}=N}} \sigma_{m,\bar{m}} z^m \bar{z}^{\bar{m}} + O(|z|^{N+1}) \tag{6.1.11}$$

into (6.1.8 and 9) leads to

$$z = Z + 2\mathrm{i} \sum_{\substack{m,\bar{m} \\ m+\bar{m}=N}} \bar{m}\sigma_{m,\bar{m}} Z^m \bar{Z}^{\bar{m}-1} + \cdots, \tag{6.1.12}$$

$$\bar{z} = \bar{Z} - 2\mathrm{i} \sum_{\substack{m,\bar{m} \\ m+\bar{m}=N}} m\sigma_{m,\bar{m}} Z^{m-1} \bar{Z}^{\bar{m}} + \cdots. \tag{6.1.13}$$

Thus

$$\mathrm{i}\omega z\bar{z} = \mathrm{i}\omega Z\bar{Z} + 2 \sum_{\substack{m,\bar{m} \\ m+\bar{m}=N}} \omega(m - \bar{m})\sigma_{m,\bar{m}} Z^m \bar{Z}^{\bar{m}} + \cdots. \tag{6.1.14}$$

Clearly, provided the integer

$$k = m - \bar{m} \neq 0, \tag{6.1.15}$$

$\sigma_{m,\bar{m}}$ can be chosen so as to eliminate the terms of order N from the transformed Hamiltonian. However, the point is that k can now take either sign because it is the difference between two positive integers. For n degrees of freedom, (6.1.14) generalises to

$$\sum_{j=1}^{n} \mathrm{i}\omega_j z_j \bar{z}_j = \sum_{j=1}^{n} \mathrm{i}\omega_j Z_j \bar{Z}_j$$

$$+2 \sum_{\substack{\mathbf{m},\bar{\mathbf{m}} \\ \Sigma_j (m_j + \bar{m}_j) = N}} \left[\sum_{j=1}^{n} \omega_j (m_j - \bar{m}_j) \right] \sigma_{\mathbf{m},\bar{\mathbf{m}}} Z_1^{m_1} \cdots Z_n^{m_n} \bar{Z}_1^{\bar{m}_1} \cdots \bar{Z}_n^{\bar{m}_n} + \cdots \tag{6.1.16}$$

and the resonance condition of Definition 6.1.1 follows. Details are considered in Exercise 6.1.5.

Let us turn to the analogous problem for area-preserving maps. We assume that the origin is a stable fixed point of the map and that its linear part has complex eigenvalues with modulus unity. Such a fixed point is said to be *elliptic*.

Definition 6.1.3 *A* Birkhoff normal form *of degree s for an area-preserving planar* map *in the neighbourhood of an elliptic fixed point is a rotation by a variable angle which is a polynomial of degree not more than* $M = [s/2] - 1$ *in the radial variable* τ *of the canonical polar coordinate system, i.e.*

$$(\theta_1, \tau_1) = (\theta + \alpha_0 + \alpha_1 \tau + \cdots + \alpha_M \tau^M, \tau), \tag{6.1.17}$$

where $q = (2\tau)^{1/2} \cos\theta$ *and* $p = (2\tau)^{1/2} \sin\theta$.

Notice the phrase 'canonical polar coordinate system' is used in Definition 6.1.3 to describe the coordinates (θ, τ) and to distinguish them from the more familiar polar coordinates (θ, r). Recall (see Exercise 1.9.5) that the former are related to the cartesian coordinates (x, y) by a symplectic transformation, while the latter are not.

Theorem 6.1.2 *If the eigenvalue of the linear part of an area-preserving map at an elliptic fixed point is not a root of unity of degree s or less, then the map can be locally reduced by a symplectic change of variables to a Birkhoff normal form of degree s plus higher order terms.*

Proof. We have already seen that the symplectic transformation generated by (6.1.7) can be written in the complex form (6.1.12). Let $N = n + 1$, $l = \bar{m} - 1$ so that (6.1.12) becomes

$$z = Z + 2\mathrm{i} \sum_{\substack{m,l \\ m+l=n}} (l+1)\sigma_{m,l+1} Z^m \bar{Z}^l + O(|Z|^{n+1}),$$

$$= Z + 2\mathrm{i}\mathrm{D}_{\bar{Z}} \mathscr{S}_{n+1}(Z, \bar{Z}) + O(|Z|^{n+1}). \tag{6.1.18}$$

Without loss of generality, we can assume the elliptic fixed point is at the origin and write the area-preserving map in complex notation as

$$z_1 = \exp(\mathrm{i}\alpha_0)z + \sum_{j=2}^{n} p_j(z, \bar{z}) + O(|z|^{n+1}), \tag{6.1.19}$$

where $p_j(z, \bar{z})$ is a homogeneous polynomial of degree j in $z, \bar{z}$. Substitution of (6.1.18) into (6.1.19) gives (see Exercise 6.1.6)

$$\begin{aligned} Z_1 = {} & \exp(\mathrm{i}\alpha_0)Z + \sum_{j=2}^{n} p_j(Z, \bar{Z}) \\ & + 2\mathrm{i}[\exp(\mathrm{i}\alpha_0)\mathrm{D}_{\bar{Z}}\mathscr{S}_{n+1}(Z, \bar{Z}) - \mathrm{D}_{\bar{Z}}\mathscr{S}_{n+1}(Z_1, \bar{Z}_1)] + O(|Z|^{n+1}). \end{aligned} \tag{6.1.20}$$

However,

$$\mathrm{D}_{\bar{Z}}\mathscr{S}_{n+1}(Z_1, \bar{Z}_1) = \mathrm{D}_{\bar{Z}}\mathscr{S}_{n+1}(\exp(\mathrm{i}\alpha_0)Z, \exp(-\mathrm{i}\alpha_0)\bar{Z}) + O(|Z|^{n+1}) \tag{6.1.21}$$

and consequently,

$$\begin{aligned} & [\exp(\mathrm{i}\alpha_0)\mathrm{D}_{\bar{Z}}\mathscr{S}_{n+1}(Z, \bar{Z}) - \mathrm{D}_{\bar{Z}}\mathscr{S}_{n+1}(Z_1, \bar{Z}_1)] \\ & = \exp(\mathrm{i}\alpha_0) \sum_{\substack{m,l \\ m+l=n}} (l+1)\sigma_{m,l+1}[1 - \exp(\mathrm{i}\alpha_0(m-l-1))]Z^m\bar{Z}^l + O(|Z|^{n+1}). \end{aligned} \tag{6.1.22}$$

These terms are $O(|Z|^n)$ and, provided

$$\exp(\mathrm{i}\alpha_0(m-l-1)) \neq 1, \tag{6.1.23}$$

the $\sigma_{m,l+1}$ can be chosen to eliminate terms from $p_n(z, \bar{z})$, without affecting $p_j(z, \bar{z})$ with $j < n$. If α_0 is not a root of unity of degree s or less, then

$$\exp(\mathrm{i}\alpha_0 k) \neq 1, \tag{6.1.24}$$

for any $k = \pm 1, \ldots, \pm s$. Thus (6.1.23) always fails to be satisfied when $m = l+1$, i.e. when $n = 2l+1$, and it may fail when $m = 0$, $l = s$, i.e. $n = s$. It follows that the area-preserving map can be reduced to the form

$$z_1 = \exp(\mathrm{i}\alpha_0)z\left(1 + \sum_{l=1}^{[s/2]-1} a_{2l+1}|z|^{2l}\right) + O(|z|^s). \tag{6.1.25}$$

Now introduce canonical polar coordinates (θ, τ) via $z = (2\tau)^{1/2}\exp(\mathrm{i}\theta)$ and define

$$\left(1 + \sum_{l=1}^{[s/2]-1} a_{2l+1}|z|^{2l}\right) = R(\tau)\exp(\mathrm{i}\Theta(\tau)). \tag{6.1.26}$$

The $(s-1)$-jet of (6.1.25) then becomes

$$(\theta_1, \tau_1) = (\theta + \alpha_0 + \Theta(\tau), \tau R(\tau)^2). \tag{6.1.27}$$

This map has Jacobian $(\mathrm{d}\tau_1/\mathrm{d}\tau) = \mathrm{d}[\tau R(\tau)^2]/\mathrm{d}\tau$, which must be identically equal to unity by the area-preserving property. We therefore conclude that $R(\tau) \equiv 1$ and the Birkhoff normal form follows by expressing Θ in powers of τ. □

If the linearisation of the map at the elliptic fixed point has eigenvalues of the form $\exp(\mathrm{i}\alpha_0)$, where α_0 is an irrational multiple of 2π, Theorem 6.1.2 states that the map can be locally reduced to the form

$$(\theta_1, \tau_1) = (\theta + \alpha_0 + \alpha_1\tau + \cdots + \alpha_M\tau^M + R_\theta(\theta, \tau), \tau + R_\tau(\theta, \tau)). \quad (6.1.28)$$

where R_θ, R_τ are $o(\tau^M)$ with $M = [s/2] - 1$, for any $M \in \mathbb{Z}^+$. The form (6.1.28) is precisely what is obtained if the area-preserving condition

$$\mathrm{Det}(\mathbf{Df}(\mathbf{x})) \equiv 1 \quad (6.1.29)$$

is imposed on the normal form

$$\theta_1 = \theta + \alpha_0 + \sum_{l=1}^{N} d_l r^{2l} + O(r^{2N+2}),$$
$$r_1 = r\left(1 + \sum_{l=1}^{N} c_l r^{2l}\right) + O(r^{2N+3}), \quad (6.1.30)$$

given (see Exercise 6.1.7(a)) by (2.5.15) for irrational $\alpha_0/2\pi$. Let (6.1.30) define a map $\mathbf{F}: (\theta, r) \to (\theta_1, r_1)$, then it can be shown that

$$\mathrm{Det}(\mathbf{Df}(\mathbf{x})) = (r_1/r)\,\mathrm{Det}(\mathbf{DF}(\theta, r)) \quad (6.1.31)$$

(see Exercise 6.1.7(b)). Thus

$$\mathrm{Det}(\mathbf{Df}(\mathbf{x})) = \left(1 + \sum_{l=1}^{N} c_l r^{2l}\right)\left(1 + \sum_{l=1}^{N} (2l+1)c_l r^{2l}\right) + O(r^{2N+2}), \quad (6.1.32)$$

and (6.1.29) implies $c_l = 0$ for $l = 1, \ldots, N$. Hence, (6.1.30) must take the form

$$(\theta_1, r_1) = \left(\theta + \alpha_0 + \sum_{l=1}^{N} d_l r^{2l} + O(r^{2N+2}), r + O(r^{2N+3})\right). \quad (6.1.33)$$

In terms of the variable $\tau = r^2/2$, (6.1.33) becomes

$$(\theta_1, \tau_1) = \left(\theta + \alpha_0 + \sum_{l=1}^{N} 2^l d_l \tau^l + o(\tau^N), \tau + o(\tau^N)\right) \quad (6.1.34)$$

which is (6.1.17) with $\alpha_l = 2^l d_l$, for $l = 1, \ldots, M = N$.

The form (6.1.28) with infinite s is often said to be obtained by a 'formal' transformation, i.e. an infinite sequence of symplectic normal form transformations is constructed but the convergence of their composition is not considered. In fact, the convergence of this composition cannot be guaranteed (see Moser, 1968, p. 28), indeed, divergence is to be expected in general (see Rüssmann, 1959). However, when $\alpha_0/2\pi$ is irrational, M in (6.1.28) may be taken to be arbitrarily large. This means that the residual θ-dependence appearing in $(R_\theta(\theta, \tau), R_\tau(\theta, \tau))$ can be removed to arbitrarily high orders in τ. This is not the case when $\alpha_0/2\pi$ is rational. If $\alpha_0 = 2\pi p/q$, then (6.1.23) fails to be satisfied when

$$m - l - 1 = kq \quad (6.1.35)$$

$k \in \mathbb{Z}$ (cf. (2.5.16)). The case $k = 0$ gives rise to the inevitable resonance terms that appear in the Birkhoff normal form, but when $k \neq 0$ additional resonance terms arise and the normal form is no longer of the Birkhoff type. In canonical polars, (θ, τ), the additional resonance terms are not functions of τ alone and therefore the map can no longer be reduced to the form $(\theta + \alpha(\tau), \tau)$ to arbitrarily high order in τ. These additional resonances are prevented from occurring at order less than or equal to s in Theorem 6.1.2 by the requirement that α_0 should not be equal to any of the appropriate roots of unity. However, an alternative use of the transformations generated in the proof of Theorem 6.1.2 would be to allow the additional resonances to occur and to give up the requirement that a *Birkhoff* normal form be obtained. This approach leads to the following corollary to Theorem 6.1.2.

Corollary *At an elliptic fixed point, an area-preserving map can be locally reduced to* normal form *by a symplectic change of coordinates.*

Since the transformation involved is symplectic, the resulting normal form, though not necessarily of Birkhoff type, is still area-preserving.

Elliptic fixed points satisfying the requirements of Theorem 6.1.2 with $s \geqslant 5$ are said to be *generic* if the constant α_1 in (6.1.17) is non-zero. This condition is also sufficient to ensure that the Birkhoff normal form (6.1.17) is an example of an *area-preserving twist map*.

Twist maps are usually taken to be defined on annuli (or equivalently cylinders) where radial (or longitudinal) and angular variables can be used as global coordinates. Here we will assume the annulus lies in the plane and is given in terms of canonical polar coordinates as $A = \{(\theta, \tau) | a \leqslant \tau \leqslant b, 0 \leqslant \theta < 2\pi\}$. An area-preserving twist map, $\mathbf{T}$, then takes the form

$$\mathbf{T}\colon (\theta, \tau) \mapsto (\theta + \alpha(\tau), \tau), \tag{6.1.36}$$

where $\mathrm{d}\alpha/\mathrm{d}\tau = \alpha'(\tau) \neq 0$ for $\tau \in [a, b]$. For the generic case, (6.1.17) has $\alpha'(\tau) = \alpha_1 \neq 0$ for all $\tau > 0$ and therefore the restriction of this map to A is an area-preserving twist map for any a, b with $b > a$. Like a rotation, the orbits of (6.1.36) lie on circles centred on the origin. It differs from a pure rotation in that the rotation number depends on the radial coordinate. Clearly, the rotation number on the invariant circle with $\tau = \tau_0$ is $\alpha(\tau_0)$. This quantity can take both rational and irrational values as τ_0 varies. When it is rational (see § 1.2) every point of the invariant circle is a periodic point. On the other hand, when it is irrational the corresponding invariant circle contains no periodic points and the orbits of (6.1.36) fill out the circle densely. What is more, $\alpha(\tau_0)$ may not lie in $[0, 2\pi)$. It is then convenient not to reduce mod 1 in the calculation of rotation number. This 'lifted' rotation number is commonly used in the topics that follow. Since $\alpha(\tau_0)$ varies continuously with τ_0, maps like (6.1.36) are topologically complicated in a subtle and very degenerate way. This complexity can be destroyed by arbitrarily small perturbations:

for example the map

$$\mathbf{T}_\varepsilon: (\theta, \tau) \mapsto (\theta + \alpha(\tau), (1-\varepsilon)\tau + \varepsilon a), \qquad \varepsilon > 0, \tag{6.1.37}$$

has no invariant circles for $a < \tau \leqslant b$. However, if the perturbations are suitably restricted (e.g. area-preserving; although, as we shall see later, weaker conditions also give interesting phenomena) then area-preserving twist maps can be unfolded to reveal the kind of complexity that is observed in the Hénon map (see § 1.9). In order to explain how this comes about, we will consider the effect of perturbations on the rational and irrational invariant circles of (6.1.36).

6.2 Rational rotation numbers and Birkhoff periodic points

6.2.1 The Poincaré–Birkhoff Theorem

Consider the following perturbation of (6.1.36)

$$\mathbf{M}_\varepsilon: (\theta, \tau) \mapsto (\theta + \alpha(\tau) + f(\theta, \tau, \varepsilon), \tau + g(\theta, \tau, \varepsilon)), \tag{6.2.1}$$

where $\alpha'(\tau) \neq 0$ and f, g are 2π-periodic in θ with $f(\theta, \tau, 0) = g(\theta, \tau, 0) \equiv 0$. We will suppose that $\mathbf{M}_\varepsilon$ is defined on an annulus $a \leqslant \tau \leqslant b$, $a < b$ and that it is area-preserving for all values of ε, i.e.

$$\int_\Gamma \tau \, d\theta = \int_{\mathbf{M}_\varepsilon \Gamma} \tau \, d\theta, \tag{6.2.2}$$

for any closed curve Γ in the annulus.

Theorem 6.2.1 (Poincaré–Birkhoff) *Given any rational number, p/q, between $\alpha(a)/2\pi$ and $\alpha(b)/2\pi$, then there are $2q$ fixed points of $\mathbf{M}_\varepsilon^q: (\theta, \tau) \mapsto (\theta_q, \tau_q)$ satisfying*

$$(\theta_q, \tau_q) = (\theta + 2\pi p, \tau), \tag{6.2.3}$$

provided that ε is sufficiently small.

This result is a weaker version of a much more substantial theorem due to Birkhoff (see § 6.5) that does not make use of a small parameter. The periodic points that are predicted here are often distinguished from other periodic points that may be present by associating them with Birkhoff's name.

It can be shown (see Exercise 6.2.1(a)) that

$$\begin{aligned} \theta_q &= \theta + q\alpha(\tau) + \tilde{f}_q(\theta, \tau, \varepsilon), \\ \tau_q &= \tau + \tilde{g}_q(\theta, \tau, \varepsilon), \end{aligned} \tag{6.2.4}$$

where $\tilde{f}_q(\theta, \tau, 0) = \tilde{g}_q(\theta, \tau, 0) \equiv 0$. If $\alpha(\tau_0)/2\pi = p/q$, $a < \tau_0 < b$, then, when $\varepsilon = 0$, $(\theta_q, \tau_q) = (\theta + 2\pi p, \tau)$ has a solution (θ, τ_0) for every $\theta \in [0, 2\pi)$. Since $\alpha'(\tau_0) \neq 0$, the Implicit Function Theorem ensures that the equation $\theta_q = \theta + 2\pi p$ has a unique solution,

$$\tau = F(\theta, \varepsilon), \tag{6.2.5}$$

such that $F(\theta, 0) \equiv \tau_0$, for sufficiently small ε. Typically, the curve $\mathscr{C}$ defined by (6.2.5) takes the star-like form shown in Figure 6.1. Clearly, points of $\mathscr{C}$ are mapped *radially* by $\mathbf{M}_\varepsilon^q$, since θ and $\theta + 2\pi p$ define the same radial direction, however, the property (6.2.2) means that $\mathscr{C}$ and $\mathbf{M}_\varepsilon^q(\mathscr{C})$ must enclose the same area. It follows that $\mathscr{C}$ and $\mathbf{M}_\varepsilon^q(\mathscr{C})$ must intersect in at least two points (see Exercise 6.2.1(b)). These points are fixed points of $\mathbf{M}_\varepsilon^q$. Moreover, if $\mathbf{x}^*$ is such a fixed point then so are $\mathbf{M}_\varepsilon^l(\mathbf{x}^*)$, for $l = 1, \ldots, q-1$, because $\mathbf{M}_\varepsilon^q \mathbf{M}_\varepsilon^l(\mathbf{x}^*) = \mathbf{M}_\varepsilon^l \mathbf{M}_\varepsilon^q(\mathbf{x}^*) = \mathbf{M}_\varepsilon^l(\mathbf{x}^*)$. Thus $\mathbf{M}_\varepsilon^q$ must have at least $2q$ fixed points. Figure 6.1 shows how these fixed points arise geometrically (see Exercise 6.2.2). It can be seen that two distinct types of periodic point must occur. For example, given that $\alpha'(\tau) > 0$, it follows (see (6.2.4)) that, for sufficiently small ε, points inside $\mathscr{C}$ rotate clockwise, while those outside $\mathscr{C}$ rotate anticlockwise under $\mathbf{M}_\varepsilon^q$. It is then apparent from Figure 6.1 that elliptic • and hyperbolic ∘ fixed points occur alternately at the points of intersection of $\mathscr{C}$ and $\mathbf{M}_\varepsilon^q(\mathscr{C})$. In fact, the corresponding periodic points lie at the heart of the island chains that we encountered in § 1.9. However, it is important not to confuse the loops in Figure 6.1 with the islands themselves. More precise information about the occurrence of island chains can be obtained by examining vector field approximations to carefully selected families of maps on the plane.

6.2.2 *Vector field approximations and island chains*

We begin by considering a theorem (see Takens, 1973a) concerning one-parameter families of area-preserving diffeomorphisms on $\mathbb{R}^2$. In the following discussion, a

Figure 6.1 Schematic representation of the generic forms for $\mathscr{C}$ and $\mathbf{M}_\varepsilon^q(\mathscr{C})$, illustrating how the fixed points of $\mathbf{M}_\varepsilon^q$ arise. The curve $\mathscr{C}$ is mapped radially and, given that $\alpha'(\tau) > 0$, $\mathbf{M}_\varepsilon^q$ moves points inside (outside) $\mathscr{C}$ in a clockwise (anticlockwise) sense. It follows that the fixed points are alternatively elliptic and hyperbolic. Notice that $\mathbf{M}_\varepsilon^q$ has infinitely many fixed points for $\varepsilon = 0$, while for $\varepsilon > 0$ only finitely many occur.

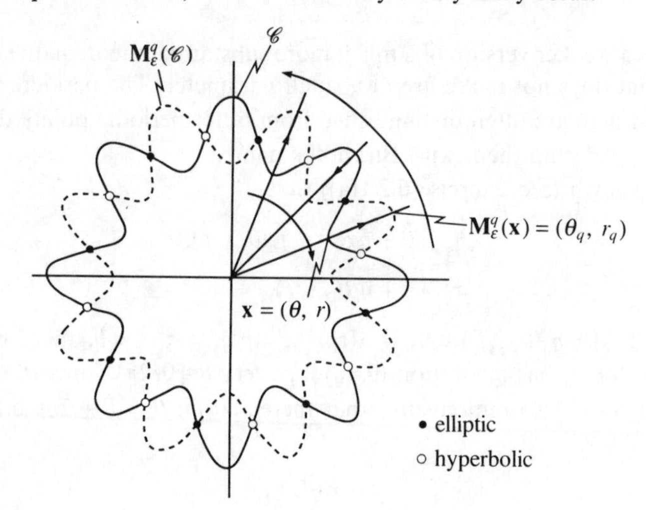

one-parameter family $\mathbf{F}: \mathbb{R}^3 \to \mathbb{R}^3$ of the form $\mathbf{F}(\varepsilon, x, y) = (\varepsilon, F_1^\varepsilon(x, y), F_2^\varepsilon(x, y))^\mathrm{T}$ is conveniently denoted by $(\varepsilon, \mathbf{F}_\varepsilon(x, y))$.

Theorem 6.2.2 *Let $\mathbf{P}: \mathbb{R}^3 \to \mathbb{R}^3$ be a C^∞-diffeomorphism of the form $\mathbf{P}(\varepsilon, x, y) = (\varepsilon, \mathbf{P}_\varepsilon(x, y))$ with $\mathbf{P}_\varepsilon$ area-preserving for all ε and $\mathbf{P}_0(0, 0) = (0, 0)^\mathrm{T}$. Suppose furthermore that $\mathrm{D}\mathbf{P}_0(0, 0)$ has eigenvalues $\exp(\pm\alpha i)$. Let $\mathbf{R}_\alpha: \mathbb{R}^2 \to \mathbb{R}^2$ be the rotation through angle α and define $\tilde{\mathbf{R}}_\alpha: \mathbb{R}^3 \to \mathbb{R}^3$ by*

$$\tilde{\mathbf{R}}_\alpha(\varepsilon, x, y) = (\varepsilon, \mathbf{R}_\alpha(x, y)). \tag{6.2.6}$$

Then there is a function $H: \mathbb{R}^3 \to \mathbb{R}$, invariant under the map $\tilde{\mathbf{R}}_\alpha: \mathbb{R}^3 \to \mathbb{R}^3$, and there is a diffeomorphism $\mathbf{\Psi}: \mathbb{R}^3 \to \mathbb{R}^3$ of the form $\mathbf{\Psi}(\varepsilon, x, y) = (\varepsilon, \mathbf{\Psi}_\varepsilon(x, y))$, where $\mathbf{\Psi}_\varepsilon$ is area-preserving and $\mathbf{\Psi}_0(0, 0) = (0, 0)^\mathrm{T}$, such that the infinity jets of $\mathbf{\Psi}^{-1} \cdot \mathbf{P} \cdot \mathbf{\Psi}$ and $\boldsymbol{\varphi}_1 \cdot \tilde{\mathbf{R}}_\alpha$ are the same in $(0, 0, 0)^\mathrm{T}$. Here $\boldsymbol{\varphi}_1: \mathbb{R}^3 \to \mathbb{R}^3$ is the time-one map of the vector field

$$\left(0, \frac{\partial H_\varepsilon}{\partial y}, -\frac{\partial H_\varepsilon}{\partial x}\right)^\mathrm{T} = (0, X^\varepsilon, Y^\varepsilon)^\mathrm{T}, \tag{6.2.7}$$

i.e. $\boldsymbol{\varphi}_1 = (\varepsilon, \boldsymbol{\varphi}_1^\varepsilon)$, where $\boldsymbol{\varphi}_1^\varepsilon$ is the flow of $(X^\varepsilon, Y^\varepsilon)^\mathrm{T}$ and $H_\varepsilon(x, y) = H(\varepsilon, x, y)$.

The invariance of H under $\tilde{\mathbf{R}}_\alpha$ means that the flow, $\boldsymbol{\varphi}_t^\varepsilon$, of the vector field $(X^\varepsilon, Y^\varepsilon)^\mathrm{T}$ satisfies $\mathbf{R}_\alpha \cdot \boldsymbol{\varphi}_t^\varepsilon = \boldsymbol{\varphi}_t^\varepsilon \cdot \mathbf{R}_\alpha$ for each ε. Thus, if $\mathbf{x}^*$ is a fixed point of $\boldsymbol{\varphi}_t^\varepsilon$, then so is $\mathbf{R}_\alpha^j(\mathbf{x}^*)$, for any $j \in \mathbb{Z}$. When $\alpha = 2\pi p/q$, $\mathbf{R}_\alpha^q$ is the identity and the number of fixed points of $\boldsymbol{\varphi}_t^\varepsilon$ is a multiple of q. These fixed points of the flow are periodic points of $\mathbf{R}_\alpha \cdot \boldsymbol{\varphi}_1^\varepsilon$ with period-q. Moreover, since the infinity jets of $\mathbf{\Psi}^{-1} \cdot \mathbf{P} \cdot \mathbf{\Psi}$ and $\mathbf{R}_\alpha \cdot \boldsymbol{\varphi}_1^\varepsilon$ agree, it can be shown that the topological structure of the periodic points of these two maps are the same, at least in the generic case and near zero (see Exercises 6.2.3,4). The function H can be simplified further by transformations which preserve $\tilde{\mathbf{R}}_\alpha$-invariance but which are not necessarily area-preserving. In the generic cases, this leads to the topological types shown in Figures 6.2–6.8 according to the value of the denominator q. Features resembling island chains are apparent in these phase portraits, particularly for $q \geqslant 5$.

The family $\mathbf{P}$ unfolds the map $\mathbf{P}_0$ which has a non-hyperbolic fixed point at the origin. The first step in the construction of the function H is to obtain the corresponding family of normal forms $\mathbf{N}(\varepsilon, x, y) = (\varepsilon, \mathbf{N}_\varepsilon(x, y))$. Clearly, a linear change of coordinates is required to bring $\mathrm{D}\mathbf{P}_0(\mathbf{0})$ into Jordan form after which $\mathbf{P}_0$ can be reduced to normal form. It is perhaps worth recalling that when $\mathrm{D}\mathbf{P}_0(\mathbf{0})$ is a rotation these calculations are best carried out in terms of the complex coordinates $(z, \bar{z})$ (see Example 2.5.2). If $\mathbf{N}$ is to be a smooth function of ε, the same resonant terms must be retained when $\varepsilon \neq 0$. This procedure allows us to construct the normal forms $\mathbf{N}_\varepsilon$ from $\mathbf{P}_\varepsilon$. The transformations involved in this stage of the calculation can be taken to be symplectic for α both irrational and rational by Theorem 6.1.2 and its corollary. The composition of all these changes of variables is the transformation $\mathbf{\Psi}_\varepsilon(x, y)$ in the statement of Theorem 6.2.2.

Figures 6.2–8 Topological types of phase portraits occurring in the one-parameter families $\varphi_t = (\varepsilon, \varphi_t^{\varepsilon})$ when $\alpha = 2\pi p/q$. The trajectories coincide with the level curves of the function $H(\varepsilon, x, y)$ which is given for each value of q. Notice that these figures are similar to those obtained in § 5.6 but without dissipation. What is more as in Chapter 5, the cases $q = 1, 2$ are distinguished by linearisations that are shears rather than rotations. In the following H_q denotes the polynomial on $\mathbb{R}^2$ which has the form $H_q(\theta, r) = r^q \sin(q\theta)$ in polar coordinates. (After Takens, 1973a.)

Figure 6.2 $q = 1$: $H(\varepsilon, x, y) = \frac{1}{2}y^2 - \frac{1}{3}x^3 - \varepsilon x$.

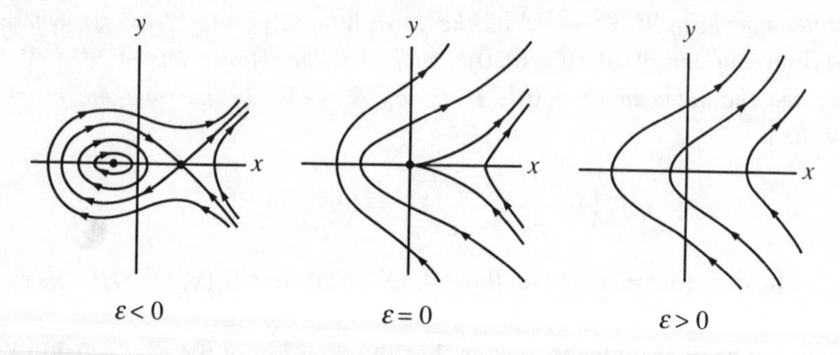

Figure 6.3 $q = 2$: $H(\varepsilon, x, y) = \frac{1}{2}y^2 + \frac{1}{4}x^4 + \varepsilon(x^2 + y^2)$.

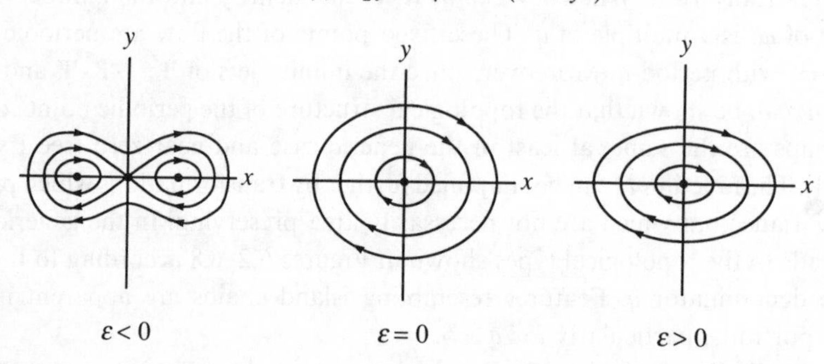

Figure 6.4 $q = 2$: $H(\varepsilon, x, y) = \frac{1}{2}y^2 - \frac{1}{4}x^4 + \varepsilon(x^2 + y^2)$.

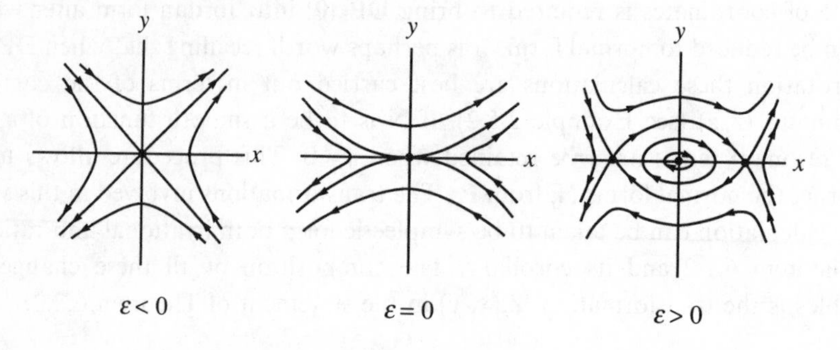

Figure 6.5 $q = 3$: $H(\varepsilon, x, y) = H_3 + \varepsilon(x^2 + y^2)$.

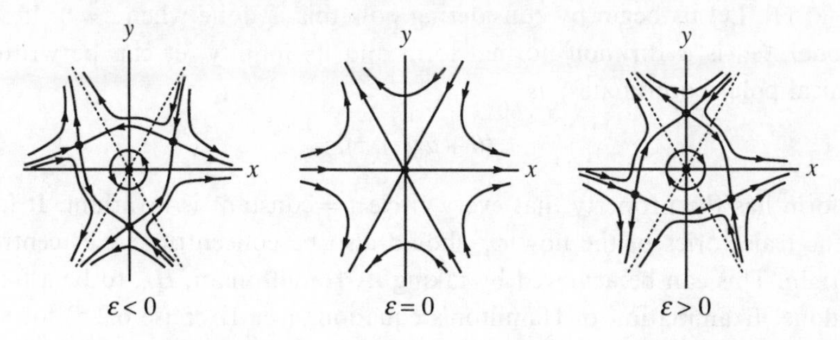

Figure 6.6 $q = 4$: $H(\varepsilon, x, y) = H_4 + \kappa(x^2 + y^2)^2 + \varepsilon(x^2 + y^2)$, $H(0, x, y)$ positive definite.

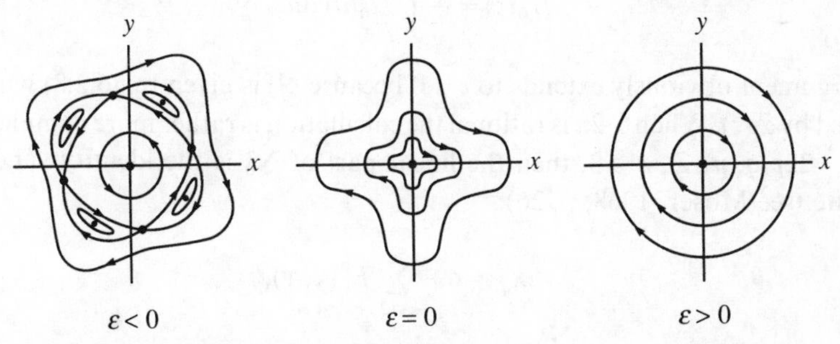

Figure 6.7 $q = 4$: $H(\varepsilon, x, y) = H_4 + \kappa(x^2 + y^2)^2 + \varepsilon(x^2 + y^2)$, $H(0, x, y)$ not definite.

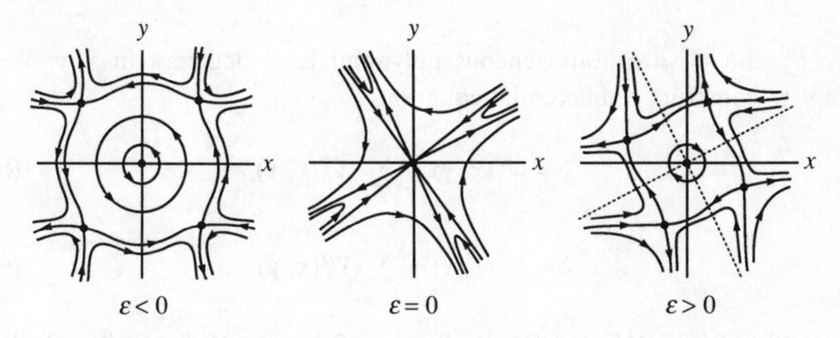

Figure 6.8 $q \geqslant 5$: $H(\varepsilon, x, y) = H_q + (x^2 + y^2)^2 + \varepsilon(x^2 + y^2)$.

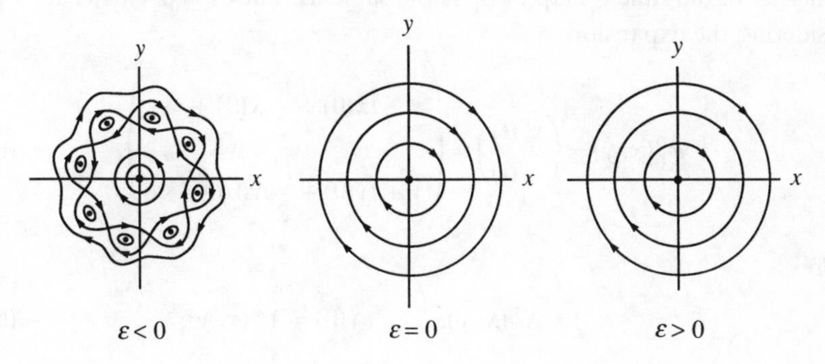

The function H is constructed from the family of normal forms $\mathbf{N} = \mathbf{\Psi}^{-1} \cdot \mathbf{P} \cdot \mathbf{\Psi} = (\varepsilon, \mathbf{N}_\varepsilon(x, y))$. Let us begin by considering how this is done when $\varepsilon = 0$. If $\alpha/2\pi$ is irrational, $\mathbf{N}_0$ is a Birkhoff normal form and its infinity jet can be written, in canonical polar coordinates, as

$$(\theta + \alpha_0(\tau), \tau). \tag{6.2.8}$$

This form has the property that every circle $\tau = \text{constant}$ is invariant. It follows that the trajectories of the flow $\boldsymbol{\varphi}_t^0$ should also be concentric circles centred on the origin. This can be achieved by taking its Hamiltonian, H_0, to be a function of τ alone. Examination of Hamilton's equations (see Exercise 6.2.5) for such a Hamiltonian then gives

$$H_0(\tau) = -\int^{\tau} \alpha_0(u)\, du. \tag{6.2.9}$$

This argument obviously extends to $\varepsilon \neq 0$ because $\mathbf{N}_\varepsilon$ is given by (6.2.8) with $\alpha_0(\tau)$ replaced by $\alpha_\varepsilon(\tau)$. When $\alpha/2\pi$ is rational the calculation is rather more complicated.

If $\alpha = 2\pi p/q$, $p \in \mathbb{Z}$, $q \geqslant 3$, then the linear part of $\mathbf{N}_0^q$ is the identity. Therefore, we write (see Moser, 1968, p. 26):

$$x_q = x + \sum_{k=2}^{\infty} F_k^0(x, y), \tag{6.2.10}$$

$$\mathbf{N}_0^q:$$

$$y_q = y + \sum_{k=2}^{\infty} G_k^0(x, y), \tag{6.2.11}$$

where F_k^0 and G_k^0 are homogeneous polynomials of degree k in x, y. We now attempt to construct a differential equation

$$\dot{x} = X^0(x, y) = \sum_{k=2}^{\infty} X_k^0(x, y), \tag{6.2.12}$$

$$\dot{y} = Y^0(x, y) = \sum_{k=2}^{\infty} Y_k^0(x, y), \tag{6.2.13}$$

where $(X_k^0, Y_k^0) \in H^k$ (H^k is defined following (2.3.1)), with flow $\boldsymbol{\varphi}_t^0$ such that the infinity jet of the time-q map of $\boldsymbol{\varphi}_t^0$ is the same as that of $\mathbf{N}_0^q$. This is achieved by considering the expansion

$$\boldsymbol{\varphi}_t^0(x, y) = \begin{pmatrix} x^0(t) \\ y^0(t) \end{pmatrix} = \begin{pmatrix} x + t\dot{x}(0) + \dfrac{t^2}{2}\ddot{x}(0) + \cdots \\ y + t\dot{y}(0) + \dfrac{t^2}{2}\ddot{y}(0) + \cdots \end{pmatrix}, \tag{6.2.14}$$

where

$$\dot{x}(0) = X^0(x, y), \qquad \dot{y}(0) = Y^0(x, y), \tag{6.2.15}$$

$$\ddot{x}(0) = \left[\frac{\partial X^0}{\partial x} X^0 + \frac{\partial X^0}{\partial y} Y^0 \right]_{(x,y)}, \tag{6.2.16}$$

$$\ddot{y}(0) = \left[\frac{\partial Y^0}{\partial x} X^0 + \frac{\partial Y^0}{\partial y} Y^0 \right]_{(x,y)}. \tag{6.2.17}$$

Observe that the terms involving t^l, $l \geqslant 2$, in (6.2.14) are at least of order 3 in x, y. Thus the coefficients of x^2, xy, y^2 in (X_2^0, Y_2^0) can be chosen so that $\boldsymbol{\varphi}_q^0$ agrees with $\mathbf{N}_0^q$ to this order in x, y, since $F_2^0(x, y)$ and $G_2^0(x, y)$ are assumed known. In general, the terms of order k in (6.2.14) take the form

$$tX_k^0 + \text{terms depending on } X_i^0, Y_i^0, i = 2, \ldots, k-1, \tag{6.2.18}$$

$$tY_k^0 + \text{terms depending on } X_i^0, Y_i^0, i = 2, \ldots, k-1, \tag{6.2.19}$$

(see Exercises 6.2.6 and 6.2.7) so that X_k^0 and Y_k^0 can be calculated given X_i^0, Y_i^0 for $i = 2, \ldots, k-1$. Hence, by induction, X^0 and Y^0 can be constructed so that $\boldsymbol{\varphi}_q^0$ and $\mathbf{N}_0^q$ agree to any desired order.

When $\varepsilon \neq 0$, (6.2.10,11) are replaced by

$$x_q = x + \sum_{k=2}^{\infty} F_k^0(x, y) + \sum_{k=0}^{\infty} \sum_{i=1}^{\infty} \varepsilon^i F_k^i(x, y), \tag{6.2.20}$$

$\mathbf{N}_\varepsilon^q$:

$$y_q = y + \sum_{k=2}^{\infty} G_k^0(x, y) + \sum_{k=0}^{\infty} \sum_{i=1}^{\infty} \varepsilon^i G_k^i(x, y), \tag{6.2.21}$$

and (6.2.12 and 13) by

$$\dot{x} = X^\varepsilon(x, y) = \sum_{k=2}^{\infty} X_k^0(x, y) + \sum_{k=0}^{\infty} \sum_{i=1}^{\infty} \varepsilon^i X_k^i(x, y), \tag{6.2.22}$$

$$\dot{y} = Y^\varepsilon(x, y) = \sum_{k=2}^{\infty} Y_k^0(x, y) + \sum_{k=0}^{\infty} \sum_{i=1}^{\infty} \varepsilon^i Y_k^i(x, y), \tag{6.2.23}$$

An inductive construction of the type described above must then be carried through for each order in ε. For example, (see Exercise 6.2.8) we find that the terms of order ε in $\boldsymbol{\varphi}_t^\varepsilon(x, y)$ take the form

$$tX_k^1 + \text{terms depending on } X_i^1, Y_i^1, i = 0, \ldots, k-1$$
$$\text{and } X_i^0, Y_i^0, i = 2, 3, \ldots, k+1, \tag{6.2.24}$$

$$tY_k^1 + \text{terms depending on } X_i^1, Y_i^1, i = 0, \ldots, k-1$$
$$\text{and } X_i^0, Y_i^0, i = 2, 3, \ldots, k+1, \tag{6.2.25}$$

Thus, given that X_i^0, Y_i^0, $i = 2, 3, \ldots, k+1$, are known from $\mathbf{N}_0$ and assuming X_i^1, Y_i^1, $i = 0, \ldots, k-1$, have already been calculated, it follows that X_k^1 and Y_k^1 can be found.

The vector field $\mathbf{X}^\varepsilon = (X^\varepsilon, Y^\varepsilon)^{\mathrm{T}}$ constructed in this way can be shown to be

Hamiltonian. This follows because the area-preserving property of $\mathbf{N}_\varepsilon$ allows us to show that

$$\operatorname{Tr} \mathbf{DX}^\varepsilon = \frac{\partial X^\varepsilon}{\partial x} + \frac{\partial Y^\varepsilon}{\partial y} = X_x^\varepsilon + Y_y^\varepsilon \equiv 0 \tag{6.2.26}$$

in a neighbourhood of $(x, y) = (0, 0)$. It can be shown (see Exercise 6.2.9) that the Jacobian determinant of the transformation $\boldsymbol{\varphi}_t^\varepsilon$ satisfies

$$\frac{\mathrm{d}}{\mathrm{d}t}\{\operatorname{Det}(\mathbf{D}\boldsymbol{\varphi}_t^\varepsilon(\mathbf{x}))\} = \operatorname{Tr} \mathbf{DX}^\varepsilon(\boldsymbol{\varphi}_t^\varepsilon(\mathbf{x})) \operatorname{Det}(\mathbf{D}\boldsymbol{\varphi}_t^\varepsilon(\mathbf{x})), \tag{6.2.27}$$

subject to the initial condition $\operatorname{Det}(\mathbf{D}\boldsymbol{\varphi}_0^\varepsilon(\mathbf{x})) = 1$. Suppose

$$\operatorname{Tr} \mathbf{DX}^\varepsilon(\mathbf{x}) \not\equiv 0, \tag{6.2.28}$$

then there is a value of $k \geqslant 1$ such that

$$\operatorname{Tr} \mathbf{DX}^\varepsilon(\mathbf{x}) = \tau_k^\varepsilon(x, y) + \cdots, \tag{6.2.29}$$

where τ_k^ε is a homogeneous polynomial of degree k in x, y. In view of (6.2.14), this means that

$$\operatorname{Tr} \mathbf{DX}^\varepsilon(\boldsymbol{\varphi}_t^\varepsilon(\mathbf{x})) = \tau_k^\varepsilon(x, y) + \cdots, \tag{6.2.30}$$

and (6.2.27) then implies

$$\operatorname{Det}(\mathbf{D}\boldsymbol{\varphi}_t^\varepsilon(\mathbf{x})) = 1 + t\tau_k^\varepsilon(x, y) + \cdots. \tag{6.2.31}$$

Thus

$$\operatorname{Det}(\mathbf{D}\boldsymbol{\varphi}_q^\varepsilon(\mathbf{x})) = 1 + q\tau_k^\varepsilon(x, y) + \cdots. \tag{6.2.32}$$

This contradicts the hypothesis that $\mathbf{N}_\varepsilon$ is area-preserving as follows. By construction, $\boldsymbol{\varphi}_q^\varepsilon$ agrees with $\mathbf{N}_\varepsilon^q$ to any order, and consequently

$$\operatorname{Det}(\mathbf{D}\boldsymbol{\varphi}_q^\varepsilon(\mathbf{x})) = \operatorname{Det}(\mathbf{DN}_\varepsilon^q(\mathbf{x})) \tag{6.2.33}$$

to order k in $|\mathbf{x}|$, but $\mathbf{N}_\varepsilon^q$ is area-preserving and therefore

$$\operatorname{Det}(\mathbf{DN}_\varepsilon^q(\mathbf{x})) \equiv 1 \tag{6.2.34}$$

in a neighbourhood of the origin. We conclude therefore that

$$\operatorname{Tr} \mathbf{DX}^\varepsilon(\mathbf{x}) = X_x^\varepsilon(x, y) + Y_y^\varepsilon(x, y) \equiv 0. \tag{6.2.35}$$

What is more, it can then be shown (see Exercise 6.2.10) that there is a formal power series $H_\varepsilon(x, y)$ such that

$$X^\varepsilon = \frac{\partial H_\varepsilon}{\partial y}, \qquad Y^\varepsilon = -\frac{\partial H_\varepsilon}{\partial x}, \tag{6.2.36}$$

and $(X^\varepsilon, Y^\varepsilon)$ is a Hamiltonian vector field. Since ε is a constant independent of t, the component $\dot{\varepsilon} = 0$ is added to (6.2.36) to obtain (6.2.7).

The above arguments lead to a smooth, one-parameter family of approximating vector fields. As in § 5.5, these vector fields are each invariant under rotations by $2\pi/q$ and this leads to an approximation for $\mathbf{N}_\varepsilon$ consisting of a composition of $\boldsymbol{\varphi}_1^\varepsilon$ with a rotation by $2\pi p/q$; i.e. (see Exercise 6.2.11)

$$\mathbf{R}_{p/q}\cdot\boldsymbol{\varphi}_1^\varepsilon = \boldsymbol{\varphi}_1^\varepsilon\cdot\mathbf{R}_{p/q} \tag{6.2.37}$$

(see Proposition 5.5.1). These results are conveniently presented in terms of the maps $\tilde{\mathbf{R}}_\alpha$, $\boldsymbol{\varphi}\colon \mathbb{R}^3 \to \mathbb{R}^3$ defined in (6.2.6) and (6.2.7), respectively.

The use of Theorem 6.2.2 to obtain information about the island chains in a family of maps on the annulus like (6.2.1) is not trivial. We can assume that the annulus is embedded in the plane, centred on the origin and that (6.2.1) is an appropriate restriction of a family, $\mathbf{P}$, of planar maps. However, in studying (6.2.1) we are interested in the effect of perturbations on the invariant circle of the twist map (6.1.36) with rotation number p/q. If this circle is given by $\tau = \tau_0$ then we know that $\alpha(\tau_0) = 2\pi p/q$; but to apply Theorem 6.2.2 we need $\alpha = \alpha(0)$ and this is not available. The weak resonance case $q \geqslant 5$ provides a clue as to how we might proceed. Figure 6.8 clearly shows an approximation to a q-fold island chain. The fixed points of the vector field in this phase portrait are period-q points of $\mathbf{R}_\alpha\cdot\boldsymbol{\varphi}_1^\varepsilon = \boldsymbol{\varphi}_1^\varepsilon\cdot\mathbf{R}_\alpha$. They form two periodic orbits, one stable the other unstable, each rotation number p/q. However, Figure 6.8 corresponds to $\alpha(0)$, rather than $\alpha(\tau_0)$, equal to $2\pi p/q$. Thus, if we are to associate the periodic orbits in this vector field approximation with those predicted by Theorem 6.2.1, we must show that $\mathbf{M} = (\varepsilon, \mathbf{M}_\varepsilon)$, where $\mathbf{M}_0$ is given by $(\theta + \alpha(\tau), \tau)$ and $\alpha(\tau_0) = 2\pi p/q$, behaves in the same way as $\mathbf{P} = (\varepsilon, \mathbf{P}_\varepsilon)$, where $D\mathbf{P}_0(\mathbf{0})$ has eigenvalues $\exp(2\pi i p/q)$. This is a non-trivial task. The desired relationship can be proved (see Chenciner, 1982), at least when $\alpha(0)$ is sufficiently close to $2\pi i p/q$, by studying the two-parameter family of maps obtained by composing $\mathbf{M}_\varepsilon$ with a rotation through $[\alpha(\tau_0) - \alpha(0) + s]$, where s is a new parameter. This new family obviously unfolds a map whose linear part at the origin is a rotation through $\alpha(\tau_0) = 2\pi p/q$. Moreover, $\mathbf{M}_\varepsilon$ can be recovered by setting $s = -[\alpha(\tau_0) - \alpha(0)]$. In fact, the new family can be shown to behave in the same way with ε for all sufficiently small s and Theorem 6.2.2 can be applied when $s = 0$. In this way, a formal link can be forged between the families $\mathbf{M}$ and $\mathbf{P}$. This same problem recurs in § 6.8 and we will postpone further discussion of it until then.

From the practical point of view, it is important to realise that Theorem 6.2.2 does not exclude the existence of periodic points with period other than q, it merely focusses attention on a chosen period q. We can emphasise this point and, at the same time, illustrate the use of the theorem, by considering the Hénon map introduced in § 1.9.

When suitably restricted, the map (1.9.40) (Hénon, 1969) defines a family of the type described in Theorem 6.2.2; namely

$$(\alpha,\ x\cos\alpha - y\sin\alpha + x^2\sin\alpha,\ x\sin\alpha + y\cos\alpha - x^2\cos\alpha)^{\mathrm{T}}. \tag{6.2.38}$$

Moreover, we can satisfy the requirement that the parameter remains small by writing $\alpha = 2\pi p/q - \varepsilon$ for various choices of p/q. Thus, for example, if we pick $p = 1$ and $q = 4$ we might expect to find something resembling Figure 6.6 or 6.7 in the numerical experiments for $\cos\alpha$ near $\cos\pi/2$. This is indeed the case as Figures 6.9 and 6.10, which are reproduced from Hénon (1969), show. This is not a coincidence: $p = 1$, $q = 5$ suggests a five-fold island chain for $\cos\alpha \lessapprox \cos 2\pi/5 \simeq 0.3090$ and Hénon's work shows such a chain for $\cos\alpha = 0.24$ (see Figure 6.11). What is more, it can be shown (see Exercises 6.2.12,13) that this five-fold island chain does seem to disappear for $\cos\alpha \simeq 0.3$. The case $q = 3$ is also visible in Hénon's paper. Observe that $\cos 2\pi/3 = -\frac{1}{2}$ and Figure 6.12(*a*)–(*c*), again reproduced from Hénon (1969), clearly reflect the behaviour predicted in Figure 6.5. Of course, Theorem 6.2.2 does not imply that the island chains occur one at a time. For example, $p = 5$ and $q = 12$ gives $\cos 2\pi p/q = -0.8660$, while $p = 3$ and $q = 7$ gives $\cos 2\pi p/q = -0.9010$. Theorem 6.2.2 would suggest (see Figure 6.8) that both seven-fold and twelve-fold island chains should be present for $\cos\alpha \lessapprox -0.9010$. Figure 12 of Hénon (1969) showing both of these island chains present at $\cos\alpha = -0.95$, is reproduced in Figure 6.13.

Of course, as we saw in § 5.7, the separatrices of the vector field approximation will not be found generically in $\mathbf{P}_\varepsilon$. Instead, we expect homoclinic tangles to occur at each hyperbolic periodic point. In the light of our discussion in § 3.7, these

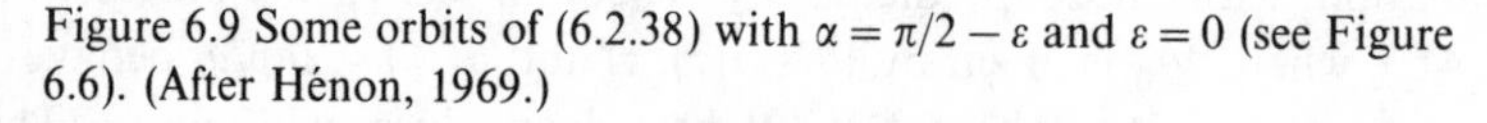

Figure 6.9 Some orbits of (6.2.38) with $\alpha = \pi/2 - \varepsilon$ and $\varepsilon = 0$ (see Figure 6.6). (After Hénon, 1969.)

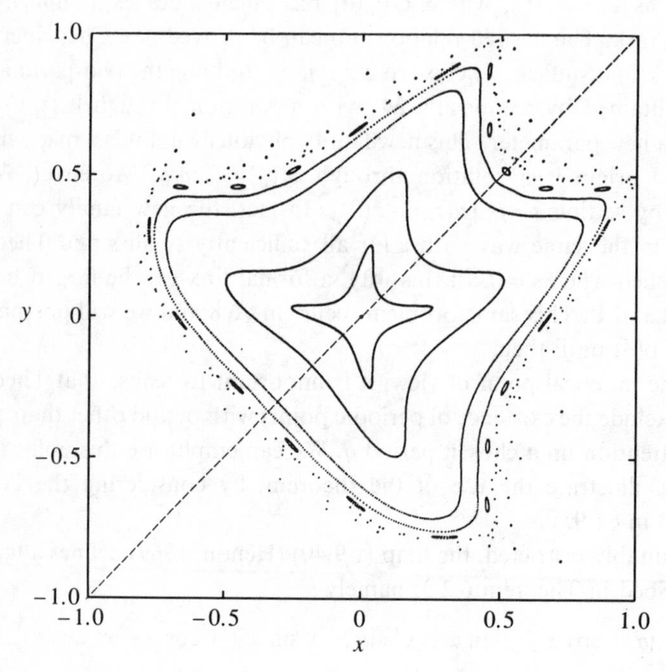

tangles are responsible for the 'two-dimensional' orbits appearing in the figures shown above.

6.3 Irrational rotation numbers and the KAM Theorem

The fact that (6.2.1) has invariant circles on which $\mathbf{M}_\varepsilon$ is conjugate to an irrational rotation is the content of the celebrated theorem of Kolmogorov, Arnold and Moser (see Kolmogorov, 1957; Arnold, 1963; Moser, 1962). Essentially, the theorem states that invariant circles of this kind occur provided that the rotation number is 'sufficiently irrational' and the perturbation, (f, g), is 'sufficiently small'. This result was first enunciated by Kolmogorov and proved by Arnold for the Poincaré maps of non-integrable, analytic Hamiltonian systems with arbitrary numbers of degrees of freedom (Arnold, 1963). An independent proof of an analogous result for sufficiently differentiable, area-preserving, planar maps was given by Moser (1962). The latter form is known as the Moser Twist Theorem and our aim in this section is to highlight some important features of its proof. Complete details of Moser's work can be found in Siegel & Moser (1971) and Sternberg (1969), while Arnold's approach to the result in higher dimensions is presented concisely in Arnold & Avez (1968).

Before embarking on our main discussion, it will be helpful to examine the

Figure 6.10 Selected orbits of (6.2.38) with $\alpha = \pi/2 - \varepsilon$ and $\varepsilon < 0$ (see Figure 6.6). In fact, $\cos\alpha = \sin\varepsilon = -0.01$. (After Hénon, 1969.)

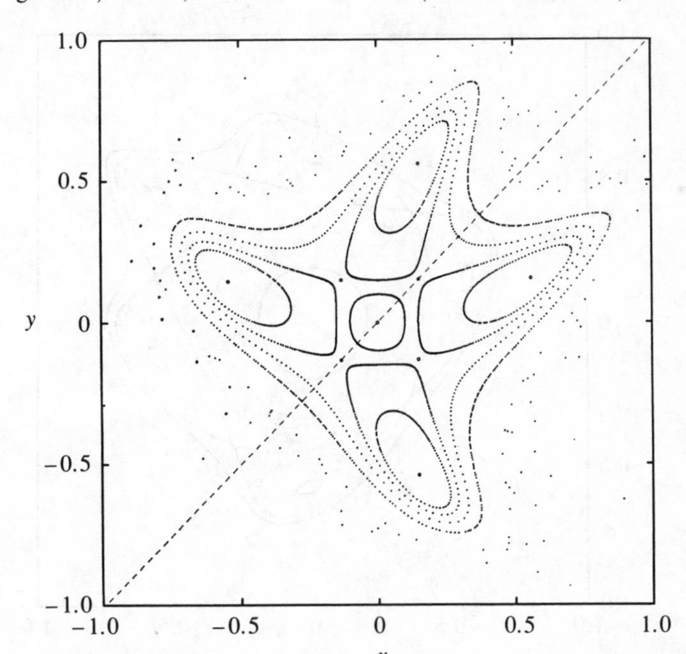

solution of the linear difference equation

$$w(x+\omega)-w(x)=h(x), \tag{6.3.1}$$

where $\omega/2\pi$ is irrational, $h(x)$ is periodic with period 2π and

$$\int_0^{2\pi} h(s)\,\mathrm{d}s = 0. \tag{6.3.2}$$

We can obtain a formal solution by expanding w and h as Fourier series. Thus, let

$$w(x)=\sum_{k\in\mathbb{Z}} \hat{w}_k \exp(ikx) \tag{6.3.3}$$

and, in view of (6.3.2),

$$h(x)=\sum_{\substack{k\in\mathbb{Z}\\ k\neq 0}} \hat{h}_k \exp(ikx). \tag{6.3.4}$$

Substitution into (6.3.1) and comparison of Fourier coefficients yields

$$\hat{w}_k = \hat{h}_k/[\exp(ik\omega)-1], \qquad k\neq 0, \tag{6.3.5}$$

and $\hat{w}_0$ arbitrary. Observe that (6.3.2) is essential if this solution is to exist. What

Figure 6.11 Let $\alpha = 2\pi/5 - \varepsilon$. In view of Figure 6.8, we might expect a five-fold island chain for ε small and negative, i.e. for $\alpha > 2\pi/5$ and, therefore, $\cos\alpha < 0.3090$. This diagram shows typical orbits of (6.2.38) with $\cos\alpha = 0.24$, i.e. $\varepsilon \simeq -0.07$. (After Hénon, 1969.)

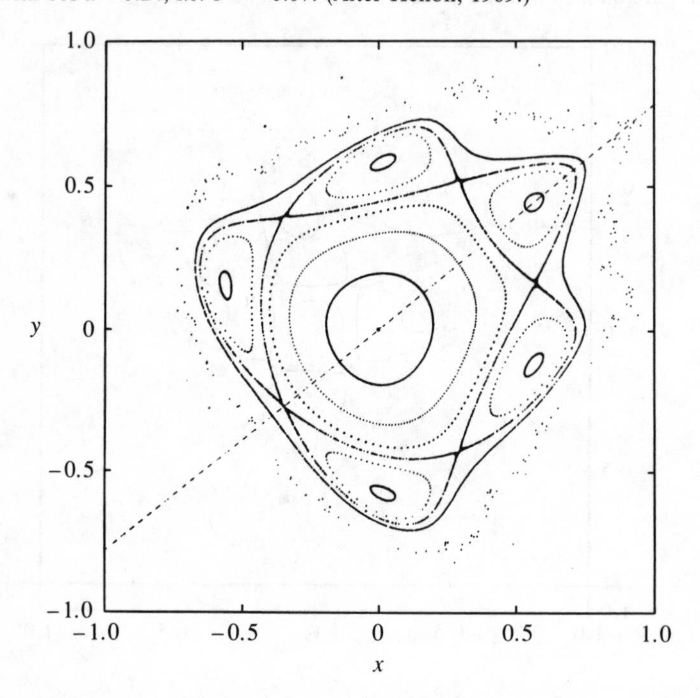

is more, if the series (6.3.3) is to converge the denominators, $[\exp(ik\omega) - 1]$, must not become too small relative to $|\hat{h}_k|$. The following lemma provides a route around this 'small-denominator problem'.

Lemma 6.3.1 *Let $h(x)$ be a real analytic function and assume that there exist c, $\mu > 0$ such that, for all $q \in \mathbb{Z}^+$ and $p \in \mathbb{Z}$,*

$$\left| q\frac{\omega}{2\pi} - p \right| \geqslant cq^{-\mu}. \tag{6.3.6}$$

Then the Fourier series (6.3.3) for $w(x)$ is convergent for all x.

Proof. Elementary manipulation shows that

$$|[\exp(ik\omega) - 1]| = 2\left|\sin\frac{k\omega}{2}\right|. \tag{6.3.7}$$

Figure 6.12 Sample orbits of (6.2.38) for three values of $\cos\alpha$ near $\cos 2\pi/3 = -0.5$: (*a*) $\cos\alpha = -0.42$; (*b*) $\cos\alpha = -0.45$; (*c*) $\cos\alpha = -0.60$. If $\alpha = 2\pi/3 - \varepsilon$, then the corresponding values of ε are: (*a*) $\varepsilon = 0.09$; (*b*) $\varepsilon = 0.06$; (*c*) $\varepsilon = -0.12$. This sequence of diagrams is consistent with the bifurcation shown in Figure 6.5. (After Hénon, 1969.)

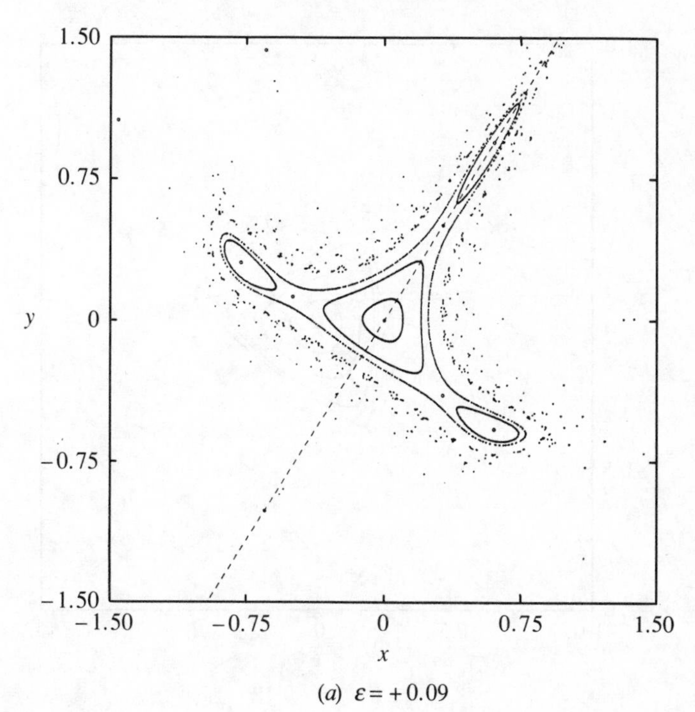

(*a*) $\varepsilon = +0.09$

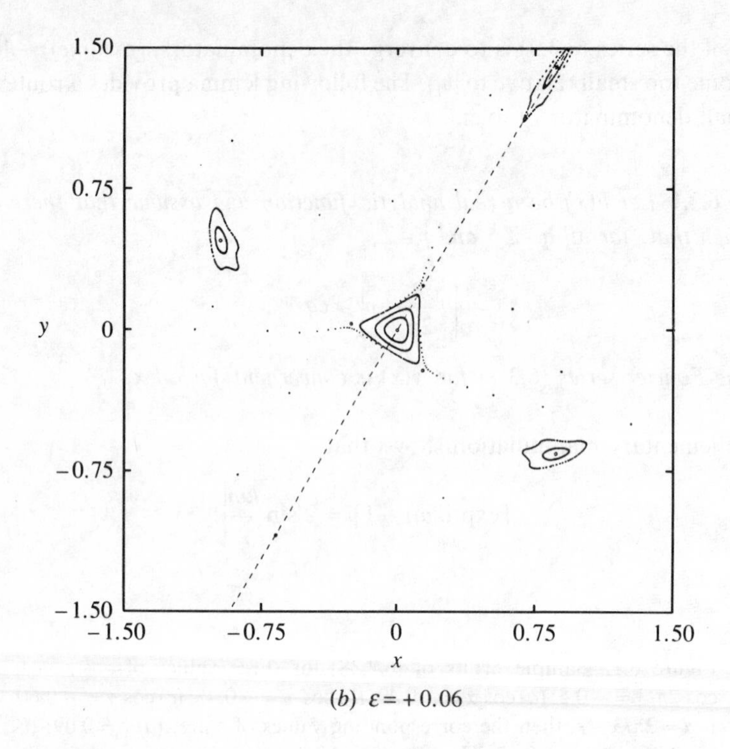

(b) $\varepsilon = +0.06$

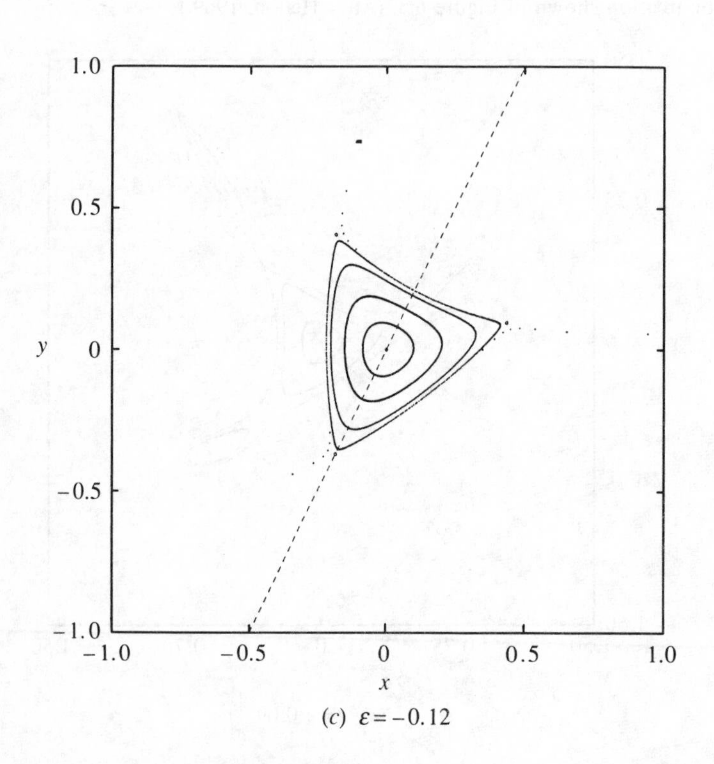

(c) $\varepsilon = -0.12$

However, for each $k \in \mathbb{Z}^+$, there is an integer p_0 such that $(2\pi p_0/k) < \omega < 2\pi(p_0+1)/k$. Thus

$$2\left|\sin\frac{k\omega}{2}\right| = 2\left|\sin\left(\frac{k\omega}{2} - \pi p_0\right)\right|, \tag{6.3.8}$$

with $0 < [(k\omega/2) - \pi p_0] < \pi/2$. Since sine is an increasing function on $(0, \pi/2)$, (6.3.6) implies

$$2\left|\sin\frac{k\omega}{2}\right| \geqslant 2\sin\left(\frac{c\pi}{k^\mu}\right) \geqslant \frac{4c}{k^\mu}. \tag{6.3.9}$$

Here we have used $(2x/\pi) \leqslant \sin(x)$, $x \in (0, \pi/2)$. Similar arguments for $-k \in \mathbb{Z}^+$ allow us to conclude that

$$|[\exp(ik\omega) - 1]|^{-1} \leqslant |k|^\mu/4c. \tag{6.3.10}$$

Since h is real analytic, it can be naturally extended to an analytic function $h(z)$, $z = x + iy$, defined on a strip $|y| < r$, for some $r > 0$. It follows from Cauchy's Theorem that

$$\frac{1}{2\pi}\int_0^{2\pi} h(x+iy)\exp[-ik(x+iy)]\,dx \tag{6.3.11}$$

Figure 6.13 Behaviour of the map (6.2.38) for $\cos\alpha = -0.95$. We can think of this value of $\cos\alpha$ in two ways: (i) $\alpha = 6\pi/7 - \varepsilon$, with $\varepsilon = -0.13$; or (ii) $\alpha = 10\pi/12 - \varepsilon$, for $\varepsilon = -0.21$. Given that Theorem 6.2.2 holds for such values of ε, the presence of both seven-fold and twelve-fold island chains is to be expected for this value of $\cos\alpha$. (After Hénon, 1969.)

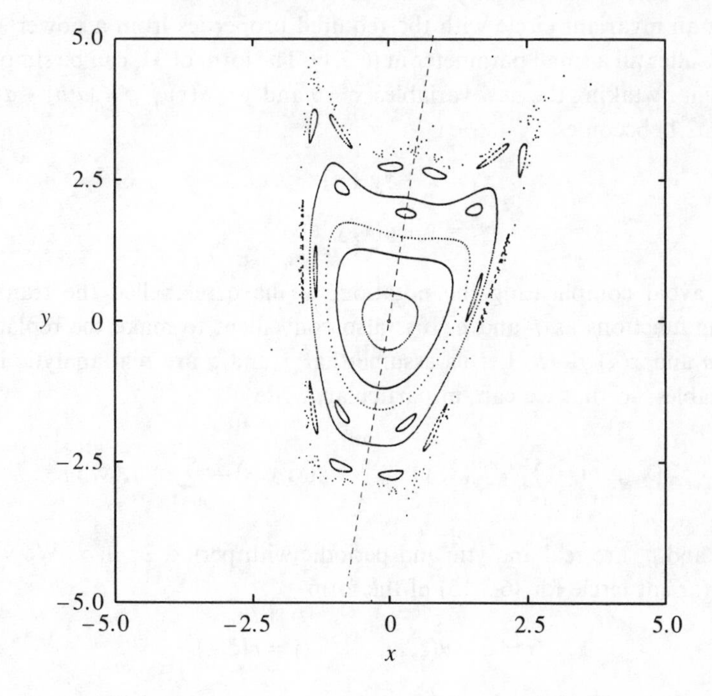

is independent of y for $|y| < r$ and hence (6.3.11) provides an alternative expression for $\hat{h}_k$. Given that $|h| < K$ in $|y| < r$, we conclude that

$$|\hat{h}_k| \leqslant K \exp(ky), \tag{6.3.12}$$

for each y with $|y| < r$. Consequently, the best upper bound is given by taking $y = -r$ (r) when $k > 0$ (<0) giving

$$|\hat{h}_k| \leqslant K \exp(-|k|r), \tag{6.3.13}$$

for some $r > 0$. Finally, observe that

$$\begin{aligned}|w(x)| &= \left|\hat{w}_0 + \sum_{\substack{k\in\mathbb{Z}\\k\neq 0}} \frac{\hat{h}_k \exp(ikx)}{[\exp(ik\omega) - 1]}\right|,\\ &\leqslant |\hat{w}_0| + \sum_{\substack{k\in\mathbb{Z}\\k\neq 0}} |\hat{h}_k|\,|[\exp(ik\omega) - 1]|^{-1},\\ &\leqslant |\hat{w}_0| + \sum_{k=1}^{\infty} \frac{K}{2c} k^{\mu} \exp(-kr),\\ &\leqslant |\hat{w}_0| + \frac{KC(\mu)}{2c} r^{-(\mu+1)},\end{aligned} \tag{6.3.14}$$

(see Exercise 6.3.1). Thus the Fourier series for w converges for all x. □

Another instructive preliminary consideration is to examine the possibility of obtaining an invariant circle with the required properties from a power series in ε, which is after all a small parameter in (6.2.1). The form of $\mathbf{M}_\varepsilon$ can be simplified to some extent by taking the new variables $x = \theta$ and $y = \alpha(\tau)/\gamma$, $\gamma = |\alpha(b) - \alpha(a)| > 0$, so that (6.2.1) becomes

$$\begin{aligned}x_1 &= x + \gamma y + f(x, y, \varepsilon)\\ y_1 &= y + g(x, y, \varepsilon).\end{aligned} \tag{6.3.15}$$

Here, to avoid complicating the notation, we have relabelled the transformed perturbing functions as f and g. It is also convenient to make the replacements $\alpha(a)/\gamma \mapsto a$ and $\alpha(b)/\gamma \mapsto b$. Let us assume that f and g are real analytic in all of their variables, so that we can, in particular, write

$$f(x, y, \varepsilon) = \sum_{n=1}^{\infty} \varepsilon^n f_n(x, y), \qquad g(x, y, \varepsilon) = \sum_{n=1}^{\infty} \varepsilon^n g_n(x, y), \tag{6.3.16}$$

where f_n and g_n are real analytic and periodic with period 2π in x. We will look for an invariant circle for (6.3.15) of the form

$$x = \xi + u(\xi, \varepsilon), \qquad y = v(\xi, \varepsilon), \tag{6.3.17}$$

on which we will assume the map takes the form

$$\xi_1 = \xi + \omega, \tag{6.3.18}$$

where $\omega/2\pi$ is irrational. Our aim is to obtain u and v from power series in ε. To this end we let

$$u(\xi, \varepsilon) = \sum_{n=1}^{\infty} \varepsilon^n u_n(\xi), \qquad v(\xi, \varepsilon) = \omega\gamma^{-1} + \sum_{n=1}^{\infty} \varepsilon^n v_n(\xi). \tag{6.3.19}$$

Here we have chosen $v(\xi, 0) = \omega\gamma^{-1}$ because (6.3.15,17,18) give $\omega = \gamma y = \gamma v(\xi, 0)$ for $\varepsilon = 0$ and $x = \xi$.

On the invariant circle (6.3.17), the map (6.3.15) becomes

$$\begin{aligned} \xi_1 + u(\xi_1, \varepsilon) &= \xi + u(\xi, \varepsilon) + \gamma v(\xi, \varepsilon) + f(\xi + u, v, \varepsilon), \\ v(\xi_1, \varepsilon) &= v(\xi, \varepsilon) + g(\xi + u, v, \varepsilon). \end{aligned} \tag{6.3.20}$$

Substituting (6.3.18) and (6.3.19) into (6.3.20) and comparing coefficients of ε^n yields

$$u_n(\xi + \omega) - u_n(\xi) - \gamma v_n(\xi) = F_n(\xi), \tag{6.3.21}$$

$$v_n(\xi + \omega) - v_n(\xi) = G_n(\xi). \tag{6.3.22}$$

Here

$$F_n(\xi) - f_n(\xi, \omega\gamma^{-1}) \tag{6.3.23}$$

and

$$G_n(\xi) - g_n(\xi, \omega\gamma^{-1}) \tag{6.3.24}$$

depend only on f_l, g_l, u_l, v_l for $l < n$ and may therefore be taken as known (see Exercise 6.3.2).

Since $\mathbf{M}_\varepsilon$ is area-preserving, $\mathbf{M}_\varepsilon(\Gamma) \cap \Gamma$ is non-empty for every closed curve Γ enclosing the origin. This property of $\mathbf{M}_\varepsilon$ can be used (see Siegel & Moser, 1971, pp. 229–30) to show that

$$\int_0^{2\pi} G_n(s)\, ds = 0, \tag{6.3.25}$$

for every n. If ω satisfies (6.3.6), then Lemma 6.3.1 assures us that (6.3.22) can be solved for $v_n(\xi)$ up to an arbitrary constant $\hat{v}_{n0} = (2\pi)^{-1} \int_0^{2\pi} v_n(s)\, ds$. By choosing this constant to be $-\gamma^{-1}\hat{F}_{n0} = -(2\pi\gamma)^{-1} \int_0^{2\pi} F_n(s)\, ds$, we can ensure that (6.3.21) has a convergent solution for $u_n(\xi)$ also. Of course, this argument assumes that $F_n(\xi)$ and $G_n(\xi)$ are real analytic functions for every n. Unfortunately, although we are able to obtain convergent Fourier series for $v_n(\xi)$ and $u_n(\xi)$, the convergence of the power series (6.3.19) for $v(\xi, \varepsilon)$ and $u(\xi, \varepsilon)$ cannot be guaranteed and the perturbation method fails. In order to get round this difficulty, a completely different approach is necessary.

In the following discussion, ε is held constant and we denote $\mathbf{M}_\varepsilon$ simply by $\mathbf{M}$. The first step of the new approach is to replace $\mathbf{M}$ by its extension into the complex domain. Thus f and g are taken to be analytic functions of the complex variables x and y in a complex region $\mathscr{D}$ where $|\text{Im } x| < r_0$, $r_0 \in (0, 1]$, and $y \in \mathbb{C}$ belongs to a complex neighbourhood of $a \leqslant y \leqslant b$. What is more, it is assumed that each curve $y = \varphi(x) = \varphi(x + 2\pi)$ intersects its own image under $\mathbf{M}$. For given ε, the functions $u(\xi, \varepsilon)$ and $v(\xi, \varepsilon)$ in (6.3.17) are then obtained from an iteration scheme whose convergence is carefully controlled. This scheme generates a sequence of maps $\mathbf{M}_i$ (where $\mathbf{M}_0$ is a restriction of $\mathbf{M}$) with the following properties:

(i) the limit of the $\mathbf{M}_i$ as $i \to \infty$ is a twist map $\mathbf{M}_\infty$;
(ii) each $\mathbf{M}_i$ is conjugate to $\mathbf{M}_0$ on a suitably restricted domain;
(iii) the limiting conjugacy between $\mathbf{M}_\infty$ and $\mathbf{M}_0$ exists.

Clearly, if $\mathbf{M}_\infty$ in item (i) has an invariant circle with rotation number $\omega/2\pi$ then the limiting conjugacy in item (iii) allows us to find a corresponding invariant circle in $\mathbf{M}$. Let us consider the construction of the sequence $\{\mathbf{M}_i\}_{i=0}^{\infty}$ a little more closely so that the role played by the domain restriction, mentioned in item (ii), can be better appreciated. For simplicity, we will consider the case $\gamma = 1$.

Observe that in the absence of the perturbation (f, g), (6.3.15) is an area-preserving twist map for which the invariant circle given by $y = \omega$ has rotation number $\omega/2\pi$ lying between a and b. It is a feature of Kolmogorov's method that we focus attention on an *irrational* value of $\omega/2\pi$ in this interval at the beginning of the calculation. The key step in the construction of the maps $\mathbf{M}_i$, $i = 1, 2, \ldots$, is given in the following lemma in which *ω is assumed to satisfy* (6.3.6). The initial map $\mathbf{M}_0$ is defined as the restriction of $\mathbf{M}$ to the domain $|\text{Im } x| < r_0, |y - \omega| < s_0$.

Lemma 6.3.2 *Let $\mathbf{M}_i$ take the form* (6.3.15) *with $f \equiv f_i$, $g \equiv g_i$ and $\gamma = 1$. Suppose $\mathbf{M}_i$ is defined on the domain $\mathscr{D}_i$: $|\text{Im } x| < r_i$, $|y - \omega| < s_i$ and that $|f_i| + |g_i| < d_i$ on $\mathscr{D}_i$. Then there exists a coordinate transformation $\mathbf{U}_i$ of the form*

$$\mathbf{U}_i\colon x = \xi + u_i(\xi, \eta), \qquad y = \eta + v_i(\xi, \eta), \tag{6.3.26}$$

such that $\mathbf{M}_{i+1} = \mathbf{U}_i^{-1}\mathbf{M}_i\mathbf{U}_i$ takes the form

$$\xi_1 = \xi + \eta + f_{i+1}(\xi, \eta), \qquad \eta_1 = \eta + g_{i+1}(\xi, \eta). \tag{6.3.27}$$

Here f_{i+1} and g_{i+1} are real analytic and satisfy

$$|f_{i+1}| + |g_{i+1}| < C'\left[(r_i - r_{i+1})^{-(2\mu+3)}\left(\frac{d_i^2}{s_i} + s_i d_i\right) + \left(\frac{s_{i+1}}{s_i}\right)^2 d_i\right] \tag{6.3.28}$$

on a domain $\mathscr{D}_{i+1}$: $|\text{Im } x| < r_{i+1}, |y - \omega| < s_{i+1}$, where $0 < r_{i+1} < r_i$ and $0 < s_{i+1} < s_i$. Moreover, the functions u_i and v_i are such that

$$|u_i| + |v_i| < \tfrac{1}{7}s_i. \tag{6.3.29}$$

In Lemma 6.3.2, $|f_i| = \sup_{(x,y)\in\mathscr{D}_i} \{|f_i(x, y)|\}$. Note also that the exponent μ occurring in (6.3.6) appears in (6.3.28).

A proof of Lemma 6.3.2 involves the careful estimation of a number of quantities in the complex plane and we refer the interested reader to Siegel & Moser (1971, pp. 235–43) for such detail. However, some important features of the lemma can be explained. For example, why is ω assumed to satisfy (6.3.6)? The answer to this question is that the transformation (6.3.26) is obtained by solving a difference equation of the form (6.3.1). This equation arises from a desire to transform $\mathbf{M}$ into a twist map $\mathbf{T}$: $\xi_1 = \xi + \eta$, $\eta_1 = \eta$. To this end, we seek a transformation, $\mathbf{U}$, of the form

$$\mathbf{U}: x = \xi + u(\xi, \eta), \qquad y = \eta + v(\xi, \eta), \tag{6.3.30}$$

such that $\mathbf{UT} = \mathbf{MU}$. It is easily shown that u and v must satisfy the *non-linear* difference equations

$$\begin{aligned} u(\xi + \eta, \eta) &= u + v + f(\xi + u, \eta + v), \\ v(\xi + \eta, \eta) &= v + g(\xi + u, \eta + v), \end{aligned} \tag{6.3.31}$$

where $u = u(\xi, \eta)$ and $v = v(\xi, \eta)$. These equations cannot be solved directly and are replaced by the linear counterparts

$$\begin{aligned} u(\xi + \omega, \eta) - u(\xi, \eta) &= v(\xi, \eta) + f(\xi, \eta), \\ v(\xi + \omega, \eta) - v(\xi, \eta) &= g(\xi, \eta) - g^*(\eta), \end{aligned} \tag{6.3.32}$$

where $g^*(\eta) = (2\pi)^{-1} \int_0^{2\pi} g(\xi, \eta)\, \mathrm{d}\xi$. Notice $\xi + \eta$ has been replaced by $\xi + \omega$ in the left hand side of (6.3.32) and the mean value, $g^*(\eta)$, of $g(\xi, \eta)$ has been subtracted from the right hand side of the second equation. The linear difference equations (6.3.32) can be solved along the same lines as (6.3.21,22), provided we impose the condition that ω satisfies (6.3.6), and their solution defines a transformation $\mathbf{U}_0$. Of course, $\mathbf{U}_0^{-1}\mathbf{M}_0\mathbf{U}_0$ is not the twist mapping, $\mathbf{T}$; indeed, it is the proof of Lemma 6.3.2 that assures us that it is closer to $\mathbf{T}$ than $\mathbf{M}_0$ is. The connection with Lemma 6.3.1 is also apparent in (6.3.28), where the factor of $(r_i - r_{i+1})^{2\mu+3}$ arises from estimates similar to those considered in Exercise 6.3.1(b) (see Siegel & Moser, 1971, pp. 238–41).

The quantities r_i, s_i, d_i, $i = 0, 1, 2, \ldots$, control the convergence of the sequence $\{\mathbf{M}_i\}_{i=0}^{\infty}$. In the construction of $\mathbf{U}_i$ it is assumed that the following inequalities are satisfied:

$$0 < r_i \leqslant 1, \qquad 0 < 3s_{i+1} < s_i < \frac{r_i - r_{i+1}}{4}, \qquad d_i < \frac{s_i}{6}, \tag{6.3.33a}$$

$$C''(r_i - r_{i+1})^{-2(\mu+1)} \frac{d_i}{s_i} < \tfrac{1}{7}. \tag{6.3.33b}$$

However, if r_i, s_i, d_i are chosen within such limits then lemma 6.3.2 assures us of the existence of the corresponding $\mathbf{U}_i$. In fact, by choosing a suitable sequence of r_i, s_i, and d_i, $i = 0, 1, \ldots$, repeated application of Lemma 6.3.2 yields a sequence $\{\mathbf{M}_i\}_{i=0}^{\infty}$ with the properties listed above. The choice

$$r_i = \frac{r_0}{2}\left(1 + \frac{1}{2^i}\right), \qquad s_i = d_i^{2/3}, \qquad d_{i+1} = C^{i+1} r_0^{-(2\mu+3)} d_i^{4/3} \tag{6.3.34}$$

(see Siegel & Moser, 1971, p. 233), where C is a constant greater than or equal to 2, not only satisfies all the necessary requirements but is also such that, for sufficiently small d_0 (see Exercise 6.3.3),

$$r_i \to \frac{r_0}{2}, \qquad s_i \to 0, \qquad d_i \to 0, \tag{6.3.35}$$

as $i \to \infty$. It follows that $\mathbf{M}_\infty$ is defined on the complex domain

$$|\mathrm{Im}\, x| < \frac{r_0}{2}, \qquad y = \omega, \tag{6.3.36}$$

where it takes the form

$$x_1 = x + y, \qquad y_1 = y, \tag{6.3.37}$$

because d_i (and hence $|f_i|$, $|g_i|$) tends to zero as $i \to \infty$. In view of (6.3.36), $\mathbf{M}_\infty$ may also be written as

$$x_1 = x + \omega, \qquad y = \omega. \tag{6.3.38}$$

For real x, (6.3.38) can be interpreted as the restriction of a twist map to a single circle with rotation number $\omega/2\pi$.

The existence of an invariant circle in $\mathbf{M}_0$ and the conjugacy between an irrational rotation and the restriction of $\mathbf{M}_0$ to that circle then depends on the convergence of the composition $\mathbf{V}_i = \mathbf{U}_0\mathbf{U}_1 \cdots \mathbf{U}_i$ as $i \to \infty$. In essence, this follows from (6.3.29) and the observation that $\mathbf{V}_i$ takes the form

$$\mathbf{V}_i\colon x = \xi + p_i(\xi, \eta), \qquad y = \eta + q_i(\xi, \eta), \tag{6.3.39}$$

where

$$\begin{aligned} p_i &= u_i + p_{i-1}(\xi + u_i, \eta + v_i), \\ q_i &= v_i + q_{i-1}(\xi + u_i, \eta + v_i), \end{aligned} \tag{6.3.40}$$

so that

$$|p_i| \leqslant |u_i| + |u_{i-1}| + \cdots + |u_0|, \qquad |q_i| \leqslant |v_i| + |v_{i-1}| + \cdots + |v_0| \tag{6.3.41}$$

(see Exercise 6.3.4). Given that $\lim_{i\to\infty} p_i$ and $\lim_{i\to\infty} q_i$ exist, they are functions only of ξ because the domain of $\mathbf{V}_\infty = \lim_{i\to\infty} \mathbf{V}_i$ is $|\mathrm{Im}\, \xi| < r_0/2$, $\eta = \omega$ (see (6.3.36)). If we write $\lim_{i\to\infty} p_i = u(\xi)$ and $\lim_{i\to\infty} q_i = v(\xi) - \omega$, then $\mathbf{V}_\infty$ takes the form

$$\mathbf{V}_\infty : x = \xi + u(\xi), \qquad y = v(\xi) \tag{6.3.42}$$

and the conjugacy $\mathbf{M}_\infty = \mathbf{V}_\infty^{-1}\mathbf{M}_0\mathbf{V}_\infty$ provides the form of the invariant circle in $\mathbf{M}$. From (6.3.38), let us write $\mathbf{M}_\infty$ as

$$\xi_1 = \xi + \omega, \qquad \eta = \omega, \tag{6.3.43}$$

then, with the aid of (6.3.42) and (6.3.15), $\mathbf{V}_\infty\mathbf{M}_\infty = \mathbf{M}_0\mathbf{V}_\infty$ becomes

$$\begin{aligned} \xi + \omega + u(\xi + \omega) &= \xi + u + v + f(\xi + u, v), \\ v(\xi + \omega) &= v + g(\xi + u, v), \end{aligned} \tag{6.3.44}$$

where $u = u(\xi)$, $v = v(\xi)$. Comparison with (6.3.20) shows that this is precisely the condition for $x = \xi + u(\xi)$, $y = v(\xi)$ to be an invariant circle for $\mathbf{M}$ on which the map behaves like an irrational rotation through ω (see Exercise 6.3.5). Thus, this new approach allows us to determine $u(\xi, \varepsilon)$ and $v(\xi, \varepsilon)$ in (6.3.17) without using an expansion in powers of ε.

The ideas outlined above allow a proof of the following version of the Moser Twist Theorem for the map (6.3.15). Details can be found in Siegel & Moser, 1971, pp. 230–5.

Theorem 6.3.1 *Given $\varepsilon > 0$ and the domain $\mathscr{D}$, defined above, there exists a δ, depending on ε and $\mathscr{D}$ but not on γ, such that for*

$$|f| + |g| < \delta \text{ on } \mathscr{D}, \tag{6.3.45}$$

the mapping $\mathbf{M}$ in (6.3.15) admits an invariant curve, $\mathscr{C}$, of the form

$$x = \xi + u(\xi), \qquad y = v(\xi), \tag{6.3.46}$$

where u, v are real analytic functions of period 2π in the complex domain $|\mathrm{Im}\, x| < r_0/2$. The parametrisation in (6.3.46) is such that $\mathbf{M}$ is given by $\xi_1 = \xi + \omega$ on $\mathscr{C}$, where $a < \omega < b$ is an irrational multiple of 2π and satisfies (6.3.6). What is more, the functions, u, v, are such that

$$|u| + |v - \omega| < \varepsilon. \tag{6.3.47}$$

The inequality (6.3.47) shows that, for sufficiently small perturbations (f, g), the invariant curve $\mathscr{C}$ is close to the circle $x = \xi$, $y = \omega$. Using (6.3.29), (6.3.41) and $s_{i+1} < s_i/3$ (see Exercise 6.3.6), it can be seen that

$$|u| + |v - \omega| < \frac{1}{7}\sum_{i=0}^{\infty} s_i < \frac{1}{7}\sum_{i=0}^{\infty}\frac{s_0}{3^i} < s_0 \tag{6.3.48}$$

and (6.3.47) follows if we choose $\delta = d_0$ so small that $s_0 = d_0^{2/3} < \varepsilon$.

It is also important to realise that Theorem 6.3.1 predicts the existence of an invariant circle, $\mathscr{C}$, for every irrational $\omega/2\pi$ satisfying (6.3.6) with $\omega \in (a, b)$. It can be shown that the subset of ω-values for which (6.3.6) does not hold has measure-zero (see Arnold, 1983, p. 114) and therefore invariant circles of almost all rotation numbers between $a/2\pi$ and $b/2\pi$ can occur.

Theorem 6.3.1 assumes that the functions f and g in (6.3.15) are real analytic. In this sense it differs from Moser's original work (Moser, 1962) in which only finite differentiability of the perturbation is required. The following statement, which appears in Moser (1968), emphasises this point and neatly summarises the conditions of the theorem. It refers directly to a map of the form (6.2.1) with $a = 1$ and $b = 2$.

Theorem 6.3.2 (Moser Twist Theorem) *Let $\alpha'(\tau) \neq 0$ and let any curve Γ surrounding $\tau = 1$ and its image $\mathbf{M}_\varepsilon(\Gamma)$ intersect each other. The functions f and g are assumed to be sufficiently often differentiable. Then, for sufficiently small ε, there exists an invariant curve $\mathscr{C}$ surrounding $\tau = 1$. More precisely, given any number ω between $\alpha(1)$ and $\alpha(2)$ incommensurable with 2π and satisfying the inequalities*

$$\left|\frac{\omega}{2\pi} - \frac{p}{q}\right| \geqslant c|q|^{-5/2}, \tag{6.3.49}$$

for all integers p, q, there exists a differentiable closed curve

$$\theta = \xi + G(\xi, \varepsilon), \qquad \tau = F(\xi, \varepsilon), \tag{6.3.50}$$

with F, G of period-2π in ξ, which is invariant under the mapping $\mathbf{M}_\varepsilon$, provided ε is sufficiently small. The image of a point on the curve (6.3.50) is obtained by replacing ξ by $\xi + \omega$.

The assumption of finite differentiability has an important effect on the proof of Lemma 6.3.1. For example, if h in (6.3.1) is N-times differentiable rather than analytic then the estimate (6.3.13) of its Fourier component $\hat{h}_k$ is replaced by

$$|\hat{h}_k| \leqslant \frac{K}{|k|^N}, \tag{6.3.51}$$

where $|\mathrm{d}^N h/\mathrm{d}x^N| \leqslant K$ (see Exercise 6.3.7). If (6.3.51) is used in conjunction with (6.3.10), we obtain

$$\left|\frac{\hat{h}_k}{\exp(ik\omega) - 1}\right| \leqslant \frac{K}{4c|k|^{N-\mu}}. \tag{6.3.52}$$

Since $\sum_{k=1}^{\infty} k^{-\nu}$ converges for $\nu > 1$, the series for $w(x) - \hat{w}_0$ converges uniformly and absolutely, by the Weierstrass M-test, provided $N - \mu > 1$. Moreover, since the partial sums form a sequence of continuous functions converging uniformly to a limit, we can conclude that $w(x)$ is continuous, i.e. $w \in C^0$. If we require $w \in C^l$,

$l > 0$, then we make similar estimates for

$$w^{(l)}(x) = \frac{d^l w}{dx^l} = \sum_{\substack{k \in \mathbb{Z} \\ k \neq 0}} \frac{(ik)^l \hat{h}_k}{[\exp(ik\omega) - 1]} \exp(ikx) \tag{6.3.53}$$

and obtain

$$|w^{(l)}(x)| \leqslant \sum_{\substack{k \in \mathbb{Z} \\ k \neq 0}} |k|^l \left| \frac{\hat{h}_k}{\exp(ik\omega) - 1} \right| \leqslant \sum_{\substack{k \in \mathbb{Z} \\ k \neq 0}} \frac{K}{4c|k|^{N-l-\mu}}. \tag{6.3.54}$$

It follows that $w^{(l)}(x)$ is continuous provided $N - l - \mu > 1$. The crucial point to note here is that $N - l > 1 + \mu$ and therefore differentiability is lost every time a difference equation of the form (6.3.1) is solved. Since such a solution is involved in the construction of each member of the sequence $\{\mathbf{M}_i\}_{i=1}^{\infty}$ in Lemma 6.3.2, it is clear that steps must be taken to counteract this loss of differentiability or all derivatives will be exhausted after a finite number of iterations. This is achieved by a smoothing procedure in which the solution to the difference equation is replaced by a sufficiently accurate, more differentiable approximation (see Moser, 1962; Rüssmann, 1970). The differentiability class of the functions f, g in (6.3.15) and the sequence of smoothing operations must be chosen in such a way that the resulting invariant circle has finite differentiability.

Obviously, the overall loss of differentiability in the calculation depends on the details of the smoothing operation, but it is also affected by the *magnitude* of the loss occurring at each solution of (6.3.1). This, in turn, depends on the precise nature of the rotation number constraint (6.3.6). In the above discussion, $\mu > 0$ so that l is at best $N - 2$. In Moser's original work (1962), the parameter μ satisfies $\mu \geqslant \frac{3}{2}$, so that $l \leqslant N - 3$. Let $\Omega_{\mu_0} = \{\omega \,|\, (6.3.6) \text{ is satisfied with } \mu \geqslant \mu_0\}$, then, since q^μ increases with μ for every $q \in \mathbb{Z}^+$, $\Omega_{0^+} \subseteq \Omega_{3/2}$. Thus, the weaker the restriction on allowed values of ω, the greater is the minimum loss of differentiability in solving the difference equation. Several versions of the restriction (6.3.6) appear in the literature. For example, the constraint

$$\left| \frac{\omega}{2\pi} - \frac{p}{q} \right| \geqslant \frac{c}{q^{2+\sigma}}. \tag{6.3.55}$$

for all $p \in \mathbb{Z}$, $q \in \mathbb{Z}^+$, with $c, \sigma > 0$, which corresponds to $\mu > 1$ in (6.3.6), is used in the work of Arnold (1963, 1983). Later work of Moser (see Moser, 1973; Guckenheimer & Holmes, 1983) has the more restrictive condition

$$\left| \frac{\omega}{2\pi} - \frac{p}{q} \right| \geqslant \frac{c}{q^{\tau}}, \tag{6.3.56}$$

for all $p \in \mathbb{Z}$, $q \in \mathbb{Z}^+$, with $c, \tau = \mu - 1 > 0$.

In the statement of Theorem 6.3.2, μ is taken equal to $\frac{3}{2}$ for definiteness and, if the original proof given by Moser (1962) is referred to, then 'sufficiently differentiable' means 'of class C^N with $N \geqslant 333$'. A more recent statement (see Moser, 1973 and Guckenheimer & Holmes, 1983) has $N \geqslant 5$, rotation numbers restricted by (6.3.56) and predicts the existence of a continuously differentiable invariant circle. Moser (1973) points out (cf. Moser, 1969; Rüssmann, 1970) that it is sufficient to take $N > 3$ and conjectures that it may be possible to obtain continuous invariant curves for $N > 2$. However, the result is certainly not true for $N = 1$ as a counterexample has been given by Takens (1971).

6.4 The Aubry–Mather Theorem

In §§6.2,3, we argued for the existence of periodic orbits and invariant circles in sufficiently small area-preserving perturbations of the planar twist map (6.1.36). The following theorem due to Aubry (Aubry & Le Daeron, 1983) and Mather (1982) (see also Katok, 1982) recognises that invariant Cantor sets may also occur. The nature of these invariant sets follows from the theory of circle homeomorphisms.

6.4.1 Invariant Cantor sets for homeomorphisms on S^1

Recall that Denjoy's Theorem (see §1.5) states that any orientation-preserving C^2-diffeomorphism, f, of the circle with irrational rotation number β is conjugate to a pure rotation through β. However, when f is not C^2 counterexamples to this result, involving invariant Cantor sets, can be constructed. These *Denjoy counterexamples* play a vital role in the Aubry–Mather result (see §6.4.2).

Suppose $f: S^1 \to S^1$ is a homeomorphism with *irrational* rotation number $\rho(f) = \beta$, and let $\omega(x)$, $x \in S^1$, be the set of accumulation points of the sequence $\{f^m(x)\}_{m=1}^{\infty}$.

Proposition 6.4.1 *The set $\omega(x)$ is independent of x. Moreover, $E = \omega(x)$ is the unique, minimal, closed invariant set of f.*

Proof. Define $I = [f^m(x), f^n(x)]$ for distinct integers m, n, $m < n$. Then each successive interval of the sequence $\{f^{-k(m-n)}(I)\}_{k=0}^{\infty}$ abuts its predecessor. Therefore, the union of the intervals either covers the whole of S^1 or the end points of the sequence converge to a fixed point of f^{m-n}. If f^{m-n} has a fixed point, f has periodic points and $\rho(f)$ would be rational which is a contradiction.

Now let $z \in \omega(x)$, i.e. there exists a sequence $\{f^{m_l}(x)\}_{l=1}^{\infty}$ with limit z. Consider the interval $I_l = [f^{m_l}(x), f^{m_{l+1}}(x)]$ and the corresponding sequence $\{f^{-k(m_l - m_{l+1})}(I_l)\}$. This sequence is of the type described above and therefore it covers S^1. Thus, given any $y \in S^1$, there is $k = k_l$ such that $y \in f^{-k_l(m_l - m_{l+1})}(I_l)$ and therefore $f^{K_l}(y) \in I_l$, where $K_l = k_l(m_l - m_{l+1})$. This argument is valid for each l and, since $f^{m_l}(x) \to z$ as $l \to \infty$, the intervals I_l collapse towards z as l increases. It follows that $f^{K_l}(y) \to z$ as $l \to \infty$ and so $z \in \omega(y)$. However, since z can be any point in

$\omega(x)$, we conclude that $\omega(x) \subseteq \omega(y)$. Reversal of the roles of x and y gives $\omega(x) = \omega(y)$ and, consequently, $\omega(x) = E$ independent of x.

If $z \in E$, then there exists a sequence $\{f^{m_l}(x)\}_{l=1}^{\infty}$, for any x, with limit z. The sequence $\{f^{m_l+1}(x)\}_{l=1}^{\infty}$ has limit $f(z)$ and therefore E is invariant under f. Moreover, E is closed because it is the set of accumulation points of $\{f^m(x)\}_{m=1}^{\infty}$. To show that E is unique and minimal, let E' be closed and invariant under f. If $x \in E'$, then the orbit of x is a subset of E'. However, E' is closed and hence contains the accumulation points, $\omega(x)$, of the orbit. Therefore $E = \omega(x) \subseteq E'$. We conclude therefore that E is minimal and that there is no closed, f-invariant set that does not contain E, i.e. E is the *unique*, minimal, closed, f-invariant set. □

Now recall the following definition of a Cantor set.

Definition 6.4.1 *A Cantor set is a set which is perfect and nowhere dense.*

A set is *perfect* if it is equal to its set of accumulation points and it is *nowhere dense* if its closure has an empty interior (see Exercise 6.4.1). With this in mind, we can prove the following result.

Proposition 6.4.2 *The set E is either the whole of S^1 or it is a Cantor set.*

Proof. For each $x \in E$ we can take $E = \omega(x)$. Consequently, there must be a convergent subsequence, $\{f^{m_l}(x)\}_{l=1}^{\infty}$, of the orbit of x with limit x itself. The rotation number $\rho(f)$ is irrational and therefore f has no periodic points. It follows that the points $f^{m_l}(x)$ are all distinct and accumulate at x. Thus x is an accumulation point of E and hence E is perfect.

Now observe that the boundary, ∂E, of E is invariant under f. This follows from the continuity of f, the invariance of E and the fact that points in ∂E are limits of sequences of points in both E and E^c. However, since E is minimal, it can have no proper subsets that are f-invariant. Therefore, either $\partial E = \varnothing$; i.e. $E = S^1$, or $\partial E = E$. In the latter case, the interior of the closed set E is empty and so E is nowhere dense. Since E is also perfect, it is a Cantor set. □

Let us now turn to the problem of constructing a homeomorphism for which E is a Cantor set. The crucial observation is that E is always perfect in Proposition 6.4.2. The two cases are distinguished by the fact that E must be nowhere dense if it is to be a Cantor set. The starting point of the construction is an irrational rotation, R_β, for which $E = S^1$. The idea is to 'splice in' intervals at the points of an orbit of this rotation in such a way as to make E nowhere dense and perfect. If $\{x_n\}_{n\in\mathbb{Z}}$ is an orbit of R_β, then the splicing is achieved by the *semi-conjugacy*, Π, illustrated in Figure 6.14. The image of the x_n under Π^{-1} is the interval $I_n = [a_n, b_n] \subseteq S^1$ for each x_n in the chosen orbit, while the restriction of Π to $(\{x_n\}_{n\in\mathbb{Z}})^c$ is a bijection. The intervals I_n are disjoint so for such a Π to exist we require that $\sum_{n=-\infty}^{\infty} l_n < 1$, where l_n is the length of $[a_n, b_n]$. Given that such a set of intervals $\{I_n\}_{n\in\mathbb{Z}}$, ordered on S^1 by $\{x_n\}_{n\in\mathbb{Z}}$, do exist (see Exercise 6.4.2), we define $f(I_n) = I_{n+1}$, for all n, with $f|I_n$ an increasing homeomorphism satisfying $f(a_n) = a_{n+1}$ and $f(b_n) = b_{n+1}$. It can be verified (see Exercise 6.4.3) that $f: S^1 \to S^1$ defined in this way is indeed a homeomorphism. Moreover, if $\bar{f}$, $\bar{R}_\beta$ are suitably chosen lifts on $\mathbb{R}$ of f and R_β, respectively, then

$$\bar{\Pi}\bar{f}(\bar{x}) = \bar{R}_\beta \bar{\Pi}(\bar{x}), \tag{6.4.1}$$

for all $\bar{x} \in \mathbb{R}$, where $\bar{\Pi}$ is a lift of the semi-conjugacy Π. It follows that

$$\rho(f) = \rho(R_\beta) = \beta, \tag{6.4.2}$$

which is irrational (see Exercise 6.4.4). However, if $\tilde{y} \in (\bigcup_{n\in\mathbb{Z}} \operatorname{int}(I_n))^c = \tilde{C}$ then $\{f^m(\tilde{y})\}_{m\in\mathbb{Z}^+} \subseteq \tilde{C}$, by construction. It follows that $E = \omega(\tilde{y})$ must be such that

Figure 6.14 Illustration of the map Π used to construct the Denjoy counterexample f from the irrational rotation R_β via the semi-conjugacy

$$\begin{array}{ccc} S^1 & \xrightarrow{f} & S^1 \\ \Pi\downarrow & & \downarrow\Pi \\ S^1 & \xrightarrow{R_\beta} & S^1 \end{array}$$

The intervals I_n lie in the same order on S^1 as do the points, x_n, of the orbit of x_0 under the irrational rotation R_β. Moreover, $l_n \to 0$ as $n \to \infty$, since $\sum_{-\infty}^{\infty} l_n < 1$. If the sequence of points $\{x_{n_k}\}_{k\in\mathbb{Z}^+}$ converges to $y \in (\bigcup_{n\in\mathbb{Z}} x_n)^c$ from above (below) as $k \to \infty$, then the sequence of intervals $\{I_{n_k}\}_{k\in\mathbb{Z}^+}$ also converges to the point $\tilde{y} = \Pi^{-1}(y)$ from above (below) as $k \to \infty$. While $C = (\bigcup_{n\in\mathbb{Z}} x_n)^c$ is neither perfect nor nowhere dense, $\tilde{C} = (\bigcup_{n\in\mathbb{Z}} \operatorname{int}(I_n))^c$ is a Cantor set (see Exercise 6.4.5).

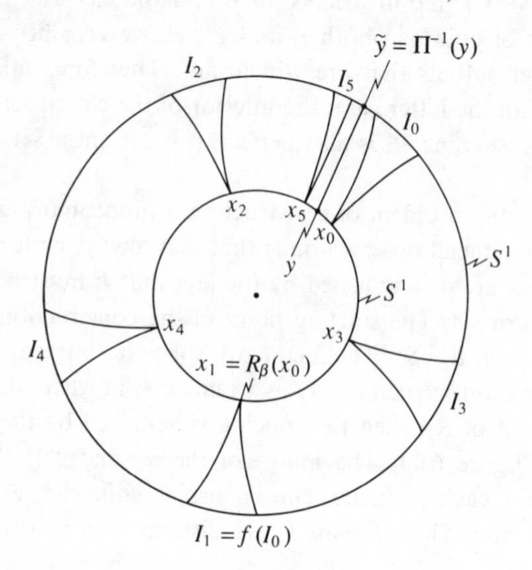

$E \cap \text{int}(I_n)$ is empty, for every n. Hence $E \neq S^1$ and must therefore be a Cantor set by Proposition 6.4.2.

This example shows that Denjoy's Theorem cannot be extended to homeomorphisms. Indeed, the definition of f on $\bigcup_{n \in \mathbb{Z}} I_n$ can be chosen (see Nitecki, 1971, p. 48) to provide a counterexample for C^1-diffeomorphisms also.

6.4.2 Twist homeomorphisms and Mather sets

The Aubry–Mather theorem refers to an area-preserving twist homeomorphism, **f**, defined on the standard annulus $A = S^1 \times [0, 1] = \{(\theta, r) | 0 \leqslant \theta < 1, 0 \leqslant r \leqslant 1\}$. In an analogous manner to that employed in § 1.2 for circle maps, we can define a lift, $\bar{\mathbf{f}}$, of any homeomorphism $\mathbf{f}: A \to A$ on the strip $\mathscr{S} = \mathbb{R} \times [0, 1] = \{(x, y) | x \in \mathbb{R}, 0 \leqslant y \leqslant 1\}$ (cf.(1.2.5)). Corresponding to Proposition 1.2.1, we have

$$\bar{\mathbf{f}}(x + 1, y) = \bar{\mathbf{f}}(x, y) + (1, 0)^{\mathrm{T}}, \tag{6.4.3}$$

for every $x \in \mathbb{R}$ and $y \in [0, 1]$. A homeomorphism **f** of A is a *twist homeomorphism* if it preserves orientation, maintains the boundary components of A and if $\bar{\mathbf{f}}(x, y) = (\bar{f}_1(x, y), \bar{f}_2(x, y))^{\mathrm{T}}$ is such that $\bar{f}_1(x, y)$ is a strictly monotone function of y for every $x \in \mathbb{R}$ (see Exercise 6.4.6).

The lift, $\bar{\mathbf{f}}$, allows us to define a rotation number for some of the invariant sets of **f**. When dealing with twist homeomorphisms it is important to use the same lift for all rotation number calculations. This requirement leads to the 'lifted' rotation numbers mentioned at the end of §6.1. Recall the expression (1.5.18) for the rotation number of a circle diffeomorphism f. Suppose that $\lim_{n \to \infty} (\bar{f}(x) - x)/n = k + \alpha$, $k \in \mathbb{Z}$, for a particular lift, $\bar{f}$, of f, then (1.5.18) gives $\rho(f) = \alpha$. However, $\bar{f}_k(x) = \bar{f}(x) - k$ is also a lift of f (see §1.2) and $\lim_{n \to \infty} (\bar{f}_k(x) - x)/n = \alpha = \rho(f)$. This example shows that the process of taking mod 1 in (1.5.18) is equivalent to using an appropriate alternative lift of f. Thus, if we are to maintain a fixed choice of lift in calculating rotation numbers, then we must not reduce mod 1 in (1.5.18). The resulting lifted rotation number, $\rho(\bar{f})$, can differ from $\rho(f)$ by an integer (k in the above example) which depends on $\bar{f}$. Now consider the restriction of **f** to $S^1 \times \{0\}$ and $S^1 \times \{1\}$. These homeomorphisms of S^1 have lifts $\bar{f}_1(x, 0)$ and $\bar{f}_1(x, 1)$, respectively, given by the chosen lift $\bar{\mathbf{f}}$ of $\mathbf{f}: A \to A$. We can calculate $\rho_0(\bar{\mathbf{f}})$ and $\rho_1(\bar{\mathbf{f}})$ by using (1.5.18) (without taking mod 1). Given that $\bar{f}_1(x, y)$ is strictly increasing in y for every $x \in \mathbb{R}$, then $\rho_1(\bar{\mathbf{f}}) > \rho_0(\bar{\mathbf{f}})$ and $[\rho_0(\bar{\mathbf{f}}), \rho_1(\bar{\mathbf{f}})]$ is called the *rotation* (*or twist*) *interval* of $\bar{\mathbf{f}}$. Use of a different lift for **f** can only change both $\rho_0(\bar{\mathbf{f}})$ and $\rho_1(\bar{\mathbf{f}})$ by the same integer amount. Thus the rotation interval of **f** is determined up to an integral translation. It follows that the rotation number of all invariants sets of **f**, when calculated with respect to $\bar{\mathbf{f}}$, lie in the interval $[\rho_0(\bar{\mathbf{f}}), \rho_1(\bar{\mathbf{f}})]$. For example, if $\mathbf{f}^q(\mathbf{x}) = \mathbf{x}$, $\mathbf{x} \in A$, then $\bar{\mathbf{f}}^q(x, y) = (x + p, y)^{\mathrm{T}}$, where p

is determined up to a multiple of q (see Exercise 6.4.7(a)). The fraction p/q, which must lie in $[\rho_0(\bar{\mathbf{f}}), \rho_1(\bar{\mathbf{f}})]$, is called the *rotation number of the periodic point* $\mathbf{x}$. In particular, this definition allows us to associate a rotation number with all of the periodic points predicted by the Poincaré–Birkhoff Theorem (see Exercise 6.5.1). These Birkhoff periodic points, as they are called, appear in the statement of the Aubry–Mather result as presented by Katok, who defines them in the following way (see Katok, 1982).

Definition 6.4.2 *A point* $\mathbf{x} \in A$ *is said to be a* Birkhoff periodic point *of type* (p, q) *if, for a lift* (x, y) *of* $\mathbf{x}$, *there is a map* $\mathbf{\Theta}: \mathbb{Z} \to \mathcal{S}$ *with the properties*:

(i) $\mathbf{\Theta}(0) = (x, y)$;

(ii) $\mathbf{\Theta} = (\xi, \eta)$, *where* ξ *is a strictly monotone function*;

(iii)
$$\mathbf{\Theta}(n + q) = \mathbf{T\Theta}(n), \qquad \mathbf{T\Theta} = (\xi + 1, \eta), \tag{6.4.4}$$

$$\mathbf{\Theta}(n + p) = \bar{\mathbf{f}}\mathbf{\Theta}(n), \tag{6.4.5}$$

for all $n \in \mathbb{Z}$.

The orbit of a Birkhoff periodic point of type (p, q) *is called a* Birkhoff periodic orbit *of type* (p, q).

It is not difficult to show (see Exercise 6.4.7(b)) that (6.4.4,5) imply $\mathbf{x}$, in Definition 6.4.2, is a periodic point with rotation number p/q. What is more, if $\mathbf{x}$ has polar coordinates (θ, r) and $\mathbf{f}^n: (\theta, r) \mapsto (\theta_n, r_n)$ then the θ_n occur in the same order on S^1 as the orbit of θ under rotation by $2\pi p/q$ (see Figure 6.15). However, Definition 6.4.2 also imposes bounds on the radial displacements, $|r_m - r_n|$, of the periodic orbit (see Katok, 1982) and therefore rotation number and ordering alone are not enough to ensure that a periodic orbit is of the Birkhoff type.

A Birkhoff periodic orbit is an example of a more general structure, called a *Mather set*, for which a rotation number can be defined.

Definition 6.4.3 *Let* $\mathbf{f}$ *be a twist homeomorphism. A closed,* $\mathbf{f}$*-invariant set* $E \subseteq A$ *is a* Mather set *if*:

(i) *E is the graph of a continuous function* Φ *defined on a closed subset K of* S^1 *taking values in* $[0, 1]$;

(ii) $\bar{\mathbf{f}}$ *preserves the order on the covering of E.*

The properties of E required by Definition 6.4.3 are illustrated in Figure 6.16. Clearly, every closed, $\mathbf{f}$-invariant subset of a Mather set is also a Mather set and, hence, every Mather set contains a minimal subset.

It can be shown (see Exercise 6.4.9) that $\mathbf{f}|E$ is topologically conjugate by order-preserving homeomorphism to a restriction of a homeomorphism of the circle to the set K. Thus the rotation number $\rho(E)$ is defined. Moreover, it then

Figure 6.15 Illustration of the relationship between the function $\mathbf{\Theta} = (\xi, \eta)$ that characterises a Birkhoff periodic point of type (p, q), $(\theta, r) \in A$, and the angular order of the orbit of the point $\theta \in S^1$ under the rotation R_α with $\alpha = 2\pi p/q$. The first component of $\mathbf{\Theta}$ reflects this ordering so that only S^1 and its covering space $\mathbb{R}$ are shown. The example chosen has $p = 2$, $q = 5$, the point θ lifts to $x + n$, $x \in [0, 1)$, $n \in \mathbb{Z}$ and $\bar{R}_\alpha^k x$, for $k = 0, 1, \ldots, 6$, is given in relation to ξ. $\bar{R}_\alpha$ is a lift of R_α. Observe that ξ is strictly monotone and that $\xi(0), \ldots, \xi(4)$ are lifts of the periodic points in order of increasing angular coordinate rather than in the order $R_\alpha^k \theta$, $k = 0, 1, \ldots, 4$. The properties of $\bar{R}_\alpha$ ensure that, for any $n \in \mathbb{Z}$, $\xi(n + 5) = \xi(n) + 1$ and $\xi(n + 2) = \bar{R}_\alpha \xi(n)$ in agreement with (6.4.4) and (6.4.5).

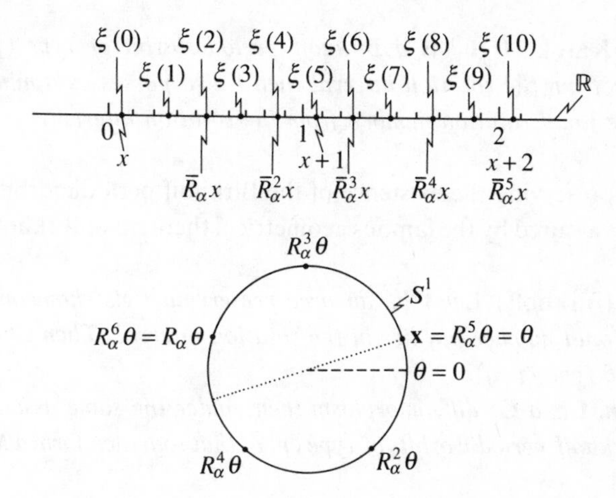

Figure 6.16 Schematic illustration of Definition 6.4.3 for a Mather set. The function Φ is defined on a closed subset K of S^1. The Mather set E and its covering $\bar{E}$ are shown. Some typical orbit points are used to indicate how $\bar{\mathbf{f}}$ preserves order on the covering of E.

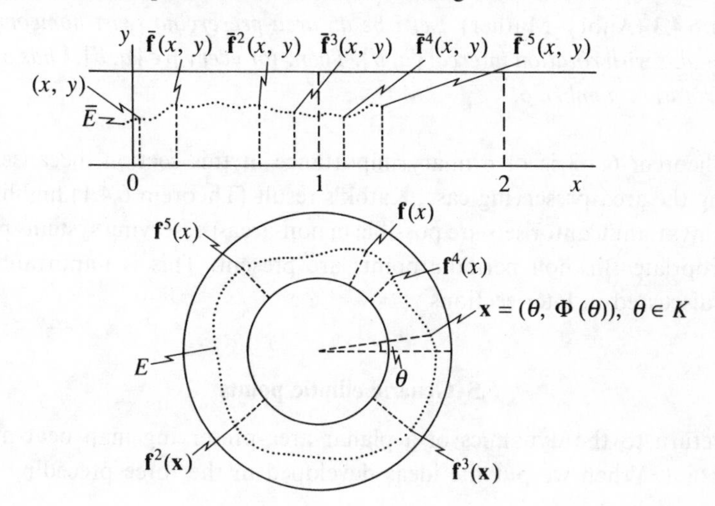

follows from the theory of circle homeomorphisms (see §6.4.1 and Nitecki, 1971, pp. 32–50) that there are exactly three types of the minimal Mather set. If $\rho(E)$ is rational, equal to p/q say, then E is a Birkhoff periodic orbit of type (p, q). If $\rho(E)$ is irrational then either:

(i) $E = S^1$ and $\mathbf{f}|E$ is conjugate to a rotation by $\rho(E)$;
(ii) E is a Cantor set and $\mathbf{f}|E$ is conjugate to the restriction of one of the Denjoy counterexamples to its invariant Cantor set (see §6.4.1).

Katok (1982) proves the following result for a twist homeomorphism $\mathbf{f}: A \to A$, whether or not it is area-preserving.

Theorem 6.4.1 (Katok) *If* $\mathbf{f}$ *has a Birkhoff periodic orbit of type* (p, q) *for any rational number from the rotation interval then* $\mathbf{f}$ *also possesses a minimal Mather set with any irrational rotation number from the rotation interval.*

When $\mathbf{f}$ is area-preserving the existence of the Birkhoff periodic orbits required in Theorem 6.4.1 is assured by the famous geometrical theorem of Birkhoff (1968a).

Theorem 6.4.2 (Birkhoff) *Let* $\mathbf{f}$ *be an area-preserving twist homeomorphism and suppose the rational number* p/q *lies in the rotation interval. Then* $\mathbf{f}$ *has a Birkhoff periodic orbit of type* (p, q).

If, in addition, $\mathbf{f}$ *is a* C^1*-diffeomorphism then, under the same assumptions,* $\mathbf{f}$ *has two different Birkhoff periodic orbits of type* (p, q) *that together form a Mather set.*

It is apparent that the second part of Theorem 6.4.2 includes Theorem 6.2.1. The latter result is clearly weaker because it assumes that the map in question is a small perturbation of the twist map (6.1.36).

Together Theorems 6.4.1 and 6.4.2 give the Aubry–Mather result.

Theorem 6.4.3 (Aubry–Mather) *Let* $\mathbf{f}$ *be an area-preserving twist homeomorphism of the annulus with rotation interval* $[a, b]$*. Then, for every* $\rho \in [a, b]$*,* $\mathbf{f}$ *has a Mather set with rotation number* ρ*.*

While Theorem 6.4.3 is of primary importance in this section, because we are discussing the area-preserving case, Katok's result (Theorem 6.4.1) highlights the fact that invariant Cantor sets are possible in non-area-preserving systems provided the appropriate Birkhoff periodic points are present. This is important for the systems discussed in later sections.

6.5 Generic elliptic points

Let us return to the dynamics of a planar area-preserving map near a generic elliptic point. When we put the ideas developed in the three preceding sections together, the picture that emerges is one of great complexity. By Theorem 6.3.2, the map (6.2.1) has an invariant circle, surrounding the origin, corresponding to almost every irrational number in $(\alpha(a)/2\pi, \alpha(b)/2\pi)$. These circles are often referred to as KAM circles. Moreover, the discussion of § 6.2 predicts the existence of an island chain associated with every rational number in this interval. Figure 6.17 attempts to illustrate this situation. Unfortunately, the best one can do is to show individual invariant circles and island chains. It is therefore important to point out that between any two of either type of structure there are infinitely many more of both of them.

Theorem 6.3.2 assures us that the restriction of the map to a KAM circle is conjugate to an irrational rotation, however, the dynamics associated with an island chain is a great deal more complicated. Each chain is associated with two types of periodic point: one hyperbolic and the other elliptic. We have already noted (see § 6.2) that homoclinic tangles will occur in the neighbourhood of the hyperbolic points giving rise to complexity of the kind found in the horseshoe map (see § 3.7). Thus, chaotic orbits can result from numerical iterations of the map starting in the vicinity of these tangles. In the neighbourhood of the elliptic points there is even more complexity. Each elliptic point of period q is an elliptic fixed point for $\mathbf{M}_\varepsilon^q$ and the results developed in §§ 6.2–3 can equally well be applied to any one of these points. We therefore conclude that there will be invariant circles and island chains of $\mathbf{M}_\varepsilon^q$ surrounding each of the elliptic periodic points in a q-fold island chain (see Figure 6.18). These q'-fold chains of $\mathbf{M}_\varepsilon^q$ will contain both

Figure 6.17 Schematic illustration of KAM circles and island chains surrounding a generic elliptic point of a planar area-preserving map. The island chain of period q is formed around two periodic orbits of that period; one elliptic and the other hyperbolic. Eight- and twelve-fold island chains are shown. Generically, homoclinic tangles occur at the hyperbolic periodic points. KAM circles $\mathscr{C}_i$, $i = 1, 2, 3$ are shown with rotation numbers $\omega_i/2\pi$, where $\omega_1 < \omega_2 < \omega_3$.

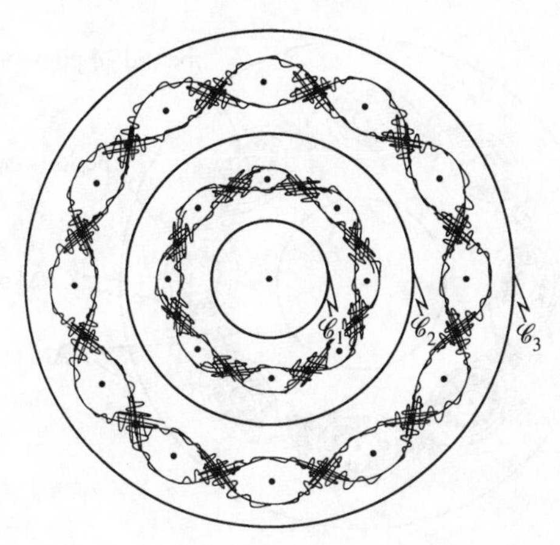

hyperbolic and elliptic periodic points of $\mathbf{M}_\varepsilon$ with period qq'. It is worth noting (see Exercise 6.5.1) that all these periodic points of $\mathbf{M}_\varepsilon$ have the same rotation number even though they have different periods. The hyperbolic points will be associated with homoclinic tangles while the elliptic points will be surrounded by further invariant circles and island chains corresponding to even higher periods, and so on. It follows that the complexity of the generic elliptic point extends to finer and finer scales. This kind of behaviour was observed in the numerical experiments of Hénon (see Figure 1.42). Of course, only those island chains with relatively low periods for which the angular separation of the stable and unstable manifolds at the saddle points is sufficiently large are detected in numerical calculations.

The KAM circles have an important effect on the nature of the dynamics because they separate the orbits of points inside them from those of points outside them. Since the KAM circles surround the elliptic point and occur in every neighbourhood of it, the dynamics of points that do *not* belong to them are severely restricted. The situation is particularly well illustrated by the so-called *standard map* (see Taylor, 1968, unpublished; Chirikov, 1979; Greene, 1979). This map, which is defined on the cylinder, takes the form

$$\theta_1 = (\theta + r_1) \bmod 1, \qquad r_1 = r - \frac{k}{2\pi}\sin 2\pi\theta, \tag{6.5.1}$$

where the r-coordinate lines lie along the generators of the cylinder and θ is the usual angular coordinate. For $k = 0$, the orbit of a point (θ, r) is confined to the circle wrapping around the cylinder in the θ-direction with r fixed. The restriction

Figure 6.18 Elliptic points of the q-fold island chain are surrounded by KAM circles and island chains of $\mathbf{M}_\varepsilon^q$. The periodic points of period q' in the latter are periodic points of $\mathbf{M}_\varepsilon$ of period qq'. The case when $q = q' = 8$ is shown. The KAM circles of $\mathbf{M}_\varepsilon^q$ are each filled out densely by every qth iterate of $\mathbf{M}_\varepsilon$. Note that homoclinic tangles occur generically at the hyperbolic points of all periods.

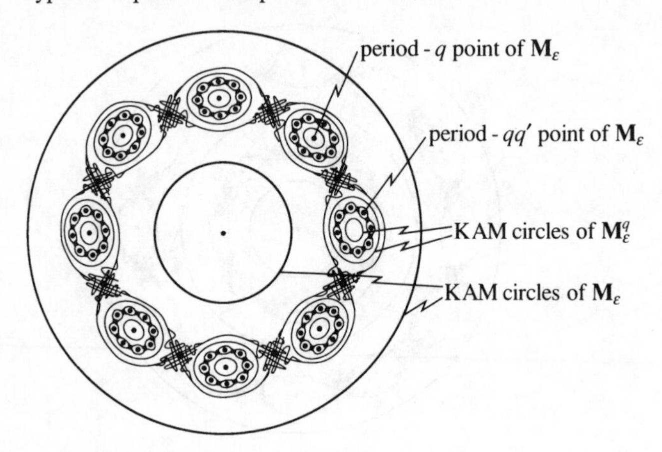

of (6.5.1) to the circle at r has rotation number equal to r. If r is rational, equal to p/q say, then every point of this circle is a periodic point of period q. On the other hand, if r is irrational the iterates of (6.5.1) fill out the circle densely. For sufficiently small $k>0$, there are, by Theorem 6.3.2, KAM circles with almost every irrational rotation number that occurs in the unperturbed map. The KAM circle with rotation number $r \in \mathbb{R}\backslash\mathbb{Q}$ goes continuously into the invariant circle of the $k=0$ map with the same rotation number as $k \to 0$. It follows that, if k is close enough to zero, then the rotation number of the KAM circles is an increasing function of r for every θ (see Figure 6.17). The periodic points lying at the heart of the island chains predicted by the Poincaré–Birkhoff Theorem also evolve continuously from the invariant circles with rational rotation number at $k=0$ (see Exercise 6.5.2 and 3). Numerically, this ordering of the island chains appears to extend to values of k of order unity. Of course, the KAM circles cannot be observed directly in numerical experiments because of the finite resolution of the computing but some of the island chains can. In fact, it is the ordering of special sequences of island chains that indicate where a given KAM circle lies.

The sequence chosen is given by the continued fraction expansion of the rotation number of the KAM circle. If α is a real number then we write

$$\alpha = a_0 + \frac{1}{\alpha_1}, \tag{6.5.2}$$

where a_0 is the largest integer less than or equal to α, i.e. $a_0 = [\alpha]$, and $\alpha_1 > 1$. We then define integers a_n, $n = 1, 2, \ldots$, inductively by

$$\alpha_n = a_n + \frac{1}{\alpha_{n+1}}, \tag{6.5.3}$$

where $a_n = [\alpha_n]$ and $\alpha_{n+1} > 1$. If α is rational this process terminates after a finite number of steps (see Exercise 6.5.4), i.e. there is an m such that $\alpha_m = a_m$ and

$$\alpha = a_0 + \cfrac{1}{a_1 + \cfrac{1}{a_2 + \cfrac{1}{\ddots \, \cfrac{}{a_{m-2} + \cfrac{1}{a_{m-1} + \cfrac{1}{a_m}}}}}}. \tag{6.5.4}$$

We then write

$$\alpha = [a_0, a_1, a_2, \ldots, a_m]. \tag{6.5.5}$$

If α is irrational then the process never terminates, i.e.

$$\alpha = a_0 + \cfrac{1}{a_1 + \cfrac{1}{a_2 + \cfrac{1}{\ddots \cfrac{}{a_{n-2} + \cfrac{1}{a_{n-1} + \cfrac{1}{\alpha_n}}}}}}, \tag{6.5.6}$$

for every $n \in \mathbb{Z}^+$, and we write

$$\alpha = [a_0, a_1, a_2, \ldots, a_{n-1}, \alpha_n], \text{ for all } n \in \mathbb{Z}^+, \tag{6.5.7}$$

or

$$\alpha = [a_0, a_1, a_2, \ldots]. \tag{6.5.8}$$

The rational number

$$p_n/q_n = [a_0, a_1, \ldots, a_n] \tag{6.5.9}$$

is called the *nth principal convergent* of α and the sequence $\{p_n/q_n\}_{n=0}^{\infty}$ is the continued fraction for α. The principal convergents for an irrational number have the following important properties.

(i) For each n, p_n, q_n are relatively prime and $q_n \geqslant 1$.

(ii) The denominators q_n form an increasing sequence

$$0 < q_1 < q_2 < \cdots < q_n < \cdots. \tag{6.5.10}$$

(iii) p_n/q_n is the 'best approximation' to α in the sense that

$$|q\alpha - p| > |q_n\alpha - p_n|, \tag{6.5.11}$$

for every $1 \leqslant q < q_n$.

(iv) For n even (odd) the nth principal convergents form a strictly increasing (decreasing) sequence converging to α.

The origin of these properties is examined in Exercises 6.5.5–6.5.7.

Let us now consider the KAM circle with an irrational rotation number, $\omega/2\pi$, satisfying (6.3.6) and suppose the ordering of the KAM circles and island chains is as described above. There will be an island chain corresponding to each nth principal convergent, p_n/q_n, of $\omega/2\pi$ based on periodic orbits of period q_n. Property (iii) assures us that, of all periodic orbits of period q_n or shorter, these orbits lie nearest to the desired KAM circle. Moreover, Property (iv) predicts that the periodic orbits corresponding to even values of n are inside the KAM circle, while those associated with odd values of n are outside of it. The appearance of these sequences of island chains in numerical experiments not only confirms the assumed ordering but can also locate the given KAM circle to quite fine tolerances (see Figure 6.19).

Of course, it is reasonable to ask how one can obtain an irrational number that satisfies (6.3.6). The answer is again provided by the theory of continued fractions. The form (6.5.6) shows that the larger the value of α_n, the closer α is to the rational p_{n-1}/q_{n-1}. However, the size of α_n is reflected in $a_n = [\alpha_n]$ and therefore we can bound α away from all its principle convergents by giving an upper bound to a_n for all n. Given that the nth principal convergent, p_n/q_n, is the best rational approximation to α with denominator less an or equal to q_n, then this constraint will, in some sense, ensure that α is bounded away from *all* rationals. The connection with (6.3.6) is provided by the fact that the following properties of an irrational number α are equivalent (see Lang, 1966, p. 24).

(i) The continued fraction for α is $[a_0, a_1, \ldots]$ and there is a constant $a > 0$ such that $a_n < a$ for all n.

Figure 6.19 The number $\alpha^* + 1$ is the 'golden mean' $[1, 1, 1, \ldots] = (1 + 5^{1/2})/2$ (see Niven, 1956; Gardner, 1961). The diagram shows some island chains converging towards the 'golden' KAM circle, with rotation number $\alpha^* = (5^{1/2} - 1)/2$, in the standard map.

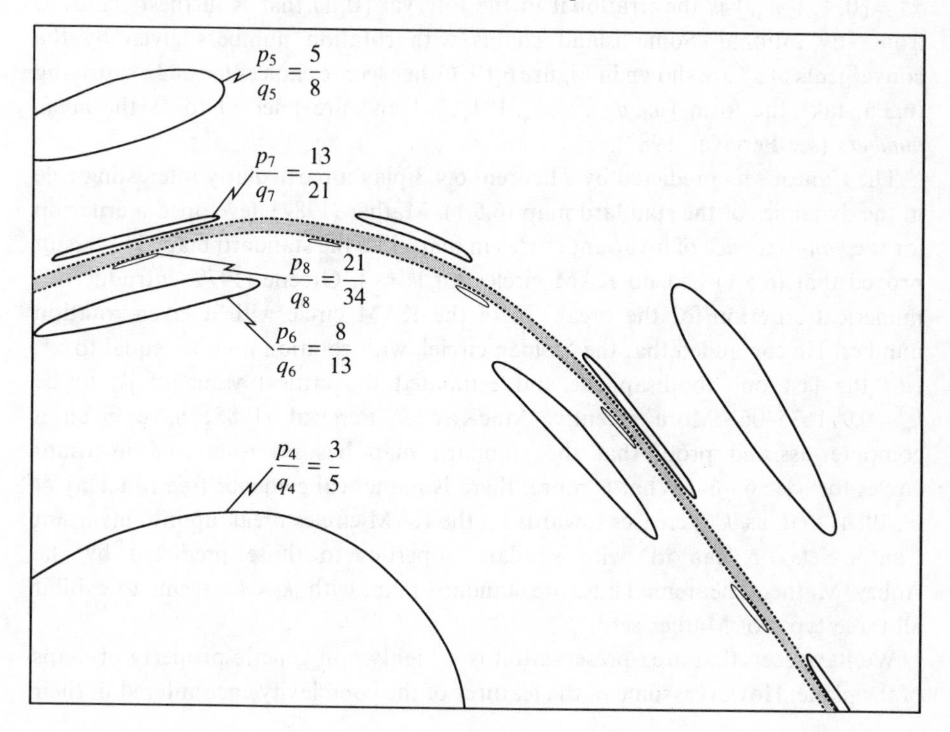

(ii) For any positive function ψ with convergent sum

$$\sum_{q=0}^{\infty} \psi(q), \tag{6.5.12}$$

the inequality

$$|q\alpha - p| < \psi(q) \tag{6.5.13}$$

has only a finite number of solutions.

If we let $\psi(q) = Cq^{-(1+\varepsilon)}$, $\varepsilon > 0$, then (6.5.12) is convergent and we conclude that

$$|q\alpha - p| \geqslant Cq^{-(1+\varepsilon)} \tag{6.5.14}$$

for all but a finite number of values of q, say q_i, $i = 1, \ldots, l$. However, for each q_i, we can obviously find a constant, C_i, such that

$$|q_i\alpha - p_i| = C_i q_i^{-(1+\varepsilon)} \tag{6.5.15}$$

and, by picking $c = \min\{C, C_1, \ldots, C_l\}$, we can ensure that

$$|q\alpha - p| \geqslant cq^{-\mu}, \tag{6.5.16}$$

for all $q \in \mathbb{Z}^+$, with $\mu = 1 + \varepsilon$. Hence (6.3.6) is satisfied if we take $\omega = 2\pi\alpha$.

Clearly, the smaller we make a in (i) above, the further the α is from any rational. However, $a_n \geqslant 1$ for all $n \geqslant 1$, because $a_n = [\alpha_n]$ and $\alpha_n > 1$. Hence, in these terms, $\alpha^* = [0, 1, 1, \ldots]$ is the irrational in the interval $[0, 1)$ that is furthest removed from any rational. Some island chains with rotation numbers given by the convergents of α^* are shown in Figure 6.19. Other secure choices for $\omega/2\pi$ satisfying (6.3.6) take the form $[a_0, a_1, \ldots, a_n, 1, 1, \ldots]$ and are referred to as the *noble numbers* (see Percival, 1982).

The Cantor sets predicted by Theorem 6.4.3 play a particularly interesting role in the dynamics of the standard map (6.5.1). Mather (1984) developed a criterion for the *non-existence* of invariant circles in maps like the standard map. This result proved that (6.5.1) had no KAM circles for $|k| > \frac{4}{3}$. Greene (1979) introduced a numerical criterion for the break up of the KAM circle with a given rotation number. He concluded that the 'golden circle', with rotation number equal to α^*, was the last one to disappear and estimated the critical value of $|k|$ to be $k_c = 0.971635406$. More recently, MacKay & Percival (1985) have given a computer-assisted proof that the standard map has no rotational invariant circles for $|k| \geqslant 63/64$. What is more, there is numerical evidence (see MacKay *et al*, 1984) that, as $|k|$ increases towards k_c, the KAM circles break up into invariant Cantor sets or 'cantori' with similar properties to those predicted by the Aubry–Mather Theorem. Thus, the standard map, with $|k| < k_c$, seems to exhibit all three types of Mather set.

We have seen that area-preservation is a highly non-generic property of maps of the plane. However, some of the features of the complexity encountered in their dynamics are structurally stable. For example, the hyperbolic periodic points and the transverse intersections of their stable and unstable manifolds should persist under sufficiently small perturbations whether the perturbations are area-preserving or not. To what extent is the complexity of area-preserving maps reflected in the dynamics of nearby planar maps? The examples described in the subsequent sections shed some light on the answer to this question and are of current research interest.

6.6 Weakly dissipative systems and Birkhoff attractors

The notation used in the literature on this topic replaces the annulus of previous sections by a cylinder C. We noted in Chapter 2 that these two domains are, topologically, completely interchangeable (see Figure 2.13). However, area-preservation for a map of the annulus requires a zero-Calabi invariant for the corresponding map of the cylinder (see MacKay *et al*, 1984). This excludes area-preserving maps that involve net translations along the cylinder. The twist maps considered here also differ from those defined in § 6.4.2. In this section we are concerned with diffeomorphisms rather than homeomorphisms. Differentiability allows the twist condition to be expressed in the form

$$\frac{\partial}{\partial y}\bar{f}_1(x, y) > \delta > 0, \tag{6.6.1}$$

where $\bar{\mathbf{f}}\colon (x, y) \mapsto (\bar{f}_1(x, y), \bar{f}_2(x, y))$ is a suitable lift of the twist map $\mathbf{f}\colon C \to C$ (see (6.4.3)). Furthermore, the cylinder C is assumed open, i.e. $C = S^1 \times \mathbb{R}$ and $\mathbf{f}$ is taken to be end-preserving. These conditions replace the closed annulus with $\mathbf{f}$-invariant bounding circles that appeared in § 6.4.2. If $\sup_{\mathbf{x}\in C}\{|\mathrm{Det}(\mathrm{D}\mathbf{f}(\mathbf{x}))|\}$ is less than one, then the twist map is said to be *dissipative*. Can complex Mather-set structures reminiscent of area-preserving maps occur in such systems?

At most one rotational invariant circle can remain in the presence of dissipation. Suppose there are two such invariant circles, then $\mathbf{f}$ would have to preserve the area of the cylindrical region, C', between them, i.e. $\mathbf{f}(C') = C'$. However, this contradicts the hypothesis that $\mathbf{f}$ is dissipative under which

$$\mathrm{Area}[\mathbf{f}(C')] < \max_{\mathbf{x}\in C}\{\mathrm{Det}(\mathrm{D}\mathbf{f}(\mathbf{x}))\} \times \mathrm{Area}[C'] < \mathrm{Area}[C'], \tag{6.6.2}$$

for all such C'.

Birkhoff periodic orbits can persist under sufficiently small dissipative perturbations. For example, let us take an area-preserving map, $\mathbf{M}_\varepsilon$, of the form (6.2.1), with $\varepsilon \neq 0$ and sufficiently small, as a starting point. Theorem 6.2.1 assures us that Birkhoff periodic points exist for this map. Let us now construct a dissipative map by composing $\mathbf{M}_\varepsilon$ with the contraction $\mathbf{C}_\delta\colon (\theta, \tau) \to (\theta, (1+\delta)\tau)$, $-1 < \delta < 0$. It is easily shown (see Exercise 6.6.1) that $(\mathbf{M}_\varepsilon \cdot \mathbf{C}_\delta)^q$ takes a similar form to (6.2.4) and the Implicit Function Theorem once again gives the existence of a radially mapped closed curve, $\mathscr{C}'$ say, near to the corresponding curve $\mathscr{C}$ for $\mathbf{M}_\varepsilon^q$; indeed $\mathscr{C}'=\mathscr{C}$ when $\delta=0$. Since the intersections of $\mathscr{C}$ and $\mathbf{M}_\varepsilon^q(\mathscr{C})$ are generically transverse, $\mathscr{C}'$ and $(\mathbf{M}_\varepsilon\cdot\mathbf{C}_\delta)^q(\mathscr{C}')$ will also intersect transversely provided δ is sufficiently small. It follows, by the same arguments as were used to prove Theorem 6.2.1, that $\mathbf{M}_\varepsilon\cdot\mathbf{C}_\delta$ has Birkhoff periodic points of period q for sufficiently small δ.

Now suppose that we have a dissipative map with the above property whose restriction to some annulus is a twist homeomorphism in the sense defined in §6.4.2. Given the existence of the necessary Birkhoff periodic orbits, Theorem 6.4.1 would allow us to conclude that Mather sets with an irrational rotation number must also occur. Since there is at most one invariant circle, at least some of these sets would have to be of Cantor type. This line of argument suggests that it may be possible to obtain an analogue of the Aubry–Mather Theorem (Theorem 6.4.3) for dissipative twist maps.

It is important to emphasise that Theorem 6.4.1 requires *both* a twist homeomorphism with a non-trivial rotation interval *and* the occurrence of Birkhoff periodic orbits of type (p, q) for every rational number in that interval. How should we go about finding such behaviour in dissipative twist maps? It is plausible that Birkhoff periodic orbits persist in dissipative twist maps that are near area-preserving maps but in order to use Theorem 6.4.1 we apparently require them to occur in a compact cylinder (i.e. a closed annulus) for which the map has a non-trivial rotation interval. A clue as to how one might proceed lies in the observation that the stable (in the sense of Liapunov) periodic points of the area-preserving map should become asymptotically stable in the presence of a dissipative perturbation. Of course, the saddle-like periodic points would persist since they are locally structurally stable. This suggests that, if Birkhoff periodic points exist, they will appear in the *attracting sets* of the dissipative twist map. Recall that a closed invariant set, A, is attracting for $\mathbf{f}$ if within any neighbourhood of A there is a trapping region T such that $A=\bigcap_{n\in\mathbb{Z}^+}\mathbf{f}^n(T)$. A closed set T is said to be a *trapping region* for $\mathbf{f}$ if $\mathbf{f}(T)\subseteq\operatorname{int}(T)$. An attracting set that contains a dense orbit of $\mathbf{f}$ is called an *attractor*. Thus we should look for attracting sets with non-trivial rotation intervals.

Birkhoff (1968b) gave an example of a dissipative twist map which has an attracting set, A, with a non-trivial rotation interval. The set A is contained in a compact subcylinder of C and separates C into two connected components. Birkhoff introduced the notion of internal and external rotation numbers of such an attracting set as follows. Lines of constant r are taken on C above and below A and vertical ($\theta=$ constant) projections of these lines onto the attracting set are considered (see Figure 6.20). The points of first intersection from above and below, respectively, give subsets of A which are called the 'top' and 'bottom' of the attracting set. These subsets are invariant under $\mathbf{f}^{-1}$ and rotation numbers, ρ_e and ρ_i, respectively, can be associated with them. This is done by using a fixed lift, $\bar{\mathbf{f}}$, of **f** and projection onto the reference circles shown in Figure 6.20 (see Exercise 6.6.5). Given that $\rho_e \geqslant \rho_i$, the rotation interval of A is given by $[\rho_i, \rho_e]$.

To sum up, we might hope to prove a generalisation of the Aubry–Mather result for dissipative twist maps when the attracting set of the whole cylinder lies in a compact subcylinder for which the map has a non-trivial rotation interval. This line of approach has been investigated by Le Calvez (1986) and Casdagli (1987a,b). In this work, the dissipative twist map **f** defined above is subject to an additional constraint. It is assumed that there exists a $K_0 \in \mathbb{R}^+$ such that, for all $K > K_0$, the compact subcylinder $T_K = S^1 \times [-K, K]$ is a trapping region for **f**. Not all maps that satisfy the twist and dissipation conditions described above satisfy this 'trapping hypothesis' (see Exercise 6.6.2), however, Casdagli has shown that a dissipative twist map **f** with this property has a unique attracting set of the type described by Birkhoff.

Definition 6.6.1 *The* Birkhoff set, $B(\mathbf{f})$, *of a dissipative twist map* **f**, *that satisfies the trapping hypothesis, is defined by* $B(\mathbf{f}) = Cl(L) \cap Cl(U)$, *where*

$$L = \{(\theta, r) \in C \mid r_{-j} < -K_0 \text{ for some } j \in \mathbb{Z}^+\}, \tag{6.6.3a}$$

$$U = \{(\theta, r) \in C \mid r_{-j} > \ K_0 \text{ for some } j \in \mathbb{Z}^+\}, \tag{6.6.3b}$$

and $\mathbf{f}^{-j}: (\theta, r) \mapsto (\theta_{-j}, r_{-j})$.

Notice that the lower (L) and upper (U) sets in (6.6.3a,b) are defined in terms of the expanding map $\mathbf{f}^{-1}$. Since **f** is end-preserving, L and U are disjoint sets. Thus $Cl(L) \cap Cl(U)$ must consist entirely of boundary points of L and U. Therefore $B(\mathbf{f})$ has no interior or, alternatively, it is its own boundary. Indeed one might look

Figure 6.20 Schematic illustration of the top and bottom of a Birkhoff attracting set. The reference circles $r = R_1$ and $r = R_2$ allow us to define rotation numbers for the restriction of **f** to these sets. Observe that the top and bottom of an attracting set are not necessarily continuous curves (cf Figure 6.21).

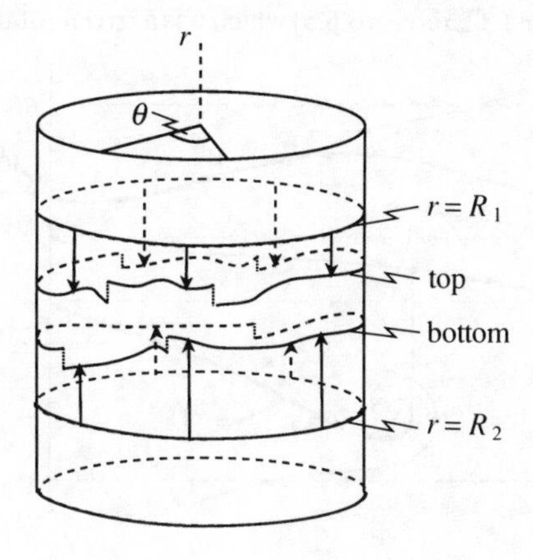

on $B(\mathbf{f})$ as the boundary of the basin of attraction of the sinks of $\mathbf{f}^{-1}$ at $r = \pm\infty$. It is important to emphasise that $B(\mathbf{f})$ is not necessarily an attractor for it may not contain a dense orbit (see Exercise 6.6.5). For this reason, we have avoided calling it a 'Birkhoff attractor', although this term is commonly used in the literature. However, it is an attracting set of the type envisaged by Birkhoff. For example, it is a non-empty, compact subset of the attracting set $\bigcap_{n \in \mathbb{Z}^+} \mathbf{f}^n(T_K)$, for any $K > K_0$. Moreover, it can be shown to separate C into two connected components (see Exercises 6.6.3 and 4(a)). In some cases, $B(\mathbf{f})$ is simply an invariant circle that winds once around the cylinder, i.e. a rotational invariant circle (see Figure 6.21 or Exercise 6.6.5). The rotation interval of the Birkhoff set is then said to be trivial since it degenerates to a single point. Notice that, in both of the examples mentioned above, there are periodic points lying outside the Birkhoff set. Thus $B(\mathbf{f})$ does not necessarily encompass all the non-trivial dynamics of $\mathbf{f}$.

When $B(\mathbf{f})$ is a rotational invariant circle, it is not surprising that it can contain Birkhoff periodic points if the rotation number is rational. However, the following result, due to Casdagli (1987a), is far from obvious.

Theorem 6.6.1 *Let the Birkhoff set $B(\mathbf{f})$ have internal and external rotation numbers ρ_i and ρ_e, $\rho_i \leqslant \rho_e$. Then $B(\mathbf{f})$ contains a Birkhoff periodic orbit of type (p, q) for all $p/q \in [\rho_i, \rho_e]$.*

Recall that in §6.4 area-preservation forced the intersection of rotational circles and their images under the twist maps considered there. This topological property

Figure 6.21 The map $\mathbf{f}$ shown is on the cylinder and therefore the lines AB and $A'B'$ are to be identified. There are three main features: an invariant circle $\mathscr{C}$, such that $\rho(\mathbf{f}|\mathscr{C}) = \alpha > 0$; a focus and a saddle point $\mathbf{x}^*$. The union of $\mathscr{C}$ and the unstable manifold of the saddle is an attracting set A for $\mathbf{f}$. The top set of A is $\mathscr{C}$, while its bottom set is a segment of the unstable manifold of $\mathbf{x}^*$. Thus $\rho_i = 0$ and $\rho_e = \alpha$, so that A has a non-trivial rotation interval. However, the Birkhoff set $B(\mathbf{f})$ is simply the invariant circle $\mathscr{C}$ (cf Exercise 6.6.5) which has a trivial rotation interval.

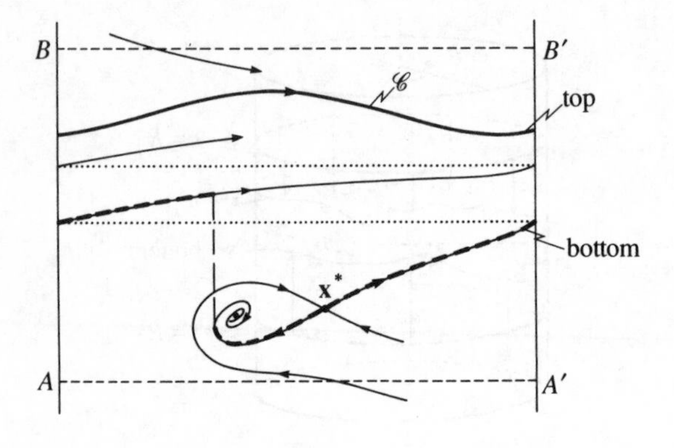

of area-preserving twist maps played a key role in the occurrence of their Birkhoff periodic points. It is perhaps worth noting that the set $B(\mathbf{f})$ has an analogous property. More precisely, it can be shown (see Exercise 6.6.4(c)) that, for every rotational circle R in C, $B(\mathbf{f}) \cap R \neq \emptyset$ implies $\mathbf{f}(R) \cap R \neq \emptyset$. Thus there is a sense in which the behaviour of $\mathbf{f} | B(\mathbf{f})$ parallels that of area-preserving twist maps.

Finally, an analogue of Theorem 6.4.3 follows from Katok's result (Theorem 6.4.1) as suggested earlier.

Theorem 6.6.2 *Let $B(\mathbf{f})$ be a Birkhoff set with non-trivial rotation interval $[\rho_i, \rho_e]$, $\rho_i < \rho_e$. Then, for every rational number $p/q \in [\rho_i, \rho_e]$, $B(\mathbf{f})$ contains a Birkhoff periodic orbit of type (p, q) and, for every irrational number $\beta \in [\rho_i, \rho_e]$, $B(\mathbf{f})$ contains a Mather set with rotation number β.*

Notice the Mather set with irrational rotation number in Theorem 6.6.2 cannot be an invariant circle, for the Birkhoff set would then be the circle itself and that would contradict the hypothesis that the rotation interval is non-trivial.

The family of maps $\mathbf{f}_{b,k,\omega}$ defined by

$$\theta_1 = (\theta + \omega + r_1) \bmod 1, \qquad r_1 = br - \frac{k}{2\pi} \sin 2\pi\theta, \tag{6.6.4}$$

$0 \leqslant b \leqslant 1$, $0 \leqslant \omega < 1$, $k \in \mathbb{R}$, exhibits a Birkhoff set with non-trivial rotation interval for certain values of the parameters. This family has the property that for $b = 0$ it is essentially the Arnold circle map discussed in § 5.2, while for $b = 1$ it gives the area-preserving standard map (6.5.1) (see Exercise 6.6.6). The Jacobian of $\mathbf{f}_{b,k,\omega}$ is identically equal to b and therefore the map is dissipative for $b < 1$. Indeed, (6.6.4) is often referred to as the *dissipative standard map*. It is easy to show that $\mathbf{f}_{b,k,\omega}$ is a twist map. Moreover, the radial component of (6.6.4) implies that

$$b - \left|\frac{k}{2\pi r}\right| \leqslant \left|\frac{r_1}{r}\right| \leqslant b + \left|\frac{k}{2\pi r}\right|. \tag{6.6.5}$$

Thus, $|r_1| < |r|$ for all $|r| > h$, where

$$h = |k| / [2\pi(1 - b)]. \tag{6.6.6}$$

It follows (see Exercise 6.6.7(a)) that $\mathbf{f}_{b,k,\omega}$ in (6.6.4) satisfies the trapping hypothesis with $K_0 = h$. Hence there is a Birkhoff set in the closed subcylinder $-h \leqslant r \leqslant h$.

If $\mathbf{f}^n_{b,k,\omega}\colon (\theta_0, r_0) \mapsto (\theta_n, r_n)$, $n \in \mathbb{Z}^+$, it can be shown (see Exercise 6.6.7(b)) that $|r_0| > h$ implies $|r_n| < |r_0|$, for all $n \in \mathbb{Z}^+$. However, $|r_n| < |r_0|$ for all n means there can be no fixed or periodic points of the map with radial coordinate r_0. Therefore, fixed or periodic points of $\mathbf{f}_{b,k,\omega}$ cannot occur outside the region $-h \leqslant r \leqslant h$. It is not difficult to show that they do occur inside this region. For example, the fixed points, (θ^*, r^*), of $\mathbf{f}_{b,k,\omega}$ are given by

$$r^* = -\omega + m, \tag{6.6.7}$$

where m is an integer for which the equation

$$\omega - m = k \sin(2\pi\theta^*)/[2\pi(1-b)], \tag{6.6.8}$$

admits solutions for θ^*. By way of illustration, for $b = 0.95$, $k = 1$ and $\omega = 0.2$, fixed points exist for $r^* = \{-2.2, -1.2, -0.2, 0.8, 1.8, 2.8\}$. All six values of $r^* \in [-h, h]$, with $h = 10/\pi = 3.18\ldots$, and each one gives rise to a pair of values for θ^* from (6.6.8). Of the pair of fixed points occurring at each value of r^*, one is of saddle-type and the other is a stable focus. The separation of the saddle and the focus with the same radial coordinate r^* decreases with increasing $|r^*|$. If b is reduced, the saddle-focus separation becomes smaller and, eventually, the saddle and focus for which $|r^*|$ is greatest coalesce and disappear in a saddle-node bifurcation. Alternatively, if b is increased, further saddles and foci are created at similar bifurcations until the area-preserving standard map is recovered when $b = 1$.

It is important to note that the trapping hypothesis does not guarantee that the Birkhoff set has a non-trivial rotation interval nor does it exclude the possibility that it occurs as a subset of the attracting set $A = \bigcap_{n \in \mathbb{Z}^+} \mathbf{f}^n(T_K)$, $K > K_0$. For example, A may consist of a union of periodic orbits along with an attracting invariant circle (see Exercise 6.6.5). Here the invariant circle is the Birkhoff set. The existence of a non-trivial Birkhoff set can be proved when the stable and unstable manifolds of the saddle point at the lowest value of $|r^*|$ intersect in such a way as to give homoclinic orbits that wind around the cylinder. More precisely (Rand, private communication), let $\mathbf{x}^*$ be the saddle point and $\mathbf{x}^\dagger$ be a homoclinic point, then the closed curve formed by taking the union of the segment of $W^u(\mathbf{x}^*)$ going from $\mathbf{x}^*$ to $\mathbf{x}^\dagger$, and the segment of $W^s(\mathbf{x}^*)$ going from $\mathbf{x}^\dagger$ to $\mathbf{x}^*$, must not be homotopic to zero (see Figure 6.22). Figure 6.23 shows that $\mathbf{f}_{b,k,\omega}$ does exhibit this kind of behaviour for certain values of the parameters.

Figure 6.22 Schematic illustration of the union of the segment of $W^u(\mathbf{x}^*)$ from $\mathbf{x}^*$ to $\mathbf{x}^\dagger$ and that of $W^s(\mathbf{x}^*)$ from $\mathbf{x}^\dagger$ to $\mathbf{x}^*$. In (*a*) the resulting closed curve $\mathscr{C}$ is not homotopic to zero but in (*b*) it is. The situation shown in (*a*) is compatible with homoclinic orbits which rotate around the annulus (or cylinder) whilst that in (*b*) is not.

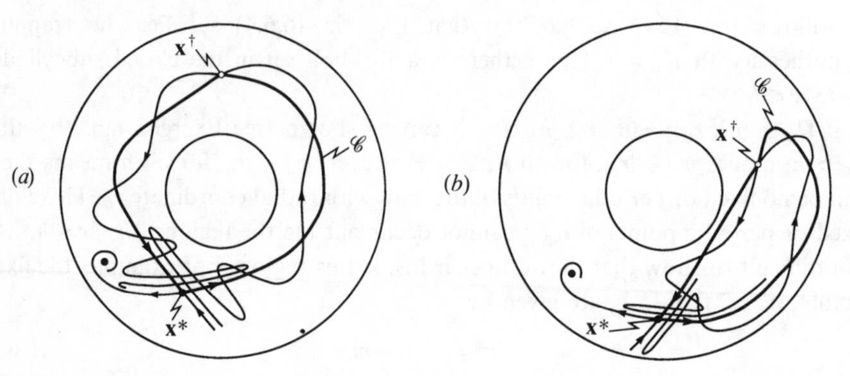

The proof of the existence of a non-trivial rotation interval makes use of the so-called *Shadowing Lemma* (see Bowen 1970, 1978; Newhouse, 1980). Let $\mathbf{f}$ be a C^l-diffeomorphism on $\mathbb{R}^n$ with a hyperbolic invariant set Λ and recall that the orbit of a point $\mathbf{x}_0 \in \mathbb{R}^n$ under $\mathbf{f}$ is the sequence of points $\{\mathbf{x}_i\}_{-\infty}^{\infty}$, where $\mathbf{x}_i = \mathbf{f}^i(\mathbf{x}_0)$, so that $\mathbf{x}_{i+1} = \mathbf{f}(\mathbf{x}_i)$.

Figure 6.23 Numerical approximations to the stable (open arrow) and unstable (closed arrow) manifolds of the saddle point with $r^* = -0.2$ in the dissipative standard map with $b = 0.95$, $k = 1$ and: (*a*) $\omega = 0.1$; (*b*) $\omega = 0.2$; (*c*) $\omega = 0.3$. The left and right hand sides of these diagrams are to be identified to form a cylinder. Observe that a homoclinic tangle exists in (*b*) but not in either (*a*) or (*c*). Note that the manifold segments plotted only show some of the early homoclinic oscillations. These segments are sufficient to determine whether or not homoclinic points occur. Longer segments of the unstable manifold are shown in Figure 6.25(*a*).

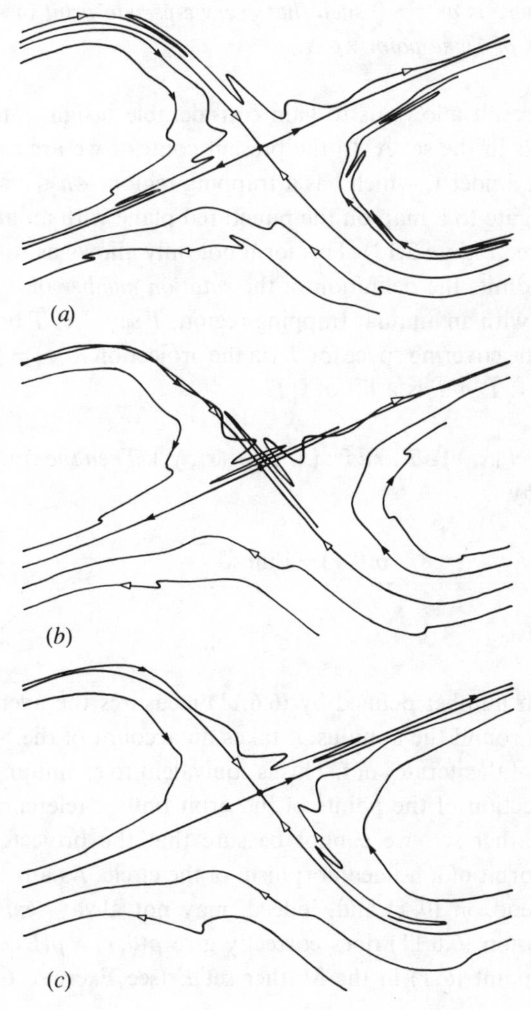

Definition 6.6.2 *An* ε-pseudo-orbit *in* Λ *is a bi-infinite sequence of points* $\{\mathbf{y}_i\}_{-\infty}^{\infty}$, $\mathbf{y}_i \in \Lambda$, *such that*

$$d(\mathbf{y}_{i+1}, \mathbf{f}(\mathbf{y}_i)) < \varepsilon, \tag{6.6.9}$$

where $d(\mathbf{x}, \mathbf{y})$ *is any suitable distance function on* $\mathbb{R}^n$.

Definition 6.6.3 *An actual orbit* $\{\mathbf{x}_i\}_{-\infty}^{\infty}$ *in* Λ *is said to be a* δ-shadow *of the pseudo-orbit* $\{\mathbf{y}_i\}_{-\infty}^{\infty}$ *if*

$$d(\mathbf{x}_i, \mathbf{y}_i) < \delta, \tag{6.6.10}$$

for all $i \in \mathbb{Z}$. *When this is the case, the actual orbit is also said* to δ-shadow *the pseudo-orbit.*

Theorem 6.6.3 (Shadowing Lemma) *Let* Λ *be a hyperbolic invariant set for* $\mathbf{f}$. *Then, for every* $\delta > 0$, *there is an* $\varepsilon > 0$ *such that every* ε*-pseudo-orbit in* Λ *is* δ*-shadowed by the actual orbit of some point* $\mathbf{x} \in \Lambda$.

This remarkable result allows us to gain considerable insight into the nature of the orbits of points in the set Λ. In the present context, we are concerned with a map of the open cylinder C which has a trapping region $-h \leqslant r \leqslant h$. Such a map is smoothly conjugate to a map on the punctured plane with an annular trapping region $1 \leqslant r \leqslant 2$ (see Figure 2.13). This form not only allows us to apply Theorem 6.6.3 but it also admits the definition of the *rotation number* of a *point*. Consider a map $\mathbf{f}: \mathbb{R}^2 \to \mathbb{R}^2$ with an annular trapping region, T say. Let $\bar{T}$ be the strip in the x, y-plane that is the covering space for T via the projection $(\theta, r) = (x \bmod 1, y) \in T$ and suppose that $\bar{\mathbf{f}}: \bar{T} \to \bar{T}$ is a lift of $\mathbf{f}|T$.

Definition 6.6.4 *Let* $(x, y) \in \bar{T}$ *and* $\bar{\mathbf{f}}^n: (x, y) \mapsto (x_n, y_n)$. *Then the* rotation number of $(\theta, r) \in T$ *is given by*

$$\rho(\theta, r) = \lim_{n \to \infty} \frac{x_n - x}{n}, \tag{6.6.11}$$

when this limit exists.

While the rotation number defined by (6.6.11) measures the average rotation of the orbit of (θ, r) around the annulus, it takes no account of the behaviour of the radial coordinate of the iterates of $\mathbf{f}$. This is equivalent to examining the behaviour of the radial projection of the points of the orbit onto a reference circle. Unless (θ, r) lies in a Mather set, we cannot be sure that the projected orbit can be interpreted as an orbit of a homeomorphism of the circle. As a result the rotation number may depend on (θ, r) and, indeed, may not always exist (see Exercise 6.6.8). The expression (6.6.11) does correctly give $\rho(\theta, r) = \rho(E)$ for the rotation number of every point (θ, r) in the Mather set E (see Exercise 6.4.9). However, care must be taken not to confuse the properties of $\rho(\theta, r)$, for general (θ, r), with those of the rotation numbers of Mather sets. For example, if the rotation number of a general point is found to be rational, one cannot conclude that its orbit under $\mathbf{f}$ is periodic. The orbit of any point with $1 \leqslant r \leqslant 2$ under the map $(\theta, r) \to ((\theta + (p/q)) \bmod 1, (1-\eta)r + \frac{3}{2}\eta)$, $0 < \eta < 1$, has rational rotation number according to Definition 6.6.4 but it only has periodic points on the circle $r = \frac{3}{2}$. In this example, $\rho(\theta, r) \in \mathbb{Q}$ arises because the forward orbit of (θ, r) approaches a periodic orbit with rational rotation number. Thus, in the contracting environment of a trapping region, rational $\rho(\theta, r)$ may be indicative of the existence of stronger recurrences. Finally, observe that $\rho(\mathbf{f}^n(\theta, r)) = \rho(\theta, r)$ for all $n \geqslant 0$ and therefore we can associate a unique rotation number with the forward orbit of the point (θ, r), given that the limit in (6.6.11) exists. We can therefore associate the rotation number with the orbit rather than with an individual point in it.

In the following discussion the parameters b, k, ω are assumed fixed at values corresponding to Figure 6.23(b), i.e. there is a homoclinic tangle of the type shown in Figure 6.22(a), associated with the saddle point with $r^* = -0.2$. The map $\mathbf{f}_{b,k,\omega}$ with these parameter values will be denoted simply by $\mathbf{f}$. In order to use Theorem 6.6.3, the hyperbolic set Λ is taken to be the closure of the set of transverse homoclinic points associated with the saddle point, $\mathbf{x}^*$, of $\mathbf{f}$. Since these points lie on the intersection of the stable and unstable manifolds of the saddle it is not difficult to check that this set is hyperbolic for $\mathbf{f}$. What is more, $W^u(\mathbf{x}^*) \subseteq T_h = S^1 \times [-h, h]$ and is invariant under $\mathbf{f}$. Thus, since every homoclinic point lies in $W^u(\mathbf{x}^*)$, Λ is a subset of the attracting set $A = \bigcap_{n \in \mathbb{Z}^+} \mathbf{f}^n(T_h)$.

Our aim is to demonstrate that there exist orbits on Λ with a range of rotation numbers. The following results play a crucial role in the argument. They can be shown to hold in the circumstances considered here (see Aronson *et al*, 1983, pp. 324, 331).

Proposition 6.6.1 *An* ε*-pseudo-orbit in* Λ *and a* δ*-shadow of it have the same rotation numbers provided* $\delta, \varepsilon < \pi$ *and one or other rotation number exists.*

Proposition 6.6.2 *If* $\{\mathbf{y}_i\}_{-\infty}^{\infty}$ *is a periodic* ε*-pseudo-orbit in* Λ*, then it has a* δ*-shadow in* Λ *that is periodic.*

For a particular choice of δ, Theorem 6.6.3 gives the existence of an actual orbit of $\mathbf{f}$ in Λ corresponding to every ε-pseudo-orbit in Λ. Proposition 6.6.1 assures us that provided $\delta, \varepsilon < \pi$ the δ-shadow has the same rotation number as the ε-pseudo-orbit. Finally, Proposition 6.6.2 allows us to take the δ-shadow to be periodic if the pseudo-orbit is periodic. Therefore, if there exists a set of periodic ε-pseudo-orbits in Λ with every rational rotation number in some interval, then there is a set of actual periodic orbits in Λ with the same property. It only remains to show that the required set of ε-pseudo-orbits exists. To do this consider the transverse homoclinic point, $\mathbf{x}_0$, shown in Figure 6.24 and focus attention on $\{\mathbf{f}^n(\mathbf{x}_0)\}_{-\infty}^{\infty}$. Observe that the lines AB and $A'B'$ in the figure are to be identified to form the cylinder C and let $N_{\varepsilon'}$ denote an ε'-neighbourhood of the saddle point $\mathbf{x}^*$ with $\varepsilon' \leqslant \varepsilon/2$. Forward iterates of $\mathbf{x}_0$ accumulate on $\mathbf{x}^*$ from the right in Figure 6.24, while backward iterates accumulate on $\mathbf{x}^*$ from the left. It follows that there is a segment of the orbit of $\mathbf{x}_0$ under $\mathbf{f}$, given by

$$B_1 = \{\mathbf{f}^{-M}(\mathbf{x}_0), \ldots, \mathbf{f}^{-1}(\mathbf{x}_0), \mathbf{x}_0, \mathbf{f}(\mathbf{x}_0), \ldots, \mathbf{f}^{M'}(\mathbf{x}_0)\}, \tag{6.6.12}$$

say, that leaves $N_{\varepsilon'}$, goes once around C – from right to left in Figure 6.24 – and returns to $N_{\varepsilon'}$ again. Let us now construct an ε-pseudo-orbit by repeating the orbit block B_1, i.e. consider $E_1 = \{\ldots B_1, B_1, B_1, B_1, \ldots\}$, and calculate its rotation number. E_1 gives rise to one revolution around C for every $N = M + M'$ iterations of $\mathbf{f}$ and therefore has rotation number $1/N$. Observe that E_1 is not an actual orbit for $\mathbf{f}$ because the true orbit through $\mathbf{x}_0$ does not satisfy $\mathbf{f}(\mathbf{f}^{M'}(\mathbf{x}_0)) = \mathbf{f}^{-M}(\mathbf{x}_0)$. However, it is an ε-pseudo-orbit for $\mathbf{f}$ because $\mathbf{f}(\mathbf{f}^{M'}(\mathbf{x}_0))$ and $\mathbf{f}^{-M}(\mathbf{x}_0)$ both lie in $N_{\varepsilon'}$. What is more, E_1 is periodic with period-N. Now let $B_0 = \{\mathbf{x}^*\}$ and take $E_2 = \{\ldots B_0, B_0, B_0, B_0, \ldots\}$. E_2 is an actual orbit for $\mathbf{f}$ which has rotation number zero and consequently there are at least two rotation numbers, $1/N$ and zero, associated with Λ. Of course, there are other possibilities. Let $uB_1 + vB_0$ denote u copies of B_1 followed by v copies of B_0. Then, $E_3 = \{\ldots uB_1 + vB_0, uB_1 + vB_0, \ldots\}$ is an ε-pseudo-orbit for $\mathbf{f}$ with rotation number $u/(uN + v)$, which is a rational number in the interval $(0, 1/N)$. Indeed, since any rational number $p/q \in (0, 1/N)$ can be written in the form $u/(uN + v)$ with $u = p$ and $v = (q - pN)$, ε-pseudo-orbits can be constructed with any rational rotation number in $(0, 1/N)$. Since all these pseudo-orbits are periodic, it follows that there are actual periodic orbits in Λ with every rational rotation number in the interval $[0, 1/N]$. Given that these orbits are of Birkhoff type, Theorem 6.4.1 then implies the existence of a Mather

Figure 6.24 Schematic illustration of the homoclinic tangle at the saddle point $\mathbf{x}^*$ in the dissipative standard map, showing the orbit block B_1. Observe that $d(\mathbf{f}^{-M}(\mathbf{x}_0), \mathbf{f}(\mathbf{f}^{M'}(\mathbf{x}_0))) < \varepsilon$ so that E_1 is an ε-pseudo-orbit of $\mathbf{f}$.

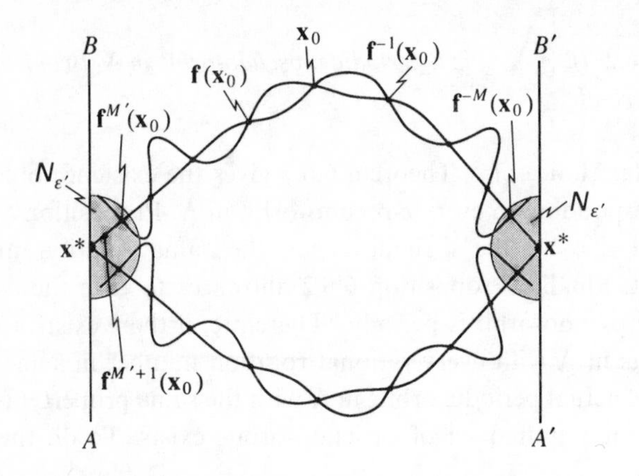

set with any *irrational* rotation number in this interval. It is therefore not surprising that ε-pseudo-orbits with irrational rotation number can be constructed. Let $\{p_i/q_i\}_0^\infty$ be a sequence of rational numbers tending to an irrational $\beta \in (0, 1/N)$, then the ε-pseudo-orbit

$$E = \left\{ \ldots \sum_{i=1}^{\infty} m_i(u_i B_1 + v_i B_0) \right\}, \tag{6.6.13}$$

has rotation number β for suitable choice of $\{m_i\}_1^\infty$ (see Exercise 6.6.8). It follows that for every $\alpha \in (0, 1/N)$ there exists an orbit of $\mathbf{f}$ with rotation number α. Notice that the choice of δ determines ε and hence a finite number of iterations of $\mathbf{f}$ are required to traverse the cylinder from $N_{\varepsilon'}$ to itself. Thus $[0, 1/N]$ is a non-trivial interval.

The arguments presented above show that the dissipative standard map $\mathbf{f} = \mathbf{f}_{b,k,\omega}$ with $b = 0.95$, $k = 1$ and $\omega = 0.2$ has a Birkhoff set, $B(\mathbf{f})$, with non-trivial rotation interval. This set separates the cylinder into two connected components (see Exercise 6.6.4) in rather a subtle fashion. We can gain some insight into the nature of $B(\mathbf{f})$ from numerical experiments. Figure 6.25(a) shows the unstable manifold of the saddle point $\mathbf{x}^*$ with $r^* = -0.2$. It is important to note that the left and right hand sides of this figure are to be identified. Observe that the loops of the homoclinic tangle that stretch off to the right, re-appear on the left and lie alongside and above the initial oscillations of $W^u(\mathbf{x}^*)$ shown in Figure 6.23(b). Of course there are infinitely many such loops, so that the unstable manifold wraps itself around the cylinder in an extremely complicated way. Figure 6.25(b) shows a sample of orbits of $\mathbf{f}$ that enter the trapping region T_h from above. They all apparently accumulate on $W^u(\mathbf{x}^*)$. Note that the plotted points appear to infiltrate the complex loop structure of $W^u(\mathbf{x}^*)$. What is more, these points must lie in U, since their reverse iterates escape from T_h. Figure 6.25(c) shows a similar sample of orbits that enter T_h from below. These orbits are ultimately drawn into the stable focus with $r^* = -0.2$; however, once again their upward motion is bounded by $W^u(\mathbf{x}^*)$. In this case, the plotted points must lie in L. It is tempting, therefore, to suggest that $B(\mathbf{f}) = Cl(L) \cap Cl(U)$ is the closure of the unstable manifold $W^u(\mathbf{x}^*)$. Further numerical experiments involving $\mathbf{f}_{b,k,\omega}$ are considered in Exercise 6.6.9.

In the above discussion shadowing arguments are applied to the homoclinic points associated with a hyperbolic fixed point to obtain an interval of rotation numbers near to zero. In §6.7, these ideas are used to obtain a rotation interval near a rational number, p/q, by considering hyperbolic periodic orbits of period q.

6.7 Birkhoff periodic orbits and Hopf bifurcations

Recent work of Aronson *et al* (1983) examines the bifurcations undergone by the invariant circle created in a non-degenerate Hopf bifurcation. Theorem 5.4.2 is a local result and holds for parameter values sufficiently close to the bifurcation point.

Figure 6.25 Results of numerical experiments on the dissipative standard map $\mathbf{f} = \mathbf{f}_{b,k,\omega}$, with $b = 0.95$, $k = 1$ and $\omega = 0.2$. Note that left and right hand sides of these diagrams are to be identified. (*a*) Approximate unstable manifold for the saddle point $\mathbf{x}^*$ with $r^* = -0.2$. The loops of the unstable manifold stretching off to the left of the diagram are drawn directly into the stable focus with $r^* = -0.2$, whilst those stretching off to the right first wind around the cylinder. (*b*) Iterates of a sample of 200 points whose orbits enter T_h from above. (*c*) Plots of orbits entering T_h from below. Observe the accumulation of points plotted in (*b*) and (*c*) on the unstable manifold shown in (*a*).

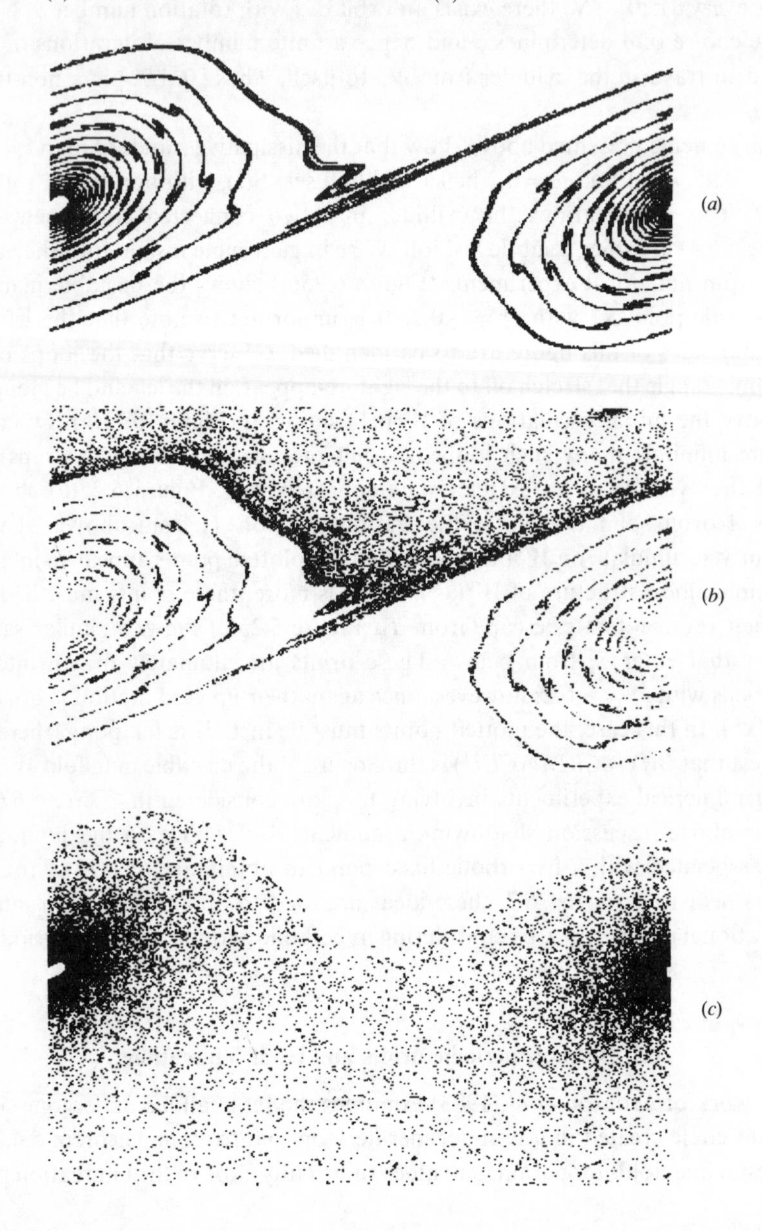

In Aronson *et al* (1983) (see also Hockett & Holmes, 1986) the fate of the invariant circle on moving out of this neighbourhood is considered. Under these circumstances, the circle may no longer be differentiable (cf. Figure 5.19). Indeed, far from the bifurcation point, it may become an attractor that is no longer a topological circle. One way in which this latter change can take place is via a bifurcation to a more complicated attracting set which has a non-trivial rotation interval. With the aid of the following definitions, Aronson *et al* give sufficient conditions for the existence of such an attractor.

Let $\mathbf{f}: \mathbb{R}^2 \to \mathbb{R}^2$ have an annular trapping region, T, associated with an attracting set $A = \bigcap_{n \in \mathbb{Z}^+}^{\infty} \mathbf{f}^n(T)$.

Definition 6.7.1 *Let $Y \subseteq T$ be a periodic orbit of* $\mathbf{f}$. *A point $\mathbf{x}^\dagger \in T$ is said to be* homoclinic to Y *if* $\mathbf{x}^\dagger \in W^s(Y, \mathbf{f}) \cap W^u(Y, \mathbf{f}) \setminus Y$.

Definition 6.7.2 *Let Y and Y' be two different periodic orbits of* $\mathbf{f}$. *A point $\mathbf{x}^\dagger$ is said to be* heteroclinic from Y *to* Y' *if* $\mathbf{x}^\dagger \in W^u(Y, \mathbf{f}) \cap W^s(Y', \mathbf{f})$.

Theorem 6.7.1 *Let $\mathbf{y} \in A$ be a periodic point of $\mathbf{f}$ of period q and rotation number p/q, where p, q are relatively prime. Suppose that $W^u(\mathbf{y}, \mathbf{f}^q)$ intersects $W^s(\mathbf{f}^k(\mathbf{y}), \mathbf{f}^q)$ transversely for some $0 < k < q$. Then there exists a non-trivial interval I containing p/q such that, for every $\alpha \in I$, there is a point with polar coordinates $(\theta, r) \in A$ with $\rho(\theta, r) = \alpha$. Furthermore, there are points in A for which the rotation number does not exist.*

This theorem gives sufficient conditions for the existence of a non-trivial rotation interval associated with a given attracting set A. It is important to note that the transverse intersection of $W^u(\mathbf{y}, \mathbf{f}^q)$ and $W^s(\mathbf{f}^k(\mathbf{y}), \mathbf{f}^q)$, with $0 < k < q$, has two implications:

(i) the existence of a point $\mathbf{x}^\dagger$ homoclinic to $Y = \{\mathbf{y}, \mathbf{f}(\mathbf{y}), \ldots, \mathbf{f}^{q-1}(\mathbf{y})\}$;
(ii) $\mathbf{x}^\dagger$ is heteroclinic from the fixed point $\mathbf{y}$ of $\mathbf{f}^q$ to the fixed point $\mathbf{y}' = \mathbf{f}^k(\mathbf{y})$ of $\mathbf{f}^q$.

The proof of Theorem 6.7.1 uses the shadowing results introduced at the end of § 6.6. Let us suppose that $\mathbf{y}, \mathbf{y}' \in Y$ satisfy $\mathbf{y}' = \mathbf{f}^k(\mathbf{y})$, $0 < k < q$. Both $\mathbf{y}$ and $\mathbf{y}'$ are fixed points of $\mathbf{f}^q$ and we will assume that $W^u(\mathbf{y}, \mathbf{f}^q) \cap W^s(\mathbf{y}', \mathbf{f}^q) \neq \varnothing$ (see Figure 6.26). Now consider a point, $\mathbf{x}_0$, in this intersection near to $\mathbf{y}$ and observe that:

(i) $\mathbf{f}^{nq}(\mathbf{x}_0) \to \mathbf{y}'$ as $n \to \infty$;
(ii) $\mathbf{f}^k(\mathbf{x}_0)$ is a point $\mathbf{x}_0'$ on $W^u(\mathbf{y}', \mathbf{f}^q)$ near to $\mathbf{y}'$;
(iii) $\mathbf{x}_0$ can be chosen to ensure that $\mathbf{x}_0'$ lies in an ε'-neighbourhood, $N_{\varepsilon'}(\mathbf{y}')$, of $\mathbf{y}'$ with $\varepsilon' \leqslant \varepsilon/2$.

Hence, there exists an $n = N$ for which the orbit segment

$$B_r = \{\mathbf{x}_0', \mathbf{f}(\mathbf{x}_0'), \ldots, \mathbf{f}^{Nq-k}(\mathbf{x}_0')\} \tag{6.7.1}$$

starts in $N_{\varepsilon'}(\mathbf{y}')$, winds a number, r, (given by a lift of $\mathbf{f}|T$) of times around the annulus and returns to $N_{\varepsilon'}(\mathbf{y}')$ once again. Clearly, the sequence $E = \{\ldots B_r, B_r, B_r, \ldots\}$ is an ε-pseudo-orbit on the closure, Λ, of the set of homoclinic points associated with Y. Moreover, it is periodic and has rotation number r/s, where $s = Nq - k$. Since $0 < k < q$, q does not divide s and therefore $r/s \neq p/q$. However, the closure of the set of points homoclinic to Y contains Y itself and Y has rotation number p/q. Thus there are two periodic orbits in Λ with different rotation numbers. As in §6.6, B_r and $B_p = Y$ can be used (cf. Exercise 6.6.8) to show that, for every α in the interval defined by p/q and r/s, there is an orbit in Λ with rotation number α. In this way, the existence of a non-trivial rotation interval is established. The construction of ε-pseudo-orbits for which the rotation number does not exist is also considered in Exercise 6.6.8.

Theorem 6.7.1 can help us to understand in more detail how an invariant circle 'expands through' a periodic orbit. We have already encountered an example where this phenomenon takes place; namely the case of strong resonance with $q = 4$ discussed in §5.6.3. The vector field approximation to the family of maps of

Figure 6.26 (a) Schematic diagram of the homoclinic tangle envisaged in Theorem 6.7.1. In the interest of clarity, the case $k = 1$, when $\mathbf{y}$ and $\mathbf{y}'$ are adjacent points of the periodic orbit, is shown. (b) When $k > 1$, the intervening periodic points introduce additional contortions in the intersecting manifolds; illustrated here for $k = 2$. In terms of $\mathbf{f}^q$, intersections of the bold curves are points that are heteroclinic from $\mathbf{y}$ to $\mathbf{y}''$.

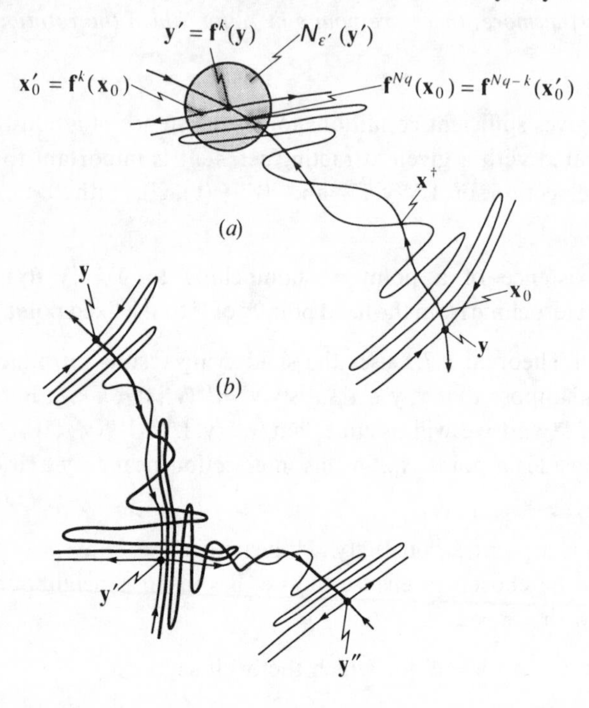

interest is shown in Figure 5.17(e)–(i). In the dynamics of the family itself, we would expect the heteroclinic saddle connections appearing in Figures 5.17(f) and (h) to be replaced by homoclinic tangles as shown in Figure 6.27. Moreover, the two possibilities illustrated in this figure should be separated by behaviour corresponding to Figure 5.17(g), in which the invariant circle is resonant with the period-4 orbit. In view of the behaviour of the vector field at the origin and at infinity in Figure 5.17, it is not unreasonable to conjecture that the conditions of Theorem 6.7.1 can be met.

In the above problem the periodic orbit arises from strong resonance terms in the normal form. However, it is not difficult to find examples near to area-preserving maps where the periodic orbits are of Birkhoff type. For example, consider the following unfolding of (3.7.1)

$$x_1 = x + y_1, \qquad y_1 = y + \varepsilon y + kx(x-1) + \mu xy, \tag{6.7.2}$$

with $k > 0$ and $(\varepsilon, \mu) \in \mathbb{R}^2$. This choice is motivated by the two-parameter unfolding

$$\dot{x} = y, \qquad \dot{y} = \varepsilon y + kx(x-1) + \mu xy, \tag{6.7.3}$$

of the Hamiltonian vector field

$$\dot{x} = y, \qquad \dot{y} = kx(x-1). \tag{6.7.4}$$

The family of vector fields (6.7.3), with $k = 1$, was studied by Bogdanov (1981a) as a preliminary to his derivation (1981b) of the versal unfolding of the cusp singularity (see §4.3.1). As a result, we will refer to (6.7.2) as the Bogdanov map. For each fixed k, the bifurcation diagram for the family of vector fields (6.7.3) is given in Figure 6.28 (see Exercise 6.7.1). The differential equation (6.7.4) takes

Figure 6.27 The homoclinic tangles that generically replace the saddle connections in: (a) Figure 5.17(f); (b) Figure 5.17(h). Observe that in (a) only the inner branches of the stable and unstable manifolds intersect while in (b) only the outer branches meet.

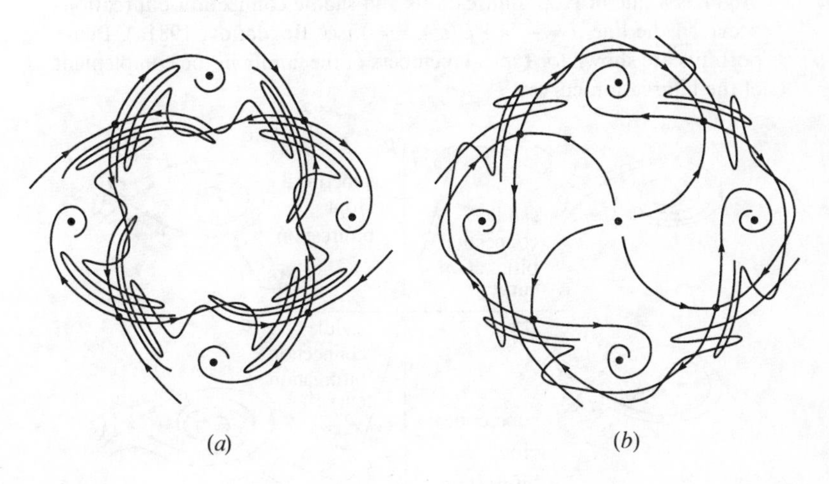

the form

$$\dot{x} = y, \qquad \dot{y} = f(x), \tag{6.7.5}$$

and it is easily verified that the discretisation

$$x_1 = x + y_1, \qquad y_1 = y + f(x), \tag{6.7.6}$$

yields an area-preserving map. This approximation to the differential equation (6.7.4) gives (3.7.1), while an analogous discretisation of (6.7.3) leads to the Bogdanov map.

In § 3.7, we found that the complexity of (3.7.1) was readily visible in numerical experiments for $k \simeq 1$. Throughout the following discussion we will take $k = 1.15$. This choice ensures that the six-fold island chain is a significant feature of the dynamics of (6.7.2). For given k, it is easily shown that the one-parameter family obtained from (6.7.2) by fixing $\mu \neq 0$ satisfies items (a)–(d) in Theorem 5.4.2. The coefficient of r^3 in the normal form of (6.7.2), with $\varepsilon = 0$ and $\mu \neq 0$, is given by (5.4.14) and a straightforward, if tedious, calculation (see Exercise 6.7.2) confirms that sub(super)critical Hopf bifurcations occur when $\mu > (<) 0$ (see Figure 6.29). As we have already noted in § 3.7, there are homoclinic tangles in the neighbourhood of the fixed point of (3.7.1) at (1, 0). These tangles remain in (6.7.2), near the origin of parameters, with the result that the saddle-connection curve in Figure 6.28 thickens into a 'strip' (see Figures 6.29 and 30).

More importantly from the point of view of the present discussion, each Birkhoff periodic orbit of (3.7.1) persists in (6.7.2) for (ε, μ) in some neighbourhood of the origin of the parameter plane. It follows that (6.7.2) may provide an example in which the interaction of the Hopf invariant circle with Birkhoff periodic orbits can be studied numerically. Some typical results are shown in Figure 6.31, where the invariant circle 'passes through' the Birkhoff periodic orbit of type (1,6).

Figure 6.28 Bifurcation diagram for the two-parameter family of vector fields obtained by holding k constant in (6.7.3). The μ-axis, excluding the origin, is a line of Hopf bifurcations and saddle connection bifurcations occur on the line $\mu = -7\varepsilon + O(\varepsilon^2)$, $\varepsilon \neq 0$ (see Bogdanov, 1981a). Phase portraits are shown for typical members of the family in the complement of the bifurcation curves.

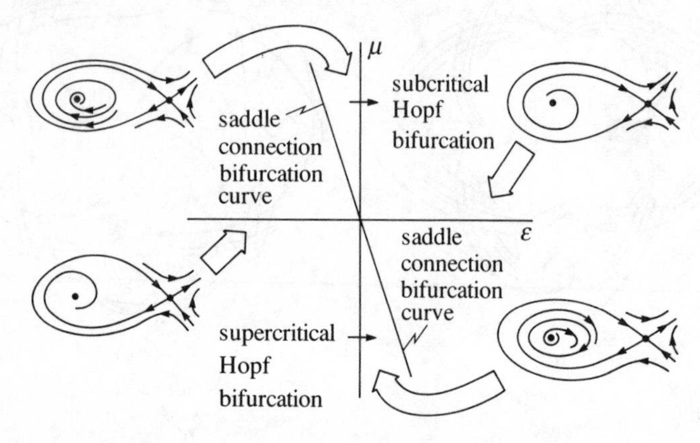

Figure 6.29 Analogues, for the Bogdanov map (6.7.2), of the bifurcations taking place in the family of vector fields (6.7.3). The parameter k is assumed to be fixed in both systems. The saddle connection bifurcation curve is replaced by a set, represented schematically by the shaded region, where homoclinic tangles occur (see Figure 6.30).

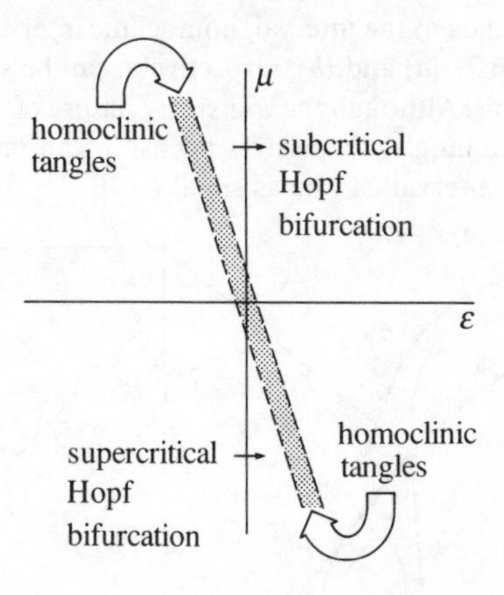

Figure 6.30 Numerical approximations to the stable and unstable manifolds of the saddle point near $(1, 0)$ in (6.7.2), as the saddle-connection strip of Figure 6.29 is traversed. The data shown is for $k = 1.15$, $\mu = -0.1$ and: (*a*) $\varepsilon = 0.0059$; (*b*) $\varepsilon = 0.015$; (*c*) $\varepsilon = 0.0256$.

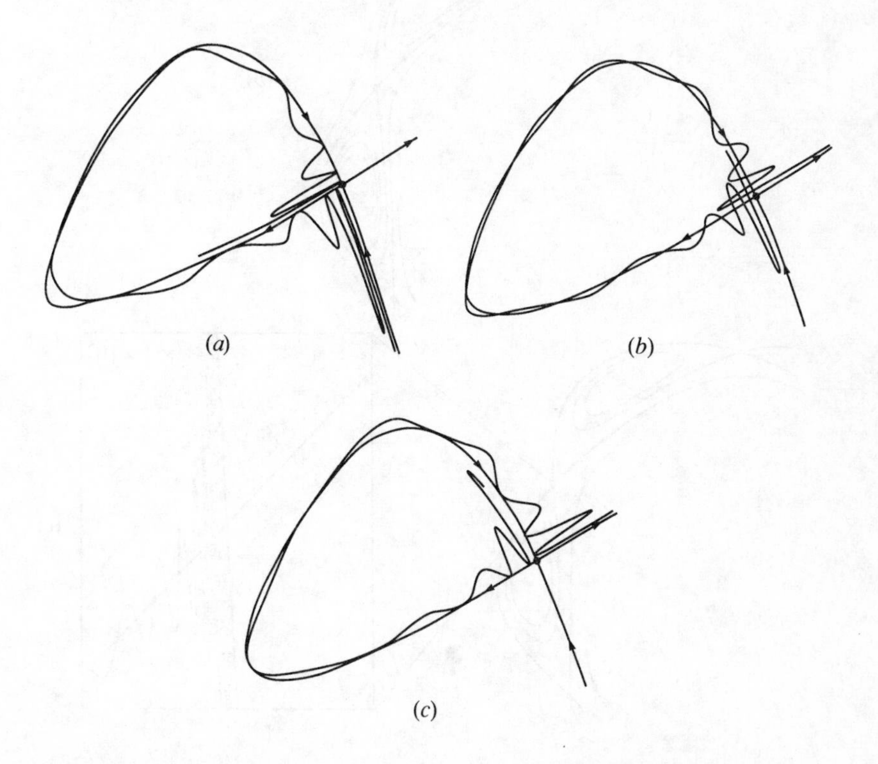

Figure 6.31 Computer-generated approximations to the stable and unstable manifolds of the (1,6)-periodic orbit in the Bogdanov map. The data shown illustrates how the manifolds behave when $k = 1.15$, $\mu = -0.1$ and ε is increased through an interval of length $\simeq 0.0009$ centred on $\varepsilon = 0.0044$. Note that in (*b*), where ε lies in the heart of the interval, there is no visible evidence of homoclinic points. In (*a*) and (*c*), where ε is approaching the ends of the interval, homoclinic intersections of the type shown in Figure 6.27 (*a*) and (*b*), respectively, can be seen at sufficiently high magnifications. Although the transverse nature of these intersections assures us that the tangles themselves persist for an interval of ε values, the extent of this interval can be as small as 10^{-11}–10^{-12}.

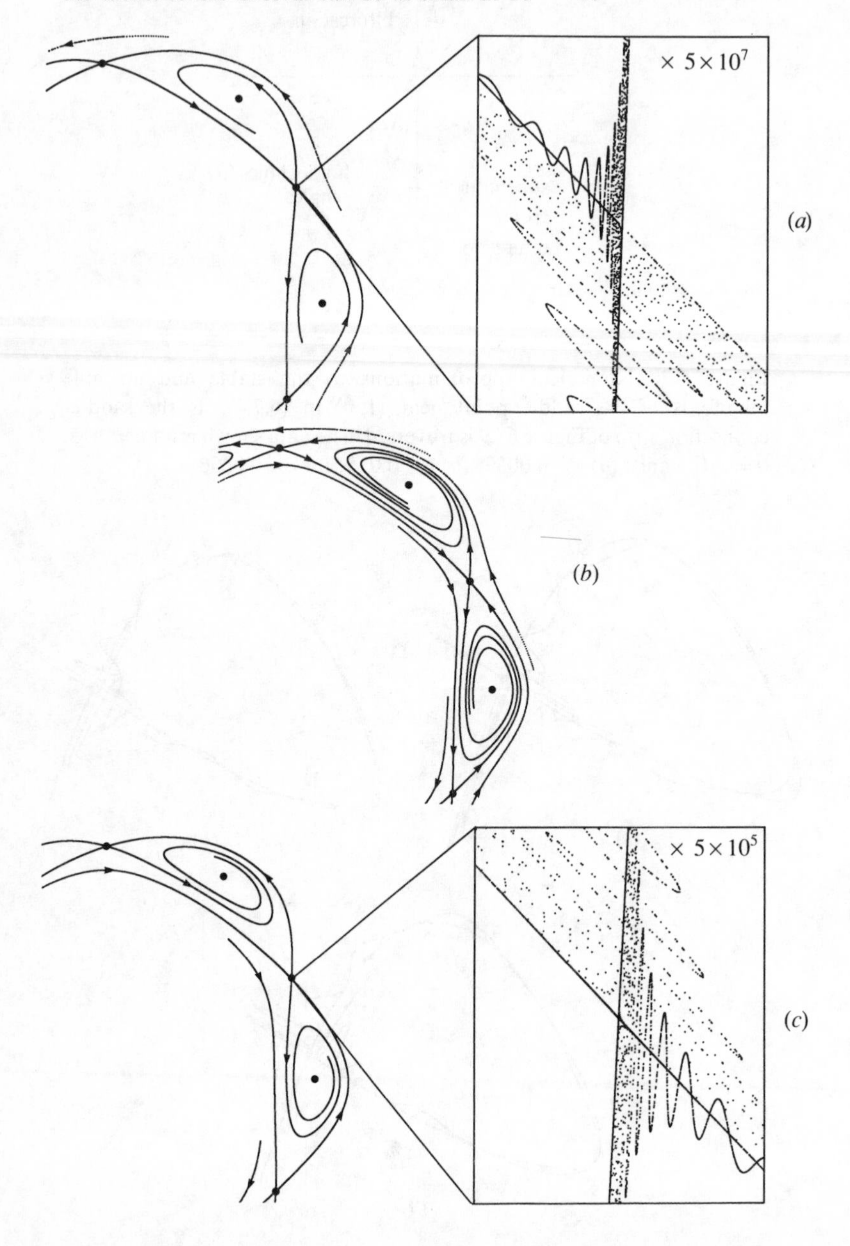

In Figure 6.31(*b*) ε and μ are chosen so that the invariant circle is resonant with the (1,6) periodic orbit. The circle is non-differentiable and the restriction of such a member of (6.7.2) to it has rotation number $p/q = \frac{1}{6}$. The situation shown in Figure 6.31(*b*) is structurally stable and therefore we expect it to persist for an interval of values of ε. However, Figures 6.31(*a*) and (*c*) show that homoclinic tangles of the form shown in Figure 6.27 develop as ε approaches the ends of this interval. In view of Theorem 6.7.1, one might reasonably ask whether or not the invariant circle becomes part of a more complicated attractor in the process of passing out of resonance with a given periodic orbit.

In order to make use of Theorem 6.7.1 to prove the existence of a non-trivial attractor, we attempt to establish that:

(i) homoclinic points of the appropriate type do occur;

(ii) there is a trapping region containing the circle and the periodic orbit.

Let us begin by considering the example involving strong resonance mentioned above. Recall (see (5.6.20)) that the equivariant unfolding of the vector field approximation for $q = 4$ is

$$\dot{\zeta} = \varepsilon\zeta + c\zeta|\zeta|^2 + \bar{\zeta}^3, \tag{6.7.7}$$

where $c, \varepsilon \in \mathbb{C}$. We wish to examine a typical family of maps near to that given by the flow of (6.7.7) at time 2π composed with a rotation through $2\pi p/q = p\pi/2$.

The polar form of (6.7.7) is

$$\dot{\theta} = v_2 + br^2 - r^2 \sin 4\theta, \qquad \dot{r} = v_1 r + ar^3 + r^3 \cos 4\theta, \tag{6.7.8}$$

where $\varepsilon = v_1 + iv_2$ and $c = a + ib$. The Euler approximation to (6.7.8) with unit step length gives the two-parameter family of maps

$$\theta_{l+1} = \theta_l + v_2 + \frac{p\pi}{2} + r_l^2(b - \sin 4\theta_l), \qquad r_{l+1} = r_l + v_1 r_l + r_l^3(a + \cos 4\theta_l). \tag{6.7.9}$$

The connection with the vector field approximation is important here, because the problem of choosing a suitable boundary for a trapping region is then seen to be closely related to the problem of finding a Liapunov function for the vector field. Specifically, it is not unreasonable to suggest that a level curve of a Liapunov function for the vector field approximation (6.7.8) will serve as a boundary to a trapping region for corresponding map (6.7.9).

Recall that the example used in § 5.6.3, to investigate the phenomena shown in Figure 5.17, had $a = -\frac{1}{2}, b = -5$. With a and b fixed at these values and $0 < v_2 \lessapprox 0.3$ homoclinic tangles like those shown in Figure 6.27 can be obtained by careful choice of v_1. The tangles occur for two disjoint intervals of v_1; one where the tangle is formed by the 'inner' branches of the manifolds of the periodic saddle points and the other where the 'outer' branches are involved. As v_1 is varied between these intervals, there appear to be values of v_1 for which no homoclinic points can be detected; apparently the invariant circle is non-differentiable and resonant with the period-4 orbit. The numerical results are shown in Figure 6.32.

When $v_1 = a = 0$, (6.7.8) is a Hamiltonian system, with Hamiltonian function

$$H(\theta, r) = -\tfrac{1}{2}v_2 r^2 - \tfrac{1}{4}br^4 + \tfrac{1}{4}r^4 \sin 4\theta \tag{6.7.10}$$

Figure 6.32 Numerical approximations to the unstable manifold of the period-4 orbit of (6.7.9). The data shown is for $a = -0.5, b = -5, v_2 = 0.3$ and v_1 increasing through an interval of length $\simeq 0.0082$ centred on $v_1 = 0.0285$. Observe that the homoclinic oscillations in (*a*) and (*c*) are similar to those shown in Figure 6.27. Note that the inverse of (6.7.9) is not available in closed form so that accurate approximations to the stable manifolds are not easy to obtain.

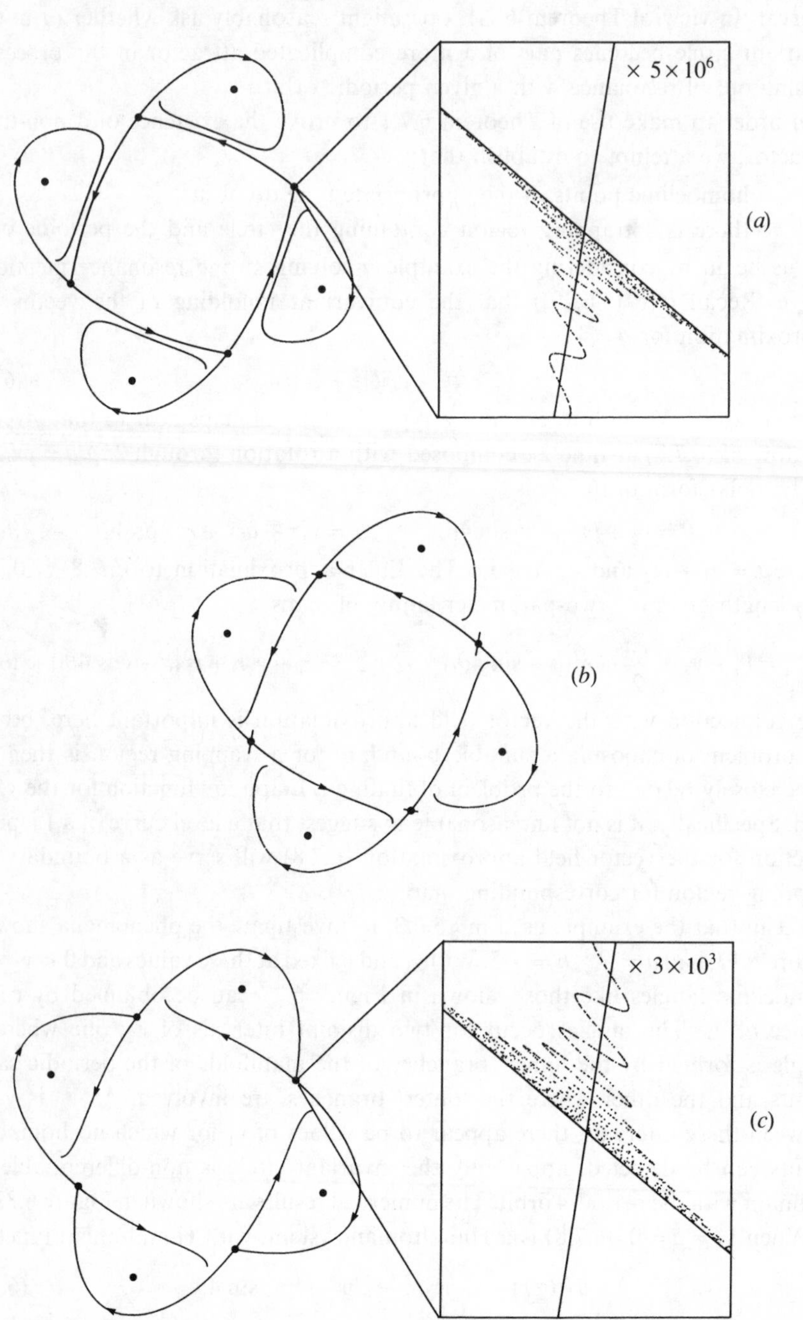

(see Exercise 6.7.5). The level curves for (6.7.10) with $b=-5$ and $v_2=0.3$. are shown in Figure 6.33. One might hope that the trajectories of (6.7.9) cross level curves like $\mathscr{C}_i$ and $\mathscr{C}_e$ as shown when v_1 and a are non-zero. Unfortunately, things are not quite so simple because not only do the positions of the fixed points change with v_1 and a but also additional rotations are induced as changes in r feedback into θ (see (6.7.9)). However, we can compensate for these effects by a small rotation of the level curve considered (see Exercises 6.7.6 and 7). Figure 6.34 shows that these ideas can be used to obtain an outer boundary to a trapping region for (6.7.9). The inner boundary is far less problematic, since the level curves of (6.7.10) are essentially circular for small r. Iteration of these boundaries provides tighter inner and outer bounds to the attractor. Figure 6.35 suggests how these iterates might collapse onto the attractor itself. Here, as in the Duffing attractor (see § 3.8 or Guckenheimer & Holmes, 1983), it is tempting to conjecture that the attractor is simply the closure of the unstable manifold of the periodic orbit. Finally, by Theorem 6.7.1, we conclude that, as the invariant circle moves out of resonance with the period-4 orbit (with both increasing and decreasing v_1), it bifurcates to an attractor which has a non-trivial rotation interval.

Let us now turn to the Bogdanov map (6.7.2) and fix $\mu=-0.1$. Numerical experiments show that ε can be chosen so that the homoclinic points required in item (i) do occur in the (1,6)-case (see Figure 6.31). As in the strong resonance example, two intervals of ε are found for which inner and outer tangles occur, respectively. Given the unstable fixed point at the origin, it is not difficult to argue

Figure 6.33 Level sets for the Hamiltonian (6.7.10) with $b=-5$, $v_2=0.3$ and H equal to: (*a*) 0.01; (*b*) -0.001; (*c*) -0.002; (*d*) -0.0035. In (*b*), (*c*) and (*d*) these sets consist of a union of two disjoint closed curves. If the curves $\mathscr{C}_i$ and $\mathscr{C}_e$ are, respectively, chosen to lie sufficiently far inside and outside the invariant circle in (6.7.8), the trajectories of this vector field, with $a=-0.5$ and v_1 small, cross $\mathscr{C}_i$ and $\mathscr{C}_e$ as shown (see Exercises 6.7.5 and 6).

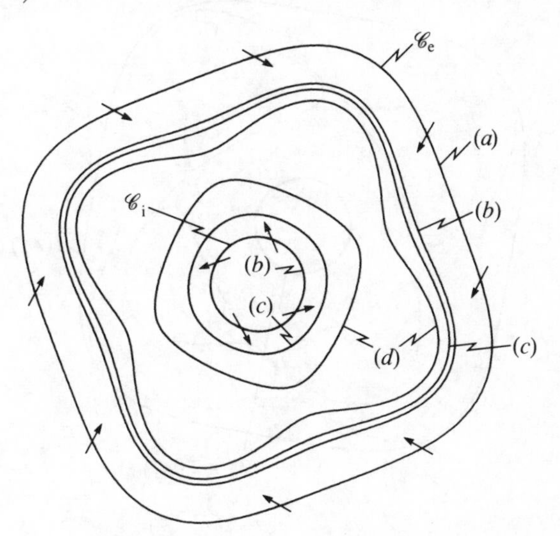

for the existence of an inner boundary for a trapping region. The outer boundary is a little more problematic. Consider stable and unstable manifolds of the saddle point near to (1,0). The behaviour of the unstable manifold shown in Figure 6.36 is typical of values of ε in the interval covered in Figure 6.31. Observe that there are no homoclinic intersections of the stable and unstable manifolds at this fixed point for these values of ε; the unstable manifold lies wholly inside the stable manifold. Hence we conjecture that there is a closed curve, $\mathscr{C}$, of the shape illustrated schematically in Figure 6.36 which acts as an outer boundary to a trapping region containing the homoclinic tangles shown in Figure 6.31. We would therefore argue that, by Theorem 6.7.1, attractors with non-trivial rotation intervals exist for the corresponding values of ε.

Of course, there is nothing special about the six-fold island chain, other than that it is easy to detect numerically. Similar bifurcations can occur when the invariant circle is associated with any dissipative island chain. One might argue that it is no longer appropriate to speak of an expanding invariant circle except in the immediate neighbourhood of the Hopf bifurcation. It is perhaps better to picture an expanding attracting set that can be recognised as an invariant circle when it is resonant with a Birkhoff periodic orbit but which bifurcates to a non-trivial attractor of Birkhoff type as it moves into or out of resonance.

Figure 6.34 Illustration of the outer boundary, $\mathscr{C}'_e$, to a trapping region for the map $\mathbf{f}$ given by (6.7.9) with $a = -0.5$, $b = -5$, $\nu_1 = 0.033$ and $\nu_2 = 0.3$. The image $\mathbf{f}(\mathscr{C}'_e)$ is also shown. The curve $\mathscr{C}'_e$ is obtained by rotating the $H = 0.01$ level set of (6.7.10) by -0.2 radians anticlockwise. Orbits of initial points $(\pm 0.3, 0)$, $(0, \pm 0.3)$, $(\pm 0.09, 0)$, $(0, \pm 0.09)$ are shown to indicate the position of the period-4 orbit. An inner boundary $\mathscr{C}'_i$ is also shown but $\mathbf{f}(\mathscr{C}'_i)$ is not resolved on this scale. Here $\mathscr{C}_i = \{(\theta, r) | H(\theta, r) = -0.001\}$ and $\mathscr{C}'_i$ are essentially circular.

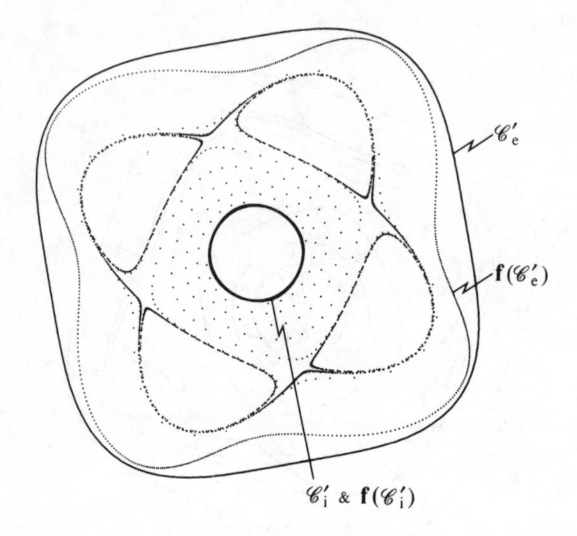

Figure 6.35 (*a*) $\mathbf{f}^{n_i}(\mathscr{C}'_i)$ and $\mathbf{f}^{n_e}(\mathscr{C}'_e)$, where $\mathbf{f}$ is the map (6.7.9) with $a = -0.5$, $b = -5$, $v_1 = 0.033$, $v_2 = 0.3$. Observe in particular how the image of $\mathscr{C}'_i$ is distorted as it approaches the attracting periodic orbit. Similar pictures can be obtained when the homoclinic tangle is as shown in Figure 6.27(*a*) (cf Exercise 6.7.7) but it is then the image of $\mathscr{C}'_e$ that suffers this kind of distortion. (*b*) Schematic diagram showing the ultimate fate of $\mathbf{f}^{n_i}(\mathscr{C}'_i)$ and as $n_i \to \infty$. As n_i becomes large, $\mathbf{f}^{n_i}(\mathscr{C}'_i)$ presses closer and closer to the inside of the unstable manifold and is sucked deeper into each focus. Meanwhile, $\mathbf{f}^{n_e}(\mathscr{C}'_e)$ collapses onto the same manifold from the outside. This suggests that the attractor is the closure of the unstable manifold of the period-4 orbit.

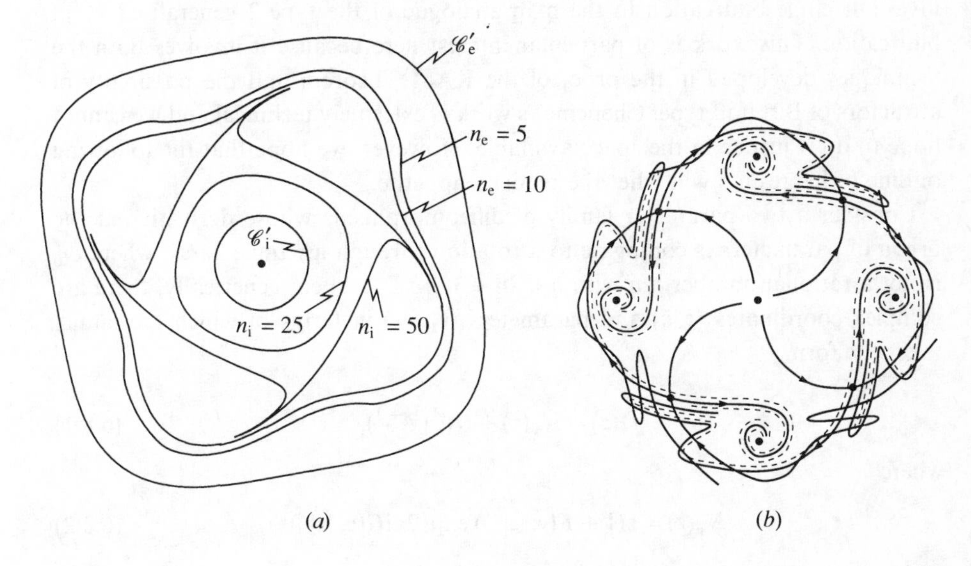

Figure 6.36 Typical behaviour of the unstable manifold of the saddle point near (1, 0) in (6.7.2) with $k = 1.15$, $\mu = -0.1$ and $0 < \varepsilon < 0.0059$. The unstable manifold lies wholly inside the stable manifold. The closed curve $\mathscr{C}$ is a schematic representation of an outer boundary to a trapping region containing the homoclinic tangles shown in Figure 6.31.

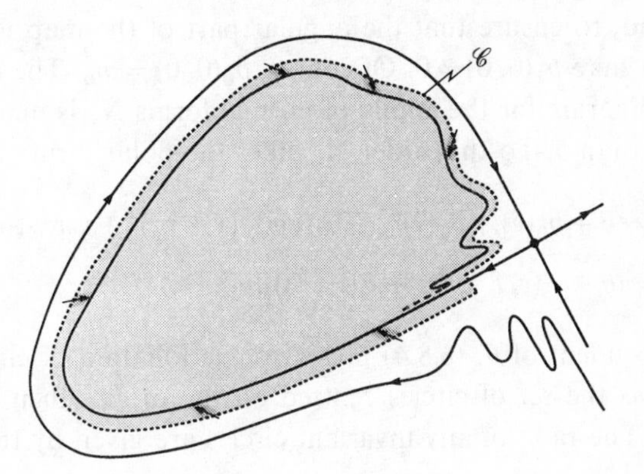

6.8 Double invariant circle bifurcations in planar maps

We have encountered double limit-cycle bifurcations for vector field problems in §§4.3.2 and 5.6. In that case the two limit cycles coalesce simultaneously over their whole length forming a single (non-hyperbolic) cycle at the bifurcation point (see Figure 4.11 or Figure 5.12). The reason for this is that if two trajectories of a flow intersect in one point, then they must coincide completely (see Exercise 1.3.2). We have already noted in § 5.7 that this is not generically the case for invariant circles of planar maps and that more complicated bifurcations can be expected. Chenciner (1985a,b and 1987) has made a detailed study of the double invariant circle bifurcation in the map analogue of the type-2 generalised Hopf bifurcation. This work is of particular interest here because it involves both the techniques developed in the proof of the KAM Theorem and the possibility of attractors of Birkhoff type. Chenciner's work is extremely technical and we cannot hope to do it justice in the space available. However, we hope that the following outline of his results will whet the reader's appetite.

Consider a two-parameter family of diffeomorphisms whose derivative at the origin of parameters is conjugate to a rotation through an angle $2\pi\omega_0$, where ω_0 is not a rational number p/q with $q < 2n+3$, $n \in \mathbb{Z}^+$. Then, generically, there are complex coordinates $(z, \bar{z})$ and parameters (v_1, v_2) in terms of which the family takes the form

$$f_{\mathbf{v}}(z) = N_{\mathbf{v}}(z) + O(|z|^{2n+2}), \tag{6.8.1}$$

where

$$N_{\mathbf{v}}(z) = z(1 + F(\mathbf{v}, |z|^2)) \exp\{2\pi i G(\mathbf{v}, |z|^2)\}, \tag{6.8.2}$$

with

$$\begin{aligned} F(\mathbf{v}, X) &= v_1 + v_2 X + a_2(\mathbf{v})X^2 + \cdots + a_n(\mathbf{v})X^n, \\ G(\mathbf{v}, X) &= b_0(\mathbf{v}) + b_1(\mathbf{v})X + \cdots + b_n(\mathbf{v})X^n. \end{aligned} \tag{6.8.3}$$

In analogy with the type-$(2, +)$ generalised Hopf bifurcation, we will assume that $a_2(0, 0) = +1$ and, to ensure that the angular part of the map is not a constant rotation, we will take $b_1(0, 0) \neq 0$. Of course, $b_0(0, 0) = \omega_0$. The main features of the bifurcation diagram for the family of normal forms $N_{\mathbf{v}}$ is unaffected by terms of order greater than 5. To this order $\mathbf{N}_{\mathbf{v}}$ takes the polar form

$$\begin{aligned} (\theta_1, r_1) &= (\theta + b_0(\mathbf{v}) + b_1(\mathbf{v})r^2 + b_2(\mathbf{v})r^4, (1 + v_1)r + v_2 r^3 + a_2(\mathbf{v})r^5), \\ &= (\theta + G(\mathbf{v}, r^2), r(1 + F(\mathbf{v}, r^2))). \end{aligned} \tag{6.8.4}$$

Since r_1 is independent of θ, (6.8.4) preserves the foliation of circles around the origin, i.e. if L_c is the set of circles centred on the origin, then for each $C \in L_c$, $\mathbf{N}_{\mathbf{v}}(C) = C' \in L_c$. The radii of any invariant circles are given by the positive roots of the equation

$$F(\mathbf{v}, r^2) = v_1 + v_2 r^2 + a_2(\mathbf{v}) r^4 = 0 \tag{6.8.5}$$

and their rotation number can then be obtained from θ_1 in (6.8.4). The analysis of the number of invariant circles occurring for given values of $\mathbf{v}$ is essentially the same as that of the type-(2, +) generalised Hopf bifurcation given in §4.3.2 (see Exercise 6.8.1). The resulting bifurcations are shown in Figure 6.37. Clearly, in this respect, the invariant circles of $\mathbf{N}_\mathbf{v}$ behave in the same way as the limit cycles in the vector field case (see Figure 4.11). However, it is perhaps worth emphasising that, while the period of a limit cycle is not important for topological equivalence of vector fields, the same is not true of the rotation number of an invariant circle of a diffeomorphism. The topological conjugacy class of a diffeomorphism is not determined solely by its invariant circles and their stability. Since the rotation numbers on the invariant circles depend on $\mathbf{v}$ within a given region of Figure 6.37, the diffeomorphisms associated with such a region do not form a single conjugacy class. However, these diffeomorphisms do have some major topological features in common and, following Chenciner, we will say that they 'look alike'. More precisely, we will say that $\mathbf{f}_\mathbf{v}$ looks like $\mathbf{N}_\mathbf{v}$ if in a $\mathbf{v}$-independent neighbourhood, Ω, of the origin, $\mathbf{f}_\mathbf{v}$ and $\mathbf{N}_\mathbf{v}$ have the same number of invariant curves and the same topological decomposition of Ω into basins of attraction and repulsion of these curves and of the origin.

Let us focus attention on the double invariant circle bifurcation curve Γ in Figure 6.37. For $\mathbf{v} \in \Gamma$, we have

$$v_2^2 = 4 v_1 a_2(\mathbf{v}) \tag{6.8.6}$$

Figure 6.37 Invariant circle bifurcations for the family of normal forms $\mathbf{N}_\mathbf{v}$. $\Gamma = \{\mathbf{v} \mid v_2^2 - 4v_1 a_2(\mathbf{v}) = 0,\ v_2 < 0\}$ is the double invariant circle bifurcation curve. The v_2-axis with $v_2 < 0\ (>0)$ is a line of supercritical (subcritical) Hopf bifurcations with increasing v_1. Two invariant circles are simultaneously present in the dynamics of $\mathbf{N}_\mathbf{v}$ when $\mathbf{v}$ lies in the shaded region of the half-plane $v_1 > 0$.

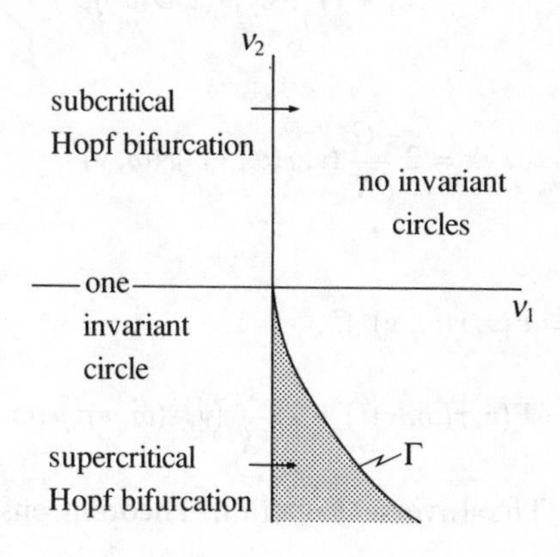

and $\mathbf{N}_\nu$ has a single non-normally hyperbolic invariant circle of radius

$$r_c^2 = -\nu_2/2a_2(\boldsymbol{\nu}). \tag{6.8.7}$$

The restriction of $\mathbf{N}_\nu$ to this circle has rotation number

$$\omega = b_0(\boldsymbol{\nu}) + b_1(\boldsymbol{\nu})r_c^2 + b_2(\boldsymbol{\nu})r_c^4. \tag{6.8.8}$$

Clearly, ω is parametrised by ν_2 and a Taylor expansion about $\nu_2 = 0$ gives

$$\omega = b_0(\mathbf{0}) + \left[\frac{\partial b_0}{\partial \nu_2}(\mathbf{0}) - \frac{b_1(\mathbf{0})}{2a_2(\mathbf{0})}\right]\nu_2 + O(\nu_2^2) \tag{6.8.9}$$

(see Exercise 6.8.2(a)). We will assume that

$$\left[\frac{\partial b_0}{\partial \nu_2}(\mathbf{0}) - \frac{b_1(\mathbf{0})}{2a_2(\mathbf{0})}\right] \neq 0, \tag{6.8.10}$$

so that ω changes monotonically with ν_2 and hence with arc length along Γ. Thus, given an appropriate (see Exercise 6.8.2) value of ω sufficiently close to ω_0, there is a unique point $\gamma_\omega \in \Gamma$ with $\boldsymbol{\nu} = \boldsymbol{\nu}_\omega$ such that $\mathbf{N}_{\boldsymbol{\nu}_\omega}$ has a single (non-normally hyperbolic) invariant circle with rotation number ω. Can we obtain similar information when $\boldsymbol{\nu} \notin \Gamma$?

To do this, Chenciner investigates $C_\omega = \{\boldsymbol{\nu} | \mathbf{N}_\nu$ has an invariant circle with rotation number $\omega\}$. Obviously, $C_\omega \cap \Gamma$ is the single point γ_ω, but consider the parameter points $\boldsymbol{\nu}$ in a neighbourhood of γ_ω. Provided $\boldsymbol{\nu}$ is sufficiently close to γ_ω, there will be a unique circle of radius $r = r(\omega, \boldsymbol{\nu})$ whose points are rotated by ω under $\mathbf{N}_\nu$. This does not mean that the circle is necessarily invariant, only that $G(\boldsymbol{\nu}, r(\omega, \boldsymbol{\nu})^2) = \omega$. Now introduce a local radial coordinate, σ, by taking

$$r = r(\omega, \boldsymbol{\nu})(1 + \sigma). \tag{6.8.11}$$

Taylor expansions of $G(\boldsymbol{\nu}, r^2)$ and $F(\boldsymbol{\nu}, r^2)$ then allow (6.8.4) to be expressed in terms of (θ, σ) (see Exercise 6.8.2(b)). The result is that $\mathbf{N}_\nu$ takes the form

$$\begin{aligned} \theta_1 &= \theta + \omega + \tau\sigma + O(\sigma^2), \\ \sigma_1 &= \varepsilon_1 + (1 + \varepsilon_2)\sigma + O(\sigma^2), \end{aligned} \tag{6.8.12}$$

where

$$\tau = 2\frac{\partial G}{\partial X}(\boldsymbol{\nu}, r(\omega, \boldsymbol{\nu})^2) r(\omega, \boldsymbol{\nu})^2 \tag{6.8.13}$$

and

$$\varepsilon_1 = F(\boldsymbol{\nu}, r(\omega, \boldsymbol{\nu})^2), \tag{6.8.14}$$

$$\varepsilon_2 = F(\boldsymbol{\nu}, r(\omega, \boldsymbol{\nu})^2) + 2\frac{\partial F}{\partial X}(\boldsymbol{\nu}, r(\omega, \boldsymbol{\nu})^2) r(\omega, \boldsymbol{\nu})^2, \tag{6.8.15}$$

are new parameters. The Inverse Function Theorem ensures that (ν_1, ν_2) and

$(\varepsilon_1, \varepsilon_2)$ are related by a local diffeomorphism at γ_ω. It follows from (6.8.14) that $\varepsilon_1 = 0$ corresponds to $r(\omega, \mathbf{v})$ being a root of (6.8.5). Moreover, given that $\varepsilon_1 = 0$, then (6.8.15) shows that this is a double root when $\varepsilon_2 = 0$. In fact, $(\varepsilon_1, \varepsilon_2) = 0$ is the point γ_ω. Furthermore, if $\varepsilon_1 = 0$, it is clear from (6.8.12), with $\sigma = 0$, that $r = r(\omega, \mathbf{v})$ is an invariant circle for $\mathbf{N_v}$ with rotation number ω. More precisely, the restriction of $\mathbf{N_v}$ to $r = r(\omega, \mathbf{v})$ is a pure rotation through angle ω. Thus, C_ω is the ε_2-coordinate line passing through γ_ω (see Figure 6.38). The radial component of the form (6.8.12) also shows that, when $\varepsilon_1 = 0$, the invariant circle $r = r(\omega, \mathbf{v})$ is stable for $\varepsilon_2 < 0$ and unstable for $\varepsilon_2 > 0$. Since $\mathbf{N_v}$ has normally hyperbolic invariant circles for $\mathbf{v} \in C_\omega \backslash \gamma_\omega$ (see Exercise 6.8.4), it follows that the branch of C_ω with $\varepsilon_2 < 0$ (>0) is contained in a 'cone' of parameter points for which $\mathbf{N_v}$ possesses an attracting (repelling) invariant circle. Of course, these circles can only be guaranteed to have rotation number ω on C_ω itself. It is important to remember that, for $\mathbf{v}$ in the region of the parameter plane under consideration, $\mathbf{N_v}$ has two invariant circles. Thus, if $\mathbf{v}$ lies in the intersection of two cones one circle is attracting and the other is repelling.

We should not expect the state of affairs described above for the normal form $\mathbf{N_v}$ to occur for the map $\mathbf{f_v}$ itself. For example, when ω is rational and $\mathbf{v} \in C_\omega$, every point of the circle $r = r(\omega, \mathbf{v})$ is a periodic point of $\mathbf{N_v}$. We have already seen that, generically, this behaviour does not persist when perturbed. In the light of §§6.2 and 6.7 we might expect Birkhoff periodic points in association with dissipative island chains to result. If we can show that dissipative island chains do occur for some value of $\mathbf{v}$, then they must persist in a neighbourhood of that value because they contain hyperbolic elements. This suggests that the single point $\mathbf{v} \in \Gamma$ in Figure 6.38 for which $\omega = p/q$ might correspond to an open set of points in the bifurcation diagram for $\mathbf{f_v}$. This is indeed the case but how do we obtain these features from (6.8.1)?

In order to detect the period-q points associated with the dissipative island

Figure 6.38 Schematic illustration of the relationship between the parameters (ν_1, ν_2) and $(\varepsilon_1, \varepsilon_2)$. C_ω corresponds to the ε_2-axis. The shape of this curve and its tangency to Γ at γ_ω is considered in Exercise 6.8.3.

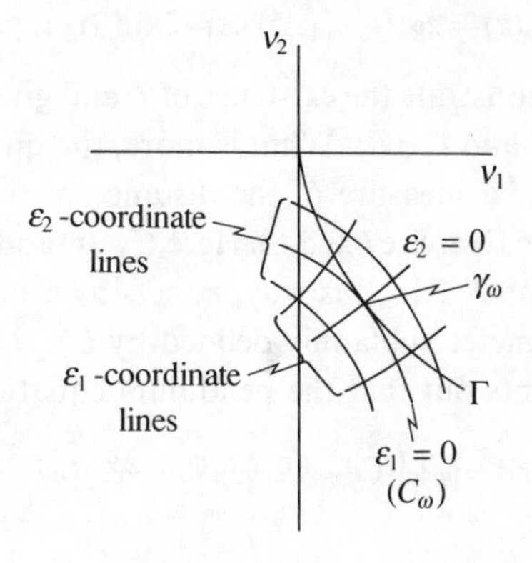

chains, Chenciner embeds the two-parameter family $\mathbf{f}_\nu$ in a three-parameter family $\mathbf{f}_{\nu,t}$ defined by the complex form

$$f_{\nu,t}(z) = \exp\{2\pi i((p/q) - \omega_0 + t)\} f_\nu(z). \tag{6.8.16}$$

Clearly, we can recover f_ν from $f_{\nu,t}$ by setting $t = t_0 = (\omega_0 - (p/q))$ but, if a local analysis of (6.8.16) at $(\nu, t) = (\mathbf{0}, 0)$ is to be appropriate, $|t_0|$ must not be too large. Under these circumstances, the bifurcations occurring in $\mathbf{f}_{\nu,t}$ near $(\nu, t) = (\mathbf{0}, 0)$ can be followed back to $\mathbf{f}_\nu$. In suitable coordinates, (6.8.16) can be written

$$f_{\nu,t}(z) = N_{\nu,t}(z) + \exp(2\pi i p/q) c(\nu, t)\bar{z}^{q-1} + O(|z|^q), \tag{6.8.17}$$

where

$$N_{\nu,t}(z) = z(1 + F(\nu, t, |z|^2)) \exp(2\pi i G(\nu, t, |z|^2), \tag{6.8.18}$$

with

$$F(\nu, t, X) = \nu_1 + \nu_2 X + a_2(\nu) X^2 + \cdots, \tag{6.8.19}$$

$$G(\nu, t, X) = \bar{\omega}(\nu) + t + b_1(\nu) X + \cdots, \tag{6.8.20}$$

Here F and G are polynomials of degree $[(q-2)/2]$,

$$\bar{\omega}(\nu) = b_0(\nu) + \frac{p}{q} - \omega_0, \tag{6.8.21}$$

and $c(\nu, t)$ can be taken to be real (see Exercise 5.6.14). It follows from (6.8.21) that $\bar{\omega}(\mathbf{0}) = p/q$ and, in contrast to $\mathbf{Df_0(0)}$, $\mathbf{Df_{0,0}(0)}$ is a rotation by $2\pi p/q$. This linear form is more appropriate to the search for q-periodic points. As in the proof of Theorem 6.2.1 (Poincaré–Birkhoff), the idea is to find a closed curve, $\mathscr{C}$, encircling the origin that maps radially under $\mathbf{f}_{\nu,t}^q$ and obtain necessary and sufficient conditions for $\mathscr{C} \cap \mathbf{f}_{\nu,t}^q(\mathscr{C}) \neq \varnothing$. The critical step in achieving this objective is to recognise the importance of the radius, $r_q = r_q(\nu, t)$, of the unique circle, centred on the origin, which is mapped radially by $\mathbf{N}_\nu$, i.e. the unique positive root of

$$\beta_q(\nu, t, r^2) = 0 \tag{6.8.22}$$

close to zero, where β_q is defined by

$$N_{\nu,t}^q(z) = z\alpha_q(\nu, t, |z|^2) \exp(2\pi i \beta_q(\nu, t, |z|^2)). \tag{6.8.23}$$

Chenciner is able to demonstrate the existence of $\mathscr{C}$ and give the required conditions for the intersection of $\mathscr{C}$ and $\mathbf{f}_{\nu,t}^q(\mathscr{C})$. What is more, the quantity $\partial\alpha_q(r_q^2)/\partial X$, where α_q appears in (6.8.23), is a measure of the distance of the parameter point (ν, t) from $\gamma_{p/q}(t)$ on $C_{p/q}(t)$ (see Exercise 6.8.5,6). Here $C_{p/q}(t)$ and $\gamma_{p/q}(t)$ are the analogues for $\mathbf{N}_{\nu,t}$ of C_ω and γ_ω for $\mathbf{N}_\nu$. In this way, r_q acts as a bifurcation parameter for the particular one-parameter subfamily defined by $C_{p/q}(t)$.

Chenciner (1982) points out that the pendulum equation

$$\ddot{\psi} = 2\delta_q \tilde{\beta}_q(r_q^2) r_q^q \sin 2\pi q\psi + 2\tilde{\alpha}_q(r_q^2) r_q^2 \dot{\psi}, \tag{6.8.24}$$

provides a vector field approximation to $\mathbf{f}^q_{\nu,t}$. The bifurcations occurring in this approximation as we move along $C_{p/q}(t)$ are shown in Figure 6.39. Observe how the stable and unstable manifolds of consecutive saddle points move together as the undamped case (*c*) is approached. For $\mathbf{f}_{\nu,t}$, these manifolds touch and then intersect transversely giving rise to a homoclinic tangle between each pair of adjacent periodic saddle points (see Figure 6.40). Since the two manifolds intersect transversely, there is an open set of parameter values near $\gamma_{p/q}(t)$ for which such tangles occur. Once Figure 6.39(*c*) is replaced by Figure 6.40 it is apparent that similar sequences of bifurcations will occur in other one-parameter families corresponding to curves near to $C_{p/q}(t)$.

Another point to note about Figures 6.39(*a*), (*b*), (*d*) and (*e*) is that in each case there is an invariant circle; possibly of finite differentiability in (*a*), (*e*) and non-differentiable in (*b*), (*d*). Furthermore, this circle is resonant with the period-*q* orbit. The following arguments confirm that this is not surprising. If $\mathbf{f}_{\nu,t}$ has an invariant circle with rotation number p/q, then, generically, its restriction to this circle will have two periodic orbits of period q: one stable and the other unstable. Thus the resonance is to be expected. More importantly, the invariant circle in question is created at a Hopf bifurcation (in the sense of Theorem 5.4.2) on the negative ν_2-axis (see Exercise 6.8.7). For $q \geqslant 5$, we have seen (see Theorem 5.4.1) that such a circle has rotation number p/q in a resonance tongue as shown in Figure 5.7. If $\tilde{C}_{p/q}(t) = \{\boldsymbol{\nu} \,|\, \mathbf{f}_{\nu,t}$ has an invariant circle with rotation number $p/q\}$, then $\tilde{C}_{p/q}(t)$ should contain the (p/q)-resonance tongue predicted by Theorem 5.4.1. However, we also expect $\tilde{C}_{p/q}(t)$ to lie near $C_{p/q}(t)$. Together, these pieces of information suggest that $\tilde{C}_{p/q}(t)$ takes the form shown in Figure 6.41. This, in turn, confirms the findings of the vector field approximation (6.8.24).

Observe that the approximation (6.8.24) only involves the invariant circle with rotation number p/q. The invariant circle of this type shown in Figure 6.39(*a*) and (*b*) ceases to exist when the homoclinic tangle of Figure 6.40 is present. An invariant circle of opposite stability but the same rotation number reappears in Figure 6.39(*d*) and (*e*). The whereabouts of any other invariant circle is not revealed.

Notice, finally, that (6.8.24) does not depend explicitly on the parameter t. This reflects the fact that similar bifurcations are to be found for all sufficiently small values of t and, in particular, for $t = t_0$. Thus, we can expect the same kind of behaviour for $\mathbf{f}_\nu$ when $\boldsymbol{\nu}$ is near $C_{p/q}$.

Chenciner also discusses the set, N, of parameter points where $\mathbf{f}_\nu$ *does* 'look like' $\mathbf{N}_\nu$. For technical reasons (see eqn (108), p. 101 of Chenciner, 1985a), $n \geqslant 15$ is required in (6.8.1) for the following results to hold. The set N is contained in the complement of an infinite number of 'bubbles' arranged in a string near to the bifurcation curve Γ of $\mathbf{N}_\nu$ (see Figure 6.42). The bubbles are pinched in at points $\tilde{\gamma}_\omega$ such that $\mathbf{f}_\nu$ looks like $\mathbf{N}_\nu$, i.e. $\mathbf{f}_\nu$ has a non-normally hyperbolic invariant circle with rotation number ω. The values of ω for which this can be shown to occur must be 'sufficiently irrational' in the same sense as is required for the KAM

Figure 6.39 Bifurcation diagram for the vector field approximation (6.8.24) with frictional coefficient $2\tilde{\alpha}(r_q^2)r_q^2$ as the single parameter. The variables ψ and $x = \psi/2\tilde{\beta}_q(r_q^2)r_q^{q/2}$ are, respectively, local angular and radial coordinates at one of the saddle points of $\mathbf{f}_{v,t}^q$. Each diagram gives the local behaviour at typical points of the stable and unstable period-q orbits. The quantities $\tilde{\alpha}_q$ and $\tilde{\beta}_q$ are given by

$$\tilde{\alpha}_q(r_q^2) \equiv \frac{\partial \alpha_q}{\partial X}(v, t, X)|_{r_q^2} \quad \text{and} \quad \tilde{\beta}_q(r_q^2) = \frac{\partial \beta_q}{\partial X}(v, t, X)|_{r_q^2}.$$

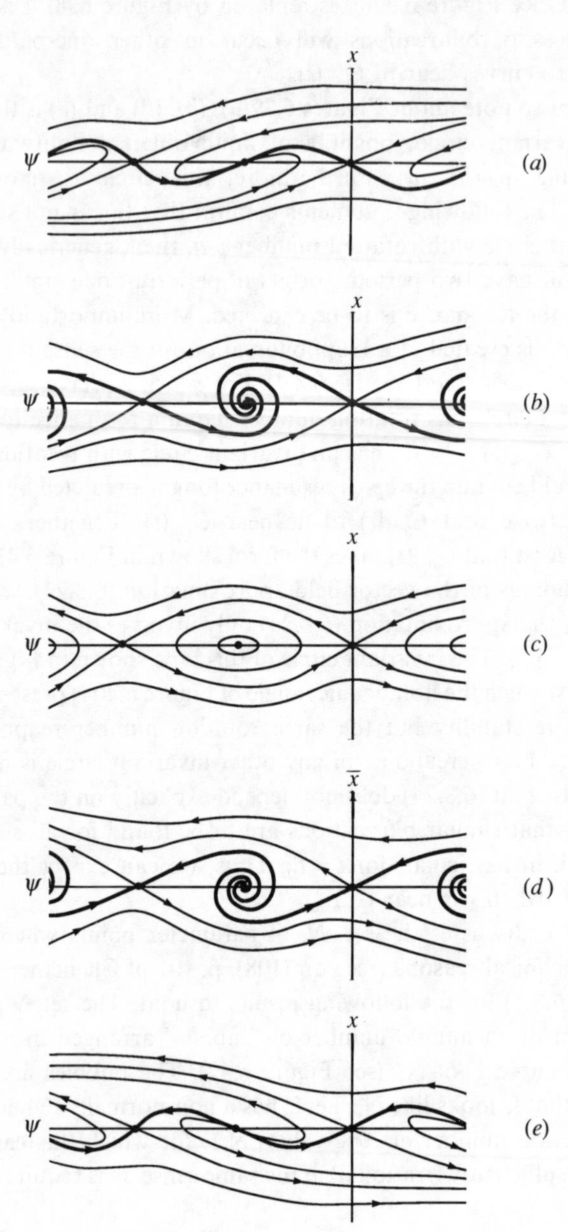

theorem, i.e.

$$|\omega-(p/q)| \geqslant C\tau_{\omega}/|q|^{2+c}, \tag{6.8.25}$$

where C and c are constant for all $p/q \in \mathbb{Q}$ and τ_{ω} is (6.8.13) evaluated at γ_{ω}. Indeed, techniques, similar to those used to prove the KAM Theorem, are required to demonstrate the existence of a unique invariant circle on which the restriction of $\mathbf{f}_{\mathbf{v}}$ is conjugate to a rotation through angle ω. It follows from (6.8.25) that the set, $\tilde{\Gamma}$, of all points $\tilde{\gamma}_{\omega}$ forms a Cantor set in the $\mathbf{v}$-plane, close to Γ (see Chenciner, 1983). Every local one-parameter subfamily going through $\tilde{\gamma}_{\omega} \in \tilde{\Gamma}$ undergoes a double invariant circle bifurcation where the two invariant curves coalesce and

Figure 6.40 Homoclinic tangle that arises generically at the periodic points of $\mathbf{f}_{\mathbf{v},t}$ in place of the saddle connection shown in Figure 6.39(c).

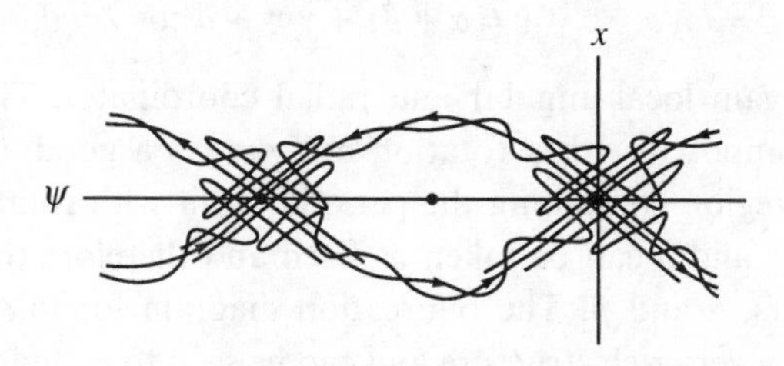

Figure 6.41 Schematic diagram showing the form of $\tilde{C}_{p/q}(t)$ based on the position of $C_{p/q}(t)$ and the assumption that the (p/q)-resonance tongue, predicted by Theorem 5.4.1, extends as far as $\Gamma(t)$. Aronson *et al.* (1983) have carried out independent investigations of the bifurcations undergone by a Hopf invariant circle for parameter values beyond the cuspoidal tip of the resonance tongue where Theorem 5.4.1 holds. They find that the circle loses differentiability and eventually foci, similar to those shown in Figure 6.39, occur followed by the development of homoclinic tangles. The work of Chenciner is further complicated by the presence of a second invariant circle. Details of the connection between this diagram and Figures 5.6 and 7 are considered in Exercises 6.8.7 and 8.

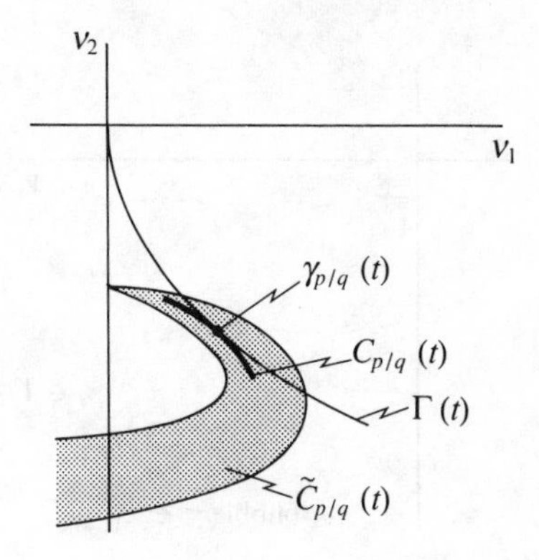

disappear in the same way as for a family of vector fields. What is more, if ω satisfies (6.8.25), it can be shown that there is a curve $\tilde{C}_\omega$, passing through $\tilde{\gamma}_\omega$, on which $\mathbf{f}_\mathbf{v}$ has the same properties as $\mathbf{N}_\mathbf{v}$ has on C_ω.

Generically the bubbles are open, so that the set of parameter points for which $\mathbf{f}_\mathbf{v}$ does not look like $\mathbf{N}_\mathbf{v}$ contains an infinite union of disjoint sets. In this region, we have already seen that $\mathbf{f}_\mathbf{v}$ can have periodic points with associated homoclinic tangles. Chenciner (1985b) has also proved that there are values of $\mathbf{v}$ for which $\mathbf{f}_\mathbf{v}$ has a Mather set of Cantor type. Consequently, it is perhaps more appropriate to redefine $\tilde{C}_\omega = \{\mathbf{v} | \mathbf{f}_\mathbf{v}$ has an Aubry–Mather set of rotation number $\omega\}$.

Our earlier discussion focussed attention on the bifurcations that occur as we move along $C_{p/q}(t)$ but Chenciner (1987) has recently looked more closely at what happens in families that do not remain near this curve. He obtains a vector field approximation to $\mathbf{f}^q_{\mathbf{v},t}$, namely

$$\dot{\theta} = \omega y, \qquad \dot{y} = \alpha + \beta y + \gamma y^2 + \delta \cos 2\pi q\theta, \tag{6.8.26}$$

where (θ, y) are again local angular and radial coordinates. The time-one flow map of (6.8.26) composed with a rotation by $2\pi p/q$ is a good approximation to $\mathbf{f}_{\mathbf{v},t}$ in an annular region containing the periodic orbit with rotation number p/q. The quantities ω, γ and δ can be taken as fixed and therefore the family (6.8.26) has two parameters, α and β. The bifurcation diagram for (6.8.26) is shown in Figure 6.43. It has a very rich structure and can be seen to include the bifurcations exhibited by (6.8.24). This approximation can be used to suggest a particularly interesting scenario for the disappearance of the two invariant circles of $\mathbf{f}_\mathbf{v}$ in certain one-parameter subfamilies that pass through the (p/q)-bubble in Figure 6.42. Consider the subfamily corresponding to the path through this bubble illustrated in Figure 6.44. The behaviour of $\mathbf{f}_\mathbf{v}$ at the labelled points is shown in Figure 6.45. Observe that the subfamily can be chosen so that, as the circles approach one

Figure 6.42 The map $\mathbf{f}_\mathbf{v}$ can be shown (see Chenciner, 1985a) to look like the normal form $\mathbf{N}_\mathbf{v}$ for values of $\mathbf{v}$ lying *outside* the string of bubbles illustrated schematically below. Each bubble is associated with a rational number p/q. The bubbles are pinched in at points $\tilde{\gamma}_\omega$ which form a Cantor set near the double invariant circle bifurcation curve of the family $\mathbf{N}_\mathbf{v}$.

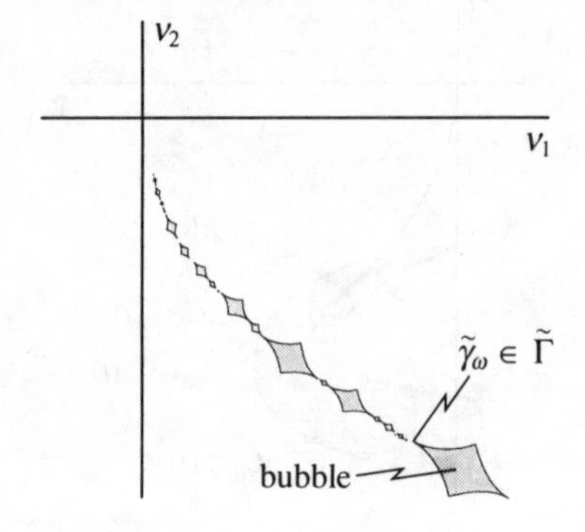

Figure 6.43 The bifurcation diagram for the two-parameter family of vector fields (6.8.26). Note that a periodic orbit with rotation number p/q exists in the region $\hat{C}_{p/q}(t)$ but: (i) there is no invariant circle below the line AOB; (ii) there is no invariant circle resonant with the periodic orbit above the line COD. The bifurcations of the invariant circle in resonance with the periodic orbit occurring in (6.8.24) can be recognised in the one-parameter subfamily obtained by setting $\alpha = 0$; i.e. changing β corresponds to moving along $C_{p/q}(t)$. However, in this approximation we can see that the homoclinic tangle suggested in Figure 6.40 also involves the second invariant circle. Allowing α to become non-zero corresponds to moving away from $C_{p/q}(t)$. In particular, the subfamily obtained by fixing $\beta = 0$ and allowing α to decrease through zero leads to Figure 6.45. (After Chenciner, 1987, p. 88.)

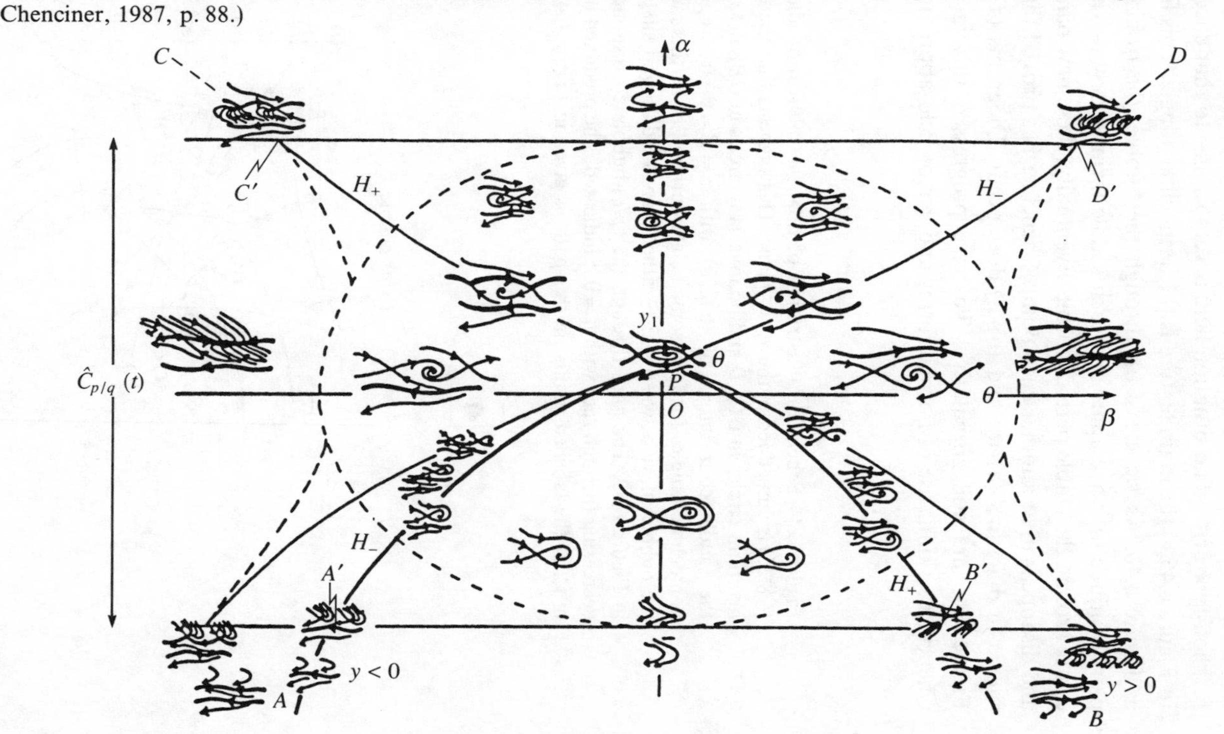

another, period-q points appear at what are cusp points in (6.8.26). Chenciner calls them 'Bogdanov points'. Each of these points develops into a saddle and a centre. For $\mathbf{f}_\nu$, the stable and unstable manifolds of the saddles are involved in a homoclinic tangle. As the centre/saddle separation increases and the invariant circles continue to shrink towards them, Chenciner conjectures that attractors of Birkhoff-type arise. The situation here is very similar to that encountered in §6.7. In Figure 6.45(e) the main feature is the homoclinic tangle. This figure is the same as Figure 6.40. As we move on through the bubble, the original saddle/centre pairs continue to move apart and each saddle forms a new pairing with the next centre around the stable periodic orbit. Eventually, these new pairs move together and disappear in at another cusp point. Chenciner has proved that the behaviour shown in diagrams (a)–(c) and (g)–(i) does occur. Diagrams (d), (e) and (f) in Figure 6.45 are still speculative. To have progressed this far is a tribute to Chenciner's skilful use of normal forms and vector field approximations.

Figure 6.44 Schematic illustration of how the bifurcation diagram of (6.8.26) is embedded in the ν_1, ν_2-plane. This suggests that there are two invariant circles in $OCQD$ and no invariant circles in $OBPA$. Given that the saddle connections of (6.8.26) with $\alpha = \beta = 0$ are replaced by homoclinic tangles for $\mathbf{f}_\nu$ in some neighbourhood of O, this means that the bubbles have a 'wasp waist' rather than the diamond shape shown in Figure 6.42. The path through the (p/q)-bubble corresponding to α decreasing through zero with $\beta = 0$ is indicated. The predicted behaviour of $\mathbf{f}_\nu$, at the labelled points on this path, is given in Figure 6.45.

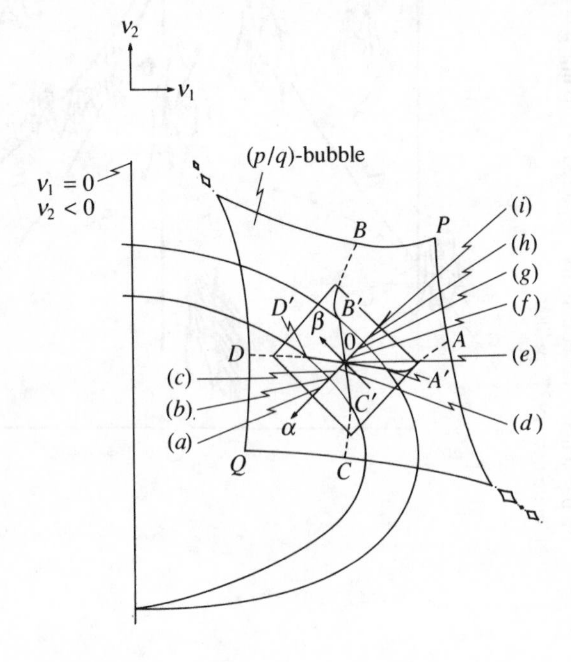

Figure 6.45 Sketches illustrating the behaviour of $\mathbf{f}_\nu$ at the labelled points of the path through the (p/q)-bubble shown in Figure 6.44. Chenciner has proved that (*a*)–(*c*) and (*g*)–(*i*) do occur but (*d*)–(*f*) and the occurrence of Birkhoff attractors is as yet only conjecture. The similarity between what is happening here and the attractors discussed in § 6.7 is blurred by the conjecture that an attractor and a repellor of Birkhoff-type occur simultaneously. Figure 6.43 shows that circumstances closer to those found in § 6.7 should be encountered in subfamilies given by varying α through zero with β equal to a small, non-zero constant. In these subfamilies the invariant circles interact separately with the periodic orbit. (After Chenciner, 1987, p. 67.)

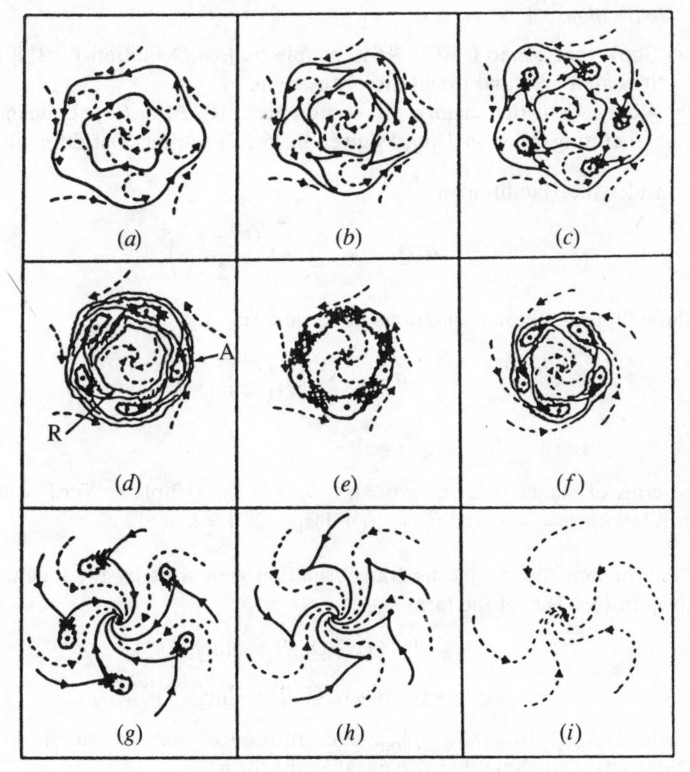

? A, R = attractor, repellor of Birkhoff type

Exercises

6.1 Introduction

6.1.1 (a) Let $\mathbf{U}: \mathbb{R}^n \to \mathbb{R}^n$ be a linear diffeomorphism. Show that

$$\begin{pmatrix} (\mathbf{U}^{\mathrm{T}})^{-1} & \mathbf{0} \\ \mathbf{0} & \mathbf{U} \end{pmatrix}: \mathbb{R}^{2n} \to \mathbb{R}^{2n} \tag{E6.1}$$

is symplectic.

(b) Consider the Hamiltonian

$$K(\mathbf{Q}, \mathbf{P}) = \mathbf{Q}^{\mathrm{T}}\mathbf{A}\mathbf{Q} + \mathbf{P}^{\mathrm{T}}\mathbf{B}\mathbf{P}, \tag{E6.2}$$

where **A** and **B** are non-singular, symmetric matrices with eigenvalues $\{\mu_i\}_1^n$ and $\{\lambda_i\}_1^n$, respectively. Assume that **B** is positive definite and prove that K can be reduced to the form

$$\begin{aligned}\bar{K}(\bar{\mathbf{q}}, \bar{\mathbf{p}}) &= K(\mathbf{Q}(\bar{\mathbf{q}}, \bar{\mathbf{p}}), \mathbf{P}(\bar{\mathbf{q}}, \bar{\mathbf{p}})), \\ &= \sum_{i=1}^{n} (\mu_i \bar{q}_i^2 + \lambda_i \bar{p}_i^2),\end{aligned} \tag{E6.3}$$

by a symplectic change of coordinates.

(c) Given that $\mu_i > 0$, $i = 1, \ldots, n$, find a further symplectic transformation $\bar{\mathbf{q}} = \bar{\mathbf{q}}(\mathbf{q}, \mathbf{p})$, $\bar{\mathbf{p}} = \bar{\mathbf{p}}(\mathbf{q}, \mathbf{p})$ such that $H(\mathbf{q}, \mathbf{p}) = \bar{K}(\bar{\mathbf{q}}, \bar{\mathbf{p}})$ takes the form (6.1.3) with $\omega_i = (\mu_i \lambda_i)^{1/2}$.

6.1.2 (a) Show that a map $\mathbf{f}: \mathbb{R}^2 \to \mathbb{R}^2$ is symplectic (see Definition 1.9.2) if and only if it is *both* area and orientation preserving.

(b) Verify that the change of coordinates, $(Q, P)^{\mathrm{T}} = \mathbf{f}(q, p)$, defined by the generating function $F_2(q, P)$ given in (6.1.7) satisfies $\mathrm{Det}(\mathrm{D}\mathbf{f}(q, p)) \equiv 1$.

6.1.3 Consider the Hamiltonian

$$H(Q, P) = \sum_{k=1}^{\infty} \alpha_k \left(\frac{Q^2 + P^2}{2} \right)^k. \tag{E6.4}$$

Show that Hamilton's equations take the form

$$\begin{aligned}\dot{\theta} &= -\sum_{k=1}^{\infty} k\alpha_k \left(\frac{r^2}{2} \right)^{k-1}, \\ \dot{r} &= 0,\end{aligned} \tag{E6.5}$$

in terms of planar polar coordinates (θ, r) in the QP-plane. Verify that (E6.5) is (6.1.2) with $\beta = -\alpha_1$ and $b_l = -(l+1)\alpha_{l+1}/2^l$, $l = 1, 2, \ldots$.

6.1.4 Confirm that the symplectic transformation generated by the function $F_2(q, P)$ given in (6.1.7) is of the form

$$\begin{aligned}p &= P + \mathrm{D}_q S_N(Q, P) + O(|(Q, P)|^N), \\ q &= Q - \mathrm{D}_P S_N(Q, P) + O(|(Q, P)|^N),\end{aligned} \tag{E6.6}$$

where $\mathrm{D}_q S_N(Q, P) = (\partial S_N / \partial q)|_{Q,P}$, etc. Introduce complex variables $z = q + \mathrm{i}p$, $Z = Q + \mathrm{i}P$ and show that (E6.6) takes the form

$$\begin{aligned}z &= Z + 2\mathrm{i}\mathrm{D}_{\bar{z}} \mathscr{S}_N(Z, \bar{Z}) + O(|Z|^N), \\ \bar{z} &= \bar{Z} - 2\mathrm{i}\mathrm{D}_z \mathscr{S}_N(Z, \bar{Z}) + O(|Z|^N),\end{aligned} \tag{E6.7}$$

where $\mathscr{S}_N$ is defined in (6.1.11). Hence deduce that

$$\mathrm{i}\omega z\bar{z} = \mathrm{i}\omega Z\bar{Z} + 2 \sum_{\substack{m,\bar{m} \\ m+\bar{m}=N}} \omega(m - \bar{m})\sigma_{m\bar{m}} Z^m \bar{Z}^{\bar{m}} + O(|Z|^{N+1}). \tag{E6.8}$$

6.1.5 Generalise the calculations of Exercise 6.1.4 to systems with n degrees of freedom. Consider the generating function $F_2(\mathbf{q}, \mathbf{P}) = \mathbf{q}^{\mathrm{T}}\mathbf{P} + S_N(\mathbf{q}, \mathbf{P})$, where $\mathbf{q} = [q_j]_{j=1}^n$, $\mathbf{P} = [P_j]_{j=1}^n$ are column vectors and $S_N(\mathbf{q}, \mathbf{P})$ is a homogeneous polynomial of degree N in the $2n$ variables q_j, P_j, $j = 1, 2, \ldots, n$. Define $p_j = \partial F_2 / \partial q_j$ and

$Q_j = \partial F_2/\partial P_j$. Introduce complex variables $z_j = q_j + \mathrm{i}p_j$, $Z_j = Q_j + \mathrm{i}P_j$ and show that

$$\sum_{j=1}^{n} \mathrm{i}\omega_j z_j \bar{z}_j = \sum_{j=1}^{n} \mathrm{i}\omega_j Z_j \bar{Z}_j$$
$$+2 \sum_{\substack{\mathbf{m},\bar{\mathbf{m}} \\ \Sigma_j (m_j + \bar{m}_j) = N}} \left[\sum_{j=1}^{n} \omega_j(m_j - \bar{m}_j)\right] \sigma_{\mathbf{m},\bar{\mathbf{m}}} Z_1^{m_1} \cdots Z_n^{m_n} \bar{Z}_1^{\bar{m}_1} \cdots Z_n^{\bar{m}_n} + O(|\mathbf{Z}|^{N+1}). \tag{E6.9}$$

6.1.6 Show that the symplectic change of variable $z \to Z$ defined in (6.1.18) transforms the area-preserving map (6.1.19) in such a way that terms of order $|z|^k$, $k < n$, remain unchanged. Verify that the change in $O(|z|^n)$-terms is given by (6.1.20).

6.1.7 (a) Consider the complex normal form

$$f(z) = \exp(\mathrm{i}\alpha_0)z\left(1 + \sum_{l=1}^{N} a_{2l+1}|z|^{2l} + O(|z|^{2N+2})\right), \tag{E6.10}$$

for which $\alpha_0/2\pi \in \mathbb{R}\backslash\mathbb{Q}$ (see (2.5.15) and (5.3.10)). Define $z = r\exp(\mathrm{i}\theta)$, $f(z) = r_1 \exp(\mathrm{i}\theta_1)$ and show that

$$\theta_1 = \theta + \alpha_0 + \sum_{l=1}^{N} d_l r^{2l} + O(r^{2N+2}),$$
$$r_1 = r\left(1 + \sum_{l=1}^{N} c_l r^{2l}\right) + O(r^{2N+3}), \tag{E6.11}$$

where $c_l, d_l \in \mathbb{R}$.

(b) Suppose that a planar map is represented in cartesian coordinates, (x, y), by $\mathbf{f}$: $(x, y) \mapsto (x_1, y_1)$ and in plane polar coordinates, (θ, r), by $\mathbf{F}$: $(\theta, r) \mapsto (\theta_1, r_1)$. Show that

$$\mathrm{Det}(\mathrm{D}\mathbf{f}(x, y)) = (r_1/r)\,\mathrm{Det}(\mathrm{D}\mathbf{F}(\theta, r)). \tag{E6.12}$$

6.2 Rational rotation numbers and Birkhoff periodic points

6.2.1 The Poincaré–Birkhoff Theorem

6.2.1 Consider the area-preserving map $\mathbf{M}_\varepsilon$ defined in (6.2.1) with $\alpha(\tau_0) = 2\pi p/q$, $\tau_0 \in (a, b)$, and assume that $p/q \in \mathbb{Q} \cap (0, 1)$ is in lowest terms.

(a) Show that $\mathbf{M}_\varepsilon^q$ can be written in the form (6.2.4) and verify that (θ, τ_0) is a periodic point of period q for every $\theta \in [0, 2\pi)$, when $\varepsilon = 0$. Use the Implicit Function Theorem to prove that, for sufficiently small ε, the equation $\theta_q = \theta + 2\pi p$ has a unique solution $\tau = F(\theta, \varepsilon)$, $\theta \in [0, 2\pi)$.

(b) Let $\mathscr{C}$ be the circle defined by $\tau = F(\theta, \varepsilon)$. Show that $\mathscr{C}$ and $\mathbf{M}_\varepsilon^q(\mathscr{C})$ must intersect in at least two points. Explain why these points of intersection of $\mathscr{C}$ and $\mathbf{M}_\varepsilon^q(\mathscr{C})$ are period-q points of $\mathbf{M}_\varepsilon$ and why, generically, there are $2q$ of them.

6.2.2 Consider the area-preserving planar map defined by

$$\mathbf{M}_\varepsilon\colon (\theta, \tau) \mapsto (\theta + \tau + \varepsilon \sin\theta, \tau + \varepsilon\sin\theta), \tag{E6.13}$$

where (θ, τ) are canonical polar coordinates. Let $p/q = \frac{1}{2}$ and obtain the radially mapped circle $\mathscr{C}$ described in Exercise 6.2.1 in the form

$$\tau = F(\theta, \varepsilon) = F_0(\theta) + F_1(\theta)\varepsilon + F_2(\theta)\varepsilon^2 + O(\varepsilon^3). \tag{E6.14}$$

Plot the quadratic approximation to $\mathscr{C}$ and $\mathbf{M}_\varepsilon^2(\mathscr{C})$ for $\varepsilon = 0.5$ and verify that the intersections of these curves are of two distinct types.

6.2.2 Vector field approximations and island chains

6.2.3 Consider area-preserving planar diffeomorphisms $\mathbf{P}_i$, $i = 1, 2$, which satisfy $\mathbf{P}_i(\mathbf{0}) = \mathbf{0}$ and:

$$\mathrm{D}\mathbf{P}_1(\mathbf{0}) = \begin{pmatrix} 1 & 1 \\ 0 & 1 \end{pmatrix}; \qquad \mathrm{D}\mathbf{P}_2(\mathbf{0}) = \begin{pmatrix} -1 & 1 \\ 0 & -1 \end{pmatrix}.$$

Write down normal forms for $\mathbf{P}_1$ and $\mathbf{P}_2$. It is a generic property of such diffeomorphisms that the lowest order terms, consistent with symmetry, have non-zero coefficients in these forms. For the generic case, show that:
(a) $y = \pm C_1 x^{3/2} + O(x^2)$ is an invariant curve for $\mathbf{P}_1$;
(b) $y = \pm C_2 x^2 + O(x^4)$ is an invariant curve for $\mathbf{P}_2$.
Verify that the Hamiltonian vector fields:

$$\mathbf{X}_1(\mathbf{x}) = (y, x^2)^{\mathrm{T}}, \qquad \mathbf{X}_2(\mathbf{x}) = (y, x^3)^{\mathrm{T}}; \tag{E6.15}$$

have invariant curves of the same form as those in (a) and (b), respectively.

6.2.4 Let $F_2(x, Y)$ be the generating function given in (6.1.7). Obtain the area-preserving diffeomorphisms generated by

$$F_2(x, Y) = xY - (x^2/2) \tag{E6.16}$$

and

$$F_2(x, Y) = xY - (x^2/2) + R(Y), \tag{E6.17}$$

where $R(Y) = \exp(-1/Y^2)$ when $Y \neq 0$, and $R(Y) = 0$ when $Y = 0$. Verify that these diffeomorphisms have the same infinity jet but have different fixed point behaviour. In what sense are both diffeomorphisms non-generic? Modify (E6.16) and (E6.17) so that the diffeomorphisms generated are generic. To which of the topological types given in Theorem 6.2.2 do these generic examples correspond?

6.2.5 Express Hamilton's equations, $\dot{x} = \partial H/\partial y$, $\dot{y} = -\partial H/\partial x$, in terms of canonical polar coordinates (θ, τ) defined by $x = (2\tau)^{1/2} \cos \theta$, $y = (2\tau)^{1/2} \sin \theta$. Hence show that, for $\alpha/2\pi$ irrational, the Hamiltonian, H_0, referred to in Theorem 6.2.2 can be written in the form

$$H_0(\tau) = -\left[\alpha\tau + \sum_{i=1}^{\infty} \frac{a_i \tau^{i+1}}{i+1}\right], \tag{E6.18}$$

where the coefficients a_i are determined by the Birkhoff normal form of $\mathbf{P}_0$.

6.2.6 Let $\varphi_t: \mathbb{R} \to \mathbb{R}$ be the flow of the differential equation $\dot{x} = X(x) = \sum_{k=2}^{\infty} X_k(x)$, where $x \in \mathbb{R}$ and $X_k(x) = a_k x^k$. Given that $\varphi_q(x) = x + \sum_{k=2}^{\infty} A_k x^k$, show that:

$$\text{(a)}\ a_2 = A_2/q; \tag{E6.19}$$

$$\text{(b)}\ A_k = \left[ta_k + \sum_{j=2}^{k-1} (t^j/j!)a_k^{(j)}\right], \text{for } k \geqslant 3, \text{ where } [\mathrm{d}^j x/\mathrm{d}t^j]_{t=0} = \sum_{l=j+1}^{\infty} a_l^{(j)} x^l; \tag{E6.20}$$

(c) $a_k^{(j)}$, $j=2,\ldots,k-1$, depends only on $a_2,\ldots,a_{k-1}$.
Explain how (a)–(c) can be used to obtain $X(x)$ (or equivalently $\varphi_t(x)$) from $\varphi_q(x)$. Illustrate your answer by finding a_3 and outlining how a_4 can be calculated.

6.2.7 Consider the flow, $\boldsymbol{\varphi}_t^0(x, y)$, of the differential equation

$$\dot{x}=\sum_{k=2}^{\infty} X_k^0(x, y), \qquad \dot{y}=\sum_{k=2}^{\infty} Y_k^0(x, y), \tag{E6.21}$$

where $(X_k^0, Y_k^0)\in H^k$ (see (6.2.12,13). Show that the terms in $\boldsymbol{\varphi}_t^0(x, y)$ that lie in H^k take the form $(tX_k^0(x, y), tY_k^0(x, y))$ plus terms depending on (X_i^0, Y_i^0), $i=2,\ldots,k-1$.

6.2.8 Extend the result of Exercise 6.2.7 to the ε-dependent system (6.2.22,23). Show that the flow, $\boldsymbol{\varphi}_t^\varepsilon(x, y)$, of this system has terms of order 1 in ε and k in x, y of the form (tX_k^1, tY_k^1) plus terms depending on $X_i^1, Y_i^1, i=0,\ldots,k-1$, and X_j^0, Y_j^0, $j=2,\ldots,k+1$.

6.2.9 Prove that the Jacobian determinant of the flow $\boldsymbol{\varphi}_t$ of the differential equation $\dot{\mathbf{x}}=\mathbf{X}(\mathbf{x})$, $\mathbf{x}\in\mathbb{R}^2$, satisfies

$$\frac{\mathrm{d}}{\mathrm{d}t}(\mathrm{Det}(\mathrm{D}\boldsymbol{\varphi}_t(\mathbf{x})))=\mathrm{Tr}\,\mathrm{D}\mathbf{X}(\boldsymbol{\varphi}_t(\mathbf{x}))\,\mathrm{Det}(\mathrm{D}\boldsymbol{\varphi}_t(\mathbf{x})). \tag{E6.22}$$

Verify (E6.22) explicitly when $X(x)=(x^2, y)^{\mathrm{T}}$.

6.2.10 Consider the vector field $\mathbf{X}^\varepsilon$ obtained by the comparison of $\boldsymbol{\varphi}_q^\varepsilon$ and $\mathbf{N}_\varepsilon^q$. Show that $\int^y X^\varepsilon(x, y)\,\mathrm{d}y=-\int^x Y^\varepsilon(x, y)\,\mathrm{d}x=H_\varepsilon(x, y)$, where H_ε satisfies Hamilton's equations (6.2.36) and is invariant under $\mathbf{R}_\alpha$.

6.2.11 Let the area-preserving, planar diffeomorphism $\mathbf{P}_\varepsilon$ be such that $\mathbf{P}_\varepsilon(\mathbf{0})=\mathbf{0}$ and $\mathrm{D}\mathbf{P}_\varepsilon(\mathbf{0})$ has eigenvalues $\exp(\pm 2\pi \mathrm{i}p/q)$, $q\geqslant 3$, $p>0$. Given that $\boldsymbol{\varphi}_t^\varepsilon$ is the flow of the vector field (6.2.36), prove that the infinity jet of the normal form, $\mathbf{N}_\varepsilon$, of $\mathbf{P}_\varepsilon$ can be written as the composition of $\boldsymbol{\varphi}_1^\varepsilon$ and a rotation through $2\pi p/q$.

6.2.12 Consider the Hénon area-preserving map

$$\begin{aligned} x_1&=x\cos\alpha-y\sin\alpha+x^2\sin\alpha,\\ y_1&=x\sin\alpha+y\cos\alpha-x^2\cos\alpha. \end{aligned} \tag{E6.23}$$

(a) Show that the two fixed points of (E6.23) lie on the line $y=x\tan(\alpha/2)$.
(b) Verify that (E6.23) is conjugate to

$$\begin{aligned} X_1&=X\cos\alpha-Y\sin\alpha+(X\cos(\alpha/2)-Y\sin(\alpha/2))^2\sin(\alpha/2),\\ Y_1&=X\sin\alpha+Y\cos\alpha-(X\cos(\alpha/2)-Y\sin(\alpha/2))^2\cos(\alpha/2), \end{aligned} \tag{E6.24}$$

by a rotation through the angle $\alpha/2$.
(c) Prove that (E6.24) is conjugate to its own inverse by using the transformation $(X, Y)\mapsto(X, -Y)$.

6.2.13 (a) Use the result in Exercise 6.2.12(c) to show that the set of periodic points of (E6.24) is symmetrically placed with respect to the line $Y=0$.
(b) Perform the following computation. Let (X_5, Y_5) be the fifth iterate of $(X_0, 0)$ under (E6.24). For each $\alpha=1.24(0.01)1.27$, plot $(Y_5^2+(X_5-X_0)^2)$ as a function of X_0 for $-0.25\leqslant X_0\leqslant 0.25$.
(c) Explain how (a) and (b) can be used to suggest that a period-5 orbit of the Hénon map appears as α increases through $2\pi/5$.

6.3 Irrational rotation numbers and the KAM Theorem

6.3.1 (a) Prove that, for $r > 0$,

$$\sum_{k=1}^{\infty} k^{\mu} \exp(-kr) \leqslant C(\mu) r^{-(\mu+1)}, \tag{E6.25}$$

where $C(\mu) = \mu^{\mu+1} \exp(-\mu) + \Gamma(\mu + 1)$.

(b) Consider the analytic continuation,

$$w(z + \omega) - w(z) = h(z), \tag{E6.26}$$

$z = x + iy$, of the difference equation (6.3.1) to the strip $|y| < r$. Assume that $|h(z)| < K$ on this domain and show that (E6.26) has a unique solution, $w(z)$, with mean value zero, which satisfies

$$|w(z)| \leqslant C_1(\mu) K (r - \rho)^{-(\mu+1)}, \tag{E6.27}$$

on the strip $|y| < \rho$, where $0 < \rho < r$.

6.3.2 Consider the coupled difference equations (6.3.21,22).

(a) Verify that $F_n(\xi) - f_n(\xi, \omega\gamma^{-1})$ and $G_n(\xi) - g_n(\xi, \omega\gamma^{-1})$ depend only on f_l, g_l, u_l, v_l for $l < n$.

(b) Assume that ω satisfies (6.3.6) and that (6.3.25) is valid. Obtain the solution to (6.3.21,22) in the form $v_n(\xi) = \sum_{k \in \mathbb{Z}} \hat{v}_{nk} \exp(ik\xi)$, $u_n(\xi) = \sum_{k \in \mathbb{Z}} \hat{u}_{nk} \exp(ik\xi)$, where

$$\begin{aligned} \hat{v}_{n0} &= -\gamma^{-1} \hat{F}_{n0}, \\ \hat{v}_{nk} &= \hat{G}_{nk} / [\exp(ik\omega) - 1], \qquad k \neq 0, \\ \hat{u}_{nk} &= \frac{\hat{F}_{nk}}{[\exp(ik\omega) - 1]} + \frac{\hat{G}_{nk}}{[\exp(ik\omega) - 1]^2}, \qquad k \neq 0, \end{aligned} \tag{E6.28}$$

and $\hat{u}_{n0}$ is arbitrary.

(c) If the solution (6.3.3) of the difference equation (6.3.1) is written in the operator form $w(x) - w^* = Lh(x)$, $w^* = (2\pi)^{-1} \int_0^{2\pi} w(x)\,dx$, show that (E6.28) corresponds to

$$\begin{aligned} v_n(\xi) + \gamma^{-1} F_n^* &= LG_n(\xi), \\ u_n(\xi) - u_n^* &= LF_n(\xi) + L^2 G_n(\xi), \end{aligned} \tag{E6.29}$$

where $F_n^* = \hat{F}_{n0}$, $u_n^* = \hat{u}_{n0}$.

6.3.3 Consider the choice of parameters r_i, s_i, d_i, $i = 0, 1, 2, \ldots$, given in (6.3.34).

(a) Let $e_i = r_0^{-3(2\mu+3)} C^{3(i+4)} d_i$ and show that

$$(d_{i+1}/d_i) < C^{-3}, \tag{E6.30}$$

if

$$0 \leqslant d_0 < r_0^{3(2\mu+3)} C^{-12}. \tag{E6.31}$$

Hence deduce that $d_i \to 0$ as $i \to \infty$ provided d_0 is taken to be sufficiently small.

(b) Given that d_0 satisfies (E6.31), prove that

$$|f_{i+1}| + |g_{i+1}| < d_{i+1}, \tag{E6.32}$$

for a sufficiently large choice of the constant C.

6.3.4 Let $\mathbf{U}_j$, $j = 0, 1, \ldots, i$, be given by (6.3.26). Prove that $\mathbf{V}_i = \mathbf{U}_0\mathbf{U}_1 \cdots \mathbf{U}_i$ takes the form

$$\mathbf{V}_i\colon x = \xi + p_i(\xi, \eta), \qquad y = \eta + q_i(\xi, \eta), \qquad \text{E6.33)}$$

where

$$\begin{aligned} p_i(\xi, \eta) &= u_i + p_{i-1}(\xi + u_i, \eta + v_i), \\ q_i(\xi, \eta) &= v_i + q_{i-1}(\xi + u_i, \eta + v_i), \end{aligned} \qquad \text{(E6.34)}$$

with $u_i = u_i(\xi, \eta)$, $v_i = v_i(\xi, \eta)$. Let $|u_i| = \sup_{(\xi,\eta)\in\mathscr{D}_{i+1}} (|u_i(\xi, \eta)|)$, $|v_i| = \sup_{(\xi,\eta)\in\mathscr{D}_{i+1}} (|v_i(\xi, \eta)|)$ and deduce that

$$|p_i| + |q_i| \leqslant \sum_{j=0}^{i} \{|u_i| + |v_i|\}, \qquad \text{(E6.35)}$$

for any non-negative integer i.

6.3.5 (a) Use the conjugacy of $\mathbf{M}_\infty$ and $\mathbf{M}_0$ to prove that $u(\xi)$ and $v(\xi)$ in (6.3.42) satisfy (6.3.44).

(b) Suppose that

$$\mathbf{M}\colon \begin{aligned} x_1 &= x + y + f(x, y), \\ y_1 &= y + g(x, y) \end{aligned} \qquad \text{(E6.36)}$$

has an invariant circle, $\mathscr{C}$, given by

$$x = \xi + \tilde{u}(\xi), \qquad y = \tilde{v}(\xi) \qquad \text{(E6.37)}$$

such that $\mathbf{M}|\mathscr{C}$ takes the form $\xi_1 = \xi + \omega$. Show that $\tilde{u}(\xi)$ and $\tilde{v}(\xi)$ also satisfy (6.3.44).

(c) Verify that $f(x, y) \equiv g(x, y) \equiv 0$ corresponds to $\tilde{u}(\xi) \equiv 0$ and $\tilde{v}(\xi) \equiv \omega$ in (E6.37) and explain the significance of the choice $\lim_{i\to\infty} p_i = u(\xi)$, $\lim_{i\to\infty} q_i = v(\xi) - \omega$ used to obtain (6.3.42).

6.3.6 Consider the solution of the coupled, linear difference equations

$$\begin{aligned} u_i(\xi + \omega, \eta) - u_i(\xi, \eta) &= v_i(\xi, \eta) + f_i(\xi, \eta), \\ v_i(\xi + \omega, \eta) - v_i(\xi, \eta) &= g_i(\xi, \eta) - g_i^*(\eta), \end{aligned} \qquad \text{(E6.38)}$$

$\xi, \eta \in \mathscr{D}_{i+1} \subseteq \mathbb{C}$ (see (6.3.32)), in which η plays the role of a parameter. Given that the mean value of $u_i(\cdot, \eta)$ is zero, the solution of (E6.38) can be written in the form

$$v_i = -f_i^* + Lg_i, \qquad \text{(E6.39)}$$

$$u_i = L(v_i + f_i), \qquad \text{(E6.40)}$$

where f_i^* is the mean value of f_i and L is an operator of the type introduced in Exercise 6.3.2(c).

Assume $f_i^* = 0$ for all i and use (E6.27) with:
(a) $\rho = r_i - [(r_i - r_{i+1})/16]$ to estimate $|v_i|$;
(b) $\rho = r_i - [(r_i - r_{i+1})/8]$ to estimate $|u_i|$.
Deduce that

$$|u_i| + |v_i| < (s_i/7). \qquad \text{(E6.41)}$$

Use (E6.30) in Exercise 6.3.3 to show that $(s_{i+1}/s_i) < \frac{1}{3}$ and hence confirm (6.3.48).

6.3.7 (a) Let $h: \mathbb{R} \to \mathbb{R}$ be a 2π-periodic function that is N-times differentiable. Show that the kth coefficient, $\hat{h}_k$, of the complex Fourier series for h satisfies $|\hat{h}_k| < K/|k|^N$, where $K = \max_{0 \leqslant x \leqslant 2\pi} \{|d^N h/dx^N|\}$.

(b) Verify that the bound obtained in (a) is consistent with that given in (6.3.13) for real analytic functions.

6.4 The Aubry–Mather Theorem

6.4.1 Invariant Cantor sets for homeomorphisms on S^1

6.4.1 Consider the 'middle-third' Cantor set constructed as follows. Trisect the interval $[0, 1]$ and define $T_1 = [0, \frac{1}{3}] \cup [\frac{2}{3}, 1]$. Now trisect the two subintervals of T_1 to form $T_2 = [0, \frac{1}{9}] \cup [\frac{2}{9}, \frac{1}{3}] \cup [\frac{2}{3}, \frac{7}{9}] \cup [\frac{8}{9}, 1]$. Continue this process to obtain the sequence

$$T_1 \supset T_2 \supset T_3 \supset T_4 \supset T_5 \supset \cdots.$$

Define $T = \bigcap_{i=1}^{\infty} T_i$. Prove that T is closed and nowhere dense. Show that every $x \in T$ can be written in the form

$$x = \sum_{j=1}^{\infty} a_j/3^j, \tag{E6.42}$$

where $a_j = 0$ or 2. Hence show that T is perfect.

6.4.2 Describe an algorithm for obtaining a circle with intervals $\{I_n | n \in \mathbb{Z}\}$ spliced into the same relative positions as the orbit $\{x_n | n \in \mathbb{Z}\}$ of an irrational rotation (see Figure 6.14). Explain how to ensure that the set $\bigcup_{n \in \mathbb{Z}} I_n$ is dense in S^1.

6.4.3 Define the semi-conjugacy $\Pi: S^1 \to S^1$ as follows. If $x \in I_n$, then $\Pi(x) = x_n$. If $x \notin I_n$, for any $n \in \mathbb{Z}$, then $\Pi(x) = y$, where y lies in the same position relative to $\{x_n\}_{n \in \mathbb{Z}}$ as x does relative to $\{I_n\}_{n \in \mathbb{Z}}$.

(a) Show that Π is well defined, continuous and injective on $S^1 \backslash \bigcup_{n \in \mathbb{Z}} I_n$.

(b) Check that the circle map f defined in the text is a homeomorphism and that $\tilde{C} = \left(\bigcup_{n \in \mathbb{Z}} \text{int}\,(I_n) \right)^c$ is the invariant set E considered in Proposition 6.4.1

6.4.4 (a) Let $\bar{\Pi}$ be a lift of the semi-conjugacy Π. Prove that: (i) $\bar{\Pi}(\bar{x} + 1) - \bar{\Pi}(\bar{x}) \equiv 1$; (ii) $\bar{\Pi}(\bar{x}) - \bar{x}$ is periodic of period 1.

(b) Show that the lift of R_β can be chosen such that

$$\bar{\Pi}\bar{f}(\bar{x}) \equiv \bar{R}_\beta \bar{\Pi}(\bar{x}), \tag{E6.43}$$

$\bar{x} \in \mathbb{R}$. Deduce that $\rho(f) = \beta$.

6.4.5 Let $f: S^1 \to S^1$ be a Denjoy counterexample as constructed in §6.4.1. With the notation used in the text, prove that: (a) every point of $C = \left(\bigcup_{n \in \mathbb{Z}} x_n \right)^c$ is an accumulation point of C but that C is not perfect; (b) C is not nowhere dense; (c) $\tilde{C} = \left(\bigcup_{n \in \mathbb{Z}} \text{int}\,(I_n) \right)^c$ is both perfect and nowhere dense.

6.4.2 Twist homeomorphisms and Mather sets

6.4.6 (a) Let $\bar{\mathbf{f}}: \bar{A} \to \bar{A}$ be a lift of $\mathbf{f}: A \to A$ where $A = \{(\theta, r) | \theta \in S^1, r \in [0, 1]\}$. Show that if $\mathbf{f}$ is a twist homeomorphism then $\bar{f}_1(x, y)$ is either a monotone increasing function of y for every $x \in \mathbb{R}$ or a monotone decreasing function of y for every $x \in \mathbb{R}$.

(b) Consider the following maps defined for $(x, y) \in \mathbb{R} \times [0, 1]$. Which of them are lifts of twist homeomorphisms of the annulus A? Explain your answers.

$$\begin{aligned}
&\text{(i)} \quad x_1 = x + \alpha, \; y_1 = y, \\
&\text{(ii)} \quad x_1 = x + y + \tfrac{1}{2} y \sin(2\pi x), \\
&\qquad\; y_1 = y + k \sin(2\pi x), \\
&\text{(iii)} \quad x_1 = x + y, \; y_1 = y^2, \\
&\text{(iv)} \quad x_1 = x + y, \; y_1 = y(1 - y),
\end{aligned} \tag{E6.44}$$

where $\alpha, k \in \mathbb{R} \backslash \{0\}$.

6.4.7 (a) Let $\mathbf{x} \in S^1 \times [0, 1] = A$ be a periodic point of period q of a map $\mathbf{f}: A \to A$. Show that, if $\bar{\mathbf{f}}: \mathbb{R} \times [0, 1] \to \mathbb{R} \times [0, 1]$, is a lift of $\mathbf{f}$ by the projection $\boldsymbol{\pi}(x, y) = \mathbf{x}$, then $\bar{\mathbf{f}}^q(x, y) = (x + p, y)^{\mathrm{T}}$, where p is determined up to an integer multiple of q. Hence verify that, up to an integer, the rotation number of $\mathbf{x}$ is independent of the choice of lift $\bar{\mathbf{f}}$.

(b) Prove that a Birkhoff periodic orbit of type (p, q) is a periodic orbit with rotation number $(p/q) + k$, $k \in \mathbb{Z}$.

6.4.8 Show that the map on A with lift $(x, y) \mapsto (x + y, y)$ has Birkhoff periodic orbits of all rational rotation numbers in $[0, 1]$.

6.4.9 (a) Show that if E is a Mather set of a twist homeomorphism $\mathbf{f}$ of the annulus A, then $\mathbf{f}|E$ is topologically conjugate to a restriction of an orientation-preserving homeomorphism of the circle.

(b) Let $\mathbf{x} \in A$, with lift (x, y), lie in a Mather set E. If $\bar{\mathbf{f}}$ is a lift of $\mathbf{f}$ and $\bar{\mathbf{f}}^n: (x, y) \mapsto (x_n, y_n)$, prove that

$$\operatorname*{Lim}_{n \to \infty} \frac{x_n - x}{n} \tag{E6.45}$$

exists and is independent of $\mathbf{x}$. This limit is defined to be the rotation number, $\rho(E)$, of the Mather set E.

(c) Let E be a minimal Mather set of $\mathbf{f}$. Prove that $\rho(E) \in \mathbb{Q}$ if and only if E is a Birkhoff periodic orbit.

6.5 Generic elliptic points

6.5.1 Suppose that $\mathbf{f}$ is an area-preserving planar map with an elliptic fixed point at the origin and a periodic point of period q at $\mathbf{x}_0$. Let $\mathbf{x}_1$ be a periodic point of $\mathbf{f}^q$, of period q', in a neighbourhood of $\mathbf{x}_0$ bounded by a KAM circle of $\mathbf{f}^q$. Show that the rotation numbers of $\mathbf{x}_0$ and $\mathbf{x}_1$ are the same.

6.5.2 Consider the standard map in the form

$$\theta_1 = (\theta + r_1) \bmod 1, \qquad r_1 = r - (k/2\pi) \sin(2\pi\theta). \tag{E6.46}$$

Investigate periodic points of the form $(0, r(k))$ which have period: (a) two; (b) three; and obtain equations defining $r(k)$. In both cases, show that $r(k)$ is a C^1-function of k.

6.5.3 Define the maps $\mathbf{I}_1$, $\mathbf{I}_2$ and $\mathbf{S}$ by

$$\mathbf{I}_1\colon (\theta, r) \mapsto (-\theta \bmod 1,\ r - (k/2\pi)\sin(2\pi\theta)), \tag{E6.47}$$

$$\mathbf{I}_2\colon (\theta, r) \mapsto ((-\theta + r) \bmod 1,\ r), \tag{E6.48}$$

$$\mathbf{S}\colon (\theta, r) \mapsto ((\theta + r - (k/2\pi)\sin(2\pi\theta)) \bmod 1,\ r - (k/2\pi)\sin(2\pi\theta)) \tag{E6.49}$$

Verify that:
(a) $\mathbf{S} = \mathbf{I}_2\mathbf{I}_1$, $\mathbf{I}_1^2 = \mathbf{I}_2^2 = \mathbf{id}$;
(b) if (θ_0, r_0) and $\mathbf{S}^N(\theta_0, r_0)$ are fixed points of $\mathbf{I}_1$ then (θ_0, r_0) is a $2N$-periodic point of $\mathbf{S}$.
(c) $\theta = 0$ is a line of fixed points for $\mathbf{I}_1$.
Hence show that $\mathbf{S}$ has period-$2N$ points at $(0, r(k))$, for sufficiently small k, where $r(k)$ depends continuously on k.

6.5.4 Prove that the algorithm used to construct the continued fraction representation of a real number α terminates after a finite number of steps if and only if α is rational.

6.5.5 (a) Consider the convergents p_0/q_0, p_1/q_1, p_2/q_2 of the continued fraction, $[a_0, a_1, a_2, \ldots]$, of a real number α. For $n = 0, 1, 2$, show that

$$\frac{p_n}{q_n} = \frac{P_n(a_0, \ldots, a_n)}{Q_n(a_0, \ldots, a_n)}, \tag{E6.50}$$

where P_n, Q_n are polynomials in $n+1$ variables given by:

$$P_0(x) = x, \quad Q_0(x) = 1; \quad P_1(x, y) = xy + 1, \quad Q_1(x, y) = y;$$

$$P_2(x, y, z) = xyz + x + z, \quad Q_2(x, y, z) = yz + 1.$$

(b) Define P_n and Q_n, $n \in \mathbb{Z}^+$, by

$$\frac{p_n}{q_n} = \frac{P_n(a_0, \ldots, a_n)}{Q_n(a_0, \ldots, a_n)}, \tag{E6.51}$$

and use the form of (6.5.4) to confirm that

$$\begin{aligned} P_n(a_0, \ldots, a_n) &= a_0 P_{n-1}(a_1, \ldots, a_n) + Q_{n-1}(a_1, \ldots, a_n), \\ Q_n(a_0, \ldots, a_n) &= P_{n-1}(a_1, \ldots, a_n). \end{aligned} \tag{E6.52}$$

(c) Prove by induction that

$$p_n = a_n p_{n-1} + p_{n-2}, \quad q_n = a_n q_{n-1} + q_{n-2}. \tag{E6.53}$$

6.5.6 Let α be the real number considered in Exercise 6.5.5. Use (E6.53) to prove that:
(a) for $n \geqslant 1$,

$$q_n p_{n-1} - p_n q_{n-1} = (-1)^n; \tag{E6.54}$$

(b) for $n \geqslant 2$,

$$q_n p_{n-2} - p_n q_{n-2} = (-1)^{n-1} a_n; \tag{E6.55}$$

(c)

$$0 < q_1 < q_2 < \cdots < q_n < \cdots. \tag{E6.56}$$

Hence deduce that for n even (odd) the set of nth principal convergents form a strictly increasing (decreasing) sequence converging to α.

6.5.7 Consider the continued fraction representation of an irrational number α. Prove that the nth principal convergent, p_n/q_n, is the best rational approximation to α in the sense that, for all rational numbers, p/q, with $1 \leqslant q < q_n$,

$$|q_n\alpha - p_n| < |q\alpha - p|. \qquad \text{(E6.57)}$$

6.6 Weakly dissipative systems and Birkhoff attractors

6.6.1 Let $\mathbf{M}_\varepsilon$ be the map considered in Exercise 6.2.2. Define $\mathbf{f}_{\varepsilon,\delta}(\theta, \tau)$ by $\mathbf{f}_{\varepsilon,\delta} = \mathbf{M}_\varepsilon \cdot \mathbf{C}_\delta$, where

$$\mathbf{C}_\delta\colon (\theta, \tau) \mapsto (\theta, (1+\delta)\tau), \qquad \text{(E6.58)}$$

$-1 < \delta < 0$. Show that $\mathbf{f}^2_{\varepsilon,\delta}$ has a radially mapped circle, $\mathscr{C}'$, of the form $\tau = F(\theta, \varepsilon, \delta)$, for ε, δ sufficiently small. Verify that $\mathscr{C}'$ evolves continuously from the radially mapped circle, $\mathscr{C}$, of $\mathbf{M}^2_\varepsilon$ at $\delta = 0$. Explain why, generically, the period-2 points of $\mathbf{f}_{\varepsilon,\delta}$ are preserved for $|\delta|$ small enough but the linear type of some of them may change.

6.6.2 Consider the map $\mathbf{f} = \mathbf{T} \cdot \mathbf{h}$ of the cylinder C given by

$$\mathbf{T}\colon (\theta, r) \mapsto ((\theta + r) \bmod 1, r), \qquad \text{(E6.59)}$$

$$\mathbf{h}\colon (\theta, r) \mapsto ((\theta + (\beta/2\pi)\cos(2\pi\theta)) \bmod 1, \alpha r(1 + \beta \sin(2\pi\theta)), \qquad \text{(E6.60)}$$

where $0 < \alpha$, $\beta < 1$. Show that, for suitably chosen α and β, $\mathbf{f}$ is a dissipative twist diffeomorphism which does not have annular trapping regions of the form $S^1 \times [-K, K]$, $K > K_0$ for some $K_0 > 0$.

6.6.3 Let $B(\mathbf{f})$ be the Birkhoff set of a dissipative twist map $\mathbf{f}\colon C \to C$ that satisfies the trapping hypothesis. Show that $B(\mathbf{f})$ is a non-empty compact subset of $\bigcap_{n \in \mathbb{Z}^+} \mathbf{f}^n(T_K)$, where $T_K = S^1 \times [-K, K]$, $K > K_0$, is a trapping region for $\mathbf{f}$.

6.6.4 Let $B(\mathbf{f})$ be the Birkhoff set defined in Exercise 6.6.3.
(a) Show that $C \backslash B(\mathbf{f})$ consists of two connected components.
(b) Let $B'(\mathbf{f})$ be any rotational $\mathbf{f}$-invariant set that separates C into two connected components. Show that $B(\mathbf{f})$ is minimal in the sense that $B(\mathbf{f}) \subseteq B'(\mathbf{f})$.
(c) Let R be a rotational circle on C such that $B(\mathbf{f}) \cap R \neq \varnothing$. Prove that $\mathbf{f}(R) \cap R \neq \varnothing$.

6.6.5 Consider the flow, $\varphi_t\colon C \to C$, illustrated in Figure E6.1. Prove that the flow φ_t and the diffeomorphism $\mathbf{f} = \varphi_1 \cdot \mathbf{R}$ have the same attracting set A. Describe the set A. Sketch its 'top' and 'bottom' and explain why these sets are $\mathbf{f}^{-1}$ invariant but not $\mathbf{f}$ invariant. Find all the rotation numbers associated with A. Locate the Birkhoff set $B(\mathbf{f})$ within A and determine its rotation interval. Under what circumstances is $B(\mathbf{f})$ a Birkhoff *attractor*.

6.6.6 Consider the map

$$\mathbf{f}_{b,k,\omega}\colon (\theta, r) \mapsto ((\theta + \omega + r_1) \bmod 1, br - (k/2\pi)\sin(2\pi\theta)), \qquad \text{(E6.61)}$$

where $(\theta, r) \in C = S^1 \times \mathbb{R}$ and $0 \leqslant b \leqslant 1$ (see (6.6.4)).
(a) Confirm that, for $0 < b < 1$, (E6.61) is a dissipative twist diffeomorphism of the cylinder.
(b) For $b = 1$, verify that (E6.61) is conjugate to the area-preserving standard map (6.5.1).
(c) Discuss the nature of (E6.61) when $b = 0$. Define the curve $\mathscr{C} \subset C$ parametrically

by

$$\theta = (\xi + \omega - (k/2\pi)\sin(2\pi\xi)) \bmod 1, \qquad r = -(k/2\pi)\sin(2\pi\xi), \quad \text{(E6.62)}$$

$\xi \in S^1$. Verify that $\mathbf{f}^n_{0,k,\omega}(\theta_0, r_0) \in \mathscr{C}$ for every $(\theta_0, r_0) \in C$ and all $n \in \mathbb{Z}^+$. Show that $\mathbf{f}_{0,k,\omega}|\mathscr{C}$ is topologically conjugate to the Arnold circle map

$$\theta \mapsto (\theta + \omega - (k/2\pi)\sin(2\pi\theta)) \bmod 1. \quad \text{(E6.63)}$$

6.6.7 (a) Prove that the dissipative standard map, $\mathbf{f}_{b,k,\omega}$, given in (6.6.4), with $0 < b < 1$, satisfies the trapping hypothesis with $K_0 = h = |k|/[2\pi(1-b)]$.
(b) Let $\mathbf{f}^n_{b,k,\omega}: (\theta_0, r_0) \mapsto (\theta_n, r_n)$, $n \in \mathbb{Z}^+$. Show explicitly that $|r_n| < |r_0|$, for all $n \in \mathbb{Z}^+$, provided $|r_0| > h$.

6.6.8 Consider the homoclinic tangle occurring at the saddle point $\mathbf{x}^*$ of the dissipative standard map shown schematically in Figure 6.24. Use the orbit blocks $B_0 = \{\mathbf{x}^*\}$ and B_1, defined in (6.6.12), to construct ε-pseudo-orbits with rotation number α for every real number α in the interval $[0, 1/N]$, where $N \in \mathbb{Z}^+$ is the length of the block B_1. Give an example of an ε-pseudo-orbit whose rotation number does not exist.

6.6.9 Consider the dissipative standard map $\mathbf{f}_{b,k,\omega}$ given in (6.6.4) with $b = 0.95$, $k = 1$. Verify numerically that the sets U and L accumulate on the unstable manifold of the saddle point at $r^* = -\omega$ when $\omega = 0.2$ and $\omega = 0$. Show that the map $\mathbf{f}_{b,k,-\omega}$ is conjugate to $\mathbf{f}_{b,k,\omega}$ and describe the Birkhoff set when ω is: (a) 0.2, (b) 0.0, (c) -0.2.

6.7 Birkhoff periodic orbits and Hopf bifurcations

6.7.1 Consider the family of differential equations

$$\dot{x} = y, \qquad \dot{y} = \varepsilon y + kx(x-1) + \mu xy, \quad \text{(E6.64)}$$

with $k > 0$. Show that, when $\mu < 0$ (>0), (E6.64) undergoes a super(sub)critical Hopf bifurcation with increasing ε at $\varepsilon = 0$. Verify numerically that the saddle-

Figure E6.1 Phase portrait for the flow, $\boldsymbol{\varphi}_t$, considered in Exercise 6.6.5. The limit cycle $\mathscr{C}$ has period τ and $\boldsymbol{\varphi}_t$ is invariant under the rotation $\mathbf{R}: (\theta, r) \mapsto ((\theta + 2/3) \bmod 1, r)$. The lines $\theta = 0$ and $\theta = 1$ are to be identified to form the cylinder C. Note that the fixed points of $\boldsymbol{\varphi}_t$ form a pair of period-3 orbits of $\mathbf{f} = \boldsymbol{\varphi}_1 \cdot \mathbf{R}$.

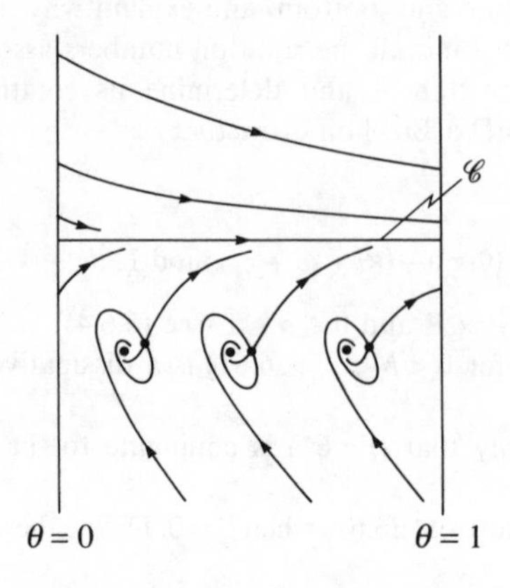

connection bifurcation curve is approximated by $\mu = -7\varepsilon$ for (ε, μ) near $(0, 0)$ when $k = 1$. Show that this asymptotic behaviour of the saddle-connection bifurcation curve for (E6.64) is independent of k.

6.7.2 Consider the Bogdanov map (6.7.2) with $k = 1$ and μ fixed at a non-zero value μ_0. Verify that the resulting subfamily, parametrised by ε, satisfies the requirements (a)—(d) of Theorem 5.4.2 and show that a super(sub)critical bifurcation is predicted if $\mu_0 < 0$ (>0).

6.7.3 Apply the Euler method, with step length $h > 0$, to the two-parameter family of differential equations (6.7.8) to obtain a family of Euler maps E_h: $\mathbb{R}^2 \times \mathbb{R}^2 \to \mathbb{R}^2$.
(a) Show that, for every h, E_h is equivalent to a family induced by E_1 via a diffeomorphism of (v_1, v_2)-space.
(b) Verify that each member of E_1 commutes with a rotation through $\pi/2$. Explain why the unstable manifolds of the period-4 points of (6.7.9) coincide with those of the fixed points of E_1.

6.7.4 (a) Consider the family of maps $\mathscr{E}_h$: $\mathbb{C} \times \mathbb{C} \to \mathbb{C}$ obtained by applying the Euler method with step length $h > 0$ to the complex differential equation (6.7.7). Prove that $\mathscr{E}_h$ is equivalent to a family that is induced by $\mathscr{E}_1$ via a real scaling of ε. Find the scale factor involved.
(b) Show that the family given by the composition of $\mathscr{E}_1$ with a rotation through angle $p\pi/2$ induces the complex normal form (5.4.6) with $q = 4$ via a local diffeomorphism at $\varepsilon = 0$.

6.7.5 Express Hamilton's equations, $\dot{x} = \partial K/\partial y$, $\dot{y} = -\partial K/\partial x$, in terms of polar coordinates, (θ, r) (defined by $x = r \cos \theta$, $y = r \sin \theta$) and the Hamiltonian $H(\theta, r) = K(r \cos \theta, r \sin \theta)$. Verify that

$$H(\theta, r) = -\tfrac{1}{2}v_2 r^2 - \tfrac{1}{4}br^4 + \tfrac{1}{4}r^4 \sin 4\theta \qquad \text{(E6.65)}$$

is a Hamiltonian for the system (6.7.8) when $v_1 = a = 0$. Assume that $b \ll -1$ and show that, for each fixed θ, $H(\theta, r)$ has a minimum at $r = r_m$, where $r_m^2 = v_2/(-b + \sin 4\theta)$ and $H(\theta, r_m) = -v_2 r_m^2/4$. Locate the fixed points for (6.7.8), determine their linear type and sketch the level sets for $H(\theta, r)$.

6.7.6 (a) Plot level curves of $H(\theta, r)$ given in (E6.65) when $b = -5$, $v_2 = 0.3$ and: (i)$H = -0.001$ (ii) $H = 0.01$.
(b) Confirm numerically that the vector field (6.7.8), with $v_1 = 0.033$ and $a = -0.5$, crosses the curves $\mathscr{C}_i$ and $\mathscr{C}_e$ as shown in Figure 6.33. Using the value of H as an indicator, verify that the region between $\mathscr{C}_i$ and $\mathscr{C}_e$ is *not* a trapping region for (6.7.9) when the homoclinic tangle shown in Figure 6.32(*c*) is present.
(c) Repeat the computations of (b) with $v_1 = 0.024$ and $a = -0.5$. Confirm that the region between $\mathscr{C}_i$ and $\mathscr{C}_e$ is *not* a trapping region for (6.7.9) with the parameter values occurring in Figure 6.32(*a*).

6.7.7 Consider the images under rotations of the curve $\mathscr{C}_e$ in Figure 6.33. Confirm numerically that an anticlockwise rotation of $\mathscr{C}_e$ by -0.2 radians yields trapping regions for (6.7.9) when v_1 is equal to: (a) 0.033; (b) 0.024. In each case, investigate the behaviour of the images of the inner and outer boundaries of the trapping region under repeated application of (6.7.9). Confirm the results illustrated in Figure 6.35 and produce the corresponding diagram when the inner homoclinic tangle is present (see Figure 6.32(*a*)).

6.8 Double invariant circle bifurcations in planar maps

6.8.1 Investigate the occurrence of invariant circles in the members of the family of truncated complex normal forms

$$N(\boldsymbol{v}, z) = z(1 + v_1 + v_2|z|^2 + a_2(\boldsymbol{v})|z|^4) \times \exp\{2\pi i(b_0(\boldsymbol{v}) + b_1(\boldsymbol{v})|z|^2 + b_2(\boldsymbol{v})|z|^4)\} \quad \text{(E6.66)}$$

(see (6.8.2) and (6.8.4)). Verify that, in the generic case, **N** exhibits the bifurcations illustrated in Figure 6.37.

6.8.2 Consider the maps, **N**, defined by (E6.66) in Exercise 6.8.1.

(a) Write down sufficient conditions for **N** to have a non-normally hyperbolic invariant circle, $\mathscr{C}$. Show that, if $|\boldsymbol{v}|$ is sufficiently small, $\mathbf{N}_{\boldsymbol{v}}|\mathscr{C}$ is a rotation through angle ω, where ω is given by (6.8.9).

(b) Assume that

$$\left[\frac{\partial b_0}{\partial v_2}(\mathbf{0}) - \frac{b_1(\mathbf{0})}{2a_2(\mathbf{0})}\right] > 0 \quad \text{(E6.67)}$$

and that ω is less than, but sufficiently close to, ω_0. Let $\gamma_\omega \in \Gamma$ be such that $\mathbf{N}_{\boldsymbol{v}_\omega}$ has an invariant circle with rotation number ω. Show that, for every $\boldsymbol{v}$ in some neighbourhood of γ_ω, there is a unique circle, $r = r(\omega, \boldsymbol{v})$, consisting of points whose angular coordinate is advanced by ω under $\mathbf{N}_{\boldsymbol{v}}$. Define the local radial coordinate, σ, by $r = r(\omega, \boldsymbol{v})(1 + \sigma)$ and verify that $\mathbf{N}_{\boldsymbol{v}}$ can be written in the form (6.8.12).

6.8.3 Assume that $N_{\boldsymbol{v}}$ is defined by (E6.66) and that ω, close to ω_0, is given. Let $\boldsymbol{v}$ be sufficiently close to γ_ω for the results of Exercise 6.8.2(b) to be valid. Prove that (6.8.14) with $\varepsilon_1 = 0$ defines a curve in the v_1, v_2-plane that is tangent to the bifurcation curve Γ at the point γ_ω.

6.8.4 Recall the discussion of normally hyperbolic invariant circles given at the end of §2.2. Explain why Σ is particularly simple for $\mathbf{N}_{\boldsymbol{v}}$ and suggest a representative of the function $\mathscr{F}: \Sigma \to \Sigma$ which allows the hyperbolic nature of the fixed point of $\mathscr{F}$ to be easily recognised. Hence show that the invariant circle with rotation number ω occurring in $\mathbf{N}_{\boldsymbol{v}}$ is normally hyperbolic for $\boldsymbol{v} \in C_\omega \backslash \gamma_\omega$, while it is not so for $\boldsymbol{v} = \boldsymbol{v}_\omega$.

6.8.5 Prove that the kth iterate of the map $f_{\boldsymbol{v},t}(z)$ defined in (6.8.17) can be written in the form

$$f^k_{\boldsymbol{v},t}(z) = N^k_{\boldsymbol{v},t}(z) + \exp(2\pi i k p/q) c_k(\boldsymbol{v}, t)\bar{z}^{q-1} + O(|z|^q), \quad \text{(E6.68)}$$

where $c_k(\boldsymbol{v}, t)$ depends on $c(\boldsymbol{v}, t)$, $b_0(\boldsymbol{v})$ and ω_0.

6.8.6 Let $N^q_{\boldsymbol{v},t}(z)$ be given by (6.8.23) and use (E6.68) to show that $f^q_{\boldsymbol{v},t}(z)$ can be written in the polar form

$$\theta_1 = \theta + p + \beta_q(\boldsymbol{v}, t, r^2) - \frac{\delta_q}{2\pi}\frac{r^{q-2}\sin(2\pi U_q)}{\alpha_q(\boldsymbol{v}, t, r^2)} + O(r^{q-1}), \quad \text{(E6.69)}$$

$$r_1 = r\alpha_q(\boldsymbol{v}, t, r^2) + \delta_q r^{q-1}\cos(2\pi U_q) + O(r^q), \quad \text{(E6.70)}$$

where $c_q(\boldsymbol{v}, t) = \delta_q \exp(2\pi i U_q)$. Define a local radial coordinate, r_L, by $r = r_q(1 + r_L)$

and verify that (E6.70) takes the form

$$r_{L_1} = r_L\left(1 + 2r_q^2 \frac{\partial \alpha_q}{\partial X}(v, t, X)|_{r_q^2}\right) + O(r_L^2). \qquad \text{(E6.71)}$$

Compare (E6.71) with the radial component of (6.8.12) and find the analogue of ε_2. Discuss the significance of this result in relation to the curve $C_{p/q}(t)$ in the $\mathbf{v}$-plane, where $\mathbf{N}_{v,t}$ has an invariant circle with rotation number p/q.

6.8.7 Show that the map $\mathbf{f}_{v,t}$, defined by the complex form (6.8.17), undergoes a supercritical Hopf bifurcation (in the sense of Theorem 5.4.2) at $v_1 = 0$, when v_1 increases through zero with $v_2 < 0$. Consider a neighbourhood of the origin in the v_1, v_2-plane. Let R be the region between the negative v_2-axis and the bifurcation curve $\Gamma(t)$ and let $\mathscr{C}$ denote the invariant circle created at the above Hopf bifurcation. Given that $\bar{v}_2 < 0$, $\bar{\omega}(0, \bar{v}_2) + t = p'/q'$ is rational and that v_1 is sufficiently small, describe the set of points in R for which the rotation number of $\mathbf{f}_{v,t}|\mathscr{C}$ is p'/q'. Justify your answer.

6.8.8 Consider the following example of (6.8.18),

$$N_{v,t}(z) = z(1 + v_1 + v_2|z|^2 + |z|^4)\exp\{2\pi i(\omega_0 + v_2 + (p/q) + t + |z|^2/2)\}, \qquad \text{(E6.72)}$$

for $t > 0$. Find $C_{p/q}(t)$ and show that it touches the double invariant circle bifurcation curve $\Gamma(t)$ at $\mathbf{v} = (4t^2/9, -4t/3)$ and terminates at $(0, -t)$. Explain how the results of Exercises 6.8.7 and 8 appear in Figure 6.41.

HINTS FOR EXERCISES

Chapter 1

1.1.1 $W_1 = \{\exp(ix)|a < x < b\}$, $W_2 = \{\exp(ix)|c < x < d\}$ such that $W_1 \cup W_2 = S^1$, $(b-a)$, $(d-c) < 2\pi$. C^∞-overlap maps.

1.1.2 $\mathbf{h}_\beta \cdot \mathbf{f} \cdot \mathbf{h}_\alpha^{-1} = (\mathbf{h}_\beta \cdot \mathbf{h}_\delta^{-1})(\mathbf{h}_\delta \cdot \mathbf{f} \cdot \mathbf{h}_\gamma^{-1})(\mathbf{h}_\gamma \cdot \mathbf{h}_\alpha^{-1})$. Composition of two C^k-maps is C^k and overlap maps $\mathbf{h}_\beta \cdot \mathbf{h}_\delta^{-1}$ and $\mathbf{h}_\gamma \cdot \mathbf{h}_\alpha^{-1}$ are C^k since $r \geqslant k$. Differentiability of $\mathbf{f}$ is independent of charts.

1.1.3 (a) Pick open subsets A, B, C, D of $\mathbb{R}^2$ such that $\{\pi(A), \pi(B), \pi(C), \pi(D)\}$ is an open covering of T^2 and restrictions of π to A, B, C, D are homeomorphisms.
(b) $W_1 = S^2 \backslash N$, N the north pole; $W_2 = S^2 \backslash S$, S the south pole, $\mathbf{h}_1(\mathbf{h}_2)$ is the stereographic projection from $N(S)$ poles. Overlap map $(r, \varphi) \mapsto (4/r, \varphi)$, $r \neq 0$.

1.2.1 Arnold, 1973, pp. 163–5.

1.2.2 (a) yes; (b) no; (c) circle map not homeomorphism.

1.2.3 Fixed points $x = 0, \frac{1}{2}$; all other points period-2.

1.2.4 Plot $y = \bar{f}^2(x)$.

1.2.5 $\bar{f}$ an orientation-reversing homeomorphism on $\mathbb{R}$ implies it is strictly decreasing. $\bar{f}(x+1) = \bar{f}(x) - 1$ as in proof of Proposition 1.2.1. Fixed points of f only at intersection of $y = \bar{f}(x)$ with $y = x$ and $y = x + 1$ (see Proposition 1.2.2).

1.3.1 Definition 1.3.1 implies $\boldsymbol{\varphi}_t$ is C^1 for all $t \in \mathbb{R}$. $\boldsymbol{\varphi}_t^{-1} = \boldsymbol{\varphi}_{-t}$.

1.3.4 $X(x) = x - x^2$.

1.3.5 (a) $\dot{x} = x^3$; (b) $\dot{x} = x$, $\dot{y} = y^2$.

1.4.1 Minimal: (a) S^1; (b) $\{x, R_{p/q}(x), \ldots, R_{p/q}^{q-1}(x)\}$, $x \in S^1$.
General: (a) S^1; (b) $S = U \cup R_{p/q}(U) \cup \cdots R_{p/q}^{q-1}(U)$, $U \subseteq S^1$, closed.

1.4.2 (a) Show Ω^c is open.

1.4.5 Separatrices connecting $n = 1, 2, 3, 4$ saddle points enclosing unstable focus. Consider Hamiltonian system with desired saddle connection and introduce dissipation in the region bounded by the separatrices, e.g. $\dot{x} = -2y(1-x^2) + \mu x B(x)$, $\dot{y} = 2x(1-y^2) + \mu y B(y)$, $\mu > 0$,

$$B(x) = \begin{cases} \exp[-x^2/(1-x^2)], & |x| < 1 \\ 0 & |x| \geqslant 1. \end{cases}$$

1.4.6 Polar coordinates in the x, $\dot{x}$-plane give

$$dr/d\theta = -\varepsilon r \sin^2\theta(1 - r^2\cos^2\theta) + O(\varepsilon^2);\ r(0) - r(2\pi) = -\varepsilon\pi r\left(1 - \frac{r^2}{4}\right) + O(\varepsilon^2).$$

Construct positively invariant set containing no fixed points.

1.5.1 (i) Similar construction to Example 1.5.1 for topological conjugacy with reference intervals [1,2] for f and [1,8] for g.

1.5.4 Conjugacy preserves fixed points.

1.5.5 Use Proposition 1.2.2. Plot $y = \bar{f}^2(x)$. Conjugacy preserves periodic points.

1.5.6 Use (1.5.11).

1.5.7 Differentiate (1.5.11) and set $\mathbf{x} = \mathbf{0}$.

1.5.8 If x^* periodic with period-q, $\rho(f) = \left(\lim_{n\to\infty}(\bar{f}^{nq}(x^*) - x^*)/nq\right) \bmod 1$.

1.6.3 Consider separatrix of the saddle which is of opposite stability to that of the node.

1.6.4 Consider the lifted flow $\bar{\varphi}_t(x, y) = (x + t, y + \alpha t)$ on $\mathbb{R}^2$. Periodic orbits are given by $\bar{\varphi}_T(x, y) = (x + m, y + n)$, $T \neq 0$, $m, n \in \mathbb{Z}$.

1.6.5 Recall that $(\boldsymbol{\varphi} \times \boldsymbol{\psi})_t(\mathbf{x}, \mathbf{y}) = (\boldsymbol{\varphi}_t(\mathbf{x}), \boldsymbol{\psi}_t(\mathbf{y}))$, $\mathbf{x} \in M$, $\mathbf{y} \in N$. Consider the lifted flows $\bar{\varphi}_t(x) = x + t$, $\bar{\varphi}'_t(x) = x + 2^{1/2}t$, $\bar{\psi}_t(x) = \bar{\psi}'_t(x) = \bar{\varphi}_t(x)$.

1.6.6 Arrowsmith & Place, 1982, § 2.3.

1.7.1 $\varphi_1(x) = xe/[xe - x + 1]$.

1.7.2 Fixed point $\mathbf{x}^*$ must lie in Σ but $\mathbf{X}(\mathbf{x}^*) = \mathbf{0}$.

1.7.3 Show that $\mathbf{P}_2(\boldsymbol{\varphi}_{\tau_0}(\mathbf{x})) = \boldsymbol{\varphi}_{\tau_0}(\mathbf{P}_1(\mathbf{x}))$, $\mathbf{x} \in S_1$.

1.7.4 Cylinder. Two limit cycles: stable $x = 0$; unstable $x = 1$.

1.7.5 (a) Möbius band; (b) Klein bottle.

1.8.2 $x(t) = C\exp(t - \cos t)$.

1.8.4 If $\mathbf{Q}(t)$ is a fundamental matrix so is $\mathbf{Q}(t + T)$ and $\mathbf{Q}(t) = \mathbf{Q}(t + T)\mathbf{Q}^{-1}(t_0 + T)\mathbf{Q}(t_0)$. All $\mathbf{P}_{\theta_0}$ are conjugate.

1.8.5 (a) $\boldsymbol{\varphi}(t, t_0) = \begin{pmatrix} \exp(\lambda_1\tau) & 0 \\ 0 & \exp(\lambda_2\tau) \end{pmatrix}$, $\tau = t - t_0$;

(b) $\mathbf{x}(t) = \begin{pmatrix} 4 - \exp(-t) \\ 2\exp(t) - 1 \end{pmatrix}$.

1.8.7 Use polar coordinates. Null solution is stable.

1.8.8 $\mathbf{x}(t) = \begin{pmatrix} \cos t & -\sin t \\ \sin t & \cos t \end{pmatrix}\mathbf{x}_0 + \begin{pmatrix} 0 \\ -\sin t \end{pmatrix}$.

1.9.1 Sketch level curves of $H(x_1, x_2)$.

1.9.3 The generic case has non-zero eigenvalues.

1.9.4 Hamilton's equations in plane polars are $\dot{r} = r^{-1}\,\partial H/\partial\theta$, $\dot{\theta} = -r^{-1}\,\partial H/\partial r$. Examine extrema of H as a function of r for various fixed values of θ.

1.9.5 (i) $\dfrac{\partial(r, \theta)}{\partial(x, y)} = \dfrac{2}{r}$; (ii) $\dfrac{\partial(\tau, \theta)}{\partial(x, y)} = 1$.

1.9.6 $\mathbf{A}=\begin{pmatrix}1 & a\\ 0 & 1\end{pmatrix}$, $\mathbf{B}=\mathbf{C}=\mathbf{0}$, $\mathbf{D}=\begin{pmatrix}1 & b\\ 0 & 1\end{pmatrix}$, a and/or $b\neq 0$.

1.9.7 If $\dot{x}=\dfrac{\partial H}{\partial y}$ and $\dot{y}=-\dfrac{\partial H}{\partial x}$ then $\dfrac{\partial \dot{x}}{\partial x}=-\dfrac{\partial \dot{y}}{\partial y}$.

1.9.8 Ψ is symplectic to order $|I|$.

Chapter 2

2.1.1 If the Jordan form of $\mathbf{L}$ is not diagonal examine the powers of blocks of the form $\lambda\mathbf{I}+\mathbf{N}$, $N_{ij}=\delta_{i,j-1}$. Observe $(\mathbf{N}^k)_{ij}=\delta_{i,j-k}$, $1\leqslant k\leqslant n-1$, $\mathbf{N}^n=\mathbf{0}$.

2.1.2 (a) $\mu=\max\{|\lambda_1|,\ldots,|\lambda_l|\}$, (b) Pick $N>3$ such that $\mu=N^{l/N}|\lambda|<1$.

2.1.3 (i) $\mathbf{A}|E^{\mathrm{u}}\colon u\mapsto\left(\dfrac{3+5^{1/2}}{2}\right)u$, orientation-preserving expansion;

$\mathbf{A}|E^{\mathrm{s}}\colon v\mapsto\left(\dfrac{3-5^{1/2}}{2}\right)v$, orientation-preserving contraction;

(ii) $\mathbf{A}|E^{\mathrm{u}}\colon u\mapsto(1+2^{1/2})u$, orientation-preserving expansion;
$\mathbf{A}|E^{\mathrm{s}}\colon v\mapsto(1-2^{1/2})v$, orientation-reversing contraction.

2.1.4 Real Jordan form of $\mathbf{A}$ is $\begin{pmatrix}-\frac{1}{2} & -\frac{1}{2} & 0\\ \frac{1}{2} & -\frac{1}{2} & 0\\ 0 & 0 & 2\end{pmatrix}$. $\mathbf{A}|E^{\mathrm{s}}$ is a rotational contraction.

2.1.6 $\dot{\mathbf{x}}=\mathbf{A}\mathbf{x}$ is linearly conjugate to $\dot{\mathbf{y}}=\mathbf{\Lambda}\mathbf{y}$, $\mathbf{\Lambda}=[\lambda_i\delta_{ij}]_{i,j=1}^3$. Show that $\dot{y}_i=\lambda_i y_i$ is topologically conjugate to $\dot{z}_i=\operatorname{sign}(\lambda_i)z_i$, $i=1,2,3$, and use Exercise 1.6.5.

2.1.7 Use Theorem 2.1.2.

2.1.8 $\dim E^{\mathrm{s}}+\dim E^{\mathrm{u}}=n$ and restrictions to E^{s} and E^{u} may be orientation-preserving or -reversing.

2.2.1 (a) $\mathbf{Df}(0,0)=\begin{pmatrix}2 & 0\\ 1 & -\frac{1}{2}\end{pmatrix}$, saddle-type with reflection;

(b) $\mathbf{Df}(0,0)=\begin{pmatrix}0 & 1\\ 2 & 0\end{pmatrix}$, expansion with reflection.

2.2.2 $\mathrm{D}\boldsymbol{\varphi}_1(\mathbf{0})=\exp(\mathrm{D}\mathbf{X}(\mathbf{0}))=\exp\begin{pmatrix}0 & 1\\ 1 & 0\end{pmatrix}$ has eigenvalues $\cosh(1)\pm\sinh(1)$. Show that $W^{\mathrm{s,u}}_{\mathbf{f}}(\mathbf{0})=W^{\mathrm{s,u}}_{\boldsymbol{\varphi}}(\mathbf{0})$, where $\boldsymbol{\varphi}$ is the flow of $\dot{x}=y$, $\dot{y}=x-x^2$. Obtain $W^{\mathrm{s}}_{\boldsymbol{\varphi}}(\mathbf{0})\cap W^{\mathrm{u}}_{\boldsymbol{\varphi}}(\mathbf{0})$ from a first integral.

2.2.3 If $\mathbf{f}^k(\mathbf{x})=\mathbf{y}$ then $\mathbf{f}^q|U$ and $\mathbf{f}^q|V$ are conjugate by $\mathbf{f}^k$.

2.2.4 If $\mathbf{y}\in W^{\mathrm{s}}(\mathbf{f}^i(\mathbf{x}^*))$ then $\lim_{n\to\infty}\mathbf{f}^{nq+k}(\mathbf{y})=\mathbf{f}^{i+k}(\mathbf{x}^*)$, $k=0,\ldots,q-1$. No, construct counterexample:
(a) for period-1, consider $\boldsymbol{\varphi}_t$ in Figure 1.16, let $\mathbf{f}=\boldsymbol{\varphi}_1$ and observe $\mathbf{y}\in A\notin W^{\mathrm{s,u}}_{\mathbf{f}}(P_0)$ but $P_0\subset L_\omega(\mathbf{y})$;
(b) for period-$q>1$, construct periodic orbit in similar manner to Exercise 2.2.5.

2.2.5 Vector field is symmetric under clockwise rotation by $\pi/2$. A fixed point $\mathbf{x}^*$ of φ_1, with topological type given by $\mathrm{D}\varphi_1(\mathbf{x}^*) = \exp(\mathrm{D}\mathbf{X}(\mathbf{x}^*))$, becomes a periodic point of $\mathbf{f} = \varphi_1 \cdot \mathbf{R}_{-\pi/2}$.

2.2.6 Let $\mathbf{x}_1 = \varphi_{\tau_0}(\mathbf{x}_0)$ and define $S_0' = \varphi_{-\tau_0}(S_1)$. Use flow box coordinates to prove $\mathbf{P}_0'\colon S_0' \to S_0'$ and $\mathbf{P}_0\colon S_0 \to S_0$ are C^1-conjugate and result then follows from Exercise 1.7.3.

2.2.7 Introduce cylindrical polar coordinates and recognise closed orbit for $r = (x_1^2 + x_2^2)^{1/2} = 1$, $z = x_3 = 0$. The Poincaré map $\varphi_{2\pi}$ defined on the plane $\theta = \text{constant}$ has a fixed point at $(r, z) = (1, 0)$. Hyperbolicity follows from $\mathrm{D}\varphi_{2\pi}(1, 0) = \exp(\mathrm{D}\mathbf{X}(1, 0))$ and Hartman's Theorem.

2.3.1 (i) Solve quadratic for y and expand square root.
(ii) Use (i) to obtain y_1 and substitute into expansion for $(1 + y_1)^{-1}$.

2.3.4 Let $\mathbf{h}_2(\mathbf{y}) = \begin{pmatrix} a_1 y_1^2 + a_2 y_1 y_2 + a_3 y_2^2 \\ b_1 y_1^2 + b_2 y_1 y_2 + b_3 y_2^2 \end{pmatrix}$, write down $\mathbf{L_A}\mathbf{h}_2(\mathbf{y})$ and show that a_i and b_i, $i = 1, 2, 3$, can be chosen such that $\mathbf{L_A}\mathbf{h}_2(\mathbf{y}) = \mathbf{X}_2(\mathbf{y})$. Find $a_1 = a_2 = a_3 = b_1 = b_2 = 1$, $b_3 = 0$.

2.3.5 Use resonance condition to show that $\begin{pmatrix} x_2^3 \\ 0 \end{pmatrix}$ is the *only* resonant term.

2.3.6 Use resonance condition and $q\lambda_1 + p\lambda_2 = 0$.

2.3.7 Use resonance condition. Normal form when $\lambda_1 = m\lambda_2$, $m > 2$, $\begin{pmatrix} \lambda_1 x_1 + c x_2^m \\ \lambda_2 x_2 \end{pmatrix}$.

2.3.8 Matrix representing $L_{\mathbf{A}}$ is triangular with repeated eigenvalue λ. Since $\lambda_1 = \lambda_2 = \lambda$, $\Lambda_{\mathbf{m},i} = \lambda$ for all $\mathbf{m}$, i.

2.3.9 Use basis $\left\{ \begin{pmatrix} 0 \\ x_1^r \end{pmatrix}, \ldots, \begin{pmatrix} 0 \\ x_1^m x_2^{r-m} \end{pmatrix}, \ldots, \begin{pmatrix} 0 \\ x_2^r \end{pmatrix}, \begin{pmatrix} x_1^r \\ 0 \end{pmatrix}, \ldots, \begin{pmatrix} x_1^m x_2^{r-m} \\ 0 \end{pmatrix}, \ldots, \begin{pmatrix} x_2^r \\ 0 \end{pmatrix} \right\}$ for H^r.

2.4.1 (1) $ad - bc \neq 0$; (2) $ad - bc = 0$, $a + d \neq 0$; (3) $ad - bc = a + d = 0$, $a^2 + b^2 + c^2 + d^2 \neq 0$. $\mathrm{cod}(S_1) = 0$; $\mathrm{cod}(S_2) = 1$; $\mathrm{cod}(S_3) = 2$. Linear vector fields satisfying (2.4.1) and (2.4.3) have codimension 1 and 2, respectively.

2.4.3 (i) $A = \frac{1}{2}(b + f)$, $B = c$, $C = 0$, $D = e/2$, $E = f$, $F = 0$; $\alpha = a + \dfrac{e}{2}$, $\beta = d$.
(ii) $A = \frac{1}{2}(b + f)$, $B = c$, $C = 0$, $D = -a$, $E = f$, $F = 0$; $\gamma = d$, $\delta = e + 2a$.

2.4.4 $c = b$, $d = 2a$.

2.4.5 Consider the types of Jordan block which give rise to non-hyperbolic linear systems. Show that each type of block satisfies a resonance condition for all $r \geqslant 2$.

2.5.2 Observe that $c > 0$ implies $f(x) > (<) - x$ for x sufficiently small and positive (negative).

2.5.3 Use (2.5.8) for complex form with $n = 2$. Observe that $f_{\mathbf{m},1} = 0$ for $m_2 = 0$ implies no $\bar{z}$-dependent terms arise. Alternatively, use (2.5.8) with $n = 1$ and a single (complex) variable z (see Exercise 2.5.2). Note $\lambda^{q+1} = \lambda$.

2.5.4 $a = \exp(3i\alpha)/[1 - \exp(4i\alpha)]$. Note $\exp(4i\alpha) \neq 1$ for $\alpha \neq 2\pi p/q$, $q = 1, 2, 3, 4$.

2.6.1 $\mathbf{M}=\begin{pmatrix}0 & -1\\ 1 & 0\end{pmatrix}$ has eigenvalues $\pm \mathrm{i}$ but $\begin{pmatrix}0 & -1\\ 1 & 0\end{pmatrix}=\exp\begin{pmatrix}0 & -\pi/2\\ \pi/2 & 0\end{pmatrix}$.

2.6.2 Note:
(i) if $\mathbf{AB}=\mathbf{BA}$ then $\exp(\mathbf{A}+\mathbf{B})=\exp(\mathbf{A})\exp(\mathbf{B})$;
(ii) $\mathbf{N}^n=\mathbf{0}$.

2.6.3 Let $\mathbf{S}^{-1}\mathbf{MS}=\mathbf{J}$, find $\ln \mathbf{J}$ from Exercises 2.6.1 and 2. $\exp(\ln \mathbf{J})=\mathbf{J}$ implies $\mathbf{L}=\mathbf{S}\ln \mathbf{J}\mathbf{S}^{-1}$.

Alternative implies state transition matrix (itself a particular fundamental matrix) $\varphi(t,0)=\mathbf{U}(t)\exp(\mathbf{C}t)$, make change of variable $\mathbf{x}=\mathbf{U}(t)\mathbf{y}$ and show that $\dot{\mathbf{y}}=\mathbf{C}\mathbf{y}$. Thus alternative implies Theorem 2.6.1 with $\mathbf{\Lambda}=\mathbf{C}$ and $\mathbf{B}(t)=\mathbf{U}(t)$.

Theorem 2.6.1 implies $\dot{\mathbf{x}}=\mathbf{A}(t)\mathbf{x}$ has solutions $\mathbf{x}(t)=\mathbf{B}(t)\exp(\mathbf{\Lambda}t)\mathbf{y}_0=\varphi(t,0)\mathbf{x}_0=\mathbf{Q}(t)\mathbf{Q}^{-1}(0)\mathbf{x}_0$ for any fundamental matrix $\mathbf{Q}(t)$. Let $\mathbf{y}_0=\mathbf{Q}^{-1}(0)\mathbf{x}_0$ to obtain the statement given in the question and $\mathbf{y}_0=\mathbf{x}_0$ for $\mathbf{B}(t)=\varphi(t,0)\exp(-\mathbf{\Lambda}t)$.

2.6.4 If $\mathbf{J}_{\mathbb{C}}$ is a Jordan block corresponding to a complex eigenvalue λ^2 of $\mathbf{M}^2$ then the complex linear transformation that reduces $\mathbf{J}_{\mathbb{C}}$ to the real Jordan form $\mathbf{J}_{\mathbb{R}}$, transforms $\ln \mathbf{J}_{\mathbb{C}}$ into a real matrix, i.e. $\mathbf{J}_{\mathbb{R}}$ has a real logarithm.

If $\mathbf{P}=\varphi(2\pi,0)$ then $\mathbf{P}^2=\varphi(4\pi,0)=\exp(4\pi\mathbf{\Lambda})$, for real $\mathbf{\Lambda}$, by the first part of the question. Show that $\mathbf{B}(t)=\varphi(t,0)\exp(-\mathbf{\Lambda}t)$ is 4π-periodic in t.

2.6.5 $\mathbf{Q}(t)=(\mathbf{x}_+ \vdots\, \mathbf{x}_-)$, $\varphi(t,0)=\mathbf{Q}(t)\mathbf{Q}(0)^{-1}$,

$$\varphi(2\pi,0)=\begin{pmatrix}\cosh 2\pi & \sinh 2\pi\\ \sinh 2\pi & \cosh 2\pi\end{pmatrix}=\exp\left\{2\pi\begin{pmatrix}0 & 1\\ 1 & 0\end{pmatrix}\right\}.$$

2.6.8 (a) $z|z|^2$, $z|z|^4$; (b) $z|z|^2$, $z|z|^4$, $\bar{z}^4\exp(2it)$.

2.6.9 For $\lambda_i=0$, $i=1,2$, (2.6.14) implies resonance only if $v=0$, i.e. all time-dependent terms can be removed. Let $x=y+\dfrac{y^2}{4}\sin(2t)$ and find $a=\frac{1}{2}$, $b=-c$.

2.7.1 $\mathbf{A}|E^{\mathrm{c}}$ given by (a) $\begin{pmatrix}0 & -\beta\\ \beta & 0\end{pmatrix}$; (b) $\begin{pmatrix}0 & 1\\ 0 & 0\end{pmatrix}$; (c) $\begin{pmatrix}0 & 0\\ 0 & 0\end{pmatrix}$. (b) gives unbounded motion.

2.7.2 Decompose $\mathbb{R}^n$ into the direct sum $E^{\mathrm{s}}\oplus E^{\mathrm{c}}\oplus E^{\mathrm{u}}$ and consider restrictions of $\exp(\mathbf{A}t)$ to E^{s} and E^{u}.

2.7.3 C^{∞}, unstable.

2.7.4 No, origin is hyperbolic node. Maximum differentiability given by $[b/a]$.

2.7.5 $a_{2j}\equiv 0$, $a_{3j}\equiv 0$, $a_{4j}=-4^j$, $a_{5j}\equiv 0$, $a_{6j}=2(6^{j+1})(1-(\frac{2}{3})^{j+1})$; $\sum_{j=0}^{\infty} a_{ij}\mu^j$ converges for $i=4$ if $\mu<\frac{1}{4}$ and for $i=6$ if $\mu<\frac{1}{6}$.

2.7.6 For $C\neq 0$ centre manifold is non-analytic.

2.7.7 Assume centre manifold given by $y=\sum_{i=0}^{\infty} a_i x^i$ and show that $a_0=a_1=0$; $a_{2k}=(k-1)!$, $a_{2k+1}=0$, $k\geqslant 1$. Hence y has zero radius of convergence.

2.7.8 Assume centre manifold E^{c} of the form $y=a_0+a_1x+a_2x^2+O(x^3)$. Show that $a_0=a_1=0$, $a_2=1$. Consider restriction of system to E^{c}.

2.7.9 Assume centre manifold of the form $x=c_0+c_1y+c_2y^2+O(y^3)$ and show that $c_0=c_1=0$, $c_2=-a$.

2.7.10 $r=1$, linearise; $r>1$, assume E^c of the form $y=a_2x^2+O(x^3)$.

2.7.11 $\mathbf{M}=\begin{pmatrix}1 & 1\\ 0 & -1\end{pmatrix}$, $y=-\alpha x^2+O(x^3)$.

2.8.1 Polar blowing-up gives: (i) saddles at $\theta=0, \pi$, $\dot{r}>(<)\,0$ for $r>0$ and $\theta=0$ (π); (ii) $\theta=0$ unstable node, $\theta=\pi$ stable node.

2.8.2 Singularities on $r=0$ circle are: (a) $\theta=0$ unstable node, $\theta=\pi/4$ saddle, $\theta=\pi/2$ unstable node, $\theta=\pi$ stable node, $\theta=5\pi/4$ saddle, $\theta=3\pi/2$ stable node; (b) $\theta=0$ unstable node, $\theta=\pi/4$, $\pi/2$ saddles, $\theta=\pi$ stable node, $\theta=5\pi/4$, $3\pi/2$ saddles.

2.8.3 Repeated blowing-up along positive y-axis gives further saddle-node singularities.

2.8.4 Division by $|u|^k$ and $|v|^k$ is necessary to prevent orientation reversal.

2.8.5 Positive x-blow-up, unstable node; negative x-blow-up, stable node. (cf. Exercise 2.8.1 with $a=-1$, $b=2$.)

2.8.6 (a) Do polar blow-up, investigate resulting singularities at $\theta=\pi/2$, $3\pi/2$ with further polar blow-ups. Obtain non-hyperbolic saddle.
(b) Polar blow-up gives six hyperbolic singularities. Obtain 'monkey' saddle.

Note that the unfoldings of the vector fields considered in this question appear in Section 5.6 (see (5.6.2) $(q=2)$ and (5.6.14) $(q=3)$). The reader may like to confirm that the underlying singularity for $q=5$ (see (5.6.34) and (5.6.35)) is a focus, while $q=4$ (see (5.6.21)) admits a variety of singularity types.

Chapter 3

3.1.1 Recall:
(i) the spectral radius, $\rho(\mathbf{A})$, of $\mathbf{A}$ is the maximum of the absolute values of the eigenvalues of $\mathbf{A}$; (ii) the spectral norm, $\sigma(\mathbf{A})$, of $\mathbf{A}$ is the positive square root of the largest eigenvalue of $\mathbf{A}^T\mathbf{A}$; (iii) $\rho(\mathbf{A})\leqslant\sigma(\mathbf{A})$; (iv) $\sigma(\mathbf{A})\leqslant\|\mathbf{A}\|$, where $\|\mathbf{A}\|=\sum_{ij}|a_{ij}|$.
(a) Let $\mathbf{M}^{-1}\mathbf{AM}=\mathbf{D}$, $\mathbf{D}=[\lambda_i\delta_{ij}]$. Consider $\det(\mathbf{M}^{-1}\mathbf{BM}-\mu\mathbf{I})$, with μ an eigenvalue of $\mathbf{B}$ that is not equal to λ_i for any $i=1,\ldots,n$, and show that $\sigma(\mathbf{D}_\mu^{-1}\mathbf{C}_1)\geqslant 1$, where $\mathbf{D}_\mu=\mathbf{D}-\mu\mathbf{I}$ and $\mathbf{C}_1=\mathbf{M}^{-1}\mathbf{CM}$. Observe $\max_i[|\lambda_i-\mu|^{-1}]\geqslant c^{-1}$ implies $\min_i[|\lambda_i-\mu|]\leqslant c$.
(b) Let $\mathbf{M}^{-1}\mathbf{AM}=\mathbf{D}+\mathbf{T}$, $\mathbf{T}=[t_{ij}]$, $t_{ij}=\begin{cases}0 & j\neq i+1\\ 1 \text{ or } 0 & j=i+1\end{cases}$. Proceed as in (a) and note that $(\mathbf{I}+\mathbf{D}_\mu^{-1}\mathbf{T})^{-1}=\mathbf{I}+\sum_{k=1}^{n-1}(-1)^k\mathbf{D}_\mu^{-k}\mathbf{T}^k$, $\sigma(\mathbf{T})=1$, $\sum_{k=1}^{n}x^k<nx^n$ for $x>1$.
(c) Observe similar results follow for $S_{\mathbf{B}}(\mathbf{A})$.

3.1.2 Hyperbolic implies structurally stable: use Exercise 3.1.1. Structural stability implies hyperbolic: observe a Jordan block $\mathbf{J}$ associated with an eigenvalue λ of absolute value unity satisfies $|\mathbf{J}^k\mathbf{x}|=|\mathbf{x}|$ for all $k\in\mathbb{Z}^+$ if $\mathbf{x}$ lies in the eigenspace of λ. The block $(1-\delta)\mathbf{J}$, $\delta>0$, can be made arbitrarily close to $\mathbf{J}$ but $|\mathbf{J}^k\mathbf{x}|\to 0$ as $k\to\infty$ for all $\mathbf{x}$. Density: note that, if $\mathbf{A}$ is non-hyperbolic with eigenvalues λ_i, then $\mathbf{A}+\delta\mathbf{I}$ has eigenvalues $\lambda_i+\delta$.

3.1.3 $\mathbf{A}$ structurally stable in S: let $\mathbf{A}, \mathbf{B}\in S$ be ε-close and apply Theorem 2.1.3 in the subspaces on which the restriction of **A** and **B** is not the identity. Hence construct a conjugacy for **A** and **B**. **A** is not structurally stable in $L(\mathbb{R}^2)$: consider $\mathbf{A} = \begin{pmatrix} 1+\varepsilon & 0 \\ 0 & \lambda \end{pmatrix}$.

3.1.4 Show that $\mathbf{A} \in O(\mathbb{R}^2)$ is a rotation so that every circle, centre $\mathbf{x} = \mathbf{0}$, is invariant under **A**. Observe $\mathbf{B} = (1-\varepsilon)\mathbf{A}$, $\varepsilon > 0$, has no invariant circles. Prove that conjugacy preserves invariant circles.

3.2.1 (a) Use the Implicit Function Theorem and Exercise 3.1.1
(b) Use Exercise 3.1.1.
(c) $\mathbf{D\tilde{X}(0)} = \mathbf{DX(0)}$.

3.2.2 (a) $|\eta| < \varepsilon/4$; (b) $|\eta| < \varepsilon/32$; (c) $|\eta| < \varepsilon/13$. Use $\mathrm{D}\varphi_{2\pi}(r_0) = \exp(2\pi \mathrm{D}X_r(r_0))$ (see Exercise 2.2.2), where $X_r = \dot{r}$ and $X_r(r_0) = 0$.

3.2.3 (a) $|\delta| < \varepsilon/(2+R)$; (b) $|\delta| < \varepsilon/(R^2 + 2R)$. The fixed point $\mathbf{x}^*$ is not hyperbolic.

3.3.1 Theorem 3.3.1 is:
(a) applicable, non-hyperbolic closed orbit;
(b) applicable, non-hyperbolic fixed point at $(x, y) = (1, 0)$;
(c) not applicable, use perturbation $(\delta B(y), 0)$, $\delta \in \mathbb{R}$.

3.3.2 (a) Theorem 3.3.3(i) fails; (b) Theorem 3.3.3(iii) fails.

$$\left.\begin{array}{l} \dot{x} = \sin(2\pi x), \dot{y} = \delta \sin(2\pi y) \\ \dot{x} = 1, \dot{y} = 2 + \delta \sin[2\pi(y - 2x)] \end{array}\right\} \text{is } \varepsilon\text{–}C^1\text{-close to } \left\{\begin{array}{l} \text{(a) for } |\delta| < \varepsilon/(1+2\pi) \\ \text{(b) for } |\delta| < \varepsilon/(1+6\pi). \end{array}\right.$$

3.3.3 Apply Theorem 3.3.1 to S_n. The ε–C^1-close perturbation given by $\dot{r} = \varepsilon + [r\cos(2\pi r)/(1+r^2)]$, $\dot{\theta} = 1$ has no limit cycles for $|r| > 1/\varepsilon$.

3.4.1 Use appropriate lifts to examine fixed and periodic points of f.
(a) infinite number of periodic points implies non-wandering set not finite.
(b) no fixed points but every orbit is dense therefore non-wandering set does not consist of fixed and periodic points.
(c) four fixed points on S^1 but none are hyperbolic.
(d) non-hyperbolic period-2 points.

3.4.2 Recall that $\boldsymbol{\pi}: \mathbb{R}^2 \to T^2$ is a local diffeomorphism and differentiate $\boldsymbol{\pi}(\mathbf{\bar{f}}^q(\mathbf{x})) = \mathbf{f}^q(\boldsymbol{\pi}(\mathbf{x}))$ with respect to **x** to show that $\mathbf{Tf}^q(\boldsymbol{\pi}(\mathbf{x}))$ (see §3.6) and $\mathbf{D\bar{f}}^q(\mathbf{x})$ are conjugate. $\mathbf{D\bar{f}}^q(\mathbf{x}) = \mathbf{A}^q$, for all **x**, and $\mathbf{A}^q$ is hyperbolic. $W^{s,u}(\mathbf{f}^i(\boldsymbol{\pi}(\mathbf{x}^*))) = \boldsymbol{\pi}(\mathbf{A}^i\mathbf{x}^* + E^{s,u})$, where $E^s(E^u)$ are the stable (unstable) eigenspaces of **A**. $W^{s,u} = \bigcup_{i=0}^{q-1} W^{s,u}(\mathbf{f}^i(\boldsymbol{\pi}(\mathbf{x}^*)))$.

3.4.3 Observe that $\boldsymbol{\pi}(\mathbf{x}^*)$, where $\mathbf{x}^* = (\mathbf{A}^q - \mathbf{I})^{-1}\mathbf{p}$, $\mathbf{p} \in \mathbb{Z}^2$, is a periodic point of **f** of period at most q. Verify that, when **A** is given by (3.4.12), $|\mathrm{Det}(\mathbf{A}^q - \mathbf{I})|$, $q \geq 2$, is an integer greater than unity, hence show that $(\mathbf{A}^q - \mathbf{I})^{-1}$ has at least one element that belongs to $Q^2 \backslash \mathbb{Z}^2$. When q is prime deduce that **f** has a period-q point.

3.4.4 Differentiable conjugacy of **f**, **g** by **h** implies $(\mathbf{\bar{h}}(\mathbf{\bar{f}}(\mathbf{x}))) = (\mathbf{\bar{g}}(\mathbf{\bar{h}}(\mathbf{x}))) + \mathbf{k}$ where $\mathbf{k} \in \mathbb{Z}^n$ and $\bar{\cdot}$ denotes a lift of $\cdot$. Differentiate with respect to **x** and set $\mathbf{x} = \mathbf{0}$. Conditions on **C** mean it is a lift of a diffeomorphism, **h**, say, on T^n. $\mathbf{CAx} = \mathbf{BCx}$ implies $\mathbf{\bar{h}}(\mathbf{\bar{f}}(\mathbf{x})) = \mathbf{\bar{g}}(\mathbf{\bar{h}}(\mathbf{x}))$; take projection $\boldsymbol{\pi}$ and show $\mathbf{h}(\mathbf{f}(\boldsymbol{\theta})) = \mathbf{g}(\mathbf{h}(\boldsymbol{\theta}))$, where $\boldsymbol{\theta} = \boldsymbol{\pi}(\mathbf{x})$.

3.4.5 $y = ((1 \pm 13^{1/2})/6)x$. Irrational slope implies stable and unstable manifolds of fixed point at $\mathbf{x} = \mathbf{0}$ wind densely around the torus without closing. Homoclinic point is given by the intersection of $y = (1 + 13^{1/2})x/6$ and $y = (1 - 13^{1/2})(x-1)/6$.

Restriction of **A** to its stable manifold is orientation-reversing while for (3.4.12) it is orientation-preserving.

3.5.1 $\mathbf{f}|P_0: (x, y) \mapsto (-5x-2, -y/5+2/5)$;
$\mathbf{f}|P_1: (x, y) \mapsto (5x-2, y/5-2/5)$.

3.5.3 (a) Note $\mathbf{g}|Q = \mathbf{f}$; use (3.5.3) and (3.5.1).
(b) Explicit form of $\mathbf{f}$ given in Exercise 3.5.1 shows x-component of $\mathbf{f}(x, y)$, $(x, y) \in P_0 \cup P_1$, is independent of y. Let $\mathbf{x} \in \bigcap_{n\in\mathbb{N}} Q^{(-n)}$ and $\mathbf{x}' \in \Lambda$ have the same x-coordinate, show that $|\mathbf{f}^n(\mathbf{x}) - \mathbf{f}^n(\mathbf{x}')| \to 0$ as $n \to \infty$. Eliminate $\mathbf{x} \in Q \backslash \bigcap_{n\in\mathbb{N}} Q^{(-n)}$.
Similarly, for $\mathbf{x} \in \bigcap_{n\in\mathbb{Z}^+} Q^{(n)}$ using $\mathbf{f}^{-1}$ in place of $\mathbf{f}$.

3.5.4 $N(d) = [\ln(2/d)/\ln 5]$.

3.5.5 Cf Propositions 3.5.3–5.

3.5.6 (b) Consider $h: \Sigma \to \Sigma$ defined by $h(\sigma)_n = \sigma_{-(n-1)}$.

3.5.7 Observe that if $\alpha^q(\sigma) = \sigma$ then σ is periodic with period-q' where $q'|q$. 335 period-12 orbits.

3.5.9 Λ is repeated within itself on all scales.

3.5.10 Fixed points: $(-1/3, 1/3)$, $(1/2, -1/2)$.
Period-2 points: $(4/13, 6/13)$, $(-6/13, -4/13)$.

3.5.11 (a) Verify for $\bigcap_{n=0}^{1} \mathbf{g}^n(P_{\sigma_n})$ and use induction. Exercise 3.5.9 gives:
(i) square of side $2/5$, centre $(x, y) = (2/5, -2/5)$;
(ii) square of side $2/5^2$, centre $(x, y) = (-8/25, -12/25)$;
(iii) square of side $2/5^3$, centre $(x, y) = (38/125, 38/125)$.
(b) $\sigma = \{\ldots \nu_{-(N-1)}, \ldots, \nu_N, \eta_{-(N-1)}, \ldots, \eta_0 \cdot \eta_1, \ldots, \eta_N, \ldots\}$.
(c) Central block of $2N$ symbols must be preserved for k shifts of binary point to the left, i.e.

$$\sigma = \{\ldots \underbrace{i, \ldots, i}_{k}, \underbrace{i, \ldots, i}_{N} \cdot \underbrace{i, \ldots, i}_{N}, \ldots\}, i = \{0, 1\}.$$

Maximum number of blocks that can be reached in k-iterations is 2^k.
(d) 'Chaotic' motion.

3.5.12 Note that if $(x, y) \in \Lambda$ is given by $\mathbf{h}(\sigma)$, then x is determined by the part of σ lying to the left of the binary point. If Λ_1 denotes the invariant Cantor set of f_1, then show that $|f_1(x) - f_1(x')| = 5|x - x'|$ for any $x \in \Lambda_1$, $x' \notin \Lambda_1$, i.e. f_1 is locally repelling at each point of Λ_1.

3.5.13 Show that σ_1 and σ_2 both represent the point $(\frac{1}{2}, 0)$. The map $\mathbf{h}$ fails to be injective at points corresponding to the dyadic fractions $(m_1/2^{n_1}, m_2/2^{n_2})$, $m_i, n_i \in \mathbb{Z}^+, i = 1, 2$ (see Arnold & Avez, 1968, p. 125). If these points are disregarded, the symbolic dynamics can be used to show that the periodic orbits are dense in T^2.

3.6.1 $\dot{\gamma}(0) = (1, 1)^{\mathrm{T}}$; $(\mathbf{g} \cdot \gamma)(0) = (2, 0)^{\mathrm{T}}$.

3.6.2 Overlap map $h_{12}(x_1) = 4/x_1$. (b) For $v \in TU_{2x_2}(=\mathbb{R})$, $\|v\|_{x_2} = 4|v|/x_2^2$ on $U_2 \backslash P_2$, where $|\cdot|$ is the Euclidean norm of $\cdot$.

3.6.3 Let $(U_i, \mathbf{h}_i)$, $i = 1, 2$, be overlapping charts and assume that $\mathbf{h}_{12} = \mathbf{h}_2\mathbf{h}_1^{-1}: U_1 \to U_2$ is C^1. Show that $\mathrm{D}(\mathbf{h}_1\mathbf{f}\mathbf{h}_1^{-1})(\mathbf{h}_1(\mathbf{x}))$ and $\mathrm{D}(\mathbf{h}_2\mathbf{f}\mathbf{h}_2^{-1})(\mathbf{h}_2(\mathbf{x}))$ are similar. If $\mathbf{f}(\mathbf{x}^*) = \mathbf{x}^*$ the eigenvalues of the tangent map $T\mathbf{f}_{\mathbf{x}^*}$ can be unambiguously defined to be those of its local representatives. This means that a fixed point on M is hyperbolic if all its local representatives are hyperbolic in the sense of Definition 2.2.1.

3.6.4 Chart T^2 with π and define $\|T\mathbf{f}_\theta^n(\mathbf{v}_\theta)\| = |\mathrm{D}\bar{\mathbf{f}}^n(\mathbf{x})\mathbf{v}_\mathbf{x}|$, $\mathbf{v}_\theta \in TT_\theta^2$ and $\mathbf{v}_\mathbf{x} = T\pi_\mathbf{x}^{-1}\mathbf{v}_\theta$. $\mathbf{A}$ has eigenvalues $\lambda_1 = (3 + 5^{1/2})/2$, $\lambda_2 = (3 - 5^{1/2})/2$. Take $\mu = |\lambda_1|^{-1} = |\lambda_2|$, $C = 2$ and $c = \frac{1}{2}$.

3.6.6 (a) $\mathbf{f}_1$ is an Anosov automorphism; Theorem 3.6.1 applies; T^2 connected implies there is only one basic set $\Omega_1 = T^2$.
(b) $\mathbf{f}_2$ has no periodic points but $\Omega = T^2$ (note $\mathbf{f}_2^{2n}(x, y) = (x, y + 2n(3^{1/2}))$ mod 1, $n \in \mathbb{Z}$); Theorem 3.6.1 is not satisfied.
(c) Four hyperbolic fixed points, $P_1, \ldots, P_4$; Theorem 3.6.1 applies; $\Omega = \{P_1, P_2, P_3, P_4\}$.

3.7.2 Recall: a Cantor set is a closed, uncountable set with empty interior such that every point is an accumulation point.

3.7.5 $k = 4\sin^2(\alpha/2)$.

3.7.6 Note $(\mathbf{f}\dot{\cdot}\Gamma)(0)$ points into the image of γ under $\mathbf{f}$ and $\mathbf{Au} \wedge \mathbf{Av} = \mathrm{Det}(\mathbf{A})(\mathbf{u} \wedge \mathbf{v})$.

3.7.7 $\mathbf{x}^* = \mathbf{h}(\sigma^*)$, $\sigma^* = (\ldots \vdots \sigma^{(q)} \vdots \cdot \sigma^{(q)} \vdots \sigma^{(q)} \vdots \ldots)$. Let $\mathbf{x}^\dagger = \mathbf{h}(\sigma^\dagger)$ where

$$\sigma^\dagger = (\ldots \vdots \sigma^{(q)} \vdots \sigma^{(q)} \vdots \sigma_1 \cdot \underbrace{0, \ldots, 0}_{k} \vdots \sigma^{(q)} \vdots \ldots),$$

where $\sigma^{(q)} = \sigma_1\sigma_2$ and σ_1 is a subblock of $\sigma^{(q)}$ containing $q - k$ symbols. Show that $\alpha^{nq}(\sigma^\dagger) \to \sigma^*$ and $\alpha^{-nq}(\sigma^\dagger) \to \alpha^k(\sigma^*)$ as $n \to \infty$. No.

3.8.1 $\mathbf{Au} \wedge \mathbf{v} + \mathbf{u} \wedge \mathbf{Av} = (\mathrm{Tr}\mathbf{A})\mathbf{u} \wedge \mathbf{v}$.

3.8.3 $M(\theta_0) = 2^{1/2}\pi\omega \sin(\omega\theta_0)\,\mathrm{sech}(\pi\omega/2)$.

3.8.5 (E3.2) is an autonomous system.

3.8.6 Transverse heteroclinic points for $b/a < \frac{1}{4}\pi\omega\,\mathrm{sech}(\pi\omega/2)$.

Chapter 4

4.1.1 (a) Terms of order $r \geqslant 3$ are not removed as in (4.1.14).
(b) Linear terms are not removed when $\mu_1 \neq 0$. Transformed system is not an unfolding of $\dot{y} = -y^2$.

4.1.2 (a) Take $X(\boldsymbol{\mu}, x)$ non-versal and $\mathbf{Y}(\nu, x)$ versal. Show $X \sim Y$ but $Y \nsim X$. For example $X(\boldsymbol{\mu}, x) = x^2$, $Y(\nu, y) = \nu + y^2$.
(b) (i) $X \sim Y$: $h(\mu_0, \mu_1, x) = x - \mu_1/2$, $\varphi(\mu_0, \mu_1) = \mu_0 - \mu_1^2/4$;
$Y \sim X$: $h(\nu_0, x) = y$, $\quad \boldsymbol{\varphi}(\nu_0) = (\nu_0, 0)$.
(ii) Let $x = x^*(\mu_0, \mu_1)$ be fixed point of $\dot{x} = X(\mu_0, \mu_1, x)$. Then
$X \sim Y$: $h(\mu_0, \mu_1, x) = x - x^*(\mu_0, \mu_1)$,
$\boldsymbol{\varphi}(\mu_0, \mu_1) = (\mu_1 - 3x^*(\mu_0, \mu_1)^2, -3x^*(\mu_0, \mu_1))$;

$$Y \sim X: h(\nu_0, \nu_1, y) = y - \frac{\nu_1}{3}, \; \varphi(\nu_0, \nu_1) = \left(\frac{2\nu_1^3}{27} + \nu_0\nu_1, \nu_0 + \nu_1 - \frac{\nu_1^2}{3}\right).$$

4.1.3 Let $x = ay$, then $\dot{y} = Y(\boldsymbol{\eta}, y)$, $Y(\mathbf{0}, y) = ay^2$, becomes $\dot{x} = X(\boldsymbol{\eta}, x) = aY(\boldsymbol{\eta}, a^{-1}x)$ so that $X(\mathbf{0}, x) = x^2$. Note for $a < 0$, $x = ay$ is an order reversing homeomorphism of $\mathbb{R}$. Use versality of $\dot{x} = v + x^2$ and transform back.

4.1.5 (a) $q(\mu, x) = \dfrac{x^2 + \mu^2 x + \mu}{\mu x^3 + x^2 + \mu^2 x + \mu}$ is not continuous at $(\mu, x) = (0, 0)$ for any choice of $q(0, 0)$.

(b) Comparison of coefficients yields $b_0(\mu) \equiv 1$ or $b_0(\mu) = -\mu^2$. Latter implies $s_1(\mu)$ not defined at $\mu = 0$ and therefore certainly not smooth on neighbourhood of $\mu = 0$. Hence $b_0(\mu) \equiv 1$, $q(\mu, x) = (1 + \mu x)^{-1}$, $s_1(\mu) \equiv 0$, $s_0(\mu) = \mu$.

(c) Let $\hat{q}(\mu, x) = \dfrac{q(\mu, x)}{(1 + \mu x)}$, $\hat{s}_0(\mu) = s_0(\mu)$, where $q(\mu, x)$ and $s_0(\mu)$ are given in (E4.7).

4.1.6 (a) Take $G(\boldsymbol{\mu}, x) = x^k$ in Mather Division Theorem. Set $\boldsymbol{\mu} = \mathbf{0}$, differentiate k times and conclude $Q(\mathbf{0}, 0) = 1/q(\mathbf{0}, 0) = g(0)$.

(b) Let $X(\boldsymbol{\mu}, x)$ be any unfolding of $\dot{x} = -x^k$ and take $F = X$ in (E4.14). Since $Q(\mathbf{0}, 0) = -1$, $X(\boldsymbol{\mu}, x)$ is equivalent to family induced by (E4.15) with $\boldsymbol{\varphi}(\boldsymbol{\mu}) = (-s_0(\boldsymbol{\mu}), \ldots, -s_{k-1}(\boldsymbol{\mu}))^{\mathrm{T}}$.

(i) Let $y = x - \dfrac{v_{k-1}}{k}$ in (E4.15).

(ii) For k odd, right hand side of (E4.15) has at least one real zero.

4.2.1 (a) (iv), (v); isoclines are tangent to each other but neither is tangent to either coordinate axis.

(b) (iv) and (v); (v).

4.2.2 If $\dot{y} = 0$ then $2yx^2 - x - (2y - 2y^3 - 1) = 0$. Take $y \neq 0$ and examine the discriminant of this quadratic equation. Equation (E4.19) arises in connection with the averaged forced Van der Pol oscillator (see Arrowsmith & Place, 1984).

4.2.3 (a) Observe distance, $\mathrm{d}(\boldsymbol{\mu}, x)$, between $\dot{x} = 0$ and $\dot{y} = 0$ isoclines satisfies $\mathrm{d}(\mathbf{0}, x) = x^2$ and use Malgrange Preparation Theorem;

(b) $(y - x^3 + \mu x, y)^{\mathrm{T}}$;

(c) $(y - x^4 + \mu x^3, y)^{\mathrm{T}}$, $(y - x^4 + 3\mu x^2 - 2\mu^2, y)^{\mathrm{T}}$, for hyperbolic points isoclines must intersect transversely – three such intersections are not possible for $\mathrm{d}(\mathbf{0}, x) = x^4$.

4.2.4 Show that $\psi(x)$ has a unique maximum on $(0, 1]$ at $x = \frac{1}{2}$.

4.2.5 (a) $\dot{P} = 0$ isocline is given by an increasing function of H bounded by γ with slope γ/β at $H = 0$. The $\dot{H} = 0$ isocline is independent of β and γ. A single fixed point arises at tangential intersection of $\dot{H} = 0$ and $\dot{P} = 0$ isoclines.

(b) Find where the non-trivial $\dot{a} = 0$ and $\dot{p} = 0$ isoclines meet in a single point. More details of both of these models appear in Arrowsmith & Place (1984).

4.2.6 (a) Evaluate $L_{\mathbf{A}} x_1^{m_1} x_2^{m_2} \mathbf{e}_i$, $i = 1, 2$, where $\mathbf{e}_i$ is the ith column of $\mathbf{I}_{m+2}$ and $\mathbf{A}$ is the coefficient matrix of the linear part of the extended vector field in (4.2.16). Use (2.3.7).

(b) Consider $L_{\mathbf{A}} x_1 \mu_k \mathbf{e}_2$ and $L_{\mathbf{A}} x_2 \mu_k \mathbf{e}_1$.

(c) Apply Theorem 2.7.2 to the 2-jet of the transformed extended system to obtain equivalent system with $\dot{x}_1 = \lambda x_1$. In absence of terms of order three and higher, $\dot{x}_2$ depends only on x_2 and $\boldsymbol{\mu}$. Complete the square on x_2 to obtain (4.2.2) with $a_2 = 0$, $b_2 = c_{23}$, $v = P(\boldsymbol{\mu}) - \dfrac{(Q(\boldsymbol{\mu}))^2}{4c_{23}}$, where

$$P(\boldsymbol{\mu}) = \sum_{j=1}^{m} a_{2j}\mu_j + O(|\boldsymbol{\mu}|^2), \qquad Q(\boldsymbol{\mu}) = \sum_{j=1}^{m} \left(b_{2j}^{(2)} - \frac{c_{22}}{\lambda} a_{1j} \right) \mu_j.$$

4.2.7 Note Proposition 4.2.2 requires only the terms $O(\kappa)$ and $O(y_2^2)$ in the second component of form (4.2.13).

4.2.8 (a) (4.2.20) fails, there is a limit cycle for all $\mu \neq 0$;
(b) (4.2.20) fails and origin not asymptotically stable at $\mu = 0$, two limit cycles for $\mu > 0$;
(c) (4.2.20) fails, limit cycle for $\mu > 0$;
(d) Not asymptotically stable for $\mu = 0$, no limit cycles for $\mu \neq 0$.

4.2.9 (E4.28) undergoes subcritical Hopf bifurcation at $\mu = 0$.

4.2.10 $\mathbf{x} = \mathbf{M}\mathbf{x}'$, $\mathbf{M} = \begin{pmatrix} 0 & 1 \\ -1 & 1 \end{pmatrix}$, reduces linear part to real Jordan form, $a_1 = -\frac{1}{8}$.

4.2.11 Use Theorem 4.2.1 and (4.2.24). (E4.30) undergoes supercritical Hopf bifurcation at $\mu = 0$. Time reversal of (E4.31) undergoes supercritical Hopf bifurcation at $\mu = 0$, hence (E4.31) undergoes subcritical bifurcation with decreasing μ.

4.2.12 Use Theorem 4.2.1 and (4.2.24) at $b = b^*$, $\mathbf{x} = \mathbf{M}\mathbf{x}'$, with $\mathbf{M} = \begin{pmatrix} 0 & a^2 \\ a & -a^2 \end{pmatrix}$, reduces the linear part of (E4.32) to real Jordan form.

4.3.1 The linear part of $\dot{\mathbf{y}}$ is in real Jordan form and Proposition 4.2.2 can be applied. Moreover $\dot{y}_2 = \dfrac{v_1}{v_2} + \dfrac{b_2}{v_2} y_2^2 + \cdots$ and fixed points occur for $v_1 < 0$ whatever the sign of v_2.

4.3.2 (a) $\dfrac{\mathrm{d}}{\mathrm{d}v_2}(\mathrm{Re}\, \lambda(v_2))|_{v_2^*} = 1 > 0$, $16a_1 = \dfrac{2a_2}{\varepsilon} < 0$, $\varepsilon = \left(\dfrac{-v_1}{b_2}\right)^{1/2}$.
(b) $\dfrac{\mathrm{d}}{\mathrm{d}(-v_1)}(\mathrm{Re}\, \lambda(v_1))|_{v_1^*} = -\dfrac{v_2}{b_2} > 0$, a_1 as in (a).

4.3.4 At non-trivial fixed point, $\mathbf{x}^* = (x^*, y^*)^{\mathrm{T}}$, $y^{*2/3} = \lambda_1(1 + x^*)/(1 - x^*)$. Tr $\mathbf{DX}(\mathbf{x}^*) = 0$ implies $\lambda_2^* = 4\lambda_1^*$, Det $\mathbf{DX}^*(\mathbf{x}^*) = 0$ implies $27\lambda_1^*\lambda_2^{*2} = 1$.

4.3.5 Transformed 2-jet is

$$\begin{pmatrix} -\dfrac{1}{2\varepsilon}\mu_1 + y_2 - \mu_1 y_1 - \frac{8}{3}\varepsilon^2 y_1^2 - 2\varepsilon y_1 y_2 - \frac{2}{3} y_2^2 \\ \frac{1}{2}\mu_1 + \frac{1}{4}\mu_2 + \varepsilon\mu_1 y_1 + \varepsilon\mu_2 y_1 + \mu_2 y_2 + 2\varepsilon^3 y_1^2 + 2\varepsilon^2 y_1 y_2 + 2\varepsilon y_2^2 \end{pmatrix}.$$

Identify $[a_{ij}]$, $[b_{ij}^{(1)}]$, $[b_{ij}^{(2)}]$, $[c_{ij}]$.
(a) $a = -\frac{5}{3}\varepsilon^2 < 0$, $b = 2\varepsilon^3 > 0$.
(b) $e_1 = \frac{1}{6}$, $e_2 = \frac{13}{12}$, $e_1 a_{22} - e_2 a_{21} = -\frac{1}{2}$. $a < 0$, $b > 0$ implies stable limit cycles.

4.3.6 Note $\lambda_2 = [xy^{-1/3}/(1 + x)]_{\mathbf{x}^*}$ for any non-trivial $\mathbf{x}^*$ of (E4.20).
(a) $\mathbf{x}^* = (\frac{1}{2}, y^*)^{\mathrm{T}}$, given by tangential intersection of non-trivial isoclines when $27\lambda_1\lambda_2^2 = 1$ in Exercise 4.2.4. Tr $\mathbf{DX}(\mathbf{x}^*) \neq 0$ provided $\lambda_2 \neq 4\lambda_1$, i.e. $\mathbf{x}^*$ not the cusp point.
(b) Tr $\mathbf{DX}(\mathbf{x}_1^*) = 0$ with Det $\mathbf{DX}(\mathbf{x}_1^*) > 0$ implies $\mathbf{x}_1^* = (x_1^*, y_1^*)^{\mathrm{T}}$ such that $x_1^* < \frac{1}{2}$. Exercise 4.2.4 gives $\psi(x_1^*) = \lambda_1\lambda_2^2$, $0 < x_1^* < \frac{1}{2}$. Solve simultaneously with Tr $\mathbf{DX}(\mathbf{x}_1^*) = 0$ and use $x_1^* = s$ as a parameter.

4.3.7 Equate (E4.42) and (E4.43) and compare coefficients of powers of $(x_1^2 + x_2^2)$. Scaling is $r' = (|a_l'|)^{1/2l} r$, $\theta' = \theta$. Take time reversal of $(l, +)$ and let $\theta \mapsto -\theta$.

4.3.8 Non-trivial zeroes of $\dot{r}$ are given by $\rho^2 + \nu_1\rho + \nu_0 = 0$, $\rho = r^2$.

4.3.9 For $\nu_2 \geqslant 0$ the double limit cycle bifurcation curve is defined for $\nu_1 < 0$ and is given by $\nu_0 = (-\nu_1\rho - \nu_2\rho^2 - \rho^3)|_{\rho_{\min}}$, where $\rho_{\min} = -\frac{\nu_2}{3} + \left(\frac{\nu_2^2}{9} - \frac{\nu_1}{3}\right)^{1/2}$. This represents a curve in the $\nu_1\nu_0$-plane which has zero slope at the origin for all $\nu_2 \geqslant 0$. However it only reduces to $\nu_0 = 2(-\nu_1/3)^{3/2}$ for $\nu_2 = 0$.

4.3.11 Double limit-cycle bifurcations occur when $P_\nu(\rho_e) = 0$, where ρ_e is a local maximum or minimum of P_ν.

4.4.3 $f(\mu, -x) = -f(\mu, x)$ implies $a_{2k}(\mu) = 0$, $k = 0, 1, 2, \ldots$. $a_3(0) = \frac{1}{6}\mathrm{D}_{xxx}f(\mathbf{0})$.

4.4.5 Let $x = y + \beta y^2 = K(y)$, $\beta = b_2(\mu)/[(\nu(\mu) - 1)(\nu(\mu) - 2)]$ and show that $\tilde{F}(\mu, y) = K^{-1}(F(\mu, K(y)))$. Note $b_3(\mu) = \frac{1}{6}\mathrm{D}_{xxx}f(\mathbf{0}) + O(\mu)$, $b_2(\mu) = \frac{1}{2}\mathrm{D}_{xx}f(\mathbf{0}) + O(\mu)$.

4.5.4 Show that $\mathrm{D}F_\rho^3(x^*) = \mathrm{D}F_\rho^3(F_\rho(x^*))$ if $F_\rho^3(x^*) = x^*$.

4.5.5 (a) Well defined: let $\pi(x) = \pi(x')$, show that $f(\pi(x)) = f(\pi(x'))$. Non-injective: $f(\pi(\frac{1}{2})) = f(\pi(0))$ but $\pi(\frac{1}{2}) \neq \pi(0)$.
(i) Let $x = 0 \cdot b_1 \ldots b_n \ldots \in \Sigma'$, define $y_n = 0 \cdot \overline{b_1 \ldots b_n}\, \overline{b_1 \ldots b_n} \ldots$, $|x - y_n| < 2^{-n}$ and $\lim_{n\to\infty} y_n = x$. Use continuity of π and $\pi(y_n)$ periodic to obtain density.
(ii) See discussion of Proposition 3.5.5.
(b) Φ continuous and onto implies (i) and (ii).
(iii) f doubles lengths of intervals on S^1. Note $\varepsilon = \frac{1}{2}$ is maximum possible separation for every $x \in [0, 1]$.

Chapter 5

5.1.1 $\lambda_2 = 1$, $\tilde{\mathbf{f}}(\mu, \mathbf{x}) = (\mathrm{sgn}(\lambda_1)2x, \mu + y + y^2)^{\mathrm{T}}$;
$\lambda_2 = -1$, $\tilde{\mathbf{f}}(\mu, \mathbf{x}) = (\mathrm{sgn}(\lambda_1)2x, (\mu - 1)y + \delta y^3)^{\mathrm{T}}$, $\delta = \pm 1$;
Reduce $\mathbf{f}$ to form $(\lambda_1 x, g(\mu, \mathbf{x}))^{\mathrm{T}}$ then $\delta = \mathrm{sgn}(-\mathrm{D}_{xxx}g^2(0, \mathbf{0}))$.

5.1.2 $\mathrm{Tr}\,\mathbf{A} = 0$, $\mathrm{Det}\,\mathbf{A} = -1$ implies $bc = 1 - d^2$ which defines a family of parabolae in bcd-space for $b \neq 0$. The 2-dimensional manifold is charted by:

$$(b, d) \mapsto (-d, b, [1 - d^2]/b, d) \text{ for } b \neq 0;$$

$$(c, d) \mapsto (-d, [1 - d^2]/c, c, d) \text{ for } c \neq 0;$$

$$(b, c) \mapsto (-d, b, c, d), \text{ with } d = \pm(1 - bc)^{1/2}, \text{ for } |bc| < 1.$$

5.1.3 For normal form, use the resonance condition $\lambda_1^{m_1}\lambda_2^{m_2} - \lambda_1 = 0$, $i = 1, 2$, with $\lambda_1 = 1$ and $\lambda_2 = -1$. For bifurcational behaviour, consider each quadrant of the $\boldsymbol{\mu}$-plane separately and use the results in §4.4.

5.1.4 (b) Let $(x', y') = (\mu x, \mu y)$ in $\mathbf{f}_1(x, y)$.

5.1.5 Introduce new coordinates $(b + c, b - c, d)$ in bcd-space. Find which points of the double cone correspond to the different possible Jordan forms for $\mathbf{A}$.

5.1.6 Consider $\mathbf{f}(\mu, \mathbf{x}) - \mathbf{x} = \mathbf{0}$ and use the Implicit Function Theorem. Give counter-examples when the hypotheses of this theorem are not satisfied.

5.2.1 (a) Use induction.
(b) $\bar{f}^k_{(\alpha,\varepsilon)}(x + 1) = \bar{f}^k_{(\alpha,\varepsilon)}(x) + 1$.
(E5.8) implies $T_{p/q}$ is well defined.

5.2.3 Introduce complex form for cos, sin and use results in Exercise 5.2.2.

5.2.4 Prove that $\bar{f}^n_{((1-\alpha),\varepsilon)}(x) = -\bar{f}^n_{(\alpha,\varepsilon)}(-x) + n$ and use (1.5.18).

5.3.1 Invariant circles exist for $|a(\mu)|^2|z|^4 + 2c(\mu)|z|^2 + (1 - (1/|\lambda(\mu)|^2)) = 0$ which has roots $|z|^2 = r_0(\mu)^2$ and $|z|^2 = r_1(\mu)^2$ where $r_1(0) > 0$. Neighbourhood for local bifurcation can always be chosen to exclude circle radius $r_1(\mu)$.

5.3.2 (a) Show that $|f_\mu(z)| = |z|(1 \pm (c + O(\mu))\varepsilon + O(\varepsilon^2))$ when $|z| = r_0(\mu) \pm \varepsilon$, $\varepsilon > 0$.
(b) Find $\arg(1 + a(\mu)r_0(\mu)^2)$.
(c) Yes, but not relevant.
(d) As μ increases through zero, fixed point at $z = 0$ changes stability and stable invariant circle surrounding it exists for $\mu > 0$.

5.3.3 (a) (i) $\mu = 0$: diagonalise, use resonance condition with $\lambda_1 = \bar{\lambda}_2 = \lambda(0) = \exp(2\pi i\beta)$ to generalise (5.3.1) to arbitrary order in $|z|$; (ii) $\mu \neq 0$: $\lambda(0) \mapsto \lambda(\mu)$, no resonant terms, invoke smoothness in μ.
(b) Require $|\lambda(\mu)|\left|\left(1 + \sum_{m=1}^{N} \tilde{a}_{2m+1}(\mu)|z|^2\right)\right| = 1$ and use roots of polynomial depend continuously on its coefficients. Let $r_0^2(\mu) = -\dfrac{b\mu}{c} + E(\mu)$ and determine order of $E(\mu)$ in μ.

5.3.4 Consider Det $\mathrm{D}_{\boldsymbol{\mu}}\boldsymbol{\lambda}(\boldsymbol{\mu})$, $\boldsymbol{\lambda}(\boldsymbol{\mu}) = (\mathrm{Re}\,\lambda(\boldsymbol{\mu}), \mathrm{Im}\,\lambda(\boldsymbol{\mu}))^{\mathrm{T}}$;

$$\lambda(\boldsymbol{\mu}) = \tfrac{1}{2}\{\mathrm{Tr}\,\mathrm{D}\mathbf{f}_{\boldsymbol{\mu}}(\mathbf{0}) + i(4\mathrm{Det}\,\mathrm{D}\mathbf{f}_{\boldsymbol{\mu}}(\mathbf{0}) - [\mathrm{Tr}\,\mathrm{D}\mathbf{f}_{\boldsymbol{\mu}}(\mathbf{0})]^2)^{1/2}\};$$

$$\lambda_0(\mathbf{v}) = (1 + v_1)\exp(2\pi i(\beta + v_2)).$$

(a) Cf derivation of (5.3.8); (b) estimate $\arg(1 + \tilde{a}_3(\mathbf{v})|z|^2 + R_1(\mathbf{v}, \bar{z}))$, $R_1(\mathbf{v}, z) = O(|z|^4)$, when $z = r_0(\mathbf{v}, 2\pi\theta)\exp(2\pi i\theta)$, $r_0(\mathbf{v}, 2\pi\theta) \ll 1$.

5.4.1 Obtain $r = m_1 + m_2$ using (5.4.3) and (5.4.4) and take minimum over m and l. Retain inevitable resonance terms up to and including $O(|z|^{q-1})$.

5.4.2 Remove factor of $\lambda(\mathbf{v})z\left(1 + \sum_{m=1}^{M} \dfrac{a_{2m+1}(\mathbf{v})}{\lambda(\mathbf{v})}|z|^{2m}\right)$, $M = [\tfrac{1}{2}(q-2)]$, in (5.4.6), substitute $z = r\exp(2\pi i\theta)$, $\tilde{f}_{\mathbf{v}}(z) = R\exp(2\pi i\Theta)$ and take logarithms. Observe that when $x, y = O(r^{q-2})$, $\ln(1 + x + iy) = \ln(1 + x + O(r^{q-1})) + iy + O(r^{q-1})$.

5.4.3 (a) Periodic points are given by $\Theta(\theta) - (\theta + p/q) = 0$, $\theta \in [0, 1)$.
(b) Tongues narrow as q increases.

5.4.4 Apply Theorem 5.4.2 and evaluate $c_3(0)$ in (5.4.14); attracting.

5.5.2 Express (5.5.10) and (E5.24) as autonomous differential equations on $\mathbb{R}^3$; corresponding flows satisfy $\mathbf{h}\hat{\boldsymbol{\varphi}}_t = \boldsymbol{\varphi}_t\mathbf{h}$ for all t, where $\mathbf{h}$ is obtained from $z = \exp(i\beta t)\zeta$. Take $t = 2\pi q$ and observe that $\hat{\boldsymbol{\varphi}}_{2\pi q} = \hat{\mathbf{P}}_0^q$, $\boldsymbol{\varphi}_{2\pi q} = \mathbf{P}_0$.

5.5.3 For resonance, use $a_{\mathbf{m}}(t) = \sum_{\nu=-\infty}^{\infty} a_{\mathbf{m},\nu}\exp(i\nu t)$ in (E5.28) and do integral over t. For symmetry consider $\exp(2\pi i/q)\tilde{Z}_r(\zeta\exp(-2\pi i/q))$.

5.5.4 (a) Assume $z_j = \lambda^j z + \sum_{r=2}^{\infty}\sum_{s=0}^{j-1} \lambda^{j-1-s}M_r(z_s, \bar{z}_s)$ and use induction, set $j = q$;
(b) Use (5.5.21). To deduce (5.5.22) equate $|z|^r$-terms in $M^q(z) = N(z)$. $|z|^r$- terms in M^q are of two types, those depending on: (i) $c_{\mathbf{m}}^{(k)}$, $k < r$; (ii) $c_{\mathbf{m}}^{(r)}$. Put former in $g_r(z, \bar{z})$ and latter give left hand side of (5.5.22).

5.5.5 Use induction to prove that $M_r(z_j, \bar{z}_j) \in \Sigma_{\mathbb{C}}$ for $j = 0, \ldots, q-1$, $r \geqslant 2$. Hence conclude that $g_r(z, \bar{z}) \in \Sigma_{\mathbb{C}}$ if $N(z)$ does. Equation (5.5.22) then determines M_r uniquely in $\Sigma_{\mathbb{C}}$. If $N(z) \notin \Sigma_{\mathbb{C}}$ then (5.5.23) implies (5.5.22) is not consistent since $g_r(z, \bar{z})$ may then contain terms $\notin \Sigma_{\mathbb{C}}$.

5.5.6 Use induction to prove that $\mathrm{d}^k z/\mathrm{d}t^k \in \Sigma_{\mathbb{C}}$ for all k and use Taylor expansion of $\varphi_t(z)$.

5.5.7 Consider the effect of $M_{\mathrm{D}\mathbf{f}(\mathbf{0})}$ on H^r (cf § 2.5) for r even and r odd.

5.5.8 Find $\mathrm{D}\boldsymbol{\varphi}_t^0(\mathbf{x})$ for $t \neq 0$ and use (2.5.6) to find $B_i = M_{\mathrm{D}\boldsymbol{\varphi}_t^0(\mathbf{x})}(H^i)$, $i \geqslant 2$. Use scaling of y and π-rotation symmetry to obtain (E5.33). Assume two normal forms exist and show that corresponding coefficients must be equal.

5.6.1 Obtain (E5.34) and (E5.35) for $\delta = +1$ and -1 separately and combine results.

5.6.2 (E5.36) and (E5.37) suggest γ_3 in Figure 5.11 has slope $-a_1^{-1}$ at $\mathbf{v} = \mathbf{0}$.

5.6.3 Apply Theorem 4.2.1 and use (4.2.24) to show that $a_1 < 0$. Implies stable limit cycles in region (d) of Figure 5.12.

5.6.4 $\zeta = \alpha\zeta'$, where $\alpha = |d|^{-1/(q-2)} \exp(\mathrm{i}[\arg(d)/q])$.

5.6.5 Observe that $(x - \mathrm{i}y)^{q-1}$ has real and imaginary parts that are linear combinations of $x^{m_1} y^{m_2}$ with $m_1 + m_2 = q - 1$.

5.6.6 Note that $r^{-1} X_r = r^{-1} X_\theta = 0$ at a non-trivial fixed point.

5.6.7 Let $\zeta^2 = w = x + \mathrm{i}y$.

5.6.8 Note $r^{-1} X_r = r^{-1} X_\theta = 0$ at a non-trivial fixed point.

5.6.9 $$\Xi(v_1, v_2) = (v_1, v_2) \begin{pmatrix} 1 - b^2 & ab \\ ab & 1 - a^2 \end{pmatrix} \begin{pmatrix} v_1 \\ v_2 \end{pmatrix}.$$

5.6.10 Note solutions to $\Xi = \left\{ \xi_2 + \dfrac{v_1}{2a}(1 - |c|^2) \right\}^2$ give (v_1, v_2) for which
$$\xi_2 + \frac{v_1}{2a}(1 - |c|^2) = \pm \Xi^{1/2}.$$

5.6.11 For given a, let $\varphi(b)$ be the difference in slopes of (E5.49) and the $\Xi = 0$-line lying in the positive quadrant. Prove that $\varphi'(b) < 0$ for $b < -(1 - a^2)^{1/2}$ and $\varphi(-(1 - a^2)^{1/2}) > 0$.

5.6.13 (c) Plot suggests (i) saddle connection in Figure 5.17(f); (ii) saddle-nodes can be created inside the Hopf invariant circle (see Figure 5.17(m)).

5.6.14 (a) $\zeta = \alpha\zeta'$ where $|\alpha|^2 = |a_1|^{-1}$, $\arg(\alpha) = q^{-1} \arg(d)$. (c) When $v_2 = -b_1 v_1$ every point of $r = v_1^{1/2}$ is a fixed point. Therefore none are hyperbolic.

5.6.15 Fixed points at
$$(\theta, r) = \begin{cases} \left(v_1^{1/2} + \dfrac{\mathrm{d}v_1}{2} \cos 5\theta_0' + O(v_1^{3/2}),\ \theta_0' + \dfrac{2p\pi}{5} \right) \\ \left(v_1^{1/2} - \dfrac{\mathrm{d}v_1}{2} \cos 5\theta_0' + O(v_1^{3/2}),\ \theta_0' + \dfrac{(2p+1)\pi}{5} \right) \end{cases}, \quad p = 0, 1, \ldots, 4,$$
where $5\theta_0' = 5\theta_0 + O(v_1^{1/2})$ and $5\theta_0$ is the principal value of $\tan^{-1}(b_1)$.

5.6.16 Note $r^{-1} X_r = r^{-1} X_\theta = 0$ at a non-trivial fixed point.

5.6.17 Show that, since $b_1 = -1$, the isoclines are related by a $\pi/10$-rotation.

5.6.18 Show that $(\pi/5, r_m)$ is a point of closest approach to the origin on the $\dot{r} = 0$ isocline. The trajectory of such a point is not easy to approximate numerically since $\dot{r} = 0$ and $\dot{\theta}$ is small. Neither $\dot{r}$ nor $\dot{\theta}$ are especially small at, for example, $(3\pi/10, r_m)$.

5.6.19 In the $q = 5$ case fixed points only appear *on* the invariant circle. The angular dependence of the isoclines for $q = 5$ is weaker than it can be for $q = 4$ because the θ-dependent terms are of lower order in r. For $q = 4$, the effect of the θ-dependent terms is determined by $|a|$. For $q = 5$, it is always limited by being of lower order in r.

5.7.1 Use Taylor expansion of $r(t)$ and $\theta(t)$ about $t = 0$. Apply Theorems 4.2.1 and 5.4.2. If $a_1 > 0$ both $\tilde{\mathbf{X}}$ and $\tilde{\mathbf{P}}$ undergo subcritical bifurcations.

5.7.2 Show that restriction of family of time-2π maps to y_2-axis generically undergoes a fold bifurcation. Obtain conditions required by Proposition 4.4.1 from hypothesis that (E5.56) undergoes saddle-node bifurcation and Proposition 4.2.2.

5.7.3 $O(|\mathbf{x}|^N)$-terms in (5.7.1) generically lead to transverse intersection of stable and unstable manifolds of saddle points. Saddle-connection bifurcation curve replaced by wedge of parameter values where homoclinic or heteroclinic tangles occur (see Figures 5.21 and 5.22).

5.7.4 No; heteroclinic tangles do not occur on the edge of the tongue for $q = 5$ since the saddle-nodes lie on the invariant circle. Periodic points occur on the invariant circle for $q = 5$ but need not do so for $q = 4$ (see Section 6.7).

Chapter 6

6.1.1 (a) Use (1.9.14); (b) let $\mathbf{U}$ simultaneously diagonalise $\mathbf{B}$ and $\mathbf{A}^{-1}$; (c) $\bar{q}_i = c_i^{-1} q_i$, $\bar{p}_i = c_i p_i$, $i = 1, \ldots, n$, with $c_i = (\mu_i/\lambda_i)^{1/4}$ is symplectic. See also Arnold, 1978, Appendix 6.

6.1.2 (a) (1.9.14) with $n = 1$ if and only if $\mathrm{Det}(\mathbf{Df}(\mathbf{x})) \equiv 1$; (b) let

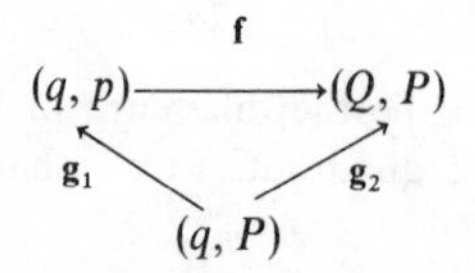

and differentiate $\mathbf{f} \cdot \mathbf{g}_1 = \mathbf{g}_2$.

6.1.4 Recall $p = \mathrm{D}_q \mathrm{F}_2$, $\mathrm{Q} = \mathrm{D}_p F_2$ and note $2N - 3 \geqslant N$ for $N \geqslant 3$.

6.1.6 $(Z + O(|Z|^n))^k (\bar{Z} + O(|Z|^n))^l = Z^k \bar{Z}^l + O(|Z|^{n+1})$ for $k + l \geqslant 2$.

6.1.7 (a) $\mathrm{Ln}(f(r \exp(i\theta))) = \ln(r_1) + i\theta_1$; (b) let $(x, y) = (r \cos\theta, r \sin\theta) = \boldsymbol{\varphi}(\theta, r)$, differentiate $\boldsymbol{\varphi}(\mathbf{F}(\theta, r)) = \mathbf{f}(x, y) = \mathbf{f}(\boldsymbol{\varphi}(\theta, r))$.

6.2.1 (a) Use induction on the form of $\mathbf{M}_\varepsilon^k$. For $\varepsilon = 0$, map restricted to $\tau = \tau_0$ is simply a rotation. Apply Implicit Function Theorem to $\theta_q = \theta + 2\pi p$ at $(\theta, \tau, \varepsilon) = (\theta_0, \tau_0, 0)$ to obtain $\mathscr{C}$ locally as $\tau = F(\theta, \varepsilon)$. Global curve $\mathscr{C}$ can be obtained from the submersion $G: S^1 \times \mathbb{R}^2 \to \mathbb{R}^2$ defined by $(\theta, \tau, \varepsilon) \to (q\alpha(\tau) + \tilde{f}_q(\theta, \tau, \varepsilon), \varepsilon)$.
(b) Show $\mathscr{C}$ and $\mathbf{M}_\varepsilon^q(\mathscr{C})$ intersect at least twice and that the fixed points are of different topological types. $\mathbf{M}_\varepsilon$ near to a rotation implies $2q$ points.

6.2.2 $\tau = \pi - \frac{1}{2}\sin(\theta)\varepsilon + \frac{1}{8}\sin(2\theta)\varepsilon^2$.

6.2.3 Generic forms:

$$\mathbf{P}_1(x, y) = (x + y + ax^2, y + ax^2)^{\mathrm{T}} + O(|\mathbf{x}|^3),$$
$$\mathbf{P}_2(x, y) = (-x + y + ax^3, -y + ax^3)^{\mathrm{T}} + O(|\mathbf{x}|^5).$$

Invariant curves – use slopes from $(0, 0)$ to (x, y) and (x, y) to: (a) $\mathbf{P}_1(x, y)$; (b) $\mathbf{P}_2^2(x, y)$; are asymptotically equal at 0.

6.2.4 (a) $(X, Y) = (x, x + y)$; (b) $(X, Y) = (x + DR(x + y), x + y)$. Modify generating functions by adding term $-Y^3/3$ to obtain behaviour in Figure 6.2.

6.2.5 H independent of θ implies $\mathrm{d}H/\mathrm{d}\tau$ independent of t. Integration of $\dot{\theta} = -\mathrm{d}H/\mathrm{d}\tau$ gives $\mathrm{d}H/\mathrm{d}\tau = -\alpha(\tau)$. $\alpha(\tau) = \alpha + \Sigma a_i \tau^i$ for Birkhoff form.

6.2.6 Taylor expand φ_t. Put $t = q$ and compare with given form. Use induction for (c).

6.2.9 Obtain $\mathrm{D}\varphi_t(\mathbf{x}) = \mathbf{I} + t\mathrm{D}\mathbf{X}(\mathbf{x}) + O(t^2)$ and $\mathrm{Det}(\mathrm{D}\varphi_t(\mathbf{x})) = 1 + t\,\mathrm{Tr}(\mathrm{D}\mathbf{X}(\mathbf{x})) + O(t^2)$.

6.2.10 Assume formal power series in x, y for X^ε and Y^ε. Use (6.2.35) to find relation between coefficients. Check that formal integrations give same function H_ε. Invariance under $\mathbf{R}_\alpha$: for $q = 2$, X^ε (H_ε) contains only odd (even) order terms; for $q \geqslant 3$, introduce complex coordinates, $z, \bar{z}$.

6.2.11 Use Proposition 5.5.1.

6.3.1 (a) Show maximum term of series occurs at $k = [\mu/r]$; split series at this term. Obtain bound on the finite sum; use the integral test on the infinite series. (b) Substitute Fourier series for analytic functions; use (6.3.9) and (6.3.13) together with (E6.25).

6.3.2 (a) Observe how coefficients of ε^n occur in (6.3.20) with (6.3.18,19) substituted. (b) Use Fourier series to solve (6.3.22) and pick $\hat{v}_{n0}$ to allow solution of (6.3.21). (c) Show that L is linear to obtain second equation of (E6.29).

6.3.3 (a) Show $e_{i+1} = e_i^{4/3}$, $0 \leqslant e_0 < 1$. Find d_{i+1}/d_i in terms of e_{i+1}/e_i. (b) Use (6.3.33) and (E6.30) in (6.3.28) to obtain

$$|f_{i+1}| + |g_{i+1}| < C'[2^{(i+2)\chi}C^{-(i+1)}(1 + d_i^{1/3}) + (d_{i+1}/d_i)^{1/3}]d_{i+1}, \qquad \chi = 2\mu + 3.$$

6.3.4 Use induction and $|p_{i-1}| = \sup(|p_{i-1}(\xi + u_i, \eta + v_i)|)$ for $(\xi, \eta) \in \mathscr{D}_{i+1}$.

6.3.5 Choice ensures that (6.3.41) matches (6.3.19) so that iteration determines $u(\xi, \varepsilon)$, $v(\xi, \varepsilon)$ in (6.3.17) without expanding in powers of ε.

6.3.6 See Siegel & Moser, 1971, pp. 237–8.

6.3.7 (a) Show that $\int_0^{2\pi} \mathrm{D}^N h(x) \exp(ikx)\, \mathrm{d}x = 2\pi \mathrm{i}^N \hat{h}_k k^N$. (b) Show (6.3.13) gives bound of the form $K'/|k|^N$, for every $N \in \mathbb{Z}^+$.

6.4.1 T is nowhere dense if and only if $Cl(T^c) = [0, 1]$. Note that: (i) the truncation of (E6.42) at $j = i$ gives the initial points of the intervals in T_i; (ii) the lengths of these intervals $\to 0$ as $i \to \infty$. Use finite truncations of (E6.42) to show that every point of T is an accumulation point.

6.4.2 Assume that $I_1, \ldots, I_n$ have been positioned on S^1 so that they are in the same relative order as $x_1, \ldots, x_n$ and with positive distance between adjacent intervals. Place I_{n+1} in the same relative position as x_{n+1} and not intersecting adjacent intervals. Density can be ensured by making the interval I_{n+1} of half the length of the gap into which it is inserted. Repeat with negative n.

6.4.3 (a) Π is well-defined if y is unique. For continuity of Π use ordering to show that inverse images of open intervals are open intervals. (*b*) f is homeomorphism on $\bigcup_{n\in\mathbb{Z}} I_n$ by construction and $f=\Pi^{-1}R_\beta\Pi$ on the complement. Obtain continuity for f as with Π above. Note $x\in\tilde{C}$ is an accumulation point for f but $x\in\mathrm{Int}(I_n)$ is not.

6.4.4 (a) Show $\bar{\Pi}(\bar{x}+1)-\bar{\Pi}(\bar{x})\equiv k\in\mathbb{Z}$ and that injectivity of Π on $\left(\bigcup_{n\in\mathbb{Z}} I_n\right)^c$ implies $k=1$. (b) Use $\bar{\Pi}-\mathrm{id}$ periodic to obtain bounds for $\bar{\Pi}(\bar{f}^k(\bar{x}))=\bar{\Pi}(\bar{x})+k\beta$.

6.4.6 (a) Consider $\varphi(x)=f_1(x,0)-f_1(x,1)$ and assume $\varphi(x_1)>0$, $\varphi(x_2)<0$. Use intermediate value theorem for continuous function φ and interpret. (b) (iii) is the only lift of a twist homeomorphism.

6.4.7 (a) Use $f^q\pi=\pi F^q$ at the lift of a periodic point for $F=\bar{f}$ and $\bar{f}'$. (b) Show $\bar{\mathbf{f}}^q(\mathbf{\Theta}(0))=(x+p,y)^{\mathrm{T}}$, $\mathbf{\Theta}(0)=(x,y)^{\mathrm{T}}$.

6.4.8 Give $\mathbf{\Theta}$ in Definition 6.4.2 for each orbit.

6.4.9 (a) Show that radial projection gives a conjugacy between $\mathbf{f}|E$ and an order-preserving homeomorphism g on K. Note K^c is a union of open intervals. Extend g to a circle homeomorphism $G: S^1\to S^1$ by defining $G|K^c$ for each $(a,b)\subseteq K^c$ to be an order-preserving homeomorphism onto $(g(a),g(b))$. (b) Extend lift component $\bar{f}_1$ of $\mathbf{f}$ to provide lift of G. Relate (E6.45) to the rotation number of $G: S^1\to S^1$. (c) $\rho(E)\in\mathbb{Q}$ if and only if G has periodic point, show that this point can be chosen in K. Periodic orbit is Birkhoff since $\bar{f}$ preserves order on the covering of E.

6.5.1 Let $\bar{\mathbf{f}}: \mathbb{R}^2\to\mathbb{R}^2$ be a lift of $\mathbf{f}|\mathbb{R}^2\backslash\{\mathbf{0}\}=S^1\times\mathbb{R}$. Show that if $\bar{\mathbf{f}}^q(\bar{\mathbf{x}}_0)=\bar{\mathbf{x}}_0+(p,0)^{\mathrm{T}}$, $p\in\mathbb{Z}$, then $\bar{\mathbf{f}}^{qq'}(\bar{\mathbf{x}}_1)=\bar{\mathbf{x}}_1+(pq',0)^{\mathrm{T}}$.

6.5.2 (a) $r\equiv\frac{1}{2}$, (b) $r=\frac{1}{3}+k/(4\pi\sqrt{3})+O(k^2)$.

6.5.3 (b) Use (a) and (θ_0,r_0), $\mathbf{S}^N(\theta_0,r_0)$ fixed points of $\mathbf{I}_1$ to show that $\mathbf{S}^{2N}(\theta_0,r_0)=(\mathbf{S}^{N-j}\mathbf{I}_2)^2(\theta_0,r_0)$, for $j=1,\ldots,N$.

(c) $\mathbf{I}_1(\theta,r)=(\theta,r)$ if and only if $2\theta=0 \bmod 1$. Given $\mathbf{S}^N: (0,r)\mapsto(\theta_N,r_N)$ with $\theta_N=0$, then (c) gives $(0,r)$, $\mathbf{S}^N(0,r)$ fixed points of $\mathbf{I}_1$; (b) gives $(0,r)$ $2N$-periodic point of $\mathbf{S}$. To obtain $r(k)$, find the form of θ_N and use Implicit Function Theorem.

6.5.4 If $\alpha\in\mathbb{Q}$ use induction to show $\alpha_n=p^{(n)}/q^{(n)}\in\mathbf{Q}$ with $q^{(n)}\in\mathbb{Z}^+$ and $q^{(n)}<q^{(n-1)}$ for each n. Conversely, termination gives $\alpha_n=a_n\in\mathbb{Z}$ which implies α_{n-1} rational, etc.

6.5.5 (a) Reduce first three convergents to explicit rational form. (b) Use (6.5.4). (c) Use (E6.52).

6.5.6 (a) Use induction; (b) Use (E6.54); (c) Note $a_n\geqslant 1$, for all $n\geqslant 1$. Monotone behaviour of odd, even convergents obtained from (E6.55) in which a_n, $q_n>0$.

6.5.7 See Lang, 1966, p. 10.

6.6.1 Let $\mathbf{f}^2_{\varepsilon,\delta}: (\theta,\tau)\mapsto(\theta_2,\tau_2)$. Find θ_2 and apply Implicit Function Theorem to $\theta=\theta_2 \bmod 2\pi$. For $\delta=0$, check that $\tau=F(\theta,\varepsilon,\delta)$ is the radially mapped circle $\mathscr{C}$ of $\mathbf{M}^2_\varepsilon$. Generically, intersections of $\mathscr{C}$ and $\mathbf{M}^2_\varepsilon(\mathscr{C})$ are transverse and will persist for sufficiently small $|\delta|\neq 0$. However, $\mathbf{M}_\varepsilon$ has elliptic periodic points and, since $\mathbf{f}_{\varepsilon,\delta}$ is not area-preserving, these will generically become foci.

6.6.2 Global diffeomorphism if and only if bijective and local diffeomorphism. Consider $\mathbf{f}(0.25,r_0)$ for any r_0 with $\alpha(1+\beta)>1$.

6.6.3 Show that $\mathbf{f}(L)=L$ and $\mathbf{f}(U)=U$, $\partial(L)$, $\partial(U)\subset T_K$ and $B(\mathbf{f})=\partial(L)\cap\partial(U)$. Use $\mathbf{f}(\partial(L))=\partial(\mathbf{f}(L))$ to obtain $\partial(L)\subset\mathbf{f}(T_K)$. Repeat for U. Obtain $B(\mathbf{f})\subset\mathbf{f}^2(T_K)$, etc. Closed and bounded by T_K implies compact. To show non-emptiness, note $S=T_K\backslash\{Cl(L)\cup Cl(U)\}$ is open and show that $\mathbf{f}(S)=S$; $\mathbf{f}$ dissipative implies S is empty. Thus $Cl(L)$ and $Cl(U)$ disconnect T_K unless $B(\mathbf{f})$ is non-empty.

6.6.4 (a) Use $Cl(L)\cup Cl(U)=C$ to obtain disconnection. Show $L=\bigcup_{n\in\mathbb{Z}^+}\mathbf{f}^n(S^1\times(-\infty,-K))$ and $U=\bigcup_{n\in\mathbb{Z}^+}\mathbf{f}^n(S^1\times(K,\infty))$, $K>K_0$. Therefore L, U are each unions of topologically connected sets with non-trivial intersection. (b) Let $B'(\mathbf{f})$ be an attracting set satisfying condition (a). Let S_1 and S_2 be the associated lower and upper connected components respectively. Show $L\subset S_1$ and $U\subset S_2$. If $B(\mathbf{f})\not\subset B'(\mathbf{f})$, suppose $\mathbf{x}\in B(\mathbf{f})\cap S_2$. Show $\mathbf{x}\notin Cl(U)$. (c) Use contradiction: assume $\mathbf{f}(R)$ is wholly below R and show there are points of $B(\mathbf{f})$ not in $Cl(L)$.

6.6.5 A is the union of $\mathscr{C}$ and the unstable manifolds of the periodic saddle points. For top and bottom of A see analogous Figure 6.21. Rotation numbers for attracting set are $\frac{2}{3}$, $(1/\tau+\frac{2}{3})$. $B(\mathbf{f})$ is $\mathscr{C}$ with trivial rotation interval. $B(\mathbf{f})$ contains a dense orbit only if τ is irrational.

6.6.6 (a) Use local diffeomorphism plus 1–1. (b) Use $r'=r+\omega$, $\theta'=\theta$. (c) $\mathbf{f}_{0,k,\omega}$ not even bijective; conjugacy exhibited by restriction to $\mathscr{C}$ of projection on the first component.

6.6.7 (a) Fix r and let $\max_\theta(r_1(\theta,r))=M_r$, then $M_r<|r|$ if $|r|>h$. Observe $\mathbf{f}_{b,k,\omega}(T_K)\subseteq T_{M_K}$ and $T_{M_K}\subseteq \mathrm{int}(T_K)$. (b) Show that $r_n=b^n r_0-(k/2\pi)\sum_{j=0}^{n-1}b^{n-1-j}\sin 2\pi\theta_j$.

6.6.8 Let $\beta\in[0,1/N]$ be irrational. If $p_i/q_i=u_i/(u_iN+v_i)\to\beta$, then consider the pseudo-orbit $\{\ldots\Sigma\, m_i(u_iB_1+v_iB_0)\}$. Choose m_k such that

$$(m_1w_1+\cdots+m_{k-1}w_{k-1})/m_k\to 0$$

as $k\to\infty$ for $w_i=u_i$ and v_i. For no rotation number, consider the pseudo-orbit $\{\ldots+m_1B_1+m_2B_0+m_3B_1+m_4B_0+\ldots\}$ where $(m_1+m_2+m_3+\cdots+m_{k-1})/m_k\to 0$ as $k\to\infty$.

6.6.9 $r'=-r$, $\theta'=-\theta$ exhibits the conjugacy.

6.7.1 The transformation $\bar{x}=x$, $\bar{y}=y/k^{1/2}$, $\bar{t}=k^{1/2}t$ reduces k to unity in (E6.64) with $\bar{\varepsilon}=\varepsilon/k^{1/2}$, $\bar{\mu}=\mu/k^{1/2}$.

6.7.2 Change to $(u,v)=(x,x/3^{1/2}-2y/3^{1/2})$ and introduce complex z, $\bar{z}$. Use (5.4.14) to obtain $c_3(0)=\mu/8$. Note a longer calculation with $k\neq 1$ shows that $c_3(0)$ is independent of k.

6.7.3 (a) $E_h\mapsto E_1$ under $r\mapsto h^{1/2}r$, $v_1\mapsto hv_1$, $v_2\mapsto hv_2$.

6.7.4 (a) $\mathscr{E}_h\mapsto\mathscr{E}_1$ under $\zeta\mapsto h^{1/2}\zeta$, $\varepsilon\mapsto h\varepsilon$. (b) Let $\varepsilon=\varepsilon_1+i\varepsilon_2$, show that inducing function is given by $(1+\varepsilon_1)=(1+v_1)\cos(2\pi v_2)$, $\varepsilon_2=(1+v_1)\sin(2\pi v_2)$ and is a local diffeomorphism at $\varepsilon=0$.

6.7.5 Check relative positions for min H at $\theta=\pi/8$, $3\pi/8$. Find saddle at $\theta=\pi/8$ and centre at $\theta=3\pi/8$. Use four-fold symmetry to complete phase portrait.

6.8.1 Consider positive roots of (6.8.5). Use (5.4.14) to check criticality of Hopf bifurcation.

6.8.2 (a) Use repeated root of (6.8.5) in (6.8.8) and expand in powers of ν_2. (b) Use the Implicit Function Theorem on (6.8.8). Substitute $r = r(\omega, \nu)(1 + \sigma)$ in (6.8.4).

6.8.4 Introduce metric distance between two circles as maximum radial displacement. Show that associated functional equation which maps circles using $\mathbf{N}_\nu$ is effectively the second component of (6.8.12).

6.8.5 Use induction on k and show that $|\mathbf{N}^k(z) + \exp(2\pi ipk/q)c_k\bar{z}^{q-1}|^2 = |\mathbf{N}^k(z)|^2$ up to order $|z|^q$.

6.8.6 Express in polar coordinates, take logarithms and separate real and imaginary parts. Introduce local coordinate r_L and consider Taylor expansion.

6.8.7 Use Theorem 5.4.2 and (5.4.14). Resonance tongue with tip at $(0, \bar{\nu}_2)$. Show that (ν_1, ν_2) near $(0, \bar{\nu}_2)$ and $(\mathrm{Re}\,\lambda, \mathrm{Im}\,\lambda)$ near $(\cos(2\pi p'/q'), \sin(2\pi p'/q'))$ are related by a local diffeomorphism (cf Figures 5.5 and 5.6).

6.8.8 Use generalisation of (6.8.5) and (6.8.8).

REFERENCES

Abraham, R. & Marsden, J. (1978) *Foundations of Mechanics*, Benjamin/Cummings.
Arnold, V. I. (1963) Proof of A. N. Kolmogorov's theorem on the preservation of quasiperiodic motions under small perturbations of the Hamiltonian, *Russ. Math. Surv.*, **18**(5), 9–36.
Arnold, V. I. (1973) *Ordinary Differential Equations*, MIT Press.
Arnold, V. I. (1978) *Mathematical Methods of Classical Mechanics*, Springer-Verlag. (Russian original, Moscow, 1974.)
Arnold, V. I. (1983) *Geometrical Methods in the Theory of Ordinary Differential Equations*, Springer-Verlag.
Arnold, V. I. & Avez, A. (1968) *Ergodic Problems of Classical Mechanics*, W. A. Benjamin.
Aronson, D. G., Chory, M. A., Hall, G. R. & McGehee, R. P. (1983) Bifurcations from an invariant circle for two-parameter families of maps of the plane; a computer assisted study, *Commun. Math. Phys.*, **83**, 303–54.
Arrowsmith, D. K. & Place, C. M. (1982) *Ordinary Differential Equations*, Chapman & Hall.
Arrowsmith, D. K. & Place, C. M. (1984) Bifurcations at a cusp singularity with applications, *Acta Applicandae Mathematicae*, **2**, 101–38.
Aubry, S. & Le Daeron, P. Y. (1983) The discrete Frenkel–Kontorova model and its extensions, *Physica*, **8D**, 381–422.
Barnett, S. (1975) *Introduction to Mathematical Control Theory*, Oxford University Press.
Birkhoff, G. D. (1968a) An extension of Poincaré's last geometrical theorem, *G. D. Birkhoff Collected Mathematical Papers*, Vol. 2, pp. 252–66, Dover.
Birkhoff, G. D. (1968b) Sur quelque courbes fermée remarquables, *G. D. Birkhoff Collected Mathematical Papers*, Vol. 2, pp. 418–43, Dover.
Bogdanov, R. I. (1981a) Bifurcation of the limit cycle of a family of plane vector fields, *Sel. Math. Sov.*, **1**(4), 373–88.
Bogdanov, R. I. (1981b) Versal deformation of a singularity of a vector field on the plane in the case of zero eigenvalues, *Sel. Math. Sov.*, **1**(4), 389–421.
Bowen, R. (1970) Markov partitions for Axiom A diffeomorphisms, *Amer. J. Math.*, **92**, 725–47.
Bowen, R. (1978) On Axiom A diffeomorphisms, *CBMS Regional Conference Series in Mathematics*, **35**, AMS Publications, Providence.
Carr, J. (1981) *Applications of Centre Manifold Theory*, Springer-Verlag.
Casdagli, M. (1987a) Periodic orbits for dissipative twist maps, *Ergod. Th. & Dynam. Sys.*, **7**, 165–73.
Casdagli, M. (1987b) Rotational chaos in dissipative systems, *Research Report* No. 103, Dept of Mathematics, University of Arizona.

Chenciner, A. (1982) Points homoclines au voisinage d'une bifurcation de Hopf dégénérée de diffeomorphismes de $\mathbb{R}^2$, *C. R. Acad. Sci.*, Paris, **294**, 269–72.
Chenciner, A. (1983) Bifurcations de diffeomorphismes de $\mathbb{R}^2$ au voisinage d'un point fixe elliptique, Les Houches, Session XXXVI, 1981, *Chaotic Behaviour of Deterministic Systems*, ed. G. Iooss, R. H. G. Helleman & R. Stora, pp. 275–348, North-Holland.
Chenciner, A. (1985a) Bifurcations de pointes fixes elliptiques: I – Courbes invariantes, *Publ. Math. I.H.E.S.*, **61**, 67–127.
Chenciner, A. (1985b) Bifurcations de points fixes elliptiques: II – Orbites periodiques et ensembles de Cantor invariants, *Inventiones Math.*, **80**, 81–106.
Chenciner, A. (1987) Bifurcations de points fixes elliptiques: III – Orbites périodiques de 'petites' périodes et élimination résonannante des couples de courbes invariantes, *Publ. Math. I.H.E.S.*, **66**, 5–91.
Chillingworth, D. R. J. (1976) *Differential Topology with a View to Applications*, Pitman.
Chirikov, B. V. (1979) A universal instability of many-dimensional oscillator systems, *Phys. Rep.*, **52**(5), 263–379.
Chow, S. N. & Hale, J. K. (1982) *Methods of Bifurcation Theory*, Springer-Verlag.
Collet, P. & Eckmann, J. P. (1980) *Iterated Maps on the Interval as Dynamical Systems*, Birkhauser, Boston.
Cvitanovic, P. (1984) *Universality in Chaos*, Adam Hilger.
Devaney, R. L. (1986) *An Introduction to Chaotic Dynamics*, Benjamin.
Dumortier, F. (1977) Singularities of vector fields on the plane, *J. Diff. Eqns.*, **23**, 53–106.
Dumortier, F. (1978) *Singularities of Vector Fields, Monografias de Matematica*, No 32, Instituto de Matematica Pura e Aplicada, Rio de Janeiro.
Eckmann, J. P. (1983) Les Houches, Session XXXVI, 1981, *Chaotic Behaviour of Deterministic Systems*, ed. G. Iooss, R. H. G. Helleman & R. Stora, pp. 455–510, North-Holland.
Gardner, M. (1961) *Second Scientific American Book of Mathematical Puzzles and Diversions*, Simon & Shuster.
Golubitsky, M. & Guillemin, V. (1973) *Stable Mappings and Their Singularities*, Springer-Verlag.
Greene, J. M. (1979) A method for determining a stochastic transition, *J. Math. Phys.*, **20**(6), 1183–201.
Guckenheimer, J. (1979) Sensitive dependence on initial conditions for one-dimensional maps, *Commun. Math. Phys.*, **70**, 133–60.
Guckenheimer, J. &. Holmes, P. (1983) *Nonlinear Oscillations, Dynamical Systems and Bifurcations of Vector Fields*, Springer-Verlag.
Guckenheimer, J., Oster, G. & Ipatchki, A. (1977) The dynamics of density dependent population models, *J. Math. Biol.*, **4**, 101–47.
Gumowski, J. & Mira, C. (1980) *Dynamique Chaotique*, Cepadues Editions.
Hale, J. K. (1969) *Ordinary Differential Equations*, Wiley.
Hardy, G. H. (1979) *An Introduction to the Theory of Numbers*, p. 125, Oxford University Press.
Hassard, B. D., Kazarinoff, N. D. & Wan, W-H. (1981) *Theory and Applications of the Hopf Bifurcation*, Cambridge University Press.
Helleman, R. H. G. (1980) Self generated chaotic behaviour in non-linear mechanics, *Fundamental Problems in Statistical Mechanics*, Vol. 5, ed. E. G. D. Cohen, pp. 165–233, North-Holland.
Hénon, M. (1969) Numerical study of quadratic area-preserving mappings, *Quart. Appl. Math.*, **27**, 291–311.
Hénon, M. (1976) A two-dimensional mapping with a strange attractor, *Commun. Math. Phys.*, **50**, 69–77.

Hénon, M. (1983) Les Houches, Session XXXVI, 1981, *Chaotic Behaviour of Deterministic Systems*, ed. G. Iooss, R. H. G. Helleman & R. Stora, pp. 57–170, North-Holland.
Hirsch, M. W. & Smale, S. (1974) *Differential Equations, Dynamical Systems and Linear Algebra*, Academic Press.
Hirsch, M. W., Pugh, C. C. & Shub, M. (1977) *Invariant Manifolds*, Springer-Verlag.
Hockett, K. & Holmes, P. (1986) Josephson's junction, annulus maps, Birkhoff attractors, horseshoes and rotation sets, *Ergod. Th. and Dynam. Sys.*, **6**, 205–39.
Iooss, G. (1979) *Bifurcations of Maps and Applications*, North Holland Math. Stud., No. 36.
Jordan, D. W. & Smith, P. (1977) *Non-linear Ordinary Differential Equations*, Oxford University Press.
Katok, A. (1982) Some remarks on Birkhoff and Mather twist map theorems, *Ergod. Th. & Dynam. Sys.*, **2**, 185–94.
Kolmogorov, A. N. (1957) General theory of dynamical systems and classical mechanics, *Proc. Int. Conf. Math.*, **1**, 315–33, North-Holland.
Lang, S. (1966) *Introduction to Diophantine Approximations*, Addison Wesley.
Le Calvez, P. (1986) Existence d'orbites quasi-periodiques dans les attracteurs de Birkhoff, *Commun. Math. Phys.*, **106**, 383–94.
Li, T-Y. & Yorke, J. A. (1975) Period three implies chaos, *Amer. Math. Monthly*, **82**, 985–92.
Lichtenberg, A. J. & Lieberman, M. A. (1982) *Regular and Stochastic Motion*, Springer-Verlag.
Lorenz, E. N. (1963) Deterministic non-periodic flow, *J. Atmos. Sci.*, **20**, 130–41.
MacKay, R. S., Meiss, J. D. & Percival, I. C. (1984) Transport in Hamiltonian Systems, *Physica*, **13D**, 55–81.
MacKay, R. S. & Percival, I. C. (1985) Converse KAM: theory and practice, *Commun. Math. Phys.*, **98**, 469–512.
Manning, A. (1974) There are no new Anosov diffeomorphisms on tori, *Amer. J. Math.*, **96**, 422–9
Marsden, J. E. & McCracken, M. (1976) *The Hopf Bifurcation and its Applications*, Springer-Verlag.
Mather, J. N. (1967) Anosov diffeomorphisms, *Bull. Amer. Math. Soc.*, **73**(6), 792–5.
Mather, J. N. (1982) Existence of quasi-periodic orbits for twist homeomorphisms of the annulus, *Topology*, **21**(4), 457–67.
Mather, J. N. (1984) Non-existence of invariant circles, *Ergod. Th. & Dynam. Sys.*, **2**, 397.
May, R. M. (1976) Simple mathematical models with very complicated dynamics, *Nature*, **261**, 459–67.
May, R. M. (1983) Nonlinear problems in ecology and resource management, Les Houches, Session XXXVI, 1981, *Chaotic Behaviour of Deterministic Systems*, ed. G. Iooss, R. H. G. Helleman & R. Stora, pp. 515–63, North-Holland.
Misiurewicz, M. (1983) Maps of an interval, Les Houches, Session XXXVI, 1981, *Chaotic Behaviour of Deterministic Systems*, ed. G. Iooss, R. H. G. Helleman & R. Stora, pp. 565–90, North-Holland.
Moser, J. K. (1962) On invariant curves of area-preserving mappings of an annulus, *Nachr. Akad. Wiss., Gottingen, Math. Phys. Kl.*, **2**, 1–20.
Moser, J. K. (1968) Lectures on Hamiltonian systems, *Mem. A. M. S.*, **81**, 1–60.
Moser, J. K. (1969) On the construction of almost periodic solutions for ordinary differential equations, *Proc. Int. Conf. on Functional Analysis and Related Topics*, pp. 60–7, Tokyo.
Moser, J. K. (1973) Stable and random motions in dynamical systems, *Ann. Math. Studies*, **77**, Princeton University Press.
Newhouse, S. E. (1979) The abundance of wild hyperbolic sets and non-smooth stable

sets for diffeomorphisms, *Publ. Math. I.H.E.S.*, **50**, 101–51.
Newhouse, S. E. (1980) Lectures on Dynamical Systems, *Prog. in Math.*, **8**, pp. 1–114, Birkhauser, Boston.
Nitecki, Z. (1971) *Differentiable Dynamics*, MIT Press.
Niven, I. (1956) *Irrational Numbers*, Math. Assoc. Amer., Menasha, Wisconsin.
Palis, J. & Takens, F. (1977) Topological equivalence of normally hyperbolic systems, *Topology*, **16**, 335–45.
Peixoto, M. M. (1962) Structural stability on two-dimensional manifolds, *Topology*, **2**, 101–21.
Percival, I. C. (1982) Chaotic boundary of a Hamiltonian map, *Physica*, **6D**, 67–77.
Percival, I. C. & Richards, D. (1982) *Introduction to Dynamics*, Cambridge University Press.
Rand, D. A. Institute of Mathematics, University of Warwick, private communication.
Robbin, J. (1972) Topological conjugacy and structural stability in discrete dynamical systems, *Bull. Amer. Math. Soc.*, **78**, 923–52.
Rössler, O. E. (1979) Continuous chaos-four prototype equations, *Bifurcation Theory and Applications in Scientific Disciplines*, ed. O. Gurel & O. E. Rössler, *Ann. N.Y. Acad. Sci.*, **316**, pp. 376–92.
Ruelle, D. (1989) *Elements of Differentiable Dynamics and Bifurcation Theory*, Academic Press.
Rüssmann, H. (1959) Uber die Existencz einer Normalform inhaltstreuer elliptischer Transformationen, *Math. Ann.*, **137**, 64–77.
Rüssmann, H. (1970) Uber invariante Kurven differenzierbarer Abbildungen eines Kreisringes, *Nachr. Akad. Wiss., Gottingen II, Math. Phys. Kl.*, 67–105.
Sarkovskii, A. N. (1964) Coexistence of cycles of a continuous map of a line into itself, *Ukrainian Math. J.*, **16**, 61–71.
Siegel, C. L. & Moser, J. K. (1971) *Lectures on Celestial Mechanics*, Springer-Verlag.
Smale, S. (1965) Diffeomorphisms with many periodic points, *Differential and Combinatorial Topology*, ed. S. Cairns, pp. 63–80, Princeton University Press.
Smale, S. (1966) Structurally stable systems are not dense, *Amer. J. Math.*, **88**, 491–6.
Smale, S. (1967) Differentiable dynamical sytems, *Bull. Amer. Math. Soc.*, **73**, 747–817.
Sotomayor, J. (1974) Generic one-parameter families of vector fields on two dimensional manifolds, *Publ. Math. I.H.E.S.*, **43**, 5–46.
Sparrow, C. (1982) *The Lorenz Equations*, Springer-Verlag.
Sternberg, S. (1969) *Celestial Mechanics Part II*, W. A. Benjamin.
Takens, F. (1971) A C^1-counterexample to Moser's Twist Theorem, *Indag. Math.*, **33**, 379–86.
Takens, F. (1973a) Introduction to global analysis, *Commun. Math. Inst., Rijksuniversiteit Utrecht*, **2**, 1–111.
Takens, F. (1973b) Unfoldings of certain singularities: Generalized Hopf bifurcations, *J. Diff. Eqns.*, **14**, 476–93.
Takens, F. (1974a) Singularities of vector fields, *Publ. Math. I.H.E.S.*, **43**, 47–100.
Takens, F. (1974b) Forced oscillations and bifurcations, *Commun. Math. Inst., Rijksuniversiteit Utrecht*, **3**, 1–59.
Taylor, J. B. (1968) unpublished.
Thompson, J. M. T. & Stewart, H. B. (1986) *Non-linear Dynamics and Chaos*, Wiley.
Van Strien, S. J. (1979) Centre manifolds are not C^∞, *Math. Z.*, **166**, 143–5.
Whitley, D. C. (1983) Discrete dynamical systems in dimensions one and two, *Bull. Lond. Math. Soc.*, **15**, 177–217.
Wiggins, S. (1988) *Global Bifurcations and Chaos*, Springer-Verlag.

INDEX